BRITISH WRITERS
OF THE
THIRTIES

British Writers of the Thirties

VALENTINE CUNNINGHAM

Oxford   New York

OXFORD UNIVERSITY PRESS

1989

Oxford University Press, Walton Street, Oxford OX2 6DP
Oxford New York Toronto
Delhi Bombay Calcutta Madras Karachi
Petaling Jaya Singapore Hong Kong Tokyo
Nairobi Dar es Salaam Cape Town
Melbourne Auckland
and associated companies in
Berlin Ibadan

Oxford is a trade mark of Oxford University Press

First published 1988
First issued as an Oxford University Press paperback 1989

British Library Cataloguing in Publication Data
Cunningham, Valentine
British writers of the thirties.—
(Oxford paperbacks)
1. English literature, 1900–1945.
Critical studies
I. Title
820.9'00912
ISBN 0-19-282655-7

Library of Congress Cataloging in Publication Data
Cunningham, Valentine.
British writers of the thirties.
Bibliography: p. Includes index.
1. English literature—20th century—History and criticism.
2. Literature and society—Great Britain—History—20th century.
3. Great Britain—Popular culture.
4. Spain—History—Civil War, 1936–1939—Literature and the war.
I. Title.
PR478.S57C86 1987 820'.9'00912 87-7630
ISBN 0-19-282655-7

Printed in Great Britain by
J. W. Arrowsmith Ltd
Bristol

The ivory towers are draughty nowadays. It's warmer in the street.

Arthur Calder-Marshall

And learn a style from a despair.

William Empson

IN MEMORIAM DISCIPVLAE MEAE
MGBAFOR NNENNA INYAMA
LITTERIS ÁNGLICIS AMANTISSIME STVDVIT
INGENIOSA HILARA HVMANA FVIT
OBIIT XXIV AETATIS ANNO

CONTENTS

ABBREVIATIONS

C	the *Criterion*
CPP	*Contemporary Poetry and Prose*
DW	*Daily Worker*
FNW	*Folios of New Writing*
L	*Listener*
LR	*Left Review*
N&D	*Night and Day*
NS	*New Statesman*
NV	*New Verse*
NW	*New Writing*
PNW	*Penguin New Writing*
S	*Spectator*
TCV	*Twentieth Century Verse*

1

Thick With One's Spittle

The worst of their poems may only be read for the light they throw on their decade; but the study of their decade will continue to throw some light on even the best of their poems.

Francis Hope

Yet if I raise the question of literary history, it is not merely to urge its importance as an intellectual discipline or to deplore the absence of methodological thinking in that area.... My argument will be that literary history is necessary less for the sake of intellect than for the sake of literature.

Geoffrey Hartman

Form and content in discourse are one, once we understand that verbal discourse is a social phenomenon—social throughout its entire range and in each and every of its factors, from the sound image to the furthest reaches of abstract meaning.... The living utterance, having taken meaning and shape at a particular historical moment in a socially specific environment, cannot fail to brush up against thousands of living dialogic threads, woven by socio-ideological consciousness around the given object of an utterance; it cannot fail to become an active participant in social dialogue. After all, the utterance arises out of this dialogue as a continuation of it and as a rejoinder to it—it does not approach the object from the sidelines.

Mikhail Mikhailovich Bakhtin

THIS book is an account of British literature in the 1930s. A literature in a history and a society. It does not, though, propose to endorse the separations between 'literature' on the one hand and 'society' on the other that are still all too conventional in literary historical discussions, not least in approaches to 'literature and society in the 1930s'. The metaphors we are accustomed to using imply severe gulfs, rather extreme differences in kind, between written texts and events 'out there' in 'life'. This is so whether we employ the old painting or stage metaphors (foreground and background), as George Eliot and Henry James did and some post-structuralists still do, or Marxism's architectural, building-site imagery (base and superstructure), or more up-to-date sounding talk of text and sub-text or text and context (as Bernard Bergonzi does in his slimline *Reading the Thirties: Texts and Contexts*). The practice of the present analysis, on the contrary, assumes no such harsh gulf. It allows there to be a gap only if it is held to be usually blurred over and perceived as practically invisible because of the busy mediations that are continually filling it. Bluntly: all texts and contexts will be thought of here as tending to lose their separate identities, collapsing purposefully into each other and existing rather as what we might call (con)texts.

The language of post-Saussurian semiotics, despite some of its more embarrassing bursts of recent notoriety, can still help us here. A period of history and its literature are, like a language, most realistically to be seen as a sign-system, or set of sign-systems, of signifying practices, composing a structural and structured whole. So that, whilst commonsense will always want to retain a notion that facts are separable from fictions, that perceptions of events may not necessarily be the same as those events, intellectual history must also acknowledge not only (as the Reader-Reception theorist H. R. Jauss has persuasively argued) that fictions alter our perceptions of facts, change history, but that fictions are themselves history, are frequently major facts of a time and its aftermath, that images and discourses themselves comprise events and sets of happenings. In other words, that what we perceive to happen is as important, as much a part of the truth, the reality of a time, as what 'actually' happens. Reality is, of course, in the first place characterized by its givenness; it is there just to be discovered, and described, to be *made out* by writers. But, as Henry James and other modernists have taught us, this discovery business is inseparable from the processes of invention, making, *making up*. Christopher Columbus was a great man, admitted the girl in the Louvre to Christopher Newman in James's novel *The American*: the discoverer 'invented America'. And like America, the world is being discovered and invented, all at once, all the time. Which is what writing and writers are all about.

Writers, I once heard Christopher Ricks allege, have very cooperative subconsciousnesses. So, it often appears, do their times. People and events, fictions and facts, have a way of fitting together; they can be detected responding to each other's pressure, moulding each other, being made together. It's impossible, then, finally to separate what is to be discovered, made out, from what is made up, invented; to distinguish (using the Yeatsian metaphor) the literary and textual dancer from the historical dance. So the dominant images of the 1930s do have a way of turning out to be intimately one with the events of that time, and vice versa. Is, for instance, school a common preoccupation of the period's literature? We soon discover that the writers have frequently only just left school and have then most promptly, as schoolmasters, gone back to school again. Do images of the frontier predominate? The writers turn out to be continually travelling, and what's more, across a world whose frontiers, recently realigned after the First World War ('Ostnia and Westland; Products of the peace which that old man provided', as a Chorus of Auden and Isherwood's *The Dog Beneath the Skin* put it in 1935, referring to Clemenceau), were almost all under threat of enemy invasion. Is gigantism an obsession? Or going up mountains and up in the air? It is then intriguing to spot how many writers are actually physically small, and also 'feel small', diminished and so excited or threatened beside the physically big, the morally and politically great, the uprisers, the triumphant. And so on. It certainly is not hard for the intellectual historian to perceive that the facts and fictions offered for his decoding comprise, in large measure, something like a connected field, a whole text, a set of diverse signs adding up, more or less, to a single semiotic. So that he must try and grapple with as much of the components of that scene, that text, as he can: to read, to interpret, and so to connect up, its public events and private lives, the books read and the books written and written about, the discoveries and inventions, the facts and the beliefs, the journeys, holidays, machines, dictators, churches, poems, songs, plays, films, novels—messy and resistant to analysis as this whole abecadarial grab-bag of a period's signs undoubtedly is.

It has to be admitted that some of the most recent adaptations of structuralist theory would resist this insistence on a necessary and busy interaction between texts and contexts. In our post-structuralist period, history and the world around the written text have suffered yet one more severe downgrading. The sceptical spirit of earlier twentieth-century rejections of real, external reference for language and for the texts made out of language—Ferdinand de Saussure's sidestepping of speech acts made by actual persons (*parole*), of history (*diachrony*), and of semantics (the *signatum* behind the sign); I. A. Richards's resistance to the idea that poems might require 'belief' from readers and his displacing of poetical statement by mere 'pseudo-statement'; the American 'New' Critics' subsequent retreat behind the boundaries of the text, a stratagem that spurned history, biography, politics, religion, that is, ideology of all sorts, in favour of a reading dealing merely with the play and interplay of the boxed-in 'words on the page'—this spirit has issued clamantly in our time in the post-structuralist concept of texts as mere plays of signifiers (*jeux des signifiants*). Geoffrey Hartman's Preface to the now famous collection of 'Yale School' essays entitled *Deconstruction and Criticism* roundly puts the case for what he calls 'the strength of the signifier vis-à-vis a signified (the "meaning") that tries to enclose it':

> Deconstruction, as it has come to be called, refuses to identify the force of literature with any concept of embodied meaning . . . We assume that, by the miracle of art, the 'presence of the word' is equivalent to the presence of meaning. But the opposite can also be urged, that the word carries with it a certain absence or indeterminacy of meaning. Literary language foregrounds language itself as something not reducible to meaning . . .

Such scepticism about the meaning of texts—borrowed from the earlier writings of Jacques Derrida with great enthusiasm by the 'Yale School' of critics—begins, of course, in a simple fading or dissolving of the sense of history as a considerable presence in writing, and of writing as an historical force, in a readjustment of the reader's priorities in favour of something queerly thought of as language as such.

> In literary studies, structures of meaning are frequently described in historical rather than in semiological or rhetorical terms. This is, in itself, a somewhat surprising occurrence, since the historical nature of literary discourse is by no means an *a priori* established fact, whereas all literature necessarily consists of linguistic and semantic elements.

That's how Paul de Man civilizedly puts his sceptical scalpel into the historically minded reader of his *Allegories of Reading*. That mild proposition is, however, the porchway of the more sweepingly radical and Derrida-incited vision of J. Hillis Miller and Hartman, in which texts draw in their horns, lose contact with history, become solipsistic enclosures, and then in a captivatingly negative extension of the process proceed to disappear implosively into a kind of linguistic black hole: in a set of deferrals, absentings, decenterings, references back. The more serious 'deconstructionists' have clung, like Harold Bloom, to vestiges of history, meagre kinds of historical siting for texts, even to actual authors. Far too many deconstructive camp-followers are content with much fainter traces of historical presences, happy to see texts as really being 'about' language and its ways, especially its tricksy ways in whatever particular text is under discussion. What interests them most is acts of textual self-mirroring, meta-texts, interpreters, and centres of interpretation within the text, particularly when self-reflection seems to suggest the negation, the undermining of orthodox 'meaning'. What is offered is the self-meaning, self-reading, if not altogether the self-writing text. What seems to overjoy most deconstructionists most is the self-destroying text. The classical texts are all now being reread as Indian rope-tricks. This is a position that cannot be ignored in any discussion of the 1930s, partly because Frank Kermode has challengingly applied it, in *The Genesis of Secrecy: On the Interpretation of Narrative*, to a representative '30s novel, Henry Green's *Party Going* (1939) but mainly because it's an argument that in fact recognizably went on in the 1930s itself as part of the intense debate between Marxists and Modernists.

At the heart of '30s critical disagreements is the contention between the opposed ideas about language and writing that issued on the one hand in *Finnegans Wake* (published complete at last, after years of piecemeal appearance, in 1939), and on the other in Moscow-based Socialist Realism. Of *Finnegans Wake* (when it was just *Work in Progress*) Samuel Beckett declared in ur-deconstructionist tones that 'It is not to be read—or rather it is not only to be read. It is to be looked at and listened to. His [Joyce's] writing is not *about* something; *it is that something itself*'. That was in *Our Exagmination Round his Factification for Incamination of Work in Progress* (1929). Beckett's chapter ('Dante . . . Bruno. Vico . . Joyce') in this piece of publicity that Joyce actually engineered for himself from twelve chosen disciples, had first appeared in *transition* magazine in Paris. Eugène Jolas was another of the Joycean Twelve. His piece, 'The Revolution of Language and James Joyce', sounded yet more ur-deconstructionist notes: 'The new artist of the word has recognised the autonomy of language and, aware of the twentieth century current towards universality, attempts to hammer out a verbal vision that destroys time and space'; 'While painting . . . has proceeded to rid itself of the descriptive, has done away with the classical perspective, has tried more and more to attain the purity of abstract idealism, and thus led us to a world of wondrous new spaces, should the art of the word remain static?'; 'The revolution of the surrealists who destroyed completely the old relationship between words and thought, remains of immense significance'; 'In his [Joyce's] super-temporal and super-spatial composition, language is being born anew before our eyes.' In other words, *Finnegans Wake* was already Exhibit Number One in the surrealistic 'Revolution of the Word' proclaimed in *transition* in June 1929—a revolution whose credo was a clear forerunner of more recent declarations. 'The Imagination in search of a fabulous world is autonomous and unconfined'; 'Narrative is not mere anecdote, but the projection of a metamorphosis of reality'; 'The expression of these concepts can be achieved only through the rhythmic "Hallucination of the Word" (Rimbaud)'; 'The "litany of words" is admitted as an independent unit'; 'The Writer expresses. He does not communicate'; 'Time is a tyranny to be abolished.' Subsequent *transition* manifestos, like 'Poetry is Vertical' (1932), made this revolution sound even more surrealistic and much less Marxist, as well as still more vatic and religiose, endowed as it now became with an explicitly cabbalistic (and Kermode-like) sense of language's secretive, hermetic properties and a wish as sharp as Roland Barthes's or Michel Foucault's to put the author out of the picture. 'The reality of depth can be conquered by a voluntary mediumistic conjuration, by a stupor which proceeds from the irrational to a world beyond a world.' 'The final disintegration of the "I" in the creative act is made possible by the use of a language which is a mantic instrument, and which does not hesitate to adopt a revolutionary attitude towards word and syntax, going even so far as to invent a hermetic language, if necessary.' And this verbal activity was envisaged in terms of collectives that were to outdo the socialistic kind: 'Poetry builds a nexus between the "I" and the "you" by leading the emotions of the sunken, telluric depth upward toward the illumination of a collective reality and a totalistic universe'; 'The synthesis of a true collectivism is made possible by a community of spirits who aim at the construction of a new mythological reality.'

It's easy to see why this sort of platform—one that sanctioned a good deal of frequently embarrassing self-conscious British word-mongering in the '30s, from *Finnegans Wake* to our egregious home-raised irrationalists such as the poets David Gascoyne, Philip O'Connor, and Dylan Thomas (who cut his early teeth in the pages of *transition*, felt happiest when his poems' words had been liberated from any conventional contact with the external world into their own fast-breeding, self-propagating world of the 'contradictory image', who thought poetry 'should work from words, from the substance of words and rhythm of substantial words set together, not towards words', and who wouldn't be averse to cobbling together poems merely out of entries in *Roget's Thesaurus*)—it's easy to see how this position irked Marxists. 'The problem that most deeply concerns artists today is that of realism', declared Marxist art critic Anthony Blunt in *Left Review*.[1] And to Marxist man of letters Randall Swingler, summarizing a debate between Realists and Surrealists held at the Group Theatre rooms (*LR*, April 1938), the answer was obvious. 'When artists have the confidence to *look* outside their own protective circle and discover something about the larger social relations, when they get to work to define the totality of the relations there perceived, then Surrealism will die a natural death, and Realism will prove to be not the label of a certain conscientious group, but the basis of real painting.' *Formalist* and *formalism* were dirty words in Marxist quarters. British Marxists had put their shirts on Comrade Radek's doctrines of Socialist Realism issued at the Moscow Writers' Congress of 1934. They didn't care for Beckett's reminder that Radek himself had fallen foul of Stalin ('ex-comrade Radek' Beckett called him in a piece on Denis Devlin in *transition* in 1938). They didn't care either for Beckett's praise of Devlin's poetic unrealism. *Extraaudenary* Beckett called it.[2] Most Marxists preferred linguistic ordinariness. They held more or less conventional and commonsensical views of language, and so of texts, as being simply mirrors on to reality. The world was antecedent to, and more important than, the word, and words had better not stand too much in the light. The plainer, the more revelatory the linguistic medium and the more like the most naturalistic of nineteenth-century fiction the novel, the better. Eugène Jolas's 'revolution of the word' was just 'mumbo-jumbo', an irrationalists' racket. Its message was 'chuck reason and intellect overboard and line up with the very discreditable emotion-mongers of Nazi "paganism".' There was, concluded G. A. Hutt's attack on *transition* (*LR*, July 1935), 'a strong near-Fascist flavour about Jolas' julep'.

Iris Murdoch's hostile summary of practices she correctly traces back to Mallarmé and Rimbaud—it's a sympathetic entering into the Marxist Sartre's views, to be found in 'The Sickness of the Language' chapter in her *Sartre*—nicely encapsulates the '30s Marxian attitude:

> Rimbaud seems to seek to achieve a dream-like plenitude wherein language disintegrates through an over-determination of meaning; the thick accumulation of exact and highly sensory imagery produces a rich blending all-enveloping confusion in the mind

of the reader. Mallarmé seeks rather to make language perform the impossible feat of simply *being* without referring at all. The reader is held by a pure incantation where-from the ordinary senses of the words have been systematically purged. Meaning has been destroyed in the one case by being crowded in, in the other by being charmed out. Characteristic of both poets is the way in which language appears to them like a meta-physical task, an angel to be wrestled with. Their attention is fixed upon language itself to the point of obsession, and their poems are thing-like, non-communicative, non-transparent to an unprecedented degree; they are independent structures, either outside the world or containing the world. Language loses its character of communicative speech. For both poets, the conclusion is silence . . . Both enterprises have, too, a touch of madness about them.

(For Rimbaud we can perhaps read Beckett, and for Mallarmé the Joyce of *Finnegans Wake*.)

Arguably, the '30s debate went in, as the current one does, for too recklessly absolutist distinctions. 'The language of poetry is not in the ordinary sense "communicative" ', as Iris Murdoch also observes—however much one's feeling (again in Iris Murdoch's words) for 'the normal power of reference possessed by words and sentences, their power to point fairly unambiguously at items in the world.' For language always faces two ways, as Ferdinand de Saussure rightly noted. Words are in one aspect mere words, signifiers; in another aspect they are bolted inextricably on to signifieds, and signifieds are mental versions of 'items in the world'. My account of the '30s continuously assumes this dual reference of words, and so also of texts. It refuses at any point to concede the extremist case to either the Jolas-ists or the Marxists of the '30s. It particularly insists that our present Jolas-ists, the self-declared deconstructionists, have only got half an argument.

In *The Genesis of Secrecy* Frank Kermode rather naughtily borrows St Paul's distinction between 'carnal' and 'spiritual' interpreters to suggest that a deconstruc-tionist reading of *Party Going* which would merely uncover a self-referential play of signifiers, of *textualité*, in that novel is better, more inward with the text (is more 'spiritual', less 'carnal'), than an older-fashioned, quasi-Marxist one which would pay attention to Henry Green's class-consciousness, his political and historical concerns, and so on. Kermode fails, of course, to break completely free from the old-fashioned presence of the author—he soon falls to comparing *Party Going* with other novels by Henry Green and to speculating (and most persuasively) about the effect of Henry Green's known deafness upon the way he writes. But even with this (reluctant?) concession, Kermode's partiality for the 'hermetic' reading of *Party Going* will evidently not pass for anything like the whole story.

Like all linguistic objects (according to my view), Green's novel looks inward; but again like all texts it is indissolubly joined to whole rosters of contexts and subtexts that are necessarily related, even if not in any simple sense determining, historical antecedents to the text.

Take an example Kermode doesn't use. About two-thirds of the way though *Party Going* (it has no chapter titles or numbers to signpost one's own going through it) rich Amabel, holed up with a group of wealthy chums in a private suite in the hotel of a London railway station, takes a bath in a bedroom whose walls are all mirrors. She's thus in a totally solipsistic enclosure. No wonder that in it 'she looked as though she were alone in the world she was so good'. But this self-admiration comes later; first she has to, as it were, create herself as the object of her own admirations:

The walls were made of looking-glass, and were clouded over with steam; from them her body was reflected in a faint pink mass. She leaned over and traced her name Amabel in that steam and that pink mass loomed up to meet her in the flesh and looked through bright at her through the letters of name. She bent down to look at her eyes in the A her name began with, and as she gazed at them steam or her breath dulled her reflection and the blue her eyes were went out or faded.

She rubbed with the palm of her hand, and now she could see all her face.

And she rubs herself dry, 'every inch of herself . . . as though she were polishing', all the time moving her toes 'as if she were moulding something', in rhythm with the polishing which makes her dry and restores her colour from red to her normal white. She dried herself so painstakingly 'it was plain that she loved her own shape and skin . . . When she came to dry her legs she hissed like grooms do': 'And as she got herself dry that steam began to go off the mirror walls so that as she got white again more and more of herself began to be reflected.' Clearly, within her mirrored enclosure and in a mirror, Amabel is engaged in writing herself. She writes and, so to say, rewrites herself: causes herself to fade out, then reproduces herself, brings herself into being again. Surely here, the deconstructionist might argue, is a perfect instance of imploding textuality, an allegory and an enactment of the notion that fiction needs no other justification than that it is only a fiction—in *Finnegans Wake*'s phrase (241: 36) *Just A Fication*; that it is only 'about' writing, about acts of writing; that it is not a transcribing of the world but rather an act of mere scribing or inscribing whose genesis is not to be sought in some historical place or thing or person or moment external to the text, 'before the letter', but in language itself. Amabel, this argument would proceed, originates, along with the alphabet, in the first letter of the alphabet, the *A* her name begins with, and so does the text that encloses her and gives her life. Writing writes not writers: at this point Amabel and the novel she's in seem handily to illustrate the Barthesian aphorism. In pursuit of which line of thought the deconstructionist might want (as Kermode does) to dwell on the other curious act of washing and drying that goes on in this novel—Miss Fellowes' lustration play with a dead pigeon. He would certainly hasten to agree with Kermode's suggestion that the odd fellow who passes so easily and multi-valently in and out and through the station hotel, and so also the novel (he may be the hotel detective, he looks like a poisoner, he commands many dialects) stands for Hermes, the patron of thieves and interpreters, and so also for the perennially, endless and uncircumscribable tricksiness of interpretation, especially the hermeneutical difficulties of this particular novel.

All well and good. But it's also clear that this novel, *and especially* in that inward-looking, self-reflexive moment in the shinily black hole (as it were) of Amabel's temporary bathroom, also looks determinedly outwards. What's more, I would argue (and will continually imply, and will always in the following discussion act as if this be the case), that moment of inscription, eloquent demonstration of fictional inwardness, can only be read and interpreted adequately, in anything approaching its possible fulness, if it's perceived to be backed by whole ranges of sets of meanings rippling outwards into the dense textures of the wider literature and history of the period in which this particular textual moment, and its immediately containing text, the novel *Party Going*, were produced. Behind Amabel's gesture (to adopt R. P. Blackmur's important vocabulary in *Language as Gesture*), her gesture of inscription, stand very many novels, postures, gestures of '30s writing and '30s living, unawareness and unrecognition of which surely condemn any reading of Amabel's gesture that sees it only as a textual business standing only unto and by itself, to be an impoverished and meagre reading, too altogether thinned of potential meanings.

In that bathroom Amabel was, we're told, 'alone in the world'. She was part of the retreat by the novel's small group of rich friends into the isolation of the station hotel, away from the crowd of frustrated ordinary travellers thronging the platforms outside; and now she had sought a further degree of isolatedness, cutting herself off even from her friends. The crowd outside is noisy, her friends are talkative, she signs her name in silence. Her enclosedness is as silent as she and her friends are inactive. Her writing and rubbing are almost the most strenuous activities these idle rich people undertake in the hotel; they're all extremely bored; *nothing* is a repeated key word for their condition. In other words, these people have achieved the 'still centre' that so many '30s authors craved in life and art, and Amabel has escaped into the stillest centre of all—where, as so many '30s authors wished to be left in peace to do, she just writes. And if that were all, her moment of enclosure and writing might seem to comprise the perfect and exemplary Jolas-ist textuality. But a centre always implies a circumference, and a still centre suggests a noisy periphery. And, of course, Amabel is just such a centre, a cynosure of the gawping world of newspaper readers ('shop girls in Northern England', we're informed, 'knew her name and what she looked like from photographs in illustrated weekly papers, in Hyderabad the colony knew the colour of her walls'). She has an audience; she's surrounded by readers; she's locked, in fact, into an inescapable relationship with a very loud circumambient mass. Which is a very 1930s relationship for a writer. And the novel that embraces her shares Amabel's ambivalent position: it too is locked in, an enclosed text; but it too is also bonded into its circumambient context.

And Amabel and her novel invite us continually to perceive and trace out such bondings. Amabel is, we're also told, like a creature in a zoo ('to be with her was for Angela as much as it might be for a director of the Zoo to be taking his okapi for walks in leading strings for other zoologists to see or, as she herself would have put it, it was being grand with grand people'). And zoos—very '30s places (as we shall see)—raise precisely the question Amabel's position raises: is she shut in for her own safety and by choice behind those steel hotel shutters (and behind all the novel's fashionable furniture made of steel bars plated in chromium) or is she a prisoner willy-nilly? The relationship of the individual to the surrounding crowd, the masses, was a peculiarly tensed one in the '30s, especially for writers, and even more so for rich writers like Henry Green. A crowd like the railway station throng, at least 30,000 strong, a mass of faces like 'lozenges', had featured earlier in Green's novel *Living*: the 30,000 people, 'nothing but faces, lozenges', at the Aston Villa football ground, the crowd Old Etonian Green had sought absorption into when he'd left Oxford to work for a year at his family's lavatory-ware factory in Bordesley Green, Birmingham.

Amabel's limbs, her polished rich-person's limbs, are magnetically sexy, even creative in a sybaritic fashion ('And all this time she dried herself she moved her toes as if she was moulding something'), but they're also shown as powerfully perverse, possessive, vengeful as she scratches dominatingly at Alex's knuckle 'with one long vermilion finger-nail'. Amabel's beautiful hands are also evil hands, and so they can't help reaching out to touch Henry Green's own striking ambivalence about his hands:

> I was to court the rich [this is Green's autobiography *Pack My Bag*, 1940] while doubting whether there should be great inequalities between incomes. I had a sense of guilt whenever I spoke to someone who did manual work. As was said in those days I had a complex and in the end it drove me to go to work in a factory with my wet podgy hands.

How those podgy hands might contact the workers' hands is never entirely clarified by Green. The rich in *Party Going* are snobbish (Amabel is 'a money snob'); they're self-obsessed, worshipping their own images captured in their walls of mirrors and their reflecting chrome furnishings (at one point in the novel Amabel and Max and all their wealthy friends are said to have the same decorators in to design their flats in identical style: doubtless, one can't help thinking, the same people as got up the dancer Tilly Losch's famous and ultra-fashionable bathroom of chrome and coloured mirror glass—designed by Paul Nash and described in *Architectural Review*, June 1933). *En masse*, the poor are frightening: will they break riotously into the hotel's quietened enclaves? But in bulk they're also pitiable, prone to get hurt themselves: last time there was a station riot a little boy called Tommy Tucker was injured, and now the lower class throng presents a large target for any bombing plane that cares to fly over ('What targets . . . what targets for a bomb'). Apocalypse, whether revolutionary upsurge or coming war—and the tones of the end-times, the end of peace, the end of writing, are strident in this end-of-decade novel—apocalypse threatens all classes. And *Party Going*, perpetually ambivalent, refuses to return any clear solutions to this threatening situation, especially the Marxists' clear solutions. The novel ends, as it began, in fog and with 'everything unexplained'. The man who seems to know the way out, the 'hotel detective', the one possible eraser of the prevailing sense of criminality, potential alleviator of the social guilt that hangs about the privileged isolations of the rich, is actually as dubious as the wealthy hotel inmates he serves. As a man of the crowd he hints at the familiar Marxist ideal of the proletarian saviour. As a social amphibian, the man of many parts and voices, he's reminiscent of Henry Green's own amphibious practice of bobbing about between nightclub and factory. But in either role he 'looks like a poisoner' and is meshed with the others in the novel's recurring talk of poisoners, the gibbet, murder ('that's murder'; 'What's murder'). And so if the potential solver of crime is also a criminal, then the sorts of solution to the social crimes of being rich and bourgeois that this 'detective' could stand for—siding with the masses, or at least engaging seriously in slumming, which were for bourgeois writers two of the most touted leftist solutions of the period—the solutions begin to look as tainted as the offences.

A guilty bourgeois author, once a determined slummer, Henry Green was now, then, socialistically speaking, going nowhere. Going nowhere! Of course not; he had got marooned with his characters. *Party Going* is about being stuck; and at a threshold, that most magical of '30s places, which appears variously all over the period as the border, the frontier, the doorstep, the railway station, the bus station, the dockside. And as it does so often in other '30s texts, here the literal threshold, that actual train station, merges vividly into the class border, the strong threshold and barrier between the bourgeois and the masses. One of the period's most urgent issues is whether the bourgeois will 'go over' or not. In *Party Going* the question of going, and going over, to Europe, in a party, Amabel's little group, blends insistently, and with all the inevitability of a well-rehearsed period problematic, into the question of going over to the working-classes with a Party—especially the Communist Party. One could go on. But even from this brief inspection of some of this text's points of contact with some of the contextual zones that my account will go on to inspect, it should already be clear that plugging a text into its contexts like this—seeing it as a (con)text—is hermeneutically very enriching. It's a gratifying process of opening out meanings: surely producing a kind of pleasure in reading a text more satisfying than the mere erotic Barthesian pleasure in the 'play of signifiers'. What's more, it's a practice, resort to which will help the reader not to close the text prematurely on the 'inexplicable' or 'irreducible'. Kermode, unable to rationalize Henry Green's oddities of style, suggests that the heavy use of demonstratives in *Party Going*, like the frequent omission of articles in *Living*, 'is a kind of grammatical assertion of the uniqueness of the text, a hint, perhaps, that it is not easily reducible to anything else'. And Kermode makes this 'stylistic eccentricity' serve his argument for textual inwardness: in his view, Green's strange way with demonstratives and articles hampers the interpreter who would resort to contexts, trying 'to see his text in relation to some larger whole: an œuvre, a genre, some organised corpus like mythology'. But, in fact, Green's odd play with such grammatical features—a double manœuvre involving their excessive presence and also their prominent absence—is not entirely unique nor short of explanatory contextual relations. Demonstratives and definite articles—actually part of a single grammatical group, *deictics*—are, as we shall see, extremely prevalent in other texts of this period, in Auden's and Dylan Thomas's poems just as much as in T. S. Eliot or D. H. Lawrence. The period's widespread deicticism was undoubtedly an effort to assert authority, knowledge, command of experience, the capacity to muster typologies (as in, say, *The Virgin and the Gipsy*, 1930), such as between-the-wars British authors seem exorbitantly to have craved. Green's perverse style is susceptible to several explanations. It might be an effort at regional dialect, or proletarian ideolect, or be an imitation of Anglo-Saxon poetry, or the peculiar language of Edmund Spenser, or whatever. But the period's effort to achieve an authoritative air by dint of deictics cannot be ignored: especially as in Green's fiction it backs on to the collapse of the deictic mode, the accompanying absence of definite articles and demonstratives, that seems entirely apt to Green's failure to sustain '30s Marxist solutions, indeed any solutions, of a definite kind.

I am arguing, then, that an active regard for the double-minded, Janus-facing postures of *Party Going*, its centripetal inwardness and centrifugal outwardness, is necessary to an apprehension of something approaching its textual fullness. And I would especially insist that readiness on our part to respond (con)textually in this way respects the strengthened alertness in the '30s to the extreme possibilities of inwardness and outwardness, the period's awareness of the voluble calls to take sides between Jolas and Marx. In the case of Green, furthermore, *Party Going*'s assorted gestures outwards, its fraught dwelling on the pull between writing and context, individual and crowd, are all extremely pertinent to the concerns of a guilt-afflicted author—one of the period's many—who was only too conscious of how a pampered body, those podgy fingers, were behind his pen, and how the figures on the page, the written results of the gestural act of writing, were touched by as well as touching, intussuscepted by as well as intussuscepting, the surrounding historical and social real. Any writer's hands will be, as W. H. Auden well knew, steeped in the stuff of text. The fascination of Auden for the Shakespearian metaphor of the dyer's hand stained by the materials of his craft bears witness to that. But, as Auden also recognized, there are other realities a writer's hands and writings touch and are touched by besides merely textual stuff. Auden's *The Orators* (1932) is, as we shall see, terrifically moved by what the homosexual poet-airman's hands might get up to. Auden's great poem 'August for the people' (August 1935) is also, and properly, worried about the touch of history. Even momentary, glancing contacts, it is suggested there, leave marks which, on any reading of the poem's allegation, it would be absurd for critics to think of simply ignoring:

> And all sway forward on the dangerous flood
> Of history, that never sleeps or dies,
> And, held one moment, burns the hand.

This kind of reality Henry Green—and again characteristically of his time—was at once compelled and disconcerted by. Everything, he wrote of his writing in *Pack My Bag*,

> must go down that one can remember, all one's tool box, one's packet of Wrigley's, coloured by its having been used in conversation or by one's having thought of whatever it may be so many times but necessarily truer to oneself for that reason and therefore unattractive no doubt, thick with one's spittle.

History masticates texts, whether authors (or readers) like to notice the chewing or not. Willy-nilly, whether among Marxists who covet and proudly finger it or among Jolas-ists who dread and seek to shun it, history's spittle lards the writing of this and every other period.

This is not the moment to multiply examples; this book is full of them. Suffice it now to assert that should one's deconstructionist eye jump from *Party Going* in search of another candidate for hermetic reading and alight on, say, Spender's poem 'Port Bou', another central text of the time, then the argument would proceed in exactly the same fashion. Here is a poem undoubtedly full of self-referential acts of reading and writing; in a major sense it is about the problems of writing and reading; its narrator (depending on which of several versions one is reading) reveals his worries about being written out ('My mind seems of paper . . .'), or even printed out (the machine-gun/sewing-machine which he imagines stitching his body with lead-bullets or metal needles blends frighteningly into a printing-machine wielding lead type). But it would be absurd to allege that this sort of concern was more important, more 'spiritual' to use Kermode's word, than the undeniably more deterring, more 'carnal', spittle-flecked questions about what the real Stephen Spender who wrote the poem was up to and really worried about in a real Spanish border-town called Port Bou during the Spanish Civil War that did really take place in real historical time. Which is what the poem is, with equal starkness, also about.

2

Vin Rouge Audenaire?

ALLOW for the context, and go on allowing for it, we must: and *Party Going* and 'Port Bou', as well as zoos, crowds, railway stations, detectives, bombers, Spain, and much else, will be heard of repeatedly in this context-minded (and (con)text-minded) discussion. But while we're still on preliminaries, we must note that the willingness to grant texts their necessary contextual affinities still leaves one dunked in certain problems. Hard questions remain. There's the old chestnut about the extent to which texts are actually *determined* by contexts. And (the less frequently broached issue), the question of how far contexts are determined by texts. How far, again, does our sense of the individual importance of a text, a *Party Going* say, depend upon the way it manages to deploy the uniquely individual as well as the generally historical, contrives to be distinctively personal as well as contextual? And is it the total domination of a text by borrowed, secondhand, contextual bits and bobs of subject-matter (as, say, Rayner Heppenstall's novel *The Blaze of Noon*, 1939) that helps give us a sense of that text's inferiority? What, further, about the apparently rogue text, the one for which few immediate period contexts, if any, can be ascertained, the writings that can't be fitted readily into any of the currently perceivable dominant categories: do we shove them roughly into the nearest available camp, wait for some new light about the immediate context (in this case the 1930s), retreat in search of some larger frame into which they might more plausibly be fitted, start modifying our theory, or what? Does for example, Kenneth Allott's haunting poem 'Lament for a Cricket Eleven' really belong next to Spender's 'I think continually of those who were truly great', which is where Robin Skelton's superb anthology *Poetry of the Thirties* places it? What about Richard Hughes's excellent read of a yarn *In Hazard* (1938): which commands attention all right as a notable fiction of the '30s, but doesn't really mesh snugly with the usual ranks of '30s travelling fictions and isn't really a politicizable sea-story of the period like Malcolm Lowry's *Ultramarine* (1933) or John Sommerfield's *They Die Young* (1930), despite its Red Army chinaman and its travelled acquaintance with Mao Tse-tung? What about this or the other score of writings in like case? (The question worried Richard Hughes himself. He knew that 'We all have to live with—and in—our times: we are all atoms of History', and he felt compelled to tackle the problem of his novel's historical relevance twice, first in a private note 'Why I wrote *In Hazard*' (1938), then in the Introduction to the novel's 1966 reprinting. His later argument, that this story of fabulous storm at sea contained a subconscious enactment of the coming storm of the Second World War, is pretty leaky. Much the same might be avowed of the hit-song *Stormy Weather*, which Auden liked to strum on the piano and Patrick Leigh Fermor found to be so widely popular on his walk across Europe at the end of 1933 and into 1934.[3] Anyway, Hughes's creaky attempt at contextualizing his book can now be weighed in Roger Poole's recent gathering of Hughes essays.[4])

Again—and this is a question certain to strike critics who prefer analysing single poems, single novels, single authors, reading them as on the whole discrete entities—

what about the parts of texts that contextual approaches do not reach and leave apparently cold and unexplained, but that non-contextualized, internalizing readings can show to be patently dense with meaning?

Even when one has attempted replies to these pressing critical questions (and this book does not), one thing remains certain: the reading situation will remain messy. For the mere act of granting, or trying to grant the context real importance is enough by itself to keep interpretation in difficulties. The contextual boundary, however defined, is always going to be most vague. Textual borders present problems enough. But an entire period of literary production doesn't even hold out the deceptive consolations of a printed text. It bears little resemblance to that whole-seeming textual object, holistic on the page, or to a finished box or 'house of fiction', neatly constrained by a book's covers. And if separate poems and novels lack the tidily unproblematic boundaries and edges that Matthew Arnold imagined when he talked of 'seeing the object as in itself it really is'—and my whole supposition about contexts declares that the borders of a text are indeed fluid, messy, not fixed: as Freud believed dreams to be (their thoughts interminable, *ohne Abschluss*, branching out on all sides, entangled complexly in the whole world of our thoughts), or as Derrida believes all textual edges to be ('the text overruns all limits assigned to it[5])—how much more fluid and difficult to trace out must a 'period' of textuality and (con)textuality be. To use two prominent '30s metaphors: a literary period is neither an Ordnance Survey map neatly contained and containing, nor an island 'entire unto itself'.[6] But, of course, for our continuing convenience as readers of history we still try hard to box up time and writing into docilely handleable packages: reigns, centuries, decades. We love the clarifications of labelling: the '90s, *Novels of the 1840s, The Thirties and After, Poetry of the Forties*. The process is always distortive: the myths of history and literature so produced are inevitably too pat. The smaller the segment of time, the patter the myth; and our segments are getting smaller. The fashion for literary decades is still rather new, reinforced by the rise of English studies at universities, where texts must be measured out in doses fit for 'examination purposes'. By the time, though, that Stephen Potter had got around in his acerbic attack on 'Inglit' in *The Muse in Chains* (1937) to protesting against 'thesis writers who take the *books published in any one decade* for their theme', it was too late. Literary decades, unknown before the self-conscious writers of the 1890s decided they were in one, had come to stay. After the self-publicizing '90s it hadn't taken long for people to realize that there had after all been earlier decades: the 'Hungry Forties', the 1840s, came into the language in 1905 according to the *OED Supplement*. And from then on the twentieth century would break inevitably into ten-yearly bits: the 1920s, and after that, as only the third self-conscious decade in literary history, the '30s. Everybody was ready for the '30s. 'What Will the "Thirties" Bring? Famous Folk Look Ahead': thus the magazine *Answers* (4 January 1930). '30s writers were well aware they were in a time knowable as 'the' 30s. What, Wyndham Lewis inquires in *Blasting and Bombardiering* (1937), distinguishes the 'Nineteen-Thirties'? Auden's 'Easily, my dear, you move, easily your head', which appeared untitled in his *Look, Stranger!* volume (1936), had already been published as 'A Bride in the '30's' in the *Listener* and in Janet Adam Smith's collection of *Listener* poems, *Poems of Tomorrow* (1935). Naturally enough, at the end of the decade, even before it was quite over, a rush of writers was to be found defining life just passed as a matter of a decade lived through: 'a low dishonest decade', Auden thought it as he sat in a New York dive on '1 September 1939'; 'the Marxist decade', according to F. R. Leavis in his 'Retrospect of a Decade' in *Scrutiny* (1940); a decade fixed absolutely as 'The Thirties'. Malcolm Muggeridge's *The Thirties* came out in 1940. 'What a decade!' exclaimed George Orwell, reviewing the book and its subject. Orwell's own retrospect, 'Inside the Whale' (1940), offered Henry Miller as the Walt Whitman of 'the nineteen-thirties'. Early in 1940 David Gascoyne wrote his 'Farewell Chorus' to the 'grim 'Thirties'; in Spring 1941 Walter Allen wrote in *Folios of New Writing* about Henry Green as 'An Artist of the Thirties'; in the Autumn 1941 issue of *Folios* its editor John Lehmann was confidently 'looking back' at what he calls 'the writers of the thirties' and 'the movement of the thirties', a critical process he'd already got well under way in his *New Writing in England* (1939) and his Pelican Book *New Writing in Europe* (1940). How easy, thereafter, for C. Day Lewis to write (in 'A Letter from Rome' in *An Italian Visit*, 1953) about 'We who "flowered" in the Thirties', for Edward Upward to initiate his *The Spiral Ascent* trilogy with *In The Thirties* (1962), for Spender to talk so readily in *The Thirties and After* (1978) of 'a literary movement . . . called "The Thirties" ' and for him to collectivize a '30s 'we' ('We (and here by "we" I mean the thirties writers)'). A decade, a movement, a group of writers: the '30s is now one of literary history's most stable and flourishing concepts.

It's still not too late, however, to observe the inevitably shaky foundations of the '30s structure, the gloriously murky outlines of the '30s text. For a start it's still unsettled—as busy debate at the end of the '70s once again showed—what precisely a decade is. Should we be dealing with the years 1930–9, or as A. J. P. Taylor has long maintained, with 1931–40? Are we to count from one to ten: as Virginia Woolf seems to have been doing when she took to thinking decannually in her 1924 paper 'Mr Bennett and Mrs Brown': 'in or about December 1910 human nature changed'? Or should we treat a decade like a person: whose first ten years run from when (s)he's aged nought to the end of his/her ninth year? The Wall Street stockmarket Crash coming at the end of 1929 and the Second World War breaking out at the end of 1939 help make it seem natural to observe the break at the year —9, in fact to treat the '20s and '30s as children. But even all this aside—though the uncertainty remains an important datum—the turn of literary events refuses totally to yield sovereignty to the turns of decades, however we decide to fix them.

This present account tends to settle for 1930–9. 1939 certainly exhibits many features of a satisfactory *terminus ad quem*. Britain and France declared war upon Germany on 3 September of that year, and so a renewed wartime era officially began. By then a great closing-down had already occurred; the literary blackout hadn't waited on the politician's fiat. *New Verse*, most prestigious of the little poetry magazines and centre of the celebration of Auden, folded up in the May of 1939. T. S. Eliot uttered his harrowed 'Last Words' as the editor of the *Criterion* in the January. To be sure, the last issue of John Lehmann's *New Writing* didn't appear until Christmas 1939, but June had seen the final number of the social-realist magazine *Fact*. The London Film Society, haunt of the trendy, the cognoscenti, the intellectuals, gave its last performance on 23 April. *World Film News*, launched by film documentarist John Grierson in 1936, also ceased in 1939. The same year W. B. Yeats and Ford Madox Ford died. It was the year important farewells to an era were published: Louis MacNeice's *Autumn Journal*, Christopher Isherwood's collection of stories *Goodbye to Berlin*, Wyndham Lewis's blush-less reneging on his earlier admiration for Hitler, *The Hitler Cult*, and on his previous anti-Semitism, *The Jews Are They Human?* (some apology!) There were notable farewells to England too: by Isherwood and Auden who departed for America in January (they arrived in New York on 26 January, the day on which Barcelona fell, bringing the Spanish Civil War to its disheartening end), and by Wyndham Lewis who, with wife and dog, scrambled unhappily aboard a ship for Canada only the day before the Second World War broke out.

But, inevitably, not all the props from the '30s stage removed themselves quite so conveniently. The *Daily Worker* (founded in 1930) and Claud Cockburn's scurrilously informative rag *The Week* were not banned until January 1941. The Left Book Club, started in 1936, and plausibly to be seen as the centre of Britain's attempts at a United Front against Fascism, dragged on towards a perhaps inevitable death. But its demise didn't actually come about until after the Second War. What's more—because it always, and upsettingly for any tight view of decades, takes time for books to go through the press—several important texts actually written in or before 1939 and so properly '30s texts did not get published until later, including such notable volumes, essential to any definition of the period, as Arthur Koestler's *Darkness at Noon*, Spender's oddly immature novel *The Backward Son*, Henry Green's doomy *Pack My Bag*, Auden's *Another Time*, heavily freighted with some of his very best poems—all of which books came out in 1940. Further still, most of the so-called '30s writers went on writing, slowing turning into post-'30s writers, into '40s and '50s writers, and so on. Certainly the mere chronological data of publication don't count for everything. (It comes as a shock to learn that William Golding, who sprang to renown as a ' '50s' novelist when *Lord of the Flies* appeared in 1954, had already published a volume of poems in 1934. A rogue text from a rogue '30s poet—who himself felt emphatically not a '30s' poet.) Still, if we're to carry on talking about the '30s, 1939 looks like the year when something, if not altogether the marketable literary package we would like, did come to an end.

For its part, the alleged *terminus a quo* of the '30s is an even messier site. Granted, there's a convenient tidiness about the death of D. H. Lawrence in 1930 as there is about the founding of the *Daily Worker* and of the Youth Hostels Association in the same year. Still handier perhaps is the case of the early '30s economic slump getting momentous and symbolic impetus in Wall Street's catastrophic troubles right at the end of 1929. Neat, too, is the publication on New Year's Day 1930 in Eliot's *Criterion* magazine of Auden's curious charade *Paid on Both Sides*: to be followed rapidly in the same year by Eliot, in his role as publisher in the house of Faber, bringing out Auden's first major volume, *Poems*. In 1930 Spender's *Twenty Poems* (published by the Oxford firm of Basil Blackwell), Evelyn Waugh's *Vile Bodies*, William Empson's *Seven Types of Ambiguity* (first significant product of the I. A. Richards stable of literary criticism that had already impressively dominated the Cambridge English Faculty), also saw the light of day. But, of course, this generation of writers had already got into print, sometimes well before 1930. C. Day Lewis's first volume of verse, *Beechen Vigil*, came out in 1925, Louis MacNeice's first volume, *Blind Fireworks*, in 1929. A clutch of Auden's poems had been published by Spender (some of them actually set and printed by him) in a tiny edition in

1928. In the same summer vacation of 1928 Spender printed, in an even rarer edition, his own *Nine Experiments* (by S. H. S.), *Being Poems Written at the Age of Eighteen*. Among ' '30s' novelists, Waugh had brought out *Decline and Fall*, and Isherwood *All the Conspirators*, in 1928; Graham Greene's *The Man Within* appeared in 1929. While he was still an undergraduate aged twenty in 1926 Henry Green had his first novel *Blindness* published. *Living*, Green's novel about Midlands factory workers that was such a considerable force in the '30s, dates from 1929. Quite properly, Samuel Hynes begins his book *The Auden Generation: Literature and Politics in England in the 1930s* in the 1920s. And, again properly, Hynes looks back not just to 1929—though that was importantly the year of I. A. Richards's *Practical Criticism* and Robert Graves's *Goodbye to All That* as well as of *Living* and the rest—but to earlier years still, when the famous young '30s authors were still at school and university. In a quest for '30s origins one soon finds oneself going back to Richards's *Science and Poetry* (1926), to Eliot's *The Waste Land* (1922), to the First World War. Instructively, it's at least that far back that C. Day Lewis's important little critical book *A Hope for Poetry* (1934) wants to look: to the Great War and its impact on Eliot, to the year 1918 when Gerard Manley Hopkins's poetry first became widely available, the year when Wilfred Owen, set to be one of the most potent of memories haunting the young writers of the '30s, was killed in the last few days of the War.

Admitting, then, the fluidity of the '30s bounds, what of the decade's contents? How is one to decode the plurality of signs, to read the multiplicity of texts within, and comprising, the larger period text, how to map this terrain—one that is like all literary-historical countries, never less than dauntingly rich in hermeneutical problematics? Rightly to *divide* the word of truth: St Paul's biblical Greek metaphor (2 Tim. 2: 15) gives us what should be the interpreter's proper ambition: to cut, or carve, the word straight, to plough orderly furrows across it. But the disparate and varying results of the various critical cuttings-up of the '30s text—all of them doubtless sincere—indicate the difficulty in achieving any one 'correct' slicing into the period's meat. The view from the Eiffel Tower tends to vary from observer to observer.[6] Reports sent back by readers embarked on their voyages of discovery and invention differ, often considerably. In fact, dissecting, arranging, revealing the nature of a period is a lot like Joyce's wrestle with the word-order of a couple of *Ulysses*' more involuted sentences. 'Perfume of embraces all him assailed. With hungered flesh obscurely, he mutely craved to adore': 'You can see for yourself', Joyce confided to his friend Frank Budgen, 'in how many different ways they might be arranged.'[7] Inevitably there are almost as many variant readings of the '30s as there are books and articles on the subject, and the present one doesn't at all seek finality. When one does try 'rightly' to divide, though, there are certain dominant readings, orthodoxies that have acquired canonical if not mythical stature, which cry out for attention. Should one, after all, follow the line of *these* cross-sections, tread, as one reads, the line of *that* furrow? Are the '30s, for instance, to be thought of as *The Auden Generation*? That's how Samuel Hynes thinks, and impressively so, about the period. His '30s—an affair of a group of chums clustered about their mutual friend, a tight and identifiable coterie comprising Auden himself, Isherwood, Spender, John Lehmann, Day Lewis, Rex Warner and Edward Upward—is the truly orthodox one. It's the one Bernard Bergonzi subscribes to. So does Richard Johnstone (in *The Will to Believe: Novelists of the Nineteen-Thirties*). Adding in Orwell, Greene and Waugh, with varying loads of significance, doesn't substantively alter the canonical Auden-centred picture. Alternatives have been offered. Frequently they pack little counter-punch. Like John Lucas's gatherum of essays, *The 1930s: A Challenge to Orthodoxy*, whose attack on the 'official' position so deftly promoted by Hynes is schooled by Arnold Rattenbury's shrill heterodoxy: 'That the Thirties were made of Auden and friends and such influences as reached them is as unlikely a notion as daft.' But seeking to replace the Auden Generation orthodoxy by merely substituting for it Rattenbury's *Left Review* crowd is much dafter. Edgell Rickword, Montagu Slater, Randall Swingler, and other fiercely uncompromising Communist Party writers belong to the literary '30s. They don't however comprise the whole, nor the central part of the decade. Again, Martin Green's account of English literary life in the mid-twentieth century, *Children of the Sun: A Narrative of 'Decadence' in England after 1918*, tries to shift the fulcrum back into the '20s and turn that odd couple of Old Etonian aesthetes, Harold Acton and Brian Howard, into the period's key men. Like the *Left Review*, though far less strongly or importantly, Howard and Acton are of course present in and about the '30s, but they both settled for harmless exile, and their kind became jokes, butts, jibed at by the likes of Evelyn Waugh and Cyril Connolly. No one reading Waugh's novel *Put Out More Flags* (1942) and making the connection between the sad Ambrose Silk and Brian Howard ('a cosmopolitan, jewish pansy') will be happy with Martin Green. At the end of the '40s, the dandy mode was specifically refuted by Auden himself. Dandyism, Auden claims in the Introduction added to Isherwood's translation of Baudelaire's *Intimate Journals* in 1947 (as a replacement for T. S. Eliot's original introductory essay), at least Baude-laire's dandyism, was a mistaken phase. Auden welcomes the 'real change of heart' by which Baudelaire shook Dandyism off.

Neither Green's nor Lucas-Rattenbury's '30s look like unsettling Hynes's. The Auden Generation remains canonical. Memorably, this is the reading of some major poets, reinforcers of the Auden Generation idea. Donald Davie's superb poem of the early '50s, 'Remembering the 'Thirties', remembers precisely Auden, quotes Auden, gently ribs Auden and Isherwood and their acquaintance, the Old Etonian traveller Peter Fleming. ' "Leave for Cape Wrath tonight!" They lounged away / On Fleming's trek or Isherwood's ascent.' And Davie's recall is also Robert Lowell's. Under the aegis of Lowell's poem 'Since 1939' Auden is handed on, as it were, to a still younger generation, as the proper stuff of the '30s:

> We missed the declaration of war,
> we were on our honeymoon train west;
> we leafed through the revolutionary thirties'
> *Poems* of Auden, till our heads fell down
> swaying with the comfortable
> ungainly gait of obsolescence . . .
>
> . . .
>
> I see another girl reading Auden's last book.
> She must be very modern,
> She dissects him in the past tense.

And these poetic memories ring true because, clearly, in the period itself Auden and his group were regarded by very many of their contemporaries as the central figures.

The sceptical Wyndham Lewis opined (in *Blasting and Bombardiering*, 1937) that Auden had grown into 'a national institution'. Dyspeptic iconoclast Tom Harrisson, who hated Oxford and Cambridge and all their works (his *Letter to Oxford* (1934) was published, he claimed, by the Hate Press), believed these universities were 'mentally led by Auden'. In her Workers' Educational Association lecture of 1940, 'The Leaning Tower', published in *Folios of New Writing* (Autumn 1940), Virginia Woolf identified the Auden 'group' as crucial: the names of Day Lewis, Auden, Spender, Isherwood, MacNeice (already Upward had got himself lost) 'adhere much more closely than the names of their predecessors'. The precise clientele of the group would shift in observers' minds. Day Lewis frankly thought 'the names of Auden, Spender and myself' comprised much of the going *Hope for Poetry*. Roy Campbell's gnashed-out composite 'MacSpaunday' evidently encompassed MacNeice as well. Geoffrey Grigson, the editor of *New Verse*, never quite decided whether he had a trinity or a quartet on his hands. He wrote variously in *New Verse* of 'the inevitable trio', 'the Circle (or the Triangle)', 'The Three and Mr MacNeice'. But whichever it was, Grigson thought that Isherwood's *Lions and Shadows* would 'always be a reference and key book of the Auden Age and the Auden Circle'.[8] A year later in his sharply Marxist survey *The Poet and Society* Philip Henderson affirmed that this was 'The Auden Age': that was the title of his book's Chapter 8.

New Verse had been devoted to promoting Auden as the chief poet of his time. It had granted him the handsomely celebratory Auden Double Number as a thirtieth birthday present in November 1937. *New Verse*'s parting shots (May 1939) included Grigson's using Auden's sonnets from China (published in *Journey to a War*, 1938) as justification for his magazine's kept-up advocacy and for that 'Auden Age' label: 'All who have believed in Auden are by this time a hundred times justified.' Even for some like F. R. Leavis who had stopped believing, Auden was still unarguably at stage-centre. Auden's was 'the representative career of the nineteen-thirties, and has a representative significance', Leavis alleged in the 1950 Retrospect he added to his *New Bearings in English Poetry*. He felt it necessary to explain why *New Bearings* had not bothered in 1932 to mention Auden. And for faithful believers Auden commanded in every way. After Harold Nicolson had heard Auden read a (never completed) poem in which Gerald Heard was playing Virgil to Auden's Dante on a tour of the Inferno of modern life, Nicolson confided his new and almost religious submissiveness to his Diary (4 August 1933):

> [The poem] is not so much a defence of communism as an attack upon all the ideas of comfort and complacency which will make communism difficult to achieve in this country. It interests me particularly as showing, at last, that I belong to an older generation. I follow Auden in his derision of patriotism, class distinctions, comfort, and all the ineptitudes of the middle-classes. But when he also derides the other soft little harmless things which make my life comfortable, I feel a chill autumn wind. . . . A man like Auden with his fierce repudiation of half-way houses and his gentle integrity makes one feel terribly discontented with one's own smug successfulness. I go to bed feeling terribly Edwardian and back-number, and yet, thank God, delighted that people like Wystan Auden should actually exist.[9]

For his closer friends and disciples Auden simply held court. His undergraduate rooms at Oxford were turned into what his friends thought of as a psychiatrist's consulting rooms in which their neuroses and poems were trundled out for verdicts and cures. Auden bossed. It may have been the case, as Auden reported in his 'Letter to Lord Byron' (*Letters from Iceland*, 1937), that 'At the *Criterion*'s verdict I was mute'. But not, we feel, for long. Auden was eager, as he says, loudly to pass on Eliot's dictates as his own:

> And through the quads dogmatic words rang clear,
> 'Good poetry is classic and austere'.

Auden obsessed his friends. 'Neurosis held him in the grip of Auden', wrote Cyril Connolly (itself a nicely Audenic line) about himself in Anthony Powell's copy of his *The Unquiet Grave*. Characteristically Day Lewis wanted Auden to accompany him on his travels up *The Magnetic Mountain* (1933). The whole of that volume is dedicated to Auden. In it the poet has not only caught the tone of Auden's early poems, their northern landscapes, their birds, their tonal toughness, but the world's W. H. Auden is turned chummily into Lewis's very own Wystan—just as R. E. Warner is throughout called just Rex:

> Then I'll hit the trail for that promising land;
> May catch up with Wystan and Rex my friend,
> Go mad in good company, find a good country.

Day Lewis wasn't short of pals (he had 'Wystan, Rex, all of you that have not fled') but in the end it's only Auden who is praised actually to the skies. *The Magnetic Mountain* offers a unique homage to him and his uniqueness:

> Look west, Wystan, lone flyer, birdman, my bully boy!

> Gain altitude, Auden, then let the base beware!
> Migrate, chaste my kestrel, you need a change of air!

The elevation of Auden could be mocked—as 'Joseph Gurnard' (G. W. Stonier) mocked it in his 'Poets' Excursion'. He has Stephen Spendlove, Louis MacNoose, Don Layman, and Co., entraining at Paddington, with Spendlove breathily admiring the trains, just like in Spender's poems. The train traverses a chracteristic '30s landscape ('Pylons! Arterial roads, semi-detached villas, Butlin's camps, ping-pong, scooters! Hurrah! But chiefly the pylons . . . "Like nude giant girls", said Stephen Spendlove with that wonderful felicity of his') on a pilgrimage to a school called St Audyn's Academy on the Height where the Beak reposes lofty on a pile of chairs to which another chair is added each day, in a building which gets higher by a floor every day.[10] The superiority of Auden's position ('The Beak likes to be above everything when he works') and the force of his personal myth were mockable. They weren't, however, even in the mocking, deniable.

The name of Auden became a touchstone of the period. It even appears, embedded as a single 'sentence' in the middle of a long footnote on page 279 of *Finnegans Wake*: 'Auden'. Tout court. Auden's was the mind that dominated other minds. It generated, declared Grigson reviewing Spender's critical book *The Destructive Element* (*NV*, June 1935): it 'really engendered' Spender's book 'as it has engendered something of a poetry revival, half of Mr Spender and nine-tenths of Mr Cecil Day Lewis'. Quickly it came about that you could refer to 'the Audenesque'— as Gavin Ewart did (his loving parody, 'Audenesque for an Initiation' appeared in

NV, December 1933), and as *New Verse* frequently did (*Lions and Shadows* is useful, No. 29, March 1938 tells us, for its 'items of Audenesque mythology'; a review by Ewart of Auden's *The Dance of Death*, 1933, is simply headed 'Audenesque'), and as Graham Greene did, quite casually in the course of a film review in *The Spectator* in 1936 ('a really dreadful woman singer murders the Audenesque charm of "You're the Top" ')—and in so doing you could evidently expect the literate reader to know instantly what was meant. And the Audenesque becomes perhaps the period's most catchable tone.

It was Auden's way with adjectives and adjectival phrases that particularly struck. Roy Fuller has recalled the 'Audenesque epithets that poets of the age caught as easily as a common cold'. Spender and Isherwood resisted the Audenesque's verbal influenza: their styles had rights of their own in the Auden circle. But good Audenesque phrases keep popping up like sore thumbs all over the work of the other friends: in, for example, Day Lewis's novel *Starting Point* (1937) ('the unobtrusive patience of the mole'), in MacNeice's *Autumn Journal* IX ('the even tenor of the usual day'), in his 'Leaving Barra' ('The routine courage of the worker,/The gay endurance of women'), in Rex Warner's 'The Tourist Looks at Spain' ('The tragic joke or abject surrender of the cracked nerve'). Warner's *The Wild Goose Chase* (1937) is a novel prone like MacNeice's play *Out of the Picture* (also 1937) to making cranky Audenic lists and collocations (Warner's are better: one captive policeman was 'the greatest living authority on the cuckoo, and the other had a recipe for removing fur from the tongue'). Early on, *The Wild Goose Chase* is to be found quite wallowing in an enthusiastic bath of Audenesque adjectives: 'the ghostly quick shadows of the screaming gulls', 'an unusual restlessness of hurry', 'the comforting exterior of attics', 'the anxious heads of eager girls'.

And loving imitation of this sort—not to mention mythographic Audenesque props like mountains, and frontiers, and mineshafts and airmen—went spurting rapidly outwards, beyond the immediate pally pale, into the pages of novelists like Eric Ambler and especially into the writings of the boys who were drenching themselves and their verse in Auden at school and university—John Cornford, Charles Madge, Bernard Spencer, Gavin Ewart—or, as they struggled with the poet's craft of an evening after work, into the texts of people like the young solicitor Roy Fuller. Madge's gauche expression (in his poem 'Letter to the Intelligentsia' in the *New Country* anthology, 1933) of the difference Auden made to him sounds extreme:

> But there waited for me in the summer morning,
> Auden, fiercely. I read, shuddered and knew.

But the influence of Auden on his time was extreme. It's a massive sign of it that Eliot himself absorbed it. His late '30s play *The Family Reunion* (1939) builds many of its lines out of Audenesque formulae ('To the chilly deck-chair and the strong cold tea'; 'The sudden solitude in a crowded desert'; 'The unspoken voice of sorrow in the ancient bedroom'; and the like). Bernard Bergonzi is right to suggest in his *Reading the Thirties* that the Audenesque appears to have been influenced early on by lines in Eliot's poem *Ash Wednesday*—'The vanished power of the usual reign', 'The infirm glory of the positive hour'. But the more important point is that those Eliotic tricks have been returned, stamped with the label 'Audenesque', and Eliot has been happy to use these remoulded, recharged devices. Strongest witness of all perhaps is John Cornford's best known poem, 'Heart of the heartless world'. Even as he scribbled this love poem amidst the turmoil of fighting for the Republic in the Civil War in Spain, in all those many excitements and distractions, Cornford—whose schoolboy poems, taken early with the Auden manner, had been sent, in the spirit of the epoch's reverence for The Master, by Cornford's English master at Stowe School to Auden himself for comment—Cornford had lines of Auden ringing in his head. 'Heart of the heartless world': that was Margot Heinemann, movingly embraced in the words of Marx's famous tribute to religion in his essay '*Zur Kritik der Hegel'schen Rechts-Philosophie*' ('the heart of a heartless world . . . the opium of the people'). But the phrase had also come via Auden's own adaptation of that tribute in his homage to a school rugger team in Ode No. II in *The Orators* (1932), dedicated 'To Gabriel Carritt, Captain of Sedbergh School XV, Spring, 1927':

> Heart of the heartless world
> Whose pulse we count upon.

Obviously, the strong presence of Auden in the '30s cannot be gainsaid. The grudging Arnold Rattenbury himself admits, after a bout of hostile huffing and puffing, that of course he and B. L. Coombes, the coalminer-author, 'spoke more about Auden than any other living author in our hours and hours of talk'. Obviously, one of the period's most distinctive literary tipples, among the most quotidian of its *vins ordinaires*, was—as in Faber's ghastly pun in the firm's advertisement in the Auden Double Number of *New Verse*—Vin Audenaire. (The joke was director Frank Morley's, who suggested it as a title for the volume that became *Look, Stranger!*: Auden thought the appalling suggestion 'brilliant'.) It is even more obviously the case, however, that the '30s are greatly straitened when they are defined only with reference to Auden and his closest contemporaries, even if we generously define this group to include writers roughly of the same age as Auden but not actually in the inner circles: people like Evelyn Waugh, Cyril Connolly, Graham Greene, Henry Green, Anthony Powell, and John Betjeman on the Oxford side, John Lehmann, Julian Bell, Kathleen Raine, Malcolm Lowry, Hugh Sykes Davies, Humphrey Jennings on the Cambridge side (most of these Cantabs were formed up in and about the magazine *Experiment*) and more maverick authors such as Orwell, V. S. Pritchett, Christopher Caudwell, and Samuel Beckett who attended neither Oxford nor Cambridge.

For a start the '30s do not comprise just a single generation, they contain at least three literary generations. Auden and his coterie may be justly thought of as somewhere in the middle. But a most distinguished older generation, in it the heroes and heroines of British Modernism, was still about in large numbers and still producing. D. H. Lawrence alone of his generation was dead, but his publications jutted aggressively into the decade: a stirring swirl of posthumously printed and reprinted poems, stories, articles, and essays. His presence was clearly not going to be easily rubbed out. Joyce's *Finnegans Wake* was startling readers piecemeal throughout the decade. Pound was still active; so were Wyndham Lewis and Yeats and Eliot. Virginia Woolf's *The Waves* came out only in 1932, *The Years* in 1937, *Between the Acts* not until 1941. E. M. Forster was still very much around, though he'd given up publishing fiction. And this generation of writers was important to the '30s not only for what they did and wrote themselves, but for their sponsorship of the young. Wyndham Lewis might fume like an angry papa, Virginia Woolf admonish like an exasperated mother, but at Faber's and in the *Criterion* Eliot assiduously sought out talent, encouraged, cajoled, tried to correct tendencies, and always read carefully what was offered him, and the Woolfs brought into their Hogarth Press John Lehmann, just down from Cambridge at the beginning of the decade, to help them pick up the coming talents. Lehmann's ride at Hogarth was often rocky, but his pals did get published, and Virginia Woolf's interest in the young men was astonishingly sustained through all her personal stresses and strains. Being avuncular came naturally to Eliot, just as acting auntie did to Virginia Woolf. Literally Julian Bell's aunt, it was simple enough for Mrs Woolf (as her letters to and about Spender, and his letters to her, not all of them published yet, clearly reveal) to adopt Spender as a nephew. As for E. M. Forster, he fell continuously and naturally into the role of uncle among his homosexual 'nephews', particularly Isherwood.

Of the very Grand Old Men, almost grandfathers to the period, Wells and Shaw were still actively writing, Arnold Bennett lasted until 1931, Galsworthy until 1933, and Kipling until 1936. And, of course, not all of the First World War poets—older brothers or youthful uncles to the Auden generation—had been killed off or silenced: Siegfried Sassoon was still going strong, Edmund Blunden would meander gently on for many more years yet, and Robert Graves, grouchy about Auden's prominence on the one hand, and the small credit granted his bossy American mistress Laura Riding on the other, wasn't just in magnificently full spate himself, pouring out numerous volumes of poetry and fiction and criticism, he was most keen to swap a gingerly fraternalism he couldn't accept for a disgruntled paternity that the Auden advocates wouldn't allow. Auden, Graves kept declaring, was a bogus and thieving poet, covertly stealing his best materials and techniques from Laura Riding and himself. (The debt to Miss Riding can be played down, but not Auden's indebtedness to those early poems of Grave's that are located among hills and crags and use the elevated vision of birds as an analogue for the uplifted poet's power. 'Rocky Acres', for instance, in the volume *Country Sentiment* (1920), is strikingly anticipatory of many key Auden motifs.)

Furthermore, there was an obvious third generation in the period: the immediate inheritors of the Auden generation, as it were the younger brothers' younger brothers: writers who came directly under the Audenesque's influence, the Roy Fullers and Bernard Spencers, the Cornfords and the Allotts, the Madges and Ewarts; authors flowering in the shade of Auden's output like David Gascoyne, Dylan Thomas, George Barker, who blended English Auden with wilder surrealistic influences from the Continent. Grigson's *New Verse* made room for some of these new-new boys but they were also finding newer homes for themselves in Julian Symons's *Twentieth Century Verse* (which was started in January 1937), or Roger Roughton's surrealistic *Contemporary Poetry and Prose* (started May 1936) or the periodical *Seven*, which began in Summer 1938 and featured alongside known names like Henry Miller more youthful contributors like Lawrence Durrell, D. S. Savage, G. S. Fraser. ' "I must call on Ruthven",' thinks the narrative voice of Julian Symons's poem 'On Liberals, 1938', a poem dedicated to Ruthven Todd, in the volume *Confusions About X* (1939) which also has poems dedicated to Herbert and Marjorie Mallalieu, and to 'R. B.' (presumably Fuller): already this generation

had other chums than the Auden group on their minds. And this particular generation gap, though narrow, was felt forcefully on both sides of it. David Gascoyne's
journal for January 1939 delightedly records that Spender had just told him that 'of
your generation there's only you and Dylan Thomas and George Barker', and 'John
Lehmann, whom I saw yesterday, said the same thing; so I feel rather encouraged'.

An informed reading of the '30s will, then, find itself inevitably focusing most
intently on the Auden generation, and probably also on the even narrower Auden
group. But wise reading must also, in the interests of truth, attempt to keep transcend-
ing these narrowing limits. And for at least two further good reasons. In the first place,
the most commonly accepted components of the Auden Generation orthodoxy,
Grigson's trio or quartet, are all poets. Isherwood and Upward, the novelists in the
inner circle, tend to get left out of the commonest reckoning, particularly Upward.
And there are multitudes of considerable novelists in the period besides those two. But
when we think of literature in the 1930s our current orthodoxies usually have us thinking first of poets. It's Auden, Spender, Day Lewis, MacNeice who spring immediately
to mind. It's poets, not novelists, that we see in Spain 'exploding like bombs'. It's *New
Verse* that we think of as the representative little magazine. And to some extent this
assumption of the centrality of poetry was fostered in the period itself. Eliot's
example, and then Auden's, were telling. Eliot published no prose fiction at all, and
Auden scarcely any. Their high-priestly regard for poetry was endorsed by the advocacies of *New Verse*. Then there was the influence of Q. D. Leavis. Her important book
Fiction and the Reading Public (1932), which firmly hitched the laden wagon of fiction
(most fiction at any rate: T. F. Powys was one of the very few living novelists who
escaped unscathed) to its well-managed despondency over popular culture, was evidently in Day Lewis's mind as he wrote *A Hope for Poetry* (his reference to 'far better
qualified writers than myself' who have established the 'narcotic and unnerving
property' of 'stimulants' such as the 'mass-produced novel', is clearly to Mrs Leavis).
Characteristically, the volume *New Signatures* (1932) in which the first attempt to
define the Auden Generation was seriously made, was an anthology of poets. Prose
had to wait a whole year to get into the follow-up volume *New Country*. Likewise, the
best of the '30s little magazines to carry prose as well as poetry, *New Writing*, has
always languished in the shadow of *New Verse*'s fame.

But even on the inside of the Auden coterie this feet-shuffling about fiction was
not to be found. Everyone knew that fiction was more popular than poetry, and thus
the obvious form to use if you wanted, as most of the Auden group did at some point
in the '30s, to speak to the people at large ('we wish poetry to be popular', Day Lewis
stated in the same volume as he deplored mass-produced novels). What's more, the
idea of the superiority of poetry over prose was a notion fostered by a classical
education at public schools and Oxford and Cambridge, and only stuffy old reactionaries actually supported Eliot when he advocated that kind of schooling (in his essay
'Modern Education and the Classics', 1932). The progressive note about Classics
was struck by Day Lewis in his *Left Review* attack on Latin as a bourgeois instrument
('An Expensive Education', Feb. 1937: he speaks as a teacher of Latin for the
previous eight years and a close friend of Rex Warner who was another Classics
master, but logic, as we've just seen, isn't his strong point). It's the note sounded too,
by Gavin Ewart in his poem about wanting to grow up and quit the academy ('To go,
to leave the classics and the buildings, / So tall and false and intricate with spires'),[11]

and by Arthur Calder-Marshall in his kaleidoscopic school novel *Dead Centre* (1935), where the Classics master loves 'the humanities' but hates boys, and the school games 'pro' reflects that studying Greek shuts boys out from contact with 'the ordinary things'. Even MacNeice, a university lecturer in the Classics, refuses in Section XIII of *Autumn Journal* (1939)—though with a quantity of irony that's hard to measure exactly—to defend the old Classical snobberies:

> We learned that a gentleman never misplaces his accents,
> That nobody knows how to speak, much less how to write
> English who has not hob-nobbed with the great-grandparents of English,
> That the boy on the Modern side is merely a parasite
> But the classical student is bred to the purple, his training in syntax
> Is also a training in thought
> And even in morals; if called to the bar or the barracks
> He always will do what he ought.
> And knowledge, besides, should be prized for the sake of knowledge:
> Oxford crowded the mantelpiece with gods—
> Scaliger, Heinsius, Dindorf, Bentley and Wilamowitz—
> As we learned our genuflexions for Honour Mods.

As another ironic allegation puts it, in the broadcast poem 'in praise of the great Greek athlete Pindar, a statue of whom stands, as we all know, in our own Piccadilly Circus', in MacNeice's *Out of the Picture*: 'Pindar is dead and that's no matter.'

Odder still, both for the keenness on poetry among the personnel around Auden and for our myth of a generation of poets, is the way novelists and their talents were particularly prized by Auden himself. His coterie was onion-like. If you patiently peeled away Day Lewis and Rex Warner, Spender and MacNeice, you would find Auden. But probe, as the intimates knew that you must, for a writer beneath Auden's skin and, as John Lehmann discovered, there you found Isherwood. And should you probe deeper yet, behind Isherwood you'd discover a still remoter guru (respectfully consulted by Isherwood, right up to his death in 1986, over all his writing): Isherwood's old friend and schoolmate Edward Upward. And, of course, Isherwood and Upward were prose fiction men. Auden's regard for Isherwood's craft couldn't be plainer. The novel, Auden informs Lord Byron, in his first Letter to him, is 'the most prodigious of the forms'; in fact Auden had wondered whether to address his epistles to Jane Austen instead of to Byron:

> . . . I don't know whether
>   You will agree, but novel writing is
> A higher art than poetry altogether
>   In my opinion, and success implies
>   Both finer character and faculties.
> Perhaps that's why real novels are as rare
> As winter thunder or a polar bear.
>
> The average poet by comparison
>   Is unobservant, immature, and lazy.
> You must admit, when all is said and done,
>   His sense of other people's very hazy,
>   His moral judgements are too often crazy,
> A slick and easy generalization
> Appeals too well to his imagination.

And this comparison's emphasis was repeated in Auden's sonnet 'The Novelist' (December 1938), written for Isherwood:

> Encased in talent like a uniform,
> The rank of every poet is well known;
> They can amaze us like a thunderstorm,
> Or die so young, or live for years alone.
>
> They can dash forward like hussars: but he
> Must struggle out of his boyish gift and learn
> How to be plain and awkward, how to be
> One after whom none think it worth to turn.
>
> For, to achieve his highest wish, he must
> Become the whole of boredom, subject to
> Vulgar complaints like love, among the Just
>
> Be just, among the Filthy filthy too,
> And in his own weak person, if he can,
> Must suffer dully all the wrongs of Man.

Auden loved flash and glitter; his poems seek precisely the verbal 'dash' that he associates here with hussars. He has no eye for the 'plain and awkward'; he prefers the beautiful boys, indeed the beautiful hussars. He feared that he himself was plain and awkward (he would deprecate his big bottom) and therefore 'One after whom none think it worth to turn' (on at least one occasion he burst into tears with Isherwood because he felt nobody loved him; his poems are obsessed by lovers' betrayals). Auden hated boredom, let alone 'the whole of boredom': that's why he was no good at quotidian political activity, 'the flat ephemeral pamphlet and the boring meeting', as he puts it in his poem 'Spain', and why his poems are so restlessly self-entertaining, hopping with irking brightness like a distracted magpie from one shining object to the next. Auden was right to see that novels must cope with dull ordinary life and boringly plain people. For his own part he prefers not to try and 'struggle out of his boyish gift'. But he respects the novelist for being able to endure all the things he himself can't cope with—including the sexually transmitted disease that's thought by some readers to be what the allusion to filthiness is all about—and in particular for this strain towards the maturity of the novel's quotidian subject.

It was the maturity of Isherwood's art, evinced in his coming to terms with a period when 'the wrongs of man' were so much to the fore, that Auden had earlier extolled, in 1935, in 'August for the people', yet another poem for Isherwood, and known later as 'Birthday Poem' or 'To A Writer on His Birthday':

> So in this hour of crisis and dismay,
> What better than your strict and adult pen
> Can warn us from the colours and the consolations,
> The showy arid works, reveal
> The squalid shadow of academy and garden,
> Make action urgent and its nature clear?
> Who gives us nearer insight to resist
> The expanding fear, the savaging disaster?

'[M]aking the necessity for action more urgent and its nature more clear' was a task, Auden felt (in his introduction to *The Poet's Tongue*, the poetry anthology he edited with John Garrett in 1935), that poetry might not actually be up to performing.

If only, then, because of Auden's own high regard for the novel, and for the elevation of Isherwood, and after him Upward, on the top of the Auden pyramid, we must in a major way keep bringing novelists into our reading of the '30s. It would anyway be quite absurd to disregard as components of the '30s scene not just Isherwood and Upward, but Joyce and Beckett, Greene and Green, Aldous Huxley, Waugh, Orwell, Powell, and Durrell, even the Powys brothers, let alone the kind of prose writer specifically encouraged by Lehmann's *New Writing* from its foundation in the Spring of 1936, the 'proletarian' authors eager to swell the ranks of the period's established novelists, men like James Hanley, Walter Greenwood, and Lewis Grassic Gibbon. But also, of course, and here we arrive at the second major reason why concentrating only on the Auden clique won't do: the myth of the Auden Generation, in choosing by and large to leave out novelists, and even if it does let in Isherwood and Upward and a tiny clutch of other prose writers, is clamantly leaving out women. There weren't many notable women poets in Britain in the 1930s. Laura Riding was anyway an American, even if we rank her at the ridiculously lofty level she induced Robert Graves to afford her. Sylvia Townsend Warner and Valentine Ackland were more modest about themselves, but then their poetry is only modest. Stevie Smith is perhaps the only woman poet of the period to write constantly strong poems and even she, characteristically, found it hard to get her poems into print. But the novel, in the 1930s as in the whole period since the form established itself in Britain, was the classic medium of the woman writer. It's not at all surprising that the mediocre poet Sylvia Townsend Warner should produce numerous readable, entertaining, and serious novels. The feminist publishing imprint Virago has trawled most impressively and fruitfully in the novel catalogues of the '30s. Storm Jameson, Stevie Smith, Sylvia Townsend Warner, Rosamond Lehmann, Naomi Mitchison, Antonia White, F. Tennyson Jesse, Winifred Holtby: these resuscitated names from the Virago Modern Classics list are all most competent novelists, and some of them are much more than that. And besides these names there are all the previously well-known women novelists, like Virginia Woolf, Elizabeth Bowen, Jean Rhys, Ivy Compton-Burnett, and Dorothy Richardson, who cannot be simply left, as most books about the 1930s leave them, out of the account.

The neglected sister—why didn't we know more about Shakespeare's sister, Virginia Woolf asked, famously and influentially in *A Room of One's Own* (1928), inspiring numerous recent feminist inquiries[12]—was finding her modern voice in the '20s and '30s, and her critical and editorial brother was still trying to stifle it. After Virginia Woolf herself the period's most important fiction-writing sister was Rosamond Lehmann, inexplicably ignored in all the standard accounts of '30s writing which all dote, though, on her brother John. It was, of course, inevitable that this emergent sisterly writing should be as obsessed with male lovers and wonder-brothers as are Ms Lehmann's *Dusty Answer* (1927), *Invitation to the Waltz* (1932), and *The Weather in the Streets* (1936). After all, were not Virginia Woolf's own novels also haunted by the loss of her brother and by the social power of men, and to a degree only apt to an interim, newly emergent female writing? What mattered was the sisters' new-found audibility. What was striking was their brothers' deaf ears.

We, for our part, now, cannot rest happy at remaining among those who still do not have ears to hear. This particular book lacks the space to do full justice to anything like all of these customarily absented authors. But at least the gap that commonly denotes their absence can be defined, they can be granted the mentions and some of the respect they deserve, and their place can be marked on the '30s map for future reference.

For all of these reasons, then, the 'Auden Generation' reading of the '30s must be seriously modified, bulked out, placed in more demanding perspectives. But what, we have then to ask, of that other dominant orthodoxy about the period, the notion that the Audenaire vintage produced a strong *red* wine, that the '30s were, in the label put on them by Eugene Lyons's title, *The Red Decade*? Is the type-casting of John Heath-Stubbs' poem 'The Poet of Bray' reliable?

> Back in the dear old thirties' days
> When politics was passion
> A harmless left-wing bard was I
> And so I grew in fashion:
> Although I never really *joined*
> The Party of the Masses
> I was most awfully chummy with
> The Proletarian classes.

How correct was Leavis's dubbing this 'the Marxist decade'? How right was his wife's comment in her scathing *Scrutiny* review 'Lady Novelists and the Lower Orders' (September 1935) that 'Communism is fashionable now', or Hugh Gordon Porteus's jibe in 1933 that 'Verse will be worn longer this season and rather red'? Fresh from advocating more or less in his *New Signatures* and *New Country* anthologies that verse should be rather red, Michael Roberts quoted Porteus in his *Critique of Poetry* (1934) and agreed that there was 'some justice' in the criticism (Porteus had blamed Auden as the reddening agent). And a certain amount of redness did indeed spill across the period like a stain, or (to use George Eliot's wonderful metaphor for the red assault of the Scarlet Woman, the city of Rome, on the myopic eyes of Protestant Dorothea in *Middlemarch*) 'like a disease of the retina'. This was the period of E. A. Osborne's 1938 anthology *In Letters of Red* (its twenty contributors included Auden, Grigson, Day Lewis, MacNeice, Upward, and Rex Warner). In the early '30s, before the freezing hand of Socialist Realism shut them up in the interests of the United Front, Agit-Prop theatre troupes toured the country exuberant under names like Red Pioneers, Red Radio, Red Flag, Red Magnets, Red Front, sending back enthusiastic reports of their dramatic agitations to *The Red Stage*, official organ of the Workers' Theatre Movement. Red flags were much waved; 'The Red Flag', the song that celebrates the people's red flag, was much sung. Characteristically of the times, perhaps, a bird at the London Zoo was given to crying 'Rot Front! Rot Front!'—at least according to Louis MacNeice's delighted report in his book *Zoo* (1938). T. S. Eliot, more seriously, has the 'young', the ones 'with fairly intelligent faces' coming on in his drama *The Rock* (1934) as a troupe of Redshirts. These Redshirts are clearly meant to be the poets of the age for they speak of 'our verse': 'Our Verse / is free / as the wind on the steppes / as the love in the heart of the factory worker.' Eliot's gloomy conclusion was that the young poets he was encouraging into print were Reds to a man.

And indeed the vocabulary of leftism does sound loudly through the '30s. The Left glossary got busily expanded: to *leftism* (OED 1920) and *leftist* (1924) the '30s added *leftish* (1934) and *leftward* (1936). The magazine *Left Review* was founded in October 1934 and ran until May 1938. It editorship was firmly in the hand of Party Communists: at first Montagu Slater, Amabel Williams-Ellis, and Tom Wintringham, a team soon afforced by Alick West; later Edgell Rickword; and in the paper's final phase Randall Swingler. Contributors did include non-Communists—after all *LR*'s business was to build the literary side of the United Front. So writers such as Herbert Read (who was an Anarchist), James Hanley, Rex Warner, John Lehmann, and W. H. Auden as well as short-lived Communists like Spender, all appeared in it. Mainly, though, the editors kept *LR* for each other and for other trusties who were, like themselves, members of the Party in more or less good standing, people like Day Lewis, Upward, Charles Donnelly, Ralph Fox, Anthony Blunt, Charles Madge, Jack Lindsay, Alec Brown, Sylvia Townsend Warner, John Strachey. When, however, in 1937, *LR* published the results of its survey of attitudes towards the Spanish Civil War in the pamphlet *Authors Take Sides on the Spanish Civil War* a much broader picture of the penetration of republican and left-wing sympathies among the period's writers was presented. Out of 149 authors listed, the pamphlet managed to reveal 127 names 'FOR' the Spanish Republic. Sixteen respondents were conceded to be 'NEUTRAL?', and only five were allowed to show their hand for Franco and the Right. We know that some massaging of responses went on (some replies weren't published, for example), but the general trend was pretty clear: a huge bulk of '30s writers was willing to place itself on the left or at least to declare itself leftish. 'I am rapidly enlisting myself as one of what Nevinson calls the great "stage army of the good" who turn up at every political meeting and travel about the country giving little talks, subscribe to things, do free articles, etc': so Spender wrote to Isherwood in October 1936, the year the Spanish Civil War broke out.[13] It was a not inconsiderable army. It stood for a consensus that the existence of the Left Book Club, founded in the same year, both reflected and helped to nourish.

Within the ranks of this broad leftishness, writers were happy to acknowledge the existence of this orthodoxy. George Orwell was one (in his 1948 piece 'Writers and Leviathan'):

> Obviously, for about fifteen years past, the dominant orthodoxy, especially among the young, has been 'left'. The key words are 'progressive', 'democratic' and 'revolutionary', while the labels which you must at all costs avoid having gummed upon you are 'bourgeois', 'reactionary' and 'Fascist'. Almost everyone nowadays, even the majority of Catholics and Conservatives, is 'progressive', or at least wishes to be thought so . . . We are all of us good democrats, anti-Fascist, anti-imperialist, contemptuous of class distinctions, impervious to colour prejudice, and so on and so forth.[14]

Young leftish poets seemed oddly eager to agree with their Enemy Wyndham Lewis's charge (one that became quickly notorious) in his *Listener* article 'Freedom That Destroys Itself', that there was a left-wing orthodoxy dominating British letters:

> A repressive 'left-wing' orthodoxy has for long existed in Great Britain. Freedom to express any view except one of a 'left-wing' tendency has been, if not disallowed, so much discouraged as to make it not worth any bright boy's while to transgress.[15]

Day Lewis, one of the bright boys in question, went out of his way at least twice to state his agreement with Lewis's allegation: in his 'Writers and Morals' essay, part of his 'Revolution in Writing' sequence published with *A Time to Dance* and *Noah and the Waters* in 1936, and again in the anthology *Anatomy of Oxford* (1938) compiled with Charles Fenby ('as Wyndham Lewis has pointed out, left-wing opinions are the orthodoxy of today'). According to Spender's article 'The Left Wing Orthodoxy' (*NV*, Autumn 1938), the allegation was not to be argued with. Attendance at yet one more gathering of writers united against fascism has clinched the case:

> When Mr Wyndham Lewis writes of the Left Wing orthodoxy of contemporary writers and intellectuals, none of them—except perhaps those who call themselves non-political even when they are taking part in protests against Fascism—should quarrel with the description. As the Queen's Hall meeting addressed by famous writers showed, most well-known British writers are now aware that the whole tradition which they represent is being challenged, that writers corresponding to them in Germany and Italy are forced into exile or prison or dulled acquiescence, that most of the famous living writers of Spain are on the Black List of those who will be shot when Franco wins. Apart from the direct threat to freedom of expression, the writer is forced to realise that the liberal assumptions of progress and freedom which form the so respectable background of most bourgeois literature today, are being challenged by the violent and destructive methods of power politics. He must submit to this challenge, reconsider the moral assumptions that flow so easily into his writing, or come out with a new set of values.

And it's clear that any reading of the '30s must accept the gist of Spender's (and Wyndham Lewis's) point. The readiness of a majority of writers to rally round to support United Front actions of intellectuals in the anti-fascist cause—usually meetings (at home or abroad) or manifestoes of the *Authors Take Sides* kind, and generally instigated by Communist Party Front committees and organizations— indeed proves the point. (E. M. Forster, recalling his own readiness to attend anti-fascist gatherings of intellectuals, called this attitude 'the Conference spirit': he told John Lehmann, in a letter, 21 December 1940, that his Penguin survey *New Writing in Europe* should have made more of the phenomenon.[16]) The present account will spend much of its time defining the nature of this leftward movement among the writers. But a close reading of that typically '30s, and typically Spenderian piece of prose, also helps underline the kind of caution over this Red-Decade or Left-Orthodoxy assumption that should be voiced early on in any discussion of the period, and should not be forgotten at any stage of it. Crudely: if leftism was so orthodox, how come there were still 'those who call themselves non-political'? If everyone who counted was already on the Left, who was producing the literature with the merely 'liberal assumptions', and where was that 'most bourgeois literature today' emanating from? And if the writer is already orthodoxly left, why has he still to submit to the 'challenge' of fascism, and 'reconsider the moral assumptions that flow so easily into his writing'? Spender is, in fact, guilty here of the wish-fulfilment that continually infected the writing of '30s Leftists: who tended to slip easily from what was, to what they wished were the case, an elision eased by their confused rhetoric of a future always coming into being. It was a rhetoric of 'more and more': of more and more intellectuals forever 'coming to realize' Marx was right (Edgell Rickword, *War and Culture*, 1936); of Lenin's conditions for a revolution forever being met 'more and more' (Ralph Bates, *Lean Men*, 1935); of the writer forever seeing 'more and more clearly that his interests are bound up with those of the working class' (Michael Roberts, in his introduction to *New Country*). And the sceptic wonders how, in the end, this 'more and more' process can possibly have kept itself up: the ranks of intellectuals and writers aren't infinite; if so many more and more of them kept going over to the left side shouldn't the time have rapidly arrived when they had all made it across? But, of course, the talk of more and more was continued because it was an exaggeration: by no means all writers were chiming in with the prevailing orthodoxy or by any means all joining the Communist Party.

Alleged consensuses must always alert our suspicion, and this one no less than others, especially when it comes gift-wrapped in Spender's shifty prose. On the one side, members of Britain's tiny Communist Party—with a mere 3,200 members in 1930, rising to about 11,700 in 1936–7 with the impact of the Spanish War, mounting on a crescendo of anti-fascist zeal to 18,000 or so in 1939, before wilting away to 9,000 upon the Hitler–Stalin pact in 1940—members of this small group were fond of 'more and more' because they wished to encourage the sense of a bandwagon, and anyway whistling in the dark does keep up one's courage. And on the other side, it was in the interests of hot reactionaries to create Red Scares by playing up the strength of the Left. Much of the most lurid language of redness and leftness comes from the Right: from tracts like *The Red Network: The Communist International at Work*, published by Duckworth in 1939, and blowing as it thought the gaff on the Left Book Club, Unity Theatre, the Film and Photo League, the Spanish Dependents' Aid Committee, the Relief Committee for Victims of Fascism, Collet's Bookshops, Prospect Tours, Marx House, and the like, as 'Organisations Affiliated to or Working in Close Cooperation With the Communist Party'; or from Harry Kemp, who collaborated in one of the more animated of Laura Riding's long roster of dull and dulling literary enterprises in this period, *The Left Heresy in Literature and Life* (1939); and especially from Wyndham Lewis and his torrent of lively and nonsensical articles and books—titles like *Left Wings Over Europe* (1936) and *Count Your Dead: They Are Alive* (1937). According to Lewis, Britain is almost altogether a Red place:

> 'Why are you English all so Red?' the poet, Mr Roy Campbell, who was in Toledo at the time of the first attack of the Madrid militia, was asked on all sides, for months. What Mr Campbell replied I do not know. He could have said: 'Because we are kept in ignorance of the true position'.

The BBC is, it turns out according to Lewis, if not quite Red, then pink; so are the Gaumont-British newsreels. The Press as a whole goes in for 'democratic bulldozing'. All this was news to Leftists. Lewis's footing was surer, perhaps, when he reported (in *The Hitler Cult*) that 'Red spittle' had covered the window of Zwemmer's Bookshop in the Charing Cross Road when his *Hitler* (1931) was displayed there. But fiction was even more convincing—like Lewis's poking fun in *The Revenge for Love* (1937) at the straw-man Percy Hardcaster, darling of the Salon Reds who are a nasty mob of indignant public-school boys (among them a 'pinkfaced and pinkminded, pugdog-like client of the Left from Brasenose'), readers all of Left Book Club stuff ('Red Dope for Leftie School-teachers').

Grimly joyous Red-baiting is not, of course, altogether absent from the pages of actual or near-Leftists. Connolly's imaginary leftist autobiography 'From Oscar to Stalin: A Progress', starring Brian Howard as Christian de Clavering the Oxford Red (he's given voice in a mock review of his book entitled 'Where Engels Fears to Tread'), is more sharply funny than anything Wyndham Lewis ever managed. And Spender's 'The Left-Wing Orthodoxy' has a wonderful paragraph which might have come from a Wyndham Lewis novel with better manners than any of the ones Lewis actually produced. At the last meeting of the Writers' Association for Intellectual Liberty that Spender attended, an organized discussion between Goronwy Rees and Day Lewis,

> there was very lively play put up by Mr John Brophy, who rather unsportingly attacked a speaker for not having carried sacks about at any period of his career, as Mr Brophy himself had done. Unfortunately for Mr Brophy, the speaker was able to trump this card, by proving that he had sprung from the working classes and would at this moment be a miner but for the extenuating circumstance (perhaps to be regretted after all!) of a scholarship at Oxford. Mr Brophy then played his Ace: he had brought a Worker to the meeting with him, who, it was assumed (if silence means consent), thoroughly approved of everything Mr Brophy had said. At this moment, another Worker, who had slipped in unawares (a *real* worker with a *real* accent) got up and said that anyhow the workers didn't want proletarian literature: what they wanted to read about was the love affairs of the upper classes. Confusion.

But Wyndham Lewis's zealously unflagging hundreds of pages, like his interest in Red spittle, betray intensities that are leading him to exaggeration. Of course many writers actually joined the Communist Party. Of course there was a huge leftist consensus. Scores of anti-fascist signatures could always be obtained at the drop of an envelope. Marxist discussion was fashionable: even *Twentieth Century Verse*, cannily scathing of consensuses, particularly Communist Party ones, could print in its June/July 1939 number a toughly Marxist piece like Roy Fuller's 'The Audience and Politics'. Notoriously, and typically, *New Writing*'s so-called non-partisan editorial policy was in practice leftist: the magazine was 'independent of any political party', but did 'not intend to open its pages to writers of reactionary or Fascist sentiments'. Even Virginia Woolf appeared as a contributor in the pages of the *Daily Worker*. Even Geoffrey Grigson professed to believe (in his *Criterion* review of *A Hope for Poetry* in January 1935), that 'all of us' had become 'left-handed'. Of course. Of course.

But in practice, if Grigson was left-handed another sort of hand kept showing itself in his *New Verse* pieces. Even Wyndham Lewis (in a letter to the *Observer*, 2 February 1936) was ready to exculpate Grigson as 'one of the few critics, among the young, who have not been chloroformed by salvationist politics. It is, after all not his fault, that, except for MacNeice, all his swans are *red*'. And MacNeice wasn't all. Another *NV* swan was Dylan Thomas, and he, Julian Symons was happy to proclaim in the first number of *TCV*, was 'not a Pylons-Pitworks-Pansy poet'.[17] There's no message from Dylan Thomas in *Authors Take Sides on the Spanish Civil War*. As soon, in fact, as one starts unpacking the left orthodoxy like this the caveats, exemptions and exceptions start to multiply. 'Engagement', Richard Hughes said in 1961, was 'the problem which so bedevilled the poets of the thirties'. His own solution? 'Well . . . *I* have kept carefully out of politics.'[18] And William Golding represents other '30s writers beside himself who were out of tune with what are commonly supposed the appropriate attitudes of a '30s author when he affirms his contemporary unease. 'I had no interest in politics, none in the USSR, none whatever in tractors. I felt the whole generous movement was wrong but knew that I could not be right. One solitary adolescent! And yet ————.'[19]

One notices that Eliot's observation of 'sympathy with Communism' on the part of the 'younger people with whom I talk' (*C*, April 1932) is soon followed in that paper's January 1933 number by his report from the USA that 'communistic theories appear to have more vogue among men of letters' there 'than they have yet reached in England'. 'Communist' was anyway a tag sometimes too promptly stuck on. Auden's poem 'A Communist to Others' is a clear case in point: Auden quickly dropped the title. Harold Nicolson was simply too slickly misconceiving to mix Auden up with communism, Day Lewis too glib in *A Hope for Poetry* with his talk of 'Communism' in his references to *New Signatures* and *New Country* ('definitely Communist forms by Auden, Charles Madge, R. E. Warner and others': elsewhere he speaks of 'such definitely revolutionary English writers as R. E. Warner'). When it came actually to joining parties (any party) there was great caution among the period's writers. *New Verse*'s 'Enquiry' of poets (results published October 1934), which included the question 'Do you take your stand with any political or politico-economic party or creed?', got some strongly negative replies. Eliot, Spender, Madge, Auden, Day Lewis all refused to answer (though Day Lewis said that 'several of the questions' were 'dealt with' in *A Hope for Poetry*). Of actual respondents, only Hugh MacDiarmid admitted he was in the Communist Party. MacNeice had no party ('In weaker moments I wish I could'); nor had Dylan Thomas and David Gascoyne, though they used the word 'revolutionary'; nor had Gavin Ewart, though he said he believed in 'Communism'. In fact, the most powerful single party bloc was formed by MacDiarmid's and Edwin Muir's support for Major C. H. Douglas's policies of Social Credit (with Eliot and Pound, as well as others on its side, Social Credit might be thought of as at least one orthodoxy of the '30s).

In the flurry of Caudwells, Day Lewises, Spenders, Upwards, and Madges who did at some stage join the Communist Party it's easy to forget that many important young '30s writers, including the most obvious of potential recruits, never actually signed up. A group of refuseniks that has Auden, MacNeice, Isherwood, Orwell, and Dylan Thomas in it, to name no others, cannot be sniffed at. Furthermore, there was in fact a slight drift away from the Party going on at this time. Graham Greene and Michael Roberts had been members in the '20s. Both left it, to move gingerly towards Christianity: Greene into Catholicism in the later '20s, Michael Roberts into Anglicanism in the later '30s.

John Strachey had announced in his influential polemic *The Coming Struggle for Power* (1932), mounting a breezy inspection of Waugh's *Decline and Fall* and *Vile Bodies*, that 'After writing these books, Mr Waugh had clearly only three alternatives open to him. He could either commit suicide, become a communist, or immure himself within the Roman Catholic Church. He chose this last (and easiest) alternative'. Too easy though this last may have seemed to observers such as Strachey, something like it was a widely taken option. T. S. Eliot and C. S. Lewis became Anglicans in the '20s; Charles Williams was already one. Waugh became a Roman Catholic in 1930, Roy Campbell converted some time later, J. R. R. Tolkien was already one. Aldous Huxley joined no actual church but upset the Leftists not only by his pacificism but by daring to talk about prayer in his fiction and in his pamphlet of 1936, *What Are You Going to Do About It*? And turning to religion commonly meant at this time a refutation of communism, a turning to the Right. Not always, of course: Greene was no Fascist; Eric Gill the Catholic artist was a keen supporter of the anti-Franco cause; so also was the Catholic writer Kate O'Brien. None the less, Kate O'Brien's *Farewell Spain* (1937) has the Republican slogan *No Pasarán* jostling the Francoist cry *Arriba, España!* Waugh is one of the few writers utterly 'AGAINST' the Spanish Republic in *Authors Take Sides*. Graham Greene equivocated enough not to reply. Michael Roberts refused to reply.

And there's no doubt where traditional literary power lay. The ranks of the '30s CP and the pages of *LR* tend to be thronged by authors who in world terms pack very little punch, however occasionally interesting they may be. There is only one Spender, one Caudwell, one Day Lewis, against a whole slate of Slaters, Foxes, Swinglers, and Amabel Williams-Ellises. The Christians were at least going over to the side of T. S. Eliot. In doing so they laid themselves open to the charge of thus also siding politically with Yeats, Wyndham Lewis, and Pound. But the reactionaries did enjoy the considerable attraction of including the period's biggest literary guns. Against them the literary Left seems by contrast often to be armed only with pea-shooters.

Orthodoxies impose unities. And perhaps the most important cautionary note to sound about the '30s is one against the too ready professions and appearances of unity. If we think of the '30s as a seamless political whole we are grossly distorting them. Even on the Left there was great disunity. The United Front was a seamed patchwork of revolutionaries, old Liberals, young liberals, pacifists, Trotskyites, Stalinists, members of the Communist, the Labour, and the Independent Labour Parties, as well as members of no party at all. To call all the Front's sympathizers Red would be exaggerating mightily. Even pink—and Arthur Calder-Marshall was prepared famously in the *NS* of 15 February 1941 to call the '30s 'The Pink Decade'—seems a mite too flushed a shade for many of them. And within the literary United Front, if that's how we may think of the sphere commanded by the Left Book Club and other associations for thinking and creative Leftist people, the same mixed shades and ragged divisions obtained. It was possible, to say the least, simply to dislike Auden's stuff ('The Auden's and Day Lewis's and so on are a positive menace', Humphrey Jennings wrote to Julian Trevelyan in 1933[20]). And politics exacerbated the usual proneness to such differences of opinion and divisive dealings within the cultural world. Some writers possessed visas allowing them the freedom of several frontiers, but in general the several coteries, the different cliques, the separate centres were quite clearly marked off from each other. The *Left Review* lot sniped at the *New Verse* gang, and the *NV* troops were very happy to keep shooting back at the *LR* people and at the *New Statesman* people and the *Criterion* people, indeed just at people, firing smartly from the hip at virtually everyone outside a tiny band of paid-up faithfuls. From time to time the *NV* loyalists would proceed to shoot each other up: in the end only Auden, and then only his poems and not his plays, escaped unpeppered. Grigson at *New Verse*, the Leavises in *Scrutiny*, Symons at *Twentieth Century Verse*: these editors were terribly hard to please. Certainly there were few broad churches among the little magazines. Or even the magazines. The shortlived *Night and Day*, London's snappy version of the *New Yorker*, whose literary editor was Graham Greene, was one attempted amalgam of disparate bedfellows. But it ran only for a handful of months, from 1 July 1937 to 23 December of the same year. To T. S. Eliot's great credit, the *Criterion* was another such open house. These were about all.

Critical toughness was the vogue. So much so that the closer a band of brothers might perhaps be thought of as likely to be, the more sharp the mutual knuckles-rapping that went on. *Scrutiny* soon lost faith with Eliot—one of Leavis's major critical influences—and with Auden—one of Leavis's earliest contributors. One wouldn't expect Geoffrey Grigson to accept a brief for George Barker ('huddles of verbiage and ignorance') or for David Gascoyne ('this self-interested cackle of the literary pullets'), even though he published their poems; one needn't want him to have liked the 'Old Jane' Edith Sitwell, nor Edmund Blunden ('the Merton field-mouse'), nor lots of the other writers he savaged, but he might have been expected to show kindness to his earliest contributors and to Auden's friends. But no, he takes especial delight in June 1935 in panning Spender's poems—'invertebrate sponginess—of sponge cake rather than bath sponge'—and in passing on someone's rumoured wisecrack (Norman Cameron's in fact) that 'Spender is the Rupert Brooke of the Depression'. In 1936 Grigson was sneering that Day Lewis was 'not even truth's pimple squeezer now'. December 1934 found Grigson in no mood for United poetic Fronts, not even with Auden's favourite chums:

> some believe that poets should keep together, show a united front and backside, and
> never attack each other in the *Spectator*, the *Listener*, and other places where they and
> their friends write, no matter what banalities they may have committed. In this sense
> *New Verse* must publish a decree absolute of divorce from any poet, new or bearded.
> Mr Spender has published a bad poem 'Vienna', Mr Day Lewis has written a bad book,
> 'A Hope for Poetry'. . . .

'Compromise', declared Grigson in January 1938, 'is a dangerous whore with seven-and-elevenpenny silk stockings', and 'A "Popular Front", like muscling-in, is a shadow in which all sorts of toad stools can grow up'. As for love on the Left, Grigson knew that 'The Berts of the *Left Review* will certainly shoot the Cyrils and the Raymonds of the *New Statesman*. We may clap at that: it will be small loss. But as the years pass they will shoot the Audens.' *LR*'s editors published several articles by Spender, but its editors and acolytes weren't going to wait for the revolution before they destroyed his ideas. In August 1935 Rickword went after *The Destructive Element* like a badly brought up guard dog—it was sincere, but also unclear, liberal, empiricist, naïve, and misconceived. In September 1937 Douglas Garman devoted most of his rough notice of *Fact* magazine's *Writing in Revolt* issue to Spender's contribution precisely, as he says, because 'He writes consciously as a Marxist'. It scarcely needs a reading of Rickword's hard-handed way with Philip Henderson's Marxizing *Literature* or with Party member Day Lewis's *Noah and the Waters* ('Who is this Noah?' Rickword scathed) to prove that the leftist Cains just couldn't keep their killing hands off the leftist Abels. Dog must, it seems, bite dog.

One could go on: to note the savagery of the Left towards leftist Orwell's *Homage to Catalonia*, or, closer to the Auden group, to observe MacNeice's public distaste for some of the Auden circle's stuff. In *I Crossed the Minch* (1938) he confided how he fell asleep in a train reading Day Lewis's gathering of leftist essays *The Mind in Chains* (1937). In the *Spectator* in the same year he was very critical of Auden and Isherwood's latest drama *On the Frontier* ('The mystical love scenes of Eric and Anna make one long for a sack to put one's head in'). So also, but anonymously in the *Listener*, was Day Lewis (*On the Frontier* 'possesses neither the vitality and invention of *The Dog Beneath the Skin* nor the deeply realized moral conflict of *The Ascent of F6*'). In 1941, incited by Spender's rough handling of the volume *Another Time*, Auden broke out in protest: 'Your passion for public criticism of your friends has always seemed to me a little odd.' Of course it was morally and critically to the good that the Group was ready to put critical standards above friendships. Better that than Isherwood's reiterated and increasingly meaningless praise in print for everything Upward published. And, clearly, the unity of the Group was getting more and more strained as the years passed. But more is involved than the mere passage of time. These tensions and differences of opinion must alert us to real distinctions and differences that existed all through the decade among these and other '30s authors. The necessary attempt to view things from up literary history's Eiffel Tower makes one try to impose unities and discern connections, to define a set of '30s tones which link authors and texts, as well as to mark out separated segments, regions, dialects on the '30s map. But caution is necessary too. It dictates attention to the kind of criticism Grigson made near the start of the period, in only the second issue of *NV* in March 1933, in response to the *New Country* anthology, Michael Roberts's second attempt to stage-manage a unified '30s literary presence:

> I condemn in this . . . book its union clamping disunion. . . . What joins these writers except paper? How, as an artist, is Auden united with Day Lewis, Day Lewis with Spender, Spender with Upward? How are any of these four linked to Michael Roberts, the editor? Roberts in a long preface 'usses' and 'ours' as though he were G.O.C. a new Salvation Army or a cardinal presiding over a Propaganda.

> Spender's article, Auden's poems and Day Lewis's 'Magnetic Mountain' prove it stupid to keep in fancy these three as triune. The three are distinct. Auden's system is being created by Auden. Spender is far from the others . . .

And so on. Extreme allegations, but containing enough wisdom to serve at least as cautionary notes held instructively in the back of our minds.

3

Destructive Elements

LIKE any other decade the '30s was a period of mixed emotions, mixed tones. This book has much to say about frivolity, jokiness, casualness, flash, energy, and exuberance. But early in any approach to the period must come an awareness of the multiplied fears and forebodings and the widespread sadness that make a constant background and foreground. Whatever style was learned or learnable, it came (in Empson's formula) 'from a despair'. 'I am writing a v sad book re Abyssinia', Evelyn Waugh wrote in April 1936.[21] And very many '30s texts, not just Waugh's, are strikingly 'v sad' ones, written out of the sort of despondent sense Spender was enduring in August 1939 that 'Being an artist . . . is just a kind of disease of suffering'.[22]

The potential grimness of his times had struck Waugh early. At the beginning of his first book, the biography *Rossetti: His Life and Works* (1928), the heavy-heartedly macabre note of all Waugh's writing sounds. The Lytton Strachey kind of biography has started a fashion for making the corpses of eminent Victorians dance merrily but, Waugh implies, the jest is already draining out of that sickly gruesome jokiness. 'We have discovered a jollier way [than old-fashioned biographers] of honouring our dead. The corpse has become the marionette. With bells on its fingers and wires on its toes it is jigged about to a "period dance" of our own piping; and who is not amused? Unfortunately, there is singularly little fun to be got out of Rossetti.' And Waugh's book ends with Hall Caine, like Waugh a 'young man' and novelist-to-be, discovering the horrible possibilities of life and art in the shape of Rossetti's melancholia, paranoia, suicidal tendencies, physical unfitness, Oblomovic inertia, and addiction to whisky and chloral. It was no joke for Hall Caine, not even a black one. And before long Waugh found himself heavy-laden with a perception of the same nauseous gruesomeness pervading life and art in his own times that he and Hall Caine had been initiated into in their opening the lid of Rossetti's troubled existence. 'The politics of Geneva, Rome, London, Paris horrify me', Waugh wrote from Jerusalem in 1935: 'They have a wall here where the Jews blub. V. sensible idea.'[23] And however determinedly loud the sound of fun, of joking, coming from some '30s texts—whether, for example, it's Auden's jesting or Waugh's own or Anthony Powell's or Rose Macaulay's (she dedicated her novel *Going Abroad* (1934) to friends who 'desired a novel of unredeemed levity') or whoever's—the sound of desperate 'blubbing', of the people wailing at the wall of their generalized distress, of the literary and politics prophets weeping for the awfulness of their vision, is never entirely quenched or absent. Frequently it drowns out everything but despair entirely.

At every phase of the 1930s there was good reason for grimly sensing and declaring a crisis, things going wrong, something up. 'The Crisis Hour is Here', blared the *Daily Worker* (20 October 1936), as it had cause of some sort to do throughout the decade. Observers of every stripe agreed: Oswald Mosley, Alec Waugh, Storm Jameson, Alick West, Martin Turnell, Christopher Caudwell, Allen

Hutt, Hilaire Belloc, John Strachey, Charles Madge, and Tom Harrisson.[24] You did not have to be a Marxist to perceive crisis. Hilaire Belloc and Martin Turnell weren't Marxists; nor was F. R. Leavis who set out to describe 'the crisis' of culture in *Mass Civilization and Minority Culture* (1930); nor T. S. Eliot who in the *Criterion* (October 1932) ascribed 'the problems of our time' to the *crise dans l'homme*. Ex-communist but well on the way to being a Christian at the time, Michael Roberts fell naturally into describing in his Introduction to *The Faber Book of Modern Verse* 'a crisis of a general kind' affecting poets: a crisis comprising theology, politics, aesthetics, 'a fractured personality or a decaying society', 'academic philosophy' and 'the deficiencies of language'. Marxism did help, though, to stoke one's sense of crisis. When in 1929 'the final economic crisis of capitalism' had begun, 'all bourgeois culture' had entered the 'throes of its final crisis': Caudwell's vision of '*permanent* crisis', set out emphatically in his *Illusion and Reality* (1937) and *Studies in a Dying Culture* (1938), was the orthodox Marxist doctrine shared widely on the Left. Caudwell was equipped intellectually to carry his sense of crisis further than most. His *Crisis in Physics* attempts, boldly if messily, to connect the loss of epistemological confidence in science, centred in Heisenberg's Uncertainty Principle, with the threat to bourgeois economics. Ralph Fox and Alick West knew less about science (Caudwell was a trained aeronautical engineer) but they could assert the results of the economic crisis in their own fields. West believes modern literary criticism has simply collapsed. Fox's *The Novel and the People* (1937) finds a 'crisis of quality' in the novel.

Many who were not Communists were prepared to agree with Caudwell and other Marxists that 'the crisis' was final, permanent, and general. The precise locale of the current manifestations of the crisis did, however, shift as the '30s went on. At first, as in the wake of 1929 British industry collapsed and unemployment ballooned, the crisis was felt to be largely economic. And the economic note is the one sounded early on in the depression landscapes of the early Auden poems—like 'Get there if you can', written in April 1930:

> Get there if you can and see the land you once were proud to own
> Though the roads have almost vanished and the expresses never run:
>
> Smokeless chimneys, damaged bridges, rotting wharves and choked canals,
> Tramlines buckled, smashed trucks lying on their side across the rails;
>
> Power-stations locked, deserted, since they drew the boiler fires;
> Pylons fallen or subsiding, trailing dead high-tension wires;
>
> Head-gears gaunt on grass-grown pit-banks, seams abandoned years ago;
> Drop a stone and listen for its splash in flooded dark below.
>
> Squeeze into the works through broken windows or through damp-sprung doors;
> See the rotted shafting, see holes gaping in the upper floors.

This adroit assimilation of more recent slump images to the Industrial Revolution's debris that had long littered the North of England and had fascinated Auden since he was a schoolboy and that had been one of his earliest themes (his 'shut gates of works' syndrome), was to continue. He kept it up in *The Orators* ('Systems run to a standstill, or like those ship-cranes along Clydebank, which have done nothing all this year'). As late as 1935, in *The Dog Beneath the Skin*, when 'the dynamos and turbines' and 'the Diesel engines like howdahed elephants' have started up again, and 'the Power House' is once more giving 'Power to the city', there is a hint of the earlier shut-downs in the 'locked sheds' that are made visible by moonlight. And the earlier industrial desolation is just glimpsable lurking about amidst the by now ascendant images of moral and spiritual decay, of Biblical Fall, Sin, judgement and Satanism in *The Ascent of F6* (1937): in the opening chorus of Act II scene v, among the toads, dock and darnel, the weasel, the wasting Dragon ('with the blast of his nostrils'), and the deliverer/destroyer on a white horse, are some smokeless chimneys, rusting implements, and 'tall constructions' that have been knocked down.

The old industrial regions of the country did, of course, stay depressed. South Wales, the old industrial north, the old industrial parts of Scotland, tended not to share in the new prosperity of the mid-'30s that boomed the places that became landmarks of the mid-'30s imagination—particularly the centres of the motor trade and other new luxury goods such as wireless sets, gramophones, and vacuum cleaners, places like Coventry, Cowley, Dagenham, Slough, and other locations west of London. J. B. Priestley began his *English Journey* (1934) on the Great West Road where 'the new industries' were—'little luxury trades', 'all glass and chromium plate and nice painted signs and coloured lights'—but he was right to 'feel there's a catch In it somewhere' and to give the impression that unemployment was rather widely diffused elsewhere in Britain. The 'catch' he detected was a regional one. Like *English Journey*, Orwell's *The Road to Wigan Pier* (1937), about the impoverished north, and James Hanley's *Grey Children: A Study in Humbug and Misery* (1935), a journey into the misery of industrialized South Wales, are also perfectly correct in suggesting that in particular regions of the country severe depression continued. For some people almost the whole decade amounted, in the phrase the National Unemployed Workers' leader Wal Hannington gave his 1940 account of the decade, to *Ten Lean Years*.

The Road to Wigan Pier's impressive regionalism can, though, be misread as a general case. Certainly it appears to be all the mainline '30s literature some historians have read who have recently taken to 'revising' what they appear to believe was a fairly uniform *Ten Lean Years* stereotype.[25] Some writers did, of course, carry on as if the economic crisis didn't much shift its scope. Those depressed bourgeois with whom Orwell closes his book, 'the private schoolmaster, the half-starved freelance journalist, the colonel's spinster daughter with £75 a year, the jobless Cambridge graduate', crop up again in private-schoolmaster Day Lewis and Charles Fenby's *Anatomy of Oxford* (1938), where an undergraduate laments that his 'expensive education' will probably be wasted and that he'll have to tout vacuum cleaners: the whole 'stage is about to collapse beneath our feet'. Some of that undergraduate's sense of crisis is because of impending war, but its economic aspects are uppermost. Which is odd, for Day Lewis had himself shown already that he knew perfectly well about our revisionist historians' much heralded economic boom. 'Already', he wrote in 1934 in *A Hope for Poetry*, 'as the slump shows signs in England of another feverish rise to another temporary boom, we note a slackening of Communist enthusiasm in poetry.' And he wasn't only chiming in with the more comfortable sections of Priestley's *English Journey*. Several central '30s authors acknowledged the boom. Spender's poem XXXII in his *Poems* (1933) begins 'From all these events, from the slump, from the war, from the boom . . .'. Discussing the European 'crisis' in his *Forward From Liberalism* (1935) Spender acknowledges that there's been an economic upturn in England, based in armaments. A year later, James Barke's novel *Major Operation* more accurately ascribed the cheerfully post-depression air of Glasgow ('BUSINESS PICKS UP IN THE SECOND CITY') to the flourishing of the luxury trades, wireless-making and such, as well as to armaments. Stuart Legg's script in the documentary film *Today We Live* (1936), narrated by Paul Rotha, described how 'Today depression is giving way to boom. Many men who were out of work have been reinstated.' The limited regionality of the new prosperity is rightly insisted on in this script: 'the crisis has left its mark. Some of the old industrial areas in the North, Scotland and Wales . . . now found themselves without a share in the revival of work.' But still there is a boom. So that Auden is only one among several '30s observers who is informed about the economic pick-up that's going on in some regions. As he tells Lord Byron: 'The prospects for the future aren't alluring; / No one believes Prosperity enduring', but still 'I read that there's a boomlet on in Birmingham.'

And as employment rose, as some houses, at least, filled with gramophones and radios and the new arterial roads throbbed with some of the people's cars, the headlines' sense of crisis got refocused: on to European affairs, at whose heart was the rampant rise of fascism, signalled most powerfully by the coming to power in 1933 of Adolf Hitler. And in world politics there would be no ready or regional abatement of crisis. Lighter variations on the crisis theme would be heard, such as the 'Simpson Crisis' that ended in the abdication of the uncrowned king Edward VIII at the end of 1936 ('a great delight to everyone', Waugh noted in his diary, 'There can seldom have been an event that has caused so much general delight and so little pain': but then Waugh was unlikely to sympathize with a strong popular undercurrent of unrest against the engineered triumph of a Conservative establishment and Church). But Mrs Simpson aside, the crisis theme was persistently deadly: rumours of war and actual wars, a new Black Death carried by planes and tanks, the rabies of modern mechanized warfare, were encroaching closer to England. The *cordon sanitaire* which the League of Nations was supposed to provide looked less and less a fence strong against infection. The smallest aeroplane could hop over the English Channel in minutes. Collective security, the wartime Allies, seemed hopelessly weak to stem the march of trouble. Japanese, Italian, and German militarism throve in Manchuria, in Abyssinia, in Europe. When Hitler pushed his troops into the demilitarized Rhineland on 7 March 1936 he proved that frontiers were a pushover to the Truly Strong Man, and got ready to push over some more. In February 1934 there was a fascist *putsch* in Austria. In July 1936 right-wing generals took arms against Spain's democratically elected leftist coalition government and began a civil war that, reinforced by Italian and German soldiers, planes, and pilots on the one side and by Soviet weapons and cosmopolitan volunteers on the other, was to last until nearly the last outbreak of fighting in the decade, the one that initiated the Second World War. And fascist violence didn't remain a cross-Channel affair. On Sunday, 4 October 1936 Sir Oswald Mosley attempted to strut his private army of Blackshirts—'decent, clean-living young men—some of them Public school products', according to the *Morning Post*—through London's Jewish regions and was only thwarted by collected Leftists ('germ-carriers of the Revolution', thought the *Morning Post*; a communist 'Organization of Terror', declared William Joyce, the future Lord Haw-Haw of the traitorous broadcasts from Hitler's Germany, in

Mosley's new *Action*). Mosley's clean-living young men, long renowned for their decent way with protesters at their monster fascist rallies, ran through the Mile End Road the following weekend smashing the windows of Jewish shops and looting their goods. Fascism, it seemed, was prospering at home: as Alan and Francis discover when they get back from their European journey in *The Dog Beneath the Skin* to discover 'The Lads of Pressan', a rightist Boys' Brigade which will 'teach Britain a lesson' in 'times of national crisis', happily ensconced in the vicarage garden.

It took little imagination to see the Second World War coming from a long way off. And it's quite mistaken of the otherwise excellent Martin Ceadel to suggest in his 1980 article 'Popular Fiction and the Next War'[26] that fears of impending war were limited to minor writers of the period. E. H. Carr entitled his 1939 study of international affairs *The Twenty Years' Crisis 1919–1939*, and the force of that notion had by 1939 been for most writers a long habituated feeling. Looking at events in countries like Germany and Austria that they had come in their travels to know intimately, in some cases being on the spot as troubles burst out, the writers had grown increasingly scared. Isherwood would report back to his friends at home the mounting horrors of Nazism. *Mr Norris Changes Trains* (1935) mingles a serious documentation of German Fascism in with its author's youthful gaieties and insouciance. The parts of what later became *Goodbye to Berlin* (1939) that appeared in *New Writing*—'The Nowaks' in the first number, Spring 1936; the first 'Berlin Diary' section in No. 3, Spring 1937; 'The Landauers', in No. 5, Spring 1938—spread their steady and unsensational reports of troubling German truths over an extended portion of the later '30s. Few intellectuals remained unalarmed. We're accustomed to regard the Spanish Civil War as a trauma that most '30s writers were put through. But the trouble was far wider than that. The whole of Europe, not to mention China, had become a nightmare. All Europe was, in Auden's words, 'troubled', it was 'crooked':

> sombre the sixteen skies of Europe
> And the Danube flood.

The particular 'hour of crisis and dismay' that Auden writes about to Isherwood was certainly not, by then, in August 1935, just made out of the Depression and capitalism's economic problems; economics had given way to more momentous dangers still, and ones encompassing the whole world. Writing his verse 'Commentary' for *Journey to a War* at the end of 1938, Auden finds that England, 'melancholy Hungary, and clever France', Madrid, Shanghai, even perhaps 'absolutely free America', are 'Now in the clutch of crisis and the bloody hour'. Nor was 1938 to be quite the end. The Munich Crisis of 1938 did, though, mightily multiply feelings of impending terror, providing the atmosphere in which Virginia Woolf and her Jewish husband Leonard would seriously discuss the necessity of suicide in the case of enemy invasion. On 27 September 1938, listening to Chamberlain on the wireless, David Gascoyne's dread overflowed into his diary:

> ... the tremendous perspectives of horror and desolation, worthy of the biblical language of the Psalms or of Dante, which have opened up before us: the no less sinister scenes that are actually taking place today, children trying-on gas masks, the hasty digging of bomb-shelters in the parks of London, soldiers disappearing into the bowels of the Maginot line, countless anguished faces of people listening hour by hour to their

wireless-sets. Are the scenes of H. G. Well's [*sic*] prophetic film [*Things to Come* (1935), based on Wells's *The Shape of Things to Come* (1933)] going to materialize so soon,— perhaps even next week? The spectacle of dense ranks of bombers zooming across the sky, of famous buildings tottering in flame and smoke, of masked stampeding crowds, the sinking of giant liners . . .[27]

At the very same time Virginia and Leonard Woolf 'sat and discussed the inevitable end of civilization'. Leonard couldn't sleep. He went out to 'Walk the streets'. 'So we clasped hands, as I understood for the last time.'[28] In 1938 the crisis was for England momentarily averted. But when the immediacy of war was once again most apparent, at the end of August 1939, Gascoyne's old tensions were renewed almost unbearably:

Beastly, bloody nightmare of a world, Another crisis. We're in for it now.
What's the use? We might as well all be struck off the register . . .

Either war within a week, or a miracle. Worse situation than last year. The world seems hypnotised—*Zero hour*.

And then, on 3 September itself:

Today . . . is the day on which the outbreak of the *Last War* has occurred.

The *last* war. The end, anticipated for so long, postponed in 1938, had actually arrived.

Every important feeling and key motif of a period enjoys its disbelievers, and there were some sceptics about the idea of Apocalypse Now. Empson's jingle 'Just a Smack at Auden' commands a kind of respect for its resistance to going apocalyptic:

What was said by Marx, boys, what did he perpend?
No good being sparks, boys, waiting for the end.
Treason of the clerks, boys, curtains that descend,
Lights becoming dark, boys, waiting for the end

(even though Empson's determined flippancy never entirely masks his own preoccupied apprehensions: 'What we had before us to write about, in the years when these poems were written', he admits in the Notes to his volume *The Gathering Storm* (1940) 'was chiefly the gathering storm of the present war'). Noel Coward's anti-Reds-and-Pinks rhyme, 'There are Bad Times Just Around the Corner', is catchy but, pandering as it does to the futile optimists, the seekers of diversion in the bourgeois amusement arcade and appeasers of all sorts, contemptibly misreading:

With a scowl and a frown
We'll keep our peckers down,
And prepare for depression and doom and dread,
We're going to unpack our troubles from our old kit bag
And wait until we drop down dead.

The voice of Lewis Grassic Gibbon—who did, though, die in 1935 when the decade of depression and doom and dread was only halfway through—is more heedable in its stern refusal to go all the way with communist crisis analysis. The British Section of

Writers' International had published its manifesto in the new *Left Review* in December 1934: 'There is a crisis of ideas in the capitalist world today not less considerable than the crisis in economics', its published Aims had said. To which Grassic Gibbon retorted (in the February 1935 issue) that although 'capitalist economics have reached the verge of collapse', 'capitalist literature, whether we like it or not, is not in decay'; 'To say that the period from 1913 to 1934 is a decadent period is just, if I may say so, bolshevik blah.' And he was right to advert to the Marxist promptness in sliding the analysis across from one kind of crisis to another, from economics, to the imagination, to the threat of war, without once letting the argument falter, as if it didn't matter what kind of crisis was at hand as long as there was at least one around to prove the overall analysis of capitalism's decay resoundingly correct. The Communists needed crises, and ever more crises, to keep their Party alive, just as Jehovah's Witnesses continually need the end of the world. Real crisis or not, they'd always be crying one up. So that when at last the big one actually struck home in April 1938 *Left Review* sounded as though it had been caught crying wolf: it had perhaps prematurely drained 'crisis' of the momentousness the allegation really did need now to be freighted with:

> Within a month the word Crisis has flared with very real meaning across Europe. Austria has been invaded and annexed, Czechoslovakia threatened, Lithuania has had to capitulate to Poland . . . The agony of Barcelona has set a new limit of horror to the record of fascist barbarity. Before this appears, a new invasion may have been attempted and war be blazing over Central Europe. This is the climax . . .

A few sceptics (and grounds for scepticism) aside, however, agreement about apocalypse was fairly general. When the Crystal Palace went up in flames at the end of November 1936 it was widely taken as an apt emblem for the times: a sort of Reichstag Fire for Britain, mirror image of what was happening to Madrid. It would be looked back on as an accurate augury of many blazing cities to come: Warsaw, London, Coventry, Plymouth, Dresden. But one man's apocalypse . . .: as William Plomer's edgily satirical poem 'Headline History' has it:

> Alleged Last Trump Blown Yesterday;
> Traffic Drowns Call to Quick and Dead;
> Cup Tie Crowd sees Heavens Ope;
> 'Not End of World', says Well-Known Red.[29]

It was a point Grigson was making in *New Verse*:

> Every active maker or sharer of an attitude is now obsessed with the quick arrival of the Doomsday of human culture. Things are all threatened, Mr Auden and the Archbishop of Canterbury would agree, but the Established Archbishop reads Berdyaev, deplores Communism, dreads a shrinking of Christendom, and puts his hope in God, while Mr Auden reads Lenin, deplores Fascism, dreads attacks on Communism, and puts his hope in Marx, Freud . . .[30]

The '30s apocalyptic, like the Christian apocalypse from which its characteristic language of destruction by fire and standing trial, facing the final test, and so on, is drawn, could be cut in several possible ways. The Liberal judge in Spender's *Trial of a Judge* (1938) looks forward with a pacifist's, a victim's horror:

Then let them turn their faces to a future
Of solemn words broken by rule,
Of spiritual words burned up with libraries,
And the triumph of injustice;
Of tyrants who send their messages of terror
Against the civilized and helpless.
O let them witness
That my fate is the angel of their fate,
The angel of Europe,
And the spirit of Europe destroyed with my defeat.

These fears are part of what Spender understands by his suggestion in *The Destructive Element* (1935) that the '30s artist inhabits a 'dream of violence'. It's the kind of nightmare that Lehmann gives to the writer Peter Rains in Vienna in his novel *Evil Was Abroad* (1938), a surrealistic collage of aeroplane noises, loudspeakers at the airport blaring 'To Berlin!' as the Chief Conspirator goes for German help, machine-guns firing on workers' tenements, house-fronts closing in on the crowd—a slightly hyped-up version in fact of the historical horrors of the Austrian experience that preoccupy Spender's long poem *Vienna* (1934). It's the sort of dream, this time about air-raids in Valencia, that Spender alludes to in his *Left Review* notice of George Barker's volume *Calamiterror* in July 1937, violent apprehensions that turned into his 'terrible daydreams' of September 1939 ('now we ourselves are next on the list'), in which Spender felt like Lear in the storm ('life filled with madness from within and without'). Such were Auden's fears in 1938, resulting from his China war experiences and contemplations: 'Yes, we are going to suffer, now':

Behind each sociable home-loving eye
The private massacres are taking place;
All Women, Jews, the Rich, the Human Race.

Reflections like these over 'the present state of public affairs' induced in T. S. Eliot 'a depression of spirits so different from any other experience of fifty years as to be a new emotion'; and with grim anticipations as to 'the immediate future' and 'the continuity of culture', feeling that he and civilization were being nailed to the cross (these are his 'Last Words'), he closed down the *Criterion*. It was the response of the 'Fifth Reader' who, just before the Final Act of Auden and Isherwood's *On the Frontier*, collapses into grief as he announces imminent 'world war': 'Oh dear! Oh dear! Oh dear!'

On the other hand, the would-be perpetrators of violence rather looked forward to dishing it out. Apocalypse pleased Hitler greatly. He and his followers relished the Final Solution, Fascism's own Last Judgement on the Jews. 'The day of the great reckoning is at hand', A. K. Chesterton warned 'the Yiddish' in *The British Union Quarterly* in 1937. At the end of *Trial of a Judge* the Fascists have gained the upperhand, to the tune of the Judge's wife's blood thirsty exclamatoriness:

And the aeriel vultures fly
Over the deserts which were cities.
Kill! Kill! Kill! Kill!

Meanwhile the defeated Reds ponder revenge: 'We shall be free', come the revolution, the anticipated apocalyptic of the Left's looked-for future, when, in the words of the Judge

all will be the same; only
Those who are now oppressed will be the oppressors,
The oppressors the oppressed . . .

And even Spender—who on evidence like that supplied by his *Trial* was as terrified of a revolutionary apocalypse as of a fascist one—had his moments of rather glib Marxist optimism when he suggested that 'the Crisis' was rather good for literature. That, at least, was his superior, armchair-Marxist line (*C*, July 1934): 'As a result of the Crisis, a time seems to have come when we are getting less interested in romantic and fictitious variations on themes drawn from the adventures of members of the middle-class in their leisure hours, and more interested in the working lives of those who are in closest contact with the industrial machine.' That is the tone of someone who was still hoping he, or at least his art, would profit by the revolution. Cyril Connolly seems to share it when he talks in his *Enemies of Promise* (1938) about writers flourishing 'in a state of political flux, on the eve of the crisis'.

Spender's later fears about apocalypse, like Gascoyne's neurotic dread and Eliot's crucifixion, are more persuasive than any talk about the advantages of the end times. They seem apter to a major aspect, and a continuing aspect, of the period's crisis feelings. In his book *Hitler* Wyndham Lewis refers to Hitler's book *Mein Kampf*. With approval he quotes Hitler: '*Wer leben will, der kämpfe also*': living is necessarily fighting. Lewis wishes to support the racist argument of Hitler's book: repellent as it might at first seem, Hitler's is 'Emergency-doctrine', suitable to the condition of emergency that the First World War has ushered in. After the War, says Lewis, 'A state of emergency came to appear for me, as for most soldiers, a permanent thing . . . I, figuratively, have never smiled again.' Nothing would ever be the same after the First War. *1066 And All That: A Memorable History of England*, which first appeared in 1930, stops at the First World War. Like many people, the book's authors W. C. Sellar and R. J. Yeatman found the apocalyptic note the natural one. 1914–18 was when, for them, 'History came to a . . .'. 'The End of History', they call it, 'A Bad Thing'. And it's quite clear that the '30s sense of the apocalyptic, its apprehension with Gascoyne that it was suffering or about to suffer 'the Last War', was rooted in and deeply instructed by the last war it had known.

Much of '30s experience and literature is undertaken in an atmosphere of danger, adventure, conflict, and violence programmed by the Great War. Paul Fussell in his *The Great War and Modern Memory* has already suggested as much. But the Great War's impact on the '30s is more widespread, more subtly pervasive than even Fussell's fine and wide-ranging survey of writing about, and influenced by it can indicate. In 1930 the Great War was a mere twelve years away. True, some of the most powerful First-War literature had by 1930 been long on the market. The poems of Wilfred Owen, the war's most important English poet, had been published, with an introduction by Siegfried Sassoon, as early as 1920. A. P. Herbert's novel about Gallipoli, *The Secret Battle*, came out in 1919. Perhaps the greatest English fictions produced under the War's impress by non-combatants had been published soon afterwards: Eliot's *The Waste Land* (1922), Virginia Woolf's *Jacob's Room* (also 1922) and *Mrs Dalloway* (1925), D. H. Lawrence's *Kangaroo* (1923). But these writings, important as they were, were not to be all. Suddenly, towards the end of the '20s, the blocked-up dam of bad memories, nightmares, trauma had burst and memoirs, volumes of letters, novels, autobiographies, and other troops- (rather than generals-)

centred books started to pour torrentially forth. R. C. Sherriff's play *Journey's End* began its mammoth West-End run of nearly 600 performances at the Savoy Theatre in January 1929 (it was first put on at the Apollo in December 1928). It became the famous spearhead of a whole battalion of War-minded books that appeared around the same time: among them Graves's major *Goodbye to All That* (1929); Siegfried Sassoon's *Memoirs—Memoirs of a Fox Hunting Man* (1928), *Memoirs of An Infantry Officer* (1930), *Sherston's Progress* (1936): published in one volume as *The Complete Memoirs of George Sherston* (1937); Edmund Blunden's *Undertones of War* (November 1928); Frederic Manning's *Her Privates We* (1930); Laurence Housman's edition of *War Letters of Fallen Englishmen* (1930); John Brophy and Eric Partridge's *Songs and Slang of the British Soldier* (1931); Herbert Read's *The End of a War* (1933), its poems already published in the *Criterion*. And so on.

In the *Criterion* of July 1930 Herbert Read ascribed this sudden 'fuss about the war' to the phenomenal success of Remarque's *Im Westen Nichts Neues* (1928) which appeared as *All Quiet on the Western Front* in 1929:

> All who had been engaged in the war, all who had lived through the war years, had for more than a decade refused to consider their experience. The mind has a faculty for dismissing the débris of its emotional conflicts until it feels strong enough to deal with them. The war, for most people, was such a conflict, and they had never 'got straight' on it. Now they feel ready for the spiritual awakening and *All Quiet* was the touch that released this particular mental spring.

All Quiet was almost certainly not the only spring. It certainly took survivors like David Jones still longer to come to anything like terms with their experiences: his *In Parenthesis* did not appear until 1937. But Read's chronology and general picture are broadly right.

So extensive, in fact, was the war-books fashion, that some latecomers were thrown on to the defensive. No one, insisted A. G. Macdonell at the start of his war and post-war book *England their England* (1933), 'need be afraid that this is a war book'. Wyndham Lewis suggests in *Blasting and Bombardiering* that he's held off until 'the war-sickness (the "post-war") is over' and 'we can look back at the War with fresh eyes' before starting his 'story of the Great War'. In their successor to *1066 And All That* called *And Now All This* (1932), Sellar and Yeatman used the war book as a suitable object of merriment. They cite, for instance, a cookery book by A. Battery Cook called *Underdones of War*.

Others found that memories of the war, despite their literary popularity and numerousness, just wouldn't dissolve into comedy. Maimed and blinded, war survivors filled hospitals. Jobless war derelicts trying to turn an honest penny thronged the streets. William Gaunt describes them in his *London Promenade* (1930):

> Ex-servicemen at every street corner, turning the handles of automatic pianos, blowing hard into trombones, and eliciting a scraping whine from one-stringed fiddles, reproaching the social order with arrays of medals and little cards—'Disabled, no pension, three children'.
> Spreading little mats before theatre queues, amongst the swiftly moving traffic, skinny veterans with shirt sleeves rolled up execute displays of agility. Hola! they shout, and stand quivering on their heads, their inverted eyes anxiously fixed on impending motor-buses. And others turned over in crab fashion with swelling veins and starting eyes, circle round an orange on the top of a match-box, which they pick up between their teeth.

Some of the uncured shell-shocked were shut away in mental asylums, like Ivor Gurney, the War still apparently going on in his head, so that he continued turning out war poems, rambling, oddly clairvoyant, the poems of a real-life Septimus Warren Smith, until he died, still in his mental home, in 1937. David Jones retired hurt from society: even in the 1950s he would still be thinking he was in a dug-out—which is what he called his room at the Monksdene Hotel, Harrow. Robert Graves shut himself away in Majorca. Others found no asylum, even though the Oxford Group made a point of offering Christ the healer of 'nervous weakness' to 'the men whose lives are still warped and ineffective from the ravages of the Great War'. Dorothy Sayers has a character in her story 'The Unsolved Puzzle of the Man With No Face' (in the volume *Lord Peter Views the Body*, 1928) blame the current run of sensational murders on such military derelicts:

Here we've been and had a war, what has left 'undreds o' men in what you might call a state of unstable ekilibrium. They've seen all their friends blown up or shot to pieces. They've been through five years of 'orrors and bloodshed, and it's given 'em what you might call a twist in the mind towards 'orrors. They may seem to forget it and go along as peacable as anybody to all outward appearance, but it's all artificial, if you get my meaning. Then, one day something 'appens to upset them—they 'as words with the wife, or the weather's extra hot, as it is today—and something goes pop inside their brains and makes raving monsters of them. It's all in the books.

Miss Sayers's narrative tone is scarcely kind to this opinion. And as a matter of fact the shell-shocked survivor was more likely to try and kill himself, like Edward Blake of the 'splendid war record' in Isherwood's novel *The Memorial* (1932), who sticks his revolver against his palate, earnestly pulls the trigger, but still 'mucks it'. Retching and bleeding, he has to run off in a taxi to his psychiatrist: 'What a joke.' Meanwhile, Elizabeth Bowen's limp Mr Ammering in *The Hotel* (1927) ('having been unable to find a job since the War, was said to be suffering from nervous depression in consequence') doesn't get around to suicide attempts, nor does he recruit his author's sympathy: like the women in the novel's Riviera hotel, Ammering's novelist appears to prefer the tougher Colonel Duperrier, 'who had also fought' but isn't fussing about it.

James Hanley, himself an ex-wartime sailor who'd had to struggle hard to survive after the War, enters vividly into the strains in the life of his Greaser Anderson (eponymous hero of one of Hanley's *Men in Darkness* stories, 1931). Out of the Navy and out of work, Anderson goes back to his old firm. 'We will not forget you', his testimonial had said—ironic echo of Laurence Binyon's lines, Remembrance Day favourites, about remembering the 'fallen' of England 'At the going down of the sun and in the morning'—but the doorkeeper throws him out ('thousands of bloody men got those things . . . they're no bloody good'). One-armed survivors man the company's lifts. Anderson knew one of them. Sacked for lateness after being up all night during his wife's confinement in childbirth he'd hung himself on his own braces: 'After fightin' the bloody war for them.' Incensed by thoughts of which, Anderson returns to the place of his rejection: 'What about us fellers fought in the war . . .'; 'I'm greaser Anderson. That's who I am. And don't you bloody well forget it.'

'This good man is a hero', a gent assures his small son who has noticed the 'poor man' with only one boot on Armistice Day, and he tells John Bullock, mutilated old soldier of Henry Williamson's *The Patriot's Progress* (1930), 'we'll see that England doesn't forget you fellows'. But ' "We are England"', said John Bullock, with a slow smile'. He remembers. So does Henry Williamson and so do all the other old soldiers, German ones included ('Good luck, Tommee', the German prisoner stretcher-bearers had said to Bullock), but the civilians seem to need reminding of Bullock's grim toll of suffering:

> John Bullock's leg, hanging by a stump of septic flesh and sinew, having been severed by a stretcher-bearer's jack-knife; the stump having been spattered with iodine and dressed; anti-tetanus serum having been injected into the flaccid mottled thigh, a tourniquet of puttee and shell-splinter turned; he was covered with a muddy blanket . . .

Henry Williamson, like other old-soldier memorialists, did his utmost to keep forgetful Civvy Street, the civilians 'at home, sitting in armchairs, and talking proudly of patriotism and heroism' (*The Wet Flanders Plain*, 1929), mindful of Bullock's plight. 'Remember!', Christopher Isherwood in *Lions and Shadows* remembers the man Lester shouting to him as the boat pulled away from the pier at Yarmouth, Isle of Wight: Lester with the gym shoes and the huge old military greatcoat, the 'ghost of the War', who bums around, sleeps in a tent and provides Isherwood with hauntingly 'matter-of-fact anecdotes' of the 'obscenities and horrors' he'd survived. Lester's recollections had soaked Isherwood in a bath of renewed war fears: 'one day, perhaps, it would be our turn—Chalmers', Weston's, Philip's, mine. Our little world which seemed so precious would burst like the tiniest soap bubble, unnoticed, uncared for—just as Lester's world had exploded, thirteen years ago.'

'Remember!': the ghostly's Lester's shout was the cry of that most famous ghost of an older generation imposing disruptive demands on the young, the ghost of Hamlet's father. And young Hamlet–Isherwood could scarcely not remember. For him, as for many boys of his age, the War had been a classic trauma. They watched their fathers and older brothers go off into danger and death with mixed horror and fascination. When the War had ended before their own turn to fight had come around, their relief was mixed with shame and guilt. The speaker in Clifford Dyment's sonnet, 'The Son', a simple and moving poem that deserves to be as well known as Cornford's 'Heart of the Heartless World', was by no means alone in having a slaughtered father to look back on with a survivor's disconsolateness:

> I found the letter in a cardboard box,
> Unfamous history. I read the words.
> The ink was frail and brown, the paper dry
> After so many years of being kept.
> The letter was a soldier's, from the Front—
> Conveyed his love, and disappointed hope
> Of getting leave. 'It's cancelled now', he wrote.
> 'My luck is at the bottom of the sea'.
>
> Outside, the sun was hot, the world looked bright.
> I heard a radio, and someone laughed.
> I did not sing or laugh or love the sun.
> Within the quiet room I thought of him,
> My father killed, and all the other men
> Whose luck was at the bottom of the sea.[31]

Isherwood's father was also killed fighting; so was Charles Madge's (he was Colonel of the Warwicks); so was Osbert Lancaster's. Auden's father went missing in another sense: absent in the RAMC for the whole duration of the War, 1914–19. Others lost brothers, one way or another. Alick West (who was himself in Germany in 1914 before going up to Balliol and so got himself interned) had an older brother, Graeme, killed fighting. Tom Driberg's medical brother Jim was captured, won an MC at Loos, went on after the War to unhappy divorce and bankruptcy and a Lester-like life in the Andes (ending up as a Colonel in the RAMC in the Second War, and an alcoholic, he died of morphine poisoning). Auden's eldest brother Bernard was called up and had actually begun his military training when the War ended. Malcolm Lowry was plunged into a curiously idolatrous relationship with his eldest brother Stuart for his having fought in the War. The death at school of Henry Green's brother inevitably got itself mixed up with the War. His family's house had in 1917 been commandeered as a convalescent home for officers; one of them 'had tried to commit suicide in his bath by cutting his wrists, so that, taken with my brother's death which had occurred a year before while he was still at school and with the lists of the dead each day in every paper, there was an atmosphere of death, and of dying'.

'This feeling my generation had in the war, of death all about us', Green goes on in *Pack My Bag*. At Auden and Isherwood's prep school, Isherwood tells us in *Exhumations* (1966), boys bereaved in the fighting, and there were lots, had to wear black crêpe armbands in a demonstration of public grief. In later life Auden still remembered Isherwood's. And not surprisingly there developed among those small boys a 'cult of the dead'. 'Did you blub much?' they would eagerly ask any boy called out of lessons to be told some male relation had died in action. One boy, jealous for this grisly second-hand glory, even pretended his father had been killed (when the pretence was exposed 'he suffered what, in our penal code, was equivalent to capital punishment. He was gorse-bushed').

'In former days'—so runs Herbert Read's epigraph to *The End of a War*, a quotation from Jean Bouvier, a French subaltern—'we used to look at life, and sometimes from a distance, at death, and still further removed from us, at eternity. Today it is from afar that we look at life, death is near us.' Death lay about the impressionable infancies of a whole generation of future authors. As the lists of the dead mounted daily there was no knowing which relative or family friend they might not include next. *The Joy of It* (1937), the autobiography of the former headmaster of Sherborne Preparatory School, Littleton Powys—MacNeice's old headmaster and continuing friend; brother of all the fiction-making Powyses—tells a weirdly compelling tale about how his school survived the War, a story high on trivial details about food shortages and stand-in lady teachers but also most compelling in its record of all the 'soldier brothers', the Old Boys and young staff who got killed. The 'born leader', the 'Captain of the School Football and the School Eight' (Colonel of his regiment at twenty-three), 'another Captain of Sherborne Football', 'the finest teacher we ever had', 'another member of the Sherborne XV', 'yet another member of the Sherborne XV', Jack Hooper of the 'brilliant dribbles', 'Jack Waldegrave, who captained Marlborough at cricket', the Scouts leader, 'the international hockey player': as Powys lists them all it's difficult not to be moved and impressed by this wholesale giving up of a generation, and a generation's values and myths, to the slaughter. Nor should these lists of officers and officer material distract us from those other lists—all those Smiths, Joneses, and Browns, whose names, clustered in large family groups by alphabetical order, fill the memorial stones and plaques in every English village and town, cathedral, churchyard, and chapel. 'The Lads of the Village' Stevie Smith called them (the poem is in her volume *Tender Only to One*, 1938):

> The lads of the village, we read in the lay,
> By medalled commanders are muddled away,
> And the picture that the poet makes is not very gay.
>
> Poet, let the red blood flow, it makes the pattern better,
> And let the tears flow, too, and grief stand that is their begetter
> And let man have his self-forged chain and hug every fetter.

In *The Memorial* Isherwood turns his novel's venomous attention on to the widow Lily, a troubling mother-figure who is caught in the very act of deploring the reading out of the names of The Fallen in alphabetical order—to her snobbish mind a lamentably undifferentiated mass of men:

> She had hardly thought of Richard, as one had to think of him, of course, turned forth over there, on the Other Side, with Frank Prewitt, Harold Stanley Peck, George Henry Swindells . . .
>
> . . . Why couldn't they have read out the officers' names first? She'd heard that the names on the Memorial were put in the same way. That was really disgraceful, because, in fifty years' time nobody would know who anybody was.

This fictional attack on his own mother Isherwood dedicated To My Father. The War certainly polarized his generation's family feelings, even if in a somewhat confused fashion. Fiction after fiction engaged in acts of diffused and confused hostility towards the older generations that were held responsible for the War. In Isherwood's novel old grandpa Vernon comes off even more badly than Lily, stealing the show, making a risible mess of the new War Memorial's unveiling as he lays his wreath. 'Awa ga ga, wa ga' he grunts, truly gaga, just the kind of old man, we're supposed to think, who sent younger men to their death. Such indignations surface again, this time merged with Auden's in the most angry passage in *The Dog Beneath the Skin*, where Dog's Skin recites the words of his wartime master, the 'very famous author', in his cups:

> Less than a hundred miles from here, young men are being blown to pieces. Listen, you can hear the guns doing it . . . Every time I hear that, I say to myself: You fired that shell. It isn't the cold general on his white horse, nor the owner of the huge factory, nor the luckless poor, but you. Yes, I and those like me. Invalid poets with a fountain pen, undersized professors in a classroom, we, the sedentary and learned, whose schooling cost the most, the least conspicuous of them all, are the assassins . . . Men are falling through the air in flames and choking slowly in the dark recesses of the sea to assuage our pride. Our pride! Who cannot work without incessant cups of tea, spend whole days weeping in our rooms, immoderately desire little girls on beaches and buy them sweets, cannot pass a mirror without staring . . .

When the Spanish Civil War broke out many of the young—as we shall see—seized on it as the chance to catch up with their fathers, their older brothers and the dead Old Boys, to wipe out their guilt over having missed the First War. There was a lot of catching up to do, and that the young should wish to is a tribute to the immense potency of First War images that the '20s and then the '30s carried in their heads. With a terrible kind of naturalness the writings of this period fell into war language, into a semiotic supplied by what had been learned from the war-time fronts. The War was in almost every writer's mental luggage.

In, for example, Julian Bell's ironic reflections on the post-war vision of the future 'workers' city' in his poem 'Bypass to Utopia' (published in the Memoir his brother Quentin edited after his death in Spain)—a vision that Julian Bell sees marred by the old tyrants and tyrannies surviving 'war embittered'—the First War images are as authentic as if the poet had actually been in the fighting rather than simply having heard tell of it

> Coldharbour and the Wilderness,
> Verdun and Somme and Passchendaele,
> The black mud of the battlefields
> Settles and waits.
>
> . . .
>
> Thrown away on the uncut wire,
> Squandered in a succession of frontal attacks;
>
> . . .
>
> Wasted and ruined in that useless clay
> Accumulated nitrates, phosphorus, steel,
> Experience, courage, wit and strength and skill,
> The granary of three hundred years, and all
> Gone but to smash and spoil as good as they.
> Judgement and foresight and the quiet mind,
> And sense and feeling watchful of the end,
> The flaming spirit of the golden town
> Gassed, shelled, defeated in the fighting line.

It's A Battlefield is how Graham Greene's novel of 1934 represents a '30s London troubled by the death sentence passed on a Communist who has stabbed a policeman to death at a political meeting: a battlefield that merely continues the War's struggles. 'I'm paid,' thinks the novel's Assistant Commissioner of Police, 'I've got to do my job. One did not question during the war why one fought; one waited till the war was over for that.' 'In this sector when barrage lifts': so begins Poem 22 in Day Lewis's *From Feathers to Iron* (1931), and it goes on to talk of 'strategy' and the 'midsummer offensive'. Whatever is happening in this Audenesque sequence of poems, whether the revolution or merely the development of Day Lewis's son in the womb, it draws confidently on a ready stock of wartime images. And Day Lewis's master's voice was even abler at assimilating a schoolboy myth of frontier combat, culled from school playgrounds and the Old English text about the Battle of Maldon, to the rhetoric of the Western Front. In, for instance, Ode V of *The Orators* ("To My Pupils"), 'Though aware of our rank and alert to obey orders': a poem whose schoolboy soldiers have all seemingly endured the wartime absence of fathers:

> At night your mother taught you to pray for our Daddy
> Far away fighting.

But this time, in these boys' engagement amidst the limestone, there'll be none of that fabled First War fraternization with the Enemy:

> Do you think that because you have heard that on Christmas Eve
> In a quiet sector they walked about on the skyline,
> Exchanged cigarettes, both learning the words for 'I love you'
> In either language,
> You can stroll across for a smoke and a chat any evening?
> Try it and see.

The youthful survivors found it only natural to conduct their debates about politics and literature in the jargon of the war just passed. They must 'take sides': 'The Writers Take Sides', declared Christina Stead.[32] *Authors Take Sides On the Spanish Civil War.* [N]early every serious writer . . . is passionately on our side', said Spender.[33] People were refused 'neutrality': 'neutrality is not enough', Day Lewis told readers of the *Daily Worker.*[34] Hence that question-mark 'Neutral?' in the *Authors Take Sides* calculations: 'it is impossible to take no side.' In periods of crisis, argued Spender in *Forward from Liberalism*, men 'cannot avoid taking sides', and 'the artist, I believe, will fight on the side of the workers'. Francis, at the end of *The Dog Beneath the Skin*, is one such would-be returner to the battlefield, who rejecting his class declares to the rightist village notables

> You are units in an immense army: most of you will die without ever knowing what your leaders are really fighting for or even that you are fighting at all. Well, I am going to be a unit in the army of the other side.

Was the War the source of the 'horrible virus of bitterness, pique and confused understanding' of modernity, wondered Richard Church in 1936:

> It is so easy to blame the war as the cause of these evil forces, these half-personal hatreds and wilfulnesses. But all life is a war, a desperate and wasteful conflict of self-consuming strength, and we get no further by pleading the excuse of special circumstances. Yet how else are we to explain the confusion and cross-purposes in the world of poetry today? I find myself believing that the trouble is due partly to the neurotic residues of the war . . .[35]

Wyndham Lewis entertained none of Church's hesitations on his way to a similar conclusion. 'All the war hens are coming home to roost and I'm damn glad they are. I like being here to see this roosting', he glowered in *Blasting and Bombardiering*. It was the War, he repeatedly argued, that had generated the whole series of warfares that preoccupied the '20s and the '30s: the Class War, the Generation War and the Sex War.

The War's veterans found it easier still than the more youthful onlookers to dramatize life in this fashion. Hitler casts his life and political ambitions in terms of *Mein Kampf*, my war. Wyndham Lewis takes up the part of *Frontkämpfer*, the front-line soldier, and rejoices in his role of The Enemy. Rightist politics take on, in the calls to Action from ex-soldier Mosley, the tones of a 1914 heroic. The politics of the Left, directed by men mindful of 1917 in Russia and 1918 in Germany, readily inhabits a Front-consciousness: *Rot Front*, Red Front. (The subtitles of the German version of Dziga Vertov's visual poem about the Soviet Five Year Plan, *Enthusiasm* or *Symphony of The Don Basin* (1931)—I've not seen the English version—came packed with references to the Plan as a *Kampf*, to workers as *Kämpfer* and *Stoss-kämpfer* (Shock-Troops) pushed into the *vorderste Linie* of industry, the very front-line.) When Cornford began his Spanish War poem 'A Letter from Aragon' with 'This is a quiet sector of a quiet front' it was like an affirmation that '30s poetry had come back to its source in the First War. Indeed, the '30s spirit had, as MacNeice puts it in *Autumn Journal*, Section VI, found 'its frontier on the Spanish front': all its frontier crossings, all the travels which will preoccupy later parts of this book, had on this view only been a way of leading it back to the Front, of reaffirming the front-consciousness with which the word *frontier* always comes fronted.

It became a convention of the times to depict the landscapes, geographical and spiritual, of the post-war, as war-devastated waste lands. What Spender's poem 'The Landscape Near An Aerodrome' is, among other things, striving for is a connection between its industrial topography and First War battle scenes, a link worked for in the phrases 'the outposts / Of work', 'the charcoaled batteries', and in its direct allusion to Wilfred Owen's 'Anthem for Doomed Youth' and its 'passing-bells for these who die as cattle':

> they hear the tolling bell.
> Reaching across the landscape of hysteria.

The most famous literary waste land of all, the one T. S. Eliot put before the world in 1922, was itself, of course, generated out of an immediate response to the War. And it was a vision that rapidly proved indispensable to the post-war sense of the city, of modern life: it became *the* touchstone, the most convenient shorthand way of defining the modern plight. 'If', declared Charles Plumb and W. H. Auden in their little Preface to *Oxford Poetry 1926*, 'it is a natural preference to inhabit a room with casements opening upon Fairyland, one at least of them should open upon the Waste Land.' The terminology came very naturally to the smart young literary person of 1927 when pondering means of engaging his poems with the social scene. But Eliot's Waste Land could equally well be as it were retrojected, so that one finds its images, its vocabulary running reinforcingly in and out of the war literature written after it about events that took place before Eliot's poem was published. War memories were being read and re-read in Eliotic terms. David Jones's *In Parenthesis* picks up from the 'A Game of Chess' section of *The Waste Land* its final 'Good night . . . sweet ladies' passage and transmutes it into the long 'Good night, bon swores' passage that characterizes 'the liturgy' of troops 'going-up' the line. From the same section of Eliot's poem Jones makes prominent theft of the arresting word 'cupidon'. Stretcher-bearing is made to take place in a version of Eliot's version of Ezekiel's valley of bones in the section of his poem that Eliot titled 'The Burial of the Dead': 'You mustn't spill the precious fragments, for perhaps these raw bones live.' Jones's text talks casually of life in 'the war landscape' as 'day by day in the Waste Land'.

As the war-scape became fixed in the post-war imagination as a Waste Land as written out by Eliot, so post-Eliotic England got fixed as a combination of Eliot and the War. Reports from the industrial regions fell into the habit of utilizing, as that Spender poem does, established elements from the War's metaphoric set. For example, Philip Gibbs's *England Speaks* (1935) has a chapter on 'The Front-Line of Industry' in which the industrial scene in the north of England, where the struggle against the depression is shown hotly waging, becomes 'The Battle-Line', the 'front-line trenches', the 'zone' of the 'industrial battle' where the 'heaviest casualties' are (hospitals are 'casualty clearing stations'). In a similar fashion, the exploring eye of old soldier J. B. Priestley (he'd been wounded twice on the Western Front, gassed, buried alive) is haunted by First-War scapes as he makes his *English Journey*—a most important journey that helped incite other explorers (Bill Brandt the photographer, George Orwell too). The people Priestley meets in the north are 'front-line troops' against muck; they're on 'active service', in the 'front-line trench'. The unemployed are 'prisoners of War'. Priestley has seen nothing like Jarrow since the War. But this is not all. What's particularly striking is that Priestley goes beyond this kind of observation in a reading and writing of the desolations of Wallsend–North Shields as an Eliotic warscape-waste land. And if the manufacturing landscape was perceivable in this way, as a battlefield rewritten by Eliot, so also was the moral and spiritual condition of those who had manufactured that landscape—a plight which was, after all, the main thrust of Eliot's poem. Evelyn Waugh's *A Handful of Dust* (1934) takes its title as it borrows its epigraph, from *The Waste Land*'s line 'I will show you fear in a handful of dust': which was itself an intensely concentrated intertextual moment at which Eliot's own sense of grim mortality was schooled not only by the classical Sybil of that epigraph to *The Waste Land*, condemned to a shrivelling death in life, but also by the madness, the ravings about urban deadliness and corruption in Tennyson's poem 'Maud' ('And my heart is a handful of dust, / And the wheels go over my head'). Waugh's novel takes the reader into many a 'troubled landscape'. Its contacts with Eliotic motifs of Grail and pilgrimage (Chapter 5 is entitled 'In Search of a City'), in short the Christianity, the Biblicism of *A Handful of Dust*, connect it forcefully with yet another of Waugh's disgusted inspections of English life, *Vile Bodies* (1930)—a title taken from St Paul's writing on the subject of the human need for radical change in the resurrection from the dead at the end of the world (God 'shall change our vile bodies . . .'). And *Vile Bodies*' ending ('Happy Ending') takes place 'in the biggest battlefield in the history of the world'. Waugh's wasteland thoughts about modern civilization have looped back to merge into the First-War images that sustained Eliot in the first place: 'a splintered tree stump', 'axles in mud' (cf. Eliot's 'Garlic and sapphires in the mud clot the bedded axle tree'), a 'French military greatcoat', a 'tin hut', 'gas masks', war poets. Back home, Doubting Hall is a hospital; on the battlefield false-religious revivalist Mrs Melrose Ape's songster Charity is being nice to a lecherous general as she's already been nice to countless other military men.

But for all the honesty of its recoil from what *A Handful of Dust* presents as the pervading savagery of civilized life, this final battle-scene in *Vile Bodies* is offering a none the less fantastic metaphor for modern life. And fantasy, of course, lay behind all of the Auden group's use of wartime images. They hadn't been in the War. Nor had most of the young bourgeois Communists. In the end their talk would have a disconcerting way of turning into actualities: there may be a doubt (in later life Philip Toynbee denied the truth of the episode) about the story mongered by Toynbee in *Friends Apart: A Memoir of Esmond Romilly and Jasper Ridley in the Thirties* that he and Romilly purchased knuckledusters at an ironmonger's shop in Drury Lane before heading for Mosley's Olympia Rally, 7 June 1934, to punch up a few Fascists; but the Spanish bullets were painfully real. This reality, though, was postponed until the middle and later '30s. Earlier on, a good deal of the language of Red Fronts, as of revolutionary frontiers to cross, remained mired in fantasy land. In particular, many of the massed squadron of aesthetic homosexuals of the period seem to have shared in a myth of their sexual adventuring as a kind of honourable even if honorary version of the War they'd missed—a more pleasurable variant, to be sure, of toughing it with Mosleyites or getting shot at in Spain.

Auden explicitly blamed his homosexuality on the absence of his father during the War years. Certainly, experience of the father who went missing in the War thrust both Auden and Isherwood into the unfortunately traumatic family pattern out of which came many other literary and intellectual homosexuals: Tom Driberg, whose aged father died in 1919 when his son was six; E. M. Forster, smothered in a male-less world by his mother and aunts; Walther Groddeck, Auden's psychoanalyst hero, whose father died young and who was continually seeking to escape the influence and memory of an oppressive mother amidst substitute fathers like Freud and Hitler; Rimbaud, whose absent father and dominating mother were noted in a *Criterion* review (October 1937) by the bisexual David Gascoyne ('These facts lead to only one conclusion, if we accept the relevance of the following passage from Freud's *Leonardo da Vinci* . . .')—a review which Auden probably read, a year before writing his sonnet 'Rimbaud' (certainly a phrase about Diaghilev quoted in a review of Nijinsky's *Diary* in the previous month's *C*—'does not want universal love but to be loved alone'—made its way eventually into Auden's 'September 1, 1939'). J. R. Ackerley, the homosexual literary editor of the *Listener* after 1935, whose father's homosexual identity had been hidden away from him in childhood, and Brian Howard, who suffered a vampire mother and was long perplexed over the hidden identity of his father (he never knew his grandfather's real name: was he Jewish?: 'Heaven knows who Daddy *really* is'), might also be thought of as fitting the pattern.

But whatever the precise connection in some cases between homosexuality and the absent soldier father, '30s homosexual life did have a way of turning into a curiously welcomed parody of the War. Wilfred Owen, claimed Day Lewis in *A Hope for Poetry* (and Owen's *Poems* had come out in a new edition with a memoir by war-survivor Edmund Blunden in 1931), Owen 'commends himself to post-war poets largely because they feel themselves to be in the same predicament'. Owen, himself mother-obsessed, probably homosexual, had translated his feelings about proletarian boys—particularly underground workers like the coal-miner he met at the evangelical Keswick Convention whose scarred back so fascinated him—into the world of soldier boys living and fighting in the coalmine-like underground of the trenches:

> I thought of some who worked dark pits
> Of war, and died.

Thus Owen's 'Miners', a poem which ends with 'us poor lads/Lost in the ground'. And the 'poor lads' of the '30s thought they were indeed taking on Owen's predicament as they plunged into their own kind of dugout, the underground lavatories of Britain's cities, where they hustled for male lovers, proletarians mainly, who were frequently (and bizarrely) men in military or quasi-military uniforms—guardsmen, policemen, sailors. In E. M. Forster's case—and he was in these respects still 'a bit of a lad' sexually speaking, even though he was by the '30s getting on rather in years—anybody in a uniform would do, even bus conductors and tramdrivers. Why 'this passion for . . . the policeman', wondered Virginia Woolf in a letter (21 December 1933) to Quentin Bell in which she'd raised the subject of the sexual tastes of Plomer, Spender, Forster, Auden and Ackerley. One answer was that it was fearfully exciting. Homosexuality being illegal, to consort sexually with policemen was to be engaged dangerously and literally with the enemy, and an enemy never entirely pacified even within the close-knit bonds of the uniformed-lovers' world. For instance, Ackerley's policeman chum Harry Daley disliked Ackerley's dog Queenie, given with a sure instinct to biting bus conductors on the bum: he was jealous perhaps because the dog had been purchased from another of Ackerley's boyfriends, an army deserter this time, a man who was finally taken into custody by none other than the policeman who was Forster's lifelong pal and Ackerley's friend, Bob Buckingham. The whole thing was as enticingly risky in fact as having a German boyfriend was outrageous to decent British opinion. And the '30s homosexuals, Spender, Auden, Isherwood, John Lehmann, Brian Howard, went in keenly for Germanic lovers, the boyfriend defiantly sought amongst the recent wartime foe. It was all a way of being, so to say, in the First War by proxy, a participation in a murky underground substitute for the uniformed world of the military father and elder brother, a wasteland place that was legally and physically dangerous, where one's pacifist conscience could be appeased in a parody of the Christmas Day 1914 fraternization with the enemy, and one's guilt over missing the War could be assuaged in submission to the father and elder-brother substitute as chastiser. For this was frequently rough trade: you could get hurt, as Auden was. The 'wound' that Auden sometimes refers to in his earlier poems—'Letter to a Wound' in *The Orators* is his most extended reference to it—was, as Isherwood reveals in *Christopher and His Kind*, a rectal fissure got in the homosexual trenches. It's never referred to, one notices, as a merely civilian injury: 'wound', the wartime vocable, is the one Auden prefers. And in wartime contexts: 'The wound is healing and we can now look back to war' is followed in *The Orators* by talk of 1914 and the British Expeditionary Force.

'Hands in perfect order': so ends *The Orators'* 'Journal of an Airman': a curiously dense and sticky *ragoût* of war-time heroics, wounds, illnesses, a dominant mother and a queer uncle, an allusion (helped along no doubt by *A Portrait of the Artist as a Young Man*'s account of Stephen Dedalus's sexually fraught distresses) to 'the last desperate appeals of the lost for help scribbled on the walls of public latrines', homo-erotic affections, 'unnatural offences against minors', and a guiltily onanistic obsession with what hands can get up to ('The true significance of my hands. "Do not imagine that you, no more than any other conqueror, escape the mark of grossness" '). 'Hands in perfect order': it is, of course, Auden's Airman's version, as he is about to crash to his death, of Wilfred Owen's claim just before he was killed, in a letter to his mother (he repeated it, and Blunden calls attention to the repetition in his 1931 edition of the *Poems*): 'My nerves are in perfect order.'

A period controlled by memories, fantasies, the language of the War, the '30s is commanded obsessively by a violence—its images, its tone, its horrors, its pleasures—that one wants to keep tracing back to the First World War. Or at least to the syndrome of pain and savagery within modernism's extended theatricals of cruelty of which the First World War is a dominatingly central act. The First World War did not turn Wyndham Lewis, T. E. Hulme, the *Blast* crowd, and the Italian Futurists into violent men. But it did express for them what their private thuggeries and their predilections for a violent criticism and an aesthetic of violence had been reaching towards. Hulme pinned Wyndham Lewis upside down against the railings in Soho Square, had Gaudier-Brzeska (the *Savage Messiah* of H. S. Ede's book about him, 1931) make him a brass knuckleduster (according to Michael Robert's *T. E. Hulme*, 1938, Hulme tamed his women with it), he translated Sorel's *Réflexions sur la Violence*, admired the Action Française, all before the War in which he and Gaudier-Brzeska were killed, just as *Blast* magazine and the Futurist manifestoes started to appear before August 1914. Nevertheless, it was the War, and its literature, that funnelled what we can only call this fascistic strain of modernism into the '20s and '30s.

The case of James Hanley is revealing. During the '30s his fiction developed a remarkable capacity for presenting working-class life. Hanley's run of notable short fictions and novels makes him by far the period's best proletarian realist. But his narratives are notable too for being steeped in a violence of talk and action extremer than even the commonly violent genre of 'proletarian' fiction cares to indulge in. Hanley's socially hammered people and his hammering fictional tone may well be more truthful to ordinary people's lives than gentler narratives are. Whatever the case, though, Hanley's violent touch, his sense of the violence of ordinary British existence, is clearly seeded by his experiences and lingering vision of the First War. His first published story was *The German Prisoner* (1930), an extraordinarily disturbing début, in which, in a grim version of the homosexuals' activities in the urinals and an aweing parody of their fondness for blond German boys, two British tommies stuck out in No-Man's-Land violate and eventually kill a beautiful blond German prisoner before being blown to bits themselves. They curse the German, bash his pretty face in, piss upon him. Then they 'mangle his body', worry it 'like mad dogs', 'with peculiar movements of the hands':

> 'PULL his bloody trousers down.'
> In complete silence O'Garra pulled out his bayonet and stuck it up the youth's anus.
> The German screamed.
> Elston laughed and said: 'I'd like to back-scuttle the bugger'.
> 'Go ahead', shouted O'Garra.
> 'I tell you what', said Elston. 'Let's stick this horse-hair up his penis.'
> So they stuck the horse-hair up his penis. Both laughed shrilly.
> A strange silence followed.
> 'Kill the bugger', screamed O'Garra.

But they don't have to kill him, for he is already dead. None the less they 'jump up and down on his corpse', trampling it into the mud; then ('Let's kill each other') they attack each other just before they're hit by a shell. When the shower of mud and fragments and the smoke have cleared

> the tortured features of O'Garra were to be seen. His eyes had been gouged out, whilst beneath his powerful frame lay the remains of Elston. For a moment only they were visible, then slowly they disappeared beneath the sea of mud which oozed over them like the restless tide of an everlasting night.

And this ooze of post-war violence didn't only engulf the '30s fictions of James Hanley: it managed to muddy over almost all of the period's imaginings. The '30s—it was a common allegation—was extraordinarily possessed by death. Nobody, of course, was so naïve as to suppose earlier periods hadn't also had their necessary morbidities. Had not Eliot famously advertised the graveside gravities of the Jacobeans? 'Webster was much possessed by death / And saw the skull beneath the skin'; 'Donne, I suppose, was such another.' The title of Auden's drama *The Dance of Death* (1933) was evidently mindful of the shocks of the medieval Black Death. But it is at least arguable that the consciousness of no period—not even the medieval and Reformation mind, soaked in the Last Things of Christian doctrine—had been so taken up with death as was the imagination of the '30s. 'Shapes of death haunt life' begins one Spender poem (No. XIX in his *Poems*, 1934), and it's typical of the period's incapacity to leave death alone. The period's writers address death, they sing about it, they offer it overtures, they dance with it, they visit funerals, they make journeys to wars, they watch dying and killing, they attend closely to its 'soft answer' its 'coercive rumour' and 'enticing echo'—the variant tones in which Death invites the world to dance with it in Auden's 'Letter to William Coldstream, Esq' in *Letters from Iceland*. Spender contemplates 'Beethoven's Death Mask'. Stevie Smith and William Plomer build a career of macabre wryness with rosters of sickly jokey poems about deaths of all sorts. Death comes constantly to the '30s pen, and in every kind of guise. 'Who's sitting next to you?', asks Auden's 'Blues' (for Hedli Anderson: who became the second Mrs MacNeice): and the answer is that 'It may be Death'. Death 'As a high-stepping blondie', or 'a G-man', or 'a doctor', or a 'real estate' salesman, or 'a teacher':

> So whether you're standing broke in the rain,
> Or playing poker or drinking champagne,
> Death's looking for you, he's already on the way,
> So look out for him tomorrow or perhaps today.

But, above all, as the First War and its attendant revolutionary upheavals had schooled the world to expect, Death would come with violence. With the War, violent death had become a kind of necessity. How Auden's phrase in his poem 'Spain', 'the necessary murder' annoyed lots of his readers. Orwell (in 'Inside the Whale') trounced the notion of that necessity. So did Edwin Muir, in a *Criterion* review.[36] But Auden was only putting tellingly what a whole generation felt. (The historian Robert Stradling has argued that Orwell himself was implicated in his own accusation.[37]) In his *Criterion* piece Edwin Muir professed to detect the same 'mysticism of violence' in Rex Warner's *Poems*: 'The necessary murder again, the fat man bumped off without regret.' One of the Reds in Spender's *Trial of a Judge* attacks the judge for disclaiming 'The necessary killing hatred'. Political 'necessity', John Cornford believed, justified firm measures such as Bela Kun's massacring of 5,000 prisoners. 'It had to be done', says strike leader Rod McGinn as he shoots his brother Jack for being a blackleg in MacNeice's (unpublished) play *Blacklegs* (1939).[38] 'Bourgeois culture has discovered that what pays is bourgeois violence', claimed Caudwell in his essay 'Pacifism and Violence' in *Studies in a Dying Culture*, so there is a 'stern necessity' laid on one to enlist for or against that violence: 'Under which banner of violence will [the pacifist] impose himself? The violence of bourgeois relations, or the violence not only to resist them but to end them?' Violence *must* attend the parturition of the new world, Caudwell says in *Illusion and Reality*: 'a new society' will be 'born only in suffering, torn by the violence of those who will do anything to arrest the birth of a world in which the freedom of the majority is based on their unfreedom'. Wherever one looks in '30s literature it seems that the necessity of violence is being proclaimed or insinuated. Willy-nilly, the whole '30s world was immersed 'in the destructive element'. The phrase came secondhand, from what has proved perhaps twentieth-century literature's mightiest footnote: the one in I. A. Richards' article 'A Background for Contemporary Poetry' in the *Criterion* (July 1925)—an article that became Chapter 6, 'Poetry and Beliefs', of Richards's little book *Science and Poetry* (1926)—where Richards cites *The Waste Land* as the quintessential post-war text, *the* modern poem. The poem, says Richards, not only expresses the post-war 'sense of desolation, of uncertainty, of futility, of the baselessness of aspirations, of the vanity of endeavour, and a thirst for a life-giving water which seems suddenly to have failed', but it also shows the way out of this chaos, 'the only solution of these difficulties', beyond the lost fixities of belief to a determined acceptance of the chaos engendered out of the modernist flux. After which odd and most challengeable assertion, Richards suddenly produced two sentences he'd adapted from Stein's advice in Conrad's *Lord Jim*: 'In the destructive element immerse. That is the way.' One must abandon oneself, as Richards was suggesting *The Waste Land* had done, to the horrors of modernity. Life is a dream, Conrad's Stein had said; it was like falling into the sea: you drown only if you try to get out of the water: 'The way is to the destructive element submit yourself, and with the exertions of your hands and feet in the water make the deep, deep sea keep you up.'

   The effect of Richards's footnote was magical. His hold over the younger writers was anyway hypnotically strong. So many of them actually read the new university subject of English Literature in their degree courses—the first generation of British writers and intellectuals of whom this is true—and the dominant critical influence of the day in the young English departments, especially at Cambridge, Richards's own university, but elsewhere too, Oxford included, was I. A. Richards. He was Empson's supervisor and mentor. He was Charles Madge's tutor. All of Humphrey Jennings's work in words and paint and film, derived, Kathleen Raine thought, from his training in Richards's kind of criticism (Jennings got a double First in English at Cambridge). Cornford believed Richards 'Absolutely first-rate'. Julian Bell exported 'practical criticism' in 'the manner of Dr Richards' to China. Auden and Garrett arranged the texts in their anthology *The Poets' Tongue* (1935) in alphabetical order and with authors unascribed (except in the Index) in pious imitation of Richards's practice of conducting practical-criticism lectures with anonymous poems. Evelyn Waugh professed to have read Richards's *Principles of Literary Criticism* twice. His annotated copy of *Principles* survives. Richards was 'our guide', declares Isherwood in *Lions and Shadows*, 'our evangelist, who revealed to us, in a succession of astounding lightning flashes, the entire expanse of the Modern World'. It's scarcely surprising that the 'destructive element' became one of the most enduring of his devotees' touchstones. For it encapsulated exactly what was—with the exception of Eliot himself who eventually demurred from Richards's use of his poem—being widely felt: that to be post-war, to be modern, was to be an inhabitant of a 'dream of violence', to be in a scene that threatened you constantly with destruction.

The destructive element: the phrase became a catchphrase. Caudwell uses it again and again in *Illusion and Reality* and *Studies in a Dying Culture* in the course of his attacks on bourgeois life and literature. Spender takes it as the title for his critical excursion into violence and death in modern literature, *The Destructive Element: A Study of Modern Writers and Beliefs* (1935). Wyndham Lewis applies it in *Men Without Art* (1934) to the 'suicide club' of Eliot, Lawrence, and Firbank, and then to himself as a man at one with the destructive element. The revolutionary author, wrote Day Lewis in his reply to Julian Bell's attack on his work (both pieces were published in the *Julian Bell* memoir, 1938), must 'express the "destructive element" which we feel at work within ourselves, the disintegration of a society organised primarily in the interests of our class'. In his playlet *Noah and the Waters* (1936)—a drama oddly derided by Edgell Rickword in *LR* ('One soon gives up trying to work it out')—Day Lewis was clearly trying to float a myth of the revolution as precisely a necessary trusting of oneself, just like the Old Testament Noah, to the fiercely dangerous but hopeful flood of apocalyptic social change, a revolutionary version of Richards's and Conrad's destructive element. 'Save us, save us from the destructive element of our will, for all we do is evil': so prays Ransom in Auden and Isherwood's *F6*—proving once again the extremely useful elasticity of the Ricardian proposition. And when John Lehmann's sister Rosamond had her central character Olivia reflect quite casually in the course of her serious-minded but still intentionally populist novel *The Weather in the Streets* (1936; Part One, Section V) on 'the destructive element' that is the life of her friend Marigold's tubercular boyfriend Timmy, it was clear the phrase was thoroughly at home by then in the English language.

The use Day Lewis makes of the destructive element is, for the '30s, perhaps the most natural one of all. For violence dominates the literature of revolution and aspirant revolutionaries. 'War is Also An Art', claims Tom Wintringham in *Left Review* (February 1936): workers and revolutionaries do war well. Spender, mindful of how the workers were violently suppressed in Vienna, hopes in *Forward From Liberalism* that in similar circumstances he 'would have had the courage to act violently myself'. 'Good! Let them have it', snaps Valentine Ackland in her poem '1937', as she updates 'the age-old threat of death' and gets behind her anti-fascist machine-gun.[39]

Demonstrations, street-fighting, police violence against workers and democrats fill leftist fictions. In novels and stories like Walter Greenwood's *Love on the Dole* (1933), A. P. Roley's *Revolt* (1933), Grassic Gibbon's *Grey Granite* (1934), Naomi Mitchison's *We Have Been Warned* (1935), John Sommerfield's *May Day* (1936), James Barke's *Major Operation* (1936), Arthur Calder-Marshall's 'I Want My Suitcase Back' (in *Fact*, No. 4, July 1937), Day Lewis's *Starting Point* (1937), the violent moment—as police batons crash down on proletarian heads and stunned workers fall among the hooves of the police horses, or as, in the Mitchison case, a wholesale right-wing terror reigns—is in the first place shocking, traumatic, intended as a revealing corollary of the industrial oppression and violence that also fill such fictions. 'Warm blood ran from his nose and his brains littered the road' (*May Day*): and the reader is meant to be appalled. But such moments are also aimed at being politically catalytic, exemplary proofs that the blood of the workers is the necessary seed of revolutionary progress. So the conversion to Communism of Day Lewis's Oxford undergraduate Anthony begins in such a *mêlée* in the General Strike of

1926, and Grassic Gibbon's Ewan Tavendale is drawn into Communism as a result of yet one more attack on a column of the unemployed—marching men, mindful of their 'time in the army, the rain and stink and that first queer time your feet slipped in a soss of blood and guts, going up to the front at Ypres':

> As the bobbies charged, the Broo men went mad though their leader tried to wave them back, Ewan saw him mishandled and knocked to the ground under the flying hooves of the horses. And then he saw the Broo folk in action, a man jumped forward with a pole in his hand with a ragged flag with letters on it, and thrust: the bobby took it in the face and went flying over his horse's rump, Ewan heard some body cheer—himself—well done, well done! Now under the charges and the pelt of the rain the column was broken, but it fought the police, with sticks, with naked hands, with the banners, broken and knocked down right and left, the police had gone mad as well, striking and striking, riding their horses up on the pavements, cursing and shouting, Ewan saw one go by, his teeth bared, bad teeth, the face of a beast, he hit out and an old, quiet-looking man went down, the hoof of the horse went plunk on his breast—
>
> And then Ewan saw the brewery lorry jammed by the pavement, full of empty bottles; and something took hold of him, whirled him about, shot him into the struggling column. For a minute the Broo men didn't hear or understand, then they caught his gestures or shout or both, yelled, and poured across the Mile and swamped the lorry in a leaping wave . . .

There's nothing pat about the welcoming of that kind of outraged, uprisingly violent indignation of the poor and oppressed, as the disjunctive, wrenched syntax—a sort of stream of violent consciousness—indicates. Considered violence it was, and nowhere perhaps more considered than in the novels of Ralph Bates, a novelist who was himself working for the cause of revolution in Spain in the early '30s. Francis Charing, the English revolutionary of Bates's *Lean Men* (1934) agonizes over his own act of 'necessary' murder, when he shoots an anarchist assailant ('I know I was in "a state of legitimate defence" as they say, but that doesn't make much difference when you look down and see half a face left on your man'). But *Lean Men* and Bates's later novel *The Olive Field* (1936) still grant the necessity of violence, and keep on depicting the violent doings that their characters keep expressing decent men's regrets over. *Lean Men* culminates in a communist rising in Barcelona and its bloody suppression; *The Olive Field* offers yet another bloodily put-down rising, this time of the Asturian miners. Revolution is shown as inevitable, the only solution to the ills of a country brutalized by landlords, overseers, and the Catholic Church. The violence of the oppressor—and Bates is fictionalizing Caudwell's argument—incites the oppressed to respond with guns, dynamite, knives. Bates has his novels take a determined header into the destructive element:

> Mudarra took out his knife and pulled off the sheath with his teeth. '. . . Going to speak?' 'Going to speak? No? Well, see how you like this.' Seizing the man's hair he thrust the point of the knife hard into the herbalist's gums against the base of the upper teeth, a harsh choking scream like an idiot's burst from his throat as the knife blade bored through the gum into the nerve and bone . . . When he turned to the herbalist again his face showed surrender, blood was pumping freely from his torn gums.

The fascination of this kind of horror, the relish for it too, are evident; but they're not confined to revolutionaries like Bates and Anarchists like Mudarra. Almost every '30s writer was, it would seem, afloat in the destructive waters—and even as some of them appeared to struggle they could be detected giving in to the dangerous excitements engulfing them. This is observable of Isherwood, even as he warns the world of Nazi horrors in, for instance, the closing 'Berlin Diary' section of *Goodbye to Berlin*. It's the aftermath of a Nazi meeting at the Sportpalast:

> All at once, the three S.A. men came face to face with a youth of seventeen or eighteen, dressed in civilian clothes, who was hurrying along in the opposite direction. I heard one of the Nazis shout: 'That's him!' and immediately all three of them flung themselves upon the young man. He uttered a scream, and tried to dodge, but they were too quick for him. In a moment they had jostled him into the shadow of a house entrance, and were standing over him, kicking him and stabbing at him with the sharp metal points of their banners . . .
>
> Another passer-by and myself were the first to reach the doorway where the young man was lying. He lay huddled crookedly in the corner, like an abandoned sack. As they picked him up, I got a sickening glimpse of his face—his left eye was poked half out, and blood poured from the wound. He wasn't dead.

Eyeless in the Bülowstrasse. Not dissimilarly, an overtly pacifist novel, Aldous Huxley's *Eyeless in Gaza* (1936), can't resist zestfully inventing a spitefully comic variant of the bombing raid to show just what pacifism is up against. In Chapter 12 Anthony Beavis and Helen Amberley are sunbathing naked on a rooftop in the South of France:

> A faint rustling caressed the half-conscious fringes of their torpor, swelled gradually, as though a shell were being brought closer and closer to the ear, and became at last a clattering roar that brutally insisted on attention. Anthony opened his eyes for just long enough to see that the aeroplane was almost immediately above them, then shut them again, dazzled by the intense blue of the sky . . .
>
> A strange yelping sound punctuated the din of the machine. Anthony opened his eyes again, and was in time to see a dark shape rushing down towards him. He uttered a cry, made a quick and automatic movement to shield his face. With a violent but dull and muddy impact the thing struck the flat roof a yard or two from where they were lying. The drops of a sharply spurted liquid were warm for an instant on their skin, and then, as the breeze swelled up out of the west, startlingly cold. There was a long second of silence. 'Christ!' Anthony whispered at last. From head to foot both of them were splashed with blood. In a red pool at their feet lay the almost shapeless carcass of a fox-terrier. The roar of the receding aeroplane had diminished to a raucous hum, and suddenly the ear found itself conscious once again of the shrill rasping of the cicadas.
>
> Anthony drew a deep breath; then, with an effort and still rather unsteadily, contrived to laugh. 'Yet another reason for disliking dogs', he said.

Huxley's malicious pleasure is clear, and perhaps not all that surprising. But even George Orwell, the apostle of decency, can't help getting right inside the many kinds of violence that he keeps rehearsing. Indeed his power consists precisely in his ability to get inside the skin, as it were, of the violent action: an act, it might be, of his own that an imperialist's duties or the white man's burden traps him into, as in 'A Hanging' (the *Adelphi*, August 1931) and 'Shooting an Elephant' (*New Writing*, Autumn 1936); or perhaps the good old English domestic poisoning, now so sadly missed, as in 'Decline of the English Murder' (*Tribune*, 1946): or, again, it could be the nightmare of coming fascist violence such as Orwell's plain man George Bowling encounters, deplores and then enters most feelingly into, in *Coming up for Air* (1939) at the West Bletchley Left Book Club's lecture against Fascism:

'Bestial atrocities . . . Hideous outbursts of sadism . . . Rubber truncheons . . . Concentration camps . . . Iniquitous persecution of the Jews . . . Back to the Dark Ages . . . European civilisation . . . Act before it is too late . . . Indignation of all decent peoples . . . Alliance of the democratic nations . . . Firm stand . . . Defence of democracy . . . Democracy . . . Fascism . . . Democracy . . . Fascism . . . Democracy . . .'

Hate, hate, hate . . .

I saw the vision that he was seeing . . . It's a picture of himself smashing people's faces in with a spanner. Fascist faces, of course. I *know* that's what he was seeing. It was what I saw myself for the second or two that I was inside him. Smash! Right in the middle! The bones cave in like an eggshell and what was a face a minute ago is just a great big blob of strawberry jam. Smash! There goes another! That's what's in his mind, waking and sleeping, and the more he thinks of it the more he likes it. And it's all o.k. because the smashed faces belong to Fascists.

The smashing-in of faces fascinated Orwell extremely. In his essay 'Politics vs Literature: an Examination of *Gulliver's Travels*' (1946) he singles out for special comment Swift's phrase 'battering the warriors' faces into mummy'. He has the chinless prisoner in *Nineteen Eighty-Four* (1949) knocked to the floor by a fearful blow to the mouth: 'Amid a stream of blood and saliva, the two halves of a dental plate fell out of his mouth'. In that novel O'Brien expresses an eternity of totalitarianism in terms of the smashed face:

Always, at every moment, there will be the thrill of victory, the sensation of trampling on an enemy who is helpless. If you want a picture of the future imagine a boot stamping on a human face—for ever.

Coming up for Air has Bowling accept this face-smashing future as inevitable: 'Quick, quick! The Fascists are coming! Spanners ready, boys! Smash others or they'll smash you.' As a kind of warning and proof of physical damages to come the RAF accidentally drops a bomb on a greengrocer's shop in Lower Binfield:

There was frightful smashed-up mess of bricks, plaster, chair-legs, bits of a varnished dresser, rags of tablecloth, piles of broken plates, and chunks of a scullery sink. A jar of marmalade had rolled across the floor, leaving a long streak of marmalade behind, and running side by side with it there was a ribbon of blood. But in among the broken crockery there was lying a leg. Just a leg, with the trouser still on it and a black boot with a Wood-Milne rubber heel.

William Plomer made memorable fun of dismemberment in some of his later poems—'The Dorking Thigh' and (one of his very best satires) 'The Flying Bum: 1944'—all, though, done very much in the spirit of the '30s. He could make a bombing raid more savagely funny even than Huxley. 'Mews Flat Mona. A Memory of the Twenties' (in his *Selected Poems*, 1940) has its heroine making a bad end as a human bomb:

She stepped from the top of an Oxford street store;
She might well have waited a split second more
For she fell like a bomb on an elderly curate
And his life was over before he could insure it.
Oh, Mona! you're exploded now!

But Bowling and Orwell—his visions brought up to date by what he had lived through in Spain—can't bring themselves to laugh over high explosives:

> *It's all going to happen.* All the things you've got at the back of your mind, the things you're terrified of, the things that you tell yourself are just a nightmare or only happen in foreign countries. The bombs, the food-queues, the rubber truncheons, the barbed wire, the coloured shirts, the slogans, the enormous faces, the machine-guns squirting out of bedroom windows. It's all going to happen. I know it . . . there's no way out.

Still, the engulfing pervasiveness of the destructive element in '30s literature was, as in Orwell's case, frequently being zestfully splashed about in even while it was being complained of. John Betjeman, the friendly bomber ('Come, friendly bombs, and fall on Slough'), is relishing, even as he mocks it, the violence of boys' adventure stories, the world of the fictional First World War heroes such as Biggles:

> How many vicious swings have gone socketing into midriffs, how many uppercuts have landed neatly on weak chins, how many lashes descended on trembling flesh, how many steel-blue eyes scanned open boyish faces, how many shifty dagoes have flinched before a steady gaze, since first I started to read the twenty-two boys' books it is now my pleasure to pass in review? . . .
>   Here is a nice little torture to try on the milksop next door, next time you and your brother have got him alone. It comes from *Adventure Down Channel* . . . Lash your man by his ankles, wrists and thighs so that he cannot move. Now heat an iron bar on a stove well in view of your victim. When the bar is red hot, remove his shoes and socks and make as though to brand him on the soles of his feet. Now the skill comes in. When you have got the red-hot bar near his feet, substitute for it a block of ice, without his seeing. Draw the ice across the soles of his feet. The nervous shock that results will cause yells of merriment.[40]

'By Jove, it was an ace trick the way you shot him over that balcony', enthuses the English schoolboy hero of Rex Warner's novel for boys, *The Kite* (1936), after he and his pals have got rid of a Viennese drug peddler by chucking him from the first floor into the street. Plomer, the flying bummer, starts a 1934 piece on Hemingway by appearing to complain of Hemingway's sadism. A common enough complaint, for Hemingway's tough talking affection for violence was widely noted in the period (the narrator of Anthony Powell's *What's Become of Waring* (1939) is writing a book called *Stendhal: and some Thoughts on Violence,* and trying to work in some trendy parallels between *Armance*'s Octave and 'that other apostle of violence, Mr Hemingway's hero of *The Sun Also Rises*'). But Plomer goes on to celebrate the 'particular significance' of Hemingway's violent nihilism: 'the nihilism of our time, the time of the War and of our permanent Crisis and of our spiritual dislocation . . . also the nihilism that so often goes with vitality.'[41] Mixed feelings again: and ones evidently shared by Stevie Smith, a poet forever stylishly working the graveyard-murderer-suicide-'The toll of the Roads'-lions eat Christians-painful case ticket, who yet complains in her curiously muddled novel *Over the Frontier* (1938) about fashionable artistic pain mongers. Pompey, the novel's spokeswoman, is upset early on by the German painter Georg Grosz ('a war neurotic'), and goes on to rant against Sam Whatshisname the visceral artist and 'laborious cruelty fan', against the little man in the Second Class Carriage on the Great Northern Line reading *The Pleasures of the Torture Chamber* (illustrated), and against 'writers, I name no names, that make a lot of money writing about pain, pain is to them an inexhaustible fount of inspiration and a regular income. Specially nowadays, when people are getting so strung up and enervated full of fatigues and unrest by reason of the motor bicycles, the electric drills and the radio, is the subject of primitive great chaotic pain a favourite to come in first and leave the field at the starting-gate.' 'Oh what pleasure can there be', Pompey demands, 'in this that stinks of the slums of hell and is the very spit of the fiend?':

> Sometimes it is because people are so dumb cluck on the sentient plane, they have a very high flash point, they need a burning fiery furnace heated seventy times seven before their cold damp insentiency can catch alight-o. These are the subhumans, the people that go to throng the Coliseum when the lions are on show with the Christians, and who go to throng the cinemas when there is some tough work in the prisons of the U.S.A., or maybe it has a political excuse, and it is photographs of atrocities by Abyssinians, or maybe it is negro-baiting across the Atlantic, or maybe it is the *Fall of St Petersburg* and again with the prison brutality that is so the most beastly of all, so shut in and without all hope.
>
> Oh this is a very desperate and bitter cruelty that exacerbates and affronts our immortality, that brings all to nothing to negation and betrayal. Avoid it.
>
> Oh in this desperate cruelty and pain wherein the mind is darkness and the flesh set in domination above the spirit, oh in this there is all the abomination of reversal and traduction in the backward going and the darkening and the death of the spirit. It is contrary to nature and abominable.

And so on: an astonishing tirade from the character who speaks for Stevie Smith, the period's chief poetess and high priestess of the literary macabre. Just so, it's astonishing to hear Wyndham Lewis, who did more than any other writer to carry over a wartime violence into art and criticism, and to make that toughness fashionable, who thought of his typewriter as a machine-gun, who declared that art was 'like' war, who sought by his published preachments to leave 'upon your retina a stain of blood', who doted on the 'hundred thousand fists' for Hitler, who loved to cite instances of Marxists sticking toothpicks into eyeballs, cutting off testicles, and sawing out tongues, it's odd to find this same Wyndham Lewis complaining in 1935 that the 'best-sellers', like the politicians, the motorists, the Press, the cinema, are 'Always *killing*!'[42]

Nevertheless, though they *are* compromised testifiers, Stevie Smith and Wyndham Lewis are, like Betjeman and Plomer, quite right to advert to the prevalent sadistic bent of the '30s. Now clearly, as classic 'Just War' theories, for example, indicate, violence is not inevitably an undifferentiatable business, a seamless garment. It's as obviously unhelpful not to bear in mind real distinctions between the violence of Left and Right, for instance, as it would be to ignore the possibility of 'just' and judicial violence. Also, it is critically not all that helpful always to lump together writers and texts simply because they deal extensively in violent matters. Undifferentiated, 'violence' can be a pretty low sort of Lowest Common Factor. That said, however, what is impressive in any *tour d'horizon* conducted from up the '30s Eiffel Tower is just how the various kinds of post-war violence do consort together in a powerful, if higgledy-piggledy Wyndham Lewis-type of heap. Scarcely any party or group escapes the tarnish of sadism; the destructive element washes over them all: whether it's the leftist inheritors of Futurism in its more acceptable Soviet forms, or their rightist opponents like Evelyn Waugh, men for or against speed, men of the car-smash, the aeroplane crash, the violent death by motor-bike; or whether it's Wyndham Lewis, the old Blaster and Bombardier, the writer who set the '30s a dominating model of critical discourse, with the assistance of his musclebound gorilla of a disciple Roy Campbell—the biffing bronco who preferred poets who came stinking of horses and sweaty from football, who liked punching people and knocking their teeth out and breaking their dentures and specs, who loudly despised the pansy Leftists who wouldn't step outside and fight and who wore underpants ('underpants (which they all wear)', he jeers in *Broken Record*, 1934).

'I've knocked the shit out of . . . bogus champs', boasts Lewis in his long poem 'One-Way Song'.[43] And Lewis's admirers, and they were numerous, tried to cultivate their master's morally wayward no-nonsense toughness. The example of Lewis hangs over Geoffrey Grigson's eager use of the critical 'billhook' in *New Verse*. In *Twentieth Century Verse* its editor Julian Symons clearly sought to emulate or even to outdo Grigson in critical harshness. Perhaps the closest he came in critical bitchiness was when he hit on 'Tuttifrutti' as a suitably tart way of dismissing the poetic entrepreneur Tambimuttu. That was a level of invention he couldn't, alas, keep up long. Whereas Grigson could produce the most scathing chop at the merest drop of a dislikeable poem (Campbell's *Flowering Rifle*, the last number of *NV* effortlessly jabs, 'sounds like a hyena, ambitious to be a lion, howling away to itself (and to Mr Edmund Blunden, who has crept out of his hole to praise it) in the middle of a lonely and extensive sewage farm'). Still, Symons couldn't be more open in his admiration for Lewis. When *NV* put on its Auden Double Number *TCV* went in for a Wyndham Lewis Double Number (November–December 1937), fawning over the master's toughness, 'that toughness essential to a man of genius who declares war on ignorance and pedantry, stupidity and folly'.

It was, perhaps, this affection for Lewis's rebarbativeness that led Symons to cultivate Dylan Thomas's reputation. Thomas didn't mind, in fact he rather liked, writing about burning babies. Certainly it's Lewis's kind of taste that helps generate *New Verse*'s love affair with Auden—the 'boy bushranger' as Dylan Thomas, spotting kindredness of spirit, labelled him. For what Grigson particularly valued in Auden was a Lewisian monstrousness. 'Auden the Monster'; 'this monster out of Birmingham': it was the old *Frontkämpfer*'s kind of vocabulary and enthusiasm that Grigson brought out for applauding Auden with. And it was probably as much from a wish to emulate Lewis's example as from any desire to play for their own sake the headmaster or the commissar that Auden and his friends went in for that souped-up gangster tone they commonly used in delivering their threats: Auden's 'It is time for the destruction of error', or 'The game is up for you and for the others', or 'It's no use raising a shout'; or MacNeice's 'It's no go the merry go round'; or Day Lewis's 'It is now or never, the hour of the knife, / The break with the past, the major operation'. The heartless Chicago-mobster voice is one of the '30s' most imitable tones. The young poets employ it effortlessly. It's one of the backbone strains of Graham Greene's 'entertainments', especially of *Brighton Rock* (1938). Eric Ambler's '30s thrillers keep providing a close rival to the transatlantic stuff in Britain's best approach to the *echt* American manner. Even in Ralph Bates's committedly communist *Lean Men* the CP toughs conduct a long stick-up episode (Chapter 9, 'Escapade') in a loving parody of gangster talk (*we'll stitch you up; mutt; short-arse; stow it; Compree?; a swipe in the clock*: Raymond Chandler could wish for no greater devotion).

Even Orwell—arch-spotter of the Blimpish, keen rebutter of Newboltian war-mongering wherever he detects it—keeps insisting that you've got to have the guts. According to Orwell, Auden is no good without guts: he's 'a sort of gutless Kipling'—a delightful phrase Orwell delights in chewing over again in 'Inside the Whale'. 'Few people', gibes Orwell in a review of Philip Henderson's *The Novel Today* (December 1936), 'have the guts to say outright that art and propaganda are the same thing'. 'No one had the guts to raise a riot', he sneers at the Burmans in his 'Shooting an Elephant' (lead piece in *NW*, No. 2, Autumn 1936). The criticism of Orwell, veteran of the notoriously ferocious Eton College wall game, had nothing if not the guts. And the reader has certainly got to have the guts to get through Orwell's manly Lewisian-Grigsonian trade in stinks and sewage and phrases like, for instance, that one about Rupert Brooke's 'Grantchester' (in 'Inside the Whale'): 'a sort of accumulated vomit from a stomach stuffed with place-names'.

Like Lewis and his imitators, the Surrealists went in on principle for breaking and cutting. It's scarcely surprising that the children of Dada, those dynamiters of form, should have seen themselves, to the growing consternation of Moscow, as political revolutionaries. Eugène Jolas's part as one of the twelve Joycean disciples in the piously exegetical *Our Exagmination . . .* is only typical of the Surrealists' destructive intentions towards the traditional languages of art and criticism. Jolas praises Joyce for his 'disintegration of words', for giving 'a body blow to the traditionalists', for a 'deformation' of vocabulary, for helping the 'exploding' and 'undermining' of 'the antique logic of words' that's been proceeding in Western Europe, according to Jolas, since 1914. Disintegrating speech, slashing syllables, the surrealist revolution that's 'destroyed completely the old relationship between words and thought', iconoclasm, 'demoralizing the old psychic processes by the destruction of logic', words 'disjoined from their traditional arrangements': with this slamming rhetoric of destruction Jolas proclaims a new birth. In Joyce, 'language is being born anew before our eyes', and tearingly, painfully. It comes as no surprise, after that, to glean (eventually) in one's toil through it that a central aspect of the myth of *Finnegans Wake* is an incident of war, the Crimean War. For Jolas's talk is war talk; reminiscent of Eliot's own 'wasteland' thoughts about the new poetic language ('The poet must become', he claimed in his 1921 essay on 'The Metaphysical Poets', 'more allusive, more indirect, in order to force, to dislocate if necessary, language into his meaning').

Besides breaking up, in an almost archetypal modernist fashion, the language, the images, the forms of their art, the Surrealists and their heirs would make a cult of doing violence to people within their fictions. Dylan Thomas's short story 'The Burning Baby' ('The bush burned out, and the face of the baby fell away with the smoking leaves') actually appeared in the first number (May 1936) of the would-be surrealist magazine *Contemporary Poetry and Prose*. The surrealist subject was evidently, in the mind of its editor Roger Roughton, strongly a matter of murders and smashing and guns, for these fill the poems he publishes as they keep constituting the political themes his paper contemplates—the Spanish Civil War, 'Fascism Murders Art', 'Lorca's Death' (and Roughton himself, one is not surprised to learn, later committed suicide). Surrealists practised the techniques of the cut-up art object, and were drawn almost naturally in their writing to the cutting-up of people. Roughton himself offers 'Final Night of the Bath', a collage composed of cut-up bits from an edition of the *Evening Standard* newspaper and featuring a *cutlet* (briefly) and a murder (extendedly).[44] *Petron* (1935), that curious little fiction by surrealist spokes-man Hugh Sykes Davies—perhaps the single major literary text produced by the English honeymoon with Surrealism in the '30s—wallows firmly in an abundance of gore as it follows Petron on a small fraction of the pointless journey he's in the course of making. Petron meets a hedger who suffers epileptic fits, vomits, chops his fingers up into thousands of shreds with a pruning-bill, and enjoys threading string through his head. He steps on to a toadstool from which a super-Beckettian idiot springs to trail after him, his lower jaw rattling bloodily on the ground. *Petron*'s sadistic impulses are wilfully morbid, not to say self-hating. Petron disguises himself to enter a city:

> With a small upholsterer's hammer, neat and thin-headed, he knocked out one of his teeth, and bent another back in such a way that his smile would convince and infer [*sic*] honesty. With a heavier instrument he knocked at his chin for half an hour until all signs of human power were effaced. With a large wooden mallet he so beat upon his brow that its force was no longer apparent, and with the same tool and a billet of wood he blunted the glance of his eyes to a proper civic quality. Finishing off the whole with a well-balanced tacking hammer, he brought himself to break his nose, which at first he had intended to leave as it was.

Meanwhile, David Gascoyne was enthusedly advocating in his *A Short Survey of Surrealism* (1935) the horrors of Salvador Dali, and taking a boyish delight in 'the phosphorescence of putrescence' and other lavatorial aspects of Dada. 'Negativism, revolt, destruction of all values, Dada was a violent protest against art, literature, morals, society. It spat in the eye of the world.' 'At the end of the gallery a young girl, dressed in white for her first communion, stood reciting obscene poems. Destruction and sacrilege . . .'

But if, disgusted by sacrilege, you reeled away into some more orthodoxly Christian company, you weren't likely to escape the destructive element quite that easily. The orthodox Christians were a violent lot in the '30s (were *still* a violent lot, you might say). How much of Dylan Thomas's and Hugh Sykes Davies's violent rhetoric was, one wonders, afforced by generations of chapel hell-fire in their Welsh blood? Graham Greene's tortured, hell-obsessed Catholic, the boy-gangster Pinkie in *Brighton Rock*, is a torturer of others, with school-dividers, razors, vitriol, and vitriolic taunts. And Greene, haunted by Hell for his own part, is repeatedly drawn towards Catholicism in the starker forms it assumes in Latin countries, places of dark forebodings, sadistically detailed statues of Christ crucified and passionate mortifications of the flesh:

> And then you go into the Cathedral for Mass—the peasants kneel in their blue dun-garees and hold out their arms, minute after minute, in the attitude of crucifixion; an old woman struggles on her knees up the stone floor towards the altar; another lies full length with her forehead on the stones. A long day's work is behind, but the morti-fication goes on. This is the atmosphere of the stigmata, and you realize suddenly that *this* is the population of heaven . . .

That description of Mexican Christians, from *The Lawless Roads* (1939), Greene's account of the journey that later fruited into the novel *The Power and the Glory*, gives the Spanish atmosphere Greene praised in his review of the film *From the Manger to the Cross*: 'as in Spanish churches, you are allowed no escape at all from physical suffering; Christ is a man beaten up, like a Nazi prisoner in the Brown House. The physical horror is never far away.'[45] It's the impressive atmosphere in which Christianity is continually approached in Greene's early Pyrenees novel *Rumour at Nightfall* (1931). In a fiction commanded by murk and shadow we're told that Spaniards 'fight under the shadow of this sense of immortality. Round corners, in the shadows cast by anonymous peaks, stood wooden crosses bearing bloodstained and contorted Christs . . . Their religion seemed to [the character Chase] not a consolation but a horror.' Magnetically awful for the English journalist Chase is the (extraordinarily good) scene where Colonel Riego coaxes the mortally wounded Roca to confess his sins as the bubbles of blood rise to his dying lips; Hell will be evaded even if you have to force repentance from people in pain:

> He put his mouth alongside Luis's and blew the froth away. Chase's stomach stirred with nausea, but the circle of watchers was quite unmoved by the physical ugliness of death. They were accustomed to it. Every Christ in every church seemed to suffer more.

Nor was this feeling for the tortured Christ confined to Roman Catholics. There runs through Anglican Charles William's spiritual thrillers a most troubling ecstasy of violence. Williams resorts repeatedly to nailed flesh, gnashings of teeth, the black pangs of hell, the clashing turmoils of evil, to murders and suicides, and to supernatural grisliness in his fictions, just as he was drawn to the torturous detail of witchhunting—the red-hot iron plates, the beheadings and stranglings (while crowds of children sing hymns), the hoistings and chairs and red-hot pincers—in his book about *Witchcraft* (1941). What, one wonders, did Williams's publisher T. S. Eliot make of Williams's fascinated business with the witch-chair ('an iron-chair, with blunt studs all over it, in which the accused was fastened, while fire was lit below the seat. This was frequently used, sometimes after earlier tortures of the more general kinds—the thumb-screwing, the leg-crushing, the scourging, the hoisting with weights')? Or of Williams's weird theological moralizings:

> A woman of Ziel in 1629 was convicted of having four times desecrated the Host and having murdered her child. She was therefore, on her way to be burned alive, torn six times by the glowing iron—four times to avenge our sacred Lord and twice to avenge her human child. It would be bad enough if she were innocent; it is worse if she was guilty. For then the horror of the whole thing becomes unbearable: the screams of the sacrilegious murderess, as she is stopped for the fourth, for the fifth, for the sixth tearing, ascend like an epitome of the nature of man.

The nature of man? It was, we know, Williams's own nature to try to tame the nature of woman by also offering punishments to the ladies who fell into his clutches as a spiritual adviser. These were less burning torments, to be sure, than the witch-hunters used, but persistent ones—examination papers on the English poets, with ruler-spankings on the hand if not enough marks were obtained; penitential copyings-out of Biblical verses, with the occasional whipping threatened. 'At bottom a darkness has always haunted me', Williams declared. And there was at least one perceptive woman who thought much the same of T. S. Eliot's writing. Stevie Smith wrote in 1958 about the 'fear', 'horrible' beauty, 'neurosis of the invasion of uncleanliness and sin' and the 'disgust' she felt afflicting the women of *Murder in the Cathedral*. A 'religious terror writer' was her verdict on Eliot.[46]

But if the other Inklings allowed themselves any inkling of Williams's darknesses they appear not to have let it trouble them much. Kindness towards women was not much in C. S. Lewis's or Tolkien's line—especially, in the case of Tolkien, kindness towards Mrs Tolkien. And anyway even if, on the face of it, these two First-War veterans seemed to have shed wartime morbidities for an apparently breezier world of beer and Beowulf, they too had their darker sides. Tolkien's scholarship and fiction promoted a feeling for fighting as a glamorous activity (think of all those fights in *The Hobbit* (1937)) as if somehow, despite the War, one could go back to the clean, even picturesque heroics of *The Fairie Queene* or *The Battle of Maldon*. But even Tolkien couldn't prevent more sombre notes about warfare invading *The Lord of the Rings*. And Lewis, saddled for most of his domestic life with a living ghost of the War, the oafish mother of his dead soldier friend Paddy Moore with whom he kept house, strongly continued his soldierly belligerences into the tutorial and class-room. He was a bullying, roaring Ulster logician, whose argumentative prac-tice reminds one that it was the Ulster regiments, disregarding orders and charging the enemy full pelt, who were most successful on the terrible first day of fighting on the Somme. Pursuing what John Wain has called the 'knock-down and drag out' methods of his own old tutor The Great Knock, Lewis is even reputed to have fought a duel with a pupil over a point in Matthew Arnold. A one-sided duel, of course, Lewis's broadsword against the disputatious pupil's rapier, in which Lewis, accord-ing to J. A. W. Bennett, 'actually drew blood—a slight nick'.

Inevitably, with sadism as dominant a strain in the period's letters as in its politics, there was a run in the '30s on groups of people seeking to appropriate the name of the Marquis de Sade for their cause. De Sade, claims *enfant maudit* David Gascoyne in his *Short Survey of Surrealism*, is the *fons et origo* of Surrealism. Geoffrey Gorer disagrees with this common assertion (to the annoyance of Gascoyne): the Surrealists 'completely caricature' de Sade; he is a political revolutionary, 'the first reasoned socialist'. And Gorer's *The Revolutionary Ideas of the Marquis de Sade* (1934) wasn't merely published by Wishart, the Communist publishers, but came with a Fore-word by Professor J. B. S. Haldane, the Left's most famous scientist. No wonder pacificism—for all those ten million votes for disarmament cast in the Peace Ballot of October 1934, and the scores of thousands of members (including Gerald Heard and Aldous Huxley) that Canon Dick Sheppard trawled into his Peace Pledge Union—was really such a non-starter among '30s intellectuals.

The older Cambridge Liberals and Bloomsburyites, their policies thrown into disarray by the War and its aftermath, simply failed to bequeath their views with any certainty of their sticking to their youthful heirs. Of course, a whole generation of schoolboys and university students went through the pacifist motions. They absorbed the anti-heroic message of the First War's literary survivors and surviving pieces of literature. They took in the equation between sadism and war made in books like Edward Glover's *War, Sadism and Pacifism* (1933)—which opens with a long quotation from Freud on the 'Destructive Instinct' within 'every living being'. They read Beverley Nichols' *Cry Havoc* (1933) and endorsed the Oxford debating Union's resolution (9 February 1933) 'That this House will in no circumstances fight for its King and Country'. But as soon as the matter became personally pressing most of the erstwhile pacifist young joined up: for the cause of violent revolution, for Spain, and eventually for the Second World War.

We know that Beverley Nichols was having his doubts even as he wrote *Cry Havoc*. By mid-decade Rex Warner's view of bourgeois pacifism was openly current enough among literary Leftists to get into *LR* in the shape of his poem 'Pacifist (July 1935)':

> On lovely levels, whining in a wash of gold
> pacifists, pale as porridge, are burying bugles,
> who feel no fight, who are learned liberals.
>
> Blow, bugle, blow! Let loose, lungs, over the meadows
> to tell these twisters that war is being waged
> cruelly in complete peace and pleasant pastures.

Julian Bell's case now looks a rather typical pacifistic rake's progress: born into the heart of Bloomsbury and Cambridge pacifism, editing *We Did Not Fight* (1935), a volume of pacifist recollections of the War, and then steadily backtracking and finally dying in Spain driving a conscience-appeasing ambulance. And indeed, how could Bell's generation be expected to resist the tidal wave of the destructive element sweeping the period along? Violence was the period's inherited destiny, one that Orwell, for instance, found himself having repeatedly to come to terms with—over Spain, and especially at the end of the decade over a patriotic war for the survival of his imperialistic homeland. It was a violent destiny he could withstand on neither occasion, as he admits in 'My Country Right and Left' (*FNW*, Autumn 1940), for

> On and off, I have been toting a rifle ever since I was ten, in preparation not only for war but for a particular kind of war, a war in which the guns rise to a frantic orgasm of sound, and at the appointed moment you clamber out of the trench, breaking your nails on the sandbags, and stumble across mud and wire into the machine-gun barrage.

No wonder Orwell's generation of writers turned so ferociously on Aldous Huxley for trying to prevent them from re-enacting the First War that was so vitally in their blood. Spender, in *Forward from Liberalism*; Spender again, in *Left Review*;[47] Christopher St John Sprigg;[48] C. Day Lewis: they all charged fiercely in. Huxley's Peace Pledge Union pamphlet *What Are You Going to Do About It?: The Case For Constructive Peace* (1936) must be one of the '30s' most persistently denigrated texts. *We Are Not Going to Do Nothing* toughed Day Lewis (1936). And Huxley's antagonists were right: his pacifism *is* cranky, an advocacy of the efficacies of prayer and private goodness amidst public horrors that cried out for more resistance than a quietist recipe for the well-heeled author who would shortly retire to vegetables and mantras in the Hollywood sunshine far from the threat of gas-chambers and bombs. Some texts are weightily impressive because of their wrongheadedness. Just as the reiterated assurances of Wyndham Lewis and other Fascists such as Henry Williamson that Hitler and the Germans have unwarlike intentions fail to cut any ice at all, so Huxley's most strongly spiritual fictions of the period, *Eyeless in Gaza* and *After Many a Summer* (1939) read like cosily wilful evasions in the hour of the dictators—as skittishly wayward as Virginia Woolf's pacifist-feminist resistance to meeting the war issue head-on in her *Three Guineas* essays (1938) is vexingly round about. And it's notable that even Isherwood, although he hankered all through the '30s for an apolitical retreat into some privatized island of peaceful sex, didn't bring himself to withdraw to pacifism and Huxley and Hollywood vedantism until right at the end of the period, in 1939–40.

4

In the Cage

'ALWAYS *killing!*' The representative voice of this age of the dictators, the gas-chamber and the aerial bombardment of civilian populations is certainly not that of the Peace Pledge Union. Much more likely a candidate is the extraordinarily popular crime story. 'People are today predominantly interested in crime', Robert Graves asserted to T. E. Lawrence in 1933, and 'Yes', his historical novel *I, Claudius* was 'a crime story'. '[T]hey like reading about murders, so I was careful not leave out any of the six or seven that I could tell about', Graves revealed in his more than customarily arrogant 'Postscript to "Goodbye to All That" ' (in *But It Still Goes On*, 1930): an odd comment on *Goodbye to All That* which was in some senses about several million murders. But Graves's animadversion serves not only to advertise the popularity of the murder-story but also to alert us to how that torrentially popular cult arrived precisely in time with the post-war destructive element. It is not accidental that Graves should go on to link the ways of detective stories with those of 'the *genre* war novels':

> These treat the already classical themes of fire-eating officer and sullen men, drunken officer and gallant men, fraternisation of mortally wounded soldiers in no-man's-land, or unjust Court-martial on the supposed coward, in the same free way as the murdered-financier, stupid-village-constable, cigarette-stub-in-the-garden, suspected-hero themes are treated by the detective writers.

It's not a random fact either that Agatha Christie, the '30s' completest detective-story author, got the idea of making her Poirot a *Belgian* detective from reflecting upon wartime refugees from 'Little Belgium': Poirot entered the detective world as, quite literally, a wartime refugee.

There are, to be sure, differences of form and technique to be kept constantly in mind between thrillers in general and the detective story in particular. T. S. Eliot was, like Auden, a serious analyst of the detective stories that he was hooked on: specially keen on Sherlock Holmes, he was eager in the earlier years of the *Criterion* not only to discuss Conan Doyle but also to analyse the 'rules' of detective fiction in general[49] and to notice all kinds of murder fiction in the review pages of his paper. And Eliot actually proposed three categories: Detectives, Thrillers and Curdlers ('A proper Curdler has something nasty about it', like James's *The Turn of the Screw*).[50] And these remain possible and helpful distinctions. They do not, though, at all obviate the clear case that the chillingly overwhelming, the ritualistically repeated feature of all of this kind of fiction is the presence in all of it of corpses, cadavers, the bodies of violently bumped-off people. When there is no actual corpse, as in Dorothy Sayers's *Gaudy Night* (1936), the absence is distinguishable as an especially notable, recherché variant on the norm. And what is staggering is how common among '30s writers this corpse-mongering was. It wasn't simply confined to the obvious groups: the producers of thrillers, the writers like Greene and Ambler who wrote extremely literate variants on the thriller and the well-known detective writers like Dorothy

Sayers and Agatha Christie. There was also the well-known Christian Charles Williams who produced a string of 'thrillers as thrilling as any' (the compliment was T. S. Eliot's). And G. D. H. and Margaret Cole, the well-known leftist economic historians, wrote a sizeable fistful of detective stories; so did C. Day Lewis (as Nicholas Blake); so did Christopher St John Sprigg (who turned himself into Christopher Caudwell the Marxist critic). Raymond Postgate, another leftist historian, wrote a detective novel about a murder trial, *Verdict of Twelve* (1940)—one of its epigraphs is Marx's famous allegation about social existence determining consciousness, a notion Postgate seeks to exemplify in his jurors' responses to the evidence they hear: they include a university Marxist, Auden imitator and *Left Review* contributor who craves 'a Marxist interpretation of the evidence'. According to the *Daily Worker* (6 February 1937), the Marxist commentator, critic, and novelist Ralph Fox left behind him a half-finished detective story when he was killed in Spain. Alec Brown, the 'proletarian' novelist, was also the author of *A Time to Kill* and *Green Lane or Murder at Moat Farm* (both churned out in 1930). A number of '30s authors would enter this particular genre a little later on: Julian Symons, who already had a crime story in the drawer in the '30s,[51] and who went on to do the best book on this phenomenon, *Bloody Murder* (1972), known in the US more politely as *Mortal Consequences*; Ruthven Todd (who as R. T. Campbell knocked off ten, or even twelve, detective stories in six months in 1945 in an effort to clear his debts); Montagu Slater (who wrote *Man With a Background of Flames*, 1954); Roy Fuller. Others wondered about having a go: in January 1940 Edwin Muir sounded Herbert Read's opinion on whether a detective story would help him and Willa Muir make any money. 'Willa and I have actually been thinking of starting a detective story, which we have been making up in our minds for several months now for our own entertainment.' Read was an obvious person to consult. He shared the enthusiasm of Eliot and the *Criterion* (on which he was Eliot's editorial assistant); he reviewed detective fiction for the shortlived periodical *Night and Day*. Nor was Read alone among serious '30s people of letters in putting thrillers before the public in their reviews: at various time in the period Dorothy Sayers, Charles Williams, and Dylan Thomas turned an honest penny this way. In 1940 Thomas and John Davenport collaborated on a detective story, the *Death of the King's Canary*, considered too libellous, alas, about the English poets they mocked and parodied in it to be publishable until much later in 1976. F. Tennyson Jesse, whose novel *A Pin to See the Peepshow* (1934) is a fictionalizing of the Thompson and Bywaters murder case, was a notable criminologist as well as an excellent novelist, famous for her *Murder and its Motives* and the editor of six volumes of the *Notable British Trial* series. The action and motifs of the thriller even spilled over into poetry: into poems like Roy Fuller's 'Poem' ('The murdered man was rumoured up again') and 'Thriller' ('The body was found in the first chapter'), both aptly received into the surrealist *Contemporary Poetry and Prose*,[52] or Fuller's 'To Murder Someone',[53] or Ruthven Todd's 'Poem' ('The tall detective on the landing-stage'),[54] or Auden's own 'Detective Story' (July 1936). It was a symptom of Eliot's decline, jeered Montagu Slater in a *LR* analysis of the current state of English letters done in the light of the new Soviet doctrine of Socialist Realism (October 1935), that he'd 'ended by writing . . . a poetic play with a detective-story title *The Murder in the Cathedral*'. (*Murder in the Cathedral*, Eliot's actual title, sounds, of course, a bit less like a '30s detective novel: though it is also the case, as Eliot told Harvey Breit in 1948, that the working title for the play was 'The Archbishop Murder Case' and that this was dropped so as not to confuse whodunit fans.)

So many thrillers and detective stories were produced and read at this time (Wyndham Lewis joyously passed on the rumour that the Cole stories were among Hitler's favourites; characteristically, Evelyn Waugh consoled himself on his travels with strings of fictions in which 'people's faces burned off with juice of tropical cactus' and the like) that having a critical attitude in the matter was almost inevitable. Eliot we've seen getting to grips with detectives in 1927 ('During the last year or two the output of detective fiction has increased rapidly'); Auden would fizz away later on the subject of what was for him a major craving and distraction;[55] Alick West tried out a few clumsy Marxist propositions (crime as capitalist disintegration; nice detective policemen a bourgeois myth to cover up the real baton-wielding sort) in his two-part essay 'The Detective Story';[56] Laura Riding plodded over the subject for forty-odd pages of her usual waffling opacity, describing the results of an investigation into 'why people are so interested' in crime, that Graves considered to be '*essential work*'.[57] And the period's habitual mockers naturally turned their mocking attentions on to this fashion. The narrative of Graves's novel *'Antigua, Penny, Puce'* (1936) warns the reader mock-solemnly not to expect this intrigue about a postage-stamp to go the usual way of thrillers ('nobody will try to forge a copy, nobody will even try to steal the original. There will be no murders because of it, and the services of Scotland Yard will not be called upon'). P. G. Wodehouse, whose fun tends reliably to map the fads he continually mocks, sends up the crime story marvellously in his story 'Strychnine in the Soup' (*Mulliner Nights*, 1933). 'Nothing in modern life', opines Mr Mulliner, 'is more remarkable than the way in which the mystery novel has gripped the public', and he tells of his nephew Cyril who 'had a greater passion for mysteries than anyone' and meets Amelia Bassett at a stage performance of *The Grey Vampire*:

'You are evidently fond of mystery plays.'
'I love them.'
'So do I. And mystery novels?'
'Oh, yes.'
'Have you read *Blood on the Banisters*?'
'Oh, yes! I thought it was better than *Severed Throats*.'
'So did I', said Cyril. 'Much better. Brighter murders, subtler detectives, crisper clues . . . better in every way.'

But, this joking aside, what *was* the effect of all these corpses, stabbed, shot at, poisoned, disfigured, headless, faceless, in vicarages and colleges and railway carriages, on beaches and headlands and golf links, just about everywhere and anywhere in fact? Its perpetrators and its fans were prone, of course, to gloss their fascination. Auden talked—it was his solution to other literary modes, notably travel literature—of the quest for the Grail as the key to detective stories' attractions. Graham Greene, purveyor of many a seedy little crime, not least in stories like 'The Case for the Defence' and 'A Little Place Off the Edgware Road' (both 1939), also went in for theological analogy:

Murder, if you are going to take it seriously at all, is a religious subject; the interest of a detective-story is the pursuit of exact truth, and if we are at times impatient with the fingerprints, the time-tables and the butler's evasions, it is because the writer, like some early theologians, is getting bogged in academic detail. How many angels can stand on the point of a needle?[58]

Dorothy Sayers clearly had similar feelings. When she sought to explain, in her book *The Mind of the Maker* (1941), the creative activity of the Holy Trinity she used as model what is evidently the activity and the fiction of the traditional detective-story writer. On such views detective stories were rather wholesome affairs; they did not minister to depravity or evil (despite Hitler's rumoured regard for G. D. H. Cole); on the contrary, they appealed, Auden suggested, to good people who weren't excited by all that violence, but who felt quite dissociated from what they were reading about.

One has one's doubts. So did Wyndham Lewis, who repeatedly linked the First War and the prevailing political destructive element with the thriller's world. Contemporary Europe has, he alleges in *Left Wings Over Europe* (1936), 'gone "Crime Club" '.

> 'How like a shilling shocker continental politics have become, full of the most impro-bable Secret Service gentlemen, sleekly embarked upon the most ruffianly missions, is not realized by the English . . .
>
> The France of Stavisky; the homeland of Capone in the grip of a *Machtpolitik*; the martial-law conditions obtaining in Germany; not to mention Communist Russia, hermetically sealed down upon its wretched inhabitants, and ruled by a permanent terrorist élite.

The crime-fiction fad is angrily savaged by Lewis in *The Roaring Queen* (his 1936 novel, not published until 1973), in which the crime novel literati are faced with the consternatingly bullet-riddled body of award-winning Donald Butterboy, and also in *The Apes of God* (1931), in which civilization, especially the French writers united with André Gide in what Lewis presents as a brotherhood of jailbirds, 'the league against Law', is shown caught up in crookery, in thieving, in 'a criminal mad-house' as defined by Edgar Wallace and Phillips Oppenheim. Jail is now, gibes Lewis, *de rigueur* 'in the best literocriminal New York circles'. 'Western training-for-war', the Great War, the 'Great Adventure', have brought about this cult of the *roman policier* in a *roman policier* world; it's part of a 'blood psychosis' helping prepare for the next gigantic blood-letting. Charles Dolphin argues so near the end of *The Roaring Queen*.

Already in 1931 Cicely Hamilton reported (in her *Modern Germanies*) the growing *Judenhetze*, Hitler's Jew-baiting, alongside the German fondness for the British detective-story, the *Kriminalroman* (which the Germans did not, she says, do very well for themselves). Reginald Southose Cheyney, the Mosleyite detective novelist (his first full-length novel, as 'Peter Cheyney', was *This Man is Dangerous*, 1936) is credited by Julian Symons with being the first to introduce torture for pleasure by the 'right' side into this kind of fiction. And Eric Ambler, forcibly enacting the connections that the Lewisian observations underline—much more forcibly and deliberately than Graham Greene's milder attempts at the same sort of linkage—essentially makes the British *Kriminalroman* grow up into an awareness of the 'real' violence its doings were forever shadowing forth.

Posing the realities of one's own current novel against the falser forms and expectations of the genre was, of course, a common enough device of the better detective writers, an aspect of their high formal self-consciousness. Agatha Christie, for example, is continually using the activities of Poirot to tease the whole *roman policier* scene with. But she lacks the political knowledge and intent with which Ambler uses the device when in *The Dark Frontier* (1936) he thrusts a middle-aged physicist into the politics of the international armaments traffic and of European revolutionism and reaction. What happens to this unlikely adventurer only vaguely corresponds to the fictional exploits of 'Conway Carruthers', special agent, whose name he is ironically granted. 'This was reality.' So is what embroils Charles Latimer, university lecturer and detective novelist, at the hands of the head of Istanbul's secret police (himself a *roman policier* fan and author of the unfinished *Clue of the Bloodstained Will*) in Ambler's *The Mask of Dimitrios* (1939). Latimer is offered a choice by policeman Haki of two versions of Dimitrios's story, the dull English murder story of convention, and the real, messy account that makes up Ambler's novel.

Ambler's simultaneous parodying and exploiting of the convention is wonderfully done, but always with the intent of addressing the realities of violence that the conventions believed themselves to have domesticated and tamed: the rubber truncheons applied to knee-caps (the '*Totschläger*, or beater-to-death of Nazi Germany') witnessed in *The Dark Frontier*; the revived arts of the third degree dwelled-on in *Uncommon Danger* (1937)—the castor-oil, kickings, lighted cigarettes, dentists' drills, rubber-hose and rubber truncheon to the jaw, of the Nazi concentration camps and Italian penal camps ('the stupid, fumbling, brutish forces of the primeval swamp'). And as the decade's horrors mount Ambler's tone grows increasingly anxious to make the point about 'mankind fighting to save itself from the primaeval ooze that welled from its own subconscious being' (*Epitaph for a Spy*, 1938), 'the insanity' (as it's put in *Journey into Fear*, 1940)

> of the sub-conscious mind running naked, of the 'throw back', of the mind which could discover the majesty of God in thunder and lightning, the roar of bombing planes, or the firing of a five hundred pound shell; the awe-inspired insanity of the primaeval swamp.

For I. A. Richards, the destructive element was explicitly a chaos, the chaos of the modernist collapse of belief, of solutions, of a religious and intellectual order that would make sense of things. And the popularity of the detective story—and here one *must* distinguish it from the mere thriller—doubtless owed a lot to its ritualized acts of determining order and significance amidst the seeming randomness of the murderer's bullet or cut-throat razor. The classic fictional detective, never frightened or appalled, never himself (and occasionally *herself*) a victim of events, never outwitted or daunted, never finally at a loss, finds a way through the chaos of events, follows a train of clues—as Theseus followed Ariadne's clew—through the labyrinth, and at the end of each unfolded plot is enabled to explain, to allocate responsibilities and blames, to isolate the evil-doer and exonerate the rest, in other words to build and uphold a firm structure of social and moral values. No wonder Christians and Marxists—both categories being devoted explainers and orderers—pack the ranks of the detective-story providers and readers.

> In the middle, not only in the middle of the way
> But all the way, in a dark wood, in a bramble,

> On the edge of a grimpen, where is no secure foothold,
> And menaced by monsters, fancy lights,
> Risking enchantment.

The lost-ness of T. S. Eliot in 'East Coker' (1940) is representative. Only a Sherlock Holmes can show him the way to safety and certainties (the 'great Grimpen Mire' was the marsh in *The Hound of the Baskervilles* 'into which one may sink . . . with no guide to point the track'). 'O Poirot, deliver us'; 'O Holmes, deliver us,' chants Auden's *The Orators*. And these detectives do just that: at least within the bounds of the fictions in which they triumph, they always grant answers to the prayer for meaning in an otherwise absurd, murderously contingent universe. Small wonder, then, that the detective becomes for the '30s the type of all solution seekers and perceivers of order.

Louis MacNeice decried William Empson's poems ('definitely the kind of poetry I don't like') for being merely a set of soluble puzzles, games for 'the detection fan, the statistician and the crossword puzzler' (*NV*, August–September 1935). It was an accusation that Empson in his 'Note on Notes' at the end of his *The Gathering Storm* volume (1940) fielded cleanly: yes, his poems were like crossword puzzles; the taste for 'obscure poetry' had indeed arisen 'about the same time as the fashion for crossword puzzles'; 'this revival of puzzle interest in poetry, an old and natural thing', was very good; what's more, his Notes would give the 'answers' just as the newspapers eventually published crossword solutions. For its part, Mass-Observation, which sought to describe the way social life in England really was, claimed it was detective work:

> Even the drab and sordid features of industrial life will take on a new interest when they become the subject of scientific observation. His squalid boarding-house will become for the observer what the entrails of the dog-fish are for the zoologist—the material of science and source of its *divina voluptas*. Not for nothing has the detective become a figure of popular admiration.

That's Charles Madge and Tom Harrisson's *Mass-Observation* (1937), chiming in with the efforts of ex-policeman George Orwell who, researching *Wigan Pier*, enquiring, recording, measuring people's houses with a tape-measure, reminded at least one unemployed man in Barnsley of detectives. Nicholas Blake, i.e. Day Lewis, found it only apt to do his novel about a Mass-Observer, *Malice in Wonderland* (1940) within the established detective-story form. Young Mr Perry of Mass-Observation and James Thistlethwaite, Oxford tailor and amateur criminologist, will each try—rather blunderingly to be sure—to collaborate with Nigel Strangeways in sleuthing out a holiday-camp villain.

But, of course, neither Poirot nor Peter Wimsey nor Strangeways nor any other detective could actually put an end to the killing. The genre which gave them life also enclosed them in an endless commitment to working on crime's unstoppable treadmill. 'What a mistake' it was, Agatha Christie admits in her *Autobiography* (1978), that Poirot had been a retired policeman at the start of his fictional career: he 'must really be well over a hundred by now'. The problem-solver was locked into the destructive element as surely as the victim and the suspects were contained—in that favourite set of detective-story situations—by the bounds of the locked room, the weekend houseparty, the snowed-in household, the Blue Train, the Orient Express or the Nile riverboat. The detective moved, like Death itself, from novel to novel, from container to container. (S)he 'found no end in wandering mazes lost', we might say. For detectives there were only shortlived ways out of the labyrinth, their skills never finally redeemed them from the genre's endless condemnation to a sequence of criminally disposed Chinese boxes, just as, customarily at least, those skills were powerless to hamper the violences of their novels being perpetrated in the first place.

The detective's real helplessness in this sense, then, makes just one more instance of that '30s sense of being *enclosed* in and by the destructive element. This was a period consciously caught *entre deux guerres*, in a *Lull Between Wars* (the title Graves and Hodge first thought of for their book about the '30s, *The Long Weekend*, 1940), disconcertingly in suspension between the last war and the advancing 'Last War'. There's a distinct aptness in David Jones's war-book *In Parenthesis* coming out in 1937, and striking one more emphatic war-note in the midst of the '30s respite from war that was no respite at all. It was easy, in fact, to feel that war's parenthesis marks had expanded so as to join up, close a circle, and so utterly smother and imprison the period and all its doings in a condition of violence.

'I was born', writes Henry Green in his *Pack My Bag*, 'a mouthbreather with a silver spoon in 1905, three years after one war and nine before another, too late for both. But not too late for the war which seems to be coming upon us now and that is a reason to put down what comes to mind before one is killed, and surely it would be asking much to pretend one had a chance to live.' He felt trapped by war and his fictions present people who can't escape. When his schoolboy hero goes blind in Green's first novel *Blindness* a pattern is set for all the subsequent novels that impose an experience of imprisonment upon their characters. The title of Green's 1943 novel about the London Auxiliary Fire Service (in which he served during the blitz) could not have been more succinctly indicative of his recurring nightmare: it's called *Caught*. In Green's *Living* (1929), the proletarians Bert and Lily endeavour to get out of Birmingham, to emigrate and make a better life than the industrial Midlands offer, but they only get as far as Liverpool. There Bert abandons Lily, and she has to return to her native city that encloses her as surely as the three 'coffin-shaped lumps of metal' at Prescott's foundry enclose the bodies of suicidal workers who jump into the molten metal pouring from the furnaces. Lily is drawn irresistibly back to her prisoning circumstances like the homing pigeons the novel keeps noticing, and with which, circling fascinatingly and irkingly round a pram Lily is pushing—she scatters them 'with a loud and raucous cry'—the novel ends. *Party Going*—the novel that had occupied Green's imagination for almost the whole of the decade (he worked on it from 1931 to 1938)—opens with a pigeon in London fog smacking 'flat into a balustrade' near the Departures entrance of a London railway terminus at the feet of Miss Fellowes, who has it wrapped up in brown paper and tied with string with the help of a railway attendant. There'll certainly be no more flights, nor even any attempted departures, for that now thoroughly contained bird: a striking emblem for the novel whose gaggle of upper-class characters finds it impossible to make their getaway from England to the Continent. They are for their part triply confined at the railway station: by the dense fog, then by the dense crowds milling dangerously about the platforms also unable to get away, and finally in the station hotel in which they're given by the management a refuge that's ominously difficult to distinguish from an imprisoning. 'We're shut in now,' observes the girl called Julia, as the steel shutters are erected across the hotel's main entrance: locked, as we saw earlier, into their own (and Henry Green's) selfishness, class-isolation, and guilts, their criminal feelings about being rich and cosseted. Julia had felt like 'a poisoner' in the London crowd; Alex thinks their private view of the mass of people on the station is like 'a view from a gibbet'. Violence murmurs intensely at them: not only in the oppressive noise of the crowd that's 'like numbers of aeroplanes flying by' and so brings bombing to their minds, but in Julia's fantasies of 'those frantic drinking hordes of awful people' breaking uncouthly in—'they'll come up here and be dirty and violent'. These bourgeois protagonists are locked into a Buñuel nightmare and one from which—importantly—the hotel's socially and morally ambivalent house detective can provide for them no way out.

Henry Green's *Party Going* makes a powerful rebuttal of detective solutions— more powerful even than (Henry) Graham Greene's passing exposure of Mr Muckerji the Mass-Observer's shortage of detection skills in *The Confidential Agent* (1939), a novel that's pervasively sceptical, as is much of Greene's writing, about the due processes of the Law, and which ends with a group of bogus 'detectives' rescuing its hero D. from the clutches of the real ones. *Party Going* also grants perhaps the '30s' most powerful fictional imaging of the destructive element as prison. Its feelings are Spender's. In *The Making of a Poem* (1955) he entitled his later reflections on the I. A. Richards' footnote and the arguments Richards sited around it, 'Inside the Cage'. For Spender, the destructive modernist belief-chaos composed a cage. So too it did for Eliot.

In the Notes Eliot supplied for lines 366–76 of *The Waste Land*—lines in which cities collapse apocalyptically, Bolshevik hordes swarm over the 'endless plains' of Europe, and a suggestion of a Christ figure remains a stubbornly hooded enigma— Eliot quotes Hermann Hesse's *Blick ins Chaos* to the effect that half of Europe, or at least half of Eastern Europe, is 'auf dem Wege zum Chaos', on the road to chaos. This wasn't the first coincidence between Eliot's mind and Hesse's. Hesse's article 'Recent German Poetry', gloomy about 'The experience of the Great War, with the collapse of all the old forms and the breakdown of moral codes and cultures hitherto valid', had featured in the very first number of the *Criterion* (October 1922) on the occasion of *The Waste Land*'s own first (noteless) appearance. And Eliot had originally intended that the 'A Game of Chess' section of *The Waste Land* should be entitled 'In the Cage'. The title was, as Stephen Spender well knew, that of a *nouvelle* of Henry James, the novelist whose fictions schooled Eliot's early poems as well as majorly preoccupying Spender's *The Destructive Element*. James's 'In the Cage' is the story of a girl post-office clerk stuck behind the metal grille, or cage, through which aristocratic customers pass their laconically disjunctive telegrams. All day long within her occupational prison, through which these chancily enigmatic messages, bits of utterance and encoded meanings flow, she strains to make sense of the tantalizing puzzles, the epistemological confusion that she's handling, in a strenuous hermeneutical effort that's utterly characteristic of the later James (the New York edition puts 'In the Cage' into the same volume as *What Maisie Knew*). The story makes an extremely important touchstone, taken up as it was by Eliot to stand as a momentous emblem for the twentieth century's wrestle for meaning in the chaos of conflicting information that, apparently, so intensely engenders scepticism.

The post-office girl believes she's really getting to know her telegraphic customers.

Like James's Maisie she keeps implying that she *knows* what's afoot. She toys with the notion of leaving her cage to work for the interior decorator Mrs Jordan. If only she could be confident that Mrs Jordan's work helps one get to know the wealthy more intimately! For their part too the '30s authors were always looking out for escape from the current chaos into certainties. And there were some prominent escape routes. Caudwell begins *Studies in a Dying Culture* with a quotation from Max Planck's *Where is Science Going?* (1933): the iconoclasts have filled first religion, then art, and now 'the temple of science with scepticism'. Caudwell's task is to explain the crisis of belief as it has touched in turn these various resorts from puzzle and chaos, and then to offer a Marxist escape route.

But getting out, on one's own behalf or society's, proved difficult, even for the Christians and Marxists. The shades of certain prison-houses were, of course, relatively easy to shrug off. The growing boy, for instance, soon outgrew the prison of school, at least in its immediate aspects. It was a convention among the writing ex-public-schoolboys in the '30s that their prep schools and public schools had been prisons, and fascist prisons at that, totalitarian places presided over by militaristic dictators of headmasters who made their pupils wield destructive weapons in the Officers' Training Corps, concentration camps run by chilling Gestapo bullies of prefects whose canes were as bad as rubber truncheons. 'The best reason I have for opposing Fascism', claimed Auden, quite in the spirit of most of the volume of school recollections *The Old School: Essays by Divers Hands*, edited by Graham Greene (1934), 'is that at school I lived in a Fascist state'. Eton was 'a humane concentration camp', according to Henry Green's *Pack My Bag*: 'I believe the whole system of government in Germany is founded on that evolved through centuries at the greater British public schools.' Spender's novel *The Backward Son* (1940) mounts an impressive analogy with the forced confessions of the Stalinist trials when the prep-schoolboy hero is made to write to his mother a letter dictated by the headmaster declaring he had lied when he'd earlier complained to her with truth about a certain loss of freedom. And Greene is even more impressive in his novel *It's A Battlefield* (1934, the year of *The Old School*) with the analogies which that prison-haunted novel makes between real prisons and school systems (' "It's just like a school", the warder said, raising his old kind eyes with an expression of reverence towards Block A'). But real prisons were, unfortunately, most unschool-like in that you couldn't just grow out of them: there were few, if any, truly confident graduates from the destructive element's more grown-up cages. '[P]risoners, cages (crowds of them) and bars.' Tom Harrisson's sharply accurate survey of a bunch of 1937 publications in the Oxford student paper *Light and Dark* (February 1938) was a critical exercise that proved to be, on several clear counts, the '30s' best piece of structural analysis, and one so painfully right that it roused intense ire from all sides ('dotty and emotive', 'ridiculous and ignorant', cried Grigson: poets could do without 'Poor Tom'). Here Harrisson put the finger right on the '30s' oppressed sense of being confined, enclosed, islanded, behind bars. He had inspected only *The Year's Poetry* (1937), *Letters from Iceland*, *NW* No. 4 and the Auden Double Number of *NV*, but his insight applies to the whole period. 'We show you man caught in the trap of his terror, destroying himself', announces the Chorus of *The Dog Beneath the Skin*. '[E]ach in the cell of himself': Auden's 'In Memory of W. B. Yeats' (February 1939) finds that imprisonment so axiomatic that it mentions it only in passing: it's not felt to be a proposition anyone will seriously argue with.

This was a period when the older prisoners and internees, like Alick West and Joe Ackerley (Ackerley was wounded and interned in Switzerland during the First War) and the political prisoners like Wilfred MacCartney, prominent Communist jailed for spying, and author of the Left Book Club's book about his imprisonment *Walls Have Mouths* (September 1936), simply formed a continuum with the prisoners of the more recent tyrannies: men like Arthur Koestler who was imprisoned by Franco's side in Spain, and later interned in England when he escaped from Second-Wartime France; or like Spender's boyfriend T. A. R. Hyndman, imprisoned for deserting in the Spanish Civil War. They stood in line with the prisoners of imperialism, such as Orwell described in 'A Hanging' (1931), people in 'the condemned cells, a row of sheds fronted with double bars, like small animal cages': 'the dirty work of Empire', Orwell calls it in 'Shooting an Elephant'—'the stinking cages of the lock-ups, the grey, cowed faces of the long-term convicts, the scarred buttocks of the men who had been flogged with bamboos'—dirty work that imprisons the tyrant in turn ('when the white man turns tyrant it is his own freedom that he destroys'). There was also the perturbing multiplication of the prisoners of the new dictators, men and women like Ernst Thaelmann—the German Communist whose plight was kept prominently before the '30s public in Britain. 'We're shut in now.' No wonder British writers were afraid for their freedom, as England and America filled up with refugees from Germany after 1933 (many of the escapees, like Walter Benjamin, were of course not lucky enough to get away), and as they watched Russian politicians and intellectuals whose words they'd hung on—men like Bukharin and Radek who'd featured so prominently for English Leftists in the dissemination of Socialist Realism after the 1934 Soviet Writers' Congress—being sucked down into the absurdist horrors of Stalin's show trials and associated purges.

There could be, about some of the '30s tradings in prison images, a whiff of the gratuitously unthought-out received idea. *The Muse in Chains, The Mind in Chains* (edited by C. Day Lewis, 1937): titles like that make easy claims. One's scarcely convinced nor much perturbed by all the juvenile clamour about school gestapos. The period's little clutch of British-Museum poems—Empson's 'Homage to the British Museum' ('People are continually asking one the way out'), MacNeice's 'The British Museum Reading Room' ('cells of knowledge'; 'The guttural sorrow of the refugees' a nicely Audenesque phrase)—seem to draw too easefully on the more strenuous prison and destructive-element connections established earlier around the emblem of the domed Reading Room in Virginia Woolf's novel published in *The Waste Land* year of 1922, *Jacob's Room*. Spender's poem 'The Prisoners' (No. XX in *Poems*, 1933)[59] is not only flabbily self-pitying, it's also empty of any real sense of prison. Who, or what, precisely Spender's prisoners are is impossible to determine; they seem little more than close neighbours of the 'pretty bird' of his own fleshly lusts (No. XXVI in 1934) that Spender was so eager to release into an emancipated hedonism:

> Passing, men are sorry for the birds in cages
> And for constricted nature hedged and lined,
> But what do they say to your pleasant bird
> Physical delight, since years tamed?

But prisons became much more real for Spender as he worked desperately to get T. A. R. Hyndman out of the Commissars' hands in Spain. They're real enough too for MacNeice when he's dwelling, movingly, on the prison-house of his Ulster childhood and his father, the Church of Ireland minister (as well as on 'the playing-fields of Marlborough'), in his *Modern Poetry: A Personal Essay* (1938). They're real, as well, for Bernard Spencer, whose poem 'Cage' decides that he is worse off than the canary, chirpy in its neatly clenched 'fists of wire': for the human prisoner can imagine being free of the enclosing iron bars of the oppressive régimes and military saviours that confine him.[60] And they're convincingly real, again, in Plomer's volume of poems *Visiting the Caves* (1936). Caves are one further kind of enclosure. In *A Passage to India* (1924) Plomer's friend E. M. Forster had notoriously presented the Marabar Caves nightmare as part of a sequence of containments, in boxes, compartments, coffins, a general 'bottling up', and no doubt Plomer was drawing on Forster's metaphorical larder. But there are also some more metonymic prisons in Plomer's volume. His 'A Prison for Sale' has a 'by-pass motorist' and a 'hiking-girl' noticing an old prison that's up for sale ('A ruin, where penance was long a cult'), but unable to make their own jailing a thing of the past:

> Hiker and motorist hurrying on
> To similar crimes, and a diet of dreams
> In separate cells.

And Plomer's prisoner in 'The Prisoner', caught in the modern destruction (he has an 'old wound'), finds refuge only in suicide:

> While flights of bombers streak his patch of sky,
> While speakers rant and save the world with books,
> While at the front the first battalions die,
> Over the edge of thought itself he looks,
> Tiptoe along a knife-edge he slowly travels,
> Hears the storm roaring, the serpent hiss,
> And the frail rope he hangs by, twisting, unravels,
> As he steps so lightly over the abyss.

It's as inescapably unfrivolous, that, as Greene's Conradian demonstration in his novel *It's A Battlefield* of how all of London is imprisoned, caged, from the political prisoner to the Assistant Commissioner of Police, all shut up into a sequence of see-through enclosures that give the illusion of freedom but are none the less totally entrapping: the 'glass cages' of busdrivers, the 'steel cage' of the motorcar, the 'little glass rooms' at Scotland Yard and in other offices, the headmaster's study with its glass door, the prison visiting arrangements—'a place the size of a telephone box . . . A look through the glass. A word through the wire.' This novel's London is, in fact, a Dantesque hellish sequence of enclosing circles, whose tube trains run in Outer and Inner Circles, whose buses go 'round and round Trafalgar Square'. 'Mrs Coney was encircled by death and crime and implacable justice.' London is an outer circle of Hell: 'This was not the worst pain,' thinks the Assistant Commissioner, looking in on his prisoner, 'hope and fear in a cell, visits from the Chaplain; he had a dim memory that someone had once mapped hell in circles, and as the searchlight swooped and touched and passed, and the bell ceased clanging for Block C to go to their cells, he thought, "this is only the outer circle".' 'At each station on the Outer Circle', that first chapter of the novel ends, 'a train stopped every two minutes.'

Mr Surrogate, the Bloomsbury Communist of *It's A Battlefield*, surprising a mouse in his bookshelves (he'll have it trapped: 'It had nibbled a corner off *The Dictatorship of the Worker*'), had been driven to reflect on 'the great Russian novelist comforted in the Siberian prison by the nightly visitations of a mouse. "I too. The prison of this world"', and his eyes filled with tears'. Less sentimentally, Arthur Koestler uncovered the nightmares of Stalin's G.P.U. prisons in *Darkness at Noon* (1940): fuller revelations about Siberia would come along later from the hand of Solzshenitsyn. And Isherwood dwells horrified in *Mr Norris Changes Trains* on the torturing to death in a Nazi barracks of the Communist called Bayer: 'It's a funny thing', says that novel's sleuthing Helen Pratt, 'his left ear was torn right off . . . simply looneys. Why, Bill, what's the matter? You're going green round the gills.' 'That's how I feel,' Bill Bradshaw understandably replies.

In the cooler *Goodbye to Berlin*, in the 'The Landauers' section, a couple of Europeans are overheard discussing the case of the Berlin Landauer family. Bernhard is dead, of 'heart failure': 'That's what the newspapers said':

> 'If you ask me', said the fat man, 'anyone's heart's liable to fail, if it gets a bullet inside it.'
> The Austrian looked very uncomfortable: 'Those Nazis . . .' he began.
> 'They mean business'. The fat man seemed rather to enjoy making his friend's flesh creep. 'You mark my words: they're going to clear the Jews right out of Germany. Right out.'
> The Austrian shook his head: 'I don't like it'.
> 'Concentration camps', said the fat man, lighting a cigar. 'They get them in there, make them sign things . . . Then their hearts fail.'[61]

Evelyn Waugh's heart clearly failed him as he contemplated the prison-house. In his travel-book *Waugh in Abyssinia* (1936) he's appalled and ironic about Abyssinian incarceration. The town of Harar had a new, modern prison all right, its showers, bunks, and dining rooms adding up to better accommodations than Addis Ababa's hotels offered, but it was only for showing to journalists: prisoners still went, permanently manacled, unfed, into Harar's old place, 'the lowest pit of human misery to which I have ever penetrated'. This was yet one more prison trauma for Waugh, whose first novel *Decline and Fall* (1928) had bizarrely conducted its long-suffering hero Paul Pennyfeather from one grimly comic enclosure to another: from Scone College, Oxford, to Llanabba Castle School whose head is Dr Fagan (Waugh evidently intends to recall the prison horrors lived through by Dickens's Fagin before his execution, in *Oliver Twist*), to Sir Wilfred Lucas-Dockery's model Black-stone Jail (where the doubts-burdened Chaplain Prendergast from Llanabba is bloodily slaughtered by a religious lunatic given too much free rein under Dockery's liberal practices), to Egdon Heath Penal Settlement ('anyone who has been to an English public school will always feel comparatively at home in prison'; the egregious Grimes from Llanabba is another of the inmates), to Dr Fagan's nursing home with the 'barred windows', and so back to Scone again. Pennyfeather can't ever escape, just as T. S. Eliot's Sibyl was forever shut into her bottle, a handful of mortal dust, and as, in the 'Du Côté de Chez Todd' section of *A Handful of Dust*, Tony Last, like Waugh himself a shocked refugee from English social savagery and marital casualness, is imprisoned by a crazily illiterate Dickens fan in the Brazilian interior, condemned to read Dickens's nightmarish tales of school harshness (*Nicholas*

Nickleby), of the murderous Sykes and executed Fagin (*Oliver Twist*), and of perpetually unavoidable prison in the head (*Little Dorrit*), over and over until he should die. 'Let us read *Little Dorrit* again. There are passages in that book I can never hear without the temptation to weep.'

The temptation to weep must have come pretty strongly to other shut-ins. To the workers, shut into the struggle for life, for example. As Lily is forced back into industrial Birmingham in Green's *Living*, so Walter Greenwood's *Love on the Dole* presents a closed economic circle, ending as it began with the knocker-up clattering along at 5.30 a.m. to rouse people for work, a circle from which you can only escape by crime or by submitting to being, as Sally becomes, a kept woman. Spender's neurotically personal thoughts about cages got considerably toughened (after he'd read some Marx) by their assimilation to the cages in which miners go down to their subterranean labours—as his poem 'A Footnote (*from Marx's Chapter on the Working Day*)' shows:

> 'So perhaps all the people are dead, and we're birds
> 'Shut in steel cages by the devil who's good,
> 'Like the miners in their pit cages
> 'And us in our chimneys to climb, as we should.'

The coal-miner's cage, of course, mightily preoccupied the period's proletarian novelists, as it did the accounts of visitors to coalmines such as in George Orwell's *The Road to Wigan Pier* ('You get into the cage . . . a steel box . . . ten men . . . like pilchards in a tin'). Orwell became obsessed by creatures in cages (the caged rats of *Nineteen Eighty-Four* are peculiarly haunting) and set out to demonstrate how the workers were confined in pig-sty conditions. Bars, wooden-barred windows, iron gratings, iron railings, the bars of metal bedsteads, are strongly prominent among the photographs in the original Left Book Club edition of *Wigan Pier*. One shot of 'A Basement in Limehouse' ('one of a whole street') shows a tripled enclosure: a barred cellar-grid, behind which is a wooden and metal barred window, through which can be glimpsed a bird-cage. 'Note the Parrot in a cage', the caption exhorts. In Bill Brandt's *The English At Home: 63 Photographs* (1936), a volume Orwell shows several signs of having known, angered images of cagings had likewise featured prominently: 'Miners Returning to Daylight', 'Ascot Enclosure Within and Without', 'Their Only Window' (three children peering through barred basement window-panes).

But the shocked excursionist, a Brandt or an Orwell, could never match the insider's habituated disgust, renewed every workday morning. 'The "bang . . . bang-bang" of the wooden droppers falling into place as the cage dropped out of sight sounded strange to Len and the hot, foetid atmosphere of the pit, after the long period of fresh air on the surface, made him choke. He tasted it thick in his mouth and retched like a man who had never been down a pit before.' That's from coal-miner Lewis Jones's novel *We Live* (1939). And going down into the earth in that cage was so fundamental to male coal-mining life—Harold Heslop even featured it in his title, *Last Cage Down* (1935)—that the sense of being caged readily expanded to embrace all aspects of working people's lives in such novels. Workers are imprisoned daily, literally and metaphorically. Len is jailed by coal-owner magistrates for alleged riotous behaviour during a strike in *We Live*: the prisoner's dock in which he stands to be condemned by a lopsided class justice is called 'his cage'. In Walter Brierley's *Sandwichman* (1937) the Labour Exchange becomes a set of cages through whose bars the unemployed receive their meagre handouts. In an eloquent passage in Brierley's *Means Test Man* (1935) the Cook family is shown penned tightly into the cage of the iniquitous Means Test: 'penned in a small space . . . like a lot of cattle . . . provided with what was thought enough for them. Thousands of harassed men, women and children were penned with them, beings with no independence, no freedom, underfed, underclothed, not trusted. Yet from behind their barrier they could look out into the real world and see folk less intelligent, less sensitive, less capable, moving about with even minds and certain feet, taking and giving, choosing and discarding, who would think the world gone mad if a man were to come and sit at their table and demand to know every secret of their domestic life. No one could know except the sensitive members of this suffering herd what the bitterest bitterness was . . .'.

Among the epigraphs to *Love on the Dole* is a letter written from prison by Rosa Luxemburg to Sophie Liebknecht ('We shall still live . . .') and a quotation from Sassoon's *Memoirs of an Infantry Officer*: 'in 1917 I was only beginning to learn that life for the majority of the population, is an unlovely struggle against unfair odds, culminating in a cheap funeral.' Thoughts from prison and from 1917: they're inevitable joint reflectors of the '30s worker's plight that go on ringing through the period. Towards the end of *Memoirs of a Fox-Hunting Man* Sassoon describes being in a 'steel hut' in France: it was 'like being inside a boiler'. And steel boilers, bins, boxes and their ilk pile up impressively in '30s writing. There may be some doubt, of course, how precisely significant it is that in Ambler's *Uncommon Danger* Kenton the veteran Spanish Civil War reporter and Zaleshoff the Soviet agent should be imprisoned and nearly suffocated in a vulcanizing tank; or that Ralph Bates's revolutionary hero in *Lean Men* should have, as the revolution collapses, to take refuge in an old boiler on the docks from which he scarcely escapes with his life. Though one cannot help thinking it's not accidental that Ambler should have his pair move from that metal bin to a modernistically chromium-steel plated apartment belonging to a vitriol-scarred old veteran of European revolutions. (For his part, prison-haunted Waugh is as fascinated by the awfulness of chrome-plated walls in *A Handful of Dust* as he is by people committing suicide by putting their heads into the metal boxes of gas ovens—as does Lord Balcairn in *Vile Bodies* and at least 'two more chaps' in *A Handful of Dust*.) But there's no doubt about the conscious meanings James Hanley attaches to the metal container.

In Hanley's 'Men in Darkness' (1931) a wartime ship's stokers are sailing under 'sealed orders', in the 'dark pits' of the stokehold, condemned in that metal box within the metal box that is a ship to feed the metal boxes of the boilers (no joy for them there, unlike for Michael Brady's 'scut' Elsie, who'd been caught having sex 'in a bloody ship's boiler with that fellow Farrell'). The ship is hit by a torpedo. The stokers fight to climb out of the stokehold. They're dashed against the furnace doors by the inrushing water. Blood gushes from one man's mouth in the hustle for the ladder. In the lifeboat, maddened Brady raves against the War and against novelists who glamorize the sea:

> 'What about bloody war in ship's bunkers ship's stokehold bleedin authors shipping on tramps as passengers damn-all to do and down to Borneo and Gulf and other places and masterpiece written in London true story of the sea. Know nothing about bloody war they don't.'[62]

Hanley's considerable '30s achievement was in story after story to bring to light the plight of the men in the dark post-war proletarian bunker, the 'bleak' place as he calls South Wales in *Grey Children: A Study in Humbug and Misery* (1937), 'dark and secretive', from which there is no escape for the miner and his family. As Henry Green's metal-workers leaped into the molten, so Hanley's Reilly, the old ship's fireman making *The Last Voyage* (1931) before his pension (much moved as he bids farewell to children and wife, 'always kissed her on the lips on sailing day': 'Nearly in tears he was, for was much in his mind, and "the heart is a terrible prison", he said to himself'), unable to bear being old and weak and clumsy, afraid of dismissal, puts an end to his life: 'Hey! Jesus Christ! HELP! HELP! Reilly's jumped in the furnace.' What helps make Hanley's *Boy* (1931) one of the period's most affective fictions is the sequence of incarcerations its youthful hero is thrust into. Pressed by a harsh father to leave school at thirteen (he was bereaved of his brother and sister in the War), he becomes a de-scaler of ships' boilers, a filthy, dangerous occupation. At one point, inside a boiler that's smelly and very hot, he weeps in consternation. 'The mortality rate amongst these scavengers of civilization was never inquired into by any of the companies concerned, though sometimes a boy was burned to death or suffocated in a boiler.' So he stows away in a ship's bunker—he's discovered half-dead under the coal—only to be oppressed by the terrible attentions of lustfully pawing seamen and to catch syphilis in a brothel. Then, very ill, delirious, crying for his mother, he's vengefully smothered by the ship's captain:

'Syph is really *for* boys. Boys like you. Boys who tell lies. Boys who do not do as they're told . . . Syph has got you.'

His body is dumped overboard: 'Lost at sea . . . Aged fifteen years.' When two welders are lowered into a ship's ballast tanks in ex-seaman John Sommerfield's novel *May Day* (1936) the indignations with which the scene is loaded clearly owe not a little to the anger of all such scenes in ex-seaman Hanley's writings:

The confined air grew heavy with fumes. Men often collapsed down here. The fumes were poisonous. In the end you died of them: the work was shortening the men's lives.

Sexuality and its attendant problems also compounded the prison in which many '30s women found themselves. The effort of Stevie Smith's heroine Pompey in *Over the Frontier* to escape from her suburbs-loving Freddy is yet one more case of the dangerous underwater metal container:

Well, imagine that it is a sunk submarine from which you will escape. But first you must stand quietly and without panic until the flood-water in the escape-room is covering your shoulders, is creeping up to your mouth, and only then when the whole of your escape-room is flooded to drowning point will you be able to shoot up through the escape-funnel, to shoot up for ever and away . . . Not until I am flooded with a dislike pointing to hatred can I escape . . . But I must escape from him . . .

In Jean Rhys's stories and novels rooms won't ever turn into escape-rooms. Her lost girls, commanded like their author (for these stories are hauntingly close to Jean Rhys's own life-story) by financial need and the craving for love and affection, shunt from one claustrophobically lonely and oppressive room to another. Theirs is a

Prufrockian world, 'Of restless nights in one-night cheap hotels', a sequence of narrowing coffin-like cells. 'All rooms are the same', thinks Sasha Jensen in *Good Morning Midnight* (1939):

> I lay down [this is Anna Morgan of *Voyage in the Dark*, 1934] and started thinking about the time I was ill in Newcastle, and the room I had there, and that story about the walls of a room getting smaller and smaller until they crush you to death. *The Iron Shroud*, it was called. It wasn't Poe's story; it was more frightening than that. 'I believe this damned room's getting smaller and smaller', I thought. And about the rows of houses outside, gimcrack, rotten-looking, and all exactly alike.

Writing to her friend Peggy Kirkaldy in 1941, Jean Rhys says she'll 'often think of your warm cool rooms', but her own lot is the horror of 'complete loneliness'—'my solitary confinement. And what it's made of me'. Sasha Jensen, always searching for the room that will be different ('Who says you can't escape from your fate? I'll escape from mine, into room number 219'), has tried rooms everywhere, Paris, London, Amsterdam, Brussels, and they've all been deadly, violent, as oppressive as the men she takes into them. Like the man René, with whom she enters the room the novel ends with: '*Je te ferai mal*', declares this 'wounded' man, scarred from ear to ear, 'I will do you harm'; and so he does, biting her until she bleeds:

> Now the room springs out at me, laughing, triumphant . . . '*T'as tu compris? Si, j'ai compris . . .*'. Four walls, a roof and a bed. *Les Hommes en Cage* . . . Exactly.

Likewise, Elizabeth Bowen's girls, though frequently younger and less experienced in sex than Jean Rhys's, tend to get caught. It's woman's recurrent fate. Irish Karen reflects in *The House in Paris* (1935) that 'Being caught is the word for having a child'. (*Voyage in the Dark* ends with Anna Morgan's distressing abortion: 'You girls', snaps a doctor, 'are too naïve to live'.) At the beginning of *The House in Paris* Henrietta, one of the several pushed-about Bowen girls, perpetually in transit, comes to a Paris sinisterly shuttered and barred. Its houses are fortified with grilles and shutters. In particular, the house she's brought to has wallpaper with 'barlike stripes':

> It is a wary business, walking about a strange house you know you are to know well . . . From what you see, there is to be no escape. Untrodden rocky canyons or virgin forests cannot be more entrapping than the inside of a house, which shows you what life is.

There are more would-be innocent stripes—on the awnings of a patisserie, on chair covers, on carpets, and such—that are also ready to shed their innocence fast, in Bowen's *The Hotel* (1927). Here the multiplicity of striping patterns becomes one more set of signs of the routinated social bondages and blindnesses that the novel's English people abroad, and particularly the neurotic 'modern girl' Sydney Warren, are held in:

> stripes prevailed, on the tight brocades of the upholstery, on the mats methodically diamond-wise on the polish; stripes were repeated innumerably in the satiny wallpaper and the lace-blinds over the door. One had the sense of being caged into this crowded emptiness.

And all this novel's stripes lead neatly into the zoo-thoughts that Elizabeth Bowen liked to indulge in. As Sydney and her female protector gobble pastries under the awning of the patisserie ('But we're not hungry!' Sydney protests), Sydney thinks of the London Zoo: '*This* seems in a grim kind of way funny . . . It's like being given tea by an aunt at the Zoo.' Mrs Kerr ripostes by wishing Sydney 'had been taken more to the Zoo': it might have made her appetites more normal.

And one of the most striking aspects of literary life in the '30s is just how usual zoo-going was, and the extent to which zoos became prevailing emblematic reference points for writers. At no other period can going to the zoo and thinking about zoos have been so common among intellectuals. For the Bowens, who went to live on the edge of Regent's Park in 1935, the London Zoo made a constant presence. Their Clarence Terrace house got into *The Death of the Heart* (1938) and so in consequence did the neighbouring zoo. ' "It's all draughts and stinks," Eddie said. "But we did think it was pretty, didn't we, Portia".' It's also a quiet emblem of the bad deal Portia is getting from her adoptive relatives. Karen, of *The House in Paris*, perceives that she must appear, to the girl she knows as Yellow Hat, 'like something on a Zoo terrace, cantering round its run not knowing it is not free, and spotted not in a way you would care to be yourself'. In the '50s Elizabeth Bowen would be struck by the similarities between New York's United Nations and 'that penguin pool in the London Zoo which was considered so modern in the 1930s'. So modern indeed was zoo architecture by the firm of Tecton—Britain's most prominent and progressive architectural outfit of the '30s, centred about the refugee Berthold Lubetkin (they built, for example, the Finsbury Health Centre)—that no self-respecting '30s book on buildings was complete without shots of their constructions at the new Dudley Zoo or Whipsnade Zoo or of London Zoo's Gorilla House or Penguin Pool. The '30s zoo was a modernist, Bauhaus-Gropius place, a set of architect's models affording excitingly dinky glimpses of what buildings for human beings might become. Vincennes, said MacNeice in (naturally) his book *Zoo* (1938), was 'one of the most up and coming example of modernism': 'in zoos', he added. But with any luck, or so many writers felt, this kind of building wouldn't stay confined to the zoo.

Tecton's new mid-'30s zoo at Dudley was phenomenally successful, with an aweing crowd of 250,000 flocking to its opening in 1937. But it was Whipsnade, opened in 1931 and designed on the innovative 'open prison' plan of Sir Peter Chalmers Mitchell (later to become still more prominent among writers as the saviour of Arthur Koestler's skin in Malaga during the Franco uprising), and to which Tecton contributed the famous Elephant House of 1935, that made more of the literary running (it was, of course, nearer to London than Dudley, as well as getting in first). Gladys Davidson's *At Whipsnade Zoo* (1934) finds a natural slot in the same series of Nelson Discovery Books for Boys and Girls as Christopher St John Sprigg's *British Airways* and John Sommerfield's *Behind the Scenes*: because zoos were as much central ikons of the period as aeroplanes and the masking disguises of actors and other role-players, but also because in the '30s zoos meant Whipsnade. It was a trip to the newly opened Whipsnade in his brother's motor-bike side-car in September 1931 that helped tip C. S. Lewis finally over into the Christian faith. He 'liked bears', he says in his autobiographical *Surprised by Joy* (1955). There can't have been anything specifically godly about Whipsnade, even its bears: still, T. S. Eliot does, in the second chorus of *The Rock*, describe the Church settling 'all the inconvenient saints, / Apostles, martyrs, in a kind of Whipsnade'. 'Tony and I', Spender thought it worth the trouble of informing Geoffrey Grigson one Sunday in around 1934, 'had a marvellous day at Whipsnade yesterday.'[63]

More sombrely—for a start, it hadn't Whipsnade's open-prison pretension—and so providing a more obvious ground-bass of the period, the Berlin Zoo rumbles through the considerable '30s literature of that city. No visitor could miss Berlin's Zoo Station. It was the Zoo Station that gave Bill Bradshaw and Mr Norris, and Anthony Powell's Chipchase and Maltravers (in his novel *Agents and Patients*, 1936), access to the city's delights and horrors. Chipchase and Maltravers reflect that the gorillas in the nearby zoo aren't too unlike the German people:

> The male gorilla was swinging on his trapeze, very slowly backwards and forwards, facing the people with an impression heavy with hatred and contempt. He had caught something of the national character of the race he found himself among, and his demeanour suggested a prussian captain of industry at his morning exercises.

Berlin's gunmen—like 'the armed *Zuhälter* or ponce' who 'fattens and flourishes' there—were, reported Wyndham Lewis in his *Hitler* (1931), 'the most oddly unlovely gunmen of the earth.' Maltravers gives a *Heil Hitler* to a couple of Brown-shirts: 'Just as well to be on the right side of them.'

The idea that some human beings might be rather fit for the zoo, brought into prominence by the First World War and its aftermath (and given memorably ironic life in David Garnett's little fictional squib *A Man in the Zoo*, 1924) had started, of course, to come naturally as soon as Darwinian assertions had insinuated that that was all everybody might once have been fit for. It can be the cheapest of sneers—a Grigson (in *New Verse*) presenting some ludicrous extracts from an enemy's utter-ances as 'Exhibition Page, or Private Zoo', or a Wyndham Lewis once again spraw-lingly immoderate in the over-ripely insulting *The Apes of God*. Wyndham Lewis habitually thought of his enemies as *Untermenschen*; it came easily to him ener-getically to satirize the English and England's literary and artistic groups as gangs of gone-ape sub-men. Only Lewis, it appears at such moments, is truly human:

> [Matthew Plunkett] yaw-yaw-yawned with the blank bellow of the great felines behind their bars at the scheduled feeding-hour (smelling the slaughtered meat for their consumption and sighting the figures of the keepers, moving to be their waiters, at their private table d'hôte, watched by herds of blanched apes, who had confined them) he roared—disparting and shutting his jaws, licking his lips, baying and, with his teeth, grinding, then again baying, while he stretched the elastic of his muscles elevating his arms with clenched fists, in heavy reproduction of the plastic of the Greek. Then he carried one of his exhausted hands to his head, and scratched it, between two sandy bushes, somewhat sun-bleached.

But slick gibing of this sort is far removed from Elizabeth Bowen's rather pained use of zoos; or Evelyn Waugh's savage reduction in *Scoop* (1938) of London's journalistic scene to a contest between *The Brute* and *The Beast* over stories about 'the Zoo Mercy Slaying'; or the equation made between the imprisoned miner in Heslop's *Last Cage Down* and a Sanger's Circus 'caged lion'; or Malcolm Lowry's confinement of Dana Hilliot in *Ultramarine* (1933) to a life on the waves of the destructive element aboard a floating zoo complete with a caged (and eventually drowned) pigeon and an epigraph from Chaucer about the discontents of a caged bird, even one in a cage of gold; or Orwell's gloomy reflections in *Wigan Pier* on King George VI's yearly camp at which boys from public schools 'mix' with slum lads, 'rather like the animals in one of those "Happy Family" cages where a dog, a cat, two ferrets, a rabbit, and three canaries preserve an armed truce while the showman's eye is on them'. In such cases zoos and caged animals couldn't be set for wider or more telling applications. And this is particularly true of Aldous Huxley's *After Many A Summer*, where the presence of actual monkeys and the monkey-like behaviour among the humans (lustful Dr Obispo is an 'ape-man'; prissily Paterian Jeremy Pordage gets a baboon's pleasure out of picking at a tiny scab on his scalp) are by no means just a wry comment on rich, irreligious Hollywood man and woman. '[W]e couple', reports an eighteenth-century earl's diary that Pordage is employed to work on, 'as I have seen the condemned Prisoners at Newgate coupling with their Trulls, between the bars of our cages.' And so Huxley's humankind as a whole is presented as caged within absurd actions and longings, in 'a kind of spatio-temporal cage', 'the cage of flesh and memory', above all in a lust for a humanity-falsifying immortality. Living for ever would be to make a mockery of even T. S. Eliot's Sibyl's fate, especially if that eternity were lived out in the style of millionaire Jo Stoyte, the Uncle Joe Stalin of this novel's business world. The novel ends with the discovery of the fifth earl, aged 201, victim of the elixir of life he had stumbled upon, caged in the basement of his Surrey mansion, a gibbering, regressed, 'foetal ape'.

And this frightening perpetuation of man's zoo-life is bolted into Huxley's pacifist, religious apprehension of the nature of the destructive element. Pordage, Obispo, the fifth earl: they're all devotees of the Marquis de Sade. The private sphere, like the wider cage, is a violent place. 'On the plane of the absence of God, men can do nothing else except destroy what they have built—destroy even while they build—build with the elements of destruction.' The baboons gibber as the 'transients', some Steinbeckian Okeys, scrape for a living and the doings of Hitler's, Mussolini's and Stalin's 'boys' are discussed ('the liquidation of intellectuals and Kulaks and old Bolsheviks . . . the hordes of slaves in prison camps'). And Barcelona falls as idealist Pete, late of the International Brigade ('Do you want us to sit still and do nothing?' he argues against mystical Mr Propter, using the words of Day Lewis and Spender against Huxley), is gunned down by jealous Stoyte, who mistakes him for Obispo slyly on the sexual job with Stoyte's 'Babe' ('the blood . . . pouring out of the two wounds, one clean and small, the other cavernous, which the bullet had made as it passed through Pete's head').

It was quite in the Huxley spirit, then, that MacNeice should connect in *Autumn Journal*, XXIII, the last days of Barcelona with Noah's Ark:

> two and a half millions
> Of people in circulation
> Condemned like the beasts in the ark
> With nothing but water around them.

And that his *Spectator* report, 'Today in Barcelona' (20 January 1939), should contemplate Barcelona's zoo: 'the Zoo is macabre—a polar bear 99 per cent dead, a kangaroo eating dead leaves.' This was also, of course, quite in the spirit that had earlier fired MacNeice in *Zoo* (1938)—a book that looks quite eccentric as the writing of a major poet (even when you know, as MacNeice did not, that the Russian Formalist Viktor Shklovsky had published an autobiographical novel called *Zoo* in Berlin in 1923), but only until it's granted its context. Like Elizabeth Bowen, MacNeice wrote as a near-neighbour of the London Zoo, but the zoos that fill

Zoo—London, Clifton, Whipsnade, Paris—are by no means just fun places for taking any old contingent day out in: they're seen by MacNeice as popular dream-worlds on the lines of the cinema; they're like games, like art-forms; and they're also prisons that edgily mirror lots of the dilemmas of the '30s.

In zoos, MacNeice finds, the private and the public plight merge tenaciously and suggestively. The London lion-house ('no attempt to suggest that this is a jolly place, no euphemism, no glossing over the fact that this is a prison') leads MacNeice on to nightmare thoughts of Ulster ('bigots, sadists, witchdoctors, morons'). MacNeice sees himself as 'the Irishman' who 'like the elephant, never forgets'. In Clifton Zoo, it's school and anti-semitism that spring to his mind: 'The charm of the Clifton Zoo, herein unlike the College where the Jewish boys are segregated in one house, lies in the eclectic assortment of animals in some of the enclosures.' The animals' watchers, the paying public, are, he reflects, themselves caged like the animals: 'The two of you in adjacent cages.' One should visit the zoo, *Zoo* ends by urging, for vital lessons about life: to hear the animals saying, '(if they ever say anything at all), "Le Zoo, c'est moi" '.

In the '30s cage it was apparently difficult to believe that the world had not—and in an unjeering, unfrivolous sense—gone quite ape, was not indeed sick and mad. Like schools and prisons and zoos, hospitals and loony bins also shut their inmates in with numerous unfortunates and undesirables. 'You'll find a lot of improvements since you were here last,' a warder says to Waugh's Paul Pennyfeather as they enter the Egdon Heath Penal Settlement: authorily grinning tones deftly scooped up and inverted as Waugh's short story 'Mr Loveday's Little Outing' opens (1935): ' "You will not find your father greatly changed", remarked Lady Moping, as the car turned into the gates of the County Asylum.' The loony-bin is a prison and the prison a loony-bin: connections undisguised and not at all unsettled by Waugh's rejoicingly wry handling of Lord Moping's confinement:

> Angela left the asylum, oppressed by a sense of injustice. Her mother was unsympathetic.
> 'Think of being locked up in a looney bin all one's life.'
> 'He attempted to hang himself in the orangery', replied Lady Moping, '*in front of the Chester-Martins.*'

The County Lunatic Asylum was Stevie Smith's name for the destructive element of Richards's belief problematic. The title for her brisk couple of lines:

> The people say that spiritism is a joke and a swizz,
> The Church that it is dangerous—not half it is.

is, starkly, 'From the County Lunatic Asylum.' The lines come with a sketch by their author of a naked, Munch-faced screaming woman, entangled in the gymnasium-like bars of the windows she's staring through. 'It's my belief that some of this gang are simply looneys', Helen Pratt had told Isherwood's Bradshaw with reference to the Nazis. It was natural in *The Dog Beneath the Skin* (II, i) for Auden and Isherwood to portray the Nazis as inhabitants of a lunatic asylum (Spender missed the point when he objected in *NW*, Autumn 1938, that 'the alarming fact about Nazi Germany is that the Nazis are *not* lunatics'). But Germany wasn't the only '30s loony-bin. The hero of Anthony Powell's novel *Venusberg* (1932), the journalist Lushington, leaves his London newspaper offices to go off to a Baltic country as special correspondent; he

> went down the stairs which were of stone like those of a prison or lunatic asylum and were, in effect, used to some considerable extent by persons of a criminal tendency or mentally deranged.

The whole earth could seem, in fact, like a mental hospital. Or, indeed, a general hospital. 'The whole earth is our hospital', Eliot alleges in 'East Coker' (Easter, 1940), the second of his *Four Quartets*, Christianly mindful of humankind's original sinfulness. The idea that the world was ill was not confined to the Christians, however. 'Our culture is sick', declared the anthropologist Malinowski in *First Year's Work 1937–8 by Mass-Observation* (1938): 'The symptoms of this are unmistakeable. Since 1914, most of the historical events have meant destruction or disintegration.' 'Europe is . . . a very sick neurotic', Hermann Hesse told the readers of the first *Criterion*. In *Studies in a Dying Culture* Caudwell agreed: 'One does not have to be a Marxist to declare that bourgeois culture is seriously ill.' At the time a kind of Marxist, Michael Roberts was disinclined in his *New Country* Preface to think otherwise than like Caudwell, accusing the ordinary man of failing to see the link between England's 'illness' and the wider 'disease'. 'What do you think about England, this country of ours where nobody is well?' asks the text of *The Orators* early in its career. T. S. Eliot warmed up to the heavy charges that he would bring against D. H. Lawrence in *After Strange Gods* (1934) by accusing the novelist of being 'a very sick soul'.[64] (Eliot later suggested that he himself had been sick when he gave the Lawrence-accusing lectures that made up *After Strange Gods*.) According to Spender in *Forward From Liberalism* the 'sick soul' was the 'European individualist'. Other writers cheerfully charged themselves with sickness. 'Priding myself, in the sun, on private feeling', confessed Charles Madge in his *New Country* poem 'Letter to the Intelligentsia', 'I knew that I was ill; it was slow in healing' (he does, it must be added, also say that 'All men are ill, I thought'). John Cowper Powys helped pad out his bloatedly Rabelaisian *Autobiography* (1934) with repeatedly gloating insistences on how low his mental and physical condition was 'in the lunatic-asylum which I kept locked up in my cat-head': his sadistic craving to smear females with marmalade; his slobberings over the legs of girls; his belly continually racked by ulcers, acid, bile ('my stomach seething with acids struck from the rock of my nerves by the rod of my vice'); his 'twitching in my nether gut', whenever he recalls a bad time in Hamburg, 'as if I were holding back the excrement of weeks':

> I can remember now exactly . . . when walking . . . through the darkening streets of some Midland factory-town, on my way from the railway station to where my host lived, how it felt to be driven to put down my bag on the kerbstone, and, thrusting my fingers down my throat in sublime disregard of the passers-by, to struggle to retch up some of the cruel burningness in my vitals. I can recall how once when I was doing this, certain artificial teeth, what nowadays they call a 'plate', fell out into the gutter, and how I groped for them in the obscurity and how gratefully I pressed them back, all muddied, into my mouth and taking up my bag hurried forward once more.

In a decidedly unamusing way, the world was a sick joke. John Garland, the bolshy airman of J. Leslie Mitchell's (i.e. Lewis Grassic Gibbon's) novel *Stained Radiance* (1930) who has deliberately smashed his hand with a bullet in order to get himself invalided out of the Air Force and back to his pregnant wife in London, visits her—undernourished, eclamptic, slobbering a brown acid out on to her rubber-sheet—in a paupers' hospital. 'A memory of a day in the Zoo at Regent's Park came to him', and as he watches her torment as she goes into labour he, inveterate blasphemer, has a vision of God as a devilish scientist, experimenting aloofly with the earth, flicking idly at his concoctions in a jar:

> And where he had flicked a finger the crabs of writhing cancer would move in agonised stomach, great tumours would root and sprout in rotting brain-cells, viviparous organs would contract and close on unborn offspring. Then God would replace the lid, make notes, walk away, and stand watching . . . And God himself was but a fleeting drift of atoms in an experiment by yet another super-scientist, who flicked his fingers in the brain of God to induce God to set the plagues and cancers in motion.

This sense of an absurdist world governed by a chain of sick-minded deities visiting random awfulnesses on a human race that had better laugh or its mind will snap is very like Evelyn Waugh's. Casual deaths, maimings, imprisonings, cruelties, and nonsenses of all sorts fill his '30s plots. Just how casually, can be seen in a comparison between Anthony Powell's *From A View to A Death* (1933) and Waugh's *A Handful of Dust* (1934): the hunting death of Powell's Zouch, the rather silly Nietzschean *Übermensch* of a painter, is made to matter humanly much more even in its frivolously gossipy context than the hunting death of Tony Last's little son in Waugh's novel, even though the boy is shown to have been abundantly loaded with parental hopes and family meanings. Waugh's writing and the boy's parents simply shy away from even trying to cope with their disaster and sorrow. And yet it's clear that the casual disasters of Waugh's fiction are a response to seriously felt experiences, to the nonsensical and hideous things Waugh charts in his '30s travel books, the all too many grimnesses he found to confirm the sick joke of his own broken marriage, that dementing personal perception of evil that helped to push him into the Catholic Church.

The world, Waugh kept finding, was mad, it stank, it imprisoned. *A Handful of Dust* began life, in fact, in the short story 'The Man Who Liked Dickens' (it became the chapter 'Du Côté de Chez Todd' later), generated out of Waugh's weird encounter with the religious loony in British Guiana whom he describes in the travel-book *Ninety-Two Days* (1934). Waugh 'thought of' his story as he was holed up in Boa Vista, a ramshackle, worn down nothing of a place, full of emaciated people at their 'final halting place before extinction', on a journey that had been for him unrelievedly unpalatable misery. Travelling certainly proved for Waugh no way of escape from sadness and heart sickness. 'Most visitors to Addis Ababa', he writes in *Waugh in Abyssinia*, 'arrive feeling ill.' And lodging at Addis's Deutsches Haus hostelry, which was opposite a tannery outside which loads of 'decomposing cows' feet' would be parked, they scarcely felt their journey to be giving them very much chance to get better.

In this prevailing atmosphere of encircling glooms and destructive cages it's only natural that dreams of escape and escapist dreams were widely indulged in. 'And each in the cell of himself is almost convinced of his freedom', Auden went on in his poem 'In Memory of W. B. Yeats'. The action of Powell's *Agents and Patients* expands outwards from its opening scene with a street escapologist doing his stuff, to embrace the wish for escape from bondage as its totalizing theme. 'This afternoon', moans Maltravers to his wife, 'I watched a man with a sword who was prodding another man who was gagged and chained and lying on the ground. That's what I feel like in our married life. I lie on the ground gagged and chained and you prod me with a sword.' In particular, the gullible Blore-Smith, a grey but wealthy Oxford graduate, who 'felt himself chained. Chained by circumstance' as he too watched the escapologist's efforts, is whisked around Europe by the knavish Chipchase and Maltravers in a quest for 'life' and freedom under the banners of Parisian sex, German movie culture, and Chipchase's psychoanalysis.

The Mind in Chains could never have been written, proclaimed C. Day Lewis in his Introduction to it,

> were it not for the widespread belief of intellectual workers that the mind is really in chains today, that these chains have been forged by a dying social system, that they can and must be broken—and in the Soviet Union have been broken; and that we can only realize our strength by joining forces with the millions of workers who have nothing to lose but their chains and have a world to win.

These last were, of course, the famous and stirring words from the end of *The Communist Manifesto*, as stirring as Christopher Caudwell's offer in *Studies in a Dying Culture* of release for the bourgeois trapped in the illusion of being free, imprisoned along with Bernard Shaw and D. H. Lawrence: the bourgeois 'animal in a cage' who will 'become free when it realises that a locked cage completely restricts its movements and that to be free it must *necessarily* unlock the door'. As stirring as being shown crowds of workers bursting locked gates and closed doors in Soviet movies, or to have (as Ralph Bates does in his *Lean Men*) 'the masses' storming Montjuich Model Prison, or to have them, as Rex Warner's novel *The Wild Goose Chase* (1937) does, breaching the concrete shell of the Fascist urbs ('Looking up, men who had spent their lives beneath concrete beheld with surprise a sky and George too was surprised to see so pure above their heads the blue, for always in the vicinity of the town the sky had been overcast before now'), or to imagine (as Rex Warner's people do at the end of his novel *The Aerodrome*, 1941) that 'at last the circle of sin', the Fascist aggression that had enclosed an English village, 'might be broken', or to look forward with writer after writer to the time when the destructive element, like Day Lewis's Noahic flood, shall have subsided and fun shall have returned: 'Today the struggle, tomorrow the poetry and the fun', as Edgell Rickword jeeringly put it in the Auden Double Number of *New Verse* in expressing what he took to be the gist of Auden's poem 'Spain'. The hard communist critics resisted the poets' common idea that the revolutionary life itself was destructively imprisoning; thoughts like that they said, should be reserved for Fascism's oppressiveness—though not, of course, as the excuse for the indulging in literary escapism, in fairy, that J. R. R. Tolkien used his observations of Fascism as the pretext for. Why should the prisoner dwell on 'jailers and prison walls' (or 'the misery of the Führer's or any other Reich') Tolkien asks in a version of his 1940 lecture 'On Fairy Stories': the jailed man wants to 'get out and go home', and if he can't he'll go in for escapist thoughts, 'fairy-stories'.[65]

Thoughts of escape included thoughts of cure. Briskly short and sharp measures appealed to Christians, Communists and Fascists alike. The fascist Air Vice-Marshal in Warner's *The Aerodrome* urges young Roy to make a 'clean' cut from his family and the human messiness of the village, 'confusion, deception, rankling hatred, low aims, indecision', the drunkenness and the mud in which the novel's opening discovers Roy, face down. Dictators had a way with corruption, even when degradation hadn't quite reached the gaudily Weimar levels that Hitler dealt so firmly with. Eliot agreed that what the world's 'hospital' needed was the surgeon's 'steel', the bleeding Christ's application of the scalpel. Rafts of Leftists chimed in here too, at least about the knife. Like Randall Swingler, in his poem 'Invitation to Society':

> Come then, this is the season for pruning
> To cut away ruthlessly the dried and involuted plumage,
> Age like a uniform, the cage of your possessions.

It's time, Swingler urges, to 'liven that centre with a real wound'.[66] Which is Day Lewis's notion too, in *The Magnetic Mountain*, 25 (1933):

> Drug nor isolation will cure this cancer:
> It is now or never, the hour of the knife,
> The break with the past, the major operation.

Major operation: Day Lewis's haunted metaphor came from one of the most vicious bits of *The Orators'* 'Journal of an Airman': 'Major operations without anaesthetics begin at noon.' It was staunchly enacted in the politicized plot of socialist James Barke's novel *Major Operation* (1936), just as it was recalled in Rex Warner's novel *The Professor* (1938): what the political situation needs, writes a girl in that novel who has been raped and driven to suicide by the Fascists, 'is a surgeon'. Christina Stead gloated, for her part, over the surgical atmosphere of the Congrès International des Ecrivains held in Paris in June 1935:

> the tough, fiery, humourless young ones, who have to give up their poetic solitudes and soft self-probings . . . take lessons from working men and use their pen as a scalpel for lifting up the living tissues, cutting through the morbid tissues, of the social anatomy

(a report derisively quoted in *The Left Heresy in Literature and Life*). And the prospect of calmly cutting away the old life became so habitual in some Marxist quarters that Cornford could easily borrow its language for his 'Sad Poem' that so heartlessly discarded the girl Ray who was his mistress and the mother of his child:

> Though parting's as cruel as the surgeon's knife,
> It's better than the ingrown canker, the rotten leaf.
> All that I know is I have got to leave.
> There's new life fighting in me to get at the air,
> And I can't stop its mouth with the rags of old love.
> Clean wounds are easiest to bear.

But harder to bear than no wounds at all. As the anally wounded Auden and the multitudes of wounded First World War veterans, of course, well knew.

Day Lewis, much taken by 'the hour of the knife' idea, presents Auden in *A Hope for Poetry* as the satirist with 'scalpel' in hand. But Auden was no Christopher Caudwell. He also looked for cures to psychoanalysts rather than to surgeons:

Publish each healer that in city lives
Or country houses at the end of drives.

Surgeon's knives couldn't reach the 'neural itch', the alleged roots in a neurotic impurity of heart, of psychosomatic illnesses like *The Orators'* great enemy, cancer. The thoughts of one John Layard—'loony Layard' as Auden came to call him—stood behind Auden's position on the psychosomatic. Auden had had some dramatic passages with this disturbed patient of the persecuted American psychologist Homer Lane, during his stay in 1928–9 in Berlin (annoyed by Auden's refusal to love him and by Lane's dying halfway through his analysis, Layard shot himself in the mouth only farcically to survive and dash in a taxi round to Auden's lodging to ask him to give him the *coup de grâce*: an episode transposed into the Edward Blake story in Isherwood's *The Memorial*). None the less, Layard's curious influence had been welcomed by Auden as a reinforcement of his own reading of Freud and of Walter Groddeck, that wild rebel in the Freudian camp. At the end of his article 'Psychology and Art Today', in *The Arts Today* (1935), a key gathering of critical essays edited by Geoffrey Grigson, Auden lists the chief healers he has discovered to date—a merrily bizarre jostle of philosophers, psychologists, critics and writers, I. A. Richards, Gerald Heard, D. H. Lawrence, Blake, Bergson, Freud, Jung, Homer Lane, Groddeck, Maud Bodkin and all. These were precisely the bourgeois healers that Caudwell berated in *Studies in a Dying Culture*: phoney curers in 'the bourgeois cage' with Adler, or Freud himself, who treated sufferers only 'if rich enough', otherwise offering 'the poorer classes', with Jung, 'the old myths, of the archetypal'. 'These', growled Caudwell, 'are the doctors who stand by the bedside of society in its most gigantic agony.' Nor was he mollified by Auden's at least sharing with himself and the would-be Communist surgeons an intense lack of pity for the individual human case ('Pity frustrates every attempted cure', Isherwood quotes a letter of Layard to Auden (Weston) in *Lions and Shadows*. 'If you find yourself beginning to pity anyone who is ill or in trouble, said Bernard, you cannot help him: you had far better abandon him altogether').

'Publish each healer.' But what if the healers are 'few' in number (in the words of Bernard Spencer's poem 'Ill') and their 'charms' 'weak' ones?[67] Already in 1936, in the last poem of his *Look, Stranger!* volume, 'Epilogue', Auden's tone has become regretful and valedictory: 'There were Freud and Groddeck at their candid studies/ Of the mind and body of man'; 'But where now are They . . . ?' As the '30s went on, the time seemed, for all the busy canvassing of so many remedies, to be getting sicker. Cure seemed far away. 'My God! Everyone's in hospital', Louis MacNeice has a character exclaim in 1939, in his unpublished play *Blacklegs*. The 'we' of Spencer's poem 'Waiting' are shut in with the 'crisis' of impending war in a sick time imagined as a hospital where the surgeon's hurtful knives have been sharpened for a fearfully imminent operation: 'like that bad dream of knives . . . Woundable as a man is.' 'This ship'—and the vessel of history and destiny was, of course, one more metal container—'is like a hospital': so Auden and Isherwood were discovering in 1938 as they sailed towards the China war:

we lie limp in our deck chairs, gazing out dully across a sea which is as boring and hopeless as an incurable spinal disease. The nurses are very attentive. All the meals are at unnatural invalid hours.

> Beneath our conversation, our eating, our thoughts, the engines throb, deep down,
> like a fever. This voyage is our illness.

Sometimes they fancy they 'are going to get well . . . But the relapse is immediate and violent. We shuffle off to the bar for our evening medicine: brandy and ginger ale.'[68] And soon those feverishly throbbing engines were overheard in the shape of the bombers Auden deplores in his China sonnets ('the sky / Throbs like a feverish forehead; pain is real').

It had been all very well for Auden to prate to Lord Byron of Layard and the immunity enjoyed by the pure in heart:

> I met a chap called Layard and he fed
> New doctrines into my receptive head.
>
> Part came from Lane, and part from D. H. Lawrence;
> 　Gide, though I didn't know it then, gave part.
> They taught me to express my deep abhorrence
> 　If I caught anyone preferring Art
> 　To Life and Love and being Pure-in-Heart.
> I lived with crooks but seldom was molested;
> The Pure-in-Heart can never be arrested.
>
> He's gay; no bludgeonings of chance can spoil it,
> 　The Pure-in-Heart loves all men on a par,
> And has no trouble with his private toilet;
> 　The Pure-in-Heart is never ill, catarrh
> 　Would be the yellow streak, the brush of tar.

But Layard did put the pistol barrel into his mouth, and Auden himself had those hauntingly painful medical experiences in the matter of his rectal 'wound'. The 'sense of guilt' organized by Christianity, Auden declared in a footnote to his article on the Oxford Group in R. H. S. Crossman's *Oxford and the Groups* (1934), was 'transferred to medicine. The hospital and the asylum became the punishment for moral offences, particularly the sexual.' Auden's own sexual 'offences' had landed him 'inside'. He swaps reminiscences with his Wound in 'Letter to a Wound' about his (metaphorically) War-time treatment:

> Haven't I ever told you about my first interview with the surgeon? He kept me waiting three quarters of an hour. It was raining outside. Cars passed or drew up squeaking by the curb. I sat in my overcoat, restlessly turning over the pages of back numbers of illustrated papers, accounts of the Battle of Jutland, jokes about special constables and conscientious objectors. A lady came down with a little girl. They put on their hats, speaking in whispers, tight-lipped. Mr Gangle would see me. A nurse was just coming out as I entered, carrying a white-enamelled bowl containing a pair of scissors, some instruments, soiled swabs of cotton wool. Mr Gangle was washing his hands. The examination on the hard leather couch under the brilliant light was soon over. Washing again as I dressed he said nothing. Then reaching for a towel turned, 'I'm afraid', he said . . .
>
> 　Outside I saw nothing, walked, not daring to think. I've lost everything, I've failed. I wish I was dead.

Nurses—like mummies and nannies—are always trouble in the '30s cosmos of Auden's writing. So, too, evidently, are surgeons. They represent the misunderstanding or the failure of Layardism. The operating theatre scene in *The Dog*

Beneath the Skin (III, iv) is intense with the fears of the Auden who found Layard wanting and ended up on the surgeon's 'hard leather couch'. This hospital's creed ('I believe', the 'famous amateur cricketer' surgeon begins, and they all intone it) is 'in the physical causation of all phenomena':

> And I believe in surgical treatment for duodenal ulcer, celebral abscess, pyloric stenosis, aneurism and all forms of endocrinic disturbance.

And the patient, Chimp Eagle, anaesthetized in this particular cage, dies of the surgery performed on his particular version of the Audenic wound ('we have before us a case of abdominal injury, caused by a bullet piercing the bowel. I intend therefore to make a five-inch incision, dividing the Rectus Abdominus . . .'). If only England had listened to the healers it tragically refused to heed:

> Lawrence, Blake and Homer Lane, once healers in our English land;
> These are dead as iron for ever; these can never hold our hand.

> Lawrence was brought down by smut-hounds, Blake went dotty as he sang,
> Homer Lane was killed in action by the Twickenham Baptist gang.

'Have things gone too far already?' this poem of April 1930 ('Get there if you can') goes on; 'Are we done for?' The '30s answer was increasingly a strong yes, especially after Spain. Auden was gloomy about cures before 1936—*The Dog Beneath the Skin* is sign enough of his despair; in the enclosing 'nightmare of the dark, all the dogs of Europe' were indeed beginning to bark—but what Auden saw and felt in Spain confirmed with some finality his sense of Europe's neurotic sickness unto death:

> On that tableland scored by rivers
> Our thoughts have bodies, the menacing shapes of our fever

> Are precise and alive. For the fears which made us respond
> To the medicine ad. and the brochure of winter cruises
> Have become invading battalions.

And the poems that Auden wrote between his visit to Spain and the end of the decade are a record of the triumph of uncured neurotic illness and the urge to destroy. Immediately after his Spanish experiences Auden penned his savagely jaunty sick-joke lyrics 'Miss Gee' (to the tune of the Blues, 'St James Infirmary'), 'Victor' (to the tune of 'Frankie and Johnny', itself a ditty celebrating murder) and 'James Honeyman'—each more brutally destructive than anything supplied by William Plomer. Auden and Isherwood journeyed to the China war, each of them growingly obsessed by 'The little natures that will make us cry', by the 'bacillus' bombers, Auden dwelling more and more on the likes of Rimbaud, Edward Lear, A. E. Housman, Voltaire, Arnold, Yeats, sexually abnormal, fearful men, neurotics, writers haunted by the irrationalities and perplexities of the European culture for which they wrote, quenched by the oppositions of family, society, and self, and finally ending up in the ultimate exile of death. One way or another, the writers visited in Auden's poems in the later '30s were all sick and in prison. Arnold 'thrust his gift in prison till it died, / And left him nothing but a jailor's voice and face'. Voltaire might have been a poet of the '30s:

. . . like a sentinel, he could not sleep. The night was full of wrong,
Earthquakes and executions. Soon he would be dead,
And still all over Europe stood the horrible nurses
Itching to boil their children. Only his verses
Perhaps could stop them: He must go on working. Overhead
The uncomplaining stars composed their lucid song.

The indifferent stars: that's the note of 'Musée des Beaux Arts', a poem also composed in this (end of) 1938—(beginning of) 1939 group. A world of suffering
enclosed in an indifferent universe: it's the hospital ship sailing on with the wounded
poets aboard, or the 'nondescript express' carrying the destructive lunatic with the
mysterious 'little case' in 'Gare du Midi', another impressively glooming member of
the same band of poems.

And if you could get out of the cage it would only be to confirm, as it were, that the
destructive element was really inescapable. When in 'Mr Loveday's Little Outing'
Mr Loveday, girl-throttler and loony-bin trusty (he's been inside for thirty-five
years), is let out after Angela's strenuously humanitarian efforts for his release, he
returns within two hours professing satisfaction that he's fulfilled the 'one little
treat' he's been promising himself all these years:

> Half a mile up the road from the asylum gates, they later discovered an abandoned
> bicycle. It was a lady's machine of some antiquity. Quite near it in the ditch lay the
> strangled body of a young woman, who, riding home to her tea, had chanced to overtake
> Mr Loveday, as he strode along, musing on his opportunities.

It's a replica of his original crime. His madness is circular, complete, and completely
imprisoning. His case is a nice emblem of the '30s dilemma. As we shall see, the
islanded writer was given to dreaming of happier isles elsewhere, holiday islands,
enclosures over the water that might prove happier places than the oppressively
islanded home, nice and relaxing 'private spheres out of a public chaos', as Auden
and Day Lewis's Preface to *Oxford Poetry 1926* had put it. They rarely found their
happy isles. The hospital ship would only disgorge its passengers at a war. At the
'Gare du Midi' Auden's lunatic 'walks out briskly to infect a city'. You took your
sicknesses with you wherever you went, turning the next place bad like yourself,
proving it no island refuge, merely another asylum of the mentally disturbed.

So another possibility struck some people. Why not stay put within the cage, make
the best of a bad job, look on the bin as truly a refuge? Like Henry James's caged girl
who changes her mind as the climax of her story impends: 'to be in the cage had
suddenly become her safety, and she was literally afraid of the alternate self who
might be waiting outside.' Or like the 'liberals' in a novel Spender planned to write
(but never did) after he'd done *The Destructive Element*, people contriving a kind of
'completely free life': it was to have been titled *The Liberal Cage*. Or, again, like the
inhabitants of Shangri-La in James Hilton's fundamental '30s text, *Lost Horizon*
(1933), the novel that not only incited a great deal in Auden and Isherwood's *The
Ascent of F6* (mountains, a lama, a frontier drama) and presented a myth of attractive
longevity for Huxley's *After Many a Summer* to undo, but also provided a magically
resonant name for hundreds of suburban domestic retreats. Shangri-La: it was a
fastness, in Hilton's vision, cut off from modernity (dance-bands, cinemas, sky-
signs) but, much more than that, cut off also from the Crisis, the world's 'reek of

dissolution', its moral 'blind-flying', and so offering for the man Conway a refuge from the bad memories of 1914–18 and a haven from the coming war ('The airman bearing loads of death to the great cities will not pass our way'). It offered a way of defeating death within the ultimate '30s parenthesis. It was a more positive version of the awful, conditioned, be-drugged bliss of Huxley's *Brave New World* (1932): the '30s equivalent of a nuclear fall-out shelter in Switzerland. And even though the status of Shangri-La was always dubious ('Are *we* in the prison or are *they*?'), the notion that such a cage might be enjoyable struck other writers of the period, the ones prepared to go along with those British Museum readers in MacNeice's poem who find refuge in the Reading Room, 'Folded up in themselves in a world which is safe and silent', living off the accumulated 'Honey and wax' in their 'hive-like dome'.

This is just what Caudwell in his essay on 'Freud' in *Studies in a Dying Culture* took the founder of psychoanalysis to task for doing:

> Freud imagines a pleasure-principle attempting to gain freedom for its pleasures within the bounds of the prison house of reality. Beyond those bounds of causality we must not stray, Freud admits, but inside their ever-contracting boundaries there appears to be true freedom.

'It is a fine fable', Caudwell adds, but a wilful self-blinding. The 'truth is, the world is not a prison house . . . in which man has been allotted by some miracle a honey cell of pleasure'; 'an animal in a cage' is by no means 'free because it does not realise it is in a cage'. Caudwell's objection was another way of stating what Huxley's *Brave New World* had presented as the cosy acceptance of a self-pleasuring bottled existence by people bred in bottles. 'Bottle of mine, it's you I've always wanted' ('Skies are blue inside of you, The weather's always fine'), sing the Sixteen Sexophonists to their customers who are 'safely inside' their drugged, escapist, soothed womb of a life. It was close, too, to what George Orwell described as the retreat of writers like Henry Miller 'Inside the Whale', into a '30s quietism, where they refuse to be concerned with the politics and the destruction going on outside or even to take notice of the bars of their cage:

> Get inside the whale—or rather, admit you are inside the whale (for you *are*, of course). Give yourself over to the world-process, stop fighting against it or pretending that you can control it: simply accept it, endure it, record it.

Orwell's direct inspiration for his point came, he says, from the chapter in Henry Miller's *Max and the White Phagocytes* (Paris, 1938) entitled 'Un Etre Etoilique' (and first published in *C*, October 1937). A celebration of Anais Nin's descent into the labyrinthine depths of her diary's womb-consciousness, this book includes Miller's response to Huxley's published horror over the 'visceral prison', the whale's belly that people in El Greco's picture 'The Dream of Philip the Second' look as if they are in. For his part Miller finds the idea of being inside a whale extremely attractive. Orwell declares that all Miller's best passages are written from the angle of 'a willing Jonah'. Orwell guesses that the Jonah story touches upon 'what is probably a very widespread fantasy': 'For the fact is that being inside a whale is a very comfortable, cosy, homelike thought . . . It is, of course, quite obvious why. The whale's belly is simply a womb big enough for an adult . . . Short of being dead, it is the final, unsurpassable stage of irresponsibility.'

Wombs and bellies, especially the whale's belly that engulfed the Biblical Jonah, certainly fascinated Orwell himself. Dorothy in his novel *A Clergyman's Daughter* (1935) loses her school-teaching job when outraged puritan parents object to her classroom explication of *Macbeth*'s lines about Macduff 'from his mother's womb/ Untimely ripp'd.' Gordon Comstock of *Keep the Aspidistra Flying* (1936) wants to sink into unrespectability 'Down in the safe soft womb of earth'. George Bowling of *Coming Up for Air* (1939) is much taken by 'the chap (I notice that to this day he turns up in the Sunday papers about once in three years) who was swallowed by a whale in the Red Sea and taken out three days later, alive but bleached white by the whale's gastric juices'.

Enjoying the cage as a womb, a home, was the import too of an astonishingly energetic, ranting *tour de force* of a text, entitled 'Ego' and produced by Miller's young friend, Lawrence Durrell. It appeared in the first issue of the magazine *Seven* (Summer 1938), and it was the only part of Durrell's novel *The Black Book* to get published in Britain before 1973. It's not at all surprising from Durrell's pen at this period. He was enormously under Miller's influence, was obviously extremely eager to please his master, and as ready as could be to develop any hint from the master's pen, even whale hints. The published correspondence between Miller and Durrell is in the later '30s full of mutual whale talk. What *is* surprising, though, since 'Ego' is even more vividly close to Orwell's discussion of the whale-syndrome than Miller's piece on Anais Nin, is that Orwell never once mentions it.

'Ego' is the theme of Durrell's text as well as its title. In it, a narrator known as Jonah ('for the purposes of simplification'), heads ('With that knowing look I always imagine the spermatozoa to wear on their faces') into a womb, the womb of Hilda the whale. (Curiously, Hilda is the name of the boring wife of Orwell's whale-intrigued George Bowling, and Bowling's mother was devoted to *Hilda's Home Companion*.) Hilda is a prison. 'I am going to be *walled in*. Womb, then, and tomb in one!'; 'My glyptic jailors wait stiffly for me to address them in their own language'; 'I *am* bricked in'. And so Jonah will undertake no more a free man's pursuits ('There by the door lie the ice-pick, crash-helmet and skates. If I had known beforehand I should never have brought them with me. It is always the way. They are quite useless. Such a thing as a motorbike is unheard of in this limited . . . world'). What's more, this womb is a surrealistically horrid place: 'I am blinded with blood . . . My boots are full of blood.' It is a continual death. 'This must be the end, the terminus. I am waiting forever in space.' Above all, it's a happy place: 'My only entertainment is in softly walking round the walls, repeating my own name and chuckling quietly. I am happy.' No wonder, for he dwells repeatedly on the womb's comforts. 'Plush walls, naturally! And a well-furnished house. All the genteel possessions of the cultured owl. Sherry on tap, Picasso on the wall over the piano . . .'; 'this limited plush world'; 'This little plush world'; 'this plush-lined niche in the forever'. In short, 'Ego' welcomes the solipsism of its Jonah. He was having a whale of a time.

And what Durrell's Jonah reminds one forcefully of, of course, is another solip-sistic hero of a Jonah fiction produced by yet another Irishman who was in exile, like Miller and Durrell, in and about Paris in the '30s: Samuel Beckett's eponymous Murphy. Like Durrell's 'Ego' and *The Black Book* as a whole, *Murphy* is also a jumble of egos and rooms and wombs and tombs, and like Durrell's work it was also too outspokenly visceral and Miller-ish to find an English publisher. Like *The Black Book* again, *Murphy* came out in Paris in 1938. Much clearer-cut, however, than *The Black Book*, Beckett's fictional volume opens, in its English version, with its hero (a 'seedy solipsist') in 'a medium-sized cage of north-western aspect commanding an unbroken view of medium-sized cages of south-eastern aspect'. Murphy is in a kind of bird-cage, in fact, a 'mew' (Beckett's punningly singular singular of *mews*) and the novel goes on to rejoice in Murphy's encirclement in the novel's meshing set of enclosures, short circuits, closed systems: the womb, the loony Mr Endon's eyes, the mind, lavatories (Murphy's Will asks for his ashes to be flushed down the Abbey Theatre's famously noisy loo, 'if possible during the performance of a piece': ' "infinite riches in a W.C.", said Neary', half recalling Marlowe's parody in *The Jew of Malta* of the Christian concept of Christ in the Virgin's womb, 'Infinite riches in a little room'), not to mention the padded cells at the Magdalen Mental Mercyseat where Murphy is employed. What the novel discloses, then, is enclosures and Murphy's wish to be enclosed. Murphy yearns above all for enclosure within the 'little world' of the M.M.M. loony-bin's 'indoor bowers of bliss'—as he supposes the 'pads', the padded cells, to be. His 'experience as a physical and rational being obliged him to call sanctuary what the psychiatrists called exile and to think of the patients not as banished from a system of benefits but as escaped from a colossal fiasco.' It was a welcoming of lunacy ('The hypomanic bounced off the walls like a bluebottle in a jar') and, ultimately, of death—self-bound in his favourite rocking-chair in his garret haven at the Mercyseat Murphy meets his end when someone turns on the gas, that 'superfine chaos'.

What Murphy called 'his mind functioned not as an instrument but as a place' (as with Milton's Satan), and once having made the mind his fiction's place Beckett was poised to proceed in subsequent novels, in the tactic we can now recognize as a classic tactic of modernistic (and post-modernist) writing, to make, as it were, the 'mind' *of the text itself* his fictional space and place. Arguably, this shift has already happened in *Murphy*. The movement had certainly already occurred—less vivaciously, to be sure, than in Beckett's texts—in the novels of Virginia Woolf and—at regrettably greater length—in the writing of Beckett's Irish master, also in European exile, James Joyce himself.

Virginia Woolf's '20s novel *Jacob's Room* had defined the trend for herself and her successors. Like Murphy, *Jacob's Room* is a fictional enclosure packed with emblematic, metatextual enclosures: Jacob's rooms in Cambridge and Bloomsbury; the British Museum Reading-Room; all the rooms and the chapel at King's College, Cambridge; the domed, bony, British Museum-like forehead of Huxtable, the don of that college whose head is capable of housing many other people's heads; and the yellow globes of light within the encircling glooms that always stand in Virginia Woolf's work for consciousness. Semi-transparent globes or envelopes of yellow light are, in fact, the more intensely precious to Mrs Woolf because they stand not just for the consciousness of the modern fictionist—as it's presented in her essay on 'Modern Fiction'—but because they remind her of her own earliest, initiatory states of consciousness. Her 'A Sketch of the Past' (1939) begins with a primal scene, one of her first memories—lying in bed in her nursery at St Ives in light refracted through yellow blinds, and hence having 'the feeling . . . of lying in a grape and seeing through a film of semi-transparent yellow'. If she were a painter, she writes, she would express this enclosedness of yellow globularity by painting 'these first impressions in yellow ... the pale yellow blind ... I should make a picture that was globular; semi-transparent. I should make a picture of curved petals ... of things that were semi-transparent; I should make curved shapes, showing the light through.'[69] And since a mind just like this, translucent, refractive, is the pre-condition of Modern Fiction; since her own novels and people are made continually to live bathed in such a momentary illumination (Orlando, lit by the sun shining through his family's stained-glass coat of arms, 'stood now in the midst of the yellow body of an heraldic leopard; Mrs Dalloway's curtains are yellow); since Proust, the novelist of memory whom she thought of as a kind of norm for the modern novel, is for her 'an envelope, thin but pliable'; and since her own fictional effort has to do like Proust's with the charting and recuperation of previously existent states of mind such as her yellow nursery memories, it's inevitable that the enclosures of *Jacob's Room* should be felt by the reader to stand metatextually, like Murphy's pads and Watt's word-games, for the novel that contains them and, even more, for the whole art of fiction in general.

Virginia Woolf's writing shows her various enclosures to be continually full of things, of images of things. They comprise a richly impressionistic phenomenology, a proud metonymy, that rivals the richness of the light that illuminates those things and gives them apprehended life. But at the same time these vivid containers are empty, or poised fragilely on the edge of emptiedness, as Jacob's rooms are full of books, notes, slippers, traces of his late presence, but fatally empty of Jacob's own person. The boy and the man Jacob remain reserved, held back, deferred, finally unknowable, for all his friends' and relations' claims to know him, and then ultimately he is killed in the Great War. Just so, Virginia Woolf's favourite moths fly into the circle of the lantern's light, there to be observed flickering brilliantly with all the intensity that (in her essay 'Modern Fiction') she granted both to consciousness and to the life-registering techniques of James Joyce's *Ulysses*, only in the moment of illumination to be burnt to death on the very source of their transitory brilliance. 'There ... is life', she says of Joyce, as she says so often of her own characters; but she's describing Joyce's graveyard scene, just as she's aware of how her own people's life converges customarily on to death. Not dissimilarly, the epistles that the women of *Jacob's Room* keep writing are frequent and fulsome, enclosing many words and much detail; but they are at the same time empty of revelation, full of mistakes and mendacities, finally incommunicative.

Brilliantly, if sometimes chaotically, *Jacob's Room* forces its mental worlds, its delvings into the impressionistic life of the mind, into becoming analogues for the life (or death) of fiction itself. Its *autology*, to use *Murphy*'s word, is not simply to do with people. The solipsism of Virginia Woolf's persons turns into the *autology* of the text. Her novel is, as her novels went on being, about the novel, about itself and about the art of the novel in general. Traditional fictional ways are discarded in such writing. At first there is the disclosure of characters whose reality is only in the head, who are self-enclosed, solipsistic, cut off from each other's knowability. But this fiction rapidly goes on to suggest a condition of writing in general, of text at large, that is knowable only in its textual self-boundedness, in its solipsism as what the French structuralists call *écriture*.

Murphy's enclosures only rise gently and unaggressively towards the full meta-textual significances of the enclosures in *Jacob's Room*. But they led Beckett on to *Watt* (written in 1942) and to the later fictions, writings that become less and less about anything but themselves, increasingly locked into an autology, a solipsism of the text. And this was a welcoming by fiction with a vengeance of the whale's belly. Fortunately, perhaps, for the reader, this belly was a location that Beckett would increasingly feel that he could mirror adequately in tinier and tinier fictional reports. Not so, as befits perhaps the Master, James Joyce. His *Finnegans Wake* is the solipsistic text writ at its largest. *Jacob's Room* advertises itself as mainly a play of signifiers, *une écriture*, as it opens self-referentially with the ink welling out of Mrs Flanders' pen-nib while a letter is being written and as it continues its demonstration with continual self-reference to drawing, painting, seeing, knowing, writing, reading. *The Waves* (1931) disports its textually self-conscious problematic by giving us Bernard the writer who has doubts. *The Years* (1937) mirrors its own fragmentariness by having people continually beginning stories that are patchily vague and inconclusive ('And then?', characters have to keep asking, in the manner of Forster's *Aspects of the Novel*, but their pressing never gets them anywhere with the continually lapsing narrators). *Watt*, deliberately circular, an affair of loop-lines of sense as of Watt's roundabout modes of travelling, enacts its hero Watt's and its own fall into incommunicativeness in a brilliant series of language disassemblings, an undoing of creativity in a pastiche of God's doings in the Book of Genesis. In Durrell's *The Black Book*, the play between the frame-narration and Gregory's diary, the novel's ego and its id, makes similar announcements of mere fictionality, of this text being merely, solipsistically, about, as it is confined within, the cage, the maze, of its own being. But happily for the reader, these fictions all turn out to be impure ones. Unable to stick only at the level of language, of the signifier, willy-nilly they engage with the world outside language, the realities, the worldy references implied by the signified. Not least, they give the reader enough bits of story, character, plot in the traditional sense, to make the intenser textual narcissicism, the verbal implosions, fairly bearable. What's more, and this is important to readers, their textual extent is brief. These texts are encompassably short enough to make them tolerable, possible reads.

Not so, however, *Finnegans Wake*. It offers the reader, and especially the common reader, no such biddable transactions, seeks no such accommodations with traditional fictional institutions and orthodoxies. The anti-novels of Woolf, Beckett, and Durrell refuse to some considerable extent, as all anti-novels do, to meet the solution-seeking expectations of detective fiction (of archetypal fiction, that is), by turning the novel's traditional 'answers' back on to the text itself. So that at the end of such novels the reader is convinced that what she has just read has been an attempt to reveal mere fictionality, that the novel, writing itself, is the agent who 'dunit' at least as much as any character did anything, and that there are few if any positive exits from the language labyrinth. But *Finnegans Wake* goes about these anti-novel activities much more rebarbatively still. When the *Wake* was accused of retreating into the deterring, deferring condition of the world's largest insoluble crossword puzzle, of compounding in other words the modernist sense of chaos, uncertainty, and non-solutions, Joyce quite simply welcomed the charge. Wyndham Lewis denigrated the *Wake's* 'Crossword Joys', its 'circumsolutions'. So Joyce responded by celebrating his puzzling text as (p. 619) Adam's 'beautiful crossmess parzel': the Christmas parcel that opens out only as a linguistic crossword puzzle, merely an exercise in endless parsing. *Finnegans Wake* would make no easy compromises.

In particular, this enormous text would concede nothing to the reader's stamina. And the trouble with its demonstrations of its own mere textuality, its devoted settling for a deconstructionist-delighting existence only as language, text, writing, bounded within the whale-huge cage of mere words, its monstrous refusal to give life to much besides the roistering game of signifiers, its abracadabra of the romping alphabetical, is that all these manifestations of linguistic and textual inside-the-whalery go on for longer than is, in practice, humanly endurable. *Finnegans Wake* is a life sentence. Who reads, who wants to read, who knows how to read this 'Work' so dauntingly endlessly 'In Progress'? The shorter and more compromising (formally, realistically, linguistically) Woolf, Durrell, and Beckett (even in *The Waves* Woolf is nowhere so devoted to protracted linguistic show as *Finnegans Wake* is, and the extremer texts of Beckett, the ones more obviously schooled by the Master had to wait until well after the Second World War), these three might on occasion make the reader believe that the textual whale is less a prison-house of language than a plush delight where fantasy, wish and desire reign joyfully—though the presence of death in Virginia Woolf's texts and the harrowingly golgothan sufferings of Beckett's people that intensify as their role as the heroes merely of *écriture* grows starker, do help banish those briefly enlivening senses of the plush. And indeed in short witty bursts—admittedly embedded deep in the multiplex strata of the verbiage of puns—even *Finnegans Wake* can also be catchily persuasive. Take the line *Say mangraphique, may say nay por daguerre* which pops up out of the overgrowth on p. 339 of the *Wake* in the manner that one grows used to such self-mirroring bits of phrasing, words, lines, popping up to act as signals of the self-regarding text, as minute surrogates for it, reflexive metatexts that function like the (more extended, not to say more metonymic, more referential and realistic) epistles in *Jacob's Room*, padded cells in *Murphy* and the diary in *The Black Book*. A witty calque on what the French observer in the Crimea is supposed to have said about the Charge of the Light Brigade ('*C'est magnifique, mais ce n'est pas la guerre*') it provides an extremely witty description of what Joyce believes his implosive whaling of a writing is up to: *Say mangraphique* (this text is just a writing, a matter of mere graphing), *may say nay por daguerre* (this text resists the old-fashioned photographic realism: it's not a Daguerro-type, a photograph, only printer's type). The sentence is a devastatingly brilliant little mirror of the whole, and one of many such in the *Wake*. But when this endless *écriture*, and this resistance to anything much but self-regarding parades of textuality, are kept up over the seventeen years it took Joyce to put *Finnegans Wake* together and for the 629 printed pages he eventually produced, the exercise appears as a selfish retreat into the hermetics of linguisticity that is not only destructive of traditional fiction's ways with the world, people, and reference generally, but also destructive of readers. *C'est magnifique* all right, but *ce n'est pas la guerre* indeed. It's a cage of language that has stopped being anything but an endurance text, a ripe case for (so to say) parole, long before the last page. It's *une folie*, less of *grandeur* than of elephantiasis, a folly that fails to convince one that textual madness, or any other kind of madness, is other than truly terrifying. To disclose, on such a scale, that reality is only a set of word-games, mere *écriture*; that fictions's verbal signs are locked in tightly upon themselves, and then to shut the reader into that mad and maddening textual enclosure for so extended a durance, is only to impress on that reader that the corseted cosmos of signifiers without a signified, words without reference beyond themselves (were such ambitions actually possible to achieve), this life on the inside of the textual whale, comprise as much of a destructive element as the inside of any other of the '30s characteristic prisons.

In 'Inside the Whale' Orwell described the mind of '30s Communism as 'a constipating little cage of lies'. The big cage—and *Finnegans Wake* is the biggest written cage the '30s ever granted—is actually, and for all that it looks much more like an affair of verbal diarrhoea than a case of constipation, no less morally and linguistically (epistemologically, hermeneutically) constipated and constipating.

5

Too Old At Forty

'Youth's on the march' says Jocker to Prushun.
Youth's the solution of every good scout.
Youth has the secret Toc H has found out.
Youth's a success.
Youth has the blessing of the *Sunday Express*.
Youth says the teacher.
Youth says the bishop.
Youth says the bumslapper.
'Strewth, says I.

THUS Auden in Ode IV of *The Orators*: lines written in October 1931. And lines that could scarcely be better placed to advertise the fact that in the '30s 'youth' had arrived. They're addressed to the youthful son of Auden's Oxford friend Rex Warner, a boy not yet two years old. They come jostling *The Orators'* other addresses to the young: Ode V to Auden's pupils at Larchfield Academy, Helensburgh, in Scotland; Ode II celebrating that school's Rugby XV and addressed to Gabriel Carritt 'Captain of Sedbergh School XV, Spring, 1927' who was another Oxford friend. Ode III is addressed to Edward Upward himself, like Auden a young school-master. *The Orators* was in fact at the centre of the youth cult that those lines announced. It's for youths, about youths, by a youth. It's a piece of English upper-class youth code (full, for instance, like those lines, of discreet homosexual slang familiar to those who had, and wanted to have, their bums slapped). On occasion so youthful do *The Orators'* show-off toughnesses and its brazen knowingnesses seem that one has to keep reminding oneself to take it seriously, remind oneself that it was taken extremely seriously in its time. Its child-celebrating composer was, indeed, already the feared and respected author of *Poems* (1928) and *Poems* (1930).

In May 1932, when *The Orators* came out Auden was just over twenty-five. The youth of this already reputable poet is striking. But many published '30s authors were only slightly older than Auden was. Several were exactly his age. A number of them were actually younger. In his time Auden was no freak of youthful genius. Of course, there were also many older writers about. In 1930 T. S. Eliot turned forty-two; Wyndham Lewis became forty-six; Virginia Woolf and James Joyce were each forty-eight; Yeats was a grand oldster of sixty-five. But in 1930 Evelyn Waugh was only twenty-seven. So were Edward Upward, Cyril Connolly, and George Orwell. Day Lewis, Graham Greene and Isherwood were twenty-six. Henry Green, Anthony Powell, Geoffrey Grigson and Rex Warner were twenty-five. Samuel Beckett, John Betjeman and William Empson were twenty-four; Auden, Christopher Caudwell and MacNeice twenty-three; Julian Bell and Kathleen Raine twenty-two; Malcolm Lowry and Stephen Spender only twenty-one. And, as if that catalogue isn't by itself enough to drive one's grey hairs with sorrow to the mirror: in 1930 Lawrence Durrell was a mere eighteen, Ruthven Todd and Dylan Thomas only sixteen, John Cornford a piffling fifteen, and David Gascoyne a mind-boggling fourteen.

No wonder Ode IV feels safe in sneering at

> Those over thirty,
> Ugly and dirty,
> What are they doing
> Except just stewing?

For Auden still had bags of time left. 'Too Old at Forty', piped young Evelyn Waugh in the *Evening Standard* (26 October 1928), *Decline and Fall* already four months on the bookstands. Unsurprisingly, Waugh found the thirty-two-year-old Henry Williamson 'quite elderly'-looking when he met him at the end of 1928. Waugh's diary entry sounds surprised that the old man was still 'capable of fun'. Of course by 1928 Waugh had been aware of being in a youth-boom for quite some time. 'What a ridiculous generation we are', he noted while he was still at Lancing College in 1919. 'In the last generation people never began to think until they were about nineteen, to say nothing of thinking about publishing books and pictures.' At the age of sixteen he was already designing book covers for his father's publishing firm of Chapman and Hall. His friend Molson was also straining to get into print. But there was a special reason why boyish publishing possibilities came so naturally to a member of the Waugh family. In 1917 his brother Alec's sensational fiction *The Loom of Youth* (its author only turning nineteen in 1917) came out, exposing games adulation and also little-boy adulation at public schools and causing intense excitement among public-school headmasters and public-schoolboys. It became the banned book all the boys wanted most to get their hands on. It made one more reason why the Sherborne School authorities decided not to have young Evelyn on the premises from which his older brother had been expelled—for the 'unnatural practices' that preoccupied his youthful fictional hit. Hence Evelyn's presence at Lancing. Hence also his knowing adaptation of Alec's title in his diary (1920):

> the extraordinary boom of youth, which everyone must have noticed . . . Every boy is writing about his school, every child about the doll's house, every baby about its bottle. The very young have gained an almost complete monopoly of bookshops, press, and picture gallery. Youth is coming into its own.

So it was. It became the fashion of the post-War period to be a child prodigy of letters, to hasten as it were to fill up the gaps blown open in the ranks of art and letters by the War. The Artists' Rifles, so to say, needed and got rawer and rawer recruits. Harold Acton's first volume of poems, *Aquarium*, was published in 1923 in only his second term at Oxford. Peter Quennell's *Masques and Poems* came out in 1922 when he was only seventeen and not yet an undergraduate. Henry Green's first novel *Blindness* (1926) was published while he was still a twenty-year old undergraduate. Graham Greene's poems *Babbling April* came out during his final undergraduate year in 1925. We've already noticed Spender printing his own and Auden's poems during the long vacation of 1928: his generation had caught their (slightly) elders' habit. Charles Madge and Tom Harrisson were only twenty-five when they got together to start up Mass-Observation in 1937, by which time Harrisson had already made a name for himself with his Hate Press's *Letter to Oxford* (1934). A note in the first number of *Contemporary Poetry and Prose* (May 1936) introduces its youthful contributors thus:

> William Empson was several years Professor of English at Tokio University . . .
>
> Gavin Ewart is at present reading English at Cambridge . . .
>
> David Gascoyne is, at 19, England's only wholehearted surrealist; he is the author of two books of poems, *The Roman Balcony* and *Man's Life is This Meat*, of a novel, *Opening Day*, and of *A Short Survey of Surrealism* . . .
>
> Roger Roughton, editor of the paper, is also 19 . . .

Gascoyne's novel *Opening Day* had appeared in 1933: he was seventeen at the time—the same age as Ewart when he wrote the poem 'Phallus in Wonderland' published in *New Verse* in that same year.

With scarcely any effort at all youths had stormed the citadels of the literary world. In fact it could scarcely be called storming, so welcome were they at the portals of publishers and literary editors. Their presence in the literary world was simply unignorable. Indeed it was physical, mappable, a case of geography. At the core of London's literary and political geography were to be found clearly located the hangouts of the young.

It's been too little observed about the '30s how small an area was actually occupied by its characteristic centres. Then, as now, England's literary and political life was largely managed from Central London and from a tiny part of Central London at that. 'Let's all go to Art's then. It's only ten minutes' walk', says one of the characters in Julian Symons's crime story *The Immaterial Murder Case*. Art's nightclub is in the same building as the Immaterialists' Exhibition—an affair loosely based on the 1937 Surrealists' Exhibition—and so not far from where the characters called the Rankes live. And ten minutes' walk would indeed take you a long way in the '30s across the landscape of arty-literary London. A vigorously punted beermug or policeman's helmet (Boat Race Night style) or a series of small arms shots (such as the youthful revolutionaries sometimes envisaged having to resort to), would almost do to carry you from any one major staging post to the next. It wasn't far from the BBC and the *Listener* down into Soho where the bohemian watering holes and eating places were, into Soho Square where the documentary film men worked and the Artists International Association held exhibitions and put on lectures, and off that into Frith Street (where the publisher Boriswood's offices were and the Chanticleer Restaurant where Communist Party members held fund-raising events), and then across into Denmark Street where the Nanking Chinese Restaurant was, haunt of socialists (in it John Cornford met his girlfriend Ray), and from there back into the Charing Cross Road where Eva Reckitt opened Collet's Bookshop in 1934 in the premises formerly occupied by Henderson's bookselling 'Bomb Shop', and in whose neighbouring cafés the Aid-Spain committees would meet and around which the Spanish fund-raising shops tended to spring up (there was a 'Spain Shop' at 71 Shaftesbury Avenue); next into Great Newport Street where the Group Theatre Rooms were and where the Left Book Club's readers and Writers Group met, then over a shade into Covent Garden to the Communist Party's headquarters in King Street, with Gollancz and the Left Book Club and other publishing houses next door in Henrietta Street (not to mention Chatto's—and the office from which Graham Greene edited *Night and Day* in 1937—and George Newnes—where Stevie Smith was employed—in other streets in this particular neighbourhood): then up again into Bloomsbury and to the British Museum area where at 38 Great Russell Street, with an entrance at the back in Willoughby Street, Harold Monro's famous Poetry Bookshop lived out its distinguished decline from 1926 until it closed down in the middle of 1935, and where of course shoals more of the key book publishers such as Faber's in Russell Square, Cape's in Bedford Square, Hogarth in Tavistock Square, also flourished; and then down a bit into Theobald's Road where the Communist Party had a Workers' Bookshop and *Left Review* for a time had its offices; on to the offices in Great Turnstile to which the *New Statesman* moved in 1933, off High Holborn (before that the *NS* had been nearer the Museum in Great Queen Street); on further down into the City's book and newspaper publishing region, where the *Daily Worker* was, and the *Morning Star* still is printed near the Faringdon Tube Station, and on a step or two beyond the station up to Clerkenwell Green and Marx House (opened in 1933) where Charles Donnelly was only one among several '30s leftist notables to give lectures. And, of course, somewhere in the middle of this brisk itinerary, only a hefted copy of, say, a hefty John Cowper Powys novel across High Holborn from the *NS* and just around the corner from Charlie Lahr's Progressive Bookshop at 68–69 Red Lion Street (haunt of H. E. Bates, James Hanley, Eric Partridge, James Boswell, Jack Lindsay, Hugh MacDiarmid) you came on a kind of epicentre: Parton Street, Red Lion Square (a street, alas, no longer there). At No. 1 Parton Street was the Arts Café, open from nine in the morning until ten at night, from which address *CPP* was edited. At No. 2 was the bookshop of Central Books and the offices of Lawrence and Wishart, the home of the Workers' Theatre Movement and the Student Labour Federation, and from April 1936 of *LR*. From No. 2 emanated all the first series of *NW* (Hogarth brought out the second series) and *Authors Take Sides on the Spanish War*, as well as Wishart's other material. And at No. 4, the centre—if such a metaphor is possible—of this epicentre, was the bookshop run by Old Wellingtonian David Archer, the home of the Parton Press (which issued Dylan Thomas's *Eighteen Poems*, George Barker's *Thirty Preliminary Poems*, Gascoyne's *Man's Life is this Meat*), briefly the address for *NV*, from May 1935 the headquarters of Artists International, the mecca in fact of the radical artistic and poetic young. Though Monro's Poetry Bookshop kept a lot of the magic it had possessed since its founding in 1913, by the early '30s the torch was being passed noticeably to Parton Street. To Archer's bookshop gravitated Dylan Thomas and John Cornford (who lodged in Parton Street in 1933 while he was, briefly, a student at the London School of Economics), and Esmond Romilly, who took refuge with David Archer when he ran away from Wellington College (5 Merton Street he calls it in his book *Out of Bounds*, 1935), there to publish the anti-war, anti-fascism mag for public schoolboys that was also titled *Out of Bounds*. In its turn this magazine tempted the young Philip Toynbee to leave Rugby School in the summer term of 1934 for a brief taste of the anti-fascist struggle at No. 4. It's very easy to debunk Parton Street now:

> What a bore all those politically affiliated young men
> at your Parton Street Bookshop were, David, in those
> forgettable thirties.

So writes George Barker in *In Memory of David Archer* (1973). But Parton Street is a necessary touchstone for any account of '30s politics and literature. Cris Clay— ex-public schoolboy, notoriously a version of camp Old Etonian Brian Howard, and narrator of Cyril Connolly's little '30s Rake's Progress 'From Oscar to Stalin: A Progress'—signs off with a knowing imitation of *Ulysses*'s dateline (Trieste–Zürich– Paris): Paris–Budapest–Parton Street. And Parton Street was, among other things a youth club. As Parton Street was in its way centrally situated on the small map of London literary and radical-bourgeois political life, so the kind of youths who frequented it were central to that life.

Indeed the stir, the excitement of the literary '30s were greatly generated by the youths such as frequented Parton Street and spread out from there across the tightly congested streets of the Central London literary scene. You could say, in fact, of much of the period's doings what John Grierson said in his 'Summary and Survey: 1935' about the Documentary Film movement: 'they are all young people'. (When the extremely important *The Film Till Now* came out in 1930 its author Paul Rotha was only twenty-two.) The '30s belonged to boys: 'The Oxford Boys Becalmed' in Edmund Wilson's debunking article of 1937;[70] the boys that Isherwood declared later in *Christopher and His Kind* (1977) were the reason for his journey to Berlin and that were the cause of his European wanderings in the '30s; the boy preachers (there's a faked-up one in Robert Graves's novel *'Antigua, Penny, Puce'*, 1936), the boy musicians (such as Yehudi Menuhin), the boy poets and novelists, even a boy gangster, Pinkie of *Brighton Rock*. In the American gangster movies and novels of the period (and even later) the cops and the crooks, however grown-up, all refer to each other as 'boys'. Boys were clearly a major preoccupation of *New Writing*. 'Four Boys Alone' was the editor John Lehmann's description of four little fictions about males in No. 3 (Spring 1937). T. C. Worsley's story 'A Boy's Love' appeared in No. 4 (Autumn 1937), G. F. Green's story 'One Boy's Town' in No. 5 (Spring 1938). Boys thickly populated the period's poems. Dylan Thomas's verse contemplated 'the boys of summer in their ruin'. Keidrych Rhys's verse had in mind 'the decayed boys the darlings of their year', now fallen to their death (in 'Landmark', *TCV*, January–February 1938). C. Day Lewis addressed boys. Addressing boys was the commonest of modes among the Auden gang. 'You'll be leaving soon and it's up to you, boys': section 10 of Day Lewis's *The Magnetic Mountain* appeared in the *Adelphi* (August 1932) as 'Address to Boys'. And even if you had objections to all this attention to boys—'Besides, I do not really like / The verses about "Up the Boys" ' was Empson's complaint in his poem 'Autumn on Nan-Yueh'—you had inevitably to go in for a bit of boy-addressing to get the point across. Empson's gibing assault on this 'Playing at the child', his energetic 'Smack at Auden' comes packed with more references to boys ('Waiting for the end, boys' is how it collars its subject right from the start) than any other single '30s poem. Almost willy-nilly the period's writing was stuck into—as Wyndham Lewis put it in yet one more of the current adaptations of Alec Waugh—the *Doom of Youth*.

Lewis's *Doom of Youth* appeared on the last day of June 1932—six weeks after Auden's *Orators*—and nastily cantankerous though it continually is it constitutes a key text of the period, with its hostile gathering of evidence for the high regard in which the gangs of youths and the youthful gangs were being held, with its charges that since the War Europe has entered 'A Vast Communal Nursery' peopled by Peter Pans ('The sickly and dismal spirit of that terrible key-book, *Peter Pan*, has sunk into every tissue of the social life of England'), and with its sharp garnering of the scattered phenomena of the post-war generation-war that stoke up Lewis's outraged case against his times: the child prodigies of all sorts (a cult imported from 'the Jewish East', Lewis glowers), the crowds of 'fairies' who are refusing to grow up into fatherhood, the novelists of school and cricket (Alec Waugh in particular), the youth journalists (Godfrey Winn especially), the fad for slimming and otherwise attaining the 'profile of youth', and so on. The *Doom* was calculatedly offensive, deliberately personal. There were whiffs of legal action from Alec Waugh and Godfrey Winn. Waugh resented the implication that he was 'a sexual pervert' and 'can compose nothing but books about schoolboys'. Chatto and Windus withdrew the *Doom* within sixty days, and eventually had to pay Alec Waugh's legal costs. Other writers, though, and often without proper acknowledgement, were prompt to build on Lewis's allegations. Cyril Connolly's *Enemies of Promise* certainly owes Lewis a lot of inspiration. Aldous Huxley's reduction to a deserved absurdity of the American West Coast wish not to age in *After Many A Summer* is almost certainly indebted to the *Doom*. It's true that Huxley's sharp essay on 'The Beauty Industry', a scathing piece about the new quest for a 'youthful appearance', appeared in *Music at Night* in 1931 (though Huxley was probably too blind to be able himself to see the actual results of cosmetics, 'health motors' and 'skin foods'). It's true too that *Brave New World*, which preceded *The Doom* by five months, spends a good deal of its time debunking the 'youthful appearance' craving of its futuristic world. But, still, to Lewis must go the credit of making the major case first, for assembling the dossiers of contemporary evidence, and for offering the first structural connections between many scattered phenomena of the time.

Personal rancour came into the *Doom* as it often did in Lewis's writings. In 1914 Lewis himself had been the youthful terrorist, the blaster of the fuddy-duddy, the fearsome arbiter of what was to be blessed and what was to be cursed in the culture of England. Promoting the youthful talent and the new direction had been his trick. Now, though many of the youths looked with admiration to his curmudgeonly example, he was himself undeniably an ex-youth. Other arbiters and providers of the new had deposed him, and he begrudged them their revolutionary impact. He had 'perused', he said, the Wyndham Lewis Double Number of *TCV* 'rather in the way a notorious bandit would a shower of *many-happy-returns* and other obliging messages, at his birthday breakfast-table in his hide-out, from the local constabulary'. He watched rather aghast as the new boys muscled in on his old pitches. He saw a generation-war everywhere. One of the many targets in *The Apes of God*—that loud and messy scattering of grapeshot into London's literary scene—and achieving prominence even amidst that rambling host of literary juvenile delinquents, the youth-stuff journalists, the chorus of twenty-four-year-old Old Etonians, the likes of Daniel Boleyn, nineteen-years-old and the author of one poem (Lewis always crowds his texts like this, busying his flattened scenes with the multivalences of cartoon, coopting the democratically numerous horrors of a Bosch canvas), are the pair of editors of *Verse of the Under-Thirties*. They're a couple who could obviously have been made out of bits of several suitable candidates, but they certainly remind one most forcefully of Auden and Spender. Siegfried Victor ('a massive young man . . . with a handsome nobly-proportioned head upon a greek museum-model') must have a deal of Spender in him; Hedgepinshot Mandeville Pickwort ('a small bleached colourless blond') may have a name that recalls Edgell Rickword, but he does look rather like Auden. (Opinions can differ wildly, obviously, on these identifications: Jeffrey Meyers (biographer of Lewis) is sure Daniel Boleyn is meant for Spender, and Pickwort only for Rickword. Lewis's technique, however, is an opportunistic cubism, or scissors-and-paste bricollage, a sticking together of all kinds of personal references into new satiric shapes.)

But for all his malice Wyndham Lewis couldn't stem the trend he described so nicely. It was time for the younger sons to have their turn. Day Lewis's *A Hope For Poetry* makes much of the transforming role of the younger son in English literary history. 'In English poetry there have been several occasions on which the younger son, fretting against parental authority, weary of routine work on the home farm, suspecting too that the soil needs a rest, has packed his bag and set out for a far country.' So prodigal sons had their uses! The old myth evidently meant a great deal to the post-war generation of younger sons. They loved reworking it in their fictions. Even variations on the corniest of English literary prodigalities, the running away to sea routine, are to be found in John Sommerfield's *They Die Young* (1930), James Hanley's *Boy* and Lowry's *Ultramarine*. For his part, Day Lewis uses the younger son syndrome to explain almost everything he likes in poetry: 'Hopkins, Owen, and Eliot are recent examples of younger sons who could not stay at home'; 'Prufrock was a youngest son'. And from a hint on Day Lewis's opening page ('The boy Keats . . . was rapt by a belle dame sans merci, and rode with her across the frontiers of fancy. We had many a hearty laugh at their antics, their wild-goose chases, but as the years went by we began to see that they had made the wilderness blossom like a rose'), Lewis's friend Rex Warner developed his long allegory of the new politics and new art of the hoped-for future, his novel *The Wild Goose Chase* (1937). Warner's fiction is an updating of the old tale of the three sons who travel far, this time 'on bicycles', towards the future's frontier. Especially is it the story of George the youngest, who has the rustiest machine but is destined for the greatest successes. As Auden, himself the youngest of three brothers, put it in a few lines of 1929:

> The youngest son, the youngest son
> Was certainly no wise one
> Yet could surprise one.

Younger sons, Day Lewis explains, have always claimed the new as the object of their searches. In *From Feathers to Iron* (section 26) he celebrated 'a new generation', and 'new faces'. And how widely this bid for newness was advertised: *New Signatures, New Country, New Bearings in English Poetry* (F. R. Leavis, 1932), *New Writing, New Stories, New Challenge, New Theatre, New Verse* (or *New Curse* as it was sometimes known). Everyone, so Wyndham Lewis complained in the *Doom*, 'wished to be, as it were, new born. To blot out the past'. The self-declared newness was not, of course, always all that new. '[N]ew knowledge and new circumstances have compelled us to think and feel in ways not expressible in the old language at all': thus Michael Roberts in his introduction to *New Signatures*, but he was claiming too much. Among the new signatures he was presenting to the world, Julian Bell, William Empson, and Richard Eberhart, for instance, now seem 'new' only in the sense that their verse hadn't been much before the public before. And Roberts's continual sniping at Eliot in that Preface only makes one doubt further his capacities for judging what newness might actually consist of in the '30s. Day Lewis's regard for Eliot as himself a younger son, though one gone a bit 'posh', as he says of Prufrock, seems far more accurate now. Still, the intention of the young was being unambiguously announced. The younger sons were packing their bags, kicking over traces, leaving home.

It's hard completely to repress a sneaking sympathy with the scorn of Roy Campbell (in *Flowering Rifle*, 1939):

> With 'New Verse' and 'New Statesman' to be new with
> Alas, it's a New Newness they could do with!

None the less the determinations to be new were often fierce ones. The motto of 'our youthful author', wrote Isherwood in a 1957 Foreword to *All the Conspirators* (1928) was 'My Generation—right or wrong!'. The novel recalled, he wrote in a revision of that Foreword:

> the days when parents were still heavies. It records a minor engagement in what Shelley calls 'the great war between the old and young'. And what a war that was! Every battle of it was fought to a finish, with no quarter asked or shown. The vanquished became love-starved old maids, taciturn bitter bachelors, chronic invalids, harmless lunatics; or they died, if they were lucky. You may call the motives of these characters trivial, but their struggle is mortal and passionate. And the author is as passionate as any of his characters.[71]

Isherwood's particular target in *All the Conspirators*, as elsewhere in his writings, is the predatory mother—whose familiar deficiencies can rarely have been so waspishly served since Jane Austen. 'I can't stay at home any longer. It's really essential that I don't see Mother for a bit': thus Philip Lindsay to his sister in *All the Conspirators*. ' "A bit!" she laughed sadly. "How many years does that mean?" ' Writing about Noel Coward, Isherwood dwelt warmly on Coward's early violence towards mothers in *The Vortex*: 'it was grand to watch the expensive perfume-bottles on mother's dressing-table being so lavishly and symbolically smashed. Coward, we said, was "all right": his bile-duct was in the right place.'[72]

Tyrannical matriarchs aren't confined to men's writing: the richly gruesome Ada Doom of the satirical novel *Cold Comfort Farm* (1932)—'the curse of Cold Comfort . . . the Dominant Grandmother Theme . . . found in all typical novels of agricultural life (and sometimes in novels of urban life, too)'—like the far less funny Mme Fisher of *The House in Paris*, are the products of women's imaginations, Stella Gibbons's and Elizabeth Bowen's. But almost to a boy, the '30s youths are to be found resisting some kind of mothering. Isherwood went on to write *The Memorial*, devastatingly hard on the war-widow Lily. Spender's *The Backward Son* has its boy hero losing all ways because of his mother: tormented at school as 'Mammy's little Geoff' because she was too publicly tender to him on the railway platform in front of his school-fellows, left in the lurch by her death at the end of a novel whose lack of final resolution reflects young Geoff's utter devastation over his loss ('he saw the stomach with the navel, and just below it a wound cut like a mouth, a wound which they had not sewn up because it was too late. In thick drops, blood surrounded by water oozed from the wound'). From *Paid on Both Sides* on (where John Nower is murdered by Seth Shaw who is incited by his widowed mother, and the Chorus ends by lamenting the defeat of every son: 'His mother and her mother won') mothers are bad news in Auden. C. Day Lewis's Theo Follett in the novel *Starting Point* (1937), a tiny little novelist, aghast at his mother's love-life, mentions Hamlet's horror over his mother Gertrude's second marriage, only to be abused for his tauntings ('despicable little beast', 'smug, spoilt, little egoist', 'impotent sneering little mannikin').

He shoots her dead; watching 'the stain, like spilt wine, growing upon the bosom of her white dress': a kind of revenge for when she struck him as a child for spilling wine over her white dress in a restaurant. But it's also a regression: for he then kills himself, putting the revolver, like a baby's dummy, into his own mouth: 'The metal tasted harsh against his teeth: he bit hard against the barrel, as though it were a comforter.'

C. S. Lewis always referred to the bossy Mrs Moore as 'My Mother'. The War had certainly landed him with an almost life-long mother problem. In *The Dog Beneath the Skin*, the mother-problem actually fuels the War: 'the universal mother' incites her sons to self-immolation. It was evidently easy for the friends and readers of E. M. Forster (moithered by his mother, his fiction haunted by the mind-dominating Mrs Wilcox of *Howards End* and Mrs Moore of *A Passage to India*), for the readers of Joyce (whose Stephen Dedalus has satanically to defy his mother and his Mother Church in order to make it as an artist), or of D. H. Lawrence (whose Paul Morel had to kill off his mother in *Sons and Lovers* in order himself to live, and whose *Fantasia of the Unconscious* is dominated by mothers), or of Henry James (whose *The Spoils of Poynton* is one of the strongest mother-warfare fictions ever), or of Walther Groddeck (scarred by his mother's rejection), to interpret most of experience as a mother problematic. T. S. Eliot's play *The Family Reunion* (1939), pivoted on the maternal strength of powerful Amy, supplied Spender with the epigraph for *The Backward Son*. (Much later on, in a review of Stansky and Abrahams's memoir of Cornford and Bell, *Journey to the Frontier* (1966), Spender would reflect feelingly on their childhoods, overshadowed by strong mothers and aunts; what a subject, he muses, their deaths, for an Auden–Isherwood play: 'the deaths of those heroes on the battlefield would have been seen as the finale of a dialogue with a chorus of artistic mothers and Bloomsbury aunts'.) Oxford had, of course, always been the Alma Mater, and Day Lewis and Fenby have an Ungrateful Son disputing with her in their *Anatomy of Oxford* ('a disgraceful piece of Fascist technique on your part', and so on), while an ungrateful son on the Evelyn Waugh plan could cut up his Alma Mater even rougher. Almost at the start of a career in which the name of his old college tutor, C. R. M. F. Cruttwell, would be given to all manner of nasty or ridiculous fictional characters—for instance, 'Mr Loveday's Little Outing' was originally 'Mr Cruttwell's Little Holiday'—Waugh ('Scaramel') published in the Oxford undergraduate paper *Cherwell* (13 June 1925) a story called 'Edward of unique achievement; a tale of blood and alcohol in an Oxford College', in which an undergraduate stabs his tutor to death: 'dripping blood, sir', a college servant reports, 'Quite slowly, pit-a-pat, as you might say'.

The woman 'England, my England' appears in Auden's 'Letter to Lord Byron'. Part V:

> The Mater, on occasions, of the free,
> Or, if you'd rather, Dura Virum Nutrix.

The harsh nurse of men. Byron will understand ('Your mother in a temper cried, "Lame Brat"'). The appearance of Mrs Ransom on the mountain at the end of *The Ascent of F6* as her son dies, embarrassing though it was even to Group Theatre's production team, came perfectly naturally to Auden and Isherwood. They couldn't be prevailed upon to alter it. Nor was Forster, called on for advice, any help.

Naturally. And this manifestation of 'the Mother as demon' (Yeats's phrase and his 'only complaint' about the play) deftly allied the crushing powers of mothers with the mythic ambivalences of Everest. Chomo-lungma is that mountain's name, goddess-mother of the world, and she's a killer. 'It is known', wrote Professor Dyhrenfurth in 'The Leica in the Himalayas' in *My Leica and I* (1937), 'that the English wish to keep Mount Everest to themselves and during six expeditions, from 1921 to 1936 they have made such big sacrifices to . . . the "Goddess-mother" . . . that their moral claim is admitted on all sides'. *F6*'s Ransom turns into Hamlet, the best-known of literature's youthful abusers of mother, as he addresses the skull, perhaps that of a European climber, and contemplates a long list of dead climbers, 'those to whom a mountain is a mother', who have been killed in that most awkward and chilling of maternal embraces.

Englishmen, Auden went on to confide to Lord Byron in that same context of the harsh mother,

> Taking them by and large, and as a nation,
> All suffer from an Oedipus fixation.

When he and Isherwood thought about Oedipus it was evidently the mother aspects of the myth that troubled and fascinated them most. Auden can talk (in 'Under boughs between our tentative endearments') about the 'ancestral curse' of fathers, but on the whole fathers are as strikingly missing from his early writings as they are from Isherwood's. This is much less true of other '30s authors: who when they think about Oedipus or Hamlet tend rather to dwell on the irksomeness of fathers.

Fathers do badly on numerous counts. They produced the wrong kind of writing (all those editors: Spender's father edited the *Westminster Review*, John Strachey's the *Spectator*, Aldous Huxley's the *Cornhill*, John Lehmann's *Punch*; Evelyn Waugh's father was with Chapman and Hall). They preached the wrong kinds of belief (Day Lewis's, MacNeice's, Rex Warner's, Geoffrey Grigson's, Randall Swingler's and Clough Williams-Ellis's fathers were all Anglican ministers; Montagu Slater's father was a Wesleyan preacher, Hugh Sykes Davies's a Methodist minister; Janet Adam Smith's father was the rather well-known Old Testament scholar and principal of Aberdeen University, George Adam Smith). Fathers dished out the wrong kind of medicine (Dr George Auden was no Dr Georg Groddeck). Fathers were too Liberal to be credible in the harsh *realpolitik* of the current world of politics (which is the burden of Spender's *Trial of a Judge*, and of Rex Warner's *The Professor*: Liberalism would no longer do). Fathers had connived in their own fathers' war ('a parcel of damned old men' is abused in Brian Howard's schoolboy poem of 1922, 'To the Young Writers and Artists Killed in the War: 1914–18', for having had 'a great Young Generation' killed off). Whatever father might do, seemingly, was wrong.

'I hold with Freud that most poets are neurotic above the average', declared MacNeice in his *Modern Poetry: A Personal Essay*, and he went on mildly to mention 'that I have met several poets who had had strained relations with, if not antagonism towards, one or both of their parents'. But his tone towards his own father, as towards old bourgeois people generally, is far less mild. Bath, he tells the reader of *Zoo*, 'has some of the horror of Oxford, collecting, as Oxford does, various types of monstrous old people who have money of their own but no ties and no work, who, like shoals of jelly-fish, float into ports like Oxford (North Oxford, that is), Bath and

Cheltenham, gathering culture or health and stinging at a moment's notice.' John
Betjeman devotes more time to the detail of these old people's lives, especially if
they're worshippers in dim and slightly cranky Anglican churches in leafy suburbs
like Camberley and Croydon and nice provincial towns like Royal Leamington Spa
(Betjeman's people are inevitably tested by the buildings they condone and own),
but the dead old lady in chintzy circumstances (in 'Death in Leamington'), say, or
the old dear in a mauve hat with two cherries on it who's about to die in 'Calvinistic
Evensong' aren't stopped from being threatening by Betjeman's jaunty tone of
celebration. Spender, too, appears to be more tolerant than MacNeice of the aged,
but, again, he isn't either. He disclaims Hamlet's intensities (in 'Lines Written
When Walking Down the Rhine'),[73]

> Our fathers enemies, yet lives no feud
> Of prompting Hamlet on the kitchen stair,

but he then goes on repeatedly to blacken his parents (see, for instance, the poem
'My parents kept me from children who were rough'), again and again to shadow-
box with his Liberal father (how he sympathizes in *Forward from Liberalism* with the
nineteenth-century radicals who 'suffered in youth from the dominating will of the
Victorian father'), and to round off the '30s in *The Backward Son* with a picture of
the emotional collapse of Mr Brand, Liberal MP and one-time romantic giant of a
dad, shrunken after the death of his wife into a soggy childish mess. 'The Ambitious
Son' of Spender's poem (*NW*, Christmas 1939) is of course Oedipus cast in the
mould of this special group of '30s sons. He rounds on the 'Old man, with hair made
of newspaper cutting / And the megaphone voice'.

   In comparison, though, with the tones and intentions of many of the '30s youths
Spender's Hamlet/Oedipus act is conducted quite mealy-mouthedly. In general
fathers were not going to be forgiven for the destructive intentions towards their
sons that the sons thought they detected. They'd taken the lessons of the slaughtered
generation that preceded them and of Wilfred Owen's poems, one of the most
powerful of which is 'The Parable of the Old Man and the Young', in which
Abraham does not hold back the knife at the story's climax, 'but slew his son, / and
half the seed of Europe, one by one'. On this re-reading of Abraham and Isaac's
relationship sons had better be vicious towards fathers before it became too late.
Poems like Gavin Ewart's 'Home', tersely summing up all his grouchy poems
against schoolmasters, Cambridge dons, and the adult world's oppressions in
general—'How awful to live in a horrible house', where 'Sadism, anaemia, anxiety
neurosis / . . . make our dear home such a sweet bed of roses'—are implicitly oedipal
in their efforts to strike back at father. More explicitly, Anthony Powell's *Agents and
Patients* has its Maltravers grappling with a film version of *Oedipus Rex* and wondering how to translate its hatreds into visual terms—'one can't photograph a passive
man subconsciously hating his father'. In the '50s Malcolm Lowry sharpened the
point of his novel *Ultramarine* still further by rechristening Hilliot's ship the *Nawab*
as the *Oedipus Tyrannus*. But for busy father killing there must be few texts at any
time, let alone in the '30s, to rival Geoffrey Parsons' 'The Inheritor' (1938), a surrealistic mix of Oedipus, *Hamlet*, and *Macbeth* in which the vengeful inheritor of the
title, a youth named Arnold, gloats over the corpse of a vivid father-Duncan-
Polonius whom he's just added to the period's long lists of detective story cadavers:

Who would have thought the old man had so much blood
So many guts, such yards and yards of guts?
Out of the bucket they trailed in raw festoons
The hungry dog licked at the clammy tubes
And Arnold worked at his un-neat dissection.

Who would have thought one old man went so far?
The sink still gurgles with eternal blood
The bread-bin overflowed its choppered bones
Elusive gristle slithered on the floor
And Arnold dealt with the part that fathered him.[74]

Seventeen-year-old David Gascoyne was just such another Arnoldian. His vividly precocious first novel, *Opening Day* (1933), has its youthful and ultra-aesthetic hero feeling utterly quenched by a father who not only wants to remarry after his first wife's death but also to stop his son's pampered life of cultural pursuits. In the end they come to fisticuffs. 'There he sat, his father, his deadliest enemy.' It's time, insists father, that the boy give up wasting time 'scribbling and strumming' on the piano: he must take up a job in an office. The father falls into a fury—'exhibiting the behaviour of a child, or of a gorilla, in a "paddy". In a voice that was black and primitive, like a typhoon forcing itself through a jagged hole in a bombarded cathedral . . .' And in the shadow of that image of First World War hostilities the two fight ('like a brawl in a pub'). The father slumps back into his chair, looking odd; the son leaves the house, free at last to go and live artistically with his aunt in the Gray's Inn Road, virtually in Bloomsbury (prowling Bloomsbury earlier in the day, he'd seen an old man looking through the bars into the Foundlings' Hospital: 'like a prisoner in the prison of old age gazing out at the freedom of Youth'). He wires his aunt: 'Everything—OK—Father—given—in.' He packs his bag, including his short story 'Solitary Confinement'. He is woken by the news of his father's decease: 'He has always had a weak heart—and when you fought with him—it was as good as killing him.' Gascoyne's toughly aggressive fiction seeks to present, in fact, the necessary realities behind Spender's milder-mannered exhortation to the young ('oh young men oh young comrades') to quit their fathers' luxurious prisons of houses:

> advance to rebel and remember what you have
> no ghost ever had, immured in his hall.
> (No. XXII, 1933)

In an extended sense that didn't only include the written but embraced every aspect of the culture, 'The Inheritor' and *Opening Day* are emblems of the Yale critic Harold Bloom's Freudianized version of history and especially literary history as a process in which each generation tries to repress the father's writing, the father's meanings and texts, even his actual presence. Day Lewis's younger son model for poetry's progress was an incitement to violence. And, of course, in such situations fathers had better look out. At least sons had better not be ignored. And nor were they in the '20s and '30s. In any case it was difficult to avoid and ignore your son when—as one voice in MacNeice's dialogue poem 'An Eclogue for Christmas' put it—'the sniggering machine-guns in the hands of the young men / Are trained on . . . Father's den'. Youth was going willy-nilly to be heard. And youth was indeed in many quarters taken extremely seriously, even—perhaps because the War had left only younger sons behind ('There are now practically only two generations,' Evelyn Waugh noted in his diary in 1920, 'the very young and the very old')—even extremely youthful youth. The newspapers eagerly spread gossip about upper-class youths. 'Rich Boy as Deck Hand', 'Cotton Broker's Son as Deckhand': thus the *Evening News* and the *Daily Mail* in 1927 signalling Malcolm Lowry's spell as a seaman ('I took my ukelele with me and tried to compose some fox-trots. I hope to go on to a university and compose some more fox-trots'). 'Red Menace in Public Schools'; 'Moscow Attempts to Corrupt Boys'; 'Mr Churchill's Nephew Vanishes from Public School'; 'Winston's "Red" Nephew': thus headlines gloating over the boy Esmond Romilly's political scrapes and japes. The Romilly brothers, of course, thought they'd been thrown right into the front-line of the class war and the international struggle against Fascism. So, doubtless, did Gavin Ewart when after his poem 'The Fourth of May' (cheerfully rude about Earl Haig, class isolativeness and nightly onanism) had appeared in the Romillies' anti-public school mag *Out of Bounds*, his old headmaster wrote to ban him from visiting Wellington College for at least three years. And it was characteristic of this youth-obsessed and obsessively youthful time that, in a sense, these boys *were* in the front-line: Esmond Romilly had become a veteran of the Spanish Civil War and the fight to save Madrid from Franco by the time he was seventeen.

Encroaching on real seriousness or not, however, this sort of thing was the life-blood of the public prints, especially the right-wing papers. Taken on at the *Express* in January 1928, Tom Driberg almost straightaway sealed his appointment to the gossip-columns with a scoop about Frank Buchman's Oxford Group. Youth news from the Randolph Hotel in Oxford and a Buchmanite houseparty in an Oxford women's college: that's what the *Express* was after and Driberg was rapidly established as an ace provider of it in his various *Express* features. Youth opinion was also much sought out. Evelyn Waugh was also hired by the *Express*, a year before Driberg was taken on there. For some reason he failed to please even that egregious rag and got the sack after three weeks. But he soon found similar lines of work elsewhere, pouring out his youthful mind on the issues of the day—'Too Young at Forty: Youth Calls to the Peter Pans of Middle-Age Who Block the Way' (*Evening Standard*, January 1929: not too be confused with his earlier *Standard* piece 'Too Old At Forty'—the awfulness of being forty was evidently something Waugh and the *Standard* could go on playing with for ages); 'The War and the Younger Generation' (*S*, April 1929); 'Was Oxford Worth While?' (*Daily Mail*, June 1929); 'What I Think Of My Elders' (*Daily Herald*, May 1930). Would the *Express*, Waugh wrote to his agent at the end of the '20s, 'take an article on the Youngest Generation's view of Religion? . . . It seems to me that it would be so nice if we could persuade them that I personify the English youth movement.' The *Mail* seems to have been persuaded. They signed up Waugh for a weekly piece at the Nabobic sum of thirty pounds a time, and the first half of 1930 saw a stream of his boyish chatter and opinionatedness (on lesbians, constipation, school military training, and so on) until even the *Mail* could no longer bear this bright juvenilia (they discontinued his contract in August 1930). And, *horribile dictu*, Waugh was by no means alone: he, Godfrey Winn and Beverley Nichols (and Cecil Beaton too) were all charged by older journalists with being part of a youth 'conspiracy' to command the daily papers. 'Mr Winn', snarled Wyndham Lewis in the *Doom*, 'is a member of a new profession . . . he "earns his living" by being a "Youth" '. Winn was a ' "Youth" hack'.

A ' "Youth"-racket' was another of Lewis's gnashed-out bad-mouthing descriptions of what was going on. And he was right. The smell of a racket was everywhere, and not only in the newspapers. It had penetrated to the offices of publishers. The young John Lehmann at Hogarth, the young Anthony Powell at Duckworth, Uncle Tom Eliot, all endeavoured to grant the youths prompt access to the book-trade. Publishers became eager to sign up youths to write instant books. They gave them money to go off and get some experience—virtually any experience—and then tell England about it quickly. The '30s travel book is virtually the creation of the '30s youth racket. It was youth hackery par excellence. Politicians too, of Left and Right, realized they'd better start appealing to youth. To be the party of youth would guarantee them the look of the future, they supposed. Massed ranks of young men and women were engaged to bare their youthful limbs in the Soviet movies to advertise the great youthfulness of Communism. Very soon the same iconography was filling the film productions of Fascism. Everybody, claimed Lewis in *The Old Gang and the New Gang* (the book he rushed out in 1932 to stand in for the suppressed *Doom*), whether Jesuits, Trotsky, Hitler, Mussolini, or Mosley, was 'after the Youthies'. And it does, for example, seem astonishing now to discover Karl Radek at the 1934 Soviet Writers' Congress taking an interest in Oxford student politics—as reported in *Problems of Soviet Literature* (1935). 'In the heart of bourgeois England, at Oxford, where the sons of the English bourgeoisie are educated, a group is taking shape which realizes that the only salvation lies in alliance with the proletariat.' But this does bear striking witness to Lewis's charge that the pervasive Politics of Youth had taken over politics as a whole. And nowhere in England, perhaps, was a more conspicuously blatant attempt made to corner the youth market than in Oswald Mosley's New Party. Mosley's paper *Action* came packed, in 1931, with the up-and-coming youthies: Isherwood, Peter Quennell, Alan Pryce-Jones, Eric Partridge. And it celebrated youth and youthiness: Quennell, for instance, wrote about the benefits of Oxford ('The Nursery Freedom of Oxford University'), Isherwood on 'The Youth Movement in the New Germany' ('They are sombre, a trifle ascetic and absolutely sincere. They will live to become brave and worthy citizens of their country. It is to be hoped that we can say as much for our own younger generation').

Isherwood's note was itself sombre and excessively pious. But then youth always takes itself solemnly. Seldom, though, can the youthful self have been so encouraged in a sense of its own importance. The inflated idea, for example, that life at your prep school and public school was as bad—and that you should take it as being as seriously endangering and traumatic and testing—as the horrors of a fascist State or a Nazi concentration camp was a conventional piece of schoolboy and youthy exaggeration. And it persisted among the gradually ageing youthies. In his 1939 jottings for an unpublished book, *The Prolific and the Devourer*, Auden is still to be found repeating the case he publicly made in 1934 in his contributions to Graham Greene's anthology of school memories *The Old School*: 'Politically a private school is an absolute dictatorship where the assistant staff play, as it were, Goering Roehm Goebbels Himmler to a headmaster Hitler.' (Was the recent assistant schoolmaster Auden feeling at all guilty, one wonders?)[75] And as late as 1950, in his autobiography *The Crest on the Silver*, Geoffrey Grigson was promoting the grim satire that his peculiar prep-school headmaster (whose cure for boys who urinated in their bath was applying vaseline to their penises and the cane to their bottoms) was a version of the Venezuelan dictator Gomez, a cartoon representation of whom he was thought to resemble: 'Gomez, dictator of the preparatory school.'

The self subjected to rigours like these must matter, and the youthies made no bones about advancing that self as a worthwhile literary topic. There were, naturally, attempts to jeer this posture of youthful self-importance off the stage. Cyril Connolly tried, in 'Where Engels Fears to Tread'. So did Alan Pryce-Jones in his marvellously lurid youth-picaresque novel *Pink Danube* (1939: 'by Arthur Pumphrey'). But still the youthful tone was maintained. The youthies' travel books—travel by and with Old Etonians and assorted other Old Boys, on the model established by Waugh and Peter Fleming—simply poured from the presses. And every other author insisted on pressing his youthful reminiscences upon the public. His prep school, his public school, his university popped up at you at every opportunity, out of every imaginable and unimaginable literary corner. Not just in obvious places, like Greene's *The Old School* reunion, or straight autobiography like Henry Green's *Pack My Bag*, and the teenage Giles and Esmond Romilly's *Out of Bounds*, or in more or less straight autobiographical novels like Isherwood's *Lions and Shadows*, but in a 'Letter to Lord Byron', in an *Autumn Journal*, in *The Road to Wigan Pier*, in the Prologue to a book that sounds like an international spy thriller (and both is and isn't), i.e. Greene's *The Lawless Roads*, or in volumes the reader might reasonably have expected to keep up their overt function as literary critical surveys, like Connolly's *Enemies of Promise* or MacNeice's *Modern Poetry*, even in a book about zoos.

Spender might have his doubts—and *The Destructive Element* was written against the moral insignificance, 'the neglect of subject matter, if not . . . the decadence of style' that this 'predominance of autobiographical themes' signified—but still '30s authors went on responding to the period feeling that they should bear witness about themselves. They were extraordinarily inclined towards personal confessions. They are always breaking out into 'My Case-Book' or 'A Personal Digression'. This was the age of the insistent personal testimony. Not for nothing was it the period when gossip columns, confessional film-fan mags, religious revivalism, political conversionism and the framed-up political Show Trial flourished. Few writers bothered even to try and avoid the currency of the fashion. *I Crossed the Minch* (MacNeice), 'I Drive a Taxi' (Herbert Hodge in *Fact*, 22), 'I am a Miner' (B. L. Coombes in *Fact*, 23), *I Was a Tramp* (John Brown), *Serving My Time* (Harry Pollitt), 'I Add My Witness' (Ralph Bates in 'Frank Pitcairn' i.e. Claud Cockburn, *Reporter in Spain*), *Let Me Tell You* (Leslie Halward), 'I Go On A Journey', 'I Revisit Old Scenes', 'I Explore an Historic House', 'I Am Driven By An Idea', 'I Am Filled With Doubts', 'I Become A Landowner' (and so the chapter headings of Henry Williamson's *The Story of A Norfolk Farm*, 1941, go on and on). The published diary flourished: Isherwood's Berlin diaries; the 'Travel Diary' in Isherwood and Auden's *Journey to A War*; J. R. Ackerley's *Hindoo Holiday: An Indian Journal* (1932); Gide's 'Pages From A Diary' in *Retouches à Mon Retour de l'URSS* (English translation, 1938); Georges Bernanos's *Les Grands Cimetières sous la lune*, or *A Diary of My Times* (1938: following up his international success with the novel *Le Journal d'un curé de campagne*, 1937). The open letter was in vogue too: all those in *Letters from Iceland*; Ewart Milne's *Letters from Ireland* (poems, 1940); Virginia Woolf's *A Letter to A*

Young Poet and Peter Quennell's *A Letter to Mrs Virginia Woolf* (both published in 1932 in a series of Hogarth Letters which appeared as a single volume in 1933); Elizabeth Bowen's 'Letter from Ireland'; Alistair Cooke's 'New York Letter' (first sign of a whole career to come of broadcast 'Letters from America'), William Empson's 'Letter from China', Stuart Gilbert's 'Paris Letter' (all these in the short-lived *N&D*); the several letters published in Julian Bell's *Memoir* (Bell ran into trouble with his open letters: the Hogarth Press turned down his scheme for an open epistle to Roger Fry); MacNeice's 'Letter to W. H. Auden' (in the Auden Double Number of *NV*); Charles Madge's poem 'Letter to the Intelligentsia'. And if the writer was a bourgeois and a young one at that, such confessions and diaries and bits of correspondence were inevitably about school.

> Sitting alone on summer evenings, wondering
> How long the isolation would go on,
> Remembering Winchester . . .

That's how Charles Madge's 'Letter to the Intelligentsia' (one has to wonder how *intelligent* an intelligentsia . . .) starts up in *New Country*. School had engrossed Madge—school and university are about all he's had time for so far in his short life (he was twenty-one in 1933)—and, notably, he assumes it'll impress you. And this is an assumption glibly shared not just by the host of reminiscers about their school-days but by the mass of university and school fictions they produced. Waugh's *Decline and Fall*, Henry Green's *Blindness*, Isherwood's *The Memorial*, Day Lewis's *Starting Point*, Arthur Calder-Marshall's *Dead Centre*, Spender's *The Backward Son*: these novels dunk you unmisgivingly into the esoterically limited world of English bourgeois education. It's an inside story that they circulate, lovingly told and retold, gloatingly commemorated.

> Commemoration. Commemoration. What does it mean? What does it mean? Not what does it mean to them, these, then. What does it mean to us, here now.

Thus opens the 'Address for a Prize-Day' in Auden's 'An English Study' in *The Orators*. And Auden was only doing once again, this time in public, what he'd many times entertained his chums with in private, that is parodying the headmaster's Founder's Day sermon at St Edmund's, Hindhead, the prep school which he and Isherwood both attended. Weston, writes Isherwood in *Lions and Shadows*

> was brilliant at doing one of Pa's sermons: how he wiped his glasses, how he coughed, how he clicked his fingers when somebody in chapel fell asleep: ('Sn Edmund's Day . . . Sn Edmund's Day . . . Whur does it *mean*? Nert—whur did it mean to *them, then, theah*? Bert—whur ders it mean to *ers, heah, nerw*?') We laughed so much that I had to lend Weston a handkerchief to dry his eyes.

The 'preparatory school atmosphere reasserted itself', Isherwood says, of his first meeting with Weston since St Edmund's. Auden was soon into his Pa routine. 'The Prep School atmosphere', Auden wrote in his Journal in 1929, 'that is what I want'. And that is what he sought, not least, and quite openly, in *The Orators*, with its Commemoration harangue, its celebration of an OTC Field Day gone serious (Ode V, 'To My Pupils': 'Though aware of our rank and alert to obey orders'), its Ode II to the surveyed members of a heroical school Rugby team:

> Success my dears—Ah!
> Rounding the curve of the drive
> Standing up, waving, cheering from car,
> The time of their life.

'The time of their life.' It was not always being invoked so successfully as by Auden, of course. The Audenesque page upon page devoted to an Oxford Rugby match in *Starting Point*, or Gavin Ewart's poem 'Characters of the First Fifteen', a direct crib of Auden's Ode that was kept out of the Wellington School magazine by the headmaster because of its reference to the 'ushers' wives', 'smoking and having the time of their lives' (*that* wasn't the kind of time of their lives these youthies were meant by their mentors to be celebrating), or the section of Calder-Marshall's *Dead Centre* devoted to 'Druce' who's just won his colours and is on his joyful way back to school with the victorious team in their charabanc in yet one more scenario indebted to Auden—all these lack the magical intensities of Auden's Ode II. But whether in the successful or the less successful versions the devotion evinced is loyal and seemingly inevitable. Such writing is happily committed to this narrow world of juvenile games, infant derring-do, and, above all, to the clinging esoteric rhetoric of this sort of childhood.

The lingo of prep school stuck. It stuck noticeably to Waugh. The first time she met him, claimed Theodora Benson in her collection of reminiscences *The First Time I . . .* (1935: yet another '30s I-formation), he was with Alan Pryce-Jones at the Ritz, and she did not catch Waugh's name:

> They were both very amusing, but whereas Alan said witty things in ordinary English, Mr E. Waugh said them in the limited though expressive vocabulary of a preparatory school for boys. I thought it nice and unaffected to perceive that this unliterary chump had an odd charm to him.

No wonder Waugh could do the Llanabba part of *Decline and Fall* so well, or that Isherwood's story 'Gems of Belgian Architecture' (written in 1927 and set at St Edmund's) is so high on that prep-school's brand of talk (*get lammed, lost his bate, quissed, got a swank on, giving him the frozen mitt*), the slang that makes *Dead Centre* just occasionally perk up (*prepper* and *cogging*—masturbation—and what not), the language that turns Day Lewis's first detective story *A Question of Proof* (1935) into an anthology of prep-school talk (*that squit Wemyss, slimy little chizzer, you silly gowk, so snoo to you, not a bad sort of beezer, we oiled out*: to solve the mystery the detective Strangeways has to really take the little ones seriously by becoming initiated into their Society of the Black Spot). It's the kind of ideolect that's really the only distinctive feature of the first episode of John Betjeman's 'Diary of Percy Progress'; entitled 'The Cag Bag and the Mags'. Progress is an Old Carshaltonion—a breed of men known to themselves as Old Cag Bags or Old Cags—and he lives by hiring out old magazines to doctors and dentists (his firm is called 'Back Numbers Ltd.'). It's a wheeze he pinched from another Old Cag whom he bumped into whilst drunkenly singing the Old Cag Song (in which youngsters rhymes with *bungsters*) along a street.[76]

Clearly, a strong stomach was needed to stand immersion into this intensely cultivated collective memory of the English bourgeois male, this act of collective regurgitation—drunk or sober—of gobbets of private language from the private schools. In 'Gems of Belgian Architecture' the boys' drunken vomiting is naturally a matter of this sort of language:

> 'Whoa, up. That's a good boy. Fetch 'em up.'
> 'Out she comes. Let 'em have it.'
> 'And the next please. Right up. Cafe and Roof Garden.'

Forgiveably, one looks to see whether Michael Roberts's tongue is in his cheek as he praises Auden for taking the subject of school so seriously. It's 'unreasonable', Roberts alleges in his review of *The Orators* in the *Adelphi* (August 1932), to condemn Auden's 'very frequent use of public school and O.T.C. imagery', 'for in addressing the public-schoolboy he is attacking the creeping timidity and platitudinous conventionality of English decadence at the crucial point'. *The Orators* is taking on the governing classes, seeking to save the middle classes from their shirked social responsibility: its 'attack' is 'important'. Can Roberts, you wonder, be serious? Can, for that matter, MacNeice really mean it about Auden's new poetic of school? 'You realize', he wrote to Auden ('Letter to W. H. Auden'), 'that one must write about what one knows ... It is an excellent thing (lie quiet Ezra, Cambridge, Gordon Square, with your pure cerebration, pure pattern, your scrap-albums of ornament torn eclectically from history) that you should have written poems about preparatory schools.' But yes, that long parenthesis notwithstanding, MacNeice does seem to mean it, for he expands his praise in his *Modern Poetry* discussion of Auden just a year later:

> Like Wordsworth, he brought into poetry a subject-matter considered alien to it by his elders, who would have thought brothels or champagne poetic but not changing-rooms or playing fields. Auden, who sees the adult in the child and the child in the adult, had the good sense to write about what he was acquainted with. And for this he had the sanction of his favourite psychologists—*maxima debetur puero reverentia*—and the sanction of the sociologists, adolescent youth being the cock-pit for striving ideologies.

Perhaps MacNeice also meant that Auden had had the good sense to write about what he, MacNeice, was best acquainted with and couldn't ever seem to get quite away from. (There exists, for example, a post-'30s draft for a radio play, 'Return to a School', in which MacNeice's Marlborough memories are given to a returning character called Greene: 'With Dr Arnold ... and the little victims and the Loom of Youth and the Squeaking Chalk and the Old School Tie and the Sporting Parson in Corpore Sano and Kipling and Newbolt ... and bags of Team Spirit and all.')[77] Certainly the implied audience for this loving re-creation must be other Old Boys from Auden's sort of school, their families, and their friends. Not forgetting, of course, the gawping proletarians that the *Mail* and the *Express* assumed to be eager for titillation by titled gossip—a large number, apparently, in the days when it was said that 'everyone' supported one side or the other in the annual Oxford and Cambridge Boat Race.

School images came naturally to the youthies in the first place because school was still near at hand in terms of years. School was near, too, because many of the writers had with almost indecent alacrity gone back into the system as schoolmasters. By choice or necessity large numbers of '30s literary persons put in time as educators of one sort or another, exercising what Evelyn Waugh said was 'the proud right of the educated Englishman to educate others.'[78] Isherwood did bits of private tutoring in England as well as Germany. Grigson taught briefly at a crammer's after Oxford. Connolly took a brief private tutorship when he went down from Oxford. Auden for a while coached one of the little Mitchisons in Latin. MacNeice became a university lecturer in Classics, first at Birmingham, then in London University. Anthony Blunt taught Art History at Cambridge, then in London. John Cowper Powys was a longtime roving lecturer-extraordinary. At the end of the decade Durrell and Bernard Spencer drifted into teaching for the British Council in Greece. But besides these relatively exotic activities within education, there was also a mass of just plain schoolmasters. Among writers at least, the '30s was definitely the era of Mr Chips. James Hilton's sentimental schoolmaster story *Goodbye Mr Chips* appeared, of course, in 1934 and became one of the period's most popular fictions. (James Hilton's work shows nicely how it's often the minor writers who express most fully the meanings of a time: nobody else can rival Hilton's '30s achievement of having planted in the English language two such definite catch-phrases: Shangri-La and Mr Chips.) Who indeed, one sometimes wants to ask, did not at some point do time as Mr Chips? Orwell did, so did Day Lewis and Auden (who 'inherited' his job as English master at Larchfield Academy from Day Lewis), so did T. H. White (who was Head of English at Stowe, 1932–6, and so overlapped with John Cornford, who left in December 1932), so did Tom Driberg (who taught at a prep school in Bournemouth before going up to Oxford), and Michael Roberts and Evelyn Waugh, and Randall Swingler, and Humphrey Jennings, and Arthur Calder-Marshall, and Samuel Beckett (who was both schoolmaster and *lecteur*), and Rex Warner, and John Betjeman (and he had been taught briefly by T. S. Eliot), and Willa Muir (the one female in the list: she assisted A. S. Neill in various of his experimental locations), and Spender (in Devon in 1940), and Bernard Spencer (at a whole host of prep schools from 1932 to 1940), and Peter Hewett and Edward Upward (who never gave up being a Mr Chips right to the end of his working life).

Writers in need of cash were in and out of the scholastic agencies like yo-yoes (which were themselves, by the way, a '30s craze). Waugh and Orwell both got work through Truman and Knightley—though Bernard Crick in his Life of Orwell has suggested that The Hawthorns School in suburban Hayes to which Orwell went in 1932 as 'head master' was too small and awful (Orwell's predecessor, a character straight out of *Decline and Fall*, had been jailed for indecent assault) to get on to Truman and Knightley's allegedly prestigious registers. Evelyn Waugh had a lower estimate of that agency than Professor Crick ('went', he says, 'to Truman and Knightley's and said that I wanted to be a schoolmaster, ever since which time I have been filling in forms to say that I was in my house-team for swimming'). Cowper Powys and Betjeman, Grigson, Connolly, Auden, and Isherwood called on the services of the rival outfit Gabbitas-Thring. 'As a result of my performance in the Tripos'—thus Isherwood in *Lions and Shadows*—'I hadn't even the necessary credentials for schoolmastering—that last refuge of the unsuccessful literary man. Still, teaching of some kind it would have to be. Parents on the look-out for a private tutor were not so particular, I was told. I put down my name on the books of . . . Messrs Gabbitas and Thring.' Others hedged their bets. Mr Joliffe, a young schoolmaster in *Dead Centre*, not liking his wife, his small house, his school's isolation, or his poor stipend, writes to both agencies 'for particulars of every small headmastership that's going'. Burned at Truman and Knightley—though Arnold House, Llandullas, did become the gloriously terrible Llanabba and did introduce Waugh to the drunken sodomite he turned into *Decline and Fall*'s marvellously awful Grimes—Waugh gave Gabbitas and Thring a try. Cyril Connolly was planning in 1927 a novel to be called *Green Endings* 'which would begin with the hero in a waiting-room of Gabbitas and Thring. A good subject for a cynically dreary introduction . . .' In 1937 Auden put Gabbitas and the whole schoolmasterly fix they stood for, to Lord Byron:

> The only thing you never turned your hand to
>   Was teaching English in a boarding school.
> Today it's a profession that seems grand to
>   Those whose alternative's an office stool;
>   For budding authors it's become the rule.
> To many an unknown genius postmen bring
> Typed notices from Rabbitarse and String.

Rather more deploring noises than Auden's could often be heard. Evelyn Waugh talked in his 'Truths About Teaching' piece about 'unpalatable' workhouse conditions and men 'degraded and lost to hope'. Schools, it was commonly said, were stiflingly dead places, symptomatic of much that was hateful and needing to be changed by the revolution, they were a set of conservative institutions naturally antagonistic to the revolutionary schoolmaster. This is what Day Lewis found as a Friend of the Soviet Union teaching in Cheltenham, and what Rex Warner discovered—his actual sacking from The Oratory (Catholic) School in Sunning had political overtones. Oppressed members of common-room: it's a recurrent note in *Dead Centre*, in *Starting Point*, in Edward Upward's story 'The Colleagues' (published in *New Country*). Characteristically, Upward's prep school master Mitchell dislikes games, fitness, Prayers, the OTC, Armistice Day, and has fantasies of refusal and escape from his school's prison of unreality. 'Nothing that happens in the school grounds has any connection with what happens . . . outside . . . Games are organised. Surplices are worn. Outmoded precautions are scrupulously taken. Nothing which a clergyman might think risky to neglect is neglected. We are servants of the parents' most contemptible misgivings. I shall be here or in places similar to this for the rest of my life.' And yet, for all these complainings, Upward's story is—again, characteristically—dense with the atmosphere it professes to find so stifling. These schoolmasterly critics accept deeply the system they express horror about, even rejoicing rather and in the blissful spirit of Christian martyrs, in the midst of their sufferings. So that in the 'Writers and Morals' section of his long essay 'Revolution in Writing' Day Lewis is proud to note that some writers are having a hard time as schoolmasters—it makes them one with the victimized manual worker. And a kind of permanent schoolmasterly analogy for the writer and his audience slips unselfconsciously to Day Lewis's pen: 'He is like a schoolmaster, who not only has to instruct and entertain his pupils, but must also be conscious that his lightest word may have a quite disproportionate effect on their development.'

In a *New Verse* piece called 'Poetry and Politics' Charles Madge makes a shrewd point about Auden. Madge is distinguishing the poets who wander about the world, observant but alien, and those who settle down. He quotes from what is possibly Auden's finest schoolmastering poem:

there exists another category: those who have got jobs, mostly as schoolmasters.
Their poems are not separate from the rest of their activity

> The clock strike ten: the tea is on the stove [;]
> And up the stair come voices that I love.

This is someone who likes his job, summing it up after a hard day's work, and laying plans for the future. His poems are a running over of the sense of activity, not a solidification away from it.[79]

Auden did like being a schoolmaster. He cared for the young enough to contribute a piece on 'Writing' to *An Outline for Boys and Girls and Their Parents*, edited by Naomi Mitchison (1932). He even became something of an expert on pedagogy—at least, he wrote on educational topics, frequently reviewing books on the subject. Day Lewis too entered into the spirit of traditional schoolmastering enough to compose in 1929 the Larchfield School Song:

> All that you gave our eager boyhood
> How can we hope to repay?
> For you've taught us to go straight
> When the ball is at our feet,
> And tackle any fate that comes our way.[80]

Thus complete with sporting metaphors, school songs could scarcely manage to be more traditional. And it's this slush of deep affection that helps make Auden's contribution to *The Old School* anthology so tonally odd.

In its manuscript version Auden entitled his piece 'The Liberal Fascist' and the contradiction of that title runs through his whole reminiscence. There's the usual fascism charge. Masters are acutely pinned as former school athletes and timid academics, 'would-be children', with 'no outside interests', so that they've become 'either lifeless prunes or else spiritual vampires, sucking their vitality from the children'. They're 'silted-up old maids, earnest young scoutmasters, or just generally dim'. And yet, Gresham's, Holt, is also glowingly presented in the approving adjectives of a school prospectus. The situation of Gresham's was *beautiful*; the authorities had *extraordinary good sense* ('virtually no bounds'); the buildings were *comfortable*; the class-rooms were warm and *well-lit*; the dorms had cubicles so that the boys weren't *unduly herded together*; fagging was *extremely light*, hot water *plentiful*, the food *quite adequate*; the education was *pretty good*, the library *magnificent*, the labs *excellently equipped*, the staff all *conscientious* (*some efficient*), the scholarship list *quite satisfactory*; about athletics the school was *extremely sensible*. The catalogue can be read as deliberate parody, an example of the sort of trap schools set to lure in the old maids and the Fascists among parents and set up here to lure in such people among Auden's readers. But set alongside Auden's rather similar tone in the similarly mixed report to Lord Byron, it seems more necessary to perceive here the interplay of real love with real dislike. On the one hand there are Rabbitarse and String, and on the other

> The Head's MA, a bishop is a patron,
>   The assistant staff is highly qualified;
> Health is the care of an experienced matron,
>   The arts are taught by ladies from outside;
>   The food is wholesome and the grounds are wide;
> The aim is training character and poise,
> With special coaching for the backward boys.

(That 'experienced matron', by the way, a more humane version perhaps of the Audenic nurse, had her first outing in the opening chorus of *The Dog Beneath the Skin.*)

Undoubtedly the schoolmasters' feelings were mixed, just as their reasons for going back to school were undoubtedly also mixed ones. But, mixed up about school and schoolmastering though they were, the fact is that they had gone back in. And once back in it's clear they found it easy not just to be *in* school but to be *of* it. So commonly indeed did '30s authors wear the authentic mantle of the schoolmaster that (according to Julian Maclaren-Ross) even non-schoolmasters of the period affected the prevailing 'housemaster's outfit . . . tweed jacket, woollen tie, grey flannels baggy at the knee'.[81] And in many ways, of course, the schoolmasterly writers had never left school: for school had never stopped being in them. The school memory, the reminder of school, analogues of schooldays, the sense of the world as school and a continuation of school (as in Gavin Ewart's poem XII: 'He wondered vaguely, Was the world a school?'), these kept on returning in the most far-flung of circumstances.

In the song that Auden composed for Raynes Park Grammar School when John Garrett, his friend and collaborator with him on the school poetry anthology *The Poet's Tongue* (1935), was headmaster there we're informed that

> Time will make its utter changes,
> Circumstance will scatter us;
> But the memories of our schooldays
> Are a living part of us.[82]

And how right Auden was! John Cornford, Esmond Romilly, George Orwell, all—as we shall see—found school memories powerfully present even amidst the distractions of their Spanish Civil War experiences. *Journey to a War* shows how Isherwood and Auden found the old school alive and well even in China: they think the proprietor of the Journey's End inn at Kinkiang is like a prep-school headmaster; at Yü-tsien a truce is declared between rival Chinese in terms they perceive as schoolish ('the examination results were published, and everybody came out equal top'). Evelyn Waugh's first travel book *Labels* (1930) is filled with this kind of recall and analogy: Ascension Day in Naples sparks reminiscences of that day's festivities 'at my school. It was the only whole holiday in the year'; Malaga Cathedral 'reminded me strongly of the chapel at Hertford College', and some early blows against Cruttwell are deftly got in ('ill at ease in his starched white surplice, biting his nails, and brooding, I have no doubt, on all the good he intended for each one of us'; 'there was a don at my college exactly like a prominent murderer'). Wherever he travelled Waugh never ceased to be reminded of school and Oxford. And it wasn't just that he kept bumping into other Old Boys; nor simply that the British had successfully exported the public-school idea so that in curiously remote corners of the globe he would come across wonky but redolent replicas that brought memories flooding back (in Uganda, for instance—in *Remote People*, 1931, this—Waugh finds a public school for sons of prominent natives, complete with prefects and crested blazers and 'colours', except that one boy yells and writhes extravagantly when he's caned: 'Apparently this part of the public-school system had not been fully assimilated'). There was more. The fact was that school and university, but school especially, had left uncomfortable tastes lurking in the mouth. The food, Waugh reports of his third-class trip back from Cape Town, 'was like that of an exceptionally good private school—large luncheons, substantial meat teas, biscuit suppers.' We 'imitated', writes Isherwood in *Lions and Shadows*, of that first reunion with Auden since school, 'Pillar cutting bread at supper: ("Here you are! Here you are! Help coming, Waters! Pang-slayers coming! Only one more moment before that terrible hunger is satisfied! Fight it down, Waters! Fight it down!")'. (A recollection that gets into Ode IV of *The Orators* where Auden tells John Bull to be patient:

> Calm, Bull, calm, news coming in time;
> News coming, Bull; calm, Bull,
> Fight it down, fight it down,
> That terrible hunger . . .)

'I wonder how you can write abt St Cyprian's', Orwell wrote to Connolly about his account of their prep school in Eastbourne in *Enemies of Promise*. 'It's all like an awful nightmare to me, & sometimes I can still taste the porridge (out of those pewter bowls, do you remember?)'[83]

And you couldn't rinse the tastes of school out of your mouth, because school on the model of the English bourgeois and upper-bourgeois classes—a chain of boarding-schools away from home, entered into at a cruelly early age, an experience of home deprivation kept up right into one's early twenties when it continued, as it frequently did, into the years spent at Oxford and Cambridge—was, bluntly to put the matter, traumatic. 'At first', Connolly recalled in *Enemies of Promise* about his days at Eastbourne, 'I was miserable there and cried night after night. My mother cried too at sending me, and I have often wondered if that incubator of persecution mania, the English private school, is worth the money that is spent on it, or the tears its pupils shed . . . It is one of the few tortures confined to the ruling classes and from which the workers are free. I have never met anybody yet who could say he had been happy there.' 'As for St Cyprian's', Orwell joined in , in 'Such, Such Were the Joys' (written by May 1947), 'for years I loathed its very name so deeply that I could not view it with enough detachment to see the significance of the things that happened to me there. In a way, it is only within the last decade that I have really thought over my schooldays, vividly though their memory has always haunted me.' '[T]hose filthy private schools', Orwell called them in another letter to Connolly (December 1938). His loathing makes the Ringwood House section of his novel *A Clergyman's Daughter* almost incoherent with a gnashing satirical rage over the charlatanry, the 'dirty swindle' of the dreary majority of private schools, 'Second-rate, third-rate, and fourth-rate'; 'grasping low-minded' places purveying nonsense and ignorance for ready cash. Neither 'Ma' of St Cyprian's, nor 'Pa' of St Edmund's, nor the ubiquitous matron, could be any substitute for home. 'So began', Geoffrey Grigson wrote in *The Crest on the Silver*, of the Plymouth prep school he was sent away to at the age of only five, 'the long purgatory of lying in bed during three periods of the year, and trying to imagine oneself in one's room at home, in the silent night of Cornwall, with the wail of the trains wiped out of the sky.' And though the pain of leaving home grew less as one grew inured to it, so that public school was less shocking in this matter than earlier private school, cruelties continued in other ways. Grigson says he found schoolmasters of all sorts 'a succession of dunces, bullies and nonentities', but the mental and physical punishments of public school were if anything worse than those of his prep school. 'If only . . . I had never seen the swishing MA robes and the red hood of the dull and ignorant! If only I had never entered the desert of a public school!' 'I do not propose to weep', he went on, 'though I record the fact that there exist six or seven people whom for years I would willingly have tortured and shot, and about whom I have only ceased to have dreams during the past six years or so [i.e. when he was around forty]. I lined them up against the back wall of the fives court, held up my hand, and dropped it as the signal to fire.' At least Grigson absorbed the lesson of prevailing school sadisms. The juvenile prep-school thuggery described by Isherwood in 'Gems of Belgian Architecture', culminating in the death-sentence horrors of a 'gorse-bushing'; the viciousnesses that pervade Spender's *The Backward Son* (of his own prep school Spender wrote in *The Old School* that his parents 'might just as well have had me educated at a brothel for flagellants'): such cruelties carried over into the organized brutalities of the public schools to make a keynote in public-school memoirs. Once L. P. Hartley's whole house at Harrow was 'whopped' as he puts it in *The Old School*:

> When my turn came the executioners, attired in running vests, were bright-eyed, sweating and exultant, but seemed rather tired. It turned out they were not.

Again and again Louis MacNeice referred to Marlborough's sadistic *pièce de résistance*. On one of these occasions, in his *I Crossed the Minch* (1938), the characters Perceval and Crowder discuss their author's schooldays. He didn't go to one of 'those co-educational bear-gardens where no one is allowed to be repressed', but to a 'Grand Old Public School', whose 'mild sadism' 'appealed to him', except when he witnessed 'a basketing'. This was when the 'bloods' punished 'moral offenders or unpopular figures' in the prep-hall. While the little 'plebs' howled, the victim was pushed up the aisles in a huge waste-paper basket. He was beaten up, his clothes were torn off, he was daubed with paint, ink, treacle, powdered chalk. Mild sadism indeed. No wonder Graham Greene in his editorial comments in *The Old School*, expressed the conviction that 'the system which this book mainly represents is doomed . . . the public school, as it exists today, will disappear'. Perhaps there was some excuse, after all, for the notion that such schooling was not unakin to the terrors of totalitarian regimes.

Graham Greene—who is one of the '30s most memorable investigators of trauma, especially the traumas of childhood—knew that schools like this traumatized boys as truly as the Saturday night rutting of his parents and his Catholic upbringing had traumatized Pinkie, the boy gangster of *Brighton Rock*. The bourgeois version of Pinkie is Anthony Farrant of Greene's novel *England Made Me* (1935), the doomed wastrel in exile, who has bought outright his father's large package of bogus values, the codes and beliefs of the school that he ran away from but cannot forget. School has spoiled Farrant for good. He's a ruined man, a con-artist who sports an Old Harrovian tie he has no right to, forced by circumstances uneasily to associate with the seedy Minty (another taste, you might say, that lingers on the palate), a man still living off boyish condensed milk and talking of tuck-boxes. Minty is yet another exile who devotes himself (like Wilson does later on in *The Heart of the Matter*) to the pages of his old school's magazine. He's another fake too, expelled from Harrow but continually endeavouring to fix up Old Harrovian dinners for men like himself and

Farrant who have been banished to Stockholm. An education that twisted you like this must have been Hell—as Greene indicated in the Prologue to *The Lawless Roads*:

> One began to believe in heaven because one believed in hell, but for a long while it was only hell one could picture with a certain intimacy—the pitchpine partitions of dormitories where everybody was never quiet at the same time; lavatories without locks: 'There, by reason of the great number of the damned, the prisoners are heaped together in their awful prison . . .'; walks in pairs up the suburban roads; no solitude anywhere, at any time.
>
>   . . .
>
> In the land of . . . stone stairs and cracked bells ringing early, one was aware of fear and hate, a kind of lawlessness—appalling cruelties could be practised without a second thought; one met for the first time characters, adult and adolescent, who bore about them the genuine quality of evil. There was Collifax, who practised torments with dividers; Mr Cranden with three grim chins, a dusty gown, a kind of demonic sensuality; from these heights evil declined towards Parlow, whose desk was filled with minute photographs—advertisements of art photos. Hell lay about them in their infancy.

'Heaven lies about us in our infancy!', according to Wordsworth. When Greene had earlier inverted Wordsworth's 'Immortality' Ode it was precisely in the case of Pinkie, the cruel boy who had graduated from the school dividers to the cut-throat razor: 'He trailed the clouds of his own glory after him: hell lay about him in his infancy. He was ready for more deaths.' No infant heaven for him: his razor-wielding hand is Satanic, toad-like: 'it lay like a cold paddock' on his girlfriend Rose's knee, neatly rebutting Robert Herrick's little prayer for a little child that it insistently and ironically recalls, 'Another Grace for a Child':

> Here a little child I stand,
> Heaving up my either hand;
> Cold as paddocks though they be,
> Here I lift them up to Thee.

Not everybody, it should go without saying, went through the boarding-school mill. Greene is theologically sound in having the hellish evils of the private system melding into the evils of Pinkie, the bully from the Council School yard. Original Sin prevails in every school: Greene knew that. But aside from these theological considerations, Greene also knew that in almost every other respect these kinds of schooling were worlds apart. *The Old School* is directed explicitly at the English snobberies that keep the private education system socially aloof. For its part, the hell of the poor child's school was mainly an economic and social hell, as Greene granted in a telling review in *Night and Day* (for which he was film critic as well as literary editor) of the documentary movie *Children at School*, produced by the Realist Film Unit (1937). On the one hand, he writes, there are the new, romantic, sunny Infant and Nursery Schools:

> Then the realistic movement, the reminder that hell too lies about us in our infancy. A small child hurries down a dreary concrete passage, while from behind a door comes a voice reciting the rich false lines—'And softly through the silence beat the bells along the Golden Road . . .' Cracks in the ceilings and the beams, damp on the walls, hideous Gothic exteriors of out-of-date schools, spiked railings, narrow windows, scarred cracked playgrounds of ancient concrete: AD 1875 in ecclesiastical numerals on a

corner-stone: the wire dustbin, the chipped basin, the hideous lavatory-seat and the grinding of trains behind the school-yard. Teachers with drawn neurotic faces flinching at the din: two classes to a room: conferences of despairing masters—the thin-lipped face, the malformed intellectual night-school skull, the shrewish voice of the cornered idealist as he reports to his colleagues—a new set of desks.[84]

This was the world of Spender's poem 'An Elementary School Class Room in a Slum', a scene of squalid circumstance, of illness: 'The stunted unlucky heir/Of twisted bones, reciting a father's gnarled disease, / His lesson from his desk.' It was horrid enough, but its horrors were rather different from what the bourgeois child was being put through.

Of all the contributors to *The Old School* only one did not share the bourgeois consensus which united all the rest. H. E. Bates, to be sure, went to Kettering Grammar School, but the style there was, he writes, imitation Public School. He enjoyed, as a day-boy, the pastiche of such schools if not their reality. The only real outsider, though, is Walter Greenwood, who attended Langy Road Council School in Salford during the First World War, a place marked by the absence of what packs the other reminiscences that Greene gathered. In Greenwood's school there was no systematic sadism, no male isolation, no homosexual infatuation, no constantly dire warning against masturbation, no games-code to take out with you to the Empire (for no colonial functionaries got trained there), and no trauma over leaving mother. Greenwood's school lacked memorability of the public school kinds. It had no famous Old Boys, no names to live up to on plaques on its walls. It was not a club, a freemasonry you were joining for life. Its identity was in fact extremely precarious: it was requisitioned during the First World War as a military hospital and the children had to work double shift at a neighbouring school. And in addition to this substantial lowness of profile, Greenwood was scarcely ever in it: he frequently 'wagged it' (played truant), and left for good at the age of thirteen. Small chance there was therefore of Greenwood seeing life in terms of cricket or even, as was the case with the leftist public-schoolboys, of cricket to be resisted or converted to socialism. His circle wasn't one where you'd appreciate John Strachey's joke about joining the Communist Party out of chagrin at not getting into the Eton Cricket XI, or where you'd have much time for Philip Toynbee's running away to Parton Street from Rugby School because he'd been jilted by the boy he fancied and couldn't face a summer of junior level cricket after a winter of glory in the First rugger XV. Nor was Greenwood's school experience such as to make Auden and Isherwood's cult of the Test, their seeing life in terms of playing-field heroics and the taking of exams, a set of anxieties he could share. 'At 6 p.m. passages of unprepared translation from dead dialects are set to all non-combatants. The papers are collected at 6.10. All who fail to obtain 99% make the supreme sacrifice. Candidates must write on three sides of the paper.' Thus *The Orators*, steeped in the emotion of the pass-list and the class-list—graded lists of the kind Cyril Connolly never stopped producing—written out of that urge to find the first-class man or book, in a world in which having or being an alpha, a beta, a gamma, and so on, mattered frightfully (as Huxley, for all the satire in *Brave New World*, never stopped feeling). 'What do you say to the idea of documentary play? Plays about living people? . . . Do you think we should cooperate more with other Left organisations? Or not? Why not? I'm sick of asking questions, anyhow. Don't goggle at me like that. Sit down and answer them. Three hours allowed. All those who fail to get 25% will be very severely punished.' That's in a typescript draft in the Berg Collection's Group Theatre Archive. Robert Medley suggested that MacNeice wrote it. Whoever did, it wouldn't have produced much of a frisson in the graduates of Langy Road.

Most of all, of course, Langy Road made Greenwood grow up fast. He left for 'full-time' work: a curious notion, for he'd already been putting in 33½ hours labour a week for a newsagent before and after his school-hours, to help his widowed mother keep the family going. For good or ill, and in common with other children of his class, he'd been kicked into adulthood while he was still a child. The bourgeois school system, on the other hand, contrived to keep adulthood from you. Astutely, in *Enemies of Promise*, Cyril Connolly takes up and develops those Wyndham Lewis allegations about English post-war Peter Panism. He was particularly tipped off by Chapter Ten of the *Doom*, entitled ' "Promise" Has Become An Institution. In The Future State Everyone Will, As A Matter Of Course, Possess "Great Promise".' Connolly was stimulated as well by being given to review the book *Antony (Viscount Knebworth): A Record of Youth* (1935), a glowing account of the youthful Old Etonian Tory MP who was killed in an air-crash at the age of twenty-nine, written by his father The Earl of Lytton, with a Foreword provided by Peter Pan's very own author J. M. Barrie. Antony Knebworth, whom Connolly knew at school, was the victim, Connolly suggested in his review, of their shared education. In him 'we see someone whose intellect (as opposed to his intelligence) remains the same as at his private school'. He was the prisoner of 'too early success'. 'I was myself a success at school', says Connolly—his success had come about, indeed, partly from cultivating school bigshots like Knebworth—'and it seems to me that only recently have I recovered my balance. Early laurels weigh like lead.' After such schooling, Antony found increasingly fewer worthwhile competitions until there was left 'just a slight dazzle from the conflagration of his early successes to remind him of the small school-universe where he had been most fully alive'.[85]

In his review Connolly hints that Antony might have been saved by consorting intellectually with his aesthetic contemporaries like Waugh or Quennell. In 1938, in *Enemies*, Connolly was inclined to disallow exceptions:

> were I to deduce anything from my feelings on leaving Eton, it might be called *The Theory of the Permanent Adolescence*. It is the theory that the experiences undergone by boys at the great public schools, their glories and disappointments, are so intense as to dominate their lives, and, to arrest their development. From these it results that the greater part of the ruling-class remains adolescent, school-minded, self-conscious, cowardly, sentimental, and, in the last analysis, homosexual. Early laurels weigh like lead and, of many of the boys whom I knew at Eton, I can say that their lives are over. Those who knew them then knew them at their best and fullest; now, in their early thirties, they are haunted ruins.

In various forms the point about the juvenile puerilities of English politics had been made before. William Plomer, for instance, had said as much in *The Old School*. But after Connolly it's not surprising to find the now highly clarified charge being repeated by the alerter ones among his old schoolfriends. By Henry Green, in *Pack My Bag*: 'everyone under the age of forty . . . influenced by what they went through at school . . . men who . . . burned themselves out . . . now amiable but defeated. For every one of these . . . a hundred whose mistake it was to do too well at games so that they were too happy to do more than relive their successes over again once they had left.' And by Orwell in *Coming Up for Air*, where George Bowling repeats the case in respect of old Porteous the retired public school master: 'funny how some of these public-school and university chaps manage to look like boys till their dying day'; 'Funny, these public-school chaps. Schoolboys all their days. Whole life revolving round the old school and their bits of Latin and Greek and poetry . . . And a curious thought struck me. *He's dead*. He's a ghost. All people like that are dead.' For, once it had been formulated so memorably, Connolly's 'Theory' had the ring of truth. It nicely theorized, for instance, the state of those Old Boys MacNeice discovered (in *Zoo*) at the annual Marlborough v. Rugby cricket match:

> The depressing thing about meeting Old Boys is that you feel that nothing ever changes. One was in the Army, monocled, tanned by India, but he took up at once where he had left off in the 'twenties. Whenever I opened my mouth he began to laugh; this was because at school I had been a professional wit, a role I have long abandoned. Another was an X-ray specialist, another a barrister, but none of that made much difference—at least at Lord's. Within this precinct they were Old Boys pure and simple. And many of them wore bowlers.

And what, we might enquire, drew MacNeice, this keen debunker of Old Boys, along to Lord's (his first visit, he protests, to the annual match)? It was what, presumably, induced Isherwood, when he quit London for the sodomite freedoms of Berlin, 14 March 1929, to sport his old school tie. For 'the first time in several years', he hastens to add. Once, clearly, you were an Old Boy . . .

Old Boyhood came to matter most, of course, in relation to the Old Boys' art. It is the artist's unpleasant lot, Connolly went on in *Enemies*, that he has to 'throw off those early experiences' and become (what the English hate) 'mature'. Connolly found it difficult. He was 'long dominated by impressions of school' which hampered this maturing. Going up to Balliol

> I was, in fact, as promising as the Emperor Tiberius retiring to Capri. I knew all about power and popularity, success and failure, beauty and time, I was familiar with the sadness of the lover, and the bleak ultimatums of the beloved. I had formed my ideas, and made my friends, and it was to be years before I could change them. I lived entirely in the past, exhausted by the emotions of adolescence, of understanding, loving, and learning. Denis' fearless intellectual justice, Robert's seventeenth-century face, mysterious in its conventionality, the scorn of Nigel, the gaiety of Freddie, the languor of Charles, were permanent symbols, like the old red-brick box, and elmy landscape which contained them. I was to continue on my useless assignment, falling in love, going to Spain, and being promising indefinitely.

A school-bounded '30s following on from a school-bounded '20s, a life locked into memories of school just over, and that landed you in the chummy company of Freddies, Denises, Nigels, and Roberts: Connolly's paragraph enacts the very Old Boyism of the early work of Auden and Co that it so forcibly recalls.

The Old Boys formed a phalanx, a Falange, rifted sometimes by inevitable differences of opinion, by estrangements of longer or shorter duration, but always—as with family likenesses that survive long separations and the tidal flows of affection—visibly present and observable. 'England *is* terribly provincial—it's all this *family* business. I know exactly why Guy Burgess went to Moscow . . . In the literary world in England, you have to know who's married to whom, and who's slept with whom and who hasn't. It's a tiny jungle': thus Auden complained to Robin Maugham (who wasn't unacquainted himself with the power of the literary family, of course) in the '50s. His wisdom applies pertinently to the '30s. The literary cousinhood of the Old Boys was only the cousinhood of the small enclave of the English bourgeoisie they mostly belonged to. John Lehmann was a cousin of Graham Greene; so was Christopher Isherwood. *All Stracheys Are Cousins*, Amabel Williams-Ellis astutely titled her memoirs; and not just all Stracheys. Evelyn Waugh was a cousin of Claud Cockburn. Julian Bell was Virginia Woolf's nephew. Cyril Connolly was distantly related to Robert Graves and Elizabeth Bowen via the Anglo-Irish Protestant Ascendancy. Orwell's Aunt Lilian Buddicom was a great friend of Auden's Aunt Henrietta. John Lehmann's sister the novelist Rosamond Lehmann married Wogan Philips (who drove an ambulance in Spain and is today the only professing Communist in the House of Lords). John Strachey's sister Amabel, the radical woman-of-letters and Strachey cousinhood observer, married Clough Williams-Ellis the leftist architect. Right-wing Roy Campbell married Communist Douglas (or Dave) Garman's sister Mary. Mary's sister Lorna was the wife of the leftist publisher Wishart. (And, naturally, neither Wishart nor Campbell was happy when the Campbells took refuge with the Wisharts after they fled from Spain during the Civil War; what Campbell referred to as 'Bolshevik Binsted' certainly wasn't to his taste in politics or in family.) Naomi Mitchison was J. B. S. Haldane's sister. The Mitford girls between them made all sorts of spectacular crossings of wires, left, right, and further right: Jessica ran off to Spain with Esmond Romilly; Diana married first Bryan Guinness (both of them were good friends of Evelyn Waugh), then Sir Oswald Mosley; Unity fell in love with Hitler and pursued him hungrily across the Third Reich, eventually shooting herself in a pathetic failed attempt at a *Liebestod*. At Cambridge Kathleen Raine married Hugh Sykes-Davies, then ran away to London with Charles Madge. Later Charles Madge married the first wife of Stephen Spender. The Powyses, for their part, travelled in droves of writing brothers, John Cowper, T. F., Llewellyn, Littleton, a formidable quartet to rival the unholy trinity of the Sitwells—the two brothers Sacheverell and Osbert and the sister Edith. The Duckworth brothers were publishers and Virginia Woolf's half-brothers. Gerald Duckworth's semi-niece, Evelyn Gardner, became the first Mrs Evelyn Waugh. But not for long. She soon ran off with John Heygate who had been sharing a basement flat with fellow old Etonian Anthony Powell ('a ramshackle oaf', Waugh abused Heygate in his letters, 'the basement boy'). Powell, who married Lord Longford's sister, lived with her at one point in a flat in Great Ormond Street over E. M. Forster. Powell's parents, not to be left out of all these literary contacts and contiguities, had a house next door to Elizabeth Bowen's house in Clarence Terrace near Regent's Park.

The Old Boys met each other (and each others' womenfolk) at all points of the social map, but the most important meeting places were the schools and ancient universities at which the cousinhood was sealed in permanent acquaintanceships and friendships, into precisely what the English language knows as the Old Boy Network. The overlap of shared educational background, assumptions, experience, makes one of the most notable features of the '30s literary world. Charles Madge, William Empson, Randall Swingler were all at Winchester; the Romilly brothers,

David Archer and Gavin Ewart (and Harold Nicolson) at Wellington; Tom Driberg, Evelyn Waugh and Max Mallowan (the archaeologist husband of Agatha Christie) at Lancing; Isherwood and Upward at Repton; Auden, Humphrey Spender, Benjamin Britten, Robert Medley, John Hayward, Donald Maclean (the spy), James Klugmann, John Pudney at Gresham's, Holt; MacNeice, T. C. Worsley, Betjeman, Anthony Blunt (another spy), Bernard Spencer, Graham Shepard (son of E. H. Shepard the illustrator of *The Wind in the Willows*) and David Mackenzie (a communist functionary who was killed in Spain) at Marlborough; and (most remarkable collection of all) Harold Acton, Brian Howard, John Lehmann, Connolly, Orwell, Alan Pryce-Jones (Pink Danuber and later editor of the *Times Literary Supplement*), Alan Clutton-Brock (the art critic), Peter Quennell, John Strachey, Henry Green, Roger Mynors (the classical scholar), Peter Fleming, Robert Byron, J. B. Morton (who became 'Beachcomber'), Anthony Powell at Eton. Many of them had overlapped also at their prep schools: Connolly, Orwell, Gavin Maxwell, Cecil Beaton at St Cyprian's; Auden and Isherwood at St Edmund's, Hindhead; Waugh and Cecil Beaton at Heath Mount prep school. Anthony Powell and Henry Green were at prep school together, so were Day Lewis and Basil Wright (the documentary film-maker). Graham Greene overlapped with Quennell at Berkhamstead, where Greene's father was the headmaster, before Quennell went on to Eton. Before Marlborough, MacNeice was at J. C. Powys's old school, Sherborne, where Littleton Powys had become headmaster. And after their public school they all continued and strengthened the cousinhood's school associations at Oxford and Cambridge. Rex Warner and Cecil Day Lewis overlapped at Wadham. MacNeice and Bernard Spencer were neighbours at Merton and Corpus. Not far away was Stephen Spender at University College. Harold Acton, David Cecil, Lord Longford, A. L. Rowse (making much of his poor origins in Cornwall at the same time as energetically altering his West Country accent to suit the new habitat), Brian Howard, Henry Green, Auden, Driberg were all Christ Church men. Richard Crossman, Goronwy Rees (another social infiltrator), Douglas Jay, and Richard Goodman were all at New College. Waugh and Calder-Marshall were both at Hertford College. Quennell, Powell, Greene and Connolly (like L. P. Hartley) were at Balliol. Isherwood and Upward carried their Repton friendship over to Cambridge with them. MacNeice's closest friend at Merton was Adrian Green-Armytage, yet another cousin of Graham Greene. At Trinity College, Cambridge, Lehmann became a good friend of Julian Bell. And so on and on in a superabundant burgeoning of a class's intimate connectedness.

The point has been amply illustrated. Wherever one dips one's investigative bucket into the '30s literary world in Britain one comes up with news of some close bond of friendship and/or relatedness that as often as not is rooted and emblematized in a school or college connection. One mustn't overstate, of course. Not all of these dramatis personae were exact contemporaries. They did not all know each other well. They did not all tread an exactly identikit career path (Orwell, notably, did not go on from Eton to Oxford or Cambridge, for instance). Attitudes, times, fashions could all change with some promptness over the space of two or three years within a school and particularly within a university. Prep schools and public schools prided themselves on their differences as well as on their likenesses, and often with good reason. Bedales was not Eton (one of the most informative parts of Robert Medley's autobiography, *Drawn From the Life*, is where he reflects on the more or less progressive and artsy-craftsy impulses that sent little Trevelyans and Nicholsons and Bones to Bedales and Gresham, and where he talks also of the marriages that took place between the alumni of these two schools). 'Red' Stowe (with T. H. White and George Rudé on the staff) had a very different tone from the conventionally militaristic Wellington. And so on. Some observers such as Martin Green have specialized in charting differences between generations of public schoolboys and Oxbridge undergraduates. The so-called 'dandy' group at Oxford gave way, they like to report, to the Auden years which yielded to the post-Auden generation. Which is to say that the earlier, camper, martini-swigging '20s was followed by the hairier, beerier later '20s and early '30s which in turn gave way to the still more earnest and Spanish-minded, Mass-observing mid-'30s, and so on. Some insiders like to dwell on the differences of tone between Balliol, New College, and Christ Church. It has become fashionable since the Anthony Blunt spy-scandal became public to talk about the distinctive role of Trinity College, Cambridge, in the '20s and '30s and to speculate about the peculiar qualities of the so-called Cambridge Apostles society. Limper, more arts-centred Oxford is often contrasted with the more scientifically minded Cambridge, which produced the magazine *Experiment* (1928–31) in which science-trained undergraduates like William Empson and Jacob Bronowski could express their turn towards poetry and the arts. And, certainly, Oxford did not produce the likes of Empson or Bronowski nor even (so far as we know) of Blunt, Philby, Burgess, and Maclean (and *Experiment*, for that matter, was no great shakes as a magazine, for all the contributions of Humphrey Jennings, Julian Trevelyan, and Empson—without whom it would have come nowhere: *Oxford Poetry* was much more lively). But none the less the broad picture is disconcertingly clear, the picture of a huge team whose members were at one time or another all nannied in much the same set of nurseries, had all imbibed the same team spirit at more or less the same snobbish, socially narrowing, moneyedly blinkered schools, who were all entitled to wear bourgeois England's Old School Tie—the insignia which even rebels, as we've seen, chose to wear, and latecoming adoptees, scholarship boys like A. L. Rowse, couldn't fasten about their necks fast enough. And, importantly, this training was hard to shuck off. Attitudes, arrogances, the class assumptions, and the tones of the born member of the ruling classes could persist despite the very best intentions of individuals. They have a repetitive way of ironically uniting the bourgeois Leftist with the bourgeois Rightist. It was very easy to revert to the typecast behaviour and talk one had been born and educated into. Class training *worked* (and still, of course, works).

At the simplest and most obvious level this class corporateness is the explanation of all those mutually dedicated texts, the flaunted public admiration and adulation the '30s writers kept offering each other and each other's writing. Auden's *Paid on Both Sides* (1930) was dedicated to Cecil Day-Lewis (he still retained the hyphen at the time); Rex Warner's *Poems* (1937) to Cecil Day Lewis. Day Lewis's *Transitional Poem* (1929) was dedicated to R. E. Warner; Auden's *Orators'* Ode No. IV to Warner's son; Poem 34 of *The Magnetic Mountain* (1933) to Warner's wife. *The Magnetic Mountain* as a whole was for Auden, so were Isherwood's *Mr Norris Changes Trains*, and Spender's 'Four Poems' in *C* (October 1930), and his *The Burning Cactus* volume of short stories (1936, a dedication shared with T. A. R.

Hyndman). Spender dedicated the second edition of *Poems* (1933, i.e. in 1934) to Isherwood (a tiff got in the way of a 1933 dedication). John Lehmann's novel *Evil Was Abroad* (1938) was also for Isherwood. Isherwood offered *All the Conspirators* (1928) to Edward Upward (as Auden did *Orator*'s Ode III); he dedicated *Goodbye to Berlin* to John and Beatrix Lehmann, and (with Auden) *On the Frontier* (1938) to Benjamin Britten. Connolly's novel *The Rock Pool* (1936) has a Dedicatory Letter to Peter Quennell. Auden's *Orators* was for Spender; Day Lewis's translation of Virgil's *Georgics* (1940) had Dedicatory Stanzas to Stephen Spender. Gavin Ewart's *Poems and Songs* (1939) were 'For Cuthbert Worsley'. Waugh's *Decline and Fall* was offered to Harold Acton. In another neck of the same sort of wood, C. S. Lewis's *The Problem of Pain* was dedicated (1940) To the Inklings (his Oxford coterie of like-minded Christians), so was Charles Williams's *The Forgiveness of Sins* (1942: Williams was as devoted an honorary Oxonian as A. L. Rowse became the epitome of All Soulishness). *The Screwtape Letters* of Lewis (1943) were for J. R. R. Tolkien.

Dedicating one's writings to one's closer friends and relations isn't uncommon—though the incestuousness of the Old Boys' small social circles is perhaps indicated by the (selected) list just made. But there was worse waiting the reader once he/she had passed these writings' dedicatory page. For the Old Boys simply did not worry about letting the familiarities of the coterie, the first name terms of their private lives, spill over into the more public domain of the published work (and revealingly and aptly enough, for the hegemony of a class is only the assumption of power by formerly small and even private groups, just as Standard English is only the public aggrandisement of a once merely regional dialect). The poems so larded with mutual dedications are liberally peopled by characters called Christopher and Dick, Wiz and Derek, Rex and Wystan. 'And everyone was called by their Christian names! So cosy!', exclaims Cris Clay campingly in Connolly's 'Where Engels Fears To Tread'. 'I sense', complained George Barker in the *NV* Auden Double Number 'a sort of general conspiratorial wink being made behind my back to a young man who sometimes has the name Christopher, sometimes Stephen, sometimes Derek and sometimes Wystan. Briefly I criticise its snobbery of clique.' And even when the clique was widened to let in some different Old Boys, or even a few non-Old Boys, it was still strongly the same old clique. For instance, when Oxford-bred Naomi Mitchison dedicated her novel *We Have Been Warned* 'To The Comrades', her list did embrace an anonymous Soviet girl aeroplane engineer, three other working women, Labour Party workers and Walter Greenwood 'Writer' and 'Labour Councillor'. But it was equally long on Mitchisons and Haldanes, on the Coles, Wystan Auden 'Schoolmaster and poet', Zita Baker 'Explorer' (she was also an Oxford don's spouse, the future wife of R. H. S. Crossman and the rumoured mistress of Tom Harrisson), on Victor Gollancz, Gerald Heard, and Rudi Messel 'Propagandist' (an old Etonian, he was, *inter al*, the wealthy backer of *Fact* magazine). *Plus ça change . . .*

Comrades, Old Boys: among writers it was more or less the same kind of gang. They worked together: Auden and Isherwood on their plays (with help from Benjamin Britten) and on their *Journey to a War*; Auden and MacNeice on their *Letters from Iceland*; Auden and Britten on their song cycles *Our Hunting Fathers* (1936) and *On this Island* (1936–7), and on documentary films (Britten and Basil Wright together drove down to the Downs School, near Malvern, to persuade schoolmaster Auden to join them in the GPO Film Unit); Britten and Randall Swingler on *Ballad of Heroes* in celebration of the British members of the International Brigade. 'It seemed always', grouched Evelyn Waugh in his review of Spender's autobiography *World Within World* (1951), 'to take at least two of them to generate any literary work however modest'. (He'd made the point spikily several times before in his wartime novel *Put Out More Flags* (1942) in the matter of Auden and Isherwood, a.k.a. Parsnip and Pimpernell. 'Parsnip'—this is a drunken Basil Seal ranting—'he has the alias of Pimpernell; he puts it about that he is a poet, two poets in fact, but there again the work betrays him'.) And these persistent collaborations were entirely in the spirit that had been laid down by the school friendships and the early university collaborations.

The annual student publication called *Oxford Poetry*, for example, was an extremely fruitful training exercise for the period's practice of regular literary chumminess. It now reads almost as a manual in how a class's cliquery can keep itself going and sustain any number of subsequent literary coteries (as well, of course, as fuel some notable family feuds). Certainly there could be no clearer revelation of the narrow social base of England's literary life than this magazine's self-generating annual turnover of eager editors and contributors.[86]

And at Oxford, *Oxford Poetry* was only one such meeting point. MacNeice edited the undergraduate paper *Sir Galahad*, for instance. His friend Bernard Spencer was one of its contributors. *Oxford Outlook*, edited by Calder-Marshall and Isaiah Berlin also published Spencer's work (Berlin was, like Spencer, at Corpus). Stephen Spender has talked in his autobiography *World Within World* about the circle of his friends after Auden had gone down from Oxford: MacNeice, Spencer, Calder-Marshall, Humphrey House. A shared education *drove* these men into *droves*. And they were ready to turn the accidents of social formation into critical dogma. 'All genuine poetry', Auden and Day Lewis asserted in their Preface to *Oxford Poetry 1927*, 'is in a sense the formation of private spheres out of public chaos'. And so eager were these two to seal such private spheres that they included in their volume a poem by Auden's old school-friend Isherwood, 'Souvenir des Vacances', even though Isherwood had never been to Oxford (the poem had to appear anonymously). And their successors as editors, secure in their cosy sense of the advantaged private sphere, were keen in their turn to hallow and celebrate such privatized social and poetic terrains. *Oxford Poetry 1932* is dedicated to 'Wystan Auden, Cecil Day Lewis, Stephen Spender'.

And, naturally, the old private references, the ingrained mythologies, the shared assumptions and jokes and beliefs of people whose sensibilities had all been honed at the same educational mills continued busily, along with the old friendships (even along with the old, and new, enmities) into later life and writing. Which was all very well for the reader inside the magic social circle. But how, one wonders, were readers outside the private (and private school) sphere expected to read the often extremely coterie-pandering writing that could emerge from such small quarters? Derek and Dick, Bill and Gabriel: were they expected to grab the common reader? Was that reading outsider even expected to enthuse over an essentially private myth the identity of whose biggest heroes—the likes of Rex and Wystan—there was available public evidence enough to work out? Gossip columns thrive on granting voyeuristic pleasures to outsiders, but were there actually enough of these to be gleaned from, for instance, Auden and MacNeice's 'Last Will and Testament' in *Letters from Iceland*?

Some of this poem's long roster of names were public property—published authors (Isherwood, Hugh M'Diarmid—as he appears in the 'Will', Wyndham Lewis, Peter Fleming, Spender, Day Lewis, Warner, Madge), politicians (Franco, Hitler, Sir Oswald Mosley), and so on. The public almost knew where it was with those. But only the authors' friends were on the inside track with 'our old friend Rupert Doone' and 'my friend Benjamin Britten' and 'our two distinguished colleagues in confidence' (they were Spender and Day Lewis) and 'Neville Coghill, fellow of Exeter, my tutor' and 'Erika, my wife' and jokes about 'Stephens' blue-black ink'. Only if you were extremely familiar with the coterie's ins and outs did you really know, even if then you were entirely clued up, why Berthold Viertel was there (he was the film producer Isherwood worked with), or Littleton Powys (MacNeice's old headmaster), or Robert Medley and Guy Burgess (both of Gresham's, Medley also of Group Theatre), or Bernard and Nora Spencer (Spencer, MacNeice's school and university chum), or John Layard and Gerald Heard, or the Isle of Wight (Auden's and Isherwood's and Upward's favourite island), or Acton, Howard, Forster, Empson, and Isaiah Berlin, or Olive and Sylvain Mangeot (friends of Isherwood, who had been secretary to André Mangeot's string quartet). Only the cognoscenti would be able to grant any meaning whatsoever to these honourable mentions, let alone pick up the calculated personal revenges: 'May the critic I. M. Parsons feel at last / A creative impulse' (Parsons savaged *The Dog Beneath the Skin* as 'shoddy' and 'half-baked'); 'to John Sparrow a quarter pound of fudge' (for his stubby-fingered attack on all modern English poetry in his book *Sense and Poetry*, 1934, especially his jeers at Auden's *The Orators* as 'a jumble of images and jottings' and his running-down of Naomi Mitchison's friendly puffing of that work). Reading such things as this 'Last Will and Testament' it's easy to sympathize with, and applaud Empson: it was his desire to undermine the 'unfortunate suggestion of writing for a clique about a good deal of recent poetry' that prompted him to add explanatory notes to his own *Poems* (1935).

The Old Boys fell naturally into private codes and esoteric references. Mostly, as in 'Last Will and Testament', knowingly, in a repletion of irritating nudges and nods to fellow-conspirators in the know: like the archness of *Mr Norris Changes Trains* about Bill Bradshaw's homosexual proclivities, or the winking and becking to the few in the secret about Auden's rectal fissure, or Waugh's kept-up denigrations of Cruttwell of Hertford, or Connolly's jokes about Wykehamists in *The Rock Pool* ('Cuckolded by a WYKEHAMIST!'). But sometimes also, to be sure, unknowingly. It's evidently beyond the ken of some of the Old Boys that there might exist readers not all that moved by MacNeice's bequeathing 'a lavatory seat with chromium gadgets' to Marlborough College or by yet more recalling of Oxford. Or—from this point of view, still worse—that there might indeed be readers who lacked all clue as to why being cuckolded by a Wykehamist might be amusing to an Old Etonian who knows several literary and political Wykehamists with lurid sexual reputations to live up to among their pals (men such as Empson, Madge, and Crossman) and an Old Etonian, what's more, who has already chuckled over Betjeman's' coterie poem 'The Wykehamist': a poem dedicated to Randolph Churchill ('but not about him') that adds the 'fleshy wants' of the young don Richard Crossman to something akin to Betjeman's own tastes in Anglican theology and architecture:

> Broad of Church and broad of mind,
> Broad before and broad behind,
> A keen ecclesiologist,
> A rather dirty Wykehamist.
>
> . . .
>
> He gives his Ovaltine a stir
> And nibbles at a 'petit beurre',
> And, satisfying fleshy wants,
> He settles down to Norman fonts.

This kind of poem was a deliberate settling for the company, and the readership of Old Boys who know about Wykehamists in general and in particular, in preference to the council school graduate who does not:

> A council scholar was shaking his fist,
> 'I can't understand it. O, but I see
> *Dum* not an adverb.' A Wykehamist
> Was refuting Joseph under a tree.

That's from Part One of Auden's 'A Happy New Year' (in *New Country*). It's a racily baroque phantasmagoria of English life. But it's mainly Old Boy life and bourgeois England that concern it. And characteristically so. On the rare occasion when Auden turns to address a non-Old Boy in his poem 'A Communist to Others' he allows scarcely any poetry-room at all to his proletarian comrades before turning to address the Captain of the First Fifteen and his sort, people whom he evidently much prefers. 'Comrades, yours fraternally', Charles Madge signs off: but it's at the end of a Wykehamist's 'Letter to the Intelligentsia' and one that has been all about the old school.

Now to complain thus is not at all to complain in general about occasional *romans à clef*, closed and narrow worlds as such, private references, portraits of friends getting into literary texts, or writing about what you know: all these are too common to get stirred up about. Rather, it's a complaint against the turning of such vast tracts of '30s literary country into a *roman à clef*, an enormous private joke, a huge private box or royal enclosure whose entrance fee was membership of the bourgeois cousin-hood. Evelyn Waugh, for once acting the renegade insider, made the objection very nicely in relation to Cyril Connolly's 'Where Engels . . .'. The spoof certainly enlivened *Press Gang*, the volume it first appeared in, Waugh agreed; but

> Unfortunately for most readers he has chosen to devote his prodigious talent to a private joke. The object of his satire has been a figure of fun to a tiny circle, but he is totally unknown to those whose adolescence does not happen to have coincided with his. A dozen readers in London, one in China, one in Gloucestershire, and possibly a handful in Spain, will revel in Mr Connolly's laying of this pathetic ghost.[87]

And this private literary demesne particularly irks because so much of what goes on there is the result of a shared adolescence: it's essentially a trivial, silly place, granting importance to juvenile feelings and experiences. ' "Rhythm. Spring. Waggle that thing / Do Do De O" they were singing': the 'Four saxophone boys' in Auden's 'Happy New Year' can't help reminding us of a lot of Auden's own singing. It's small wonder that Wyndham Lewis charged England with having become a nursery or that he should have turned his heaviest guns on the kind of juvenility encouraged by the Sitwells and others old enough to know better. The Shaw-Finnian family apartment in *The Apes of God* isn't a house but a nursery. 'This is God's own Peterpaniest family'. Ageing fast—in an 'Age-hell', in fact—they're still making mud pies at forty, showing off their toys (barbaric fragments, a Hottentot tricycle, Pacific pudenda), surrounded by a chorus of youths with the air 'of an impossibly early undergraduate life—as if just turned out in spick and span, passionless, lisping rows by Eton for Oxford Colleges and Inns'.

And the Old Boys' school and university world that so preoccupies their writings does frequently tap the still more infantile nursery world. And this is not just a matter of nannies in Auden poems or Waugh fictions, it's a case of characters out of nursery stories and rhymes getting turned into touchstones of the imagination and of an attempted political analysis.

> Attractions for the coming week
> Are Masters Wet, Dim, Drip and Bleak:
> Master Wet will show his pet,
> Master Drip will crack his whip,
> Master Bleak will speak in Greek,
> Master Dim will sing a hymn.

Thus Ode IV of *The Orators*: featuring a nursery group that's meant to characterize the efforts by the English press, bishops and whatnot 'to amuse' the public. It is perhaps making a serious point about the period's youth-politics; England *is* a nursery. But Auden's feeling for his own nursery is equally, if not more intensely, evident. In the same vein Auden's favourite Scissor Man out of Dr Heinrich Hoffman's *Struwwelpeter* gets into *The Dog Beneath the Skin* as an instance of Fascist terrors soon to come to Europe ('The bolt is sliding in its groove': part of the third part, slightly cut, of 'The Witnesses' that Auden transferred to the play). And *Struwwelpeter*'s Johnny-Head-in-Air runs like a talisman through all the period's thoughts about airmen and heroes come to grief. It was a way of avoiding being altogether serious about serious things. And a way of coping with grown-ups in an ungrown-up way that Old Boys from the same kind of nursery as Auden would understand. They'd all been shaped by the same sort of nursery reading which they hadn't, it seems, grown out of and which they allowed to go on defining for them the shape of the adult world.

And so Geoffrey Grigson particularly singles out the scissor-man stanza from 'The Witnesses' in his piece in the *New Verse* Double Number as a powerful illustration of Auden's crossing to 'the dangerous side' of the '30s frontier. And Empson is only half trying it on when in *Some Versions of Pastoral* (1935) he responds to the 1934 Soviet Writers' Congress dicta about socialist realism by suggesting that Pastoral is the truly proletarian literature and, on this tenuous theoretical basis, goes on to discuss *Alice in Wonderland*. And, not to be outdone, in his Introduction to *The Oxford Book of Light Verse* (1938) Auden argues that light verse is truly popular and therefore, by the sort of extension that Marxist theory would smile on no more enthusiastically than on Empson's cheeky dodge, prolet-arian. By this flashily trick device Edward Lear and nonsense verse were instantly allowed to appear more communist than one would ever have imagined. After this you're not surprised to discover Auden unable to shed a frivolous tone even about what you'd suppose to have been even for him still more urgently serious matters. What's the solution to the zestily characterized English death in Auden's dramatic piece *The Dance of Death* (1933)? Of course, Marxism. But Marxism curiously turned to in an oddly jokey version, as the Chorus sings, to the tune of Mendelssohn's Wedding March:

> Oh Mr Marx, you've gathered
> All the material facts
> You know the economic
> Reasons for our acts.

Still commanded by nursery light-heartedness the Old Boys found it difficult to come to terms with the adult world other than frivolously. And any excuse that they might make, or that we might be tempted to offer on their behalf along the lines that they had discovered Freud and that Freud had shown them the powerfully trau-matizing role of their nursery experiences which had then better be taken seriously, sounds like the convenient let-out it is. *Alice in Wonderland* is not proletarian lite-rature; a concentration camp's bloodiness is more than even *Struwwelpeter*'s grisly nursery threatens you with. The youthies were emotional cripples still enmeshed in the memories of boyish, prep-school doings, childish violences, boyish admirations for bombing-planes and for the ready-fisted cock of the school and of school-fiction—the terrors and enthusiasms of schoolboy violence that Isherwood renders so well in 'Gems of Belgian Architecture', or that Auden invokes so brightly in 'Address for a Prize Day' in *The Orators* ('Draw up a list of rotters and slackers . . . All these have got to die without issue. Unless my memory fails me there's a stoke hole under the floor of this hall, the Black Hole we called it in my day. New boys were always put in it . . . Quick, guard that door. Stop that man'). And this sort of lingering, even if it wasn't quite a permanent, adolescence hampered them from arriving at more mature thoughts about politics, about the horrors of life in the '30s, and the seriously proposed solutions to those horrors.

The prep-school atmosphere can certainly be allowed to have had *some* things in common with Fascism, even with tight-lipped Anglo-Saxon heroics such as many of the Old Boys had been introduced to in their university English courses:

> What have we all been doing to have made from Fear
> That laconic war-bitten captain addressing them now
> 'Heart and head shall be keener, mood the more
>   As our might lessens':
> To have caused their shout 'We will fight till we lie down beside
>   The Lord we have loved'?

Putting words from the Anglo-Saxon poem *The Battle of Maldon* into Ode V of *The Orators* ('To My Pupils') makes a nice connection. But insisting on the symmetries thus suggested, as Isherwood did in 'Notes on Auden's Early Poetry' (in the Auden Double Number of *New Verse*, and then in *Exhumations*)—'The saga-world is a schoolboy world, with its feuds, its practical jokes, its dark threats conveyed in puns and riddles and understatements'; 'it is impossible to say whether the characters [of *Paid on Both Sides*] are really epic heroes or only members of a school OTC'—and, again, in his *Exhumations* introduction to 'Gems of Belgian Architecture' ('a kind of hybrid language composed of saga phraseology and schoolboy slang'), insisting destroys a passingly suggestive likeness by claiming altogether too much for it. The importance of the schoolboy subject is not proved; only the observer's capacity to analyse the adult world grows more and more doubtful looking. After all, it was stretching the idea of practical jokes disconcertingly far to arm *The Orators'* vengefully violent airman with them. Isherwood's practical-joking saga heroes look even odder than they appear on their own when they're placed beside his practical-joking T. E. Lawrence (who, as Isherwood described him, 'had a giggling laugh, played practical jokes and interspersed his conversation with schoolboy slang', and just like a schoolboy enjoyed 'mechanical dodges and gadgets'). Even more peculiar is that most startling of all Auden's critical suggestions, in his later essay 'The Joker in the Pack', where the category of practical joker is stretched to embrace Shakespeare's most malevolent undoer of goodness, *Othello*'s Iago.[88] No ironies are, it seems, intended, only the youthies' persisting high estimate of the moral portentousness of the schoolboy world. But, clearly, if you can't spot the difference between schoolboys' antics, however terribly violent and malevolent they might be, and what Iago or T. E. Lawrence got up to or what went on at the actual Battle of Maldon, you're scarcely prepared to cope with the sharp disconnection between tough talk in the dorms and the Spanish Civil War or Adolf Hitler. Even Graham Greene's Pinkie, impressive illustration of the theology of Original Sin though he may be, is finally unconvincing. Boy Gangsters are for Musicals only. And the practical realization that violent practical jokes and thuggish japes among boys were really different in scale from what Franco's Moors and Stalin's OGPU and the pilots of the Stukas and the Capronis were perpetrating in Spain came, as we shall see, as an unacceptable set of shocks to many of the youthies—not least to Auden and Isherwood.

The unconsidered and immature heartlessness of the schoolboy calculation and impression was not confined to one or two writers. It was as common to Evelyn Waugh's plots, to the reportage of Bill Bradshaw in *Mr Norris*, to Geoffrey Grigson's critical passions (egged on by Auden and MacNeice's bequest 'to Geoffrey Grigson of *New Verse* / A strop for his sharp tongue before he talks'), as it was to Auden's verse. The 'boy bushranger', Dylan Thomas labelled Auden. 'Schoolboy jokes and undergraduate humour. A cruel handshake': that was Herbert Read's estimate of Auden's writing. (Both those opinions are in the Auden Double Number.) But Auden is only characteristic of the '30s Peter Pan author in whom adulthood and childhood are curiously merged, the author for whom boy-scouting ('is pure scoutmaster' jibed Orwell in 'Inside the Whale') stretched its satisfying meanings all the way from an OTC field-day through Maldon to the Airman's craving to fight over again the First World War and the revolutionary's longing for a punch-up.

Auden is typical, in fact, of the period's youthful scoutmaster-schoolmaster writers who are oddly indistinguishable from the schoolboys they should have already stopped being. Out there on field-day, scouting with his pupils, Auden's weirdly ungrown-up grown-up's 'I' gets completely submerged in the children's *we* ('aware of *our* rank'; '*we* must say goodbye. /*We* entrain at once . . . *we* shall see in the morning'). When a schoolmaster of this type actually appears as a character separate from the narratorial voice of the poem, as one does in Part I of Auden's 'A Happy

New Year', he's armed—as we might expect—with a boyish water-pistol and talks pure prep-school Hrotswitha:

> 'Just let them wait till the dark nights come',
>   A voice whispered suddenly close at my side.
> 'Walk ahead till I poke you twice with my thumb
>   And don't look round. We shall have to hide.
>   Remember how Lord Kitchener died.
> A certain person whom I shall not name
> Would shoot if he knew that we knew his game.'

The schoolmaster plays the narrator's game with him; they're two of a kind; and it's a schoolboy's sport that they're about. And the schoolmaster-poet is just as inhumanly casual about violence, as pitiless as any schoolboy: whether he's condemning other members of the school staff to the Black Hole in 'Address for a Prize Day', or threatening Beethameer (Beaverbrook/Rothermere) with 'the thrashing you richly deserve'; whether it's schoolmaster Day Lewis handing Beaverbrook another classroom warning, 'As for you, Bimbo, take off that false face!', or schoolmaster Auden coolly giving frigid old Miss Gee cancer to prove John Layard correct about psychosomatic illnesses, and then turning her over to the surgeon for turning into surgical meat:

> And Mr Rose the surgeon
> He cut Miss Gee in half.

If school violence was at all akin to Fascism, that left the thuggishly schoolboyish master in a morally quite difficult position.

Bernard Bergonzi has pointed out (it's one of the strongest points made in his *Reading the Thirties*) how often the schoolmastering Old Boys fell, in their writing, into a pedagogue's hectoring. When the Old Boys are gathered together, it's not only striking how many beaks there are on stage, but also how their collective tone is of someone bawling out the Lower Fourth.

> You're a fool if you think your system will give you cricket much longer. Haven't you realized? *Cricket doesn't pay*. If you want cricket you'd better join us. We're out for a decent life for ourselves and our successors, not for a paper profit. What good has all your pre-war profit-snatching done for us?

> You must have a conversion. And you are no more likely to have it by reading Marx than to experience a religious conversion by reading theological text-books. The certainty of new life must be your starting-point. Not jealousy, not pity, not a knowledge of economics; not hate even, or love; but certainty of new life. You may give all you have to the poor or the funds of the CPGB; you may work in boys' clubs or throw bombs; you may sell *Daily Workers*, study dialectical materialism, foment strikes, lose your job, go to prison or go mad. But you'd be better growing sweet peas outside a bungalow if you are doing all this from any other motive but the compulsion of new life. For revolutionary works without faith are vain.

> It's no use pretending you are splendidly or redeemingly or even interestingly doomed. If you are doomed at all, and it is still possible for you not to be one of those who are doomed, you are doomed like a factory which excludes the latest machinery or like a migratory bird which fails to migrate. Don't flatter yourself that history will die or hibernate with you; history will be as vigorous as ever but it will have gone to live elsewhere. No, you are not a martyr, you are not a conqueror, you recognise that . . .

These schoolmasterly voices are all taken from *New Country*. Astonishingly, for they could be the voice of one person, they are the voices of three separate schoolmaster writers: the first is editor Roberts's in his Preface, the second is Day Lewis's in his 'Letter to A Young Revolutionary', the third is Edward Upward's in his story 'Sunday'. And as if these reprimanding, cajoling pedagogic tones were not enough, *New Country* also contains the Bimbo section of Day Lewis's *Magnetic Mountain* and Auden's water-pistol packing pedagogue.

In his review of *The Mind in Chains* Evelyn Waugh alleged that there was a natural connection between the massed schoolmasterly authoritarianism of the volume's thoughts on socialism and criticism, and totalitarianism. His article is entitled 'For Schoolboys Only':

> It is not surprising to find that of the twelve socialists who have compiled *The Mind in Chains* the leading four are schoolmasters amd ex-schoolmasters, and two others lecturers. There is a natural connexion between the teaching profession and a taste for totalitarian government; prolonged association with the immature—fanatical urchins competing for caps and blazers of distinguishing colours—the dangerous pleasures of over-simple exposition, the scars of the endless losing battle for order and uniformity which rages in every class-room, dispose even the most independent minds to shirt-dipping and saluting.[89]

Waugh would make exactly the same point, and in almost so many words, in *Robbery Under Law* (1939), in a context, again, of socialist schoolteachers. It's clear by then, if it wasn't clear in 1937, that he felt his case to apply much more strongly to the Left than to the Right. Still, an ex-schoolmaster himself and a strongly authoritarian Rightist, Waugh did know an authoritarian when he saw one. And his point can be taken generally, and be given renewed general application. Those schoolmasterly voices from *New Country*: whose tones are they masters of? The speakers purport to be Leftists, but their tones could come from almost any hectoring quarter: the Bishop in his pulpit, the roused Housemaster in his study, the senior officer dressing down his troops, the District Officer putting his native subordinates in their place. As the Old Boys, even the leftist ones, are enclosed within so much of the given snobbish order of their own class, so the tone of command that springs to the lips even of the leftist schoolmaster is that of the pastors and masters whose authority and authoritarianism they mistakenly assume themselves to have resisted and sloughed off. As is often the case in the '30s the tones of Left and Right, of the people who are usually derided and the people who are generally cherished, blur and mingle confusingly and with ease.

How did the Old Boys—manifestly 'caught', as Allen Tate put it in the Auden Double Number, 'in a juvenile and provincial point of view'—get away with it? Americans outside the English world of the Old Boys, critics like Allen Tate and Edmund Wilson, clearly couldn't understand the praise the Oxford Boys were being afforded. The answer, in part at least, was the class ramp. The Old Boys were the beneficiaries of the racket that they and their friends, their schools and their colleges, their families, their mentors, their uncles and their sugar daddies comprised. The Old Boys were a connection: they had connections. And bright boys were passed along from the right schools to the right literary groups at Oxford and Cambridge to the right London literary editors. These 'moneyed young beasts who glide so grace-fully from Eton to Cambridge and from Cambridge to the literary reviews': so Orwell's Gordon Comstock bad-mouthed his better-off contemporaries in *Keep the Aspidistra Flying*. He was exaggerating, but pointfully. 'Money and culture! . . . Money for the right kind of education, money for influential friends, money for leisure and peace of mind, money for trips to Italy.' Comstock was right. On just the pattern he berated, Francis Birrell, a distant cousin of Auden's, got Sacheverell Sitwell to give Auden dinner in a London restaurant in 1927, and Sitwell got Auden to send his poems to T. S. Eliot. John Cornford's poems were sent by his Stowe English master to Auden. Arnold Rattenbury discovered that his reputation for selling the *Daily Worker* in a school dormitory had already preceded him to Cambridge. Gavin Ewart, still at Wellington, had tea in Cambridge with Grigson and a gang of *New Verse* poets and before long his own poems were joining theirs in the pages of the magazine. John Lehmann has talked of 'my key position' at the Hogarth Press, where he was able to promote *New Signatures* and *New Country*, Isherwood's *The Memorial* and Upward's *Journey to the Border*. '*The Listener* in the 1930s was one of our main outlets', Auden said of the connection that began when Janet Adam Smith, wife of the Old Boys' very own anthologist Michael Roberts, was in the literary editor's chair. Anthony Powell at Duckworth's introduced his chum Evelyn Waugh to the firm: which turned down *Decline and Fall*, but remained the publisher of Waugh's travel books. Duckworth's rejected Henry Green's *Blindness* as well, but Powell was not deterred and kept trying on behalf of his old school and university friends. Duckworth's did take Robert Byron's *The Station*.

A lot of books so promoted within the connection would undoubtedly have found outlets anyway. No one can doubt that Auden and many of the rest would have found publishers in the end. But the connection certainly helped by giving juvenile authors prompt and early access to the public prints, and also in helping sell to older members of publishing houses some of the juvenile excesses that they consented to publish. It helped too—as part of the same process—by what the toughs of *Scrutiny* (a group furiously given to its own group-politicking and with its own strong links to the main bands of literary Old Boys, but continuously hostile to 'the Group') jeered at as 'Clique-Pufferey'. One finds Auden, for instance, writing to commend Naomi Mitchison for her review of his *Poems* (1930): 'Any reviewer who tells people to buy the book has said the right thing.' And it wasn't long before he was boosting Isherwood's *The Memorial* and wanting her to do the same:

> How can you be so cold about it. If it isn't at once recognised as a masterpiece, I give up hope of any taste in this country. By far the best novel outside Lawrence since the war.

'PS', he added, and the manuscript shows the 'PS' to be heavily underlined, 'NB, Remember to boom the Memorial.'[90]

The Old Boys, even the Leavisian ones, openly espoused group loyalties. Cliques, it was widely supposed, were inevitable. Grigson thought so. 'Mr Squire had his cricket eleven. Mr Squire had his *Mercury* pages for gossip about old books, and fine books, and first editions. In other words, *The London Mercury* pimped, with much skill, for the prejudices of the bellelettrist and the middleman.' But R. A. Scott-James was a classic Liberal and killed the *Mercury* by his open-handedness. 'There is nothing so blinding as trying to see both sides, and there is only one way of avoiding cliques—to have your own cliques. Only you must'—and these were Grigson's own defiant last words in the last number of *New Verse*—'make sure that your clique is a clique of the best and truest and most lively writers of the time'. *Scrutiny* agreed, but Leavis also argued that 'the very circumstances that make the Group essential enhance its disadvantages and dangers': i.e. a clique might puff the wrong people. *Scrutiny* illustrated the danger by espousing the poems of Ronald Bottrall, at the same time as insisting with increasing loudness that the reputations of the Old Boys around Auden were being propped only by clique assiduousness. This accounted, Leavis thought, for the reputations of MacNeice, Madge, Warner, Day Lewis ('The Old Boy may have gone Left, but he remains true at heart to the Old School'), Spender (a career 'of literary frustration and dissipation'), and finally Auden himself. The present, Leavis wrote about *The Ascent of F6*, 'is the time when the young talent needs as never before the support of the group, and when the group can, as never before, escape all contact with serious critical standards'.

Scrutiny was too sweepingly dismissive, but it had a major point. Readers will argue over which authors were boomed beyond their deserts. But there's no denying that the booming went on, and is the only reason for some reputations. The glaring example is Edward Upward. Minimally productive, he'd be scarcely noticeable without his chums' repeated advertisements for his merits and importance. And the Old Boy racket is revealed at its most blatant over Upward's and Isherwood's Cambridge collaboration on the much bruited 'Mortmere'. Mortmere is the most over-rated piece of childishness in the whole of English literature. It's mainly of interest, though only at the level of gossip, because of the suggestion sometimes made that Auden's early work owed it a lot. Auden did put a handful of Mortmere characters—Bob, Miss Belmairs, Moxon—into Ode IV of *The Orators*, but only as a makeshift way of avoiding libel in the published version: 'Bob and Miss Belmairs spooning in Spain' substituted for 'Robert' (Graves) and 'Laura' (Riding); Moxon 'Dreaming of Nuns' was standing in for 'Eliot'. Unlike the Brontës' Angria and Gondal, though, Mortmere itself hardly survives at all. And the little we have of it—the achingly dull descriptions in *Lions and Shadows*, and Upward's story 'The Railway Accident'—are so thoroughly prep-school as instantly to clobber any claim for the importance or seriousness of Mortmere. And yet John Lehmann professed to think 'The Railway Accident' 'the most brilliant piece of surrealist prose to have been written in English (if you don't count the *Alice* stories). It is the culmination, and justification, of the joint Mortmere fantasies so wittily described by his friend Christopher Isherwood in . . . *Lions and Shadows*.' Isherwood, who never scrupled over puffing his old collaborator's work, called Upward 'a distinguished British prose-writer' ('not yet', he had to admit, 'as widely popular as his admirers would wish'), with 'an extraordinary technique'. 'The Railway Accident', he went on in his Foreword to its first publication in *New Directions in Prose and Poetry* (1949), is to be read seriously as 'a satire', 'a nightmare, about the English' and 'a touchstone of sanity' in 'our neurotic epoch'.[91] But can he, one thinks, reading the story, actually be describing 'The Railway Accident'? Only an Old Boy, and one as close to Upward as Isherwood had been since childhood, could so elevate this juvenile romp with water-pistol and pea-shooters, full of scragging and cries of 'beastly rotters', with its casual relish for violence and the crashing of vehicles that, even though they're done up in undergraduate surrealist purple prose, are still only characteristic of the taste of schoolboy monsters.

It's no accident that 'The Railway Accident' should be so particularly packed with the flamboyances of the prep-schoolboy's private code. Without it 'The Railway Accident' would scarcely exist:

> Shreeve was at the partition door, his hands curved to a megaphone. He shouted in padre's affected slang: 'Here, cheese that row you fellows'.
> 'Bligging spak a flunka blicking spug!' A few laughed. The lower panel split noisily. 'All together, ram the fliggering backer to bitching hell'. But already some other interest deflected them. 'Chuck over and hand back me Tin Lizzy. Look at this you gowks, Sandy's gone and cut his bum. That's his own funeral for shoving it through a closed window. Bleeding like a pig, is he? Here, take my neb-wipe.'

In a revealing moment, Shreeve's overcoat is described as 'reminiscent of frosty afternoons on the touchline'. The whole affair is redolent of the playing-field. And not at all uncentral to that location is the master Wygrave's homophiliac pederasty. The private language of the prep school is shot through with the homosexuality that's rife there. 'As surely as Wygrave's vice is branded on his face' (to quote Upward), juvenile homosexuality was deeply implanted into the Old Boys' writings, and it helped to define and motor the Old Boy racket.

As presented in *The Apes of God*, the youthful literary racket is also a sexual racket. Wyndham Lewis's youthies are made to serve the lecheries of the older generation. 'The aged drink-puffed lips pressed the baby-red and the breath of old carouse and the aridness of cigarettes blew round the astute juvenile nostrils'. Sexually, Lewis's Apes are mixed—heterosexuals, bisexuals, male and female homosexuals—a glorious Weimar debauch-full on the rampage around Bloomsbury. When he was shooting at fewer targets, however, as in the *Doom of Youth*, Lewis recognized that sexually the youth racket was predominantly male homosexual. And this was no merely right-wing slur. The recurrent motif of the false bottom that's meant to dish the literary Left in Lewis's Spanish novel *Revenge for Love* (1937) chimes in with the point that Communist Alec Brown makes in his novel *Daughters of Albion* (1935) about Gregory Bumbe—who 'belonged to that great brotherhood of monstrosities which plays such an important part in the cultured life of the last days of bourgeois England'. Everyone knew what Auden's bumslapper was after (even if they did not know that Jockers and Prushuns were homosexual slang terms).

Lewis and Brown were right. Sexual alliances actually formed at prep school and sexual habits picked up there cemented the Old Boys together just as they helped keep them in the Permanent Adolescence. The 'pylon boys', as Connolly called them, had their own private corner in communizing solidarity: 'The Homintern', he labelled it. Auden and Isherwood's long-standing sexual liaison, kept up intermittently from prep school, is the best example of this perverse variant on the Comintern: but their affair was only characteristic of the period. The shared male bed lay behind many of the coterie's dedications. Spender, for instance, repeatedly addresses poems to T. A. R. Hyndman, his boyfriend/secretary who typed out *Forward from Liberalism*, who shares the dedication of *The Burning Cactus* with Auden, and is offered the dedicatory poem that prefaces *Trial of a Judge*. The friends that Forster's *Abinger Harvest* (1936) is dedicated to are all homosexuals: William Plomer, Joe Ackerley, PC Bob Buckingham, Isherwood. One of the two dedicatees of Forster's next volume of essays after that, *Two Cheers for Democracy* (1951), is Jack

Sprott, who'd been a lover of Maynard Keynes and beneficiary of an allowance from Forster while he was a Cambridge student in the '20s. And affection for boys and among the boys led naturally to jobs for the boys. Mrs Leavis noticed it at the time, in her sharply accurate attack on England's 'closed literary society run on Civil Service lines' for the benefit of boys from the public schools and the ancient universities:

> you see how it is that these elegant unemployables get into the higher journalism, and even the academic world, and how reputations are made—you have only to get the right people, whom you already know or can get introductions to, to write the right kind of thing about you in the right places. The odious spoilt little boys of Mr Connolly's and so many other writers' schooldays . . . move in a body up to the universities to become inane pretentious young men, and, still essentially unchanged, from there move into the literary quarters vacated by the last batch of their kind . . . Mr Connolly and his set expected to succeed Rupert Brooke's, and are now seeing to it that the literary preserves are kept exclusively for their friends. We who are in the habit of asking how such evidently unqualified reviewers as fill the literary weeklies ever got into the profession need ask no longer. They turn out to have been 'the most fashionable boy in the school', or to have had a feline charm or a sensual mouth and long eye-lashes.[92]

Among Auden's schoolboy pashes was the younger John Pudney. Auden lectured him and read his poems to him at Gresham's. And when the young Pudney left school Auden looked him up in his Carnaby Street attic. 'When he came round', Pudney wrote later in *Thank Goodness for Cake*, 'there were no concessions to love. It was just meat he was after.' And when Pudney got a job at the BBC he it was who commissioned Auden and his homosexual friend Benjamin Britten, also from the same school, to collaborate in their programme about Hadrian's Wall. Later yet, Britten stood godfather to Pudney's son. Small instances, but, in accumulation helping hands like this did amount to what Wyndham Lewis called 'the intense "outcast" *esprit de corps* of the pathic'. The Old Boys' ready *we* frequently indicated not just a class but a sexual closed-circle. 'He's *one of us*', Auden snapped to Isherwood of the immigration officer who refused Isherwood's friend Heinz entry into England in 1934. Unlike him, but very much like the public school and Cambridge youthies recruited in the '30s as Soviet spies, and who make such vividly pointful epitomes of the way the bourgeois schoolboys tended to cluster in guarded coteries bonded by shared private codes, covert languages and publicly inadmissible passions, the literary homosexuals did not let each other down. E. M. Forster, claimed Joe Ackerley, 'always helped his friends'. So did they all. Spender got Tony Hyndman a half-time job on *Left Review*, thanks 'to the kindness and generosity of Derek', and promptly wrote to Isherwood urging contributions to the paper: 'if you can send a story or an article to *Left Review*, several people will be even more grateful than is usual in such cases.'[93] Homosexual John Lehmann's tastes, as editor and publisher, ran prominently among his homosexual writer friends: Plomer, Forster, Isherwood, Spender, John Hampson (John Hampson Simpson, the ex-kitchen worker and book-thief who was a close friend also of Forster's). The interests of Joe's boys were particularly well looked after at the *Listener* after Ackerley became literary and arts editor in 1935 (Auden's 'one of our main outlets' has in relation to Ackerley's reign the 'one of us' ring about it). Isherwood was one of Ackerley's most called-on reviewers, as Forster was one of Ackerley's most reliable and distinguished contributors. Just so, when Ackerley was a BBC Talks Editor earlier in his career, he had roped in Forster to do broadcast book reviews and the policemen and other proletarian chums to mount 'Conversations in the Train', or to describe 'The Day's Work'.

Mrs Leavis wasn't the only outsider to the magic homosexual circle to complain. In his review in *The Tablet* of Connolly's *Enemies of Promise*, Waugh noted Connolly's preference for 'epicene' writers—Petronius, Gide, Firbank, Wilde. Only 'virile' authors, Waugh supposed, were able to get on 'alone, without collaboration', but 'outcasts', 'epicene authors like to huddle together and imagine plots and betrayals'. And Waugh detected in this the source of Connolly's view of literary history 'as a series of "movements", sappings, bombings and encirclements, of party racketeering and jerrymandering'.[94] As he often did, Waugh went a bit too far here; as often, his extremist opinion is not altogether unconvincing.

Observing the homosexual nature of much '30s cliquery is not, of course, the same thing as complaining about friendliness and helping hands as such, nor is it to deplore homosexual writing in all of its manifestations. A love poem is a love poem. Clearly, a poem like Auden's justly famous 'Lay your sleeping head, my love' does not depend for its power—its expression of feelings all lovers can share, the extremely enterable-into language of guilt and wariness and passion—on our knowing or not knowing that it was addressed to a boy rather to a girl. Acquaintance with the sex of the particular beloved is irrelevant. (Which maybe, of course, helps heterosexuals to like it: more flamboyantly homosexual love passages are sometimes difficult for non-homosexuals to read happily.) No, what is more in question here is the politics of '30s homosexual writing, the way literary homosexuals conducted their literary affairs—admittedly in a legal context not at all of their own making. And one is, in the first place, complaining about the way homosexual liaisons and sympathies reinforced the Old Boys' proneness to clique-puffery and blunted critical standards. It was hardly literary merit alone that secured Brian Howard and T. A. R. Hyndman their niches in Lehmann and Spender's anthology *Poems for Spain*. In the second place, however, one is complaining about something that may be more intrinsic to homosexual writing and not just in the '30s. At least one must declare a worry about the prevalence in the period's literature of a number of specialized attitudes that the prevalent homosexuality of writers tended to normalize. It is arguable—and Wyndham Lewis did so argue—that the going homosexuality was not just a symptom of the youthies' current immaturity but a factor that helped stabilize that immaturity, helped keep the youthies juvenile. Mrs Leavis thought this too. 'It is no use looking for growth or development or any addition to literature in such an adolescent hot-house', she wrote after quoting MacNeice on Auden's 'not unfriendly contempt for the female sex'. And Stephen Spender, at least in some moods, also agreed. 'I stifle in a bugger's world', he confided to Grigson in 1934, envious of Grigson's approaching fatherhood.[95] And he not only married in an effort to suppress the homosexual side of himself but in his diary, 10 September 1939, he turned ruthlessly on his earlier homosexuality as evidence of a defectively ersatz humanity, a 'failure to be a complete man'.

Without doubt, the widespread homosexuality helped entrench some immaturely lopsided views. Much of the period's writing about the proletariat is vitiated by the bourgeois bugger's specialist regard. Spender's attitude to the Spanish Civil War soldiery—like Owen's towards the youthful combatants of the First World War—is deeply mixed up in homosexual affections (the fighting and the Commissars put his ex-boyfriend specially in danger, and 'the boy lying dead under the olive tree' in the poem 'Ultima Ratio Regum' is no simple object of general humanitarian sympathy: 'He was a better target for a kiss'). Forster's feelings about English justice and civil liberty, as expressed at the Congrès International des Ecrivains in Paris in 1935, focused lengthily on the suppression by the Lancashire police of James Hanley's *Boy*, a novel (as we've seen) much preoccupied by homoeroticism among sailors: why, he complains, 'any policeman in any provincial town'—and Forster knew so many policemen in ways the law did not smile on—might take it into his head to persecute a book or an author ('Next time it will be the author'). As for women, they frequently count for little in the Old Boy world of the all-male school and college. 'There is something peculiarly horrible about the idea of women pilots', Auden's Airman notes in his Journal in *The Orators*. Misogyny was rife in the writing of this period. C. S. Lewis was of the opinion that women's minds were intrinsically inferior to men's. Tolkien openly neglected his wife for the company of his university chums. Charles Williams was as eager to chastise his female followers as Charles Kingsley his wife a century earlier. Evelyn Waugh (his rocky first marriage merely underscoring his early homosexual aversions) was relentlessly hostile to women in *A Handful of Dust* as the committed destroyers of men. *Waugh in Abyssinia* opens with the Shavianly insulting 'Intelligent Woman's Guide to the Ethiopian Question'. In Arthur Calder-Marshall's *Dead Centre* women are presented as lying in wait for innocent prep-schoolboys, much as they'd been lying in wait for innocently lustful young men in the numerous Victorian and Edwardian novels also written by and for the male products of private schools and Oxbridge colleges. Proletarian, and pregnant, the girl Ada drives boys to despair, and to running away from school, with her accusations about paternity and her demands for postal orders. Marriage has cruelly trapped Calder-Marshall's young housemaster Mr Joliffe. Sex with women is even more sordid than this in Graham Greene's fictions. And even outside the immediate grimnesses of the bedroom in Greene's fiction there's the likes of *Brighton Rock*'s Ida Arnold, breasty, boozed-up, interfering, stupid, and superstitious. Greene even has the support of his Church in his novel's distaste for Ida: she's not only an example of ignorant womankind in general, but a theologically inept Protestant in particular. Given all of which, it's not surprising to find that when in 1939 Auden and Isherwood drew up for *Vogue* magazine a list of the ten most promising British writers of the day, they included no women at all. So no wonder the women readers and writers were beginning to be incensed. Stevie Smith was resigned to her poems being locked out of the male pale ('Your old battle axe on the *New Statesman* won't have me', she wrote to Naomi Mitchison, 'nor John Leighman [*sic*], nor Spender, nor Ian Somebody on *The Spectator*, nor Ackerley on *The Listener*. Only *Punch* will sometimes.'[96] Virginia Woolf, though, was fiercer. Addressing the Conference of the Workers' Educational Association in 1940 (an address published as 'The Leaning Tower' in *Folios of New Writing* II), Mrs Woolf asserted that 'in future we'—and she meant not just workers, and all people like herself who had not been through the male preserves of private and public schooling, but all reading women—'we are not going to leave writing to be done for us by a small class of well-to-do young men who have only a pinch, a thimbleful of experience to give us'. (And since she was using women's metaphors—the pinch (of salt, or flour), the thimbleful—she was actually implying that the Old Boys had probably not even that much to offer: they knew little about women's ancient work of cooking and sewing.)

And if women came off badly in much '30s writing so, inevitably, did the presentation of the family. Family life flourishes, one notices, among the works of proletarian authors and in the so-called proletarian fictions. Not so among the texts of the bourgeois Old Boys. Which is why among the few heterosexual male authors in that class of writer there is a certain assertiveness about their marriedness. MacNeice's *Poems* (1935) are pointfully dedicated 'To My Wife'. The Leavises' well-publicized deference to each other's minds and opinions was undoubtedly meant as a major protest against the homosexuality current among literary people. Robert Graves's notorious uxoriousness flowered spectacularly into his crackpot theorizing about Woman as Muse, the poet's commanding Goddess-Mistress. But even the determinedly womanizing and married found it hard to steer their texts clear of the prevailing hostility to women and marriage. Greene's continual dedication of his early novels to his wife Vivienne and Waugh's dedication of *Scoop* (1938) to his second wife Laura are no sign that womankind in general is going to be spared. (It wasn't only Graves who found that the attractions of the White Goddess had a dismaying way of fading into the less charming ways of her sister the Black Goddess.) As for the period's crowd of homosexuals, their cues were taken from Forster (and D. H. Lawrence): aggressively dominant mothers, absent or poor quality fathers, derisory wives. A whole generation of writers refused to countenance the normal family in their work, as they refused in their lives to acquire wives and become fathers. In literature, as in their life, only uncles—and Victorian and Edwardian families had unmarried uncles in plenty—were tolerable. And many of these, wicked uncles in modern fairy-stories, were homosexuals. An uncle of Driberg's died in a male brothel in Paris. Auden had a bachelor uncle (whose sexual tastes are not certainly known, to be sure). Isherwood soon winkled money for his Berlin stay out of his homosexual uncle. Like the honorary homosexual uncles the youthies made up to—Forster, Somerset Maugham (approached by the BBC in 1932 about a series of 'letters to unknown listeners', Virginia Woolf suggested 'Stephen Spender to an Uncle about Everything')—the uncle was in these circles the one respectable kind of adult relation. If he happened to be your mother's brother you could even pretend you were to him the precious *swustersunu* of the *Battle of Maldon*. And so Isherwood's Victor Page is shown early on in *All the Conspirators* with his uncle the Colonel, and the youthful Nigel Strangeways, detective hero of C. Day Lewis (Nicholas Blake), is the nephew of the Assistant Commissioner of Police (and a particularly apt solver of the mystery in Day Lewis's first detective story, *A Question of Proof*, the one set in the prep-school, involving as it does the murder of the headmaster's own nephew), and Auden's Airman, obsessed by uncles, discovers his uncle to be 'my real ancestor' (and one satisfyingly disliked by 'My mother'). This discovery was one that seemed to Day Lewis in *A Hope for Poetry* to provide 'one of the great moments' of *The Orators*.

Determined for their own part never to be fathers, destined in their turn only to be homosexual uncles, these authors would only enter parodic marriages of convenience. Auden married Erika Mann to gain her a passport out of Germany. John Hampson entered into a similar non-relationship with the Austrian actress Teresa

Ghiese. David Gascoyne spent some months waiting for a German girl called Ingrid to show up in Paris to marry him for his passport. These marriages were typical of the Old Boys' relish for the practical joke. And here, as elsewhere, the joke wasn't just a private one to the youthful coterie (who knew these marriages weren't serious), but doubly *recherché* in that only the homosexual core of the clique knew precisely why they weren't serious.

In the same way, the private language of the private school was doubly privatized when it turned into homosexual jargon:

> We wish the cottage at Piccadilly Circus kept
> For a certain novelist, to write thereon
> The spiritual cries at which he's so adept.

'Cottage' was a homosexuals' word for a public lavatory where sexual liaisons could be effected. As if uncracking the Rex and Wystan code weren't enough, the coterie's camp talk—at its most extended in MacNeice's long 'Hetty to Nancy' chapter (XII) of *Letters from Iceland* and his 'Hetty to Maisie' chapter (XVI) of *I Crossed the Minch* with its self-referential jokes about Our Memoirs from Greenland—involves one in a still more deeply excluding world of Hetty, Nancy, Maisie, and Stephanie.

But the secretiveness of the homosexual code was never, for all its frequently lighthearted pretences, just a joke. Forster could not fully speak his mind about *Boy* and policemen for real fear of what the police might do if they knew all about him and his friends.

> I've lately had a confidential warning
> That Isherwood is publishing next season
> A book about us all. I call that treason.
> I must be quick if I'm to get my oar in
> Before his revelations bring the law in.

Thus Auden to Lord Byron. But in the '30s Isherwood never did tell all 'about us all'. In his *Listener* review of *T. E. Lawrence by His Friends* in 1937 (it's in *Exhumations*) he accused T. E. Lawrence's *Seven Pillars of Wisdom* of failing 'because it was not absolutely frank. Lawrence had been unable to bring himself to record the full history of the "deep cleavage" in his own nature'. Isherwood was right about Lawrence's failure of nerve: Lawrence could never match James Hanley's boldness ('God almighty', he wrote to Hanley about *Boy* in 1931, 'you leave nothing unsaid or undone, do you? I can't understand how you find brave men to publish you!'). But Isherwood was less than frank himself, indeed he turned into an almost excessively clamant Gay Liberationist in the '70s partly from guilty hostility towards his own earlier concealments and unoutspokenness. His problem was not, though, unique. Forster's *Maurice* and other homosexual texts were passed from hand to hand among his friends, but remained unpublished in his lifetime. In his *Goldsworthy Lowes Dickinson* (1934) Forster carefully played down his subject's homosexuality and shoe fetishism, leaving the coterie—and it became a cult book for Auden, Sassoon, Isherwood—to read the true story between the lines and from the references to Edward Carpenter, Gerald Heard, Joe (probably Ackerley), German youths in shorts, and Dickinson's habit of patting his bottom with both hands during lectures (*'Si ce n'est pas vrai'*, Forster added, *'C'est bien vu*; one endorses the gesture': Isherwood and Auden and the rest knew why). John Lehmann's *Evil Was Abroad* is sexually a pretty hectic novel, but about a very unspecific relationship between the Englishman Peter and the Austrian Rudi. Peter regrets not having confessed his 'different life', not having 'seen, smelt, rubbed himself in Rudi's world' enough. But the precise nature of these hinted yearnings remains irresolutely smudged within the pages of the novel. The making of proper amends to Rudi, as it were, had to wait until 1976 when Lehmann's wildly unreticent confessional novel, the pantingly naif, determinedly explicit *In the Purely Pagan Sense*, was emboldened to rush in where *Christopher and His Kind* had not eschewed to tread just a year before. Spender complained (in *Fact*, No. 20) about the 'thinness' in *Evil Was Abroad*'s account of 'the relationship of the young Englishman and the young Austrian'. But for his own part Spender could only manage to deploy the occasional male pronoun in his love poems (as in No. XVIII in *Poems*, 1933, 'How strangely this sun reminds me of my love!'). His *World Within World*, like Lehmann's volumes of auto-biography, conceals the intimacies of life with the boyfriend behind the dim half-truth of 'my secretary'. It wasn't until the mid-1960s that Spender publicly confided that the line of his most famous of poems 'The Express', about the railway engine 'gliding like a queen', carried in part a memory of 'an Oxford queen called M— gliding down the High when I was an undergraduate'. And, of course, where coyness about speaking out has turned into a furtiveness commanded by fear, there is inevit-ably a drop in literary power. For 'Good novels', as one of Orwell's most striking axioms in 'Inside the Whale' has it—and one wants to extend the observation to all writing, all texts, and not only texts in the '30s—'Good novels are written by people who are *not frightened*.'

6

High Failure

IN *The Death of the King's Canary*—written in 1940, but like many another text of intense interest to the 1930s not published until much later (in 1976) because of fears of prosecution, this time for libel—Dylan Thomas and John Davenport have the Prime Minister choose the Poet Laureate. Out of all the possible contenders the PM selects Hilary Byrd, Old Etonian and Cambridge graduate, born in 1907, manifestly one of the Old Boys, and an Airman. He gets the post on the strength of his volume *A Time to Laugh*, which is all about a young man looking for a Leader. 'The Leader, when found, proved to be living at the top of a mountain.' Byrd's volume ends with a sonnet ('It was frightful,' the Premier thought, 'but it would do'):

MANMOUNTAIN

This was my test: not by the easier route
But through the gentians and the rocks to hurl
My adolescence; the glacier bruised my foot
and I laughed despite at the icy wind's up-curl.
My goal was there, poised on the peak's white winter
As an eagle's eyrie breasting the burning blast,
And though the old world round me cruelly splinter
Not was for me in my pride to be downcast.

'Mountains', he said, 'are only high in space:
Make the Andes your molehill and below
Map in the valleys lofty continents,
Plan power-houses for your island race.
Take a divining rod, and boldly throw
Alpenstock down; use plainsman arguments.'

The Prime Minister's (and Dylan Thomas's and John Davenport's) eye for the characteristic concatenations of '30s motifs and emblems—airmen, mountaineers, mountains, eagles, leaders, aerialism, and so on—is sharp. Nobody reading '30s writing can fail to notice the wide deployment of this metaphoric set. What's not noticed enough is the connection of these images with prevailing debates about heroism.

The First World War seriously jolted the ancient equation between the fighting man and heroism. 'This book', announced Wilfred Owen in the 'Preface' to his *Collected Poems*, 'is not about heroes. English poetry is not yet fit to speak of them.' It was a declaration that the post-war generation was inclined to accept as an axiom of the times. Heroic thoughts might do for officers, wrote Siegfried Sassoon in *Memoirs of a Fox-Hunting Man*, but they wouldn't do for the common soldiery. 'Perhaps the most significant thing about the War', proposed T. S. Eliot in the *Criterion* (January 1930: the same number that contains Auden's *Paid on Both Sides*), 'is its *insignificance*':

It is easy to convince people of the horrors, and of the harm that war does; but it is at least as important to convince them that it does no good, and has no grandeur, and that 'the sense of glory' has other, and only other, means of expression.

The 'many-ribboned hero', as MacNeice's *Autumn Journal*, XVIII, describes him, 'With half a lung or leg waits his turn to die.' The voice of anger and disillusionment, the voice of Owen and Sassoon, was readily assumed to be the necessary and authentic voice of the First World War's fighting man. The trenches scene before Act III, Sc. ii of Auden and Isherwood's *On the Frontier* (1938) is only a replay of what had become standard First World War material—grouses about officers and food, and mud and death, songs that include a version of 'Mademoiselle from Armentières', fraternization between the two sides' entrenched soldiery, a shared downbeat sense of waste and distaste:

> We're sick of the rain and the lice and the smell,
> We're sick of the noise and the shot and the shell,
> And the whole bloody war can go to hell!

This anti-heroic message of the First World War is the one that Christopher Caudwell accepts and builds on in his very important essay, 'T. E. Lawrence: A Study in Heroism', published posthumously in *Studies in a Dying Culture* after Caudwell had himself been killed fighting in a war, in Spain:

> Although the leading powers of the world directed during the four years of the Great War all their material, scientific and emotional resources to violent action, this unprecedented struggle produced no bourgeois master of action. The Great War had no hero . . . In the twentieth-century millions of deaths and mountains of guns, tanks and ships are not enough to make a bourgeois hero. The best thing they achieved was a might-have-been, the pathetic figure of T. E. Lawrence.

It is, however, one of the most important facts about the 1930s that this widely observed, world-scale collapse of the idea of heroism, this breakdown of the idea of the greatness of a life of action, this loss of 'the sense of glory', did not last very long. Characteristically, Eliot was not in 1930 discarding 'the sense of glory', only directing his readers' attention away from the scene of warfare where once it was thought specially to reside. And Caudwell himself couldn't rest, nor would he let his readers rest, with the failures of T. E. Lawrence that he announced. In the same essay he turns to Lenin. Lenin is an authentic super-man; the only one such to have emerged from the War. But Caudwell also knows that Lenin is not the only contender for a renewed super-manhood: in fact a whole gang of claimants to heroic stature has stepped vociferously into the post-war vacuum of glory. Against Lenin, and his successor Stalin, there are on offer, Caudwell recognizes, fascist alternatives—Mosley, Mussolini, Hitler—who are also appealing to the frustrated sense of glory and heroism in a western world that has been morally and physically enfeebled by the War and its aftermath: rival leaders, with rival brands of the heroic, who are all calling on 'the men of 1918' to arise and grow strong again, to join bands and armies that will again grow fit and tough, to become fighters of whom it shall again be allowable that they are truly admirable and glorious. Caudwell bluntly dismissed the fascist demagogues as charlatan heroes. As a Communist he must. But Caudwell's own preference isn't specially relevant to the most important point about his essay. That is, that here, in a central '30s text, and right at the heart of the major events of the decade, politics had been forcefully translated into a question of heroics. Once again, it had become impossible for the intellectual simply to deny the possibility of there being a great life of action, or just to ignore the question of the morality of the life of action. The search for Eliot's 'sense of glory', brushed aside for a brief time, was once more on. Living had again become a question of strength and weakness, of whether to be 'manly' or not, of choosing the right heroic leader.

'Under Which King, Bezonian?' is the title of Leavis's famous policy discussion in *Scrutiny* (1932): which course, which politics, and which set of leaders will *Scrutiny* choose to follow? The '30s writers are continually asking 'Under Which King?' Were they to go left under Lenin, or rightwards under Mosley, Hitler, Franco, Mussolini or with T. E. Lawrence's bourgeois admirers such as Winston Churchill? And what about the claims of God? Should one get embroiled in the debate between God and Monarch that T. S. Eliot's Thomas à Becket is engulfed by in *Murder in the Cathedral*, and try for Eliot's so-called 'Third Way' between Fascism and Communism, under 'Christ the Tiger', the Christians' *Man* (in the words of Dorothy Sayers's radio play of 1941) *Born to be King*? You could even try, as Leavis did in his 'Bezonian' article, to wriggle out of an answer by ridiculing the question and heaping scorn on Left and Right (Leavis didn't care either for the Marxists' critical efforts, especially Trotsky's, or for those of Eliot and the *Criterion*). But even if the question was annoying—'The Marxist challenge . . . seems to us as heroic as Ancient Pistol's and to point to as real alternatives', Leavis wrote—it had become one that couldn't be avoided. 'Under which king, Bezonian?', as Shakespeare's Ancient Pistol originally puts it: 'Speak, or die.'

In other words, before the memories of the last unheroic war had faded, the old kinds of pre-war claims were once more being widely canvassed. On occasion, Stephen Spender—who engages in a debate with Owen's anti-heroism that outlasts the decade—can be heard doubting whether Owen's point still holds: the times, he came to think, might have become ripe for heroes again, especially in Spain. The Communists were soon conceding no doubts at all. 'The great writer Ralph Fox has fallen heroically in the anti-Fascist fight as political commissar of the Anglo-Irish Company of the 14th Brigade': thus André Marty, reported in the *Daily Worker* (7 January 1937). This kind of response was, as we shall see, multiplied many times during the Spanish War. But heroes weren't just spottable in Spain. They were appearing all over the place. In Vienna (Section III of Spender's long poem *Vienna*, 1934, about the suppression of the workers' democracy there, is entitled 'The Death of Heroes'); in the Soviet Union (the novel, Ralph Fox suggested in *The Novel and the People*, 1937, had only to look to people like Reichstag Trial hero Dimitrov to allay any defeatist worries about the death of the novel's old interest in heroes); within the annals of the Christian Church (Part III of Evelyn Waugh's life of *Edmund Campion*, 1935, is entitled 'The Hero', and Waugh was still talking to his agent in 1939 about writing a 'book of [Christian] Heroes'); even in the shape of the flushed and wordily self-anatomizing solipsism of Dylan Thomas's all agog, writing, masturbating, defecating body ('My hero bares his nerves', 1933). And, of course, there were current heroes in plenty up mountains and in the air.

According to Spender, this new bout of heroic myth-mongering in and around '30s writing was only natural because the 'poet'—and the suggestion comes in a *Fact* review of André Malraux's Spanish War novel *Days of Hope*, November 1938—'is always to some extent a frustrated man of action'. The writer would always be engaged, Spender thought, in his own kind of running debate about the active life, in a tussle between mere sitting, thinking and writing, and more heroic styles of living, being and doing. Spender wasn't alone. Hence the widely felt tug of the communist call to revolutionary activity, the magnetic force of a writer like Malraux, the centrality of Spain and Ralph Bates's kind of hero. '[A]h, this was going to be great, this was life and the fulfilment of the will, Action'; 'Ah, the Deed, the joy of unfettered defiant Action': thus Mudarra in Bates's *The Olive Field* (1936), as the dynamiting of a dam is prepared. 'A spot of action would be excellent', declares Francis Charing in Bates's *Lean Men* (1935), as he responds to the Communist International's commission to Spanish revolutionary work: and it's almost an echo of the response of Graham Greene's Francis Chase in *Rumour at Nightfall* (1931): 'it was more than the hope of a good story which lightened his heart. It was the sense of action, something at last to break the monotony of endless days.' T. S. Eliot anxious to promote the idea that suffering, martyrdom, the *patience* of a Thomas à Becket are serious versions of action, are a valid Christian heroism, does so in *Murder in the Cathedral* by blurring patients into agents, suffering into action: 'both', 'agent' and 'patient', 'are fixed In an eternal action, an eternal patience'.

At all events, seeking heroes, people to admire as glorious, or looking for heroic action, for life on a grander scale, seem to have returned naturally and quickly in the dispirited trough of the post-war diminishment, in that prevailing sense of moral shrunkenness which accompanied actual physical thinning—the undernourishment and ill-health, starvation even, of the run-down (and 'flu-ridden) and increasingly economically depressed populations of Western and Eastern Europe. There were great famines in Soviet Russia and less great famines, but still famines, in the West. This was an age whose problems simply dwarfed the politicians, as Auden declares in Ode IV of *The Orators*:

> O yes, MacDonald's a giant,
> President Hoover's a giant.
> Baldwin and Briand are giants—
> Haven't they told us?
> But why have they sold us?
> They said they were winners,
> They were only beginners.
> Pygmies, poor dears,
> Beside the Giant Sloths and the Giant Despairs.

The good old English hero, the 'swaggering bully' with the 'meaty neck'—'the John Bull of the good old days'—had been finished off, declared Auden in Part II of 'Letter to Lord Byron', by the War ('He passed away at Ypres and Passchendaele'), leaving only the modern little man behind. But the little man, 'our hero' now cut down in size, hasn't forgotten bigger things; on the contrary his present diminution only makes him more mindful of them:

> Turn to the work of Disney or of Strube;
> There stands our hero in his threadbare seams;
> The bowler hat who straphangs in the tube,
> And kicks the tyrant only in his dreams,
> Trading on pathos, dreading all extremes,
> The little Mickey with the hidden grudge;
> Which is the better, I leave you to judge.

. . .

'I am like you', he says, 'and you, and you,
  I love my life, I love the home-fires, have
To keep them burning. Heroes never do.
  Heroes are sent by ogres to the grave.
  I may not be courageous but I save.
I am the one who somehow turns the corner,
I may perhaps be fortunate Jack Horner.

'I am the ogre's private secretary;
  I've felt his stature and his powers, learned
To give his ogreship the raspberry
  Only when his gigantic back is turned.
  One day, who knows, I'll do as I have yearned.
The short man, all his fingers on the door,
With repartee shall send him to the floor'.

What's perhaps most striking here is that Auden is presenting the issue of heroism as a matter of size, of scale, of contrasting dimensions. It's an affair of pygmies versus giants, of 'little Mickey', little Jack Horner, 'The short man', against the ogre with the 'gigantic back'. Auden is carefully adopting the rhetoric of nursery rhymes, little people's poetry, to grapple with big, grown-up problems. And the urgency of his anxieties shows through in a way they don't always manage to do when he's rummaging in his nursery memories for images. Auden is clearly worried about the strategies small people may adopt to defeat the great. His concerns are utterly characteristic of the period's renewed quest for heroes, and its new questioning about heroics. The '30s mind and imagination are, in such contexts, continually measuring, sizing up, wondering who is big enough to cope, who big enough to admire, and what makes a person, an ideology, a political system imposingly grand enough to compel the public to look up to it, to confess (in the words of the 1934 Cole Porter song whose charm Graham Greene found so Audenesque) that 'You're the Top'.

Looking up is one of the '30s' most prevailing attitudes. Looking up was what the new photography invited you to do. The shot taken from below was the period's 'new angle', according (for example) to *Amateur Photography* (edited by one Charles Duncan, 1935). Looking up was perforce the angle of small children, of the schoolboy so many writers found it hard—as we've just seen—to stop being. Spender's little hero Geoffrey in *The Backward Son* finds his father, as they play together, 'this free, romantic-looking giant'. 'Little beasts we were', wrote Henry Green in his *FNW* piece, 'A Private School in 1914' (Spring 1940), 'so dirty . . . and so small, as comes out in the photographs, that on looking back to what one remembers we see it all much larger than it was. The main classroom seems immense, the masters giants and of course our headmaster, and I remember when visiting him some years later I was struck by how small he was, a heroic man, colossal figure.' No wonder the small reader's hero, as described by George Orwell in the pages of 'Boys' Weeklies' (1940) was 'a superman . . . a sort of human gorilla; in the Tarzan type of story he is sometimes actually a giant, eight or ten feet high'.

Again, you looked up—especially if you were a 'short man', or a weedy writer like E. M. Forster, or a 'squat, spruce' author like Christopher Isherwood—at big proletarians. It was large workers that the '30s homosexuals liked to go to bed with, factory operatives whose bodies were developed by physical labour, or workers in jobs with height stipulations (any policeman, for instance, was, like the members of the guards regiments, bound to be big). The period's proletarian novels are all peopled, as we shall see, by large working-men—Big Jims in plenty, Big Jock, Big Joe, Big Tom, Big John, Big Ezra. 'Yes, why do we all, seeing a communist, feel small?' asked a C. Day Lewis sonnet in *Left Review* in November 1934. Day Lewis's Communist in that poem is a 'rock', absorbed into the height of a futuristic sky-scraper that he's building:

> There fall
> From him shadows of what he is building; bold and tall—
> For his sun has barely mastered the misted horizon—they seem.
> Indeed he casts a shadow, as among the dead will some
> Living one. It is the future walking to meet us all.

For feeling thus small in the presence of a big communist worker and the big communist future Day Lewis was mercilessly ragged, especially by Wyndham Lewis in *The Revenge for Love*. Several times in that novel the bourgeois Socialists Tristy and Gillian Phipps are said to 'feel small' when they meet 'real' Communists. Inevitably they are said to do so 'in the words of the poet'. Feeling small, especially in the presence of Communists, wasn't calculated to please big-muscled Roy Campbell either ('Mr Bullfighter Campbell . . . Big swelling muscles holding up highly coloured bladders of hot air', in the words of Geoffrey Grigson):

> Day Lewis to the Communist 'feels small'
> But nothing's made me feel so steep and tall:
> With me such things are easy to determine
> Who never felt this reverence for vermin
> And all I know is, communists or germs,
> He fares the best who never comes to terms!

Day Lewis's sonnet may sound like an off-moment's version of Cole Porter's 'You're the Top' ('But if, baby, I'm the bottom/You're the top!'). It is in fact a deliberate and seriously intended pastiche of the Gerard Manley Hopkins sonnet we know as 'The Soldier': 'Yes, why do we all seeing of a soldier, bless him?' And it draws on Hopkins's own repeatedly expressed homoerotic feelings for large members of the lower orders—especially soldiers and sailors, with the occasional farrier or plough-man or sturdy beggar thrown in. These feelings were shared by Auden, Plomer, Joe Ackerley, Brian Howard (a slight man, with a physically abnormally tiny heart, who had a lengthy affair with German Toni: 'so funny, guzzling [opium] down and roaring about sausages and Hitler and being the perfect great swinging captain-of-the-eleven contrast'), and by Isherwood (whose German boyfriend Heinz was photo-graphed from the 'new angle' by Humphrey Spender: a photograph that was pub-lished in the new *Photography Year Book*, 1935, and that has been rightly called a veritable ikon of the period). Such feelings were endorsed too by Uncle Morgan: weak, thin, physically inept Forster (who couldn't drive a car, didn't know how cricket worked, never had to work for a living, and who was always losing things and breaking his limbs by falling over), Forster whose recurrent desire, he said, was to be 'hurt' by a 'strong young man of the lower classes', and whose coy, closet homo-sexual stories were fond of pawing at the attractions of the huge and uncouth male lover. There is, for instance, 'The Obelisk', written in 1939, which is about school-master Ernest ('very very small', 'like a doll') and his wife whose imagination craves the embrace of some cinematic Sheik. In the story both get what they fancy in the arms of a couple of butch sailors whom they meet on a touristic visit to an obelisk: there's Stanhope, a dreamboat from Hollywood, enjoyed by Hilda on the grass, and, for little Ernest, there's big 'Tiny', a giant of a man with 'a huge paw' of a hand, who jokes hintfully after his return from being up at the obelisk with Ernest about penises and obelisks ('ever seed a bigger one? . . . a bigger obolokist, I mean').[97] And even big Spender, descriptions of whom always stress his shambling bulk and his huge excitements ('the tall chap with a leg like a flying buttress', in George Barker's Sonnet 'To Stephen Spender' in Tambimuttu's *Poetry in Wartime*, 1942; 'He had the trick of enlarging the distinctly general into the gigantically particular, so that public events became a part of the huge dramatic stage on which he lived', according to T. C. Worsley's *Behind the Battle*, 1939), even Spender felt shrunken, a little boy again, beside his guardsman boyfriend T. A. R. Hyndman (he's called Jimmy Younger in Spender's *World Within World*). Even as he writes that autobiography, Spender says he thrills once more in recalling a soldier carrying him in his arms to an air-raid shelter during the First World War: 'he held me to his heart with a simplicity which my parents with their fears for health and morals, and their view that any uninhibited feeling was dangerous, could scarcely show.'

You looked up, then, to the man with the bigger body (a creature out of Walt Whitman and Whitman-admiring Edward Carpenter—who combined democratic sentiments with simple-lifery and a liking for what E. M. Forster called 'toughs'—by thuggery-fancying D. H. Lawrence), a figure fitter, harder, stronger, more active, speedier (because he could run faster, or even because he was skilled with one of the period's obsessively doted-on aids to speed, a motorbike, a racing-car, or a plane), an altogether healthier man than yourself or any of the other Prufrock-like men of inaction you tended to associate with on the pansy bourgeois left. This gloriously fantastic hero—the *ace* or *pacer* or *racer* of John Pudney's volume of poems *Open the Sky* (1934) to whom 'Today we are devoted'—was frequently also more sun-tanned, as well as more muscled. Heinz is not only above his photographer, he's oiled and glistening as the sunlight catches the drops of water he's been fondly splashed with. A working heritage, sport, exercise, a lack of fear in the great outdoors, and some-times (as in Spender's cherishing descriptions of life in '20s and '30s Germany and Austria) an idle life of walking, swimming and sun-bathing enforced on European youth by Depression unemployment ('nude at bathing places . . . crucifying suns': *Vienna*): all these have combined to grant this heroic male type his admirable body.

And all through this period the Western populations trailed in search of this admirable physique: putting on shorts, taking all their clothes off in the new nudist camps, going in for collective physical jerks, joining the Youth Hostels Association (founded 28 May 1930), taking to the sun and the outdoors (in England and Scotland as well as abroad), rambling, swimming, hiking, bicycling, eager for health from vitamins, orange-juice, vegetarian diets, lots of roughage, and milk (free to British schoolchildren after 1934; purchasable by older youths at the new chromium-plated milk-bars). 'Hike all day on a Slab of Vitamalt!': Orwell's Gordon Comstock, in *Keep the Aspidistra Flying*, hates the continual incitements from the advertisers' copy-writers to get fit and stay well. So does Rose Macaulay's heroine Hero in *Going*

Abroad (1934): she simply loathes Sanatogen, '*And* ovaltine and bovril and milk'. Dublin's neon Bovril sign disquiets Beckett's Belacqua in *More Pricks than Kicks* (1934) by its ironic recall of heraldic patterns in a trivializing counterpoint with Christmas-season thoughts ('The lemon of faith, annunciating the series . . . A shy ooze of gules, carmine of solicitation, lifting the skirts of green that the prophecy might be fulfilled', and so on). In Beckett's *Watt* the advertisers' claims are mocked, in a wash of '30s memories, by Mr Graves's enthusiasm for Bando: 'From being a moody, listless constipated man, covered with squames . . . my breath fetid . . . I became, after four years of Bando, vivacious, restless, a popular nudist, regular in my daily health, almost a father and a lover of boiled potatoes.' And gnash people well might, for Health Centres (led by the pioneering one at Peckham, 1934–5, or the one Tecton built at Finsbury 1937–8: a 'megaphone for health' that was designed to catch day-long sunshine), swimming baths, sanatoria, as well as boys (and girls) shiningly healthy in shorts, in bathing suits, walking, leaping, running, diving, jumping, all busily crowded on to your attention. No newspaper, no photograph album, no book on architecture, no current register of what *tout le monde* was up to, felt complete unless it was laden with such motifs.

The sun was in fashion—even down to its obliquer manifestations in the period's craze for photography (light-pictures), in modernist architecture's zest for motifs adapted from the sun-worshipping Aztecs and for gleaming Riviera whiteness and light-welcoming windows (the newest houses were even given names like the 'Sun House' that E. Maxwell Fry built in Frognal Way, Hampstead), in Sunbeam cars and motorbikes (the motorbikes were pictured in adverts surrounded by sunlit bathing belles), in the Derby Winner of 1937 named Mid-Day Sun (after Noel Coward's song about mad dogs and Englishmen), in sunshine emblems such as the bursting sun-ray patterns on smart two-tone shoes, on the latest domestic furnishings, on everyman's wireless cabinet or suburban garden gate. And, moreover, the fashionable were out in the sun.

For the '30s was emphatically the era of the body. People had become reduced, confined to their bodies in the writing of the great moderns—defecating bodies in Joyce, sexual objects in Lawrence and Joyce, fragmented imagistic bits of bodies in the poetry of T. S. Eliot, grotesquely animalized gargoyle versions of bodies in the purple patches of Wyndham Lewis (Matthew Plunkett, in *The Apes of God*, 'roared —disparting and shutting his jaws, licking his lips, baying and, with his teeth, grinding, then again baying, while he stretched the elastic of his muscles elevating his arms with clenched fists, in heavy reproduction of the plastic of the Greek. Then he carried one of his exhausted hands to his head, and scratched it, between two sandy bushes, somewhat sun-bleached': *The Wild Body*—Lewis's title of 1927—indeed). It was a trend that Aldous Huxley was able, as so often, to pinpoint exactly. His Anthony Beavis in Chapter XI of *Eyeless in Gaza* is spelling out this tendency in Chapter XI of the book *Elements of Sociology* that he's writing. Old-fashioned Personality, Beavis argues, has given way to mere bodies. 'My digestion or metabolism' is the individual's last refuge in a world whose propagandists can actually alter the individual's mind:

> '*Cogito, ergo Rothermere est.* But *caco, ergo sum.*
> 'And here, I suspect, lies the reason for that insistence, during recent years, on the rights of the body. From the Boy Scouts to the fashionable sodomites, and from

Elizabeth Arden to D. H. Lawrence (one of the most powerful personality-smashers, incidentally: there are no 'characters' in his books). Always and everywhere the body. Now the body possesses one enormous merit; it is indubitably *there*. Whereas the personality, as a mental structure, may be all in bits—gnawed down to Hamlet's heap of sawdust. Only the rather stupid and insentient, nowadays, have strong and sharply defined personalities. Only the barbarians among us "know what they are". The civilized are conscious of "what they may be", and so are incapable of knowing what, for practical, social purposes, they actually are—have forgotten how to select a personality out of their total atomic experience. In the swamp and welter of this uncertainty the body stands firm like the Rock of Ages.

> *Jesu, pro me perforatus,*
> *Condar intra tuum latus.*

Even faith hankers for warm caverns of perforated flesh. How much more wildly urgent must be the demands of a scepticism that has ceased to believe even in its own personality! *Condar intra* MEUM *latus*! It is the only place of refuge left to us.'

'Let me hide myself in ME!' Beavis was theorizing the anti-intellectual, unremittingly physical condition of *Brave New World*. The anti-theological, materialist, anti-platonic implications of this extremely widespread bodiliness attracted as they tormented T. E. Lawrence. His autobiographical text *The Mint* (written in the '20s and first published in a tiny, imperfect American edition in 1936) conducts a busy polemic against the parsons who accept the biblical antithesis of flesh and spirit and it sticks up for the ordinary, inarticulate aircraftsmen's rich physical eloquence— 'their ceaseless extravagances of body flow in part from verbal poverty and relieve just those emotions which sophisticated man purges, uttered or unuttered, into phrase'. And the preference's spiritedly popular manifestations struck Auden no less forcibly than its wider implications had impressed Huxley and Aircraftsman Shaw. 'Came summer like a flood', Auden reported in the first of *The Orators'* Odes:

> Sunday meant lakes for many, a browner body,
> Beauty from burning.

Lord Byron had better be informed ('Letter to Lord Byron', Part II):

> We've always had a penchant for field sports,
>   But what do you think has grown up in our towns?
> A passion for the open air and shorts;
>   The sun is one of our emotive nouns.
> Go down by chara' to the Sussex Downs,
> Watch the manoeuvres of the week-end hikers
> Massed on parade with Kodaks or with Leicas.

'Massed on parade', 'manoeuvres': the wartime connection was obvious to Auden. Byron, though, will be interested in the current quest for fitness by means of 'salads and the swimming pool' not only because nothing of the '30s is held by Auden not to be of interest to the older poet, but also because Byron himself was a tough guy, tougher even in Auden's view than sporty Roy Campbell, who entitled his mendacious and boastful autobiography *Broken Record* (1934):

> A poet, swimmer, peer, and man of action,
>   —It beats Roy Campbell's record by a mile—
> You offer every possible attraction.

In fact this set of fads was powerfully prevalent enough to penetrate even the deliberately solipsistic crust of *Finnegans Wake*:

> Let the love ladleliked at the eye girde your gastricks in the gym. Nor must you omit to screw the lid firmly on that jazz jiggery and kick starts. Bumping races on the flat and point to point over obstacles. Ridewheeling that acclivisciously up windy Rutland Rise . . . Then breretonbiking on the free with your airs of go-be-dee and your heels upon the handlebars. Berrboel brazenness! No, before your corselage rib is decartilaged, that is to mean if you have visceral ptosis, my point is, making allowances for the fads of your weak abdominal wall and your liver asprewl, vinvin, vinvin, or should you feel, in shorts, as though you needed healthy physicking exorcise to flush your kidneys, you understand, and move that twelffinger bowel and threadworm inhibitating it, lassy, and perspire freely, lict your lector in the lobby and why out you go by the ostiary on to the dirt track and skip! Be a sportive. (p. 437)

(Coprophiliac in life and art, convinced of the generative power of the degenerative manifestations of the body in defecation, Joyce rather naturally takes, here amidst the unstoppable verbal diarrhoea of the *Wake*, to the prospect that sport and exercise can assist constipation. Dirt tracks are both bodily passages and the arenas of the speedway stars. As usual in Joyce's texts defecation and copulation merge lasciviously. Shorts fill him with all manner of unspeakable lusts.)

People in shorts were everywhere, not just in *Finnegans Wake*. 'There are hikers on all the roads': so the poem 'Pindar is Dead' declares (it's read over the radio in Act II of MacNeice's drama *Out of the Picture*, 1937). By no means every observer found them as welcome a presence as James Joyce did. William Empson essentially debunks that girl who so memorably 'cleans her teeth into the lake' (in his poem 'Camping Out': originally in *New Signatures*) with a characteristically high-falutin side-step, away from the toothpaste that she spits into the waters, to the stars that the soap-pattern reminds him of—a pastiche of the metaphysical poets' way of reaching at every instance for thoughts of the transcendental. The hastily got-up shorts that Barbara Greene wears on her walk through West Africa with her cousin Graham Greene (*Land Benighted*, 1938) make a wry leitmotif for her brisk and peppy account, but their awfulness becomes a major source of irritation within the cousins' usually placid relationship. When the would-be *Übermensch* painter Zouch of Anthony Powell's *From A View to a Death* encounters his literary acquaintance Fischbein and his 'dwarfish' bespectacled girlfriend among a gang of singing hikers trespassing on Mr Passenger's land he's embarrassedly hostile as he senses free board-and-lodging at Passenger Court slipping away. 'Was it for this, he wondered, that people spoke reverently of the duty to preserve rural England? Was there no power to protect these lovely regions from defilement by Fischbein and his filthy loves?' Powell relishes the aesthetic nastiness of the shorts wearers as well as the situation's social prickliness (the girl says she 'would have been farther if her shorts had not been so tight'). And one of the gruesomest Wyndham Lewis bodies outside the pages of Wyndham Lewis is a girl hiker, one of the many crowding the road to Loch Lomond, as described in James Barke's novel *Major Operation*:

> Clad only in a shirt, abbreviated shorts, thick hand-knitted socks turned down over a pair of heavy infantry boots, she carried on her back a Herbergan rucksack that in its enormous bulk suggested a weight of not less than fifty-six pounds. And yet the

enormous rucksack was balanced by the girl's enormous thighs which, thick as Belfast hams, were sunburned a painful lobster red. As the weight on her back and shoulders inclined her spine forward, her hips were thrust backwards adding a grotesque and slightly Rabelaisian touch to her ensemble.

'My God! What a horror', gasps one of the onlookers, in the tones of George Orwell's ordinary decent citizen on the bus in *The Road to Wigan Pier* recoiling from a grisly shorts-wearing duo, or Rose Macaulay's Hero in *Going Abroad*: ' "Good Lord", said Hero, disgusted', after Ted had rhapsodized about walks with knapsacks and hunks of bread and cheese, 'and nothing around you but hills and the sky'. Like Orwell, James Barke had no time for any sort of sun-worshipper, nudist, or hiker, even socialist ones. One of *Major Operation*'s activists is annoyed that summer distracts comrades from political work. 'Well: the hikers will soon be leaving us: we'll need to postpone the revolution till they come back.'

But hikers weren't easily deterred, even in the Communist Party. MacNeice's 'Pindar is Dead' goes on to talk of swimmers ('The swimming baths are filled for Easter Monday') and of climbers: 'There are climbers on all the hills'; 'With oiled boots and ropes they are tackling Snowdon.' And the '30s looked up not just to the Charles Atlas types with their built-up or building-up bodies, but to the man literally on high, up the mountain—just as the mountaineer himself looked up, as he climbed. It was the prevalence of mountaineering images in the period's writing that of course made Dylan Thomas and John Davenport have their spoof poem 'Manmountain' win for its author the Laureateship. Early in the decade, in 1933, Day Lewis had announced that the way to the future was up: up towards *The Magnetic Mountain*. Michael Roberts—'Michael the Mountaineer' Grigson dubbed him, inheritor of 'Snowdonia . . . with our love' in Auden and MacNeice's 'Last Will and Testament', who gave his name to the mountain-climbing hero of Ruthven Todd's novel *Over the Mountain* (1939), 'Michael the Mountaineer' as he's called, and also to Auden and Isherwood's mountaineer hero in *The Ascent of F6*, Michael Ransom— Michael Roberts kept on celebrating climbers and climbing, in his poems and his criticism. In the *Alpine Journal* (1940) he published an extremely interesting paper that he'd read to the Alpine Club in May 1939 on 'The Poetry and Humour of Mountaineering' which forces together his own activities as poet, mathematician, philosopher, critic and climber, and seeks pervasive analogies between writing and climbing. The new poetic and political country that Roberts had looked forward to early in the decade was distinctly to his taste as a mountaineer. His *New Country* anthology includes several of his own hilly poems, 'Kangchenjunga', 'St Gervais', 'Hymn to the Sun', 'Sirius B', as well as four poems from Day Lewis's *The Magnetic Mountain*. Charles Madge's poem 'Blocking the Pass', which is included in Roberts' later anthology *The Faber Book of Modern Verse*, actually gives us a mountaineering giant. He's called Grant: 'And alone, on a tall stone, stood Grant.' The father in Spender's *The Backward Son* is not only gigantic, but actually a mountain. He plays mountain-climbing with his sons on Hampstead Heath and represents grown-up achievements, the 'unscalable heights' of ambition: 'not only the mountaineer, but also'—especially since he has such a craggy head—'the mountain'. Of all the careers open to Auden and Isherwood's neurotic hero, Michael Ransom, their version in many ways of T. E. Lawrence, it's immensely significant that they turned him into a mountaineer. They needed to make it clear that looking up was what they were about.

Spender had been fixed precisely in the act of looking up within the pages of *New Signatures* itself: his part in that very first airing of a collective '30s poetic consciousness was a hero-seeking one:

> I think continually of those who were truly great.

And where are these greats? They're to be found by looking up, for their reputation was achieved where their memory is now most aptly enacted—up on high:

> Near the snow, near the sun, in the highest fields.

And one reason, evidently, for the spate of travelling north that goes on in '30s writing—Auden and MacNeice's journey to Iceland, MacNeice's Hebridean travels, Elizabeth Bowen's novel *To the North* (1932), and so on—is that northwards lie mountains, fells (Day Lewis's 'northern fell'), Audenic moors and rocks, and that in such regions you are nearer to the top of the world, to the very 'highest fields', the North Pole that attracts Spender in his poem 'Polar Exploration' (in *The Still Centre*, 1939):

> Our single purpose was to walk through snow
> With faces swung to their prodigious North
> Like compass iron.

'North, north, north' urges the opening line of Auden's soundtrack poem for the GPO Film Unit's documentary movie *Night Mail*. The north is magnetic, as alluring as Day Lewis's Magnetic Mountain ('So it's me for the mountain'), or like the roof of the world in Michael Roberts's Himalayan poems. The north is indeed a Shangri-La. It sustained Malcolm Lowry's private cult of Norway and of the Norwegian novelist Nordahl Grieg's *The Ship Sails On*, a novel peculiarly influential upon Lowry's own *Ultramarine*. It infiltrated Graham Greene's own myth-like friendship with Grieg—which led Greene to Norway and, following Grieg, to Estonia (see Greene's celebration of Grieg in *Ways of Escape*, 1980). The north is a major ingredient in the period's fascination for the coolly mysterious filmstar from Sweden, the 'Swedish Iceberg', Greta Garbo. A mythic north compelled the grown-up Auden as it had Auden the little boy, both conscious of their Icelandic roots. As Auden tells Lord Byron ('Letter', Part IV):

> With northern myths my little brain was laden,
>   With deeds of Thor and Loki and such scenes;
> My favourite tale was Andersen's *Ice Maiden*.

Northwards is where the 'young men' have gone in Spender's long poem 'The Uncreating Chaos': 'to the Pole, up Everest'.[98] The north is the zone for inevitably admirable heroic effort, a place whose axiomatic moral worth undergirds texts as apparently diverse in tone as Orwell's *The Road to Wigan Pier* and Auden's 'Letter to Lord Byron', both full of scorning contrasts between Surrey and 'the north', and of related insistences on the rebarbatively un-Wordsworthian northernscape:

> Tramlines and slagheaps, pieces of machinery,
> That was, and still is, my ideal scenery.

In Spender's early poem 'The Port' (1933), 'Northwards' is where 'the sea exerts his huge mandate', a place for working men in 'furnace' and 'yards'; 'Southwards' is where the 'Well-fed, well-lit, well-spoken' merchants loll about 'In their fat gardens'. The morality and the politics of this geographical contrast couldn't be plainer. It's the contrast that continually distinguishes the soft escape southwards in the period's travel literature, the pursuit of pampered D. H. Lawrentian climes in the sun, from grittier engagements with Icelandic, Norwegian, and Orwellian worthiness.

Spender ends his 'I think continually of those who were truly great' with this:

> Born of the sun, they travelled a short while towards the sun
> And left the vivid air signed with their honour.

Signatures in the air. This was Day Lewis's conceit too. In fact the title *New Signatures* must have been derived from Lewis's poem 'Letter to W. H. Auden' with which his volume *From Feathers to Iron* ends:

> But I, who saw the sapling, prophesied
> A growth superlative and branches writing
> On heaven a new signature.

And how better to write in the sky than in an aeroplane? 'When I . . . found', said one of the witnesses before the Parliamentary Select Committee on Sky-Writing (1932) 'that on three out of four nights in the week . . . the whole of the sky would be covered like Piccadilly Circus with advertisements, I felt that really something must be done and I wrote my letter to "The Times".' Writers frequently groused about this outbreak of capitalist scriptures in the air ('the streets are paved with gold but why', demands a satirized Lord Leverhulme in heaven, in MacNeice's *I Crossed the Minch*, 'Are there no sky-signs in the sky'), but the combination of the heroically lofty position ('We must up and find / What trade-routes are above', declares *From Feathers to Iron*) with their own craft of writing fascinated '30s writers just as Virginia Woolf's Mrs Dalloway had, on a famous fictional occasion in the '20s, been fascinated.

Airmindedness and being *airminded*, are characteristic concepts of the period. The OED Supplement traces *airminded* to 1928. Elizabeth Bowen uses *airmindedness* in *To the North* (ch. XVIII) of the altered sense of time and space that comes on Emmeline after her flight to Paris. A 'good day' for T. H. White (in his *England Have My Bones*, 1936) is when he's been 'air-minded till lunch, country-minded at a tennis party till supper, and beer-minded . . . till they closed'. In China Empson was reluctantly air-minded ('Autumn on Nan-Yueh (with the exiled universities of Peking)'):

> I have flown here, part of the way,
> Being air-minded where I must.

There had never been such an air-minded time in England (in a single day at the aerodrome where T. H. White learns to fly, as described in *England Have My Bones*, a long distance record-breaking plane covered in places-names flies in; so does the Prince of Wales; a writer of film scenarios and his film-star wife are having their flying lessons; and the lead actor in a London 'flying play' is photographed for publicity purposes in the cockpit of a grounded plane). Nor had English literature ever been so air-minded. Never before (or since) have poetry and the novel been so obsessed by the action, the clutter, the machinery, the terminology, the termini of travelling by air. Flying would help sell any piece of literature from humble detective stories (*Peril at End House*, 1932, and *Death in the Air*, 1935, by Agatha Christie; *Death of an Airman*, by Christopher St John Sprigg—not yet 'Christopher Caudwell', 1934; *Thou Shell of Death*, by Nicholas Blake/Day Lewis, 1936) to the most respectable of novels, poems, memoirs and travel books. 'Open the Sky!' indeed: the cry of John Pudney's poem echoed not only in its own title 'Open the Sky', and the title of its containing volume, *Open the Sky*, but from innumerable '30s texts.

And so the man in the aeroplane, the birdman, the maker whose Daedalian wings have indeed been transformed from feathers into something more man-made and iron-like, enjoyed a rich background of airminded expectancy among readers against which to slip easily into position, for the would-be sky-writers among '30s authors, as the quintessential man of action who was demanding to be looked up to. And where else should the post-war hero be found but up in the air, in the sun: the place where so many First War soldiers, like Oswald Mosley, Ralph Bates, Herbert Read, Wilfred Owen (and in all probability Henry Williamson) had craved to be? The air was the only location where it was, after the War, widely supposed that heroics had survived the general disillusionments consequent upon wartime active service—up there among the aviators of the Royal Flying Corps, the RFC, in the cleaner, freer element, heroically superior to the troglodyte infantry shut into their dark and grim trenches below, lapsed heroes bogged down in the mud and the messy ruins of once glorious suppositions about war. Henry Williamson's friend Captain Douglas Bell— to whom Williamson dedicated *A Fox Under My Cloak* (1955), his novel about the Battle of Loos—'was so sick of trenches (and trench mortars)', that he eagerly joined the RFC. In his anonymously published journal, introduced by Williamson and, in an interesting collaboration, apparently much revised by Williamson himself, *A Soldier's Diary of the Great War* (1929), Bell confided that as an infantryman one 'could but bury one's dirty self in a filthy smelly dug-out, and pretend one liked the cold and damp; but in the RFC one has the privilege of cleanliness, warmth and comfort after one's work is done. Having been so long an infantry man I can appreciate the RFC. Besides . . . The PBI [Poor Bloody Infantry] are slaves, in effect'.

Where a First-War heroic carried on at all uninterruptedly through the anti-heroic '20s and into the '30s, it was in air-warfare fictions, and those largely for small boys ('the air is the thing now', declared Betjeman in his 1938 review of boys' books), stories by W. E. Johns and Percy F. Westerman and their tribe, the sort of tale published by Christopher Caudwell's brother T. Stanhope Sprigg in his magazine *Air Stories* (founded 1935). Flying had remained, as Auden put it in a *Listener* review, unquestionably 'heroic travel'.[99]

In an age when the renewed quest for 'heroes' had become so keen (so much so that Mr A. of *The Ascent of F6* complains that he and his wife 'are bored by the exploits of amazing heroes'), 'heroic' airmen were especially numerous. 'Popular heroes, particularly in the air, are two a penny nowadays', declares Sir John Strangeways in *Thou Shell of Death*. And so it was only natural that T. E. Lawrence should fetch up in the Air Force. The only place for a would-be military hero was among the flying soldiery (just as, in the Air Force, Lawrence keenly rode motorbikes and ended up working on speedboats and taking part in air races). And it was equally natural that when it came to the writer as man of action ('Scholar and man of action:

an unusual mixture, eh?', muses press-baron Stagmantle to Auden and Isherwood's Ransom) it was four writing flyers who kept on fascinating the '30s imagination: T. E. Lawrence himself; André Malraux (whose 'violent life', claimed Spender, 'is an essential part of his creativeness as a writer', and whose photograph, 'André Malraux by His Machine', figured in the very first number of *New Writing* to carry photographs—snaps chosen and arranged by Humphrey Spender); Antoine de Saint-Exupéry; and Lauro de Bosis, the young Italian poet lost on his privately undertaken anti-Fascist flying mission dropping protest leaflets over Mussolini's Rome.[100] The obituarist of John Cornford in the *Cambridge Review* was reminded of de Bosis by Cornford's death in Spain. John Lehmann took de Bosis as the type of the writer who turns political activist (*LR*, January 1937). And even as late as 1943 Louis MacNeice was to be found doing a BBC radio broadcast about him.

Julian Bell, who was a huge and beefy youth never short of meat and greens within the prosperous enclaves of Bloomsbury and Cambridge, found all the looking-up going on an occasion only for a sneer against scrawny, homosexual authors:

> The idealised hero; the Proletarian, the Worker, of the cartoons and posters, ruggedly Grecian, stripped to the waist, muscular, with hammer or axe in hand; lean and brawny, smashing, kicking, humiliating the round fat capitalist in top-hat and trousers [Bell was thinking of cartoons like James Boswell's in *Left Review*, October 1935, where a large worker is shaking a tiny Mussolini, whilst a still tinier Hitler is skulking off the edge of the paper in dismay] . . . a wish-fulfilment, after all, of the scrawny, scraggy, embittered little proletarian intellectual who has pushed his way to the top by using his wits rather than his muscles.

Bell himself never made it quite 'to the top' either by his wits or his muscles, and he was wrong to suppose that such feelings for large workers were confined to scrawny proletarian intellectuals. But he was right about the element of the fantastic hereabouts. William Coldstream claimed that he and Auden found John Grierson's GPO Film Unit documentaries about the proletariat ('in which workers appeared undressed to the waist, covered in sweat, while the voice of a background narrator described in heroic tones what they were doing') richly absurd productions: they would chortle, according to Spender, over voice-over lines like 'Ever on the alert, this worker lubricates his tool with soap'.[101] But wish-fulfilment wasn't banishable by a few shared giggles. And what's clearly noticeable is how so many central '30s writers are, in life, outside the fitness-mountaineering-speed-flying heroic that absorbs so much of their writing.

Spender was big ('there's six-feet six of Spender for a start': 'Letter to Lord Byron', Part V), but in his Spanish poems he was a self-professed coward ('But I am the coward of cowards'), and in real life so unphysical that he even got duffed over at a poetry reading after the Second World War by the thuggish but also, by then, drunk and disabled Roy Campbell. Spender, Lehmann, Isherwood all took off their clothes in the Germanic sun, admittedly; but Auden remained a troglodytic pale-face, choosing to work indoors by day, even in the summer on the Baltic, with his curtains drawn fast ('A small room best, the curtains drawn, the light on', he told Lord Byron). There were, to be sure, lots of mountaineers amongst and around the prominent authors—Michael Roberts, his wife Janet Adam Smith, Mr and Mrs I. A. Richards, Robert Graves (who was taken climbing as a schoolboy by George Mallory the famous mountaineer who taught at Charterhouse and perished later on Everest), Ralph Bates, Jacob Bronowski (who displeased Graves and so got debunked *qua* climber in Graves's poem 'Dream of a Climber'), Michael Spender (who went on the 1935 Everest Reconnaissance Expedition), John Auden, whose ascent of K2, second highest mountain in the world, gave Auden the idea of *F6*. But one can hardly grace Auden and MacNeice's Iceland trip with the notion of real arduousness—in fact, 'Hetty to Nancy' makes constant mockery of the schoolmistress (actually a Bryanston schoolmaster) who keeps proposing a stepping-up of the amount of roughing-it that they're about ('Greenhalge's passion for hardship'). And hikes in the gentler German mountains did not turn Spender, nor Auden, nor Isherwood into mountaineers. Whilst working on *F6* in Portugal Auden and Isherwood (Auden emerging from behind the usual drawn curtains) went scrambling about in the Sintra hills: an attempt to 'get themselves into the mood' that, according to Isherwood in *Christopher and His Kind*, dissolved into 'laughter, lost footings, slitherings and screams'. As Auden explained, yet again, to Lord Byron (Part I):

> Parnassus after all is not a mountain,
>   Reserved for A1 climbers such as you;
> It's got a park, it's got a public fountain.
>   The most I ask is leave to share a pew
>   With Bradford or with Cottam, that will do:
> To pasture my few silly sheep with Dyer
> And picnic on the lower slopes with Prior.

Carefully offensive to 'mountain snobs' ('Wordsworthian fruits'), Auden rubbed in the point (in Part II): he's 'very fond of mountains' but he likes 'to travel through them in a car' ('I like to walk, but not to walk too far'). And, again, in the same sort of way, although there is a curiously large number of literary airmen around in the '30s, they are either not British (William Faulkner, de Bosis, Malraux, de Saint-Exupéry) or, if they are British, they are marginal to the Auden group. There's David Garnett, and T. H. White, the Irish novelist Francis Stuart, and T. E. Lawrence, the novelist Nevil Shute (who founded his own aeroplane company, Airspeed Ltd., in 1931), Christopher Caudwell, and Lewis Grassic Gibbon (if being a clerk in the Royal Air Force counts), but not Auden or Isherwood, or MacNeice or Day Lewis, nor even, despite his most suggestive name, Edward Upward. Spender's early verse makes a hero of his Oxford friend, the flying Marston, but Spender himself never became an airman.

In other words, the British authors at the heart of the Auden generation are, in practice, mere lookers-on at the period's heroics. They're a troupe of middle-class voyeurs of the big, the tough, the butch, the airborne, able to achieve heroic stature for themselves only by pretence, by proxy, in metaphor and other literary figures. Such loftiness as they attain is mainly in the head—like the 'mental mountains' Auden ascends when he unrolls the 'map of all my youth' for Lord Byron (Part IV). There's even something bogus about T. E. Lawrence's persistent efforts to acquire the extra stature he lacked and craved—though, of course, he did make more determined efforts in the physical arena of '30s heroics than any one of his admirers in the Auden group. Grounded, as it were, by government suspicion and decree, he still consorted with big machines, working as a groundsman and organizer of the 1929

Schneider Cup Competition, helping the RAF develop speedboats, and speeding about on great motorbikes, the huge Brough Superiors, on one of which (was it the one Bernard Shaw gave him?) he eventually had his fatal crash. Motorbikes featured in Isherwood's undergraduate concept of the 'Test' for those who, like Isherwood, had missed the First War, the 'Test of your courage, of your maturity, of your sexual prowess', a dream of a testing in 'romantic, heroic, dangerous, epic' action. At Cambridge the 'I' of *Lions and Shadows* acquires a 1924 AJS motorbike on which he achieves occasional Futurist speed thrills ('glorious . . . quite fast . . . exhilaration'). But he cannot start the bike properly, scares himself horribly at about 55 miles an hour, and after a near crash pushes the machine ignominiously back to Cambridge, acting the part of a man with a broken-down vehicle. A bogus hero indeed, and Isherwood is ruefully sending his failed heroic aspirations up. Isherwood was no more a racing motor-cyclist than his friend the novelist John (Hampson) Simpson who was sometimes confused with his brother Jimmy Hampson the champion motorbike racer, so that his novel *Saturday Night at the Greyhound* (1931) was hailed in France as the work of another T. E. Lawrence. How little like T. E. Lawrence Isherwood (or Hampson) was. And how much more bogus still was Isherwood's stalking about China in an enormous pair of Braggadocio boots, striving to look like the heroic journalist, an Ernest Hemingway or a Robert Capa. (And how letting down Grigson's *NV* review of *Journey to a War* must then have been (May 1939): 'Good as Isherwood's prose travel-diary is, as a description of the bigness of China and the War . . . it is really rather small beside the moral and imaginative weight and tenderness of Auden's poems.') And how, for that matter, unheroically comic is the cushioned airman of Davis and Sprigg's *Fly With Me: An Elementary Textbook in the Art of Piloting* (1932: Sprigg was, of course, Caudwell):

> If you are short in the leg, you must, of course, make sure you can reach the rudder-bar comfortably, even when it is right over to one side or the other. It may need to be adjusted for you, especially when you are piled up on cushions.

Dubious accoutrements—big bikes, big boots, a pile of cushions—apart, there was only fantasy left for Auden and Co.: Auden casting himself, and allowing himself to be cast by others as kestrel and birdman, airman and solo-flyer; or Spender and Day Lewis fictionalizing themselves and their chums as North Pole explorers or magnetic mountaineers ('One imagined his hands on the tiller steering the ship to the North', Spender recalled of Isherwood in his 1980 Preface to Lee Bartlett's edition of *Letters to Christopher*); or Isherwood choosing to regard China as a real enactment of the Test. 'Bother the Test' was Forster's waspish response to that. Forster was well placed to recognize a fantasy when he saw one, and well enough used to camp giggles not to allow them to cover such a multitude of self-deceptions.

In *Lions and Shadows* Isherwood's I-narrator, trying to be lovable and ruthless all at once, spares the reader no wry detail of the lengths fantasizing would take him in preparation for the Test.

> I built up the daydream of an heroic school career in which the central figure, the dream I, was an austere young prefect, called upon unexpectedly to captain a 'bad' house, surrounded by sneering critics and open enemies, fighting slackness, moral rottenness, grimly repressing his own romantic feelings towards a younger boy, and finally triumphing over all his obstacles, passing the test, emerging—a Man. Need I

confess any more? How, in dark corners of bookshops, I furtively turned the pages of adventure stories designed for boys of twelve years old? No illustration was too crudely coloured, no yarn too steep for my consuming guilty appetite. How, behind locked doors, I exercised with a chest expander, bought after nightfall, with precautions such as a murderer might observe in purchasing his weapon? I went out of my way to tell the shopman that it was for my younger brother.

'I Will Make You Taller': along with 'Tragic Widow Weds Lone Flyer', that is one of the advertisements at the modern end of Peter Fleetwood-Hesketh's long fold-out depiction of architecture's historical 'Street of Taste' published in Betjeman's *Ghastly Good Taste* (1932). But what man, merely by taking thought, as the Bible has it, can add a cubit to his stature? The satisfactions of fantasy remained stubbornly untranslateable into reality. Perhaps this made them all the more satisfactory. As MacNeice astutely observed in his poem 'The Glacier' (*Poems*, 1935), 'those who gaze get higher than those who climb'. (MacNeice steadily refused to accept that the would-be heroes of 'Eclogue from Iceland' (in *Letters from Iceland*), who 'out of bravado or to divert ennui' drove 'fast cars' or climbed 'foreign mountains', had done anything really to recuperate the idea of heroism, especially when the First War 'hero / With his ribbons and his empty-pinned-up sleeve' was still cadging for coppers in the street. Perhaps, this Eclogue suggests, heroism petered out long before the War, with Grettir, the 'doomed tough' and haunter of crags, 'The last of the Saga heroes'.)

And meanwhile Isherwood stayed unexpanded, as did all those physically or morally weedy authors, by contrast with their proletarian Heinzes, or the tough Malrauxes and big Hemingways, or the sturdy Stalinists who would never let them stop 'feeling small'. Truly weak men, they would have a perpetually hard time becoming Truly Strong. And that is perhaps why they found the idea of T. E. Lawrence so seizable. For Lawrence was one of their sort: 'The little fellow who is labelled for posterity as Lawrence of Arabia', as E. M. Forster describes him in his *Abinger Harvest* essay, relishing Lawrence's 'little unhusked body'. And he was not only tiny ('one's first impression was of his smallness' wrote Isherwood in his 1937 *Listener* piece about him), but an intellectual and writer who was worried about buggery, about whether he had an abnormal sexuality. Auden's Airman's circle was exclusively male. It did indeed find 'something peculiarly horrible about the idea of women pilots'. Very few women pilots are mentioned in '30s texts. Outside of Soviet women flyers, and Moira who is part of a trio making a round-the-world record attempt in J. Llewellyn Rhys's novel *The World Owes Me A Living* (1939), and Judy, the woman who learns to fly in Rhys's Flying Club novel *The Flying Shadow* (1936) and crashes to her death at the end of it, and Aunt Ada Doom in *Cold Comfort Farm* who dons an astonishingly modern 'flying kit of black leather' to fly off to Paris in—and even then, like Nina of Waugh's *Vile Bodies*, and the women of Agatha Christie's *Death in the Air*, and the 'society women' J. Llewellyn Rhys is frequently down on for wanting to take part in demeaningly silly record attempts, she's only a passenger—I can think of only one other '30s fictional flying woman togged up as the Audenic 'helmeted airman'. And though Mrs Rattery of Waugh's *A Handful of Dust* is fine to look at in her leather helmet and overalls, she's also a 'denationalized' American, not English, just one more of the cosmopolitan 'joke-women' the novel is deploring. One shouldn't try reading too much into this: Waugh despises pilots of both sexes. None the less it remains true that there were in the West in the '30s comparatively few actual women pilots. The rash of aristocratic lady flyers in the '20s described so nicely by Ronald Blythe in *The Age of Illusion* (1963) had died down. Which is why Amy Johnson, Jean Batten, the Soviet lady flyers and the Misses Audrey Sale-Barker and Joan Page whose long-distance flight is exploited as a running comic sub-text in Gandar Dower's *Amateur Adventure* (1934; Penguin Travel and Adventure Series, 1939), all enjoyed so much publicity. The air and its texts had indeed become largely male preserves. Agatha Christie's air-based stories are as rare in their airmindedness as Stella Gibbons' *Cold Comfort Farm* or Elizabeth Bowen's *To the North*. *Air Stories*'s notice in the *Writers and Artists' Year Book* (I quote from the 1937 edition) declared that it required 'strong plots, plenty of flying action and brisk dialogue'. Soppy, girlish stuff would not do: 'Love interest is not wanted'. And naturally enough on this women-drained site male homosexuality flourished: T. E. Lawrence had as sexual company from the pages of fiction Edward Blake, the suicidal airman of Isherwood's *The Memorial*, and Auden's Airman, and from real life Forster's close friends Air Commodore L. E. O. Charlton, First-War veteran and sometime Chief of the Air Staff, who was the author of air books for boys as well as being air-correspondent of the *NS*, and Tom Whichelo his ex-aircraftsman boyfriend, and also those abortive valiants of the RAF in the Second War, Harold Acton (who went into the RAF full of admiration for 'those legendary heroes' of His Majesty's flying services) and Brian Howard.

All in all, one's not surprised that the neurotic T. E. Lawrence was taken, then, as a representative, an emblem, of current anxieties, at least in and around the Auden group. Lawrence suffered, Isherwood claimed, 'in his own person, the neurotic ills of an entire generation' (a claim echoed by Auden's lines on Isherwood himself as novelist: 'in his own person, if he can, / Must suffer dully all the wrongs of man'). For Auden, Lawrence's life comprised an allegory: 'an allegory of the transformation of the Truly Weak Man into the Truly Strong Man, an answer to the question "How shall the self-conscious man be saved"?' T. E. Lawrence might, Auden was suggesting, provide a model, a way out, a solution for writers and intellectuals like himself and Isherwood, men of inaction, thrust embarrassingly into a period when action was being demanded of them. At least, the review of Basil Liddell Hart's *T. E. Lawrence* in which Auden makes this claim for Lawrence, indicates the kind of debate he and his friends indulged. 'Thinking of Lawrence', Auden is reminded of a statement, which he read, he believed, 'in McDougall's *Abnormal Psychology*' (though he may have got it from Isherwood, who says that *he* read it independently among McDougall's sources), a statement 'made by a man after he had cut the throats of his wife and family. "No, I am not the truly strong man. The truly strong man lounges about in bars and does nothing at all".'[102]

Heroic stature on that sort of scale is claimed of the fearless sailor Gerhart Meyer (in Auden's poem 'It was Easter as I walked in the public gardens'):

> From the sea, the truly strong man.

It's what 'Prince Alpha' is made to disclaim (in 'The Witnesses', Part II)—in an echo of Eliot's Prufrock declaring that he's not Prince Hamlet (and an echo too of Eliot's death-wishing Sybil):

'Children have heard of my every action
It gives me no sort of satisfaction
and why?
Let me get this as clear as I possibly can
No, I am not the truly strong man,
O let me die.'

Still, such strength is implied of Shakespeare—alone among the listed writers at the end of 'Letter to Lord Byron' to be granted anything very positive in answer to the question 'Are poets saved?': 'Shakespeare is lounging grandly at the bar'. But who among his Old Boy successors, consciously small-time achievers, would claim to equal Shakespeare? T. E. Lawrence was a far less considerable writer. But some consolation was to be derived, it's evident, from the notion that if the manifestly small, boyish, homosexual, neurotically self-conscious man could become as 'big', as 'Truly Strong' as T. E. Lawrence had contrived to be, there might be some personal redemption possible for the members of Auden and Isherwood's generation of small, boyish, neurotically self-conscious homosexuals. And that, obviously, is why Ransom, the man of action in *F6*, destined according to his mother 'to be the truly strong', is made extremely slight in physique ('Eight stone six . . . short and blue-eyed'). He has to start, at least, by physically resembling T. E. Lawrence. And a whole '30s run of claimants on true strength are given Lawrence's size of body. Isherwood reads it back on to Sherlock Holmes in his *Exhumations* essay on *The Speckled Band*: 'the well-bred muscle man, straightening the poker which that vulgar show-off, Dr Roylott, has twisted: "I am not quite so bulky, but if he had remained I might have shown him that my grip was not much more feeble than his own".' Captain W. E. Johns's First-War airman hero Biggles is not big at all, despite his name, but 'a slight, fair-haired, good-looking lad'; his features 'as delicate as those of a girl, as were his hands'. Just like Nicholas Blake's legendary airman hero in *Thou Shell of Death*, Colonel Fergus O'Brien, the 'little airman', 'almost fey', with an 'almost girlish voice' and 'delicate hands' ('a woman's hand' shows in his domestic neatness). (It's a stress on hands, incidentally, that the Biggles books share with a lot of '30s writing about boys, homosexuals, and heroes. Worry about what mastur-batory, even cowardly, hands are tempted to get up to—'A cold bath every morning. Never to funk . . . Whenever temptation is felt . . . Hands, in the name of my Uncle, I command you, or . . .'—afflicts Auden's Airman right to the end. James Hanley dwells on the 'tiny hands', the 'thin white hands' of his 'neish', 'girlish' hero in *Boy*, and on the 'little' hands of a dirty youngster he describes dragging a cart in *Broken Water: An Autobiographical Excursion* (1937), and on the 'hands of little nippers', in *Grey Children: A Study in Humbug and Misery* (1937). The bogus public-school man of action in Graham Greene's unfinished fiction 'The Other Side of the Border' is a sinisterly gone-wrong Kurtz-like figure who is actually called Hands.)

Michael Ransom and the Airman of *The Orators* also resemble T. E. Lawrence in that they turn out in the end to be neurotic failures. *The Orators*, Auden wrote to Naomi Mitchison, 'is my memorial to Lawrence; i.e. the theme is the failure of the romantic conception of personality; that what it inevitably leads to is Part 4'. And Part IV of *The Orators*, he said, is a description of 'The effect of Hero's failure on the emotional life'.[103] Auden did suggest in his Liddell Hart review that T. E. Lawrence was as much the sucessfully self-conscious, reasoning man of action as Lenin, but that was a cheeky suggestion whose nerve broke when Auden came to his own

Airman and to Ransom. It cannot ever be argued of those two that they keep up their bid for heroic stature for very long. Very speedily they join the abundant ranks of failing and failed '30s heroes: the Johnnies-Head-in-Air who quickly and inevitably come to grief. Johnny Head-in-Air, the 'skyward man' of *Struwwelpeter*, fascinated Day Lewis, who wrote a poem about him called 'Johnny Head-in-Air', and John Pudney (whose RAF poem 'For Johnny'—'Do not despair / For Johnny-head-in-air'—became one of the most famous of Second War poems), as well as attracting Auden and Walther Groddeck—who read a paper on *Struwwelpeter* at the 1930 Dresden Psychotherapy Congress. And in their failure, Auden's Airman and Ransom fall into step with the maimed heroes, like Lord Byron, who go limping; with the flyers who fall, like the Icarus of Auden's poem 'Musée des Beaux Arts' and de Bosis's poem 'Icaro' or like Auden's Dancer-as-Pilot in *The Dance of Death* or the dead airmen in Keidrych Rhys's poems 'Spin' and 'Landmark' ('I am he killed / On his first cross country solo'; 'draped in their sheets / I see the decoyed boys but stained')[104] or the hawk-watching aristocrat of the first one of Spender's *Poems* (1933):

> This aristocrat, superb of all instinct,
> With death close linked
> Had passed the enormous cloud, almost had won
> War on the sun;
> Till now, like Icarus mid-ocean-drowned,
> Hands, wings, are found.

The mere act of falling Icarus-like—the fate Joyce prevented his youthful artist from having to endure by turning the boy into his more successful father Dedalus—is sometimes felt to bestow a gratuitous heroic stature—a heroism *malgré lui*—as it does in the case of the villainous murderer in Day Lewis/Nicholas Blake's *Thou Shell of Death* who falls to his death during an aeroplane chase ('He fell with arms and legs sprawling, like a dummy figure. Down and down and down. For years he seemed to be falling. They had lost sight of him altogether a few seconds before there appeared on the sea's face a tiny white splash, as though someone had thrown a very small pebble'). But it's hard, even when armed with thoughts of the usual fall of the tragic hero, to snatch much heroic consolation from the Lawrence or the Icarus story. These fallers from the air are all cases of *High Failure* (the title of the book by long-distance flyer John Grierson that Auden reviewed in 1936, indulging in a characteristic '30s reverie on the dimensions of flying, its connection with the 'highbrow' vision that makes all else 'undersize').[105] And the oxymoronic stress in that title on the loftiness of the experience cannot erase the emphatic presence at its end of the fact of failing. Which is what '30s writing keeps returning to. As T. E. Lawrence flinched (he 'let the man-of-action down', alleged Wyndham Lewis in *Blasting and Bombardiering*, by turning his head away, 'a schoolgirlish touch', when he executed Arab boys), so Roger Garland of Isherwood's unwritten novel *The North-West Passage* (described in *Lions and Shadows*) is 'the neurotic hero', the Truly Weak Man (he sounds not unlike Eric of *The Memorial*) who can 'tell funny stories and imitate Cortot and Casals and hold forth about Hindemith and Stravinsky and Delius; but he funked the high dive in his prep-school swimming-bath'. He's also not unlike Day Lewis's airman genius Fergus O'Connor who's been 'a bit of a nervous wreck' since his last crash, and is resigned to the death he's been threatened with; or Garland's *alter ego* Tommy (who resembles Edward of *The Memorial* and his failed motorcyclist author) who crashes T. E. Lawrence-like to his death 'on his appallingly powerful motor-bicycle'. Isherwood, his characters and his many personae, are persistent failers of the Test. In this, again, they're not unlike the long-distance airman of Bunny Garnett's curious little novel that T. E. Lawrence so admired, *The Grasshoppers Come* (1931), who fails to make his destination; or Cyril Connolly's novelist hero of *The Rock Pool* who sees himself as a 'cashiered aviator', an 'inverted Samson' and who resolves to 'practise failure'; or Ruthven Todd's Michael the Mountaineer, who learns to accept his comedown fate as 'the voyager who was known . . . to fail', after an uplifted experience of being 'the greatest climber in the world' ('I was a giant, and I wore seven-league boots . . . I was nearly a mountain myself').

All these characters are on an elevating journey that will end, as Elizabeth Bowen's *To the North* or J. Llewellyn Rhys's *The Flying Shadow* ends, in a terrific comedown, a spectacular crash. They're of the same party as the mountaineers in Michael Roberts's 'Elegy for the Fallen Climbers', as Lamp, Gunn, and Shawcross of *F6*, as the men in Ransom's litany for the famous mountaineers now dead, and the 'Heroes' of Auden's poem 'From scars where kestrels hover' who could not 'resist the temptations / To skyline operations' and went out to their doom, and the 'we' of The Orators who 'forgot His will' 'on the snow-line'. They're all heroes who come to grief in the mountains, in the air, up on high. Hubris has always had a way—as Day Lewis's poem 'A Warning to those who live on Mountains' urges—of bringing you down to earth. Satanic pride leads inevitably to a fall (unless you're Stephen Dedalus). And trading in hubristic downfalls became the small author's belatedly honest acknowledgement that he'd never really make it on high, that for all the mutual elevation going on in the coterie—Auden, according to Isherwood in *Journey to a War*, is 'the truly strong' man because he slept through an air-raid—weakness rather than strength was his probable portion.

This reluctant acceptance of lowliness, cowardly baseness, heroic failure, may be—and despite the deterring air of people making too much out of a bad job that hangs about much of the discussion by Auden and Co. of T. E. Lawrence's broken career—may be the best thing about the '30s business of heroics. For looking up, going up, even sending yourself up as Isherwood likes to do (Auden may be offered as a candidate for truly strong man, but Isherwood rejoicingly describes himself and Auden in China in comically unheroic postures, rocking back and forth on their seats in railway carriages in an effort to prevent constipation, reciting passages from an imaginary travel book they call *With Fleming to the Front* as a antidote to the Old Etonian's crisp cool which threatens to put their shambling boyish messiness in its place), all of these loftiness considerations and postures are inseparable from the period's politics. What injects the business of '30s heroics with its importance, what keeps up its attraction to the writers and ultimately also grants it so much of its deep ambivalence, is that bigness entered directly into the rivalry between Socialism and Fascism. The going questions of the day were, which of the two was morally the bigger, and which would prevail physically, be militarily the stronger?

The biggest thing about the Soviet Union, according to much leftist polemic of the '30s, was that it was very big. Already socialism covered such a large series of countries as to make its presence no longer ignorable. Already the Soviet Union was *The Socialist Sixth of the World*. That was the title Hewlett Johnson, the so-called Red Dean of Canterbury, gave his book about the Soviet Union (1939). One of Dziga Vertov's films was *A Sixth of the Earth* (1926). So big was the place that it sanctioned any amount of would-be tough hero-seeking. It excited British writers not least because of the enormous reputed sales of books there. Everything about it was Promethean. John Lehmann entitled his Soviet travel book *Prometheus and the Bolsheviks* (1937). The Soviet Union ministered mightily to gigantism. Robert Byron in *First Russia Then Tibet* (1933) described the Kremlin as a Brobdingnag, a 'Caesaro-papist fantasia'. A breathless rhetoric of the giant (giant factories, giant tanks, giant refineries), of the enormous, the huge, of towers and towering, runs through the reports from Moscow by John Lehmann in *NW* (Spring 1936) and by the architect Clough Williams-Ellis in *LR* (November 1937). Lehmann takes especial delight in *Prometheus* in a children's May procession whose 'star-piece . . . was undoubtedly an enormous model boat on wheels, filled with children brandishing red flags round an outsize painted head of Stalin'. The ship he sails on is decorated on the anniversary of the October Revolution with a 'huge Soviet star' of red electric light bulbs flashing on and off around a blue and green hammer and sickle. E. M. Forster in his report from the 1937 Paris Exhibition showed himself quite fascinated by the roof of the Soviet Pavilion, a height dominated by a woman 'of enormous size . . . beside her gigantic mate'.[106] The rhetoric of the utopian Soviet architecture, as of the revolutionary Soviet poetry and film was all flight, height, and aerialism. Mayakovsky wrote about flying proletarians; Raizman made a film called *Flyers*, Dovzhenko one titled *Aerograd*. El Lissitsky's *Russland, Die Rekonstruktion der Architektur in der Sowjetunion* (Vienna, 1930) proposed filling the Soviet Union with elevated constructions, colossal sports stadia, giant complexes with huge aerial brackets for spectators to sit in, government offices that would outdo New York's sky-scrapers, a set of elevated service-roads in Moscow to be fed by elevators that Lissitsky dubbed 'sky-hooks'. Such schemes hardly got to the launching-pad in the Soviet Union, and in England only achieved concreteness in the miniaturized architecture—scarcely more than architect's models—of the new zoos. And even among zoo architecture the London Penguin Pond's uplifted, wheeling, aerial ramps were unique. Still, in the land of the Soviets the metaphors at least could soar unrestrained.

And amidst the rash of mere metaphors how the human giants, spectacularly lofty achievers of all kinds, did really flourish under eastern socialism: explorers, mountaineers, bold men like the flyers who in 1934 rescued a stranded expedition of Soviet Arctic explorers and are celebrated in Michael Roberts's poem 'Chelyuskin'. The poem was evidently incited by Alec Brown's translation of *The Voyage of the 'Chelyuskin'* (1935): a book that was 'proof', according to Amabel Williams-Ellis's review of it (*LR*, November 1935), 'of how crude is the often repeated fascist view that war is necessary to demonstrate and exercise the heroic qualities in mankind'. No wonder the British Workers Theatre Movement troupes of the early '30s would sing 'The Soviet Airman's Song' (was this the same, one wonders, as the 'Red Airman's Song' played in an Unemployed Demonstration in Barke's *Major Operation*?):

> Our engines roaring, roaring to the battle;
> High in the air above the clouds we speed;

> Our bombs are ready, our machine-guns rattle
> 'Gainst the world's imperialistic greed.
>
> Fly higher, higher, and higher . . .[107]

On Top of the World was the title of L. Brontman's Left Book Club volume (1938) celebrating the 1937 Soviet Expedition to the North Pole. Even more on top were the Soviet airmen who flew over the Pole and who are celebrated in *On Top*: 'We were very pleased that we, the northernmost Soviet citizens, heard the hum of their engines and can thus confirm that their red-winged aeroplane passed over the North Pole.'

Central to this mythography of Soviet superiority was a democratization of the heroic. Any Soviet citizen, it was constantly being pointed out, could become a parachutist—parachute-jumping from towers was a favourite Soviet pastime much reported by British visitors—and flying had been opened to all sorts and conditions of men and women. The West had its T. E. Lawrence, and he was possessed by enthusiasm for the coming of an uplifted, flying democracy. But, objected Ralph Fox in his *Left Review* article 'Lawrence, Twentieth Century Hero' (July 1935), Lawrence was a special case; it was only in the East that the air was being 'conquered by the common people of the Soviet Union, who have found the formula which reconciles the genius and the plain man'. Lehmann (in a passage of *Prometheus* that talks about T. E. Lawrence) 'pictured the Soviet masses in a few years air-minded and parachute-minded to an extent unapproached and unapproachable by any other country'. No wonder modern epics could be composed in the Soviet Union. The epic struggles of the working-class heroes of Soviet fiction and film were frequently urged on *LR* readers as shaming by contrast the collapse into neurotic pygmy individualism of Western writing. Dimitrov, hero of the Reichstag Trial, the Bulgarian who came to stand for heroic communist manhood in the midst of fascist adversity, was commonly urged upon the imagination of British readers. He was zealously pressed upon Western authors hesitant about the possibility of the great human subject by Ralph Fox in *The Novel and the People*. (In Harold Heslop's novel *Last Cage Down*, Dimitrov's name sends thrills of revolutionary hope through the miners, 'Dimitrov! Herculean in his simple bravery . . . So they stood in silent tribute to a MAN'. In Lewis Jones's novel *We Live*, 1939, communist miner Len spurs himself on in revolutionary activity with thoughts 'fixed on Dimitrov'.) Writers were warned that only right-wing Soviet authors wanted 'to get away from Magnitogorsk, from Kuznetsktroy, to "great art", which depicts the small deeds of small people': thus Karl Radek in the course of his famous attack on Joyce's *Ulysses* at the Soviet Writers' Congress in 1934. Joyce's 'basic feature', according to Radek, 'is the conviction that there is nothing big in life—no big events, no big people, no big ideas; and the writer can give a picture of life by just taking "any given hero on any given day", and reproducing him with exactitude. A heap of dung, crawling with worms, photographed by a cinema apparatus through a microscope—such is Joyce's work'. Soviet authors must not rest, exhorted Leonov at the same Congress, until they'd met 'the needs of the epoch with its "giant race of men"'. Soviet art, the Congress decided, was moved by 'the big passions of collective labour, the heroic significance of Socialist construction, the mighty emotions of the new man'.[108] And the British 'proletarian' novelists dutifully did their imitative best to keep up. In

Heslop's *Last Cage Down*, Big Joe Frost stirs his massed comrades with stories of the gigantic Soviet industrial achievement at the Dnieperstroy Hydro-Electric scheme:

> They *saw* a host of men and women pouring out their simple toil upon a gigantic construction, which, when completed, was going to change the entire nature of the Ukraine . . . They scrambled about the huge structure of steel and concrete watching men trampling great masses of concrete into the mould, building socialism with simple tools—feet, hands, concrete and steel . . . 'stupendous and magnificent'.

The dimensions of the Soviet leaders had to match all this gigantic effort. So they were hailed as indeed the 'truly great', in a succession of 'the truly great'. Behind them all there towered, in the phrase of the communist surrealist poet Roger Roughton, 'Man Mountain Marx'. Then came Lenin, who was to be celebrated, Clough Williams-Ellis reported to the *Left Review* (November 1937), by a 'colossal statue of chromium steel' to be built on top of 'the vast new Palace of the Soviets' that was itself 'designed to be the tallest building in the world'. Williams-Ellis had forgotten the statue's exact measurements, 'but I know one could play badminton in its boots'. And now, bigger than them all, there was Stalin. Stalin the proxy airman ('we elected Stalin an honorary pilot of our plane', claimed the new Soviet holders of the world non-stop flight record in 1936: success had attended their taking the 'Stalin route'). And Stalin the highest mountain in the Soviet Union.

The heroic ascent of Pik Stalin, a recently discovered mountain, that had earlier defeated (of course) a mixed Soviet-German expedition, was accomplished by the arctic explorer and airman Nikolai Gorbounov and described in the extraordinary book by Michael Romm, *The Ascent of Mount Stalin*, translated into English by the Communist and 'proletarian novelist' Alec Brown (1936). A photograph of Gorbounov as a 'helmeted airman' is the book's frontispiece. And the Soviet heights-heroic is the dominant motif of an account that is full of all that's highest: the 'highest glacio-meteorological permanent observatory in the world'; 'the highest motor-road in the world'; and, of course, 'the new mountain, the highest in the USSR'. And the surrounding mountainous Soviet landscape is made into a *paysage moralisé*, nothing less than an allegory of the Socialist new country, a terrain now dominated by the new giant, Mount Stalin. The mountainous topography becomes an emblem of Soviet politics, with Mount Stalin 'a head above all its neighbours'—that is above the lesser heights named for Stalin's predecessors. Geography itself is bent into supplying a myth of history, a text to be read as showing the historical pathway leading heroically upwards to Stalin. He's a summit approached via Mounts Lenin, Pravda, OGPU, Molotov, Voroshilov, Revolutionary Military Soviet, Mounts Marx, Engels, and Ordzhonikidze, and so on. It was a pathway whose mountainous signposts would of course demand careful repainting as the great Treason Trials proceeded: already the mountainous map had no room for any hills or even slopes in memory of, for example, Trotsky or Radek or Bukharin.

Leftist writers were soaked in a rhetoric of the slumbering giant of the people awakening from repression (there's an instructively sleeping Spanish giant, a pathetic freakshow specimen in Ralph Bates's novel of abortive Spanish revolution, *The Olive Field*), of Prometheus breaking free from bondage (compare *The Mind in Chains* volume of socialist essays edited by Day Lewis, 1937), of a socialist Gulliver towering over the rival Lilliputians (in *World Within World* Spender gloats over

Barcelona's huge revolutionary posters: 'as though some proletarian Gulliver had attached labels to his Lilliput, announcing the liquidation of the bourgeois Lilliputians and his intention to build a new city for a race of Gullivers'). Characteristically, to Day Lewis in the poem 'Learning to Talk', the future is a matter of growing into tallness, of leaping and flying upwards. 'We are growing', and so are the coming generations ('See this small one . . . he'll be the tall one').[109] Spender's famous pylons of 1933, his 'nude, giant girls that have no secret', are, of course, striding into a socialist future that dwarfs the bourgeois present:

> But far above and far as sight endures
> Like whips of anger
> With lightning's danger
> There runs the quick perspective of the future.
>
> This dwarfs our emerald country by its trek
> So tall with prophecy.

And this was a pervasive rhetoric that found its apogee in the Soviet Union's land of giants. (In Alexander Ptushko's 1935 film *A New Gulliver* there's a class war going on in Lilliput, and Gulliver naturally sides with the oppressed.) Day Lewis's celebratory poem 'On the Twentieth Anniversary of Soviet Power' naturally assembles a whole plethora of the going heroic images: flying, cragginess, northernness, upliftedness of all sorts, moral loftiness:

> We have seen new cities, arts and sciences,
> A real freedom, a justice that flouts not nature,
> Springing like corn exuberant from the rich heart
> Of a happier people. We have seen their hopes take off
> From solid ground and confidently fly
> Out to the mineral north, the unmapped future.
> USSR! The workers of every land
> And all who believe man's virtue inexhaustible
> Greet you today: you are their health, their home,
> The vision's proof, the lifting of despair.
> Red Star, be steadfast above this treacherous age!
> We look to you, we salute you.[110]

Only by joining the Soviet masses in scaling the Marxist–Leninist heights would punily hesitant individualists be saved: that's the burden of Tom Wintringham's stinging response to Day Lewis's 'feeling small' sonnet (it was printed on the same page of *LR* as Day Lewis's poem), 'Speaking Concretely: A Reply to C. Day Lewis':

> Marx for your map, Lenin theodolite—
> This is a thing Smolny's October showed—
> Crag-contour pioneered, valley and peak's height
> Known: all is ready? No, steel wire must be
> Inseparable from concrete, you from me,
> We from durable millions. Then there's a road!

Soviet history will teach Day Lewis how not to stay feeling small. And it was evidently amidst the plenitude of heroic Soviet ikons that an airman like Christopher St John Sprigg found it congenial and meaningful to join the Stalinist Communist

Party. Stalin the mountainous flyer could be seen as the personal apotheosis of Sprigg's gently heroizing book *Great Flights* (1935) ('There has been nothing like the great flights since the historic days when Magellan and Columbus were exploring the unknown New World. In the same way the great long-distance pilots have been exploring the possibilities of the new world of the air'). Nor is it surprising that the Stalinist Day Lewis should keep on celebrating an heroic long-distance flight such as that undertaken by the Australians Parer and McIntosh who took eight months to judder home in an old DH 9 in the 1919 England–Australia air-race. 'Their completion of the course in the face of so many difficulties was a triumph of pluck and perseverance', wrote Sprigg in *Great Flights*. Day Lewis thought highly enough of the exploit to adapt it for his own airman Fergus O'Brien in *Thou Shell of Death* ('his solo flight to Australia in an obsolete machine, flying one day and every other day tying the pieces together after the crack-up'). Lewis's extended poetic rhapsody on Parer and McIntosh's journey in 'A Time to Dance' (those 'haughty champions' he calls them) reminded cool observers of the imperialistic heroics of Kipling and Henry Newbolt. Mountaineer Michael Roberts, however, in his Introduction to *The Faber Book of Modern Verse*, hailed Lewis as providing in that poem a 'credible' myth for 'the modern reader'.

Cooler observers, even on the Left, did have their worries about this kind of inflationary mythography. We 'have reached the end of the Communist film', complained Graham Greene of the 1937 version of the film *Lenin in October*: 'It is to be all "Heroes and Hero-Worship" now: the old films are to be re-made for the new leaders: no more anonymous mothers will run in the van of the workers against the Winter Palace. The USSR is to produce Fascist films from now on.'[111] And from the start there was little uniqueness about the shape that leftist heroizing assumed. The Soviet Union enjoyed no monopoly on sky-scrapers or giants. When they got to America at the end of the decade, Auden and Britten had no trouble scouting out a sky-scraper sized American hero, 'the outsize lumberman' Paul Bunyan, for their 1941 operatic collaboration of that name, their joint celebration, in effect, of their newly acquired capitalist haven

> He grew so fast, by the time he was eight,
> He was tall as the Empire State.
>
> The length of his stride's a historical fact;
> 3.7 miles to be exact.
>
> When he wanted a snapshot to send to his friends,
> They found they had to use a telephoto lens.

And whatever the version, in whatever nation, dictators all came big. The unappetizing dictator-hero of Ruthven Todd's poem 'Dictator' isn't assigned a party, he's simply 'the tall man' from the hill country (a 'Napoleon', his shadow is a 'romantic size' and his 'brittle eyes could well outstare the eagle').[112] Herbert Read prefaced his attack on 'The Cult of Leadership', in *The Politics of the Unpolitical* (1943), with a declaration by William James against bigness as such: 'I am against bigness and greatness in all their forms.' And wisely so. Big Brother's face on that wall-poster in Orwell's *Nineteen Eighty-Four* is 'enormous, more than a metre wide'. He's the apotheosis of those fears of George Bowling in *Coming up for Air* about the coming totalitarian future with the 'posters with enormous faces'. Winston Smith 'looked up . . . at the portrait of Big Brother. The Colossus that bestrode the world! The rock against which the hordes of Asia dashed themselves in vain!' And Big Brother was perturbingly big in all of his manifestations. As far as oppressive sizings went, there was little to choose between the giant leader figures of Left or Right. At the 1937 Paris Exhibition the art works, especially the sculptures, of the German and Soviet pavilions were practically interchangeable in their heroic gigantizing: as E. M. Forster observed, 'the art-stuff on the walls' of the Soviet Pavilion 'might just as well hang on the walls of the German Pavilion opposite'. According to MacNeice's *Spectator* report 'Today in Barcelona', written right at the end of Spanish War, Franco was nothing less than a 'Goliath', threatening the little David of a reduced Republican enclave. 'For me', said Geoffrey Grigson in *Authors Takes Sides on the Spanish War* , 'Hitler, Mussolini and Franco are man-eating mass-giants.' He added that they 'issu[ed] from mediocrity and obscenity'. This was the kind of reflection Wyndham Lewis resorted to when he eventually tried cutting his quondam hero down to size in his book *The Hitler Cult* (1939). Hitler's Wagnerian 'cult of the Kolossal'—Wagner's music is 'mob-music', 'barbaric mass-music'—leads him to assume the 'vast face of the *Massenmensch*', but inside the Great Dictator there's really only a small Charlie Chaplin figure; a little man blown up, to be sure, by dint of floodlights and megaphonic amplification, but in actuality a weed and a food-crank (he 'would discuss a tomato salad, sip a glass of Horlick's, and nibble a rusk'), who cried out to be mothered by 'big masculine chaps, like Major Yeats-Brown'. But this was late denigration: in his *Hitler* (1931), Wyndham Lewis had been pleased to dwell on a monster-meeting at the Berlin Sport-Palast, a 'gigantic assembly of twenty-thousand people', exerting something like 'the physical presence of one immense, indignant thought'. He'd been pleased enough then to speak up for the Nazis' 'hefty young street-fighting warriors'.

Lewis's relish in the early '30s for the impressive toughness and bigness of Fascists and Fascism was only the response that Fascists everywhere sought to cultivate. Oswald Mosley's *British Union Quarterly* rhapsodized in its first number (1937) over Hitler's precipitously mountainous home at 'Berchtesgaden: An Idyll of New Germany'. 'The house dominates the valley as the Führer dominates the town.' Henry Williamson responded with enthusiasm in his *Goodbye West Country* (also 1937) to the multiplied bignesses manifested at one of the Nürnberg Rallies (or Party Days): the banners 'each 200 ft high and about 80 wide, great red roller-blinds, with swastikas'. Mussolini prided himself on his sunlit, shirt-less skiing. He had himself photographed heroically posed in the saddles of horses and motorbikes and in the cockpits of aeroplanes. He was widely bruited as being physically hardened and fit. So was Sir Oswald Mosley. *Oswald Mosley: Portrait of a Leader,* by A. K. Chesterton (1936), offers the heroic Olympic swordsman, soldier, airman, fist-fighter ('a young giant' by the age of fourteen), the 'virile' leader with the 'iron resolve', as the one who can stem Britain's enfeebled modern decline. His mass fascist movement, proceeding by means of huge rallies like the notorious one held at London's Olympia in 1934 (the 'mightiest indoor rally ever held in Britain'), would, Chesterton thought, get rid of 'unheroic' pacifists, 'dwarfish' homosexuals ('she-men' and 'he-women'), and banish a modern poetry that's not 'the language of heroes'. Henry Williamson was still using this heroic language of scale for his fascist leader Birkin, i.e. Mosley, in his novel of 1965, *The Phoenix Generation*. In 1933 Williamson dedicated a specially bound copy of his book *The Gold Falcon* to T. E. Lawrence, 'by whose taken thought the author added a cubit to his stature'. Mosley himself expanded continually on the 'great things' he expected from a 'great people':

> For this shall be the epic generation which scales again the heights of Time and History to see once more the immortal lights—the lights of sacrifice and high endeavour summoning through ordeal the soul of humanity to the sublime and eternal. The alternatives of our age are heroism or oblivion. There are no lesser paths in the history of Great Nations.

This was the sort of sloganeering the conscious imitators of Hitler's Aryan cult built themselves up on. They too might achieve the heroic physique of the athletes Leni Riefenstahl celebrated in her two-part film of the 1936 Berlin Olympic Games, *Olympia* (1938): especially the unclothed athletes, German remoulds of Greek originals, with whom the movie begins. Leni Riefenstahl herself appeared in these opening sequences as an emblem and apotheosis of naked uplifted Aryan physical beauty of the sort she'd time and again demonstrated as star of Arnold Fanck's ciné-dramas of snow and ice, pole and mountain, ski and aeroplane, and also of her own mountain film *The Blue Light*. It was a career celebrated in her lavishly illustrated *Kampf im Schnee und Eis* (1933), a book dense with shots from below, helmeted airmen and scenes for a real-life *Ascent of F6*. The combination was why Hitler chose her. He craved the heroic film treatments she specialized in. He knew how easy it would be for Riefenstahl's blond Germanic Olympians (though even she could not avoid showing the multi-medal winning American negro Jesse Owens) to be transformed into a conquering army without even breaking ranks. And, of course, no doting on the Truly Strong—however ostensibly un-fascist the doter—could escape these sorts of dubious association.[113]

The Left did put in claims upon the sun as its particular prerogative. Rex Warner's 'Hymn' (1933) advises the young that they 'must lie in the sun and walk erect and proud, / Learn diving': 'Light has been let in'; 'come out into the sun'. 'Lean down, sun', urged Randall Swingler in the very first number of *LR* (October 1934):

> Help us descry
> The fact,
> And knowing, fight not faint.

In *LR*'s fourth issue (January 1935) John Lehmann praises a new Soviet sanatorium: 'They have come from all corners of the enormous continent, the many republics, to be healed here . . . in the sun.' In *Prometheus and the Bolsheviks* Lehmann was pleased to discover in the Soviet Union 'deeply tanned young men and boys of a startling, wild beauty'. They proved for him that the taste for bronzed boys stimulated under Germanic and capitalist suns would be best met where the sun-tan had been democratized:

> Health, sex and sun—*Kurorts* [health resorts] for the people: such an idea was unimaginable in the old days, and its realization under the Soviets is one of the most positive and persuasive features of the Revolution. No one can resist the feeling of optimism they give, the impression of nature and science directly employed in the service of general human happiness. Nice and San Remo, not for the few who are rich enough to pay journey and hotel fees, but for all who work and who really need them.

And the efforts the Nazis made to end the German nudist movement, coupled with Wyndham Lewis's repeated rightist aggression towards *Naktkultur* as a symptom of the *négritude* that both he and Dr Goebbels abhorred, might have appeared to endorse Lehmann's leftist coopting of the sun. Lewis complained that in France, in August, Europeans were degenerating into a primitive savagery: 'burned to a negro mahogany . . . a veritable primitive negro community'. 'One Way Song' (1933) chimed in with that Lewis jeer from *Doom of Youth*:

> At the crisis of his boredom he confessed
> That he worked best without his pants and vest—
> And had often in the labyrinth at Antibes
> Lain sun-cooked side by side with other sheep,
> To make himself of Whiteness antipathetic,
> And meet the wishes of a kohl-lidded sapphic.

These are the tones too of *Paleface* (1929), Lewis's polemic against D. H. Lawrence and other sun-worshippers who want to turn civilized whitemen into swarthy negroes, sunburned dancers to the tune of negro music, that is jazz, 'the true hot Black stuff'.

But just as Auden's resolutely pale face, shut in behind his tightly drawn daytime curtains, didn't shut him and his writing off from the heroic cult of which the sun was an important part, so Lewis's rightist abhorrence of sunburn could not by itself rinse the sun-cult clean of its worrying German dimensions. The English varieties of nudism were consciously Germanic imports. The pioneer British sunbathers looked enviously towards Germany, in their magazines, and in books like John Langdon-Davies's *The Future of Nakedness* (1929)—full of praise for the German progress in 'sun bathing and open air nakedness in the sunlight'. One of Britain's pioneer nudist colonies was actually called *Spielplatz*. And when Germany turned Nazi its sun-cult rhetoric was absorbed into Nazi rhetoric. At midnight one 20 June *Sonnenwende*, Sun-Festival, Gregor Ziemer heard a Nazi youth leader exhort his boys: 'At this hour when the earth is consecrating itself to the sun, we have only one thought. Our sun is Adolf Hitler. We, too, consecrate our lives to the Sun, Adolf Hitler' (*Education for Death*, 1942). In *A Solitary War* (1966) the still loyal Mosleyite Henry Williamson dwelt on the way Hitler's work programmes had transformed pale and hopeless slum-dwellers into people with clear eyes, 'broad and easy rhythm', and a sun-tan. Williamson in fact continually assimilated his fascistic yearnings to his personal cult of the sun as the natural domain of the First War's survivor and of heroic leader types like Birkin/Mosley of *The Phoenix Generation*: Birkin 'the prophet . . . to lead his people back to sun'. The name of this hero, taken from D. H. Lawrence; his association with the D. H. Lawrentian emblem of the phoenix; the title of the sequence of Williamson novels in which he features, *A Chronicle of Ancient Sunlight*, all bear witness to the intensity of Williamson's cultic fascination for the sun. In imitation of Richard Jefferies Williamson linked the sun with 'a great dandelion' and in *Dandelion Days* (1930) he has a hero growing up in sunshine, and among enthusiasm for birds, into a life imaged as a growing dandelion. As 'man and writer I would like to be as the sun . . . to draw all life to oneself': so Williamson, rhapsodizing in the Preface to the US edition of *The Labouring Life* (1932). He was repeatedly saying such things. Unrepentant Fascism and sun worship went in his case unabatedly hand in hand. *Dandelion Days* is succeeded by *The Gold Falcon* (1933), as it were, and that by *As the Sun Shines* (1941) and that by *The Sun in the Sands* (1945), and so on.

Isherwood devotes a whole section of *Goodbye to Berlin* to the Baltic holiday resort Rügen Island, a place for enjoying sex in the sun with tanned German boys. What we now know is that the German code-name for Spain, the cover word for Hitler's military intervention in the Spanish Civil War, was 'Rügen Island'. Some cynic had spotted that Spain was another kind of holiday in the sun where the Aryan boys could put their new physical strengths to violent use. Like D. H. Lawrence's pagan cult of the sun, the German *Naktkultur* easily turned its nasty and brutish side uppermost. And what was true for Germany was applicable also in Japan's case, the land of the rising sun whose flag reminded '30s observers of a spot of blood and that became the place of the *Menacing Sun*—as Mona Gardner's title of 1939 had it.

Of course the ambivalences liberally scattered hereabouts were not lost upon '30s writers. Day Lewis, in his pre-communist polemic *A Hope for Poetry*, openly admitted them: the English 'revolutionary' poets, he said,

> are apt to produce work which makes the neutral reader wonder whether it is aimed to win him for the communist or the fascist state. Here again the influence of D. H. Lawrence assists to confuse the issue. We find, for instance, in Auden's preoccupation with the search for 'the truly strong man', Lawrence's evangel of spiritual submission to the great individual . . . And though this does not necessarily contradict communist theory, it is likely in practice to give a fascist rather than a communist tone to poetry.

Graham Greene was equally candid about the boy scoutery that Isherwood appeared to find so exciting in the New Germany. In March 1936 Greene found himself preferring *Merlusse*, a film about grubby French boys, to *The Day of the Great Adventure*, a Polish exercise in nordic Baden Powell-ism. The Boy Scouts' 'handsome brutish faces', their Aryan righteousness and flag-waving purity, were just Hitlerite, Greene thought: no wonder their film won prizes in Mussolini's Venice and Stalin's Moscow. In a *Left Review* piece entitled 'Heroism? Adventure? Glory?' (November 1934) T. Lincoln debunked Peter Fleming's heroic adventurism as ur-fascist: Fleming's writing showed how 'the British ruling-class' could indeed 'find, in their own ranks and in the middle classes, "troop leaders" as callously brutal as those who serve Hitler'. Eventually, it struck even Spender, doter on German male flesh though he was, that 'after all there was something exclusive about' a Germany 'which included the bronzed, the athletic, the good-looking and the smart, but shut out the old, the intellectual and the ugly'. By 1938 it's only the fascist, Jew-killing, nigger-baiting (i.e. Wyndham Lewis-like) Blacks in Spender's *Trial of a Judge* who are claiming the sun for their element:

> let the nordic
> Sunhaired head be matched against cloud drifts
> And the whip hand crack the lightning
> Canine and eye teeth laugh in the sun's face.

Isherwood, too, was finally driven to grant that there was something amiss with communist Rudi and Uncle Peter and their fetishistic get-up of big boots, shorts, 'heroic semi-nudity', and the collection of boys' photos, taken from the new angle, 'tilted upwards, from beneath, so that they looked like epic giants'. When Hitler comes to power Herr Issyvoo thinks of 'poor Rudi': his 'make-believe, story-book game has become earnest; the Nazis will play it with him'. The allegation was an extension of the one Isherwood's narrating self very honestly, if not altogether disarmingly, directed at himself in *Lions and Shadows*, just after the chest-expander confession:

> It is so very easy . . . to sneer at all this homosexual romanticism. But the rulers of Fascist states do not sneer—they profoundly understand and make use of just these phantasies and longings. I wonder how, at this period, I should have reacted to the preaching of an English Fascist leader clever enough to serve up his 'message' in a suitably disguised and palatable form? He would have converted me, I think, inside half an hour.

Lions and Shadows came out only in March 1938; the second 'Berlin Diary' in which Rudi features didn't appear until 1939 (in *Goodbye to Berlin*.) Isherwood was painfully backward in making public the ambivalences of his heroizings. '30s homosexuals, strongly involved with well-sunned German boys, found voicing worries over airmen and mountaineers rather easier. In his *Left Review* piece 'Fable and Reportage' (November 1936), Spender openly suggested that Auden and Isherwood had not tried to conceal 'that their hero', mountaineer Ransom of *F6*, 'is a Fascist type'. And, of course, '30s heroics were most obviously ambivalent in their airborne versions.

John Sommerfield's novel *May Day* (1936) made a mild enough set of demurrals. It is laced together not least by headlines from a capitalist press preoccupied by a round-the-world air race: THOMPSON HAS FORCED LANDING; FORCED LANDING IN SIBERIA; WORLD FLIERS BEAT RECORDS; THOMPSON FLIES ON. Heroic flying on Sommerfield's model is a ploy of the bosses' newspapers to distract workers from their terrible oppressions. Just so, in the same novel, the police use their gyroplanes to survey and control the May Day demonstrations of the workers: these gyroplanes are seen as part of the modern apparatus of social repression. But there were more formidable considerations even than these. Perhaps the most famous '30s account of the World War's Royal Flying Corps, Cecil Lewis's *Sagittarius Rising* (1936) (nothing to do with Cecil Day Lewis, of course) praises the author's father for his anti-democratic opinions, his assaults on the unnaturalness of Socialism, his favouring of power for those with 'superiority of position'. These quasi-fascist rumblings prompted the *NS*'s London Diarist (25 July 1936) into the generalization that RAF pilots were inevitably 'men who wear the old school tie' and are 'extreme nationalists'. Almost every week, the Diarist observed, the editor of *The Aeroplane* was saying that the Air Force was just waiting for the order to 'bomb the Bolshies'. A fortnight later *The Aeroplane*'s editor C. G. Grey protested that he was merely on the side of civilized nations like Germany and Italy against 'Oriental barbarism'; but he got chapter and verse quoted at him about the RAF getting in 'some useful bombing practice and . . . a little air-fighting' against 'live targets', in the matter of the Metro-Vickers Trial. Mosley supporter C. G. Grey hadn't a leg to stand on; *The Aeroplane* was regularly jammed with his anti-semitic, pro-German rantings (and Grey is one of the ripest of nasty fascist exhibits in Richard Griffiths' superb *Fellow Travellers of the Right: British Enthusiasts for Nazi Germany 1933–39*). And on Grey's side of the political divide Cecil Lewis's thoughts were rather widely picked up as good fascist stuff. The reviewer of *Sagittarius Rising* in Mosley's revived, now overtly fascist paper, *Action*, liked them immensely. 'Blackbird', as *Action*'s speed and air correspondent signed himself, was himself a veteran of the Royal Flying Corps. He'd already busily established in his column links between fascist policies, the fascist countries, and air-speed records and the building of bigger and better aeroplanes—just as elsewhere in its pages *Action* (edited for a time in 1937 by the ex-RFC pilot and sub-editor of *The Aeroplane*, Geoffrey Dorman) tried hard to connect Fascism with physical fitness, especially at the Berlin Olympics. The paper's line was completely consistent. Germany stages the Olympics, with results especially cheering to Aryans once the negro presence has been discounted (Germans, disgusted with the Americans for sending negro competitors over, 'have asked' *Action*'s Special Correspondent 'what the world would have said if the Germans had trained special Jewish athletes to win them gold medals'); Italy and Germany produce record-breaking planes and racing-cars; the Hindenburg air-ship 'is tremendous'; the 'Flying Hamburger' train is 'a colossal achievement'. Meanwhile, Blackbird sneered, a 'brace of Comrades' who tried to fly a stolen plane from Portsmouth aerodrome to Spain 'promptly crashed' because they were too 'extraordinarily stupid'. Even if 'they had taken off by a miracle without accident I imagine it would be far more likely that they would have arrived in Scotland than in Spain'.

And Blackbird's implication that Fascism is the aptest political home for airmen was not short of justification. Both Robert Skidelsky (in his *Oswald Mosley*) and Richard Griffiths (in *Fellow Travellers of the Right*) have demonstrated the extensive links between Fascism in Britain and the world of aviation. The link was not confined to Britain. The ranks Blackbird flew in included Gabriele d'Annunzio ('Poet, hero, and cad', Forster called him) and Charles Lindbergh as well as Henry Williamson, Oswald Mosley and Cecil Lewis. Instructively, airman fictionist Nevil Shute's thriller *So Disdained* (1928) is about a traitorous English flyer in Soviet pay who is prevented from handing his aerial spying photographs to Russian agents by the combined forces of the British fist and the much admired 'straight' young men of Italian Fascism: Shute was out to rescue any misguided airman from pro-Soviet sympathies. And flying was an activity that paid-up Fascists found it only natural to go in for. The British Union of Fascists started up its own flying clubs in 1934. Mosley promised 'every suitable boy' air training. He favoured 'the creation of a British Air Force second to none in the world'. It was a vision and a programme writers were not all immunized against.

Francis Stuart, the Irish writer and man of action, who was extremely keen on cars, boxing, horse-racing, shooting and who later spent the Second World War in Germany collaborating in radio work with the Nazi powers, learned to fly and celebrated the air and flying in a number of novels—*Pigeon Irish* and *The Coloured Dome* (both 1932), *Try the Sky* (1933)—and in *Things to Live for: Notes for an Autobiography* (1934), whose eighth chapter is actually entitled 'Night Flight'. The cult of the flier is at the centre of Henry Williamson's self-projection as a Fascist and a literary man of action. A veteran soldier of the First World War identified closely with the RFC—as *A Soldier's Diary of the Great War* and other texts reveal—he remained a speed-merchant after Blackbird's own heart. He raced at Brooklands, and drove motorbikes and fast cars on the public highways. His *Goodbye West Country*, for instance, is packed with tales of his speedings about England and

Germany in his open roadster, the Silver Eagle. Like other man of action of the '30s—like T. E. Lawrence, airman and motorcyclist, or the helmeted airman and climber Gorbounov of *The Ascent of Mount Stalin*, or Michael of Ruthven Todd's *Over the Mountain* (who climbs in 'a leather motorcyclist's jacket', 'a motor-cyclist's leather helmet', 'fleecy-lined gauntlets' and goggles)—Williamson keeps revealing the close fit between the various branches of heroic pursuit. In *Sun in the Sands*, for instance, his autobiographical volume written in 1934 (published in 1947), he describes 'whacking' along the Great West Road on his motorbike in 'my old flying helmet, new ill-fitting goggles . . . ancient flying leather-coat, my field-boots and yellow breeches of cavalry twill'. Flyer, motor-cyclist, climber, soldier: the kit was interchangeable. And as far as Williamson is concerned, the helmeted airman/ climber/motorcyclist is a Fascist. For him, Hitler, the German airmen flying over Nürnberg, and Williamson's friend Aircraftsman Shaw, are all unarguably part of the same fascist force. In short, Williamson's vision of things was the classic fascist package: the heroic union of speed-merchants and tough guys craved by all fascist leaders and actually achieved, as Robert Skidelsky shows, by Mosley—who made much of the jockey, the (woman) speedway star, the boxer, in his Union. Sir Malcolm Campbell's motor-car *Bluebird* broke the world land-speed record sporting fascist colours: precisely the sort of demonstration Mosley needed.

T. E. Lawerence is adulated, all through Williamson's *Goodbye West Country*, as the Hitler-like leader that Britain needs. Indeed that book's admirations for Hitler are thoroughly blurred into Williamson's regard for T. E. Lawrence, 'our nearest approach to Hitler'. Nor is that the end of the fascistic tendencies crowding into this particular corner of the picture. Together, for example, T. E. Lawrence and Williamson enthused over the wartime flying novel of another RFC veteran V. M. Yeates, *Winged Victory* (1934). Lawrence thought it 'Admirable, wholly admirable: an imperishable pleasure'. Williamson actually encouraged Yeates, an old school-friend, into print and wrote an introduction to the 1934 reprint of *Winged Victory* as well as a new preface for the 1961 reissue of it. Yeates, a perennially sick man, died at the end of 1934 in a sanatorium, of disappointment over his lack of literary success combined with tuberculosis that had begun in war strain ('technically known as Flying Sickness D'), but not before he'd witnessed the death of Antony Knebworth, Cyril Connolly's original Permanent Adolescent, Catholic convert, Tory MP, enthusiast for Hitler's Germany and Mussolini's Italy, and the flyer who was killed in a plane crash at Hendon aerodrome in 1932 because of dutiful sub-fascist devotion to his Squadron Leader ('in duty bound' wrote his father and memorialist, 'was closely watching the leader in front of him' and so hit the ground). 'I am reminded of flying here', Yeates wrote to Williamson from the Colindale Hospital, 'as I am over against the Hendon aerodrome. I saw the misfortunate Knebworth go up in a cloud of smoke yestere'en'.

And the possibility of such apocalypses seems actually to have been what particu-larly enticed the fascistic speed merchants, even when—perhaps especially when— their own life was put vertiginously at risk along with the lives of others. Like every Hitlerite Samson, they would as readily smash up themselves as the world. The *Letters* of T. E. Lawrence (they appeared, edited by David Garnett, in 1938) and the passages of *The Mint* that thrill with the excitements of travelling at speed are also highly excited by the prospects of imminent suicidal death. 'Many men would take the death-sentence without a whimper to escape the life-sentence which fate carries in her other hand. When a plane shoots downward out of control, its crew cramp themselves fearfully into their seats for minutes like years, expecting the crash: but the smoothness of that long dive continues to their graves. Only for survivors is there an after-pain' (*The Mint*, I, 4). T. H. White's crypto-fascist *England Have My Bones* (1936), which elevates the killing of small creatures to the status of a *soi-disant* art, an 'ecstacy of creation', ends on a craving for violent endings. White crashes his Bentley and thinks of spinning aeroplanes: he would like his 'to be a violent death'. In fact, 'the best thing' for the suburbanized world he deplores might be just such an end, 'a new war, in which, as quickly as possible, two-thirds of the population might be wiped out'. Contemplating the effects of the London Blitz, the hero of Williamson's long *roman fleuve, A Chronicle of Ancient Sunlight*, and declared *alter ego* Phillip Maddison would be not unpleased with 'the catharsis of high explosive'.

Henry Williamson's suggested title for Yeates's book was 'A Test to Destruction'. That was the title he himself later used for the 1918 volume (published 1960) of *A Chronicle of Ancient Sunlight*. So it seems not unapt that T. E. Lawrence should have crashed to his death on his way back from sending Williamson a telegram that, apparently, agreed to discuss his suggestion that Lawrence meet Hitler in order to get Fascism back on to a track more acceptable to Britain and British Fascists like Williamson, or (as Williamson's book *Genius of Friendship: 'T. E. Lawrence'*, 1941, reports) that Lawrence unite ex-Service men in a campaign against war and Jews—'the old fearful thought of Europe (usury-based)'—which came to much the same thing. Joint instrument of personal and public apocalypse, the aeroplane was the machine that most precisely apotheosized Fascism's yearnings for dangerous dynamism and the test to destruction. Mussolini had inevitably to be passed off by his publicists as an airman—whether he could actually fly or not. The arrival of Hitler by air at the beginning of Leni Riefenstahl's Nürnberg Parteitag film *Triumph des Willens* (1935) is a masterpiece of heroic suspense. By such means Hitler was devotedly mythicized as the latest of history's triumphant *Übermenschen*.

For his part, W. B. Yeats—who greatly admired Francis Stuart's novels *The Coloured Dome* and *Pigeon Irish* (the specially flying one)—yearned explicitly in language derived from Friedrich Nietzsche for the style, status, and power of the *Übermensch*. The Captains of his Blueshirt, fascistic 'Three Songs to the Same Tune' (1934) are aristocratic, lofty rulers come out of their Great Houses and on to the streets, Musso-muscularly to 'hammer down' the disagreeable mob and restore fullness of Nietzschean supremacy and spirituality 'up there at the top'. Yeats's poems of 1936 and 1937, 'An Acre of Grass', 'What Then?', 'Beautiful Lofty Things', are evidently stimulated by his re-reading of Nietzsche in those years, and by his renewed meditations on Nietzsche's visions of lofting genius, 'We aeronauts of the intellect'—'whose minds appear to be but loosely linked to their character and temperament, like winged beings which easily separate themselves from them, and then rise far above them'. And even though the poem 'What Then?' is about the ultimately unsatisfying nature of poetic elevatedness, its tone manages, like the much earlier celebration of an Irish aristocrat from one of the Great Houses the poem 'An Irish Airman Foresees His Death', to invest the sense of flying's disasters, dangers, and failures—the risks of authority—with intensely romantic appeal.

Wyndham Lewis also admired the flying *Übermensch*. One is not surprised to find this master of the purple patch devoting one of his best prose splashes to Antoine de Saint-Exupéry and the pilots of the French Sahara postal service. Busily gleaning North African material for his travel book *Filibusters in Barbary* (1932) Hitler's most fervent English literary admirer had in Fez bumped into some film-people. They're a gang of Americans, Jews, cosmopolites who get themselves jeered at for failing in their ambition to be heroes—especially the producer, a Sheikh-faking he-man whose lieutenants, especially the chest-inflating male star, wear aertex vests, 'or some similar make of sport-suggestive gentleman's hose'. Lewis labels them all, sneeringly, 'midget people', 'a diminutive, pthisic, gutter-people', 'an important subdivision of the *Untermensch*', mere apers of the eccentricities and brutalities of 'the full-grown, "normal", master-people by whom they were surrounded'. But high above these *Untermenschen* and their kind soar Saint-Exupéry and his teams of flyers, the true *Übermenschen*, of course (though Lewis, to be fair, does not actually use that word), the 'novel race of the upper-airs', the 'Aristocracy of the Air', who have redefined life for those stuck on lower planes:

> . . . now, higher even than the mountains, we have to take into our conspectus that new, very solitary, not by any means numerous, people, who for all practical purposes live in those superior altitudes. So when we are speaking of the nomads of the Rio de Oro, the fact that there are *other* nomads higher up cannot be ignored. Indeed, if they were forgotten, they would at every moment make themselves felt, and disturb the symmetries of the scene, recorded however ably. De Saint-Exupéry has expressed this *air-status*, as it may be called, with great effectiveness—in his extraordinary book, *Vol de Nuit*.

These 'extraordinary Air Men', as Lewis calls them, heroes of an 'air-epic' such as Saint-Exupéry both writes about and acts out, live in the realm of the 'superman'. They are 'superhuman'. Lewis has in fact entered in this passage into a Nietzschean rhapsody. It was an efflatus that came enthusiastically endorsed by Saint-Exupéry's own reflections. 'Flying', he wrote, 'produces a new race of men.' It was endorsed too by Brian Howard, whose excited little story 'Today!' has Howard rising Nietzscheanly into the future and the sun ('I am the Sun, and I am Dionysos') in a silver aeroplane, leaving even 'Nietzsche' (the name of a speedy chauffeur-mechanic) far below.[114] It was also endorsed by the totalitarian Air Vice-Marshal who's the villain of Rex Warner's political allegory *The Aerodrome* (1941): his air-force reaches out to crush the messily human disorder of an English village, its pub, church, and Hall, in the utopian name of 'a quite different order from that of the mass of mankind', 'what we in this Force are in process of becoming, a new and more adequate race of men'. In such texts the worrying link between flying and totalitarian aggression is as clear as in RFC-veteran Ralph Bates's *The Olive Field* where captured revolutionaries are tortured by being swung from their genitals in the punishment known as the Trimotor, the three-engined flight.

Übermenschen: on Left and Right the heroic of bigness, of lofty superiority, had to do with power. The leader is naturally sited above the led. He's always bigger and loftier than they are. It may be because he's above them on his podium and has acquired a bigger voice than theirs by means of electric amplification—a set of assets carefully debunked by Amber Blanco White in her Left Book Club text *The New Propaganda* (1939). He might, like the Old Testament King Saul, actually be head and shoulders above all the rest of the world: 'The tall unwounded leader' of Auden's poem 'From scars where kestrels hover'. He might inhabit heights, like 'The leader' of that same Auden poem, up on the fells

> looking over
> Into the happy valley,
> Orchard and curving river,

and like Rex Warner's Air Vice-Marshal who turns out to have been a youthful mountaineer, a lover of the Audenic north, the 'difficult country' where those 'whose wrists enjoy the chafing leash / Can plunder high nests'. But though this kind of clambering about is elevated enough, and obviously tuned to the primitive feeling that power always lies in bigness and loftiness—the notion that's terrifyingly tapped in the horrifying surrealist night-attack by the rightist police forces in Rex Warner's *The Wild Goose Chase* who approach mounted on motorized colossi, thirty-feet high, whose necks and heads are made up of bits of countless human beings grafted together—it's an elevation less impressively flexible than that further ancient demand, that the powerful man be able to elevate himself at will.

In his marvellous and haunting *Masse und Macht* (1960)—translated as *Crowds and Power*—the novelist Elias Canetti points to the rising hydraulic throne of the Byzantine Emperors as the supreme device for manifesting supremacy with. In it the Emperors could rise and descend as they willed; they commanded the process of elevation, of becoming big, then small, then big again. It was a power Richard Hughes ascribed to T. E. Lawrence, the hero who, like Christ on the pinnacle of the Temple, mounted high, and then deliberately yielded to the temptation to cast himself down. Hughes's clear implication is that he would have risen again if death on his huge motorbike had not prematurely intervened to avert the process.[115] But, T. E. Lawrence aside, how could the modern hero get to rise, descend to earth, and rise again? By becoming a bird, of course, or a birdman:

> Now to be with you, elate, unshared,
> My kestrel joy, O hoverer in wind,
> Over the quarry furiously at rest
> Chaired on shoulders of shouting wind.

> Where's that unique one, wind and wing married,
> Aloft in contact of earth and ether . . .

That's Day Lewis, conscious once again of Gerard Manley Hopkins, and contemplating Auden (at the beginning of *The Magnetic Mountain*). Day Lewis goes on to exhort Auden to outfly all other rival occupiers of his air-space, sky-scrapers, power-house chimneys, fireworks, sky-writing planes, philosophers in 'captive balloons':

> Look west, Wystan, lone flyer, birdman, my bully boy!

> Gain altitude, Auden, then let the base beware!
> Migrate, chaste my kestrel, you need a change of air!

Birds are certainly prominent in the '30s. There are all those metaphorical birds in their textual cages; Hitler as eagle (Berchtesgaden was the *Adlernest*, the eagle's nest; Britain was to be defeated by the Luftwaffe's *Adlerangriff*, the eagle attack); Mosley's Blackbird; Sir Malcolm Campbell's various racing cars and speedboats named Bluebird; Henry Williamson's motorcar Silver Eagle; the *Rote Falken* (Red Falcons), Austria's leftist children's group celebrated by John Lehmann in *Down River: A Danubian Study* (1939); the German hikers known as *Wandervögel* or Wander-Birds; the hostile critic Sparrow whom Auden several times gibes at ('Sparrow fails to understand their grammar', 'Letter to Lord Byron', IV). Birds flew busily in and about the nature-lovers' titles, for instance, Henry Williamson's *The Lone Swallows* (1922), *The Peregrine's Saga* (1923) and *The Gold Falcon, or The Haggard of Love,* and Sylvia Townsend Warner's and Valentine Ackland's joint volume of poems *Whether a Dove or Seagull* (1934). In a poem notably full, even for Auden, of birds (an eagle, a robin, a dove, gulls, ravens), his 'Spain' had the volunteers for liberty clinging, in a notable misprint in the original version, 'like birds to the long expresses that lurch/Through the unjust lands'. The printer's turning of *burrs* into *birds* was apt enough to this ornithologically minded epoch. Several writers were actually ornithologists. Rex Warner was. His bird poems are not only his best verses, but they shade over into his fictions: 'Egyptian Kites' into *The Kite* (1936), No. 8 in Blackwell's Tales of Action; 'Wild Goose' (it appeared in *NV*, June 1935) into *The Wild Goose Chase*—whose hero George is also an ornithologist. Tom Harrisson, too, was an ornithologist. Geoffrey Grigson's critical venom embraced a fine line in bird metaphors. Wallace Stevens is a 'Stuffed Goldfinch'. Hugh MacDiarmid is no 'bulbul', only 'a moulting maundering chiff-chaff'. The lesser Audenists are, according to this bird-watching critic, 'minimal finches' ('Mr Auden and Mr Spender and the several young minimal finches who twitter behind them').

It is not, however, minimal finches, nor even ordinary blackbirds, let alone robins, doves, gulls, and ravens that really incite and excite Day Lewis and Auden; it's high-flying, swift-swooping birds of prey, kestrels, Hitlerian and Williamsonian eagles, birds reminiscent of the modern military aeroplane. It's the commanding vision of the 'hawk . . . or the helmeted airman' that Auden again and again seeks: what the Germans intended when they thought of their airforce as Eagles and called their squadrons in Spain the Condor Legion.

From on high, of course, you can command experience, in every sense. It's from above that the fascist boot stamps down on to the human face in Orwell's haunted vision of totalitarian power. Up, in a plane, you could, if you so wished, drop bombs on the helpless populations of Abyssinia, Barcelona, Madrid, Guernica, Shanghai, and so on, in order to terrify them into submitting to your authority. The aerial position was then natural and ideal for that recognizably totalitarian strain of giving orders, hectoring, advising, preaching sermons, being schoolmasterly that we noticed in the last chapter—especially if you wished not just to threaten but to terrorize your victims, to force them to yield to you, to cow them. From the uplifted position of the airman, the early Auden can be extremely dictatorial in tone, edgily imperative in mood. Take the very well-known Poem XXIX of *Poems* (1930):

> Consider this and in our time
> As the hawk sees it or the helmeted airman:
> The clouds rift suddenly—look there
> At cigarette-end smouldering on a border
> At the first garden party of the year.
> Pass on admire, admire the view of the massif
> Through plate-glass windows of the Sport Hotel;

Join there the insufficient units
Dangerous, easy, in furs, in uniform
And constellated at reserve tables
Supplied with feelings by an efficient band
Relayed elsewhere to farmers and their dogs
Sitting in kitchens in the stormy fens.

Consider, look, Pass on, admire, Join: we're told precisely what to see, where to go, what to do with ourselves. Auden is pushing us, and especially our eyes, our perceptions, about. We must look just where he points. He's full of demonstrative, pointing, deictic words. Consider *this*, look *there*, join *there*. What we're told to look at looks a bit dotty; it's a perverse rag-bag, the result of Auden's gluttony for the oddest concatenations of things. Auden likes making surveys like this, accumulating lists, fixing up groups of people and things. He loves grabbing this and that, here and there, like a greedy child or a crazed cleric amok at a church jumble sale. But then, we have to recall, dictators are capricious, and Auden is after all the one in the commandingly lofty position. And his vision is so convincingly precise. It cuts straight through the mess and blur of the chaotic post-war crisis: 'The clouds rift suddenly' at his say-so and he can see clearly and, if we follow his gaze, so might we. Auden's position of vision and authority subverts what he and Day Lewis, schooled by I. A. Richards's 'destructive element' footnote, had defined in *Oxford Poetry 1927* as 'the chaos of values which is the substance of our environment', a blurry scene in which 'no universalized system—political, religious or metaphysical—has been bequeathed to us'.

The abundance of the airman's definite articles—they're always deictic—is noticeable. They've often irritated readers of Auden, as of T. S. Eliot. (They came eventually to vex Auden himself: in the 1965 Foreword to his *Poems* he castigated his 'slovenly verbal habits' of the '30s, and labelled his 'addiction' to definite articles a 'disease'. 'German usage' he called them: he meant grammar, not ideology.) In Poem XXIX, though, the definite articles seem far less a merely gratuitous device, a language tick, than they seem in the texts of Dylan Thomas. Here the definiteness of the articles lays claim to, and appears to manifest, precision of vision. So do the adjectives. Auden works his adjectives hard here, as he always does at his best. Like definite articles they're a major component of the Audenesque. And since adjectives describe what and how things are and seem, a command over adjectives is a command over experience and over the knowledge of things. It signals epistemological and hermeneutical mastery. And so it's not accidental that in this example of the uplifted vision the adjectives arrest the reader by their extraordinary precision, their sharp grasp of processes caught thus in the flux of life by the poet (those past participles: *helmeted, constellated*). Auden's adjectives make impressive arbiters too. With them Auden informs us about the weather (*stormy* fens). He picks out for us a decent band (this one's *efficient*). He can measure sufficiencies as well as efficiencies (*insufficient* units). He can judge morals (*insufficient*) and character (*Dangerous, easy*) at a glance. Politics, character, topography: the panoptic sweep and the incisive depth of the analysis are dazzlingly persuasive. Auden *knows*. And he knows because airmen and hawk-men are in the position to know.

Is he, though, too casually knowing? Auden knew that his curiosity incited in him a frequently jokey acquisitiveness. In one of those marvellously self-revelatory metatextual moments of his which occasionally crop up in his writing ('A phrase goes packed with meaning like a van', for instance, or 'Rummaging into his living, the poet fetches / The images out that hurt and connect') Auden did connect joking and curiosity. The 'average' man should know, he argues in his Introduction to *The Poet's Tongue* (1935), 'that whenever . . . he makes a good joke he is creating poetry, that one of the motives behind poetry is curiosity, the wish to know . . .'. Auden's busy-ness about adjectives, of course, like his brisk cataloguing (in that same Introduction: 'Memorable speech then . . . Everything that we remember no matter how trivial: the mark on the wall, the joke at luncheon, word games, these, like the dance of a stoat or the raven's gamble, are equally the subject of poetry'), could sometimes get perversely cocky. Take those lines from one of the choruses of *The Dog Beneath the Skin*:

> From the accosting sickness and
> Love's fascinating biassed hand,
> The lovely grievance and the false address,
> From con-man and coiner protect and bless.

Accosting is wonderfully accosting. But how much meaning do *fascinating* and *lovely* bear? Semantically they've become etiolated, thin, nearly empty. So in one sense Auden is doing an important job with them, recovering blanked-out adjectives like these for poetry, reclaiming important pieces of vocabulary from the white margins of meaninglessness to which overuse has banished them. Many readers have felt the power of Auden's poem 'Lay your sleeping head' to reside in the flat adjective *human*: 'Human on my faithless arm'. But isn't there also a danger of adjectival hubris here? Is it not possible to believe, and mistakenly, that your power over adjectives is so great that you can make them do almost anything, can make a play with almost any adjective whatever if the fancy takes you? Occasionally one feels that Auden's airman power is indeed verging on hubris in his language of description. The con-man and coiner come to our mind, as they did to Auden's pen. His efforts to prove himself the master of description go too far. He overreaches himself in his endeavours to colonize the empire of adjectives, to take over entirely Edward Lear's land. 'And Children swarmed to him like settlers. He became a land', Auden wrote of Lear. And Auden was one of those settlers, for where else but in Lear's limericks were such adjectives to be found: '*impulsive* old person of Stroud', '*luminous* person of Barnes', '*incipient* old man at a casement', '*intrinsic* old man of Peru'? The airman learns from his nursery reading, the boyish allegiance; but often he seems to have learned too well. So that even a great poem like 'August for the people', so nonchalantly powerful in its management of the Audenesque's adjectival business ('effusive welcome of the pier'; 'complicated apparatus of amusement'; 'solitary vitality of tramps'; 'the tigerish blazer and the dove-like shoe'; 'The stuccoed suburb and expensive school'; 'To private joking in a panelled room'; 'strict and adult'; 'acid and austere') can't quite manage to fill up the challengingly emptied semantic compartments of '*tremendous* statements' and '*beautiful* loneliness'.

F. R. Leavis thought that Auden's knowingness eventually ran away with him. 'He has made a technique out of irresponsibility, and his most serious work exhibits a shameless opportunism in the passage from phrase to phrase and from item to item—the use of a kind of bluff. That poised knowledgeableness, that impressive command of the modern scene, points to the conditions in which his talent has lost itself.' Before one is in a position to agree to that wholeheartedly one must, I think, acknowledge the dazzling excitements of what Grigson praised as 'the fragments, the sensations, the brilliant summaries here and there in an image, the brilliant buffoonery': Auden is 'so fidgety and inquisitive, so interested in things and ideas, so human and generous'.[116] Curiously, Auden almost fits the bill of the omnicomprehensive hero as defined by Caudwell in his essay on T. E. Lawrence: 'The hero understands geography, war, politics, and cities, and new techniques are instrumental to him.' One must concede the continually serious side of a richly various deictic effort to reveal and describe Britain: 'We would show you at first an English village'; 'What is it you see?'; 'We show you man caught in the trap of his terror' (all in *The Dog Beneath*); 'Let the eye of the traveller consider this country and weep' (*F6*);

> North, north, north,
> To the country of the Clyde and the Firth of Forth.
> This is the night mail crossing the border.

And so on. And it's not just a matter of a political demonstration of the British plight, it's also a confident rhetoric of pointing out, of showing and telling, and inviting readers to see and consider and know, that sharply rebuts and reverses the epistemological doubts about man's ability to descry and describe that settled heavily about modernist fiction with the works of Henry James and Joseph Conrad.

The birdman poet, in fact, persuades us that he enjoys what Roland Barthes craved from the Eiffel Tower, the complete knowledge that T. E. Lawrence believed the aeroplane to have granted mankind ('Today we know the whole earth', he claimed in *Now and Then* in 1932), the lofty photographer's commanding sweep from up the tower or the high building that became fashionable at the end of the '20s (Moholy-Nagy's views From the Radio Tower, Berlin, for instance). 'It is hard to conceal the true state of a country from a flying-man': the airman John Grierson's reflections in *Through Russia By Air* (1933) on why the Russians were reluctant to let him overfly their land might serve as a telling epigraph for Auden's powerful airman's panopticism. No wonder he was much imitated, especially by anybody with a parallel or rival claim on real knowledge. The airliner in Spender's 1933 poem 'The Landscape Near an Aerodrome' grants its passengers new acquaintance with the industrial 'landscape of hysteria'—just as railway travel took Dickens's Mr Dombey into regions of urban horror in London and Birmingham previously quite new to him:

> Here where industry shows a fraying edge.
> Here they may see what is being done.

Privileged by the air-liner or no, however, Spender's descriptions never achieved Auden's precision. Characteristically of his methods of composition, he fiddled about with the adjectives at the end of this poem, hesitating over 'the charcoaled batteries / And imaged towers'. They became 'those batteries / And charcoaled towers', but in neither version did they struggle out of their adjectival inchoateness. Still, Spender was aware what he was straining for. In *Trial of a Judge* he turns the Communist Brother into a 'helmeted airman'. This dying Red refuses to be 'a hero'—'Is the eye heroic . . . Or is the mind heroic . . . Because it has travelled further North than explorers?' But he does claim the airman's vision of the future:

> As the helmeted airman regards
> Through the glazed focus of height
> The bistre silent city abandoned like a leaf
> With veins in microscopic detail beneath him,
> So from my towered pause of death,
> O sweet carrier of life, my riveted eye looks
> Thirty years forward when your child is grown.
> Imagine . . .

It was because Communists continually insisted that Marxism gave them unique knowledge, helped them to *realize* what *really* happened in concrete *reality*, that they were so eager to get into airman Auden's seat. 'I know', was Edgell Rickword's proud claim. The airman Malraux knew too, Rickword argued in *New Writing* (Autumn 1938), because he too was a Communist:

> contact such as Malraux had established with the people's movement brings the whole of the world situation into focus. For the first time in history we possess the whole sphere of human activity under our eyes. A crowd fired on in India, a leader of the unemployed arrested in England, the fall in the franc and the outcry about the Moscow trials, all this makes sense and is not just a chaotic jumble of brutality and idealism.

This confident loftiness of vision is even allowed, and in a manner still more obviously pilfered from the Audenesque, to a couple of vaticinatory seventeenth-century Levellers, lively precursors of twentieth-century Communists, in Jack Lindsay's historical novel *1649: A Novel of a Year* (1938). 'Has there ever been such a moment before in the world's history?', the Leveller Lockyer asks of the novel's central character Ralph Lydcot:

> They sat there, both tired out, sipping their wine; and the hugeness of the moment in which they were acting their part stole over them like a gigantic shadow, the wings of time . . . lifting them up into strange and rare regions; setting them on a giddy crag of vision from which they saw the mass movements of men suddenly coherent and understandable, a map of man, a landscape of time, perilous and engrossing, terrific as a burst of storm, yet clear as the printed page of a book.

It was a 'moment of uplifted clarity'. Not, to be sure, as convincing in Lindsay's or Rickword's communist version as in Auden's omniscient original. But the effort was the same in all cases. Not just to pose as the all-knowing, far-seeing heir of the uplifted Romantic vision, as the enactor of the dreams of a Wordsworth, a Tennyson, a Marx, a Nietzsche, a Hardy, a Yeats brought right up to date (what John Cowper Powys's Darnley rhapsodizes over in *Wolf Solent*, 1929, as 'To fly over land and sea till you realize the *roundness* of the earth'), to declare a comprehension of past, present and future, to lay hold commandingly on experience. But also to impose this vision on your readers. In other words, to dictate, to bully, to behave nastily—even if it was with the exuberantly surrealistic monstrousness of Auden's Airman:

> Secret catalysts introduced into the city reservoirs convert the entire drinking supply into tepid urine. Adulterated milk drawn by order of the military from consumptive

gentlewomen is only procurable by those who are fortunate enough to possess attractive daughters. The factories, structurally altered, reduce all raw products to an irritant filter-passing dust. Eyeballs of ravished virgins, black puddings made from the blood of the saints, sucking children already flyblown, are exposed for sale at famine prices. For those who desire an honourable release, typhoid lice, three in a box, price twopence, are peddled in the streets by starving corner boys . . . epidemics of lupus, halitosis, and superfluous hair.

Not everyone approved of bullying, even among Auden's chums. MacNeice's Grettir, in 'Eclogue from Iceland', deplores the transformation of the heroic outlaw, 'the doomed tough', 'the man of will and muscle', into a mere bully just because he's 'exalted' to 'the curule chair' of magisterial dignity. But there were plenty of bully-boys to applaud Auden's monstrousness and uphold the practice of authoritarian imposition of will and/or vision. Day Lewis, lesser poet, stationed at a lower echelon of the coterie, Mount Lenin to Auden's Mount Stalin, seems fully to have recognized and approved of the bullying's most malicious aspects: 'Look west, Wystan . . . my bully boy.' Just so, Isherwood offers no rebuttal, in his *Listener* appraisal of T. E. Lawrence, to 'the terrifying little despot who had once ordered two thousand prisoners to be executed to avenge the death of a mutilated Arab woman'. If this were heroism it was the heroism of Mussolini's son who found bombing Abyssinian horsemen from his air-borne vantage-point 'most entertaining'. Even Caudwell acknowledges that his true Leninistic hero shares with the charlatan crypto-fascist hero his 'power over men'. The lessons in violence so well absorbed by John Cornford were available in numerous classrooms it seems: so whether one was listening at a particular moment to Lenin or T. E. Lawrence or W. H. Auden the instructions and examples all came to much the same set of advices. Power over other people was the issue.

Power over people, exercised from a position of privileged vantage, was extremely widespread in '30s writing and art. The corollary of looking up was, of course, looking down. It wasn't confined to Auden and his coterie. Looking down from aeroplanes was its commonest manifestation—in Waugh's *Vile Bodies*, in Bowen's *To the North*, in travel books galore (in Greene's *Lawless Roads*, for instance). And from the airman's position people appear less than human. In the new photography of Moholy-Nagy, Rodchenko, Bayer, that adopts the 'looking down' angle—a viewpoint assiduously adopted by Humphrey Spender whose camera will look up at, say, the Tyne Bridge from below, and then look down from the upper level of the bridge at the lower one (1936)—the elevated camera characteristically sees streets empty of people, or sees people as dehumanized puppets, tokens, manipulable objects, dolls, midgets. And people so dehumanized are ripe for less than human consideration. This downward viewpoint is strikingly analogous to the privileged distance, the disconnected outsidership claimed (as we shall see) by Isherwood's camera-I, and to the aloofness of some of Edward Upward's narration. One of Humphrey Spender's photographs, 'Berlin, Lützoplatz 1933', has Isherwood looking down on the Berlin street outside his apartment window. It was put with some aptness on the cover of the first edition of *Goodbye to Berlin*.[117] The attitudes cognate with the viewpoint arguably infiltrate the whole philosophy and practice of the documentary mode as such, as well as offering some sort of key to the willingness to do dehumanizing deeds in the name of historical knowledge and certainty that tarnishes the history of

Marxist–Leninist politics. It's the godlike position of Hardy's Immortals and of Cowper Powys's aeroplaning vision that makes for unpleasant thoughts on the part of those authors and uncomfortable reading for morally scrupulous readers. Mountaineer Michael Roberts tried hard to make it sound cosy and human. In his Preface to *New Country* he sought to illustrate what he intended by 'Social communism' with a mountaineering anecdote about a dozen males, schoolboys and undergraduates, climbing in the Jura in bad weather, who came to accept each other as a group:

> I don't think I had any love or personal feeling for them at all: we were, for the moment, part of something a little bigger than ourselves. Impatience and fatigue and personal delight and suffering disappeared, and I remember only, at the end of each day's work, standing at nightfall on the last spur of the ridge, counting the tiny figures moving down the slope in sight of food and warmth again: nine, ten, eleven, black dots against the snow, and knowing that again the party was complete, uninjured, tired and content.

But it's obvious that among less humane leaders looking down on people reduced to small dots, the situation and the tone could speedily get less genial. Michael the Mountaineer in fact speculates in Ruthven Todd's *Over the Mountain* about the link between mountaineers' uplifted visions and Fascism:

> I wondered idly if mountains had anything to do with Colonel Roscoe's being head of the secret police. Perhaps, I thought, he had looked down from some rocky gable and had decided that he was nearly God, looking at the people like ants, and had become possessed of a desire to rule their lives . . . Anyhow, it was a damned queer country.

A damned queer country: the words recall the title of a 1934 Grigson prose-poem, 'A Queer Country', about a curiously Nazified Iceland, full of blonde people and hatred, where still-born infants are 'fed' to 'The geysir'. A country as queer as the Audenesque perhaps.[118] And that aerial terrorism is implicit in the verbal weaponry of the Audenesque is the strong suggestion of Laurence Whistler's poem 'Flight', which nicely displays the close fit between the Audenesque's tones and the activities of birds acting the part of fighter planes:

> By day, the returning terror of swifts, the scream
> Of the loop over leaf, the power-dive over the thatch.[119]

So much so, that even Auden, the Airman himself, came in the end worriedly to devalue the airman's role. But not completely until some time after the decisive experience of Spain. In the event, Auden's 'Spain' (1937), was to be his last great panoptic effort as the uplifted observer. As the ultimate airman stunt—monstrously disconnected from the tormented Spanish earth below, arbitrating dictator-like in the matter of which murders are 'necessary' ones—this poem seems not only to have worried his contemporaries, people like Spender and Rickword and Orwell, but eventually Auden himself. His longer-term response to Spain, where he'd seen what airmen's and dictators' monstrous powers could in reality amount to, combined with his reaction to his readers' criticisms of 'Spain', clearly brought home to him the realization that his airman-fancying must stop.

In *On the Frontier,* written in 1937 after Auden's visit to Spain, the militaristic songs of the Westland students and the Ostnian Air Force cadets in Act II, sc. i, which are sung simultaneously, are also convergent in sentiment. They're like bad militaristic verse from any period, but especially like bad leftist poems written in the Spanish Republican cause. The Westland boys praise brave hearts and stout arms and celebrate 'the sun on our weapons . . . gleaming'. The monarchist Ostnians sing a song roughly of the Red Airman sort:

> Wheel the plane out from its shed,
> Though it prove my funeral bed!
> I'm so young. No matter, I
> Will save my country ere I die!
>
> . . .
>
> Far from Mother, far from crowds,
> I must fight among the clouds
> Where the searchlights mow the sky,
> I must fight and I must die!

Disillusions with airmen of all stripes couldn't be clearer. And it was confirmed in the journey to China, in 1938, where the sometime bullying airman poet found the distantly dictatorial, aloofly disconnected attentions of real airmen nothing but frighteningly monstrous:

> Engines bear them through the sky: they're free
> And isolated like the very rich;
> Remote like savants, they can only see
> The breathing city as a target which
>
> Requires their skill; will never see how flying
> Is the creation of ideas they hate . . .

Or again:

> Yes, we are going to suffer, now; the sky
> Throbs like a feverish forehead; pain is real;
> The groping searchlights suddenly reveal
> The little natures that will make us cry.

Little natures. This sonnet, No. XIV in a set of twenty-seven sonnets, and so the prominently central poem in the China sequence, evidently perturbed Auden. He took several shots at it, jumpily dodging about before he could bring himself finally to admit the depth of the airman's fall into moral meagreness: 'The little natures that can make us cry'; 'The little natures that can make cities cry'; 'The little natures that will make us cry'. At last the admission was out. Because the once loftily heroic airman not only *can* but in practice certainly *will* hurt anyone and everyone ('All women, Jews, the Rich, the Human Race': the human beings, including the poet, forced by aerial power into the little black dot position), he himself is now to be thought of as morally shrunken. It's a diminishment as startling as the one Spender's Napoleon—who is an emblem, probably, for Stalin and the poet's Communism—is observed to undergo in 'Napoleon in 1814' (in *The Still Centre*, 1939). Napoleon has been 'The Man of Destiny', envisaged in a plethora of heroic '30s images: he was a 'sun', a figure of 'superhuman' 'stature', a leader who 'stood / Upon your armies like a voyaging rock', with a 'great rhetoric', a phenomenal, legend-recovering firework:

> In you
> The Caesars tamed by dying, fired again
> Their lives in the unlegendary sky
> With all the vulgar violence of Today.

But Napoleon is eventually observed shrivelling, by the victims he's triumphed over. He shrinks as they grow again.

> The statesmen you had overthrown
> Sprouted again in their gold leaves
> And watched you shrivel back into a man.

Just so, Auden grows up as his airman-concept takes its moral tumble. He becomes an adult, serious at last about the exercise of totalitarian fire-power in a way that would have been quite beyond the more flashily naïf enthuser over military planes:

> All Siskins to be replaced by Bulldogs.
> 3 Short Gurnards for reconnaissance.
> 2 Vickers 163 for troop conveyance.
> 3 Gloster SS 19 for fast fighting.

He had at long last seen what might be the matter with T. E. Lawrence in a Bristol Fighter:

> A Bristol Fighter which flew overhead
> Swooped down as the pilot leaned out from his seat
> 'It's Lawrence of Arabia', somebody said
> And a typist tittered 'Isn't he sweet!'

But he wasn't at all sweet. 'His enthusiasm for mechanical dodges and gadgets', wrote Isherwood of T. E. Lawrence in his *Listener* piece, 'had an adolescent quality'. It was a quality of innocence that experience was now compelling Auden publicly to abandon.

Most of Auden's China sonnets were written in 1938. Five were published in December of that year in *The New Republic*. Most had to wait until 1939 before they appeared in *Journey to a War*. And 1938–9 is late in the decade. By then fears of aerial bombardment were commonplace, sponsored especially by what had taken place in Spain. John Langdon-Davies' *Air Raid* and J. B. S. Haldane's Left Book Club book *A.R.P.* both came out in 1938. Both men had been in Spain. The realization that England was as open to the sky as any bombed city in Europe or China inspired the panicky digging of air-raid trenches during the 'phoney war' period of September 1938 and the distribution of 38 million gas masks. It encouraged people to expect instant annihilation from the sky when war was finally declared in September 1939. 'What targets', someone remarks, 'what targets for a bomb', as they contemplate the crowd on the railway platform in Henry Green's *Party Going*. The noise of that crowd is 'like numbers of aeroplanes flying by', and so is full of immense menace. Even the once innocent-seeming sky-writing plane now contained explicit menace, especially to the agent in Graham Greene's *The Confidential Agent* (1939), who comes from a European country (it's very like Spain) and has seen what aeroplanes can do:

A pale winter sun shone, and the scarlet buses stood motionless all down Oxford Street: there was a traffic block. What a mark, he thought, for enemy planes. It was always about this time that they came over. But the sky was empty—or nearly empty. One winking glittering little plane turned and dived on the pale clear sky, drawing in little puffy clouds, a slogan: 'Keep Warm with Ovo'. He reached Bloomsbury . . .

So much for the affectionately viewed sky-writer over Bloomsbury in Virginia Woolf's *Mrs Dalloway* ('the strange high singing of some aeroplane overhead was what she loved') or for the poet 'writing / On heaven a new signature'.

These fears of the later '30s were, of course, developments of the fears that some observers had long cherished. As early as 1932 Graham Greene's Dr Czanner in *Stamboul Train* had belaboured old-fashioned ideas of national security: 'The aeroplane doesn't know a frontier.' 'The world will never be the same again', someone is reported as saying in Virginia Woolf's *The Years* (1937) upon hearing that the English Channel has been flown for the first time. And through the '30s many writers, good and bad (and not just the popular writers: *pace* Martin Ceadel's suggestion to that effect in his important 'Popular Fiction and the Next War' essay), lots of writers dwelt gloomily—and sometimes with relish—on coming aerial destruction.[120] At first, and with a strong mindfulness of First-War horrors, it was gas bombs that terrified and fascinated. The idea of an 'arsenic fog' that 'rots you into black filth if ever you breathe a drop of it' being dropped from enemy planes fed what almost amounts to a genre, the inter-war years Gas Bomb Novel (it included T. H. White's *Earth Stopped, or, Mr Marx's Sporting Tour,* 1934), especially after H. G. Wells's *The Shape of Things to Come* (1933) had granted this nightmare luridly definitive and heavyweight backing (mustard gas 'ate into the skin, inflamed the eyes: it turned the muscles into decaying tissue. It became a creeping disease of the body, enfeebling every function, choking, suffocating', and so on and on). A fate like this was far worse than merely being splashed with the blood of Huxley's dead dog. But worse even than gas was the high explosive that gradually emerged as the real threat to the city. Gas masks offered no defence against TNT. 'Only a little while', glooms Orwell's Comstock, 'before the aeroplanes come. Zoom-bang! A few tons of TNT to send our civilisation back to hell where it belongs.' That reflection came in 1936 and there is, of course, a great deal else beside bombers in *Keep the Aspidistra Flying.* By contrast, *Coming Up for Air* (1939), written after Orwell's Spanish experiences, has little in it, from first to last, that's not defined by the fear of impending bombers. By 1938–9 the bombing plane laden with high explosive challenged in most thinking people's minds the survival of all urban civilization.

The young mother in Lehmann's apocalyptic Introduction to *Down River* (October 1939) 'sees the swarms of deadly bombing planes suddenly gather overhead', threatening her child, her new flat and the electric light, the gas cooker, and the new health care of post-war Europe. 'Our language', sing the Fascists in Spender's *Trial of a Judge* (1938), 'Will be the bomber's drum on the sky's skin.' 'To Berlin! To Berlin!', exult the aerodrome loudspeakers as the chief Rightist conspirator leaves on the morning plane in Lehmann's *Evil Was Abroad* (1938). Spender has his Judge's fascistically hero-mongering wife exultant in the prospect of urban apocalypse made from the air:

> And the aeriel vultures fly
> Over the deserts which were cities.

In Section XXIII of *Autumn Journal* (1938) MacNeice hears 'the new valkyries ride':

> Droning over from Majorca
> To maim or blind or kill.

The din of this neo-Wagnerian music in the sky was simply deafening in the texts of 1938–9. Poems stirred by bombing-plane horrors simply poured out: William Plomer's 'The Japanese Invasion of China', Jacob Bronowski's 'Bomber', Spender's 'The Bombed Happiness', Herbert Read's 'Bombing Casualties', Grigson's 'The Bombers', George Barker's 'Elegy on Spain'.[121]

Barker's 'Elegy' is 'dedicated to the photograph of a child killed in an air raid on Barcelona':

> So close a moment that long open eye,
> Fly the flag low, and fold over those hands
> Cramped to a gun: gather the child's remains
> Staining the wall and cluttering the drains;
> Troop down the red to the black and the brown;
> Go homeward with tears to water the ground.

It's not surprising that Edwin Muir 'was very much moved, astonished (. . . as one is astonished by some beautiful natural spectacle) and impressed' by this poem. It was impossible not be moved and horrified by poems like F. L. Lucas's still more savage 'Proud Motherhood (Madrid, AD 1937)' (it appeared in *Poems for Spain*):

> Jose's an imp of three.
> Dolores' pride.
> 'One day', she dreamed, 'he'll be
> Known far and wide'.
>
> Kind providence fulfils
> Dolores' guess:
> Her darling's portrait thrills
> The foreign press.
>
> Though there's no wreath of bay
> About his hair:
> That's just the curious way
> Bomb splinters tear.

No wonder, either, that the perception of the airman as the agent of this 'indifferent death' (Forster's phrase in his 'T. E. Lawrence' essay), killed off the flying heroic of the '30s as surely as the flying-bomb, Dresden and the Atomic Bomb bitched (again in Forster's words, in a letter to Isherwood) the Second War's revived, Battle-of-Britain-and-After 'Romance of the Air—war's last beauty-parlour'. Rex Warner's *Aerodrome* and Orwell's Airstrip One (in *Nineteen Eighty-Four*) stand for a flying scene shorn of anything heroically admirable. 'This new mode of destruction, with aeroplanes', puts a stop, in George Buchanan's novel *Entanglement* (1938), to the old heroism of the long-distance flyer. And reluctant as Spender was in 1942 to deny the bravery of the Allied airmen (and of poets too: both 'require a bullet's eye of courage / To fly through this age'), his poem 'To Poets and Airmen' (in Tambimuttu's *Poetry in Wartime*) could not recover his lost sense of heroism in loftiness. Bombing, and poetry that supports it, are, he says, obscenities ('The expletive word') that will improve into the old exuberances only as slowly as the frozen smiles of the people who dig victims from bomb ruins will thaw out. Those smiles were 'frozen at the North Pole': the old stamping ground of Spender's 'truly great' is now as unpleasing as the airmen's element now is unheroic. As for John Lehmann, he came to reflect ironically on his own late intervention in the leftist-air heroic—that portion of a Soviet woman flyer's diary, 'An Airwoman Over Mayday' that he published in *NW* (Christmas 1939), accompanied by photos of Russian planes over Moscow and 'Soviet Air Women from the Caucasus'. It appeared, Lehmann later reminded himself, just as Soviet flyers were preparing to rain bombs on Poland.

So Auden's revisionary China sonnets appeared in a great cloud, indeed came rather at the end of a considerable queue, of airmen-damning texts. It took Auden, evidently, a long time to fight clear of a cult that he himself had done so much to initiate. Revealingly, he only got around to a gas-bomb fiction, and that a disconcertingly spry one, the poem 'James Honeyman'—about the specialist in toxic gases whose poisons are dropped by enemy bombers on his own Great West Road villa ('Oh kiss me, Mother, kiss me, / And tuck me up in bed / For Daddy's invention / Is going to choke me dead!')—in August 1937 when already the national gas-bomb obsession was fading and being displaced by the more realistic concern with high explosive. By August 1937 Auden had been back from Spain for some months, but he was slow in allowing himself to catch on about what airmen really had in their bomb-racks. For some others, at least, there wasn't Auden's problem of disengagement because they had never endorsed the Audenesque bigness-loftedness enthusiasm in the first place. We've noticed in passing the gibes and doubts of some writers. These objectors and critics fall noticeably into two broad categories: the jesters and the outsiders.

For funny men—and these tended to be more Right than Left—the components of the heroism business made just further occasions for merriment. Determinedly cuddly, and on the whole a-political, Betjeman chortles mockingly through the artsy-craftsy naturist voice he parodies in 'Group Life: Letchworth':

> Wouldn't it be jolly now,
> To take our Aertex panters off
> And have a jolly tumble in
> The jolly, jolly sun.[122]

But then we wouldn't expect the author of 'Slough'—

> Come, friendly bombs, and fall on Slough
> It isn't fit for humans now,
> There isn't grass to graze a cow
> Swarm over death—

to be too deeply engaged with the politics of the aeroplane.[123] Evelyn Waugh, too, was inclined early on to settle for the merely satiric possibilities (later in the period, of course, his Catholicism egging on his fascistic streak, he would become a fervent admirer of Mussolini and the heroic road-building engineers in Abyssinia: the kind of Italian who enjoyed bombing medieval warriors on horseback from multi-engined Capronis). Waugh set off, he tells us in his travel book *Labels* (1930) for Paris from Croydon aerodrome. He'd been 'up' before, he reveals (at Oxford an ex-RFC pilot

gave joy-rides in an old Avro biplane from Port Meadow: his looping the loop, at
fifteen shillings a time, had resulted, it was rumoured, in three cases of conversion to
Roman Catholicism). And Waugh (not yet a Catholic) cares little still for the airman's view of scenery (like 'a large scale map'). In fact, on the journey to Paris he's
'sick into the little brown paper bag provided for me': the paper bag that helped to
measure the nauseated debunking of an Audenesque landscape in *Vile Bodies* (published earlier in 1930 than *Labels*). In that novel, Nina and Ginger, a very dim
airman, fly off on their honeymoon:

> Ginger looked out of the aeroplane: 'I say, Nina' he shouted, 'when you were young did
> you ever have to learn a thing out of a poetry book about: "*This sceptre'd isle, this earth of
> majesty, this something or other Eden*"? D'you know what I mean?—"*this happy breed of
> men, this little world, this precious stone set in the silver sea . . .
>
> This blessed plot, this earth, this realm, this England
> This nurse, this teeming womb of royal kings
> Feared by their breed and famous by their birth . . .*"
> I forget how it goes on. Something about a stubborn Jew. But you know the thing I
> mean?'
> 'It comes in a play.'
> 'No, a blue poetry book.'
> 'I acted in it.'
> 'Well, they may have put it into a play since. It was in a blue poetry book when I
> learned it. Anyway, you know what I mean?'
> 'Yes, why?'
> 'Well, I mean to say, don't you feel somehow, up in the air like this and looking down
> and seeing everything underneath. I mean, don't you have a sort of feeling rather like
> that, if you see what I mean?'
> Nina looked down and saw inclined at an odd angle a horizon of straggling red
> suburb; arterial roads dotted with little cars; factories, some of them working, others
> empty and decaying; a disused canal; some distant hills sown with bungalows; wireless
> masts and overhead power cables; men and women were indiscernible except as tiny
> spots; they were marrying and shopping and making money and having children. The
> scene lurched and tilted again as the aeroplane struck a current of air.
> 'I think I'm going to be sick', said Nina.
> 'Poor little girl', said Ginger. 'That's what the paper bags are for.'

The paper bags feature again in Anthony Powell's *Agents and Patients* (1936) when,
in yet another instance of Powell's curious literary *pas de deux* with Waugh, hapless
Blore-Smith and contriving Chipchase also fly from Croydon aerodrome, that most
frequented of '30s literary places, to Paris, debunking the aerial view ('What are
those mauve patches on the water?' 'Cloud shadows.' 'Don't they look strange on the
blue water?' 'Like bruises on a body'), and running into bad weather:

> for a time [Blore-Smith] clutched a paper bag, wondering whether he would be able to
> hold out until they reached their destination. Chipchase had folded up *The Occult
> Review* and sat staring in front of him. His face looked like grey marble. He, too, toyed
> with a paper bag for some minutes but after a time he folded it up and returned it to its
> pocket on the wall. Several other passengers were less fortunate.

On this occasion Powell is simply taking opportunistic advantage of Waugh's established joke, just as his neo-Huxleyan satire *From A View To A Death* (1933) picked

up the contemporary phenomena of hikers and *Übermensch*—cravings to help sell a comic romp. But Waugh's case is less a matter of occasional mirth-making, rather it's indicative of a mirth-maker's congenital deficiency of analysis. Turning air-travel into a sick joke and a joke about sickness in *Vile Bodies*—and it's a novel eager to resist all restlessly speedy modes of modern living and transportation, including the car, motorbike, sea-liner, and air-ship; a fiction that even puts the Futurist T. P. Marinetti into a footnote as a kind of progenitor of its frenzied party world—is a graver matter insofar as it indicates a frivolity about the tendency of the new air technology to put people helplessly into the 'tiny dot' position. Waugh was signalling thus early his coming incapacity to see steady and whole the airborne fascist adventure in Abyssinia.

Not dissimilarly, P. G. Wodehouse's dictator Spode seems to have tickled his author so much that he could go on thinking (when he was overtaken by the German invasion of France in 1940) that he could simply joke his way around the issue of Nazism in the notorious broadcasts he made for his captors. The awful Spode, terror of *The Code of the Woosters* (1938) is very big ('About seven feet in height . . . as if nature had intended to make a gorilla, and had changed its mind at the last moment') and leads the Black Shorts movement:

> 'By the way [expostulates Bertie Wooster], when you say "shorts", you mean "shirts", of course.'
> 'No [says Gussie Fink-Nottle]. By the time Spode formed his association, there were no shirts left. He and his adherents wear black shorts.'
> 'Footer bags, you mean?'
> 'Yes.'
> 'How perfectly foul.'
> 'Yes.'
> 'Bare knees?'
> 'Bare knees.'
> 'Golly!'

Among the non-jokers in the '30s writing pack it was sustained liminality that steered some of them clear of the prevailing heroics. Graham Greene, tensed between Catholicism and socialism, calls the bluff of all heroically fantasizing types. Emptily boastful Hands in the unfinished 'The Other Side of the Border' (the fragment was eventually published in *Nineteen Stories*, 1947), a man eager to 'lead men' ('A leader's got to keep fit'; 'A man's sometimes kept—for the biggest things. Like Hitler'), and clearly being set up to be a yet hollower version of Conrad's Kurtz, is devastated when his talk of African treasure is believed and an actual and unwanted expedition is thrust upon him. Just so, young Tony Farrell in Greene's *The Bear Fell Free* (1935: a story only sold in a tiny limited edition) is taken up on his joke about flying the Atlantic and goes along with the old Battalion talk about 'the spirit of adventure' from his ex-RFC backer Carter ('he pictured himself in heroic situations, overwhelmed by ticker tape on Broadway'). But this is no way to recover a heroic smashed by the First War. 'It had been a bad joke', Farrell realizes, turning back in mid-Irish Channel. But it's a bad joke ending in a vividly grim crash:

> The plane met the water at a hundred and twenty miles an hour; the wheel smashed upwards at Farrell's head, screwed his neck, broke it without killing; the side of the cockpit caved in and cut through his skull to the brain, his knees were struck upwards, the broken bones jabbed through the broken neck.

Farrell's mother had seen him off in a plethora of First War names—she was like 'one of the innumerable generals in the peace procession. Look! There's Haig, Foch, French, Joffre, Allenby, Gough, Pétain, Sarrail, Plumer. A general with shattered nerves stumbling at the gate'—and guilty Carter, who commits suicide, weeps again as when he'd seen 'the dead face in the mud, under his boots. Wept when Conway stuck on the wire. Companionship of the trenches.' Unlike the Auden coterie's reluctant acceptance of the fall of the airborne, mountain-climbing hero at the end of his adventure, Greene's hero is fallen right from the start.

So, too, with Robert Graves. Exiled, the masked man apart, the most recessive, lonely and curmudgeonly poet of the period, Graves had personally lived through and seen through simplistic war-mongering heroism and was hardly likely to buy its revived airman counterparts, even if other old soldiers such as Henry Williamson or Wyndham Lewis could be persuaded to do so. Graves's poem 'The Clipped Stater (To Thomas Edward Shaw)' indicated how little he could believe in the efficacy of his friend T. E. Lawrence's disguising role as Aircraftsman. But Lawrence's dive downwards did apotheosize the movement which many of Graves's best poems actually celebrate. Nobody else in the '30s provided so many poems of dejection. 'Down', which has a man upstairs, sustained still by dreams of flying, of sun and towering, who is now 'sinking' in sickness, and condemned Icarus-fashion to 'falling', 'toppling', drowning and downing generally, dates from about 1920–1, and speaks out of Graves's deep despair and depressed vastation after the War. But the theme continues memorably and strongly in his '30s poems, in 'Song, Lift-Boy' (in which the narrator tells of rising in life from lift-boy to lift-man and finally cutting 'the cords of the lift and down we went', when he was threatened with damnation talk by Old Eagle); in 'Ogres and Pygmies' (about the modern declivities that have succeeded the old giant ages: 'The thundering text, the snivelling commentary'); in the related 'Being Tall' (in which the poet measures his height against his midget critics, but in the end accepts the 'smallness' of 'love'—a cryptic reference perhaps to how Laura Riding was, as a Blues singer might put it, 'bringing him all on down'); in 'To Walk on Hills' (in which elevated hill-walking and its associated Audenic panoptic vision of the moment—'A view of three shires and the sea!'—results in wearied legs, muteness, and 'head at last brought low'); and, sexily wry, in the erotolalic address to an eager phallus, 'Down, Wanton, Down!'

It's no surprise, given all these down-casting poems, to find Graves adding long-distance flights to the impressive list of things he was able crisply to 'see through'. Harold Dormer in *'Antigua, Penny, Puce'* crashes halfway on his record-breaking round-the-world flight. Later he takes his woman on a 'non-stop flight from San Francisco to Tierra del Fuego. They got killed, somewhere in the Andes, but they were together, and in love, and going at two hundred and fifty at the time, so the general effect was gay rather than gloomy.' Cynicism about woman and love-affairs combines breezily with cynicism about heroic flying.

For his part, Lewis Grassic Gibbon, a dyspeptically marginal Communist with few illusions about his Leftism (and in appearance an extraordinarily close look-alike, as photographs of him testify, of T. E. Lawrence), sweeps into the pervasively ironic world of his shambling anti-novel *Stained Radiance* (published under his own name of J. Leslie Mitchell in 1930) some airmen who stink ('The proletarirats smell bad this morning') and air-crashes devoid of all hint of heroism. A burning body in a crashed plane is just 'a blackened, rounded object, eyeless, faceless, moving slowly, lethargically, above a reddened carbonization'. Ruthless realism of this sort would have spared *The Orators* its rather surrealistic truck with the glory of the air-crash.

And, of course, there's Orwell, perpetual rogue elephant, who adds leftist fitness crankery to the fads about which he would be sturdily unillusioned even if they were pressed on him as the essence of progressive practice:

> there is the horrible—the really disquieting—prevalence of cranks wherever Socialists are gathered together. One sometimes gets the impression that the mere words 'Socialism' and 'Communism' draw towards them with magnetic force every fruit-juice drinker, nudist, sandal-wearer, sex-maniac, Quaker, 'Nature Cure' quack, pacifist, and feminist in England. One day this summer I was riding through Letchworth when the bus stopped and two dreadful-looking old men got on to it. They were both about sixty, both very short, pink, and chubby, and both hatless. One of them was obscenely bald, the other had long grey hair bobbed in the Lloyd George style. They were dressed in pistachio-coloured shirts and khaki shorts into which their huge bottoms were crammed so tightly that you could study every dimple. Their appearance created a mild stir of horror on top of the bus. The man next to me, a commercial traveller I should say, glanced at me, at them, and back again at me, and murmured 'Socialists', as who should say, 'Red Indians'. He was probably right—the ILP were holding their summer school at Letchworth.

'It would help enormously', *The Road to Wigan Pier* (1937) concluded, 'if the smell of crankishness which still clings to the Socialist movement could be dispelled. If only the sandals and the pistachio-coloured shirts could be put in a pile and burnt, and every vegetarian, teetotaller, and creeping Jesus sent home to Welwyn Garden City to do his yoga exercises quietly!' (For their part, by the way, many nudists resented the idea that nudism was for 'cranky artists and authors, self-expressionists and parlour bolsheviks'. They wanted 'people with jobs in the City . . . who like to wear the old school tie', said a George C. Foster in the Summer 1934 number of *Sunbathing Review*: 'Once let the impression get abroad that the nudist movement is a spiritual home for faddists, pacifists . . . and "Left-Wingers" generally, and the recreation of nudism is damned in this country and will deserve to be.')

Orwell was clear that the shorts-wearers were Peter Pans. To his horror, George Bowling in *Coming Up for Air* (1939), yet one more version of Orwell's ordinary bloke on the bus, discovers his old home infested by the shorts, sandals, nudist, vegetarian crowd (including Professor Woad: a gibe at hiking Professor Joad). Bowling meets an aged devotee ('something vaguely queer about his appearance'): 'He was wearing shorts and sandals and one of those celanese shirts open at the neck . . . I could see that he was one of those old men who've never grown up. They're always either health-food cranks or else they have something to do with the Boy Scouts—in either case they're great ones for Nature and the open air.' When a bomb drops by accident on Lower Binfield, it not only indicates Orwell's apocalyptic gloom at the end of a decade fearful with anticipations of TNT-to-come. It also enacts his novel's perception that the late childhood of the loony simple-lifers of Upper Binfield has been intimately intussuscepted into the now decisively soured childish heroic of the bombing plane.

There are several reasons for Auden's delayed assumption of responsibility in the matter of airmen and airman-poetics. Sexual preferences kept him and Isherwood,

Spender and Lehmann fudgy over what we might label the Rügen-Island syndrome. But more fundamentally, the lingering juvenility of which the homosexuality was arguably a part was much to blame. Biggles was kids' stuff. As huge proles looked heroic to undersized bourgeois authors, so airmen were heroes of the small reader. Auden's brightly gadget-prone enthusiasm for bombers in the early poems was an obvious carry-over from prep school. And small boys cannot perhaps be expected to spot the difference between a Bristol Fighter and the practical jokes with which Auden arms his Airman. But if Graham Greene could, shouldn't Auden have been able to? The dialogue of the film *Anne-Marie*, wrote Greene, 'jangles agreeably with gadgets' (its scenario was written by Saint-Exupéry), but its subject is 'the terrible games of men':

> The terrible games are the record flights: long distance, height, endurance, we have all heard of those, but there is no end to the crazy and childish ingenuity of these professional record-breakers.[124]

It will be recalled that another adult, Cyril Connolly, was encouraged in his Theory of Permanent Adolescence by the example of Antony Knebworth, and that Antony, type of all those permanently schoolboyish thugs and crypto-fascists in charge of England, was a fascist-sympathizing airman. And yet another adult observer, this time Virginia Woolf, in the first chapter of her volume of feminist reflections *Three Guineas*, uses Antony Knebworth—she doesn't name him, but it's clear that she means Knebworth by her talk of 'an airman'—to illustrate her strategical polemic against war as a vice of the male sex, the sex whose predominance in the business of modern mechanized warfare and also in the ranks of fascist politics had for a long time not escaped her attention.

And Auden himself wasn't so slow as sometimes might appear in growing up into adult perceptions of this sort—perceptions that weren't foreign, as we've seen, to his close friend and Iceland collaborator Louis MacNeice. It is, after all, the lunatics that Auden and Isherwood in *The Dog Beneath the Skin* (1935) have building 'a great big plane for our Leader'. In *The Ascent of F6* (1937), Mr A responds to news of Edward Lamp's death in the mountains with a rejection of the heroic in the recuperated tones—and amid recollections—of First War disillusionment:

> If you had seen a dead man, you would not
> Think it so beautiful to lie and rot;
> I've watched men writhing on the dug-out floor
> Cursing the land for which they went to war;
> The joker cut off half-way through his story,
> The coward blown involuntary to glory,
> The steel butt smashing at the eyes that beg,
> The stupid clutching at the shattered leg,
> The twitching scarecrows on the rusty wire;
> I've smelt Adonis stinking in the mire,
> The puddle stolid round his golden curls,
> Far from his precious mater and the girls;
> I've heard the gas-case gargle, green as grass,
> And, in the guns, Death's lasting animus.
> Do you think it would comfort Lamp to know
> The British Public mourns him so?

> I tell you, he'd give his rarest flower
> Merely to breathe for one more hour!
> What is this expedition? He has died
> To satisfy our smug suburban pride . . .

And feelings not unlike these had actually been enjoying some sort of covert life in Auden's work well before he went to Spain. As early as November 1933, only a year after *The Orators* was published, in the original version of the poem 'Here on the cropped grass of the narrow ridge I stand'—another one of Auden's sweepingly commanding surveys of Britain: 'Aloof as an admiral on the old rocks, / England below me'—Auden had turned guiltily away from the examples of both D. H. Lawrence and T. E. Lawrence, from the heroically worshippable body and the heroically lofty airman:

> Guilty, I look towards the Nottinghamshire mines
> Where one we quoted in the restaurants received
> His first perceptions of the human flame
> Smoky in us.
> We were to follow leaders; well, we have:
> The little runt with Chaplains and a stock
> Or the loony airman.
> We were to trust our instincts; and they come
> Like corrupt clergymen filthy from their holes
> Deformed and imbecile, randy to shed
> Real blood at last.

The poem ends with the poet as Wilfred Owen ('My will effective and my nerves in order'), openly referring to Owen (' "The poetry is in the pity", Wilfred said'), and overhearing other 'bones of the war' admitting the 'foolish' and 'farcical' elements of their sacrifice and discountenancing heroism as a mere pose:

> 'Unable to endure ourselves, we sought relief
> In the insouciance of the soldier, the heroic sexual pose
> Playing at fathers to impress the little ladies.'

But, for all such thoughts, Auden would go on playing for some while after this, trying to pretend that 'Real blood' was still just the old stage blood, a prep-school stage property. He was doing so, notably in his review of the Liddell Hart life of T. E. Lawrence and in his poem 'Spain', and arguably too in that run of cynically sprightly poems about violence and death written in the aftermath of Spain, 'Miss Gee', 'Victor', and 'James Honeyman': until his longer reflections on Spain and his China experiences made the airman finally, inescapably, only horrific and not in the least heroic.

Characteristically, Auden tried to keep having things both ways as long as possible. He wanted, in the same vein, to keep both T. E. Lawrence and Lenin as heroes of 'us, egotistical underlings'. Christopher Caudwell, intellectually much tougher, wouldn't have this. T. E. Lawrence, he argued, lacked Lenin's revolutionary consciousness and self-consciousness. The implication of Caudwell's 'T. E. Lawrence: A Study in Heroism'—and it shows in this, as in much else, the extent to which it intends a rebuking response to the Auden review of Liddell Hart's *Life*—Caudwell's clear implication is that the admirers of T. E. Lawrence are themselves short on consciousness and self-consciousness. One can only agree. It was a deficiency quite made up by the end of the decade: as events in Spain and China and elsewhere gradually cleared up the ambiguities of all that toying with the idea of heroism, the toughly active life, that had gone on in and around Auden; as the Hitler–Stalin Non-Aggression Treaty of August 1939 confirmed the monolithic likeness of those two gigantically totalitarian leaders; as what Spender called the Airman's 'dream of violence' realized itself in a long bloodletting replay of the First World War for which events in Spain and China were only curtain-raisers. 'The dead in wars are not heroes', Spender was driven to concede in his *New Statesman* article 'Heroes in Spain' (1 May 1937), 'they are freezing or rotting lumps of insanity'. The airman, like the idea of heroes in Flanders or Spain, had really turned out to be like a loony flying a machine built for him by those loonies in *The Dog Beneath the Skin*'s loony-bin. It was Spender's way of acknowledging that maybe after all Groddeck was right. 'Illness', Groddeck wrote in a 1925 essay on 'The Meaning of Illness' ('Der Sinn der Krankheit'), in words which we should clearly feel to underpin all Auden's sense of neurosis, but whose implications took a long time in the practical realizing, 'illness is the expression of the wish to be small'. You had, perhaps, to be ill to 'feel small' and to crave heroes on the '30s scale.

7

Going Over

It's the class-struggle in a final stage
That makes the intellectuals rage.

So Gavin Ewart in his characteristically cocky libretto for an Auden–Isherwood aping jazz opera, *The Village Dragon*. Leftist authors in the '30s were prone to justify and define their position by quoting that famous passage in Marx and Engels' *The Communist Manifesto*:

> Finally, as the class struggle nears its decisive stage, disintegration of the ruling class and the old order of society, becomes so active, so acute, that a small part of the ruling class breaks away to make common cause with the revolutionary class, the class which holds the future in its hands. Just as in former days, part of the nobility went over to the bourgeoisie, so now part of the bourgeoisie goes over to the proletariat. Especially does this happen in the case of the bourgeois ideologists, who have achieved theoretical understanding of the historical movement as a whole.

C. Day Lewis quotes the passage approvingly (these are Marx's 'decisive words') in his essay 'Writers and Morals' (the second part of 'Revolution in Writing'). A shortened version of it forms the epigraph to his *Noah and the Waters* (1936). The then Leftist A. L. Rowse, referring to the *Manifesto* as one of the great classics of political thought, quotes the passage in a 1929 *Criterion* essay on Communism.[125] It forms the epigraph to John Cornford's student article 'Left?'[126] Jim Wingfield, a rich and promiscuous Communist, justifies his Party membership at the end of Alec Brown's novel *Daughters of Albion* (1935) with a paraphrase of it ('Marx, you know, said . . .'). At the end of James Barke's novel *Major Operation*, Big Jock Mackelvie uses it in his funeral oration in praise of George Anderson, bourgeois recruit to Socialism, crushed to death by police horses defending Mackelvie's prone body from a like fate ('Comrades . . . in the Manifesto Marx and Engels have written . . .'). In *Illusion and Reality* Christopher Caudwell quotes the passage only to scathe over those who don't go on from there, those who merely drag their 'bourgeois consciousness' with them into the proletariat. Small wonder that in his Open Letter of 1938 to C. Day Lewis, 'The Proletariat and Poetry', Julian Bell should call it 'the rather hackneyed passage from the Communist Manifesto which prefaces your *Noah and the Waters*'.[127]

Going Over is the key metaphor here.[128] It's an intimate part of the widespread feeling among '30s authors of being travellers, on the road, making some literal or metaphorical journey (or both), of being involved in a pilgrimage to socialism and Moscow, it might be, or to Christ and the Church. In other words, the sense of being in transit or transition, on 'The Road These Times Must Take' (the title of Day Lewis's 'feeling small' sonnet), particularly into new poetic and political country. England was 'a transitional society' ('Sleep quietly, Marx and Freud, / The figureheads of our transition': MacNeice's *Autumn Journal*, XXIV). Leftists were particularly convinced about transitoriness. The crisis had generated a 'time of transition', in Arthur Calder-Marshall's phrase,[129] a transitional time that demanded an art of transition, the 'Transitional Poem' in fact (that was Day Lewis's title of 1929). Bourgeois writers would be judged according to how they coped with the problem of transition. Unable to make a 'destructive analysis and synthesis of bourgeois culture . . . a revolutionary task', Auden, Day Lewis and Spender had, snapped Caudwell, tried 'to skip this essential transition and therefore' fallen 'back into the dying world'.[130]

The problematic of the transition to the side of the worker is perhaps most vividly put by Edward Upward in his 1938 nouvelle, *Journey to the Border*.[131] In this story a tutor to a rich man's son decides, after a very Kafkan phantasmagoria of weird events at a country show, to 'get in touch with the workers' movement'. He discusses the move with his *alter ego*, who exhorts him:

> '. . . you will temporarily have to make a complete break with your former thoughts and feelings. You will have to move out of the region of thinking and feeling altogether, to cross over the frontier into effective action. For a short time you will be in an unfamiliar country. You will have taken your so-called "plunge in the dark"; but it will not be in the dark for very long. Out of action your thinking and your feeling will be born again. A new thinking and a new feeling.'

Capitalism offers 'no future for poetry or for anything worth while':

> 'Only the workers can save the things you value and love. All that is gentle, generous, lovely, innocent, free, they will fight to save. And in the end they will win. There will be a time of harshness and of bitter struggle, but out of it will come flowers; splendour and joy will come back to the world. And life will be better than it has ever been yet in the world's history.'

'How soon', comes the tutor's ready response, 'can I join the workers' movement?' It's the reaction of the lonely, bored, and frightened hero of Upward's *New Country* short story called 'Sunday' who decides to side with History against the bourgeoisie: 'He will look for history . . . in the places where those people are'.

Both men have readily absorbed their own author's polemic about the need to 'cross over the frontier into effective action', into 'unfamiliar country', the territory of the politicized workers, where political and literary theory are held to be one with Communist Party practice, in order to salvage any good writing from the crisis. As Upward notoriously put it in his 'Sketch for a Marxist Interpretation of Literature' in *The Mind in Chains*, the bourgeois author 'must change his practical life, must go over to the progressive side of the conflict, to the side whose practice is destined to be successful; not until he has done this will it be possible for his writing to give a true picture of the world'. It might be thought that the fate of the writer who fails to go over (especially since 'He will at best write something in the style of the later work of Lawrence or Joyce') is not so terrible after all, but Upward insisted contrariwise:

> A writer today who wishes to produce the best work that he is capable of producing, must first of all become a socialist in his practical life, must go over to the progressive side of the class conflict . . . unless he has in his everyday life taken the side of the workers, he cannot, no matter how talented he may be, write a good book, cannot tell the truth about reality.

Like other authors who became active in politics—like Auden, edgy about 'the expending of powers / On the flat ephemeral pamphlet and the boring meeting'; like

Edgell Rickword and John Cornford, whose poetry was squeezed aside by political activism; like C. Day Lewis and Stephen Spender who eventually felt themselves getting too caught up in committees and rallies for the good health of their writing— Upward had already experienced a problem over reconciling the demands of writing and political work. But he was still prepared to heap scorn upon those authors who hesitated on the brink of commitment. So were Rickword and Caudwell. Upward rebuked Virginia Woolf for suggesting an irresoluble conflict ('this does not mean that they would have had to spend all their time in committee meetings or in door-to-door canvassing or in composing propaganda leaflets. They could have taken part in ordinary political work and they could have written poems and novels as well').[132] But he also had to admit that politics might destructively claim all a poet's energies. This was to become later on the lengthily developed theme of his trilogy, *The Spiral Ascent.* In 1941 he granted that the 'younger writers who did become active undoubtedly found that they had less energy to spare for imaginative writing. Others who had been at one period strongly attracted to socialism were deterred from activity by this example'. In 1937, in *The Mind in Chains*, Upward was even franker: the writer

> ... is aware that [practical socialism] will involve him in extra work other than imaginative writing, and that this work will come upon him at a time when, having abandoned his former style of writing, he most needs to give all his energy to creating a new style. He is aware also that this work may in certain circumstances stop him writing altogether, that he may be required to sacrifice life itself in the cause of the workers ... He must be told frankly that joining the workers' movement does mean giving less time to imaginative writing, but that unless he join it his writing will become increasingly false, worthless as literature. Going over to socialism may prevent him, but failing to go over *must* prevent him from writing a good book.

Spender, always full of waverings and self-contradictions, recognized the dilemmas inherent in trying to proletarianize yourself at the same time as he was still straining to do it:

> If it were simply a matter of the poet or novelist who goes over to the working class, producing work that is as imbued in the life of the masses as, say, the Chapter in *Capital* called 'The Working Day' [voices from which Spender built into his poem 'A Footnote (from *Marx's Chapter on The Working Day*)', in *The Still Centre*], or an epic of the class-struggle in Spain, one could only say that here was a subject which for moral profundity, heroic magnificence, seriousness, made all other subjects seem either reflections of trivial fashions of the day or of a purely personal interest.
>
> Unfortunately, however, what is far too liable to happen is that the writer, overwhelmed by his new subject and environment, finds that he is disqualified by the weight of his past environment, for which he has now lost all respect, from writing: *e.g.,* Edgell Rickword. Poets ... join the communist party, they deliberately cut themselves off from the roots of their own sensibility, which derive from a life they have come to despise, and then they either stop writing or they produce stuff in which new and undigested material is imposed on a medium which was adapted to quite different material. The result is something effete, disappointing to the writer and to his comrades.[133]

Literary suicide, Spender calls it. But he still believed there were 'many reasons why writers should be communists today'. The risk had to be taken. And so Cyril

Connolly in *Enemies of Promise* quotes Upward's assertions from *The Mind in Chains*, questioning their hard logic and excessive Party-consciousness, but not discounting the importance of trying to cross the class frontier. You must still go over: 'I think a writer "goes over" when he has a moment of conviction that his future is bound up with that of the working classes.' What's more, many people have had this experience in Spain: crossing that literal frontier was (as we shall see), part and parcel of crossing the metaphorical class-border. Self-exile into Spain eased the transition into internalized, class exile away from the class of one's birth, family, and schooling.

Future tenses dominate the rhetoric of the going-over texts, as, to a classic extent, in Upward's 'Sunday':

> He will look for history . . . He will go back to his lodgings for lunch. He will read the newspaper, but not for more than a quarter of an hour. He will look out of the window . . . he will no longer be paralysed . . . He will go to the small club behind the Geisha Café. He will ask whether there is a meeting tonight . . . He will have to prove himself . . . It will take time . . . He will at least have made a start.

And for anyone to imitate the hero of Upward's 'Sunday', to make that Kierkegaardian leap of faith into an alien class and class-alienation, was to stake everything on the Communist Party's optimistic future tenses being right. 'Be the future as it might be, and no doubt that complete success was distant still . . .': the end of Warner's *The Wild Goose Chase* can't help sounding cagey and worried amid its expressions of revolutionary satisfaction. Upward's ' "plunge in the dark" ' deterred and frightened as well as invited and excited. No wonder Alick West argued in his critical book *Crisis and Criticism* (1937) that being challenged to go over made a peculiarly serious crisis for the bourgeois, particularly in his aspect as bourgeois individual, as the privately shut-in, lonely, and egotistic individualist, the aspect of selfhood most characteristic of the writer since the Renaissance:

> The difficulty of the change is the resistance of bourgeois habits. While the development intensifies the desire to abandon individualism for a consciously social life, it also intensifies the conflict as to how that desire is to be satisfied. There is a desire to feel, think and say 'we', instead of 'I'; but who are the 'we'? Bourgeoisie or workers?

This pronominal reflection was West's sharpest critical point. It was a crucial observation that the wider crisis was forcing a crisis of identity. 'Personality, character, self, "I", have become problematical because "we" have become problematical . . .'.[134]

The '30s make a kind of apotheosis of Romanticist individualism in a literature of self-regard that got fuelled by the new psychology and philosophy and the mass of great writers—Conrad, James, Lawrence, Proust, Joyce, Woolf, T. S. Eliot, to name no others—attendant in their train. This is the period when, as we have already seen, autobiography, the published letter and diary, the self-declarative title with *I* in it, the eagerly egotistic self-explanation ('Why *I* write'), flourish as never before. All bourgeois poems, lamented Charles Madge in *Left Review* (February 1936), now 'begin with "I" and all novels are autobiographies'. 'I', wrote Havelock Ellis in 1936, professedly opting out, 'have not the faintest wish to compete with the so-called "autobiographies", which are appearing every day.' But still, his waiver was done in a first-person narrative. 'They have been great egotists', declared Virginia Woolf of her 'Leaning Tower' authors:

When everything is rocking round one, the only person who remains comparatively stable is oneself. When all faces are changing and obscured, the only face one can see clearly is one's own . . . No other ten years can have produced so much autobiography as the ten years between 1930 and 1940. No one, whatever his class or obscurity, seems to have reached the age of thirty without writing his autobiography.

Virginia Woolf had gibed like this before, in *The Years*, when Peggy meets a young poet prating (as Auden and Co. so often did) of his Uncle:

She had heard it all before, I, I, I—he went on. It was like a vulture's beak pecking, or a vacuum-cleaner sucking, or a telephone bell ringing. I, I, I. But he couldn't help it, not with that nerve-drawn egotist's face, she thought, glancing at him. He could not free himself, could not detach himself. He was bound on the wheel with tight iron hoops. He had to expose, had to exhibit. But why let him? she thought, as he went on talking. For what do I care about his 'I, I, I'? Or his poetry?

'I'm tired', Peggy apologizes (a doctor, she has been up all night):

The fire went out of his face when she said 'I'. That's done it—now he'll go, she thought. He can't be 'you'—he must be 'I'. She smiled. For up he got and off he went.

Caudwell's contempt for this egocentricity just about matches Peggy's: the novel, he complains in *Studies in a Dying Culture*, has collapsed through the efforts of Joyce and of Proust, plunging down into the abysms of Gertrude Stein: 'complete "me-ness" ' now reigns. It was a sort of 'self-abuse': 'And that', as Tom Harrisson raged in his *Letter to Oxford*, 'means self-analysis, self-consciousness, self-selfedness, self, self, self.' E. M. Forster even suggested the condition was Satanic: when 'Satan' appeared to him in the Italian Pavilion at the 1937 Paris Exhibition ('The Last Parade'), the Devil had 'only one remark to make: "I, I, I" '. Too many writers were dwelling discomfortingly—to use Malcolm Lowry's marvellous smart-aleck formula in *Ultramarine*—in 'introverted commas'.

And the 'I' so dotingly attended to kept manifesting itself as a problematic egotism. Actually, self-consciousness, at least to this degree, is obviously itself part of the problem, and anyway quickly generates a sense of problem. 'The unselfconscious man', said Auden in his 1947 Introduction to Isherwood's translation of Baudelaire's *Intimate Journals* (1930), 'can rest in his natural individuality, in the fact that he is what others are not—but, once he becomes self-conscious, this is not enough.' The '30s readily picked up and generalized Prufrock's self-unease. The apparent social problems of C. Day Lewis's novel *Starting Point*, 'are simply a special kind of personal problem', thought T. L. Hodgkin (in *LR*, November 1937), and he recalled Prufrock. ' "Shall I part my hair behind? Do I dare to eat a peach?" is a fair expression of the state of mind of the main characters . . . Even when the refrain is altered the tune remains the same: "Shall I join the Church of Rome? Do I dare go to Spain?" ' What 'horribly self-observing, self-questioning, motive-rummaging, flagellating states of mind', Hodgkin adds.

The revolutionary McGinn in MacNeice's unpublished play *Blacklegs* (1939) rants against the text's egotistic Professor:

Dissecting, correcting, peering, poking, analysing, revising, fixing and falsifying—there you sit all day long goggling and giggling into your microscope. And there's one thing you always see . . . yourself. A little selfish self-deceiving bourgeois playing his tricks in the middle of a blob of scum.

Stevie Smith's 'Analysand' (in *A Good Time Was Had By All*, 1937) is equally contemptuous:

> But is it surprising Reader do you think?
> Would you expect to find him in the pink
> Who's solely occupied with his own mental stink?

And the '30s selves everywhere so prominently displayed are frequently all too aware of their debility:

> Here am I, here are you:
> But what does it mean? What are we going to do?

Every stanza of Auden's 'It's no use raising a shout' (*Poems*, 1930), ends with that refrain. A soberer version, perhaps, of the popular song repeatedly sung in Orwell's novel *A Clergyman's Daughter*:

> There *they* go—in *their joy*—
> 'Appy *girl*—lucky *boy*—
> But '*ere am* I-I-I.
> Broken—'a-a'arted!

'[W]ho am I?' wonders Dorothy in that novel when she's lost her memory: '*Who was she?*' 'What's "I"?' asks Maggie in Woolf's *The Years*:

> 'Yes', said Sara. 'What's "I"?' She held her sister tight by the skirt, whether she wanted
> to prevent her from going, or whether she wanted to argue the question.
> 'What's "I"?' She repeated.
> But there was a rustling outside the door and her mother came in.

Neville and Susan's 'Who am I?' is equally fruitless in *The Waves* (1931), the novel which works Mrs Woolf's problematizing of the self to its most strenuous pitch. In *The Waves* selves lose their hard edges, go fluid, shift and dissolve and, in the case of Percival, altogether disappear (he's rubbed out by death in India as Jacob Flanders was killed in the First War in *Jacob's Room*). 'My sense of self almost perishes', Neville complains. 'The danger of losing one's individuality is . . . greater in modern times than it has ever been': Auden's foreboding (in that Introduction to the Isherwood Baudelaire) could have found almost classic support from Virginia Woolf—classic exemplum, of course, for the Marxist charge that the identity problem was firmly in *bourgeois* culture's bailiwick. Christopher Caudwell's portrayal of the excruciated bourgeois self in *Illusion and Reality* couldn't it be grimmer: 'final incompleteness of the bourgeois vision . . . ravages apparent in modern consciousness . . . pangs of this dismemberment . . . chaotic and intoxicated confusion of all *sincere* modern bourgeois art, decomposing and whirling about in a flux of perplexed agony.' And it's not accidental that *The Waves*' anguished questions about identity should have so thoroughly infiltrated the fictions of younger bourgeois authors. 'I did not know who I was', declares Roy, youthful narrator of Rex Warner's *The Aerodrome*. 'Who am I then?', he keeps wondering. The boy Geoffrey in Spender's *The Backward Son* meditates on self-hood while in bed (his author not unmindful, perhaps, of that most memorable scene in *The Waves* which has Rhoda in bed, as her room bends about her, conscious of losing the boundaries of her self—'Out of me now my mind can pour'—stretching her toes to touch the bottom rail of the bed to 'assure myself . . . of something hard'). 'Supposing I am nothing?', Geoffrey thinks:

> Being I means not being Palmer or Daddy or Mummy or Hilary or Christopher. But they also are 'I' to themselves. However unhappy they may be, they all say 'I am glad that I am I'. It would be terrifying to be another. Being I is home.
>
> I shall never know what it is to be them. They must all have their excuses, all be afraid, as I am . . .

And like the writer Bernard in *The Waves* or the letter-writers in *Jacob's Room* the lad strains to capture his identity in an act of writing to his mother:

> So on one of [the sheets of notepaper] he wrote 'Dear Mummy, I . . .' and then crossed this heavily out.

The writing fizzles out—like Bernard's stories and the stories that fill *The Years*— and the headmaster makes the boy write home a lying, misleading letter (not unlike the mendacious letters of *Jacob's Room*). He meditates a way of achieving the fixity of writing; he will print out with his Christmas printing-set *My Life, by Geoffrey Brand. Collected Works. Kiyudkoo, or Who Am I? A Play in Three Acts*: 'In clear, hard, cold print, they will have to recognise me.' His usual defeatist woolliness quickly supervenes.

Very frequently these tormentedly puzzling selves of the '30s settled into split-mindedness. The bourgeois consciousness's dilemma led inevitably, according to Caudwell, to 'the schizophrenic vision of Joyce'. It was scarcely a funny condition, and Norman Cameron's mockery in 'Nostalgia for Death' is unusually jolly about it:

> Psychologists discovered that Miss B
> Suffered from a split personality.
> She had B-1, B-2, 3, 4 and 5,
> All of them struggling in one body alive.
> B-1 got tipsy and B-2 felt ill,
> B-3 got pregnant, B-4 paid the bill.
> Well, that's enough of that. What about me?
> I have, at least, N-1, N-2, N-3.
> N-1's a glutton, N-2 is a miser,
> N-3 is different, but not much wiser.
> Well, that's enough of that. What of N-0?
> That is the N I'd really like to know.[135]

It's a footnote, perhaps, to Spender's 'An "I" can never be great man' (*Poems*, 1933) in which a reconciliation between 'great I' and 'I being', 'I loving', 'I angry', 'I excreting', will only be achieved in the naughting of death:

> The 'great I' is an unfortunate intruder
> Quarrelling with 'I tiring' and 'I sleeping'
> And all those other 'I's who long for 'We dying'.

This more sombre tone characterizes the split men who throng the '30s against a richly dismaying background provided by Yeats's masks, by Eliot's insistence on the veiled impersonality of poets, and by Eliot's re-animation in Prufrock of the Conradian Double, the self accompanied by its Secret Sharer other half ('Let us go then, you and I'). Wyndham Lewis's 'The Split-Man' of *The Apes of God* (1931) actually started life in Eliot's *Criterion* in February 1924. 'We are, I know not how, double in ourselves': Montaigne's ancient sentiment inspired Auden's *The Double Man* title (1941: published as *New Year Letter*, a title not of Auden's choosing, in England). William Plomer entitled his autobiography *Double Lives* (1943). '[W]hat was I, leading this life divided', wonders the narrator of Spender's story 'Two Deaths' (*The Burning Cactus*, 1936). Eleanor Parrish of *The Years* 'seemed able to divide herself into two'. The young Evelyn Waugh was toying (late in 1920) with 'the study of a man with two characters' as a scheme for a first novel. Graham Greene's laboured first (published) novel *The Man Within* (1929) actually had such a divided hero:

> He was, he knew, embarrassingly made up of two persons, the sentimental, bullying, desiring child and another more stern critic. If someone believed in me—but he did not believe in himself. Always while one part of him spoke, another part stood on one side and wondered, 'Is this I who am speaking? Can I really exist like this?'

Rumour at Nightfall (1931), Greene's third novel, armed with an epigraph from his favourite poet Traherne about 'second selves', engineers a kind of wholeness for the woman Eulalia ('What am I? I am my father and my mother. If I have any virtue it is my father's . . . And if I have sinned, it is my mother's sin') in a short-lived love with the man Crane ('she freed herself from the past and faced the future with joy. That joy was not her father's nor her mother's, it was her own; an individual, precarious joy'). Healing for the divided self came noticeably less readily to MacNeice. He goes in for debating poems (eclogues), and dramatic conflict (his Group Theatre play *Out of the Picture* was acted in 1937). He repeatedly presented himself as the narrator tensed in divisive self-debate: in *Zoo* between Reader and Writer, and in *I Crossed the Minch* variously between Head ('in the air') and Foot ('on the floor'), between the traditionalist Crowder in his plus-fours and the trendier bourgeois Leftist Percival (with his *Daily Worker*, his copy of Kafka in German, and his cocaine), and between 'Me' ('My sympathies are Left. On paper and in the soul. But not in my heart or my guts') and my/his Guardian Angel (urging harder socialist attitudes upon me/him). In the *Criterion* (July 1938) Ruthven Todd praised those *Minch* dialogues as a mode of scrupulous doubting, but was still worrying over MacNeice's fraught self-division: 'I felt . . . that Mr MacNeice was not quite happy to be writing a personal book. He seems to be avoiding himself.'

It was an honest debate Todd could not imagine Stephen Spender conducting. Mistakenly; for just such an argument goes on in Part II of his political polemic and confession *Forward From Liberalism*, where Spender engages at length with the voice of self-doubt over Marxism ('Answer the doubts which are implied in . . .'). But Spender seeks Alick West's solution: the problematic 'I', the hesitantly divided self exposed in *Forward*'s Part II (entitled 'The Inner Journey'), once armed with a Marxist's understanding of historical events (the historical Part I of the book is called 'Journey Through Time'), will seek a solution, wholeness, by going forward, over to the communist side, to join in the workers' 'We'. '[I]n disciplined political action, the will makes the individual merge his individuality in the purpose of the whole movement': so Spender declared in a *LR* piece (July 1937) that toughly scrutinized the personal element in the writings of a number of contemporary poets—Rex Warner (too 'arrogantly individualist'), George Barker (not as 'absorbed in the private crisis of his own personality' as was frequently alleged), and Auden (his 'Spain' rigorously excludes 'the element of personal experience and direct emotional response').

Sticking over this point of the personal helped place one's distance from hard Socialist commitment. MacNeice joined 'the Comrades', he said, 'in their hatred of the *status quo*', but was unable to 'sink' his ego: 'I had a certain hankering to sink my ego, but . . .'. John Cowper Powys, far more of a mystic anarchist than he was a fringe Marxist, was determined precisely to hold on to his old self: 'Not all the heroic appeals of the noblest cause in the World can turn the unconquerable Faustian "I", the old Homeric and Biblical "I", into a permanent cog in an impersonal machine.' Christopher Caudwell, who devotes a largish section of *Illusion and Reality* to praising 'the social ego', insists that the degree to which bourgeois authors submerged their 'I' was a test of effective writing:

> They must work with the proletariat somehow, and this necessarily involves their accepting the obligations of united action. This is educative and has had, for example, a considerable effect on Spender and Day Lewis. In some cases, it may even extend to their joining the party of the proletariat—the Communist Party—but the extreme reluctance of most of these artists to take this step is symptomatic.

The notion, West's, Spender's, Caudwell's, that losing the 'I' in the Workers' 'We'—abandoning the concept of Ego, in fact, as a lie, as what Koestler's Rubashev in *Darkness at Noon* (a novel in which the I–We debate runs very strong) has christened the 'grammatical fiction'—the idea that this self-abandonment was the only salvation for the shattered modernist self, was the veriest orthodoxy of the Communist Party and of *Left Review*. It was received wisdom about existence in the Soviet Union. Randall Swingler ('the needs are communal, a giving away of the self'); John Lehmann's friend 'the Georgian poet' ('and on the scaffolding of new [Soviet] buildings men are conscious of "we" before "I" '); T. A. Jackson (the Soviet Union had, he thought, put an end to the fractionalization of individuals); Maurice Hindus (the Soviet Five Year Plan 'envisages the recasting of society into a new world. Man's "I" is no longer the centre of things. It is an organic part of the aggregate—or, as the Russians say, of the mass'): all agreed on which side redemption lay in the contemporary war of the pronouns.[136] Day Lewis's novel *Starting Point* ends with Anthony Neale having achieved the happiness of a whole selfhood in communist solidarity; *I* in *we*, *he* in *they*:

> Appleton, Grove, Sinclair, Morris: all the comrades: in streets, in the country, in prison, in factories, in the little room above the tobacconist's shop and the May-Day Demonstration—they were with him and would be with him as long as he lived. He was bound to them by the steel cables of action, the filaments of belief . . . Of these he was one. With these he was one.

'I', thinks James Seton in Sommerfield's *May Day* as the great workers' march gets on the move, 'I sink my identity into the calm quietness of this waiting crowd, I am part of it, sharer in its strength . . . and the solution of my conflicts is bound up with the fate of this mass.'

The 'frantic assertion of the ego' belongs to the fascist past that has to be overthrown in Warner's *Wild Goose Chase*, in part because it led to a withdrawn

solipsism of the self, and of the self of art—to the condition of Rudolph, forbidden to
look out of the window, concentrating in his poems on the furniture of his room and
the backs of his books, and of David sinking into the madness of a heightened
egotism:

> it was I myself who was being tortured, I myself who was the torturer. How previously
> I had been enraptured by the intellectual consideration of such a union of subject and
> object! How terrible was the reality! My self, which I had considered inviolate, was
> dissipated over the walls and floor of my room, and a reflection, a puppet, something
> that would show in a mirror was the mere toy of a system of Government which, late in
> the day, I found that I had never understood.

'[C]ould you give me back my self?', David pleads. And he is offered outdoor work in
the real world of the revolution. 'Come over to our side. Catalogue the flowers. Get
some fresh air. You'll be all right.'

'I am I': according to Rickword, André Malraux won the right thus to assert his
personal integrity only through 'contact' with 'the people's movement' ('the presence of the people'), a going beyond the treacherously romantic egotism of T. E.
Lawrence ('Egotism is not a philosophical fallacy, but treachery or a dreary pose').
So the stakes could not be higher. No wonder the end result of going over was
imagined as the discovery of new capacities for accelerated and rhythmic group
movement, for running and dancing in time (as Rickword put it) 'with the rhythmic
movement of the social process'.[137]

Section III of MacNeice's *Autumn Journal* ends with MacNeice (even MacNeice)
aspiring to:

> . . . look up and outwards
> And may my feet follow my wider glance
> First no doubt to stumble, then to walk with the others
> And in the end—with time and luck—to dance.

'Break from your trance: start dancing now', the last lines of Day Lewis's 'Magnetic
Mountain' exhort. Now is indeed *A Time to Dance*: 'together', as Lewis's 'Address to
Death' has it, 'in the rhythm of comrades'.[138] 'O communists', Gavin Ewart's
'Political Poem' declares, 'we believe you have got it'; it accepts the necessity of
joining the enlightened workers, 'in an atmosphere of dancing, / Sensitive to you as
the powerful throb of accordians'. As Lehmann's Peter Rains finds himself doing at
the end of *Evil Was Abroad*, when he answers his friend Rudi's 'appeals from across'
the class 'frontier Peter had half longed, half feared to cross'—imperatives to unite
with the oppressed, to 'learn as they felt', 'to belong to them entirely'. 'He began to
run with the people running . . .' The proposition of the loss of self in the socialist
plurality undoubtedly held extreme allure for bourgeois intellectuals worried about
their personal wholeness. Communism's 'increasing attraction for the bourgeois',
Auden wrote in a review of Bertrand Russell's *Education and the Social Order*, 'lies in
its demand for self-surrender for those individuals who, isolated, feel themselves
emotionally at sea. Does Mr Russell never contemplate the possibility that intellectual curiosity is neurotic, a compensation for those isolated from a social group,
sexually starved, or physically weak?'[139] But the trouble is that while it was easy to
talk in general terms about ditching the old bourgeois ego for the future workers'

community, it was more difficult to describe what going over meant in detail and in practice for bourgeois authors, and therefore difficult for aspirant writers not to sound flatulently glib about a case they had by and large yet to prove or find proven. As Christina Stead does, in her *Left Review* rhapsody (August 1935) over the young writers, 'the tough, fiery, humourless, young ones, who have to give up their poetic solitudes and soft-sell probings to study worldly subjects' and to 'take lessons from working men', and who have sacrificed their personal poetic futures, conscious that the 'great writers' will only come after them. Or as Spender does, in his poem 'The Funeral' (*Poems*, 1933), about the death of a communist driving-belt maker. Being a Communist, Spender appears to be claiming, means you can laugh at socialist funerals. Indeed you must be glad when you bury a comrade because you and he aren't individuals any more and any show of grief would be a terrible bourgeois lapse into the old dismally personal feelings Socialism is there to help you grow out of:

> They think how one life hums, revolves and toils,
> One cog in a golden and singing hive:
> Like spark from fire, its task happily achieved,
> It falls away quietly.
>
> No more are they haunted by the individual grief
> Nor the crocodile tears of European genius,
> The decline of a culture
> Mourned by scholars who dream of the ghosts of Greek boys.

It hardly seems a strong recommendation of going over that it turns you into a bit of machinery (a *cog*), a worker bee, an inhabitant of a *Brave New World* termitary (even if the hive is a *singing hive*), a place where you're not a person so much as a bright and happy *spark*. Nor is it either attractive or convincing to allege that Communists stay inhumanly dry-eyed when friends and loved ones die. Auden's tone falters, too, as soon as he tries in a poem to go over to the workers. In fact, so far from manifesting the respite and salvation for the self so widely canvassed by the Left, his address to the problem only plunges his pronouns still more miringly into confusion. His poem 'A Communist to Others' appeared in the magazine *Twentieth Century* in September 1932, and in *New Country* (1933). 'Comrades who when the sirens roar', it begins. Was Auden, then, a Communist? Well, no; he never joined the Party, and his hesitation is mirrored in his doubts over whom precisely he's supposed to be going over towards. In Auden's notebook the poem opens merely with 'All you who when the sirens roar'. These addressees then got clarified as 'Comrades' in the copy Auden sent to Isherwood, and in the *Twentieth Century* and *New Country* versions. But not for long: in the *Look, Stranger!* (1936) reprint they had become generalized again as 'Brothers'. Ditherings like this usually indicate, as they do in 'Spain', that it is dawning on Auden that something is up: the poet is tumbling to his own impostures (and one must insist that it is the dilemmas of Auden himself that are in question here: Edward Mendelson's suggestion in his study of the *Early Auden* that this poem is a piece of 'ventriloquism', whose voice is not Auden's but that of a Communist telling Auden what he needs to learn, is unconvincing).

> Comrades, who when the sirens roar
> From office shop and factory pour
> 'Neath evening sky;

By cops directed to the fug
Of talkie-houses for a drug
Or down canals to find a hug
        Until you die.

We know, remember, what it is
That keeps you celebrating this
        Sad ceremonial;
We know the terrifying brink
From which in dreams you nightly shrink.
'I shall be sacked without', you think,
'A testimonial'.

We cannot put on airs with you
The fears that hurt you hurt us too
Only we say
That like all nightmares these are fake
If you would help us we could make
Our eyes to open, and awake
        Shall find night day.

On you our interests are set
Your sorrow we shall not forget
        While we consider
Those who in every country town
For centuries have done you brown,
But you shall see them tumble down
Both horse and rider.

He's tired already of his working-class comrades. Perhaps because he cannot keep up the pressure of knowing what their fear of the sack is like. His use of the bourgeois word *testimonial* has already rumbled his game: and after all, this flaunted bit of knowledge was got from books rather than from life ('I might be sacked without a testimonial', thought the hero of Edward Upward's 'Sunday', only a handful of pages earlier in *New Country*). How odd, too, the tone is! Auden asks for help ('If you would help us') without managing to sound other than condescending ('We know, remember . . .'). Working-class life is made so emptily banal, a matter of cinema-mindlessness and canal-side cuddling, that it cannot possibly attract an intellectual. And it doesn't. Perhaps that's why the pronouns wobble so disconcertingly. 'If you would help us we could make / Our eyes to open . . .' A more committed, Alick West version might read: '*you* could make / Our eyes to open'. It sounds instead as though Auden is only just stopping himself saying that 'we could make / *Your* eyes to open' (help you to see your foolishness about movies and girls). Slyly, in fact, Auden's pronouns are working to suggest that there is a lot wrong with the workers whom his current socialist flirtation nevertheless has him appealing to. The half-heartedness of the appeal emerges still move vividly when after this brief and half-cock step towards going over, Auden promptly lapses back to 'consider' the workers' enemies, who happen to be Auden's own more congenially accustomed targets. In fact, it's taken him less than four stanzas to run right through his proletarian material and out of his going-over imperatives. And for the next seventeen stanzas in the poem's original version (only twelve stanzas in *Look, Stranger!*) he zestfully grouches away at a clutch of bourgeois horrids sited far to the reactionary side of the class border: the beautiful captain of the school Eleven and Fifteen, the mystic, the Liberal Cambridge intellectuals aloof from the poor, the runaway poet (hints of Robert Graves), bankers and Cambridge Apostles. His poem is back on accustomed lines, back among bourgeois family business. To be sure, Auden rages and curses in his finest, most terrible King Lear strain:

> Let fever sweat them till they tremble
> Cramp rack their limbs till they resemble
> > Cartoons by Goya:
> Their daughters sterile be in rut,
> May cancer rot their herring gut,
> The circular madness on them shut,
> > Or paranoia.
>
> Their splendid people, their wiseacres,
> Professors, agents, magic-makers,
> > Their poets and apostles,
> Their bankers and their brokers too,
> And ironmasters shall turn blue
> Shall fade away like morning dew
> > With club-room fossils.

But he's been away from his putative comrades for a very long time. No wonder he sounds both hectoring and sheepish when he elects eventually to return:

> Comrades to whom our thoughts return,
> Brothers for whom our bowels yearn
> > When words are over;
> Remember that in each direction
> Love outside our own election
> Holds us in unseen connection:
> O trust that ever.

'When words are over': it is curiously revealing that Auden should thus suggest he only has time for the workers at the end of poems or outside his poetry, his writing, his word-craft altogether: implying that writing, like most of this poem, will continue to be a bourgeois business. Arresting, too, is this last stanza's wash of Biblical talk: 'bowels yearn', 'Love', 'election', 'trust'. The connection between Auden and the working-class that's posited here is a matter of the 'unseen' (it certainly slipped very speedily out of his poem). The comrades in the factories, shops, and offices must, it seems, take Auden's support on trust. It becomes a matter of their having faith in him. In fact, Auden offers himself to them as the Christian God offers Himself to believers: 'Whom', as St Paul puts it, 'having not seen ye love.' Which is not at all what Marxists meant by their talk of joining the workers in concrete, Marxist living. And this recessiveness, disappearance even, of the poet is based in a still unsolved problematic of identity in this poem's final mess of pronouns. It's a mess Mendelson's ventriloquial suggestion is doubtless meant to cope with—and doesn't—and a confusion instructively absent from this poem's mirror version by Gavin Ewart (who frequently trod in Auden's footsteps in the '30s), his 'Though what I think is hardly news'. At the end of his poem Ewart refers to his addresses to the neurotic bourgeoisie and his pronouns, like his representation of his and his class's plight, have the sureness precisely missing from Auden's stanza. 'We send our hastily scribbled notes', Ewart declares:

> To be deciphered by our kind,
> The men who live in the tortured mind,
>   To show them their condition,
> To ask their help, to make them see
> That these things, said again by me,
>   Are worth the repetition.

No doubt Auden would also like to be still holding public negotiations with the bourgeoisie. Certainly he has not yet progressed to the *I-We* stage with the workers: with them he's still only on *We-You* terms. His sense of self has yet to emerge from its fearful, generic bourgeois *We*, into a more resolute, if still worried, socialistic *I*. He evidently has a very long way to travel before his final stanza's final attempt at Alick West's *we* ('Holds *us* in unseen connection') will carry any sort of conviction. In fact, Allen Tate's cynicism (in the Auden Double Number of *New Verse*) seems only too well founded:

> The well-brought up young men discovered that people work in factories and mines, and they want to know more about these people. But it seems to me that instead of finding out about them, they write poems calling them Comrades from a distance.

He might well have had in mind Wyndham Lewis's waspishness towards well-heeled Leftists. One of Lewis's most sustained '30s notes, here's how it sounded in *Men Without Art*:

> I have heard every sort of person, from clergyman to ex-society women (who are too old to be social butterflies any longer and have 'embraced' Communism, as they can no longer embrace or be embraced by anything more corporeal and satisfactory) hold forth for hours to the effect that no person should be tolerated who does not 'work with his hands'—and their listeners have always been too gentlemanly or ladylike to exclaim, 'But hold on old girl (or old boy) when did *you* ever lie on your belly in a mine, or do any machine-minding except punishing a typewriter—bought with twelve good guineas you had not earned yourself, but which came to you in dividends inherited from some hard-fisted old skinflint or other!'

It was easy, though, to crow over the dilemmas and glibness of Auden and his friends. They were running up against a genuine snag. For all the insistence on what Day Lewis called (in his essay 'Writers and Morals') 'contact with the workers', actual contact with the masses was exceedingly difficult to contrive. Going over, even in its loosest sense of a cross-border reconnaissance, a spying-sortie to see what the proletarian regions beyond the social frontier were like, was a tricky undertaking. For almost as rigidly as in the nineteenth century, Britain was still divided into what Disraeli had labelled the Two Nations, the rich and the poor. And they lived apart: the rich in pastoral rurality, as in the more prosperous Midlands and South, or in the smart quarters of cities; the poor shoved away in slum ghettoes in London and other great cities, or, more crucially distanced still from the political and cultural centre around which the writers and publishers congregated, in the depressed industrial provinces—away among the slack mines of South Wales, up among the dying mills of Lancashire and the mines of Yorkshire and Durham, around the slowed-down dockyards of the Tyne, the Wear, and the Clyde. To get there bourgeois writers had to make special journeys, across town, or across country. They had to leave the country cottage they had opted for, the smart or even great houses of their parents and friends, the quadrangles of Oxford and Cambridge and of the public schools and prep schools (frequently rurally located) where they customarily studied and taught. Often it was a long road. For a southerner like Orwell *The Road to Wigan Pier* was geographically and spiritually as difficult and elongated as the 'Factory Road' that the southerner Charles Dickens took for the first time in 1838. Orwell found the Black Country as horrifying ('the real ugliness of industrialism—an ugliness so frightful and so arresting that you are obliged, as it were, to come to terms with it') as Dickens had done ('miles of cinder-paths and blazing furnaces and roaring steam engines, and such a mass of dirt gloom and misery as I never before witnessed'). Both men had to take trains to the north, Dickens in 1854, Orwell in 1936; both remained outsiders, explorers peering at alien climes from behind railway-carriage windows. The most memorable vision of factories in Dickens's novel *Hard Times* is from the train: 'The lights in the great factories, which looked, when they were illuminated, like fairy palaces—or the travellers by express train said so.' And perhaps the most grimly telling impression of proletarian northern existence in *The Road to Wigan Pier* is this:

> At the back of one of the houses a young woman was kneeling on the stones, poking a stick up the leaden waste-pipe which ran from the sink inside and which I suppose was blocked. I had time to see everything about her—her sacking apron, her clumsy clogs, her arms reddened by the cold . . . She had a round pale face, the usual exhausted face of the slum girl who is twenty-five and looks forty, thanks to miscarriages and drudgery; and it wore, for the second in which I saw it, the most desolate, hopeless expression I have ever seen. It struck me then that we are mistaken when we say that 'It isn't the same for them as it would be for us', and that people bred in the slums can imagine nothing but the slums. For what I saw in her face was not the ignorant suffering of an animal. She knew well enough what was happening to her—understood as well as I did how dreadful a destiny it was to be kneeling there in the bitter cold, on the slimy stones of a slum backyard, poking a stick up a foul drain-pipe.

And this is presented as the view from a train: the train that 'bore me away, through the monstrous scenery of slag-heaps, chimneys, piled scrap-iron, foul canals, paths of cindery mud criss-crossed by the prints of clogs'.[140] The *Two Nations* were in Orwell's experience as emphatically divided in 1930 as in 1840, and must be shown to be so. Even being a regional insider, like the Yorkshireman, J. B. Priestley, or the Liverpudlian, James Hanley, did not achieve you insider status with all of Britain's proletariat: there was plenty of strange terrain left to be covered by Priestley on his *English Journey* and Hanley had to undertake a special trip to the 'special area' of South Wales for his *Grey Children: A Study in Humbug and Misery*.

The '30s saw lots of these internal voyages of national exploration. The impoverished industrial regions were exciting as much interest from government inspectors, do-gooders and writers as the industrial regions of the 1840s had. James Hanley recorded a certain John Jones's hostility to the crush of investigators:

> the men down here, in fact all the people down here, have grown very, very sensitive about the enormous number of people who come down here from London and Oxford

and Cambridge, making enquiries, inspecting places, descending underground, questioning women about their cooking, asking men strings of questions about this and that and the other . . . they object to all these people coming down and asking questions. That's all. They're not animals in a zoo. That's what it is.

Misery, alleged Hanley, had become 'a marketable commodity': there was 'something approaching a gold-rush in order to exploit it'. Certainly England was being discovered again on as massive a scale as in the 1840s.

'30s writing was obsessed by the topography of England. When people have been reduced to their bodies, as we've observed the modern period reducing them, then relationships, ideas, selfhood all tend to become, of course, merely physical positions and writing about them only a sort of charting or mapping. For in those circumstances life becomes, as D. H. Lawrence's preposition-laden writing characteristically presents it, highly position-conscious. (One of the best things in Paul Fussell's ideas-packed *Abroad* is his discussion of D. H. Lawrence's (pre)positionality.) So '30s literature is greedy to possess sites, places, landscapes. Poems eagerly announce their designs on geography: 'Do You Believe in Geography?' (Ruthven Todd), 'Poem on Geography' (George Barker), 'Landscapes' (Charles Madge), 'The Landscape Near An Aerodrome' and 'The Landscape' (Spender).[141] Allusions to maps and mapping fill the poems of the youngest poets, such as H. B. Mallalieu ('The Philologist' and 'Follow the Map') and Ruthven Todd ('It was Easier': 'Now over the Map that took ten million years'), as they pack the writing of their slightly older predecessors, Auden, Spender, MacNeice.[142] One of the best of all the short stories to appear in *New Writing* (No. 4, Autumn, 1937) was V. S. Pritchett's tale 'Many Are Disappointed' about four cyclists on holiday from their office and disappointed in their search for the pub and the Roman Road that's on their much disported map. Watching and mapping and traversing landscapes couldn't be more fundamental to '30s literature's sense of itself, to the '30s writers' typical envisaging of their art and their politics as being on the road, on the way, into or across new country. Far from being startling, in its time Auden's presentation of his self as a landscape, to be mapped in his verse—'I see the map of all my youth unroll, / The mental mountains and the psychic creeks, / The towns . . ., / The various parishes'—looked like a merely conventional approach to self-description.

These literary mapmakers seem now, of course, a lot like their immediate predecessors, the Georgian ramblers and walkers, the Hilaire Bellocs and the Richard Jefferies, who obviously engendered the brisk progressions from country pub to country pub that characterized the holidays of C. S. Lewis and his friends, and that came increasingly to attract C. Day Lewis. But the younger Day Lewis would not concede any such likeness (in *A Hope for Poetry*):

> The Georgian poets, a sadly pedestrian rabble, flocked along the roads their fathers had built, pointing out to each other the beauty spots and ostentatiously drinking small-beer in a desperate effort to prove their virility. The winds blew, the floods came: for a moment a few of them showed on the crest of the seventh great wave; then they were rolled under and nothing marks their graves.

The modern destructive element had been too much for them. By contrast, Day Lewis considers the journey of the 'younger sons who could not stay at home' but must 'set out for a far country', 'across the frontiers of fancy', on 'their wild-goose chases', to be wiser, politically more purposive, and likely to amount to something new and important for poetry. And the tune of Day Lewis's *A Hope for Poetry* is the one Rex Warner's *The Wild Goose Chase* dances to. And superficially, to be sure, the '30s road-hogs are far niftier movers than the Georgians: they're more prepared to take to bicycles ('Far On Bicycles' is the opening chapter of *The Wild Goose Chase*), to travel by train right up to the railhead, and to contemplate zooming along the new arterial roads in motor cars and on motorcycles. In the event, though, they seem possessed of rather old-fashioned romantic feelings about travelling across the surface of the earth towards enticingly distant goals, encouraged onwards by W. H. Auden's feeling (expressed in 1946 in his Introduction to Henry James's *The American Scene*) that travel should be seen in terms of the quest for 'The Great Good Place', the 'Grail or the New Jerusalem'.

The period's press of topographers reaped a frequently mocking response. 'Geography indeed!' snapped J. R. R. Tolkien about Charles Williams's fetishistic interest in maps as human bodies (and vice versa). It was obviously easy for Cyril Connolly to compile, out of current topographical orts and fragments, the gibingly parodic verses he has his Christian de Clavering admire in 'Where Engels Fears to Tread':

> It was new. It was vigorous. It was real. It was chic!
>   Come on Percy, my pillion-proud, be
>   camber conscious
>   Cleave to the crown of the road.

Easy, too, for Spender to sneer at Philip Henderson's *Poetry and Politics* as 'a conducted tour through well-charted New Country: the Cambridge critics, the Marxist critics, the Georgians, the Waste Land, the New Writers'—mindful perhaps of V. S. Pritchett's earlier hope as he opened Upward's *Journey to the Border* that it would not be 'yet another conducted tour over the well-worn macadam of New Country'.[143]

As one of Gavin Ewart's clever little pastiches, the poem 'Journey' put it, this was 'an over-charted countryside'. But that didn't stop Ewart charting this highly poeticized landscape once again, albeit jokily:

> To flood my carburettor? Well, 'Ambition'.
> 'Love of Praise' might give me an ignition.
> Are my tyres tested? I should say, Yes.
> I can hold the road, once started, more or less.
> Where do I want to go to? Let me see the map.
> All those roads are Auden's, old chap.
> I've been over them once, following his tracks;
> The private paths are Eliot's, stony and complex;
> There's the main road; too crowded, rather dull.
> The hay-carts, Morris minors, of any kind of fool.
> Besides the surface is too smooth for me.
> That broadening path is Yeats's, as you see.
> The railway on the left belongs to Lewis.
> The red cross marks a worm. It might be Powys.
> There's a big clock there. Belongs to Pound.
> You ought to see his time-table, exact and quite profound.
> The road through the forest was made by Joyce;

> The best in the district but not my choice.
> Lawrence lived in that subterranean passage;
> But that's not the way I shall go to deliver my message.
> I shall start by shaking the mud off my wheels;
> It's an over-charted countryside. I shall take my meals
> At different people's houses. I shan't stay long.
> My bike's rather jerky, by no means 'like a gong'.
> I've no official number-plate, I dislike the police,
> I foresee difficulties from the inhabitants of the place;
> If my watch stops I shall go by the sun;
> But it's too late to back out of it, it's got to be done.[144]

Mocking, though, didn't deter Auden, nor Orwell, nor the others, from their keen relish for apprehending topographies. Nor did the serious objection (from the Roman Catholic George Every, speaking up in the *Criterion*, October 1935, for the Platonic, Christian party, expectably hostile to Huxley-esque bodiliness), the objection that topography composes a merely surface set of accidentals that actually obscure realities, put many ardent map-makers off. The 'world of geography', declared Brother Every, impressively

> is the world of railroads, films, trousers and gingerbeer bottles, spread out in a network over the earth by the economic energy of industrialism. The real China, the real Europe, the real America, lie below that network, and only in and behind their history can we find the universal and lasting elements in human experience.

But Orwell, for example, a firmly anti-Christian humanist didn't care at all for this line of thought. Repeatedly and openly centring his values in thinginess and bodiliness, in an affection for the material surface of things, he tested authors by their capacity for being, as he himself liked to be, *ventre à terre*: Shakespeare 'did have curiosity, he loved the surface of the earth and the process of life' ('Lear, Tolstoy and the Fool'); 'the most essential thing in Swift is his inability to believe that life—ordinary life on the solid earth, and not some rationalized, deodorized version of it—could be made worth living' ('Politics *vs*. Literature'); 'So long as I remain alive and well I shall continue to . . . love the surface of the earth, and to take a pleasure in solid objects and scraps of useless information' ('Why I Write').

Solid objects, scraps of useless information: Orwell might well be describing the early Auden. And in both of their cases it's characteristic of the period that their discovery of England in a set of fictive, metaphoric engagements should be part also of a metonymic presentation in documentary and reportage of a literally discovered and discoverable England. Tramping along metaphoric roads into the strange imaginative country of Wigan Pier, hop-pickers and down-and-outs, is energized by the literal tramping into the actually strange territories of the lower classes that Orwell undertook. And he went in company. His fellow-travellers were not always precisely like-minded; in the end, though, there appear to be many similarities between them.

H. V. Morton's *In Search of England* (1927) caught one predominating tone. On this occasion he was questing for the 'village that symbolizes England': the village 'and the English country-side' that 'are the germs of all we are and all we have become'. Industry he skirts; the towns he prefers to investigate are the unabrasively nice old ones, Bath, Durham, York, Lincoln, Stratford. As he scoops in Zomerset,

Tintagel, Glastonbury and fox-hunting, the terrible sounds of 'Ye Olde England' can already be heard, just off-stage, knocking together its thatched wayside stall where plastic pixies, reproduction beer-mugs, relics of Shakespeare and corn-dollies would soon be on sale. It's true that Morton's subsequent searches (*In Search of Scotland*, which came out in 1929, and *In Search of Wales*, 1932) began to notice that Great Britain didn't only comprise the beauties of Scott's landscapes or of Snowdonia. In Wales, in particular, he did not 'fight shy of the black country in the South'; he went down a coalmine and some of the mining sections of the Welsh volume were actually published in the Labour newspaper, the *Daily Herald*, accompanied by photographs of radical intent by James Jarché. But though Morton paid tribute to the miners, bewailed their scant pay and their grim living conditions, and praised their bookishness, he couldn't get rid of his irksomely condescending sentimentalities of tone and his sense that mining valleys and industrialized locations were a freakishly intrusive aberration from a more permanent rural normality ('beautiful valleys now deformed'; 'black towns . . . as though Sheffield had climbed up into the Scottish Highlands'). His *In Scotland Again* (1933) returned with relief to the old touristic Scotland, its only industrial touch a romantic trip on a trawler. Morton's enduring emphasis had evidently not changed much since it first grew out of the Georgian world of the 'Squire-archy', of J. C. Squire and William H. Davies the 'Super-Tramp'. Morton's clear preferences are those of other soldiers like Henry Williamson and Edmund Blunden, who returned from the First War determined to preserve the rural England they'd known. Books like Squire-devotee Williamson's *The Village Book* (1930), Blunden's *English Villages* (1931: in Collins's Britain in Pictures series), Blunden's *The Face of England* (1932: introduced by Squire), A. G. Macdonell's *England their England* (1933: dedicated to Squire), the ruralizing publisher Batsford's *The Legacy of England* (1935: including Blunden on 'The Landscape'), are characteristic attempts to hang on to the world of Richard Jefferies, Thomas Hardy, and Edward Thomas. These texts were rewriting *The Wind in the Willows* twenty-five or so years on and the more insistently, in that the car-mad, Ur-Futurist Mr Toad seemed set now really to rule the roads. (*The Wind in the Willows* was itself vividly recuperated for the '30s when it reappeared for the first time with Ernest H. Shepard's line-illustrations in 1931.) These were the circles overcome with tremendous relief that village cricket had survived the War. Siegfried Sassoon, whose *Memoirs of a Fox-Hunting Man* (1928) lengthily celebrated a village cricket match (descriptions of cricket, like cricket itself, do rather go on), and who supposed in the trenches that he was fighting for Sussex and the world of Hardy's novels, records in *Memoirs of an Infantry Officer* (1930) his war-time day-dreams 'of an England where there was no war on and the village cricket ground was still being mown by a man who didn't know that he would one day join "the Buffs", migrate to Mesopotamia, and march to Bagdad'. Edmund Blunden came gladly back to an England where cricket was still possible (his book *Cricket Country* was published in 1944), much in the spirit of the returned village-cricketing soldiers of the simplelifer Hugh de Selincourt's novel *Cricket Match* (1922). Even Donald Cameron, A. G. Macdonell's dyspeptic ex-soldier Scot trying hard to make out the puzzling English, is made lengthily to celebrate the enjoyments of village cricket ('rural England is the real England'; the *echt* behind the *kitsch*—'as if Mr. Cochran [the theatrical impresario] had, with his spectacular genius, brought Ye Olde Englyshe

Village straight down by special train from the London Pavilion, complete with synthetic cobwebs (from the Wigan factory), hand-made socks for ye gaffers . . . and aluminium Eezi-Milk stools for the dairymaids').

Much of the '30s exploration of England persisted in linking exploring England with discovering rurality, and accepted the kind of connection and progression implied by Cecil Roberts' writings—*Pilgrim Cottage* (1933), *Gone Rustic* (1934), *Gone Rambling* (1935)—as utterly natural.[145] The link came from the Romantics and had been passed on by Hardy and the Georgians through (to some extent) D. H. Lawrence and had been mightily reinvigorated by the fictions of Mary Webb and the Powys brothers. BBC broadcasters (seventeen of the extraordinarily prolific guidebook writer and friend of Henry Williamson, S. P. B. Mais's wireless talks, including one about 'The Mary Webb Country' were published in 1932 as *This Unknown Island*), revivers of Merrie-England, zestful mongerers of Our Heritage, determined hikers such as C. E. M. Joad (author of *The Horrors of the Countryside*, 1931, and *A Charter for Ramblers*, 1934)—all of them with one walking-boot planted in the National Trust (founded 1895) and the other in the more recent Council for the Preservation of Rural England (founded 1926)—conspired to sell the nation a picture of a non-industrial England that they were at once seeking deftly to stage-manage and to condone. The Shell Guides, whose first general editor was John Betjeman, reflected this Mortonian spirit. They're sometimes praised for architectural democracy, interest in vernacular buildings, agreeable welcoming of nonconformist chapels, and the like, as if John Nash's volume about *Bucks* (1936), which even puts in a warm word for Slough, were representative. But the usual tone is much more aggressively and squirearchically up-market and blood-sporting Anglican. Naturally, the rural counties with their soothing 'Views of Castles, Seats of the Nobility . . . Picturesque Scenery . . . Churches, Antiquities, &c' crowd the Shell Guides list—Betjeman's *Cornwall* (1934) and *Devon* (1936), Robert Byron's *Wiltshire* (1935), *Derbyshire* (Christopher Hobhouse, 1935), *Somerset* (C. H. B. and Peter Quennell, 1938), and so on. Much of this kind of thing was journalistically soft-centred, even when as in Philip Gibbs's *England Speaks* (1935) and H. V. Morton's *Searches*, some attention was paid to industrialized parts of the country. Most of it—including the best polemics like the superb one Clough William-Ellis edited, *Britain and the Beast* (1938), with contributions from Keynes, Forster, Joad, A. G. Street, G. M. Trevelyan, S. P. B. Mais—was neurotically hostile to the spreading rash of the town, the encroachments of industrial, mechanical man by ribbon development along arterial roads. Forster's 'Havoc' in *Britain and the Beast* repeats his old *Howards End* (1910) worries, but intensified now over the unstoppable horror of the Great West Road and the greedy spread of aerodromes.

It could hardly fail to strike one that a lot of this ruralism was in political terms decidedly unprogressive. The typical ramblers preferred passing their nights when out on-the-road in a Youth Hostel, rather than in the kips and spikes, the workhouses George Orwell was interested in visiting. The British cult of Mary Webb and her gloriously menacing Salop only took off after her *Precious Bane* (1924) had been publicly commended from Downing Street by Conservative Prime Minister Stanley Baldwin ('Her sensitivity is so acute and her power over words so sure and swift that one who reads some passages in Whitehall has almost the physical sense of being in Shropshire cornfields'). In T. H. White's Socialism-baiting novel *Earth Stopped, or*

Mr. Marx's Sporting Tour (1934), the Communist John Marx's deficiencies include being no good on a horse: but he learns 'to be human as he grows acquainted with country life'. 'He was beginning to discover people, as apart from the proletariat.' The Communist, or so White's great testament to the country pursuits of hunting, shooting and fishing, *England Have My Bones* (1936), professes to believe, will be saved by learning to light fires, fish and shoot ('How safe would Karl Marx have been, I wonder, walking in a line of guns'). More extremely, the Fascist Henry Williamson suggested, in 1937 (in *Goodbye West Country*), that Hitler's Germany was full of the fresh air the English countryside was now deplorably short of: 'Everywhere I saw faces that looked to be breathing extra oxygen.' He was saddened by T. E. Lawrence's 'exhaustion' through 'lack of that which gave Hitler life and strength, his genius dulled-out by the nitrogen' of the poisonous English atmosphere. (Williamson never, of course, paused much to reflect on the poisons emitted by his own car and motor-bike exhausts, let alone on the poisonous gases of Hitler's killing chambers.) Edmund Blunden, one of the handful of British writers openly to declare for Franco in reply to the *Left Review* questionnaire, *Authors Take Sides on the Spanish War* ('it was necessary that somebody like Franco should arise') was warm towards 'the ideas' of Germany and Italy, and as late as July 1939 was to be found in the *Anglo German Review* praising the Williamsonian 'freshness' of Hitler's country. That Chief of good scouts, Lord Baden-Powell praised the message of *Britain and the Beast* by dispraising England's untidiness: 'Litter louts don't exist' in Hitler's Germany, he wrote (not, at any rate, for long, he might have added). T. S. Eliot was far cagier about his real feelings on the situation in Germany and Spain, but he couldn't disguise his writing's heated conservatism, an important component of which was the *Criterion*'s unbluntable sympathy for the right-wing agrarian movement of the Southern United States. (It was among a Southern audience at the University of Virginia in 1933 that Eliot felt safe in confiding his hostility towards cosmopolis and to typical cosmopolitans like Jews, in the lectures that got published as *After Strange Gods*, 1934.) The *Criterion* became a kind of house journal for the spokesmen of post-war British ruralism: Williamson himself, H. J. Massingham, H. M. Tomlinson, G. M. Trevelyan, John Betjeman and T. F. Powys. Agriculture, Eliot argued (*Criterion*, October 1931), was 'the foundation for the Good Life in any society; it is in fact the normal life'. It wasn't healthy, he insinuated (*Criterion*, October 1934), 'that the mind of the whole nation should become urban'. 'What is fundamentally wrong', he continued (October 1938), 'is the *urbanization of mind*'— which even tramping 'the countryside with a book of British Birds and a cake of chocolate in a rucksack' wouldn't save you from.

Rightwingers naturally assumed the land was for them as they were for the land. That last observation of Eliot's comes from a passage warmly praising Viscount Lymington's panicky plea for agriculture, *Famine in England* (1938), a book speedily reissued by the Right Book Club. For their part, the group of culture critics coalescing around F. R. Leavis at Cambridge (they hardened, after *Scrutiny* began to appear in May 1932, as the *Scrutiny* group), not only took their cudgels to mass urban industrial society and the Soviet literature and film dismayingly truckling to it, but also advocated as a counterweight the novels of T. F. Powys. Powys forms a central and positive reference point in Q. D. Leavis's *Fiction and the Reading Public* (1932). In the Minority Pamphlets Series (1930) in which F. R. Leavis's *Mass*

Civilisation and Minority Culture comes as No. 1, Nos. 2 and 3 are T. F. Powy's fiction *Uriah on the Hill* and William Hunter's critical booklet *The Novels and Stories of T. F. Powys*. Here was indeed a scurrying retreat from the 'technologico-Benthamite' society that wouldn't just drive the Scrutineers back into George Bourne's *Old Wheelwright's Shop*. Their search for a lost 'organic community' would take them perilously close to the mythic and escapist fairyland of J. R. R. Tolkien (another Franco sympathizer), who was at pains to state his preference for Beowulf and Hobbitry over 'electric street-lamps of mass-produced pattern', the 'mass-production robot factories' of Cowley, Bletchley railway station, and the 'municipal swimming-bath'.[146]

Not surprisingly, left-wingers evinced a certain discomfort over many aspects of this renewed ruralism. 'Cricketolatry' was part of the prep- and public-school culture many of them despised. A 'false quantity or a missed catch were' to William Plomer's Old Etonian prep-school headmaster, 'sins equal to theft, "impurity" or any other of a long catalogue of transgressions'. Rickword seizes on it as the symptom of a fascist tendency when 'a publisher observes that a very dull novel about an English village, which he only published because he was short of a title in his list, has a most marvellous press and sells steadily' (*Left Review*, October 1934). MacNeice has an 'Ode' (in *Poems*, 1936) in which a plane flies over a cricket match ('High above the bat-chock and the white umpires'); it's an augury of war; as far as he's concerned, the old cricketing world won't last. Auden's *The Dance of Death* is as scathing in its light-touch Betjemanic manner about ruralizing simple-lifers as about any other of the faddists it brings on stage

> How happy are we
> In our country colony
> We play games
> We call each other by our christian names
> Sitting by streams
> We have sweet dreams
> You can take it as true
> That Voltaire knew
> We cultivate our gardens when we're feeling blue
> Lying close to the soil
> Our hearts strike oil
> We live day and night
> In the inner light
> We contemplate our navels till we've second sight
> Gosh, it's all right
> In our country colony.

Orwell in *Coming Up for Air* finds Upper Binfield's Woodland City ('te-hee! Nature!') and its nature-worshipping food-cranks, Miss Helena Thurloe the novelist, Professor Woad the psychic researcher, and the shorts-clad old man who gloatingly tells him about them, grislier still. Auden hated the idea (in the words of F. C. Boden's poem published in 1938 in John Mulgan's Left Book Club *Poems of Freedom* for which Auden wrote an Introduction) that 'Beauty never visits mining places', and he specifically overturns the Shell Guides' preference for southern landscapes and the architecture of great houses in his 'Letter to Lord Byron'. He will not endorse Wordsworth; he 'can't read Jefferies on the Wiltshire Downs'; he won't promote views of Surrey; he opts for the industrial north:

> Slattern the tenements on sombre hills,
> And gaunt in valleys the square-windowed mills
> That, since the Georgian house, in my conjecture
> Remains our finest native architecture.

Industrialized landscape are 'the most lovely country' that he knows:

> Clearer than Scafell Pike, my heart has stamped on
> The view from Birmingham to Wolverhampton.
>
> Tramlines and slagheaps, pieces of machinery,
> That was, and still is, my ideal scenery.

Louis MacNeice can't, he says in *I Crossed the Minch*, 'bear' Morris-dancing. The 'folk-fancier', it's indicated in his poem 'On Those Islands' (1938), is a spoiler of the real Gaelic life.[147] Anything Druidic MacNeice savagely debunks. His old prep-schoolmaster Littleton Powys urged him (*Minch*) to visit Stonehenge:

> The last time he had been there he had been with John Cowper, and J. C. had put his 'shaggy old head' in a pool of water on the altar stone and worshipped. I was so stirred by Littleton Powys' enthusiasm that I drove my car at sixty m.p.h. to Stonehenge to see if I should be moved to worship also. I was not. Looking across the downs I saw them black with cars and in the middle a set of child's toy bricks . . . One had to pay for entry and enter through a stile worked by a slot machine. The man I was with found it extremely funny as we tried to picture John Cowper Powys putting his head on the altar stone. We decided that it would be a very good joke to give a party one night and come to Stonehenge and break it up.

One of the best passages in Cyril Connolly's published Journal occurs amidst tripper-bashing thoughts incited by a charabanc-thronged trip to Lulworth Cove in 1929:

> I thought of all the ardent bicyclists, all the accomplished couplet-eers, the pipe-smoking, beer-swilling young men on reading-parties, the brass-rubbing, Balliol-playing Morris dancers, the Innisfreeites, the Buchan-Baldwin-Masefield and Drinkwatermen, the Squires and Shanks and grim Dartmoor realists, the tramp lovers, and of course of Mary Webb—of everyone striding down the primrose path—Wordsworthian primroses—or down other green rides of this set subject to the glorious goal of an O. M. 'The country habit has me by the heart', I chanted to the Lulworth trippers and in a schoolmaster's voice of mincing horror '*procul o procul este profani*' . . .[148]

But, interestingly, it still wasn't easy to predict the politics of the deplorers of the revived rural-Englandism. For instance, Stella Gibbons's *Cold Comfort Farm*, which parodies and undoes the dark extravagances of Mary Webb and John Cowper Powys (its practice of starring passages to help the common reader distinguish sentences that are 'Literature' from those that are 'sheer flapdoodle', seems especially directed at Powys's habit of the purple patch), is not noticeably political at all. For his part, Henry Green, who was famously drawn to the workers in his family's Midlands factory and wrote so powerfully about his life among them in his novel

Living, but remained at best a Tory radical, appears to be mounting an extended parody of the Mary Webb manner in the 'Picture Postcardism' chapter of his first novel *Blindness* ('One big bramble twined so amorously round' Joan, the unfrocked vicar's daughter's 'skirt that her legs could not tear themselves free, and she had to bend to tear him off superbly. All over at the contact of her hands he trembled, and trembled there for some time'). And egregiously right-wing Evelyn Waugh can make even the most annoyingly posturing among his travel books delightful by his zestfully rabid assaults on the taste for picturesque English 'oldeness'. The sturdy English detestation for quaintness, he rants in *Labels*,

> has developed naturally in self-defence against arts and crafts, and the preservation of rural England, and the preservation of ancient monuments, and the transplantation of Tudor cottages, and the collection of pewter and old oak, and the reformed public house, and Ye Old Inne and The Kynde Dragon and Ye Cheshire Cheese, Broadway, Stratford-on-Avon, folk-dancing, Nativity plays, reformed dress, free love in a cottage, glee singing, the Lyric, Hammersmith, Belloc, Ditchling, Wessex-worship, village signs, local customs, heraldry, madrigals, wassail, regional cookery, Devonshire teas, letters to *The Times* about saving timbered alms-houses from destruction, the preservation of the Welsh language, etc.

(In 1934 Waugh was looking for a cottage but not, he stipulated, one utterly sunk in rustick-ness like Beverley Nichols's 'famous cottage' celebrated in *Down the Garden Path* (1932) and *A Thatched Roof* (1933): 'preferably thatch but not beams'.)

And, more interestingly still, there's evidence of a more than sneaking sympathy with the back-to-the-village movement on the part of the period's Left. At times it does appear that John Cowper Powys even wants to combine his supercharged West-Country mysticism with a pronounced sympathy with the Soviet Union. Clough Williams-Ellis was an out-and-out admirer of the Soviet Union. Cyril Joad, the defier of gamekeepers and of landowners' barbed-wire, defender of 'the interest of the people in the English countryside and their consequent claim upon it' as being 'more important than the interest and the claim of farmers, landowners, or sportsmen', was continuing a very English strain of Digger, or Cobbetian radicalism. Communists, exhorted to organize 'United Front rambles' mainly because they gave good opportunity for recruiting (*Daily Worker*, 30 July 1936), could easily find their Progressive Rambling Club speaking with Joad's own voice: 'Let's All Go to the Country—It's Ours!' (*DW*, 8 April 1939). What exactly happened on Kinder Scout, Derbyshire, during the so-called Mass Trespass organized on Sunday, 24 April 1932 to demonstrate for the ramblers' access to private land is greatly disputed, and the Communist British Workers' Sports Federation has been accused of mere political opportunism during that and subsequent actions. But there's no denying the strong feeling for the people's countryside within many Leftist formations.

It's evident that the village often stands for the 'real' England that English Leftists are imaginatively drawn to and want to preserve or restore. 'Our village', its mess and chaos, the freedom it grants humans to be contingently disordered within the due bounds imposed by rector, and squire, is what Rex Warner's *The Aerodrome* sees threatened by the insanely sanitary pressures of that novel's totalitarian air-force. It's face down in the mud of an old-fashioned English village that Roy, Warner's main character, is discovered sprawling as the novel opens. With his belly close to the old village earth, he wants Orwellianly to remain.

Noteworthily, the 'Other Town' that Isherwood and Upward invented as a youthful radicals' alternative to the real bourgeois world of Cambridge and the dons, came into its own when they realized it should be 'a village, somewhere among the enormous downs'. And how curiously the description in *Lions and Shadows* of Mortmere and its inhabitants—the Revd Welken, Gunball who suffers from Suffolk Ulcers, Dr Mears, Sergeant Claptree landlord of the Skull and Trumpet Inn, Gustave Shreeve headmaster of Frisbald College, Alison Kemp the village whore, Ensign Battersea helper at the Skull and Trumpet, Gaspard Farfox private detective—suggests affinities with the people and places of Mary Webb's Shropshire and the grislier reaches of John Cowper Powys's overheated West Country. For that matter, the major surviving piece of Mortmere fiction, Upward's 'The Railway Accident' (written in 1928), not only gives colour, by its quirkily schoolboy cleverness, its jumpy trickeries of tone, its zany eagerness to trawl in intriguing aggregations of odd facts and things, to the suggestion that Auden schooled the Audenesque upon Upward's style. But its handling of the village of Mortmere theme keeps reminding one of the village of Pressan Ambo in Auden and Isherwood's *The Dog Beneath the Skin*. At first glance, the ills, the neuroses, the looniness of European politics appear to be the main concern of *The Dog*; closer scrutiny reveals the play to be far more animated over the intrusion of fascistic militarism into this 'English village' (the militarized Lads of Pressan; the Vicar who preaches 'a sermon on Bolshevism and the Devil', originally published as 'Sermon by an Armament Manufacturer'; and so on). This is, though, a sense of political hellishness that is entirely one with the play's opening Chorus's shocked vision of rural England suburbanized by an invading population of urban outsiders:

> Brought in charabanc and saloon along arterial roads;
> Tourists to whom the Tudor cafés
> Offer Bovril and buns upon Breton ware
> With leather work as a sideline: Filling stations
> Supplying petrol from rustic pumps.

If this be Leftism, it is a Leftism charged with old Deserted Village nostalgias:

> I see barns falling, fences broken,
> Pasture not ploughland, weeds not wheat.
> The great houses remain but only half are inhabited,
> Dusty the gunrooms and the stable clocks stationary.
> Some have been turned into prep schools where the diet
>   is in the hands of an experienced matron,
> Others into club-houses for the golf-bore and the top-hole.
> Those who sang in the inns at evening have departed;
>   they saw their hope in another country,
> Their children have entered the service of the suburban
>   areas; they have become typists, mannequins and
>   factory operatives; they desired a different rhythm of life.

And we now know (Edward Mendelson's *Early Auden* tells us) that much of the phrasing of this Opening Chorus was lifted direct from Anthony Collett's *The Changing Face of England* (1926), reissued in *The Travellers' Library* in 1932, one of the many inter-war books on England's towns and countryside that smack of implicit

Fascism. Collett's grumbles include the decline of English heroism; 'frontiersman' and 'sea-rover' are dying off as the blond Anglo-Saxon type of Englishman is ousted from the cities by a new race of dark, alien (and Jewish?) racial types. When Spender accused Auden of 'National-Socialist' tendencies hereabouts, Auden admitted them and spoke of Fascism's success in making a 'national call' to ordinary people. By National Socialism both Auden and Spender can only have meant Nazism (even if Edward Mendelson cannot stomach that grim fact).

A similar vein of reactionary *Brave New World*-type dystopian dejection over the modern machine world and a craving for liberation into a braver pastoral old world, informs Day Lewis's 'Letter to a Young Revolutionary', which, after Michael Roberts's Preface, is granted the important place of opening piece in the *New Country* anthology:

> The country at last. And a poor enough outlook it is. Stunted crops, derelict barns, mills deserted to rats, good land given over to sheep and golfers. Somebody has run away. In the rectory the rector is reading Jeans or practising string tricks for the next village entertainment. Listen to the children, as we walk past the school, chanting in unison the kings of England and the capitals of Europe, their birthright of natural wisdom exchanged for a mess of knowledge. And the parents? The backbone of the country? The marrow seems to have been drained off. Can these dry bones live? Can they live on the tinned foods, cheap cigarettes, votes, synthetic pearls, jazz records and standardised clothing which the town gives them back, as a 'civilised' trader gives savages beads for gold? They damn well can't, and you know it. And it's up to you, if you want to see the country sound again, to put its heart back in the right place, even though it means what the progress-mongers call 'putting the clock back'. You must break up the superficial vision of the motorist and restore the slow, instinctive, absorbent vision of the countryman. Not exile mind, intellectual consciousness; but stop it trespassing in other fields. The land must be a land of milk and honey, not a string of hotels and 'beauty spots'. Can your revolution do something about all this? If not, I've no use for it.

In his enthusiastic review of *Britain and the Beast* (*LR*, August 1937) Day Lewis asserted that 'Socialism alone' could 'rescue Britain from the Beast': 'We Marxists declare that the English tradition has now passed into our hands.' This sorted well with Michael Roberts's own *New Country* tone towards readers who 'stand for the accepted order': 'remember that the Union Jack, the British Grenadiers, and cricket are not your private property. They are ours. Your proper emblem is a balance sheet. You're a fool if you think your system will give you cricket much longer. Haven't you realised? *Cricket doesn't pay.* If you want cricket you'd better join us.'

Just so Day Lewis, never far from Georgian ruralities, appeals in *The Magnetic Mountain* to cyclists and hikers, day excursionists:

> You that love England, who have an ear for her music,
> The slow movement of clouds in benediction,
> Clear arias of light thrilling over her uplands,
> Over the chords of summer sustained peacefully;
> Ceaseless the leaves' counterpoint in a west wind lively,
> Blossom and river rippling loveliest allegro,
> And the storms of wood strings brass at year's finale:
> Listen. Can you not hear the entrance of a new theme.

This clientele likes 'Watching birds or playing cricket with schoolboys': the glorious future Day Lewis anticipated is, indeed, as he says, 'a country vision'.

Marxist cricket! It's a very English aspiration. It's not out of keeping with Orwell's patriotic Leftism (Orwell reviewed Blunden's *Cricket Country* warmly in 1944: those Leftists are wrong, he argued, who attack cricket as snobbish). Nor is it a completely daft aspiration, especially when one considers how Marxism, a European import, got itself rapidly assimilated among English Communists like Edgell Rickword to the English radical tradition of Bunyan, Blake, Cruikshank and Cobbett (a tradition and a practice still kept vigorously alive by E. P. Thompson and Raymond Williams). Nevertheless, it can't help wearing at least something of the irony, the sense of self-delusive wish, that hangs heavy upon the rural rhetoric with which the period's homosexuals (many of them Leftists, of course) tricked out their essentially urban enjoyments: the lingo of 'cottage', for a public lavatory frequented by them (Auden and MacNeice, in their 'Last Will and Testament', 'wish the cottage at Piccadilly Circus kept / For a certain novelist'); lilies (Spender's poem 'The Port' shows 'the lily boys' flaunting 'their bright lips'; Virginia Woolf smiles in a letter of 21 December 1933 over 'the Lilies of the Valley', Spender, Plomer, Auden, Ackerley, all at that time denizens of Maida Vale); pansies ('Perennial pansy, hardiest of the blooms', according to Plomer's 'The Playboy of the Demi-World: 1938'; 'Go on', hard-mouthed Ruby rags Naylor in Cyril Connolly's novel *The Rock Pool*, 'you old English pansy. You're something out of a cottage garden'); and other flowers ('you male magnolia': Ruby again; 'S.W.', as Ackerley used to address William Plomer: Sweet William).

In country matters, fantasies certainly flourished. And '30s poetry came so curiously full of country matters. Joad once said that hiking was now 'the shortest cut out of Manchester' (in the nineteenth century it had been gin). He might have mentioned poetry, which, for all its obligatory Marxian nodding towards people in towns, and its repeated efforts at being Futuristically enamoured of the machine, remained in the '30s surprisingly unurbanized. MacNeice's poems 'Belfast' and 'Birmingham' are relatively rare in their dedicated urban interests. Spender's 'The Express' is far more characteristic in its effecting a quick escape from the city: 'Beyond the town there lies the open country'. Just so, that Chorus in *The Dog Beneath the Skin* that begins at the London stations, 'Paddington. King's Cross. Euston. Liverpool Street', speeds the reader in a mere five lines beyond the slums, 'the washing and the privies', 'To a clear run through open country'. 'After they have tired of the brilliance of cities', Spender's poem XXIV (1933) suggests, the old bourgeois power-seekers may learn the poets' new 'strange language'—and what this consists of is, in fact, a rural language of a new 'country', beyond factories, cathedrals and banks, a place of Spring, the tiger, plants, new roots, gushing waters, snow. '30s poets seem frequently to have preferred to leave the big industrialized urban conglomerations to the novelists and the other prose writers.

Not that the prose-writers were, as a class, immunized against the poets' drift towards nostalgias for Georgics and affinities for the Georgians. Indeed, it's very revealing that George Orwell, popularly remembered best, as regards his '30s work, for being the period's sternest comer-to-grips with the industrialized urban scene, actually produces text after text, including *Wigan Pier*, that is commanded by cravings for an unvexed pastoral world. And it is, of course, tricky to sustain warmly *ventre à terre* feelings and not have fields rather than pavements in mind. Even amidst the ruined countrysides of *Wigan Pier* Orwell is contriving wishful pastorals. If the fields have been ravaged, he'll find pastoral by working-class firesides. Even Orwell, then, steeliest of leftist observers and discoverers of England, isn't all that far from the dewier aspirations of Day Lewis and the rest of the pansy and pansy-loving crowd that he despised so much.

What troubles and outrages Orwell's writing is the way pastorals of every kind have been moved out of reach. His journeys across the surface of England are overwhelmed by the regretful difficulty of achieving pastoral consolations amidst the continuous erosion of pastoral places. That England as a place and a set of class systems vexes the seeker after pastoralia is the continual theme of Orwell's fictions as well as the *donnée* of his travelling investigations. It's not true rural yearnings that annoy him, it's the social and historical forces that hamper their fulfilment or fob the yearners off with bogus diversions and sentimental substitutes (hiking, simple-life-isms, Joadified Mortonizings).

In *A Clergyman's Daughter* (1935), Dorothy Hart finds her pastoral retreat away from the tight repressednesses of her religion and class when she repeats her author's own 'down and out' experiences as a hop-picker in Kent: 'happy, with an unreasonable happiness', singing joyfully in the sun in 'the smell of hops and woodsmoke', 'nesting' in straw, shaking off repressions in traditional, sub-pagan rituals, getting roaring drunk in the village pub and bawling songs along the village street. But it's all only a temporary business, characterized by amnesia and perplexing loss of identity, a kind of madness. Awful petit-bourgeois normality soon gets her back in its clutches. Gordon Comstock and his girl Rosemary find a measure of bucolic joy out beyond Thornton Common in *Keep the Aspidistra Flying* ('She waded through a bed of drifted beech-leaves that rustled about her, knee-deep, like a weightless red-gold sea') only to have their pastoral blissfulness steadily eroded by a series of sickening failures by rural England to meet their demands on it. 'They had visions of a cosy bar-parlour, with an oak settle and perhaps a stuffed pike in a glass case on the wall', but the odious tastes of stockbrokers have vulgarized everything and prices are inflated in the branch of the catering trade that Belloc and C. S. Lewis and the rest so favoured. A nastily expensive hotel breaks Comstock's budget. And George Bowling in *Coming Up For Air* fails to find even momentary rural respite on his yearning return to Lower Binfield (on whose green, incidentally, Orwell 'sprawled' in *Down and Out in Paris and London*: its air 'like sweet-briar after the spike's mingled stenches of sweat, soap and drains'):

> Of course I knew that even in Lower Binfield life would have changed. But the place itself wouldn't have. There'd still be the beech-woods round Binfield House, and the towpath down by Burford Weir, and the horse-trough in the market-place.

Alas, new houses, football grounds, a gramophone works, and a bomb-factory have supervened; there's no horse-trough now; Binfield House has become a loony-bin; and the local naturists make a poor substitute for a natural life formerly conducted in real touch with nature.

Southern, rural, nice England no longer provided any refuge from or alternative to the litanies of modern horror that Orwell keeps producing: 'gas-pipe' chairs, aspirins, tinned food, contraceptives, gramophones, bombs, aeroplanes, rubber-truncheons. The end-of-season cold drives the hoppers back to London. Comstock's idea of open-air orgasm with Rosemary comes dismayingly to grief because he's taken no french letters with him; gramophones have engulfed Lower Binfield and a bomb smashes down on it. There are 'no more pastorals': at least, they're getting scarce as the fields and villages are encroached on:

> Sing us no more idylls, no more pastorals,
> No more epics of the English earth;
> The country is a dwindling annexe to the factory,
> Squalid as an after-birth.

So MacNeice, in *Autumn Journal*, XVIII. No wonder *The Road to Wigan Pier* implicitly disclaims H. V. Morton. 'Who could resist a glimpse of Wigan?', the 'green England' seeker had asked:

I admit frankly that I, too, shared the common idea of Wigan. I admit frankly that I came here to write an impression of unrelieved gloom—of dreary streets and stagnant canals and white-faced Wigonians dragging their weary steps along dull streets haunted by the horror of the place in which they are condemned to live.

'This is nonsense', Morton bluffly asserted. 'I would not mind spending a holiday in Wigan.' And, sure enough, he discovered Wigan to be saved by its adjacent greenery:

Within five minutes of notorious Wigan we were in the depth of the country. On either side were fields in which men were making hay; old bridges spanned streams; there were high bridges, delicious little woods, and valleys.
  'This is all Wigan!' said the Town Clerk with a smile.

By contrast Wigan doesn't smile much on Orwell; nor is *The Road to Wigan Pier* a smiling account. 'Mr Orwell . . . liked Wigan very much', he says; but 'the people not the scenery'. Not for him, on this occasion, any treading in the glosing ruralist Morton's footsteps. He preferred the trail J. B. Priestley helped blaze. He doesn't, however, say so. Unlike the documentarist photographer Bill Brandt, who admitted that it was Priestley's *English Journey* that drove him and his camera northwards when he was gathering material for *The English at Home* (1936), Orwell never acknowledged his indebtedness (in fact he professed to despise Priestley's attempts at the common touch: 'And Priestley twists his proletarian awl / Cobbling at shoes that Mill and Rousseau wore / And still the wretched tool contrives to bore'), so that *The Road to Wigan Pier* is sometimes wrongly taken as a wonderfully unique achievement. Victor Gollancz, Orwell's publisher, knew his man's down-and-out proclivities, of course, and that's presumably why he commissioned Orwell to do the book. But when Orwell actually took the northern road he was not contriving completely unaided his swing away from Mortonism (and the southern Mortonian ruralities of *Down and Out* or of the article 'The Spike', an early version of parts of that book: 'Nobby and I set out for Croydon. It was a quiet road, there were no cars passing, the blossom covered the chestnut trees like great wax candles. Everything was so quiet and smelt so clean . . .').

Priestley's *English Journey*, catching up as well as firing the spirit of the new urban industrial working-class realism that was manifesting itself in the new photographic and cinematic documentary movement, had already gone northwards. What

Orwell's famous vignette of the young woman poking the wastepipe owed to Bill Brandt's photograph (in *The English at Home*) of a young woman kneeling to wash a Bethnal Green doorstep, must remain conjecture. But there's small doubt that Priestley's awed account of Shotton in County Durham, a ghastly volcanic village, where life is lived apocalyptically, Pompeii-like in the shadow of a perpetually smoking tip, greatly fed Orwell's description of the Wigan region's landscape as a lunar dereliction of slag-heaps. Orwell's factuality, his impressive use of statistics of costs and wages, owes something perhaps to Priestley's notes on the financial condition of the Lancashire poor (as well, of course, as being mightily stimulated by the tactics of the still more seminal Jack London's *The People of the Abyss*, 1903). The spirit of Priestley's deridingly ironic question 'Who wants to know about coal?', fires Orwell's insistence on taking you 'Down the Mine'. 'I know very well that if your supply of coal depended on my walking several miles to a pithead, descending in a cage for half a mile, walking again to the dwindling tunnel where I had to work, then slogging away for about seven hours in that hell, all for something like two pounds a week, your grates would be empty': that could be Orwell; it is Priestley. Both writers share the point that although these northern industrial locations are astoundingly ugly ('the beastliest towns and villages in the country'; that very Orwellian adjective is, in fact, Priestley's), the people there are human; they suffer just like the gentle reader would in their depressing circumstances. Both writers also insist that the south, the region of the bourgeois, ignores the industrial proletarians because they are so conveniently far away from the comfortably-off reader's home, from London, power, the press. 'The railway and motor-coach companies do not run popular excursions to mining districts. Pitmen are not familiar figures in the streets of our large cities. The mining communities are remote, hidden away, mysterious. If there had been several working collieries in London itself, modern English history would have been quite different.' That's Priestley again. What Orwell adds is the much more powerfully demotic vigour with which he makes the accusation:

> In a Lancashire cotton-town you could probably go for months on end without once hearing an 'educated' accent, whereas there can hardly be a town in the South of England where you could throw a brick without hitting the niece of a bishop.

And the vulgarity of the Old Etonian's prose style was no accident. It was, rather, one of the period's sharpest reminders of what the going-over programme required by way of stylistic and tonal adaptability.

8

Notes from the Underground

'THE road from Mandalay to Wigan is a long one.' And if entering the 'strange country' of the industrial, working-man's North was shocking to Priestley, the son of a Yorkshire schoolmaster, how much more traumatic it was for the southern-bred old Etonian. No wonder 'going over' was contemplated by Orwell and his kind as a sort of exile, a self-alienation from most of what was familiar to one's own region and class. One of the most moving passages, among many such, in *The Road to Wigan Pier* is where Orwell 'faces the fact' that 'to abolish class-distinctions means abolishing a part of yourself':

> Here am I, a typical member of the middle class. It is easy for me to say that I want to get rid of class-distinctions, but nearly everything I think and do is a result of class-distinctions. All my notions—notions of good and evil, of pleasant and unpleasant, of funny and serious, of ugly and beautiful—are essentially *middle-class* notions; my taste in books and food and clothes, my sense of honour, my table manners, my turns of speech, my accent, even the characteristic movements of my body, are the products of a special kind of upbringing and a special niche about half-way up the social hierarchy. When I grasp this I grasp that it is no use clapping a proletarian on the back and telling him that he is as good a man as I am; if I want real contact with him, I have got to make an effort for which very likely I am unprepared. For to get outside the class-racket I have got to suppress not merely my private snobbishness, but most of my other tastes and prejudices as well. I have got to alter myself so completely that at the end I should hardly be recognisable as the same person.

And this dramatic proposition of self-imposed exile by no means shed all of its terrors in the contemplation of its historical necessity and moral rectitude—the sort of consolation envisaged by Spender in his poem 'Exiles from their Land, History their Domicile' (in *The Still Centre*). 'The piece-meal, tentative nature of our poetry', wrote Day Lewis in reply to Julian Bell's 'Open Letter' to him, 'is due to the fact that, while as parasites we are instinctively moved to attach ourselves to the working class, our experience and tradition up to the present are purely bourgeois. In such a transitional stage, a groping, unbalanced, exile poetry is inevitable.' None the less, 'Learn the migrant's trust', the Chorus of Day Lewis's *Noah and the Waters* advised.

It was a faith that came hard, but it might pay dividends in the acquiring of what Henry Green praised as 'the idiom of the time'. He was writing about C. M. Doughty's *Arabia Deserta*, but he was thinking too about the writing service-man of the Second War, especially in the light of his own venture in the later '20s into the unknown proletarian regions of Bordesley, Birmingham, where he had migrated, as it were, to work in his family's foundry:

> Now that we are at war, is not the advantage for writers, and for those who read them, that they will be forced, by the need they have to fight, to go out into territories, it may well be at home, which they would never otherwise have visited, and that they will be forced, by way of their own selves, towards a style which, by the impact of a life strange to them and by their honest acceptance of this, will be pure as Doughty's was, so that they will reach each one his own style that shall be his monument?[149]

Walter Allen had earlier vouched in *FNW*, as both a Birmingham man and a one-time foundry employee, for the authenticity of the factory detail and the working man's dialogue in Green's second novel *Living*.[150] Going over, even if the exile was only temporary, like Green's, could sometimes evidently pay off. But the more that taking this long road into exile in new country was promised as the route to salvation for one's art and to personal integration, the more that moral and political pressure was borne in upon bourgeois Liberals and Leftists to undertake the journey, the more dangerously freighted with emotion going over became.

Consider the tone of this account, in the article 'Cambridge Socialism, 1933–1936' written 'By a Group of Contemporaries' for the John Cornford *Memoir*, of Hunger Marchers arriving in Cambridge from the North in 1934: Wigan Pier, so to speak, on its way to London:

> A big demonstration was organised to welcome the marchers at Girton, where refreshments were to be handed out by the girl students. Most of the demonstrators had little personal knowledge of the working class, and of the militant working-class almost none. It was a thrilling moment for them when their demonstration met the tired, shabby, cheerful column whose progress on the road they had followed day by day. Then the students and unemployed formed up together, and marched back down the long hill into Cambridge. At first, some of the students were a bit shy and self-conscious, wondering whether they had a right to be there, wondering whether it would be cheek to buy packets of cigarettes for the men. Gradually they began to enjoy it, singing *Pie in the Sky* and *Solidarity for Ever* and the rest of the marchers' songs. Going through the town, shouting 'Down with the Means Test!' you would see some student you knew slightly, standing on the pavement, staring, a little frightened, at the broken boots and the old mackintoshes. The phrases about the power of the workers and the right to a better life suddenly meant something quite concrete and real. It was significant that no attempt was made to break up this demonstration. A good bivouac had been provided in the Corn Exchange, and a crude medical service. The next evening an indoor meeting was held by the marchers, the biggest and most enthusiastic that Cambridge had seen for many a day. For the first time many students saw the militant workers in action, heard a new rough kind of speech-making from the contingent's leader, Wilf Jobling (since killed in Spain), and felt that men like these could lead them. For the first time many of them openly declared their solidarity with the working-class, and had a chance to prove it next day when they acted as an advance-guard and police defence squad on the way to Saffron Walden, where it was feared that the police would break up the march. Finally, students travelled up to the meeting of the marchers in Hyde Park and joined in the demonstration—the first of many such 'solidarity' contingents.

According to Kathleen Raine's volume of autobiography *The Land Unknown*, Charles Madge, who 'had never seen the working-class' until that day, was one of those who went over on this occasion. So, too, was Donald Maclean the future spy, according to the recollections of his friend Robert Cecil (*Encounter*, April 1978). And, without doubt, it was morally and socially beneficial for upper-middle-class boys and girls to have their narrow social boundaries overthrown like this. But even the CP jargon of the Cambridge account can't entirely theorize away the kinds of personal need that were being met at this 'thrilling moment'. The cult of the worker for such worshippers was never far from the search for the lost authoritative father-figure (Madge's father, Colonel of the Royal Warwickshire Regiment, had been killed, as we've already observed, in the War). It was never very distant, either, from the quest for the family warmth that the bourgeois child was robbed of by being parked at boarding schools. In Spender's case the need for a substitute, caring male from the proletariat that was met by his friend T. A. R. Hyndman (the Jimmy Younger of *World Within World*) was explicitly homosexual:

> I was in love, as it were, with his background, his soldiering, his working-class home. Nothing moved me more than to hear him tell stories of the Cardiff streets of Tiger Bay, of his uncle who was in the Salvation Army and who asked for his trumpet to be buried beside him in the grave, so that when he awoke on the day of judgement he might blow a great blast of hallelujah on it. When Jimmy talked of such things, I was perhaps nearer poetry than talking to most of my fellow poets. At such moments, too, I was very close to certain emotions awakened in childhood by the workers.

And Spender went on to recall the homely soldier who carried him to an air-raid shelter as a small child with a simple warmth his parents always fought shy of manifesting. Spender, though, only came around to frankness about his motives and desires when the '30s were well over. (When Cyril Connolly wrote in the *NS* early in 1937 about 'the typically English band of psychological revolutionaries, people who adopt left-wing political formulas because they hate their fathers or were unhappy at their public schools or insulted at the Customs or lectured about sex', Spender was extremely hurt, and wrote to Isherwood about his gloom over this public allegation that bourgeois Leftists might have odd private motivations: Connolly had been cruel, he thought, 'utterly destructive—and stupid'; his piece was 'a careless and profound and inexcusable cruelty'.[151]) But Orwell, the archetypically homeless, fatherless, boarding-schoolboy, went in for public frankness at the time, readily admitting (in *Wigan Pier*, ch. 7) that what he yearned for was to get inside the kind of pastoral of family life that he'd glimpsed among the workers:

> I have often been struck by the peculiar easy completeness, the perfect symmetry as it were, of a working-class interior at its best. Especially on winter evenings after tea, when the fire glows in the open range and dances mirrored in the steel fender, when Father, in shirt-sleeves, sits in the rocking chair at one side of the fire reading the racing finals, and Mother sits on the other with her sewing, and the children are happy with a pennorth of mint humbugs, and the dog lolls roasting himself on the rag mat—it is a good place to be in, provided that you can be not only in it but sufficiently *of* it to be taken for granted.

Orwell doesn't mind admitting he would dearly like to recover the lost Eden of childhood signified by those mint humbugs ('It always seemed to be summer when I looked back', says George Bowling, after a long rhapsody about the sweeties of his boyhood: Paradise Mixture, Farthing Everlastings, Caraway Comfits, Hundreds and Thousands, Penny Monsters), a Paradise that in *Wigan Pier* Orwell was at pains to associate with the working-class. And what now reads most tellingly in *The Road to Wigan Pier* is its strategically unmisgiving frankness about its author's emotional problems. The book's autobiographically confessional Part Two seems most necessarily to buttress the more prosaically factual Part One. If bourgeois authors are going to go over, Orwell implies, they had best be as frank as can be about their expectations, their guilts and their fears.

The most honest, and most reader-enraging lid Orwell lifted in *Wigan Pier* was his claim that he was brought up, like all the English bourgeoisie, to believe that '*The lower classes smell*'. Nothing illustrated better, for him, the distance between the

well-brought up public-schoolboy and the proletarian than this drummed-in fear
that the workers stank (he could even quote a passage from Somerset Maugham to
prove the fear's wide existence: 'stink . . . makes social intercourse difficult to persons of sensitive nostril'). And nothing else illuminates so well the anxiety of '30s
bourgeois intellectuals about what going over might entail than the panic caused on
the Left by Orwell's allegation. Victor Gollancz supplied, quite unusually, a Foreword to *Wigan Pier* for the benefit of its original Left Book Club readers (January
1937) in which whilst he praised the book's Part I as a marvellously 'terrible record',
he carefully dispraised Part II as characteristic of an unreconstructed 'lower middle-
class' snob. In particular, from among 'over a hundred' passages he wanted to
quarrel with, he singled out 'this quaint idea' about stink for specially guarded
hedging. Storm Jameson, in her important 1937 essay 'Documents', was equally
tough and even brisker. Part II of *Wigan Pier* presents irrelevant neuroses she wants
no truck with. It's the fleshing out of utterly unimportant grammatical fictions about
the writer's self. If a writer goes over to find out 'for his own sake, for some fancied
spiritual advantage to be got from the experience, he had better stay at home: his
presence in Wigan or Hoxton is either irrelevant or impudent':

> The impulse that made him [the would-be 'proletarian' writer] want to know is decent
> and defensible. If he happens to have been born and brought up in Kensington the
> chances are that he has never lifted the blind of his own kitchen at six in the morning,
> with thoughts in his mind of tumbled bed-clothes, dirty grates, and the ring of rust on
> the stove. But there is something very wrong when he has to contort himself into knots
> in order to get to know a worker, man or woman. What is wrong is in him, and he
> cannot blame on to his upbringing what is really a failure of his own will; it is still
> clenched on his idea of himself, given to him by that upbringing but now to be cast off
> as the first condition of growth . . . If, as a child, he had escaped from the nursery and
> been found in some Hoxton backyard he would have been bathed and disinfected and
> made conscious of having run an awful danger, much as though he had been visiting
> savages. The mental attitude persists. Breeding will out!
>   The first thing a socialist writer has to realize is that there is no value in the emotions,
> the spiritual writhings, started in him by the sight, smell, and touch of poverty. The
> emotions are no doubt unavoidable. There is no need to record them. Let him go and
> pour them down the drain.[152]

(Interestingly, in his own review of Koestler's *Spanish Testament*, Orwell dismissed
that book's earlier part as mere Left Book Club propaganda, but praised the personal
prison-diary as of 'the greatest psychological interest', a 'most honest and unusual
document'.)

In practice Victor Gollancz agreed completely with Storm Jameson. In May 1937
the Left Book Club reissued *The Road* as Part I only. Part II had been poured down
the drain. Mealy-mouthing, socialist romanticizing about proletarians had won. As
for Storm Jameson, Yorkshire born, closer to working people, she simply under-
estimated Orwell's problem. 'Meanwhile', Orwell had rounded on Somerset
Maugham, '*do* the lower classes smell?' And, curiously, he fudged the reply. But he
had to; for, of course, they smelled terrible to him, and his peculiarly sensitive
nostrils were the only sniffing instruments he had to rely on.

Persistently, experiences in Orwell's writing get defined by their smell, especially
if it's a bad one. The natives in his novel *Burmese Days* (1934) have a 'feral reek';

workers' houses are like pigsties; poverty is 'spiritual halitosis'. Keen of nostril, like all his main characters, Orwell goes to Wigan nose twitchingly foremost. 'Smells seem to occupy the major portion of the book', complained Harry Pollitt in his *DW* review of *Wigan Pier*.[153] He was right. The bedroom of Orwell's lodgings 'stank like a ferrets' cage'; 'The smell of the kitchen was dreadful'; 'stinking dust-bins' litter the industrial towns; 'smell upstairs almost unbearable' (Barnsley); 'the smell, the dominant and essential thing, is indescribable'; 'The dirt and congestion of these places is such that you cannot well imagine it unless you have tested it with your own eyes and more particularly your nose'; 'the stinking back streets of Leeds and Sheffield'; 'the stench!' (Sheffield). Foul smells disgust him ('All I knew was that it was *lower-class* sweat that I was smelling, and the thought of it made me sick'). He knew that an obsession with smells didn't necessarily make your writing great. Lionel Brittan's novel *Hunger and Love*, he complained in a wireless discussion on 'The Proletarian Writer', 'goes on and on about the intolerable conditions of working-class life, the fact that the roof leaks and the sink smells and all the rest of it'. But you can't, Orwell added, 'found a literature on the fact that the sink smells. As a convention it isn't likely to last so long as the siege of Troy.'[154] None the less, Orwell's own writing is repeatedly magnetized by just such stinks—in the same way as he is continually dabbling and paddling in bad memories of school porridge or ersatz food. Masochistically, Dorothy Hare the Clergyman's Daughter forces herself to drink from the communion chalice gruesome Miss Mayfill's mouth ('surprisingly large, loose, and wet') has been at ('The underlip, pendulous with age, slobbered forward, exposing a strip of gum and a row of false teeth as yellow as the keys of an old piano. On the upper lip was a fringe of dark, dewy moustache. It was not an appetizing mouth'). As a boy Orwell had deemed it obligatory to 'take a swig' from a bottle of beer being passed round among some 'shepherds and pig-men' on a train. 'I cannot describe the horror I felt as that bottle worked its way towards me. If I drank from it after all those lower-class male mouths I felt certain I should vomit.' And, in particular, Orwell feels compelled to keep returning to where odious smells are most certain to emanate: in the noxious pongs of the lavatory.

Notions Orwell hates are those that smell of lavatories: smelly little orthodoxies. His enemies smell like lavatories, too, and also love lavatorial stenches (they're bum suckers, arse lickers, arse kissers). Private schools are a dirty swindle. Tolstoy 'will do dirt on Shakespeare if he can'. But Orwell himself is magnetically drawn to the stink of loos—and not in the lighthearted spirit of Auden, who bohemianly associated Communism with being unwashed ('Not even clean': 'A Communist to Others'), and gibed at his left-wing friends for being fussy about the condition of public lavatory-seats—drawn to describe people's bottoms, to go 'on the bum' (as tramping is described), and to poke about in drains, sewers, cesspits. Going into the details of inter-party polemics, he declares in *Homage to Catalonia*, 'is like diving into a cesspool'. Now 'for a header into the sewage', he'll say, bracing up to yet another malodorous proposition he relishes exposing.

And it's always a header, a diving down: for the fascinating horror of smells accumulates downwards, in sinks, underground among the sewers, in subterranean places ('On the day when there was a full chamber-pot under the breakfast table I decided to leave . . . It was . . . the feeling of stagnant meaningless decay, of having got down into some subterranean place'; 'going into the dark doorway of that common lodging-house seemed to me like going down into some dreadful subterranean place—a sewer full of rats, for instance'). Distress for Orwell comes from nosing about *Down and Out*, 'Down the Mine', 'down' in Spain, 'down hopping', 'Dahn in Kent', in *Lower* Binsfield, among the *lower* classes. Going over is also, for Orwell—as it had been for Jack London in *The People of the Abyss*, the great master who had preceded Orwell in the descent into the social abyss, down to the bottom of society, among the denizens of spikes and the hop-pickers—a going down: a dreadful, fascinated tearing off and digging away of surfaces and upper layers in an act of penetration to the inevitably noisome filths and stinks below:

> Very early in life you acquired the idea that there was something subtly repulsive about a working-class body; you would not get nearer to it than you could help. You watched a great sweaty navvy walking down the road with his pick over his shoulder; you looked at his discoloured shirt and his corduroy trousers stiff with the dirt of a decade; you thought of those nests and layers of greasy rags below, and, under all, the unwashed body brown all over (that was how I used to imagine it), with its strong, bacon-like reek.

'I wanted to submerge myself, to get right down among the oppressed': and Orwell's guiltily atoning dive down from the upper-middle classes, from social superiority and the policies of British Imperialism, from the position of 'man's dominion over man' that he'd endorsed in the Burmese Police, would mightily offend his horrified bourgeois nostrils. But true Socialism consisted of rubbing your nose in miner's muck—whence 'the mingy little beasts' of London Socialists 'would have fled holding their noses'. In fact, by refusing to face up to themselves and their neuroses— precisely fears such as their obsessions about stinking proletarians—bourgeois Socialists have made modern Socialism smell. So that among them the old principles of justice and liberty have got buried 'like a diamond under a mountain of dung'. Inevitably 'the job of the Socialist is to get it out'. Whatever way Orwell looked at it Socialism came down to grubbing in shit.

What irked the bourgeois members of the Left Book Club was Orwell's loud insistence that there were such deep problems involved in his going over. His awkward confessions and allegations readily insinuated that lots of the nicely brought up public-schoolboys, insulated from the fearful workers since childbirth ('I mustn't play with you anymore, / My mother says you're common', as an unfinished poem of Orwell put it of his childhood friend, the plumber's daughter; or, 'My parents kept me from children who were rough', as that Spender poem complained), might really be holding their noses even though they seemed like good comrades. Perhaps, Orwell's story discomfitingly suggested, all bourgeois Socialists were really cringing and flinching disgustedly whenever they moved among workers, as overwhelmed by nausea at the dingy squalidness of seedy proletarian circumstances as, say, the narrative of Greene's *Brighton Rock*. Orwell's convincing demonstration in his tormented olfactory prose that the insulating preferences and hostilities dinned-in by bourgeois education persisted only too energetically even in a lower-upper-middle-class Socialist embarrassed because it was uncomfortably close to right-wingers' sneers ('Invariably they side with filth and famine'; 'What matters most to them is—"Does it Pong?"': Roy Campbell) and because it outrageously flouted orthodox going-over doctrine. 'Alter your Life', advised Spender's poem 'The Uncreating Chaos', quoting Rimbaud, as quoted by André Breton. And the further left the advice came from the more no-nonsense was its tone. '[N]othing on earth can prevent' poets 'from getting beneath the skin' of 'proletarian life—if they want to', brusquely asserted Montagu Slater against what he took to be dithering by Day Lewis.[155] '*Solvitur ambulando*', he repeated, in a review of *The Mind in Chains*: 'The solution is in living'; or, all you've got to do is walk over.[156] The *Memoir* of Cornford would, its compilers believed, 'bring confidence' to working-class readers because it demonstrated 'how an honest thinker, brought up in conditions in which contact with working people is impossible, nevertheless decides to ally himself with the working-class movement'. Even crusty Caudwell conceded (in *Illusion and Reality*) that there were lots of bourgeois authors ready to 'announce themselves as prepared to merge with the proletariat, to accept its theory and its organisation, in . . . concrete living'.

But, clearly, some alterations of 'concrete living' came easier than others. Auden, for instance, wrote to Naomi Mitchison in November 1936 that he was 'looking for a W.E.A. [Workers' Educational Association] job in Yorkshire next year'.[157] This scheme was displaced by his decision to go to Spain, a journey that was the period's most prevalent life-altering option. Another one, taken to with obvious enthusiasm by many bourgeois Leftists, was sexual libertarianism. The revolutionary young writer, declared Day Lewis in his 'Writing and Morals' essay, 'is generally free from those sexual conflicts with which orthodox Christian morality tormented forbears for centuries', and Lewis zealously notched up one-night-stands, busily accumulated mistresses. Most Leftists appear not to have worried about Aldous Huxley's link between a totalitarian state and guilt-free fornication—gibingly emblematized in the eager unzipping of clothing by *Brave New World's* Lenina; 'those Zippers', Huxley wrote to a friend, 'the symbol of the New World, its crest', had caused surprisingly little fuss.[158] Huxley was not the only intellectual shocked. Indeed, casualness about sex in the Soviet Union appears to excite Naomi Mitchison almost as much as its power-stations do. Donald Maclean in her novel *We Have Been Warned* (1935) finds Russian Marfa of the 'thick, brown breasts' ('useful, he thought, not the same as the white chickenny breasts he'd seen on postcards') gratifyingly responsive. 'You take—them pants—off me, Scottish comrade', she urges, 'You got—rubber goods—comrade?': he hasn't—not, evidently being a reader of his authoress's pamphlet *Comments on Birth Control* (1930)—but she doesn't care. Fornication had, in MacNeice's words, 'become a virtue' in such circles: 'the pattern of every night shot through with the pounding and jingling of bedsteads'. Much of Naomi Mitchison's volume of memoirs *You May Well Ask* (1979) is taken up with a righteously indignant account of the troubles she had getting those rubber goods (and others like them) past cautious publishers. Anybody who was offended must have been, she believes, a Tory. But the trouble is that, in truth, rubber goods have no particular politics and spicing up your sex-life didn't at all depend on being a Socialist: indeed, what Mrs Mitchison kept praising was (as Mrs Leavis nicely put it in that powerful 'Lady Novelists and the Lower Orders' essay in *Scrutiny*) 'the vulgarest idea of a good time—dancing to the gramophone, too much alcohol, and universal petting'.

More fundamental changes in sexual behaviour came harder. The 'most significant' give-away about Comrade X, old Etonian CP member and author of *Marxism for Infants* is, Orwell declares in *Wigan Pier*, that he 'invariably marries into his own class'. Orwell was right. To be sure there is a hint of proletarian origins about Alan

Sebrill's Elsie in Upward's trilogy *The Spiral Ascent* and so a possibility that Upward may have transgressed Orwell's little rule. But that would be most rare. All Orwell's own women, for instance, were as bourgeois by origin as Comrade X's, even though Eileen O'Shaughnessy, his first wife, was actually born in the north of England. Miss O'Shaughnessy—daughter of a well-off civil servant, graduate of St Hugh's College, Oxford—got married to Orwell only weeks before he wrote his sneer against Comrade X. Just as instructively, even John Cornford, who lived whilst an undergraduate with a Welsh miner's daughter and had a child by her, ditched the proletarian woman (to the accompaniment of that savagely heartless 'Sad Poem' about 'Clean wounds' being 'easiest to bear') for the more familiarly middle-class Margot Heinemann (reading English at Newnham College and a member of Professor Cornford's chamber music ensemble). Class would out. And it did so even in apparently simpler matters like trying to mix with the workers in pubs. Day Lewis advised his 'Young Revolutionary' (in *New Country*) to spend time 'Investigating the temper of the people'. Lenin did this. 'Pubs', Lewis added, a bit lamely, but quite in keeping with his growing conviction that 'Communism meant the village pub', 'Pubs are good starting-places'. Some young revolutionaries, of course, didn't even get to the starting-line: Spender confides in *World Within World* that he was too shy to enter a public house on his own. But even inside the pub, things could go badly awry, as Orwell's *Keep the Aspidistra Flying* snarlingly reveals.

Gordon Comstock's wealthy socialist chum Ravelston, editor of the periodical *Antichrist* and publisher of Comstock's poems, takes Gordon occasionally to pubs: doing so is 'part of a lifelong attempt to escape from his own class, and become, as it were, an honorary member of the proletariat'.

'Pubs are genuinely proletarian', Ravelston knows. Alas, when he enters one the smell revolts him ('A sour cloud of beer seemed to hang about it'). Ravelston can't help catching sight of 'a well-filled spittoon', or reflecting that the beer 'had been sucked up from some beetle-ridden cellar through yards of slimy tube, and that the glasses had never been washed in their lives, only rinsed in beery water'. The beer tastes horrid too:

> Ravelston walked self-consciously to the bar. People began staring at him again as soon as he stood up. The navvy, still leaning against the bar over his untouched pot of beer, gazed at him with quiet insolence. Ravelston resolved that he would drink no more of this filthy common ale.
>
> 'Two double whiskies, would you, please?' he said apologetically.
> The grim landlady stared. 'What?' she said.
> 'Two double whiskies, please.'
> 'No whisky 'ere. We don't sell spirits. Beer 'ouse, we are.'
> The navvy smiled flickeringly under his moustache. '—ignorant toff!' he was thinking. 'Asking for whisky in a—beer 'ouse!' Ravelston's pale face flushed slightly. He had not known till this moment that some of the poorer pubs cannot afford a spirit licence.

According to Orwell's brother-in-law, Orwell himself was uncomfortable in the working-class pub in Bramley, near Leeds, when he stayed with his sister: 'That bloody brother-in-law of yours', said the landlord, 'gives me the willies.' One of the most embarrassing moments in Humphrey Spender's efforts to get inside working-class Bolton on behalf of Mass-Observation came when he was prevented photographing in a pub by the landlord, and he reverted to haughty class stereotype, insisted on his 'legal rights' to take photographs and got abused for posh tones ('You needn't talk to me in that semi-educated way').

Comstock and Ravelston, Orwell and Spender were slumming. Like all slummers, they were 'going over', but with their bags perpetually packed for the journey back. Striving to transcend class differences, they also granted their stubborn reality, their inevitability, for the historical moment at least. Slummers live with their feet, and spirit, in two worlds, they lead double lives: not a happy condition however expertly its protagonists learn to manipulate it. But slumming, dodging between classes, back and forth across the social No-Man's-Land, was the best that most bourgeois '30s intellectuals could manage. Philip Toynbee describes (in *Friends Apart*) moving uneasily from house-parties at Castle Howard and the Bonham-Carters' place in Wiltshire to the cottage of an unemployed Communist miner in Tonypandy: his suitcase, hiding his white-tie and tails, lurked like an accusing albatross under the bed he shared with a seventy-year-old pensioner. Auden, at Oxford, usually (it's claimed) relaxed by frequenting dog-races, boxing-matches, and speedways, but when Isherwood visited Weston (in *Lions and Shadows*) they fetched up at the Christ Church Essay Society. Henry Green settled for longer than most of his contemporaries into manual work at his father's factory, devoting months to finding out 'for myself how by far the greater number live in England'. But he was never more than a short-term visitor and the boss's son, and *Living*—one of the most memorable novels of proletarian life in the period—remains for all its hold on inside information an outsider's story. Indeed, some of its most striking moments are the ironic contrasts achieved in young Dupret's passages between working class Bridesley and his own people's moneyed world. Even Orwell—especially Orwell perhaps—was never less than consciously a migrant with a return ticket, or who could get one from his aunt in Paris, his mother, sister, friends. He was always a visitor, however much an assiduous and sympathetic one, to the Parisian lower-depths, the spikes and hop-fields, the northern coal-face. The case of Christopher Caudwell, who after his conversion to Marxism actually went to live, and do his writing and his political work, in Poplar, in the proletarian East End of London, is the admirable and notable exception that proves the general '30s rule. Caudwell, at least, earned the right to keep harping on at the Auden group for their failures and reservations: 'We ask that you should *really* live in the new world and not leave your soul in the past.'

The problem did not, of course, lie only with the bourgeois. The British class system is built on mutual recognition of differences, on the trained capacity of people on each side of the class frontier to read the semiotic constituted by accents, clothes, names. Both sides have traditionally agreed, what's more, that it's offensive to change social stations. Bourgeois goers-over found the working-man cannily distrustful of his intentions. T. E. Lawrence's proletarian comrades (in *The Mint*) not only recognized his toff-tones, they kept urging him to use the masterfulness they recognized against uppity NCOs: 'Give the ignorant shit-bag a fucking great gob of your toffology.' Even in the International Brigades, as we shall see, the British working-men seem to have been glad to have had Cornford and Fox and their ilk fighting alongside *as* gentlemen and graduates and such. 'Me and another young lad from Reading University': seaman, railwayman, and Communist George Leeson's happy recall of his comrade's bourgeois status (in *The Road to Spain*) is typical of the working-class Brigaders' memoirs.[159]

It was much easier to get on equal terms with the working-man among foreigners. Abroad, the key clues as to class, particularly accent, got faded and blurred, and didn't matter terribly: at least if you steered clear of the Empire and other foreign situations where the English bourgeois was still expected to play the game, put up a good show and generally to behave masterfully. No wonder '30s authors (in huge numbers) put in time abroad and sought out unrespectable foreign situations; no wonder they so blissfully welcomed the democratic unity of revolutionary Spain, from which bourgeois clothes were at first banished and where British accents all registered impartially, at least on the ears of the international comrades. How different from at home. Humphrey Spender devotedly sank himself into the Stepney slums and working-class Bolton, bringing back memorable and compassionate photographic documents, but he never got over his sense of 'the class distinction, the fact that I was somebody from another planet, intruding on another kind of life'. At the very heart of Virginia Woolf's efforts to get on terms with working women, during a meeting in Newcastle in 1913 of the Women's Cooperative Guild, she came gloomily to realize her social distance, the 'impassability' of the class barrier. 'One could', she confesses in her Introductory Letter to *Life as we Have Known It, By Cooperative Working Women* (1931), 'not be Mrs Giles [of Durham] because one's body had never stood at the wash-tub: one's hands had never wrung and scrubbed and chopped up whatever the meat may be that makes up a miner's supper', and so 'Our sympathy is fictitious, not real'. Relationships would eventually improve, 'but only when we are dead'.[160] 'For some months', Orwell complains in *Wigan Pier*

> I lived entirely in coal-miners' houses. I ate my meals with the family, I washed at the kitchen sink, I shared bedrooms with miners, drank beer with them, played darts with them, talked to them by the hour together. But though I was among them, and I hope and trust they did not find me a nuisance, I was not one of them, and they knew it even better than I did. However much you like them, however interesting you find their conversation, there is always that accursed itch of class-difference, like the pea under the princess's mattress. It is not a question of dislike or distaste, only of *difference*, but it is enough to make real intimacy impossible. Even with miners who described themselves as Communists I found that it needed tactful manoeuvrings to prevent them from calling me 'sir'; and all of them, except in moments of great animation, softened their northern accents for my benefit. I liked them and hoped they liked me; but I went among them as a foreigner, and both of us were aware of it. Whichever way you turn this curse of class-difference confronts you like a wall of stone. Or rather it is not so much like a stone wall as the plate-glass pane of an aquarium; it is so easy to pretend that it isn't there, and so impossible to get through it.

One possible stratagem for confronting this dilemma, a most despairing ploy, was to attempt disguise. 'I would go', Orwell decided, 'suitably disguised to Limehouse and Whitechapel and such places'—and after all, wasn't it by a simple change of clothes that Jack London had achieved equality of social footing with the working class ('Presto! In the twinkling of an eye, so to say, I had become one of them. My frayed and out-at-elbows jacket was the badge and advertisement of my class, which was their class')?

The author who wanted to be in 'creative communion' with 'the active life' of his times, must, according to Ralph Fox in *The Novel and the People* (1937), practise disguise; must act like history's tyrants who 'mingled at night-time with their subjects, carefully disguised as common men'. The Mass-Observers loved disguise: they posed as drunks in order to observe the sex-life going on among Blackpool's sand-dunes; Walter Hood, an unemployed miner from Ruskin College, put on fox-furs to go Observing as a woman in Bolton pubs; Tom Harrisson pretended to come 'from another dialect area only a few miles away' so as to explain away his 'so-called Oxford accent'; Woodrow Wyatt once impersonated Tom Harrisson in an interview with a journalist. Harrisson denied that his observers were class spies—they were striving to become 'unobserved observers'—but he might have been describing secret agents when he talked of the typical Mass-Observer as 'a participant, invisibly controlled and disciplined from outside, reporting continuously to head-quarters'.[161] Certainly Day Lewis saw the Communist as one of history's spies (at the end of *Starting Point*): 'history had called them out of the ranks and given them her secret orders: they were the spies she sent forward into a hostile country, a land whose promise they alone could fully realize.' Some such enthusiasm for the spy's career no doubt undergirds the fascination for spy stories felt by Left-inclined Eric Ambler and Graham Greene—not to mention Auden and Isherwood and hosts of others. Lenin's phrase, 'To go hungry, work illegally and be anonymous' makes a key reference-point. Day Lewis quotes it in his 'Writers and Morals' piece, and again in his poem 'On the Twentieth Anniversary of Soviet Power' (in *In Letters of Red*). Auden ends his poem 'Our Hunting Fathers' (*Look, Stranger!*) with a variant on it. Its spirit determines the revolutionary 'Interim' of Randall Swingler's poem of that name: 'Life / Goes underground and does illegal work, / For whose effect but history has an eye.' Auden works the quotation into his *T. E. Lawrence* review (written around the time of 'Our Hunting Fathers' early in 1934): 'The self must first learn to be indifferent: as Lenin said, "To go hungry, work illegally and be anonymous": Lawrence's enlistment in the Air Force and Rimbaud's adoption of a trading career are essentially similar'.

And the concatenation of names there is immensely revealing: Rimbaud, the homosexual poet in exile and working as 'Rimbaud the Trader'; T. E. Lawrence, the fallen officer hero who merged himself in the ranks under an assumed name; Lenin the great revolutionary exile and mole (Day Lewis speaks Audenesquely of his Communists having 'the unobtrusive patience of the mole'). The romantic glamour of the going underground proposition was, it appears, matched only by the personal, political and poetical imperatives at work in it. The artist, declared Auden in 'Letter to Lord Byron' (Part III) 'must keep hidden' the nature of his trade: 'like a secret agent.'

And disguise of one sort or another was practised extraordinarily widely in the '30s. The problem of voice, in Britain the most fundamental index of class, was the most difficult to overcome. 'I sometimes had trouble to make my accent understood or to understand theirs,' confesses Green in *Pack My Bag*. Orwell might announce breezily at the end of *Wigan Pier* that 'we have nothing to lose but our aitches', but he couldn't drop his so smartly:

> At the start it was not easy. It meant masquerading and I have no talent for acting. I cannot . . . disguise my accent, at any rate not for more than a very few minutes. I imagined—notice the frightful class-consciousness of the Englishman—that I should be spotted as a 'gentleman' the moment I opened my mouth.

And he was, of course; which didn't stop him persisting (according to William Empson) in a badly fake and self-exposing Cockney accent. Mocking exposure didn't stop Isherwood either. He tells, in *Lions and Shadows* how, envious of a friend's skills as social amphibian, and anxious to discover 'the English' for the sake of his art of fiction, he developed a 'disguise-language', a 'slight Cockney twang', to help him at least to be a social spy. It was a grotesquely comic move for he knew his Isle of Wight boatmen friends saw instantly through it; none the less, he couldn't stop using it.

Trying to change one's speech—and the slanginess of both Empson's and Orwell's prose may owe something to this effort—was frequently accompanied by changes of costume. Wyndham Lewis alleges a connection in 'One-Way Song' (section XXVII):

> Great democrats they are, demotic tags
> Sprout from their mouths, they affect in public rags
> Almost, or homespun—sweatshirts and apache caps.

Rayner Heppenstall likewise gibed at 'New Country Singers' who stand 'drinks in proletpubs' or, worse, 'wear peaked caps, cultivate an accent, smoke very foul tobacco . . . and cease to wash'.[162] Changing clothes was an easier, even if finally no more successful a ploy than altering one's accent. 'I got hold of the right kind of clothes and dirtied them in appropriate places. I am a difficult person to disguise, being abnormally tall, but I did at least know what a tramp looks like'—and he did, for this is Orwell speaking, remember *The People of the Abyss*. None the less, as Orwell went on in *Wigan Pier* to confide, 'You can become a tramp simply by putting on the right clothes and going to the nearest casual ward, but you can't become a navvy or a coal-miner.' Old clothes made the bourgeois writer no more of a working-man than the Mass-Observer's fox-furs made him a woman. Class shone brightly through bohemian garb, which anyway tightened your links with the tradition of unkempt upper-class Englishmen. Ravelston doesn't sport a hat, and tries hard to look impoverished, but

> You could tell him at a glance for a rich young man. He wore the uniform of the moneyed intelligentsia; an old tweed coat—but it was one of those coats which have been made by a good tailor and grow more aristocratic as they grow older—very loose grey flannel bags, a grey pullover, much-worn brown shoes. He made a point of going everywhere, even to fashionable houses and expensive restaurants, in these clothes, just to show his contempt for upper-class conventions; he did not fully realize that it is only the upper classes who can do these things.

Wearing old clothes wasn't much: but it did perhaps help normalize you a little bit (Graham Greene frequently comments in his film reviews on the abnormal newness of the clothes worn in movies). Similarly with the wide-spread tinkering with their names indulged in by '30s authors. In a review of the Twentieth Century volume of the *New Cambridge Bibliography of English Literature* (1973) Christopher Ricks talked of the shock of seeing the full names of authors we've got used to in some shorter version: 'there are a great many writers who would ring quite differently in our ears had they used the first of their Christian names.'[163] Tinkering with given names is of course a normal part of the budding writer's growth in self-awareness and assertion of identity, and a deal of '30s name-juggling must be thought of as simply the regular process: Stephen Harold Spender becomes Stephen; Theodore Philip Toynbee becomes Philip; Frederick Louis MacNeice drops Frederick; and probably mainly because their names sound nicer to them that way. Julian Symons, who had an ear for what gave poets' names panache (he'd changed his own name from Julius) advised Roy Fuller to stop appearing as R. B. Fuller: 'Unless it is Reginald, you ought to use your first name.' Documentary film-maker Paul Rotha adopted the Rotha as an art student because his Slade professor told him success would never attend someone boringly named Paul Thompson. Commonly, though, in this period the choice of name reflects a deliberate effort to disguise a bourgeois identity. Stubby proley names became hotly fashionable: Evelyn Strachey became John; Humphrey Slater became Hugh; Gabriel Carritt turned into Bill; Reginald Ernest Warner tried being R. E., then settled for Rex. So many bourgeois Christian names got dropped, it was like jackpot-time on a socially alert fruit-machine. Rudolf John Lehmann jettisoned the Rudolf, Christopher St John Sprigg ditched the St John (always posh-sounding in English ears) and became Christopher Caudwell. Rupert John Cornford (named for his mother's friend, the poet Rupert Brooke) had early on turned into plain John (his close friends say there was no political reason for this), but still felt uneasy enough to appear in print on one occasion as the very proletarian-sounding Dai Barton. Ralph Winston Fox, born in 1900 and doubtless named after the Boer War hero Winston Churchill, quickly shed that embarrassing association to become plain Ralph. There was nothing exclusively bourgeois about the name of the science and maths student William Edward Roberts, but he still wanted to affirm a decisive shift in identity when he changed his politics and joined the Communist Party at Trinity College, Cambridge, in the 1920s. So he started calling himself Michael after the Russian scientist-poet Mikhail Lomosonov. Hence *the* Michael Roberts.

Hyphenated surnames obviously stigmatized one as bourgeois. So Cecil Day-Lewis (as such the dedicatee of Auden's *Paid on Both Sides*) became C. Day Lewis—the hyphen only creeping back in after the Second World War. (As Day-Lewis he was Poet Laureate; his son Sean Day-Lewis's biography is emphatically *C. Day-Lewis: An English Literary Life*; concern about the Cecil seems, though, to have lasted.) The Arthur Calder-Marshall of *Oxford Poetry 1928* and *1929* appears suddenly to have metamorphosed into plain Arthur Marshall in the *1930* number (though he retained the hyphen for his later '30s appearances). Isherwood didn't know for ages precisely what he wanted to be called in place of his given names. Was he to be, as he first appeared in public in *Public School Verse*, C. W. B. Isherwood ('the name of a least-likely-to-be-read author, if ever I heard one')? Or Ch. Isherwood, as the original title page of his translation of Baudelaire's *Intimate Journals* had him (the critics laid into *Charles* Isherwood's translating)? Or Bill Bradshaw (the narrator of *Mr Norris Changes Trains*)? Or Chris Isherwood (Herr Issyvoo) of *Goodbye to Berlin*? The final outcome mattered less than Isherwood's resolution not to continue bearing his 'ponderous double-barrelled name', Christopher William Bradshaw-Isherwood. Not dissimilarly, Orwell was happy with almost any name but his own. 'What I profoundly wanted, at that time, was to find some way of getting out of the respectable world altogether. I meditated upon it a great deal, I even planned parts of it in detail; how one could sell everything, give everything away, change one's name . . .' Out on the bum he called himself Edward or P. S. Burton. But he wasn't particular when it came to naming the author of *Down and Out in Paris and London*, as long as he wasn't to remain Eric Blair. He let Gollancz choose between P. S. Burton, Kenneth Miles, H. Lewis Allways, and George Orwell. Gollancz's choice, George Orwell, would do as well as any other pseudonym. A disguise of name—almost any disguise—was what mattered: a feeling apparently shared by some other writers with bourgeois backgrounds who wrote about the proletariat. There is as little obvious class distinction between Henry Green and Henry Yorke as between George Orwell and Eric Blair, or, for that matter (except for an enhanced hint of Scottishness), between Lewis Grassic Gibbon and J. Leslie Mitchell.

Unsympathetic observers found all this rather feeble disguising (much of it at the level of Christopher Isherwood eating lots of sweets in order to ruin his good bourgeois teeth, or Randall Swingler—according to Julian Maclaren-Ross—arranging his face 'to look like the face of a working man without the features being all that rough-hewn') merely snigger-worthy. Cyril Connolly chortles hugely over his Christian de Clavering who turns himself into plain Cris Clay. It was all too close to a game. It was like Agatha Runcible's confusion between Parties and parties in *Vile Bodies* ('she had heard someone say something about an Independent Labour Party, and was furious that she had not been asked'). Virginia Woolf scathingly accused Auden and Spender of merely dressing up. The 'mechanic's blue dungarees' assumed by the Communist Brother in *Trial of a Judge* (they are the famous *mono azul* widely sported in the Civil War in Spain and gear highly fancied by Spender) were, of course, affectedly dressing-up kit.[164] For his part, Auden thought highly enough of charades to subtitle *Paid on Both Sides* 'A Charade', and to allege in the Group Theatre programme for *Sweeney Agonistes* and *The Dance of Death* that 'the country house charade' was among 'the most living drama of today'. In Ode III of *The Orators* he promises Edward Upward 'Charades and ragging' over Christmas. No doubt Auden was in earnest about charades, but his attempts at social disguising, shot through with a marring frivolity, sound like just the kind of charades one needn't take seriously. Weston, in *Lions and Shadows*, is a great hat wearer:

> There was an opera hat—belonging to the period when he decided that poets ought to dress like bank directors, in morning cut-aways and striped trousers or evening swallow-tails. There was a workman's cap, with a shiny black peak, which he bought while he was living in Berlin, and which had, in the end, to be burnt, because he was sick into it one evening in a cinema. There was, and occasionally still is, a panama with a black ribbon—representing, I think, Weston's conception of himself as a lunatic clergyman; always a favourite role. Also, most insidious of all, there exists, somewhere in the background, a schoolmaster's mortar-board. He has never actually dared to show me this: but I have seen him wearing it in several photographs.

All this, and especially the workman's cap, seems only as gamily boyish as most of the spying going on in the early Auden poetry:

> Our hopes were set still on the spies' career,
> Prizing the glasses and the old felt hat,
> And all the secrets we discovered were
> Extraordinary and false.

It's 'false beard' time in 'August for the people' (as it was in 'Having abdicated with comparative ease') and nobody can take false beards seriously, which is a drawback for Waugh's *Vile Bodies*, whose Audenesque but seriously intended Father Rothschild totes one in his suitcase. (All 'beards look false nowadays', Paul Perry of Day Lewis's *Malice in Wonderland*, 1940, is made knowingly to think.) Many, Auden goes on in 'August for the People', 'many wore wigs'; and it is his wig that helps make Isherwood's Arthur Norris in *Mr Norris Changes Trains* so deflatedly absurd as a spy. It's as difficult to accept his conspiratorial credentials as it is those of Isherwood and Upward (Chalmers), conspiring, according to *Lions and Shadows*, with Mortmere against Cambridge ('venturing, like spies, into an enemy stronghold') or (in Upward's case, in 'The Colleagues', in *New Country*) against prep schools ('Imagining myself a spy of manners'). *All the Conspirators* they were, and a jolly implausible lot too. Francis in *The Dog Beneath the Skin* makes yet another disguised '30s venturer into the zones of the socially inferior ('As a dog, I learnt with what a mixture of fear, bullying, and condescending kindness you treat those whom you consider your inferiors, but on whom you are dependent for your pleasures. It's an awful shock to start seeing people from underneath') but the insistent note of charade ('Doggie! Young men wore me at charades to arouse in others undisguised human amusement and desire. Talking about charades . . .') seems only too apt to the play's cabaret lightness. The problematic of coming to grips with rifts between the classes keeps resisting its reduction to a jokey business of funny hats, dog-skins, or stiff collars. According to an altogether earnest bit of Connolly's *Enemies of Promise*, you were likely to recognize a writer who 'goes over' by his 'disinclination to wear a hat or a stiff collar'. At candid moments like that, '30s writers do indeed seem to be 'talking about charades', and to deserve the fierce hostilities of the politically more committed: of a John Cornford, for example, turning in his poem 'Keep Culture out of Cambridge' on Eliot and Auden ('The Kestrel joy and the change of heart'), on D. H. Lawrence ('The dark, mysterious urge of the blood') and the Surrealists ('The donkeys shitting on Dali's food'):

> There's none of these fashions have come to stay,
> And there's nobody here got time to play.
> All we've brought are our party cards
> Which are no bloody good for your bloody charades.

And yet any disguising, even unserious charades, must participate in the worry inevitably attendant on keeping up false appearances. 'The spy goes in disguise, trying to appear what he is not', runs the epigraph of Eric Ambler's *Epitaph for a Spy* (1938). Morally and personally, such pretending puts its protagonists in a potentially horrific position. Astutely, Aldous Huxley perceived the strains implicit in women's practice of painting their faces (still a new enough practice to be widely commented on in the early '30s). 'In Paris, where this over-painting is most pronounced', he complained in 'The Beauty Industry' (*Music At Night*, 1931), 'many women have ceased to look human at all. Whitewashed and ruddled, they seem to be wearing masks . . . the outward and visible signs of some emotional or instinctive disharmony, accepted as a chronic condition of being.' MacNeice stuck up (in *C*, October 1936) not only for Yeats's unfashionable snobberies ('There are more things in heaven and earth than are dreamt of in the philosophy of *The New Statesman and Nation*') but for his masking. 'Let us pay homage to Mr. Yeats and his mask.' Defiantly, MacNeice, the poet with the repeatedly split self, defended Yeats the masked poet: 'Who else would dare to say today that to have a full life one must wear a mask?' Who else, indeed. Earlier (in *C*, April 1933), Auden had dwelt on the tensions of school-masterly mask-wearing:

> a profession where adults are expected, perhaps inevitably, to profess official opinions on every subject of importance, to lead the private life of a clergyman, where a mask is essential, sets up a strain that only the long holidays of which other professions are often so jealous, safeguard from developing into a nervous breakdown.

This motif of people and things trying to appear what they are not pervades the '30s, generating pervasive consternation. Orwell's George Bowling, eating a horrid frank-furter ('pop! The thing burst in my mouth like a rotten pear. A sort of horrible soft stuff was oozing all over my tongue . . . It was *fish*! A sausage, a thing calling itself a frankfurter, filled with fish!') remembers reading

> about those food-factories in Germany where everything's made out of something else. Ersatz, they call it. I remembered reading that *they* were making sausages out of fish, and fish, no doubt, out of something different. It gave me the feeling that I'd bitten into the modern world and discovered what it was really made of. That's the way we're going nowadays. Everything slick and streamlined, everything made out of something else . . . when you come down to brass tacks and get your teeth into something solid, a sausage for instance, that's what you get. Rotten fish in a rubber skin. Bombs of filth bursting inside your mouth.

It's not only food that's ersatz. Bowling discovers Lower Binfield to be full of 'fake-picturesque houses', and Upper Binfield of 'sham-Tudor'—the 'bogus Tudor' of Betjeman's 'Slough'. And design in the '30s was liberally seeded by such disguise motifs. Railway trains, motorcars, chairs and houses could all come 'streamlined'. ('There is no sense in designing a streamlined house', grouched the *Criterion*'s Roger Hinks, 'for a house is rooted to the ground and was never meant to stir an inch from the spot where it was first built.') Buildings and soft furnishings were to be jazzed-up in the zig-zag of 'jazz-modern'. Buildings were made to look like ocean-liners ('Sub-urb factories, / Compact as a liner admired from the littered beach'—as Bernard Spencer described them in his poem 'Suburb Factories';[165] Greene's confidential agent describes the Southcrawl Lido as being like a ship, its rooms like cabins—'there was even a porthole instead of a window'). Factories looked like Aztec temples. Last week's houses were tricked out in fake 'Tudorbethan' dress. And what was felt to be wrong about all this was not simply that too much modernist design was too machined looking, too shaped by an inhumanly machine-age (though there was that: for instance, the ghastly architect Otto Friedrich Silenus in Waugh's *Decline and Fall*, product of Constructivism and the Bauhaus, offends by aiming to eliminate 'the human element from the consideration of form. The only perfect building must be the factory, because that is built to house machines, not men'). More troubling still was the act of disguising that was so frequently involved. Houses are not machines, despite Le Corbusier's dictum 'La maison est une machine à habiter'. (It's interesting how Le Corbusier, like Lenin and Trotsky and Stalin, a man dis-guised by a pseudonym—he was really Charles Edouard Jeanneret—presided, like the Soviet Trio, over so much '30s disguising.) Waugh's hostility to Corbusier was intense. His *Country Life* tirade, 'A Call to the Orders' (February 1938), lambasted the 'Corbusier plague' of masquerade architecture: 'Villas like sewage farms, mansions like half-submerged Channel steamers, offices like vast bee-hives and cucumber frames.' He finds it destructively bogus, almost as much a failure of morals as a jarring collapse of style, to plant into the serenely traditional English of Margot Best-Chetwynde's Tudor heritage Silenus' ferro-concrete and aluminium or (in *A Handful of Dust*) to overlay even Hetton Abbey's unfashionable Victorian Gothic with chromium plating. And what upsets Waugh in the matter of Great Houses becomes generalized in the period's writers' contempt for the suburbs, key locations of pretentious Tudorbethan, the 'bogus modern' villa, the bungalow's pseudo-ruralism. 'Bungaloid growth', 'ribbon' development ('ribbon and rash') became dirty words not least because in their hasty despoiling of the real countryside these '30s phenomena made irritatingly sham claims to a rurality they were in fact helping to ruin.

The life of suburbs and garden cities grated because it was neither fish nor frankfurter, neither completely town nor country. It served a delusion, a 'widespread day-dream', as Orwell called it in his long '40s essay 'The English People', of being a 'feudal landowner', a 'country gentleman': 'The manor house with its park and its walled garden reappears in reduced form in the stockbroker's week-end cottage, in the suburban villa with its lawn and herbaceous border; perhaps even in the potted nasturtiums on the window-sill of the Bayswater flat.' 'Every little owner of every little bungalow in every roadside ribbon thinks he is living in Merrie England because he has those "roses round the door" and because he has sweet-william and Michaelmas-daisies in his front garden. An amazing conception.' Thus Thomas Sharp in just one of his polemics against what he labelled the 'Town-Country' illusion. Garden cities particularly, gibes Sharp, try to rebut their urbanism by refusing to call a street a street (it must be Way, Avenue, Gardens, Grove, Drive, Green, Ridge, Hill, Dale, Lane: designations with 'the proper country flavour').[166] But the pretending was widespread, and attacked from all sides, in general and in particular. Eliot laments in *The Rock* that life lived 'dispersed on ribbon roads' is without real Christian community. Spender's *Forward from Liberalism* finds it '*laissez faire* run mad, a huge inflation of Tudor villas on arterial roads, wireless sets, tin cars, golf clubs—the paradise of the bourgeoisie'. Betjeman lays repeatedly into the invented bogies he holds responsible: Tudor Bungalettes Ltd., the Beautisite Bill-Posting Co., The Take It or Leave It Building Society (in 'Percy's Progress', *N & D*, 26 August 1937), The Cosyville Bungalow Co., The Easicatch Building Society (in *C*, October 1938).

It's easy, of course, to catch in that *Criterion* article a distinct note of old rentier snootiness towards the small-time mortgagee:

> It is going to take nothing short of a revolution to make Jones, who has been paying
> fifteen shillings a week for the last twenty years in order to become the owner of his
> jerry-built, semi-detached embryo slum, give up his rights. He is not going to sacrifice
> the standard roses on his front lawn, for any road widening scheme, if he can help it.

Similar snobberies animate Betjeman's 'Slough' with its harsh gibes about tinned food and the half-a-crown mortgage: all a long step away from the tennis courts and rhododendrons of the Surrey homesteads Betjeman preferred. Only the likes of J. B. Priestley, who entered readily into the feelings of working-class and petit-bourgeois folk rising up the mortgage ladder with their eye on a little car of their own, would defend Jones. Priestley's article on 'Houses' in the *Saturday Review* (11 June 1927) sticks up boldly for the hopes and aspirations invested in Tudorbethan, 'those new houses' which are 'a kind of signpost pointing to a sunlit main road of life' for people from houses with no 'proper sink in' nor a separate bathroom. Otherwise Jones had to endure the withering scorn of the well-settled bourgeois or tasteful bohemian, in whose minds existed a lurid map of horrific new England, of the offending disguised 'Town-Country' places—the suburbs and garden cities, Betjeman's 'Villadom', the boom towns and new towns of the Midlands and South East of England, the prospering places where motor-cars are king. Betjeman is one of the chief anti-cartographers:

> Surrey is all one suburb, so is most of Bucks: the town of London stretches out to Maidenhead, St. Albans, Hertford, Southend, Tonbridge, Brighton and Haslemere. At Slough, the most valuable market gardening land in England is being turned into factory sites and housing estates. In Sussex, the rich land below the downs is being covered with houses . . .

'Come, friendly bombs, and fall on Slough'; and not just on Slough. There are also Betjeman's favourite targets Welwyn and Letchworth (Betjeman excludes from his *Collected Poems* his poem 'The Outer Suburbs' with its slights against kitchenette, suburban stained-glass, 'Drage-way drawing-room' and 'Jacobethan' bedroom, and his poem 'The Garden City'—'Men of Welwyn! Men of Worth! / The Health Reform is growing': two of the liveliest of his sustained jeers at life in Oakleigh Park, Rosslyn Avenue, and Orchard Way). And then there's Rickmansworth (G. E. Trevelyan, 'On Garden Cities', in *Red Rags: Essays of Hate from Oxford*, 1933, is particularly fierce against Rickmansworth men); and Hayes (*A City of Sound*, according to its own publicity: headquarters of the Marconi Company and home of HMV gramophone records); and Dagenham where the Ford Car Company was (Orwell's *Coming Up For Air* throws Hayes and Dagenham in with Slough and Lower Binfield); and car and plane-building Coventry (birthplace of Cyril Connolly, deprecated in *Enemies of Promise* as 'ill-famed Coventry . . . a mother of bicycles'; home town of pathetic Lucia in Greene's *England Made Me:* 'all Coventry was in her gesture' as she threw herself at Farrant); and Cowley (the dread Oxford suburb where Morris motorcars were made, Betjeman's 'Motopolis'); and Croydon (Durrell's Gracie of *The Black Book* is a 'Croydon Juliet'); and Elmers End ('The road opens like a throat at Elmers End. I huddle nervously and press down my foot and Bang! down into the suburban country': *The Black Book* again); and Colindale (Francis Charing in Bates's *Lean Men* abandons his gramophone-making firm, the 'grand normality of a Little Palace in Colindale and prospects of a small car', for Spain); and Stevie Smith's Hadley Wood and Palmers Green (which was where Stevie Smith grew up and went to school; featured in her poem 'Suburb', and written about by her under various guises as Syler's Green or Bottle Green, the foci of her very mixed-up feelings, somewhere between affection for childhood scenes—'Syler's Green, Syler's Green, dear suburb of my infancy'—and the venom shown in her novel *Over the Frontier* towards 'beastly overrated bungaloid' Surrey-Sussex); and Hindhead and Maidenhead ('you may . . . drive beyond Hindhead anyhow', informs MacNeice in 'Sunday Morning'; suburbanites motor 'To Hindhead, or Maidenhead' on the Sabbath in Eliot's *The Rock*; Maidenhead is made to stand for all the bourgeois comforts the workers are deprived of in 'The Mansion', a Hunger March story by Gilbert Bradbury in *Storm*, the Revolutionary Fiction Magazine, April 1933; Betjeman's Sloughites 'often go' to Maidenhead); and Berkhamstead (fruitful base of Greene's vision of Metroland).

A measure of Orwell's patriotic mellowing in *The Lion and the Unicorn* is his capacity to see hope, Englishness, even in these new zones of 'the modern world, the technicians and the higher-paid skilled workers, the airmen and their mechanics, the radio experts, film producers, popular journalists and industrial chemists'—in the world, in short, of London's Western Avenue, of what Betjeman labelled 'the Great Worst Road', where the industrial estates of West London were sited, the temples (as they've been called) to the new gods of consumer durables, pharmaceuticals, gramophones, cosmetics, car accessories, vacuum cleaners, the 'new red cities of Greater London', as Orwell labels them, born out of light-industry by arterial road, 'Slough, Dagenham, Barnet, Letchworth, Hayes'. In general, though, in the '30s the sense of an obscene layering, an offensive masking of old country by new town—the petrol pump in the wilderness (Auden's photograph of one such features in *Letters from Iceland*), the pink bungalow and café at Stonehenge (abused in Clough William-Ellis's *England and The Octopus*, 1924)—overwhelmed even Leftists' sympathies for the ordinary people who inhabited suburbia. Swathed in bogusness, suburbanites were spared the warmth afforded industrial proletarians, the authentic poor.

> He [Anthony Hands of Greene's 'The Other Side of the Border'] walked out past the rhododendrons and the forgotten graves into Metroland. Denton sprawled in red villas up the hillside, but there remained in the long High Street, between the estate agents, the cafés and the two super-cinemas, dwindling signs of the old market town—there was a crusader's helmet in the church . . . You couldn't live in a place like this: it was somewhere to which you returned for sleep and rissoles by the 7.50 or the 8.52 . . . He stared into the photographer's window: yellowing photographs peered out of the diamonded Elizabethan pane—a genuine pane, but you couldn't believe it because of the Tudor Café across the street.

Orwell's '40s unheatedness, his coming around to Priestley's democratic enthusiasm for suburban mortgagees, is not at all typical of his own and other writers' '30s response. Nor is Elizabeth Bowen's capacity to stay resigned and calm amidst the scarred landscapes of new housing estates, white factories, and the discomforts and incivilities of her story 'Attractive Modern Homes'. Her dismay is evident; a disinheritance is going on (her story 'The Disinherited' precisely goes into it); old great houses and estates are being diluted into fraudulent petty bourgeois utopias ('They swerved north a little at Uxbridge and spun into London by the great empty by-pass of Western Avenue. Small new shops stood distracted among the buttercups; in the distance aerial glassy, white factories were beginning to go up among forlorn may trees, branch lines and rusty girders: here and there one was starting to build Jerusalem . . . Skirting the rotten ribs of the White City . . .'). *To the North* makes only a seemly squirm of protest: as though the novelist's strongest passions have been spent rather on the decline of the Irish great house. But still, for all the politeness of its concessions and the deft manœuvrings of its adjectives, a passage like this is joining in the wider lamenting chorus:

> When a great house has been destroyed by fire—left with walls bleached and ghastly and windows gaping with the cold sky—the master has not, perhaps, the heart or the

money to rebuild. Trees that were its companions are cut down and the estate sold up to the speculator. Villas spring up in red rows, each a home for someone, enticing brave little shops, radiant picture palaces: perhaps a park is left round the lake, where couples go boating. Lovers' lanes in asphalt replace the lonely green rides; the obelisk having no approaches is taken away. After dark—where once there was silence a tree's shadow drawn slowly across the grass by the moon, or no moon, an exhalation of darkness— rows of windows come out like lanterns in pink and orange; boxed in bright light hundreds of lives repeat their pattern; wireless picks up a tune from street to street. Shops stream light on the pavements, upon the commotions of late shopping: big buses swarm to the kerb, small cars dart home to the garage, bicycling children flit through the birdless dark. Bright façades of cinemas reflect on to ingoing faces the expectation of pleasure: lovers laugh, gates click, doors swing, lights go on upstairs, couples lie down in honest beds. Life here is liveable, kindly and sometimes gay; there is not a ghost of space or silence; the great house with its dominance and its radiation of avenues is forgotten. When spring is sweet in the air, snowdrops under the paling, when blue autumn blurs the trim streets' perspective or the low sun in winter dazzles the windows' gold—something touches the heart, someone, disturbed, pauses, hand on a villa gate. But not to ask: What was here?

Unhappy the person fobbed off with the petty joys of diluted urbanism amidst a destroyed rurality, the person who can't penetrate the lie, see through the hollowness of the disguise. Unhappiest of all, the ones who can't see through even the sharpest (and most frequently cited) example of this kind of fraudulent horror, Sussex's Peacehaven. Graham Greene plumbs the depths of Pinkie and Rose's spiritual plight (in *Brighton Rock*) by the Peacehaven test:

'It's lovely', Rose said, 'being out here—in the country with you'. Little tarred bungalows with tin roofs paraded backwards, gardens scratched in the chalk, dry flower-beds like Saxon emblems carved on the downs. Notices said: 'Pull in Here', 'Mazawattee Tea', 'Genuine Antiques'; and, hundreds of feet below, the pale-green sea washed into the scarred and shabby side of England. Peacehaven itself dwindled out against the downs: half-made streets turned into grass tracks.

The sense of disguising, on this and on other models, as an immoral faking, obviously reached out via the anxious writer's general feeling of class guilt over his class's privileges, to touch those guilty disguisings that were actually criminal: the converging world of the spy, the masked agent or double-agent that so many '30s Leftists fantasized themselves as, and that some (like Guy Burgess and Anthony Blunt) actually were, and the world of the homosexual that so many '30s Leftists and writers (including Burgess and Blunt) also were. 'Going underground' could include the agent's illegal double life, and the overtly law-abiding homosexual's lawbreaking excursions into the underground cottage, as well as the Orwellian dive into society's lower depths. No wonder Lenin and T. E. Lawrence got blurred in Auden's mind. It was easy in practice for criminal and uncriminal underground activities to merge. Tom Driberg went down and out—he spent time one university vacation as a pavement-artist, dressed in ragged clothes got from a homosexual pick-up on the Embankment—as part of the same movement that took him into sexual cottage work, for which he sometimes took the pseudonym Hardy. It's hardly a surprise that Orwell's Dorothy Hare (and thus Orwell himself?) should profess a feeling of guilt about living down and out under an assumed name:

Ellen Millborough, after Millborough in Suffolk. It seemed a queer thing to have to do, to use a false name; dishonest—criminal, almost.

The guilty knowledge that a despisable bourgeois self lurked behind the disguised voice, the altered name, or the hastily dirtied-up clothing—slummer's guilt stoked up by the criminal associations inevitable to having something to hide—was further exacerbated when there was writing to hide. When Christopher St John Sprigg became Christopher Caudwell he wasn't merely despising the bourgeois world of all St Johns, he was also rejecting his career as the popular author of books (mainly for boys) on the air (*The Airship, British Airways, Great Flights*) and of crime stories. St John Sprigg was responsible for *Crime in Kensington, Fatality in Fleet Street, The Perfect Alibi, Death of an Airman, Death of a Queen, The Corpse With the Sunburned Face*. And the Marxist Caudwell was able zestily to spurn detective stories as pap, as much a drug as religion or jazz (they take, he gibed in *Illusion and Reality*, 'vulgarised values and outraged instincts and soothe both in an ideal wish-fulfilment world') because they belonged to his suppressed *alter ego*: he was to be known from now on as Christopher Caudwell author of the (very slightly) more serious novel *This My Hand* (1936) (a novel which toys with the *Doppelgänger* conventions of the neurotically innocent-guilty, the bond between respectable hangman and outcast murder, the good man's underlying criminal self: 'We are all the innocent-guilty. Our very interest in a murder is due to our feeling that the murderer is bound to us . . . He is atoning for our wicked thoughts'). Caudwell's harshness was the hard Marxist's orthodox line.

C. Day Lewis shared Caudwell's contempt for 'the general anaesthetic of "popular" fiction' ('Writers and Morals'). An 'ever-increasing flood of false art has been turned upon the workers' in the present decay of capitalism; the 'workers' responses to the emotional effect of genuine art' have been weakened by 'dope-fiction', he declared. That was in *Left Review* in July 1935. In March 1935 the first of Lewis's long chain of detective novels *A Question of Proof* had come out. No wonder it was by 'Nicholas Blake': Day Lewis had taken the same evasive action as the character called Frank Cairnes in Nicholas Blake's *The Beast Must Die* (1938) who wrote detective fiction under the assumed name of Felix Lane because he was 'unable to convince myself that detective fiction is a serious branch of literature'. And the earnest Marxist poet Day Lewis fronted dope peddling Nicholas Blake all through his '30s Communist phase. The masking worked, superficially at least, to everyone's advantage: Day Lewis made money, the Party had a mole toiling for it under the lawn of popular writing, and Lewis appeased his conscience by working higher literary references and mildly left-wing demurrals into books that reached far more readers than his poetry ever would. But in the end Day Lewis would learn that there was no escaping Henry Miller's point. Miller warned Lawrence Durrell off writing bad popular stuff as 'Charles Norden': in the end, he wrote in a letter, 'You can't write good *and* bad books.' The deliberately indulged-in badness would, he implied, sooner or later take over entirely.

Nicholas Blake survived Day Lewis's Communism. The pseudonym didn't, though, escape detection. Geoffrey Grigson winkled Lewis out as

> Hiding his twopenny self in Nicholas Blake
> A thriller writer on the literary make.

Trying to hide was pretty futile, as Blair-Orwell (and Colonel T. E. Lawrence-Aircraftsman Shaw) kept discovering. Exposure might, of course, be delayed as it was for the Soviet spies recruited in the '30s—Philby, Maclean, Burgess, Blunt. For Orwell it came sooner rather than later. The 'Tramp Major' of Eric Blair's article 'The Spike' instantly spotted 'a gentleman'; so did the police sergeant who let Orwell off with a mere caution when he had tried hard by means of illegal public drunkenness and begging to get sent to prison to see what jailing was like; so did working-class writer Jack Common who found Orwell clearly 'a public school man', with 'perfect manners' and none of a real tramp's 'desperation'. A toff was a toff. How the ordinary seamen rub in that harsh social fact for little rich boy Dana Hilliot in Lowry's *Ultramarine.* 'No rich man', Orwell wrote feelingly of Ravelston, 'ever succeeds in disguising himself.' Embarrassed, George Orwell deferred the Tramp Major's immediate recognition of him as a gent—talk of his being a journalist is made to intervene—when he incorporated Eric Blair's narrative 'The Spike' into *Down and Out.* Orwell even tried to rub out the incident, absolutely to disown the futile disguise of that particularly disappointing moment in a moment of complete un-selfing: 'Which of you is Blank?', the Major is made to enquire; and Orwell disingenuously adds, 'I forget what name I had given.'

Some literary goers-over may have found their promised land of new selfhood: Cornford and Caudwell did, perhaps; though they were killed in Spain before they, or we, could properly tell. Most literary goers-over, though, were like Orwell and failed to find the land of personal integration they'd been promised. Ironically the disguises they felt driven to assume in order successfully to manage the transition only confirmed them as split men. Typically, the renewed Orwell had to keep going on vying antagonistically with the old Blair as vividly as new Nicholas Blake (or C. Day Lewis) with unrenewed Cecil Day-Lewis. The stubbornly complex problem of the bourgeois self just would not let itself be resolved in a simple commitment to going over. The end of the '30s found authors like Isherwood still casting about among what seem shakier and shakier expedients. In *Journey to a War* (1939) he discovers yet more versions of himself to try out, this time Y. Hsiao Wu and Mr Y. (to Auden's Au Dung and Mr Au): coming out as his truest self, 'Christopher', would have to wait until, perhaps, Isherwood's finally explicit 'coming out' as a surfaced homosexual in *Christopher and His Kind* (1977). Irresolution in the case of Lewis Grassic Gibbon couldn't have been clearer: in 1934 the name of J. Leslie Mitchell and Lewis Grassic Gibbon both appear on the title page of *Nine Against the Unknown.* It was the sort of continuing conflict that Orwell cast in April 1934 (in a poem by 'Eric Blair' in the *Adelphi*; reprinted in *The Best Poems of 1934*), in terms, familiar to the period, of the suburbanizing encroachments of the HMV gramophone factory that was sited on the outskirts of Hayes (where Orwell briefly taught, and wrote *Burmese Days*, upset by the smoke of the factory, the same works that also gets into *Coming Up for Air*, and by the proprietor of the nasty little school he worked at, who was employed by HMV). 'On a Ruined Farm near the His Master's Voice Gramophone Factory' has its narrating self awkwardly tensed between two 'warring worlds'—on the one hand 'the lichened gate', on the other the mechanical and modernistic ('where steel and concrete soar/In dizzy, geometric towers'). And he's an 'alien still', on the border between 'two countries', able neither completely to go over from one to the other, nor to retreat: 'Between two countries, both-ways torn'. He finds himself 'cursed', he declares, 'with double doubts'. Evidently going in for what Jack London labelled the 'double life', assuming what T. S. Eliot was to call 'a double part' (in 'Little Gidding', 1942), was more worrisome in practice than might appear from an unconsidered reading of those writers' and their '30s successors' pages.

Some '30s writers decided, willy-nilly perhaps, to accept their 'double doubts' as intensely fruitful for their art. In his contribution to the Auden Double Number of *New Verse* ('Oxford to Communism') Spender took up Caudwell's attack on half-hearted fellow-travellers:

> From the point of view of the working-class movement the ultimate criticism of Auden and the poets associated with him is that we haven't deliberately and consciously transferred ourselves to the working class. The subject of his poetry is the struggle, but the struggle seen, as it were, by someone who whilst living in one camp, sympathizes with the other; a struggle in fact which while existing externally is also taking place within the mind of the poet himself, who remains a bourgeois. This argument is put very forcibly by Christopher Caudwell . . . Whilst accepting its validity as a critical attitude, may we not say that the position of the writer who sees the conflict as something which is at once subjective to himself and having its external reality in the world—the position outlined in Auden's *Spain*—is one of the most creative, realistic and valid positions for the artist in our time?

This was, of course, precisely to rebut those sterner Marxists like Storm Jameson, to grant tremendous importance—importance most readers would now agree in allowing—to the anxious self-scrutiny of Part II of *Wigan Pier*. But in doing so, Spender was admitting defeat, settling for whatever creativity there might be in the analysis of torn, split, disguising selfhood, shelving the longterm solution of the social problem for the temporary display of the personal one. You 'do not solve the class problem by making friends with tramps', as Orwell said: 'At most you get rid of some of your own class-prejudice by doing so.' What's more, settling for a split personality can never be less than decidedly unsettling. Going in disguise is all right as a game of 'Let's pretend', as a temporary strategem, or as a consciously poetic device. Poetry has gone in characteristically for a world of disguise, of 'as if'— for what is metaphor but a technique for seeing things disguised as other things? But a self-world made entirely out of metaphor, of 'as if', is in the end immaturely childish, and if persisted in dangerously solipsistic. Resorting to it, as '30s authors could readily observe in T. S. Eliot's desperate pleas for impersonality, for the collapse of the self into the fictional, masked world, is a sign of personal despair. All those writers in the tradition of what Eliot calls the 'double part', from James Hogg's *Confessions of a Justified Sinner*, through Poe, Dostoyevsky, *Jekyll and Hyde*, Henry James, Oscar Wilde, Conrad, T. S. Eliot himself, acknowledged the proximity between the self acknowledged to be split or doubled, and evil and madness. It's 'the schizophrenic route' that Henry Miller, alluding to Dostoyevsky's *The Double*, detected as implicit in Durrell's 'pseudonyming': 'Not accepting oneself in toto. Not integrating'. And Orwell's apprehension of Dorothy Hare's falsifying of her name as criminal was amply supported by this tradition of the Double or *Doppelgänger*, the dark, other, underground, secret-sharer self, in which, of course, Orwell's own reading, as well as his practice, amply versed him. Orwell thought Secret Sharer-obsessed Conrad to be 'one of the best writers of this century'; he was planning a long essay on him just before he died. 'I've always been very pro-Wilde', Orwell admitted in 1947; 'I particularly like *Dorian Gray*.' His affection for Wilde's *Doppelgänger* tale went right back to Eton, where (as we learn from *Enemies of Promise*) he lent his own copy of the book to Cyril Connolly.

The masses, commented Harry Kemp in the monstrously right-wing *The Left Heresy in Literature and Life*, about a passage from *Lions and Shadows* where the I-narrator dithers unhappily, excluded from the company of the Isle of Wight holiday-makers, 'do not suffer from the neuroses of split personality' such as afflict '30s Leftists. And, very plainly, on some notable '30s literary occasions where a character goes over or purports to go over to the proletariat madness is in question. A loss of sanity is strongly implicit in Dorothy Hare's loss of memory and of identity that takes her down and out in *A Clergyman's Daughter*. The loss is clamant in the ex-public-schoolboy poet Gordon Comstock's choosing to sink into the depths of unemployment and poverty, the paradoxically attractive warmth and revoltingly stinking awfulness of the slums:

> Under ground, under ground! Down in the safe soft womb of earth . . . It was all bound up in his mind with the thought of being *under ground*. He liked to think about the lost people, the under ground people, tramps, beggars, criminals, prostitutes. . . . Down there in Lambeth, in winter, in the murky streets where the sepia-shadowed faces of tea-drunkards drifted through the mist, you have a *submerged* feeling. Down here you had no contact with money or with culture . . . You were just part of the slum . . . All his habits had deteriorated rapidly. He never shaved more than three times a week nowadays, and only washed the parts that showed . . . He never made his bed properly . . . and never washed his few crocks till all of them had been used twice over. There was a film of dust on everything. In the fender there was always a greasy frying-pan and a couple of plates coated with the remnants of fried eggs . . . he was letting himself go to pieces . . . Better to sink than rise. Down, down into the ghost-kingdom, the shadowy world where shame, effort, decency do not exist!

For his part Edward Upward has the tutor of *Journey to the Border* approach Socialism across a war-torn divide where sanity and insanity disconcertingly converge. The tutor discusses the madness his psycho-political crisis has induced with—inevitably—his double, who tells him:

> '. . . You can call your condition insanity if you like . . . There is room for doubt. Between sanity and insanity there is a sort of no-man's-land. The opposing forces in the deep trenches on either side wear conspicuous uniforms and can be easily distinguished from one another, but no responsible psychologist would be prepared to state categorically to which side belonged an individual crawling in the intervening mud among shell-holes and wire-snags. Your condition could be best described, in clinical language, as "on the border".'
> 'Then I was very nearly insane?'
> '. . . Put it that way if you choose. But you would be equally justified in saying that you had been very nearly sane.'

Arguably, all converts and would-be converts, goers-over of all sorts, have a period of feeling dismayingly dislocated like that as they effect or contemplate their transition. Revealingly the literature that comes directly out of the social no-man's-land in its sexual manifestation as the zone for the bourgeois inverts' transactions with proletarian or German lovers is tremulous with the doubts and worries as well as the excitements associated with what E. M. Forster expressed as the longing to press himself against the corduroys of working men. Forster's fictional fragment called 'Stonebreaking', in which a frankly lustful working man encourages a mute and shy bourgeois male to help him get off his belt (a First War relic) as a prelude to further undressing, is as touchingly fearful in miniature as Forster's novel *Maurice* is at greater length.[167] And in its more directly political manifestation, entering this class border-zone could be just as unsettling. Upward's *The Spiral Ascent*, Orwell's *Wigan Pier*, Spender's autobiography, what we know of Day Lewis's life, to name no other evidence, show bourgeois authors putting themselves through intense misery in their goings over, or attempts at going over, their efforts at transforming their old bourgeois selves and lives into something proletarianly other. The struggle may have been briefly bettering for their art—it certainly produced an intriguing phase of their writing. But the attempts they made to become something else, to turn themselves into *faux*-proletarians, to sink their worrying self into the engrossing social mass, did not last. Sooner or later most of them broke cover and fled—hurt, wounded, but relieved no longer to have to pretend—back to the old isolative status of the more or less bourgeois writer. At the end of September 1939 Spender confided to his diary that 'the views of some Communist writers that today one can only write about the workers and from their point of view seems to me not only nonsense, but also inhibitive and destructive to literature . . . The important thing is to write about what one knows and realize it as fully as possible.'[168] In August 1939 he'd written to Isherwood that he was 'fed up with politics and any kind of public life'. Politics and politicians had forced him 'into taking a totally false position in which I am joining the CP, making public speeches, sitting on committees, etc.' 'I realize that what I most wanted out of life was to write my own stuff and to have a satisfactory relation with Inez. I only imagined I wanted a lot of other things.' Now his wife had left him and the public life had turned to ashes. Spender's expression of his plight was characteristically awash with messy emotionalism. Auden's words, at least, were more coolly put. And his unpublished text *The Prolific and the Devourer* (now in *The English Auden* volume) speaks tellingly for many of the period's damaged survivors:

> The voice of the Tempter: 'Unless you take part in the class struggle, you cannot become a major writer.'
>
> Works of art are created by individuals working alone.

The hostility of those 1939 jottings and the poem 'Another Time' (October 1939) has almost entirely erased any regret over many now seemingly misdirected and wasted years:

> So many have forgotten how
> To say I Am, and would be
> Lost, if they could, in history.

9

Movements of Masses

ONCE he started looking up to big proles, the little boyish author would have in the end to come to terms with the proletariat *en masse* in its biggest aggregation as 'the masses'. It was the masses who comprised the allegedly saving but also worrying populace that the divided bourgeois individualist was invited to go over, or down to. The '30s required of the bourgeois author some sort of response to the masses. At no previous era, not even in the Victorian 'age of great cities', had people been so conscious that modern industrialized, urbanized life was mass-life. Man had become Mass-man, *Massenmensch*, 'The Man' (to use Edgar Allan Poe's title) 'Of the Crowd'. Inescapably, the post-First-War sensibility had to grasp that it was in an age of mass-production, mass-demonstrations, mass-meetings, mass sporting occasions, mass-communications, mass-armies, a time when things would be done in, and to, and for crowds. 'There is one fact, which, whether for good or ill, is of utmost importance in the public life of Europe at the present moment. This fact is the accession of the masses to complete social power.' Thus the opening of the first chapter, entitled 'The Coming of the Masses', of Ortega y Gasset's *The Revolt of the Masses* (published in 1930 in Spanish, and in 1932 in English). 'The multitude has become visible', he declared. 'What', he went on to ask, 'is he like, this mass-man who today dominates public life, political and non-political?' The question haunted every serious person.

Crowds, of course, have no given self. Intrinsically, they possess only immense potentiality. Mere numerosity is politically neutral, neither necessarily left nor right. Values and tendencies have to be imparted to crowds; their meanings await instructors, interpreters, readers. Joe Gates and Aaron Connolly, for instance, go, in Henry Green's *Living*, to a football match to watch Aston Villa, just as their author did on Saturday afternoons when his Bordesley Green factory labours were over for the week:

> As time gets nearer, so more rattles are let off, part of the crowd begins singing. The drunk man, who has a great voice, roars and shouts and near him hundreds of faces are turned to look at him. The band packs up, it moves off, then over at further corner the whole vast crowd that begins roaring, the Villa Team comes out, then everyone is shouting. On face of the two mounds great swaying, like corn before wind, is made down towards the ground, frantic excitement, Gates wailed and sobbed for now his voice had left him. The Villa, the Villa, come on the Villa. Mr Connolly stood like transfixed with passion and 30,000 people waved and shrieked and swayed and clamoured at eleven men who play the best football in the world. These took no notice of the crowd, no notice.

That crowd—it's part of one of the best soccer match descriptions in English fiction—is an obviously good place to be. The warmth of the tone announces that this is a text that's on the same side as those people, as ordinary persons. This is a writing that incites you to believe it's on the inside of the working-class condition. In terms of literary politics and the political alignments of writers, the passage offers as political a set of signals as any author could contrive. In terms of class alignments in literature it's rubbing shoulders with the fictions of a J. B. Priestley (and, of course, it bears comparison with the northern football-crowd scene early in Chapter One of J. B. Priestley's novel *The Good Companions*, 1929). But in party political terms this crowd is only ur-political, a massive political potentia, but for the moment so politically multivalent and omni-directional (like the multi-headed hydra that's a favourite conventional slur-metaphor for crowds) that it's a political blank. And in their unawakened political state of a torpid plenum of meanings, Green's Villa supporters recall other '30s crowds: those hikers, for instance, 'massed' in Auden's 'Letter to Lord Byron', or the thousands of Londoners doing the Lambeth Walk in London's parks in August 1938, a bout of 'mass-dancing' that fascinated the Mass-Observers Charles Madge and Tom Harrisson. Less neutralized thoughts about crowds were, though, also common in the '30s. When Walter Allen thought about Aston Villa, he chose to play up its positive moral health, stressing its history of high, chapel-based seriousness.[169] And for his part, Rex Warner presented football in Part I, Chapter 15 of *The Wild Goose Chase* as a politicized emblem of totalitarian control, a diversion for the masses condoned and arranged by the Government—who've fixed the score in advance, allow the Cons to advance against their opponents the Pros in an armoured car, and have lots of the Red-shirted Pros killed. In neither Allen's nor Warner's reading of it was football neutral (Warner's episode had duly appeared first in *NW*, No. 2). And by the end of the decade Green himself had inclined to social doubts about his football crowd; at least, a mass of 30,000 people is experienced as unpleasingly threatening by Green's cast of rich characters as they watch it thronging the railway station in *Party Going*, 'thousands of Smiths, thousands of Alberts, hundreds of Marys, woven tight as any office carpet':

> as she watched she saw this crowd was in some way different . . . in one section under the window it seemed to be swaying like branches rock in a light wind and, paying greater attention, she seemed to hear a continuous murmur coming from it . . . she flung her window up . . . there was a shriek from somewhere in the crowd, it was all on a vast scale . . . she had not realized what this crowd was, just seeing it through glass. It went on chanting WE WANT TRAINS, WE WANT TRAINS from that one section which surged to and fro and again that same woman shrieked, two or three men were shouting against that chant but she could not distinguish words. She thought how strange it was when hundreds of people turn their heads all in one direction, their faces so much lighter than their dark hats, lozenges, lozenges, lozenges.

' "It's terrifying", Julia said, "I didn't know there were so many people in the world".' The beauty of crowds certainly lay in the eye of the beholder.

'30s responders to the masses called variously on the variously possible readings of the text of the crowd. The Left was committed, more or less, to the communal idea as a good. As we've seen, Socialists believed in plural pronouns, the social 'we', especially the Party 'we' in solidarity with the working-classes. Leftist fiction is full of the Folk: of groups, crowds, multitudes, of large throngs purposefully united in dances, singing, strikes, demonstrations, marches, meetings:

> And now you were all thudding into step, and beyond the drum saw Royal Mile, flashing with trams, thick with bobbies: and here out from the wynds came the Ecclesgriegs men and the fisher-chaps from Kirrieben . . . Communionists like Big Jim

might blether damned stite but they tried to win you your rights for you. And all the march spat on its hands again and gripped the banners and fell in line . . .

Boom-roomr, wee Jake'll brain that damn drum if he isn't careful, God, how folk stare! A new song ebbing down the damp column, you'd aye thought it daft to sing afore this, a lot of faeces, who was an outcast? But damn't, man, now—

Arise, ye outcasts and ye hounded,
Arise, ye slaves of want and fear—

And what the hell else were you, all of you? Singing, you'd never sung so before, all your mates about you, marching as one, you forgot all the chave and trauchle of things, the sting of your feet, nothing could stop you.

A million moving dots of blackness, clustering together, breaking apart, drifting through the streets like wind-driven autumn leaves rustling along the dried-up ruts of woodland cart-tracks.

The dark mass flows through the streets, meanders like some caterpillar crawling across a map of London, its hand a mile away from its tail, its red spots the colours of banners, its solidity the cohesiveness of men and women marching in fours.

Drum-beats . . . marching feet . . . rhythm . . . slogans . . .

Dagenham and West Ham are marching; they have picked up other contingents on the way, they grow in strength all the time . . .

Starvelings arisen from their slumbers. Criminals of want on the march: two hundred and fifty thousand of them: a quarter of a million. Marching from every point of the city . . .

. . . An army with banners. And what banners! Elaborate designs of trade union branches. The Hammer and Sickle of the Communists, the white initials of the Independent Labour Party. Portraits of Lenin and Marx and John MacLean. Hundreds of banners bearing inscriptions: N.U.W.M. Branches: Anti-War Committees: League Against Imperialism: Scottish Socialist Party: Guild of Youth: Young Communist League: Scottish Republican Party: Relief Committee for the Victims of German Fascism. The British Section of the International Socialist Party. And many other leagues, parties and committees including the important United Front Committee and the League against War and Fascism.

There were miles of banners, flags and slogan-boards. It was like ten May Day processions . . .

The National Hunger March had been transformed by the intensification of events into a mass United Front demonstration against the danger of War and Fascism.

These welcomed crowds are from Grassic Gibbon's *Grey Granite*, Sommerfield's *May Day* and James Barke's *Major Operation*. And Leftist literature of that sort is, of course, preoccupied with the going-over drama of the isolated individual, often a bourgeois intellectual, finding his or her true self among the people, or at least in the People's Party. Leftists were expected to be clubbable, to congregate, to join in. Their slogans and organizations proclaimed the goal of communality: United Front, Popular Front, Group Theatre, Unity Theatre, the Left Book Club ('the Club *is* a United Front' insisted their first *DW* advertisement).[170] For Leftist artists and authors 'going over' included, in practice, being willing to enact an awakened group sensitivity in public by appearing on a round of United Front platforms, attending International Writers' Conferences in Paris (1935), London (1936), Republican Spain (1937), throwing in their weight with Artists Against War and Fascism, or the Artists International Association, or the British Section of the International Association of Writers for the Defence of Culture, or the Left Book Club Writers and

Readers Group ('In these days of totalitarian wars and devastating intrigues', wrote Julius Lipton of this last Group in *LR*, April 1938, 'the writer who is isolated no longer glorifies in his isolation. Rather does he feel as one ignored by the mass of the people, as well as by his enlightened fellow writers').

So a positive rhetoric of the masses (not to mention *New Masses*) flourished on the Left. *The Red Stage, Organ of the Workers' Theatre Movement* (1931, 1932) was eager to spread 'mass spirit', 'mass effort', 'mass enthusiasm' by its actors' use of 'mass slogans'. 'Our Theatre Awakens The Masses', runs one of its headlines. The leftist writer would take the leftist politician's 'broad masses' for his province (in Warner's *The Wild Goose Chase* the great peasant Joe is described knowingly as having a 'broad mass'). Socialism and socialist writers turned undismayed to those ready aggregations of people that had tended so frequently to frighten the non-leftist imagination, the crowding inhabitants of the world's great cities. As subject and object of revolutionary hopes the urban masses were diligently smiled on. Knowing no better at the time John Cowper Powys and John Sommerfield at first reeled away from the apocalyptic shocks of New York: Sommerfield in his pre-socialist novel *They Die Young* (1930), in which Manhattan smites young Christopher heavily ('shadowed by abysses, they were crushed by architecture . . . People swarmed. They were not individuals, they were masses, corpuscles in a blood stream: they seeped among buildings cast up to the sky by their masses') and Cowper Powys in *After My Fashion* (his 'missing' post-First War American novel, not published until 1980), where Richard Storm feels 'of no more weight than a floating straw borne on the tide of great irresistible forces' in New York's 'vortex of ferocious energies', its iron and steel hugeness, its 'tornado' of people, 'where the crowds of men and women scourged by economic necessity seemed to dehumanize themselves and become just one more mechanically moving element, paralleled to the iron and steel and stone and marble, to the steam and electricity, whose forces, brutal and insistent, pounded upon it, hammered upon it, resisted it or drove it relentlessly forward'. But Powys was, of course, already in two minds, torn between nostalgic English ruralism ('no fields or lanes in Manhattan') and excited Soviet-incited Futurism. And Storm, an aspirant poet moved by Russian 'experiments', can also be aroused by the intense crowd excitements of New York, the energies—'an unconquerable vitality, a ferocious joyousness and daring'—evinced by the massive chaos of 'this reckless, gay, aggressive crowd'. And by 1931 Powys was enthusing unreservedly over James Hanley (in his Preface to Hanley's *Men in Darkness*) as one of the 'school of young modern writers' which has thrown in 'at least its aesthetic lot with *the masses*', in 'pursuit of the hitherto undiscovered aesthetic possibilities of life in the mass'. This, Powys guesses, may be an instance of 'the Future' itself communicating with artists and 'throwing out electric waves of a great actual change about to take place, whose outward manifestation may be already visible above the horizon in Russia'. This was the accepting mood of Socialist city celebrators: of Sommerfield on London in *May Day*, of James Barke on Glasgow in *Major Operation*, of John Lehmann. In the first number of *New Writing* Lehmann was reporting eagerly from Paris about 'the new revolutionary unity of the workers' and 'the rhythm' granted the city by 'these masses . . . that swarm over the surrounding hills and the factory suburbs':

To the poet, who sits in one corner and lifts his head from his paper to watch the incessant flow of faces before the door, the exultation of the afternoon, fed by the pictures of yesterday's demonstrations before him, seems to deepen with the deepening night, to rise as a challenge to the insolence of the hoardings and the sky-signs, and all these masses suddenly to be transformed by the confidence of victory.

Even more emphatically, Lehmann found Moscow to be the home of 'the dense masses', overflowing 'with the eager life of thousands', who flood 'from the factory and office' (a little echo from Auden's 'A Communist to Others'), 'like an enormous river'. In his *Down River* (1939) Lehmann praised Vienna's huge housing blocks built by socialist town-planners (the Karl Marx Hof, Matteoti Hof and Goethe Hof), as 'historical landmarks in the solution of mass-housing'. Their lecture rooms and libraries were proof of an admirably 'strong emphasis . . . on raising the cultural level of the masses'. It was urban building materials that fell patly into Edgell Rickword's metaphors when he praised Malraux's Spanish-War novel *L'Espoir* (*NW*, Autumn 1938): it was full of 'the presence of the people'; its individuals existed only as 'the steel rods that bind together a concrete mass'.

Noticeably, among mass-minded Leftists, metaphors for mass-life ranged opportunistically from the common organic ones (floods, tornadoes, tides, corpuscles, swarms, herds—the artist is 'as much a member of the common herd as a riveter or a glassblower', insisted Paul Rotha in his book *Documentary Film*, 1936), to inorganic ones, like Rickword's reinforced concrete, or the machined metal of Edgar Foxall's 1933 poem 'A Note on Working-Class Solidarity':

> There will be no festivities when we lay down these tools,
> For we are the massed grooves of grease-smooth systems.[171]

What mattered, though, organic or inorganic, was to be in the mass movement. 'The only way' that the Levellers can succeed, according to Jack Lindsay's mass-minded novel about revolution, *1649* (1938), is to create 'such a mass-movement among the soldiers that . . .' (etcetera, etcetera). And how devotedly '30s writing massages the idea and the phraseology of the movement of masses:

> Come then, you who couldn't stick it,
>   lovers of cricket, underpaid journalists,
>   lovers of Nature, hikers, O touring cyclists,
> now you must be men and women, and there is a chance.
> Now you can join us, now all together sing All power
> not tomorrow but now in this hour, All Power
> to Lovers of Life, to Workers, to the Hammer, the Sickle, the Blood.
>
> Come then, companions. This is the spring of blood,
> heart's hey-day, movement of masses, beginning of good

Thus the ending of Rex Warner's 'Hymn' (1933; in the version in Warner's *Poems*, 1937). Despite protests to the Hendon Borough Council's Public Libraries Committee that 'Hymn' was 'a gushing welcome to a coming revolution' ('A better ground of complaint', retorted Harry Kemp gleefully in *The Left Heresy in Literature and Life*, 'would have been that it was blasphemously shabby writing'), the poem's refrain, that last couplet, really caught on. It was taken for the epigraph of Day Lewis's *The Magnetic Mountains* (1933) (a poem whose revolutionary appeal was directed to the same hiking-cycling clientele). Day Lewis quoted it again in his 'Letter to a Young Revolutionary' (*New Country*):

it is interesting to see how our generation, sick to death of protestant democratic liberalism and the intolerable burden of the individual conscience, are turning to the old or the new champions of order and authority, the Roman Catholic Church or Communism. They seem to thirst for communal contact, to be carried away in the movement of masses; they have been insulated too long.

Variants on the phrase run repeatedly through Warner's *The Wild Goose Chase:* 'the organised movement of masses', 'the inevitable movement of the mass', 'the revolutionary movement of masses'. It had become well-known enough to be parodied by Cyril Connolly in 'Where Engels Fears to Tread': Christian de Clavering is particularly 'haunted' by a youthful quatrain he came across in the Parton Street bookshop:

> M is for Marx
> and Movement of Masses
> and Massing of Asses
> And Clashing of Classes.

Available widely enough, also, to be slipped casually by John Mair into a *New Statesman* review of a couple of proletarian novels, Lewis Jones's *We Live* and Frank Griffin's *October Day:* 'they describe the movement of masses rather than the interplay of personality.'[172] Unfortunately, however, for the Leftist's peace of mind, the Left wasn't alone in fancying the movement of masses. Evelyn Waugh raved in the *Catholic Herald* in June 1938 about the 'inspiring spectacle' of crowds of Roman Catholics at the Budapest Eucharistic Congress—'great masses of people', worshipping at a vast Mass presided over by 'Cardinal Goma, attended by heroes of the Alcazar Siege'. Fascism—spilt, skewed Socialism—certainly took crowds for its province. Hitler's apotheosis as a version of the Wagnerian Siegfried, completely dominating the anonymous masses of his subserviently enthusiastic followers, was carefully stage-managed at the Nürnberg Nazi Parteitag of 1934 in order to be filmed by Leni Riefenstahl. Her tediously bloated cinematic record *Triumph of the Will*, drawing considerably on Fritz Lang's epic silent-film version of *The Nibelungen*, was made in collaboration with Walter Ruttmann, famous for his earlier cinematic celebration of urban mass-life, *Berlin: die Sinfonie der Grosstadt* (1927). Success for fascist dictators, as for fascist artists, lay precisely in their being masters of the urban crowd. Sir Oswald Mosley, noted Harold Nicolson in his diary in 1931, 'conceives of great mass meetings with loud speakers—50,000 people at a time'. And A. K. Chesterton praised just that aspect of his leader in *Oswald Mosley: Portrait of a Leader*: Mosley's importance in the 'mass Fascist movements' was confirmed by *his* Party Rally of 1934, at Olympia, the 'mightiest indoor rally ever held in Britain'.

It was easier to discern the fascist potentials of crowds and crowd-masters if you perceived modern war's mass aspect. Rex Warner's presentation of Fascism in the case made for it by the obese and cruelly barbaric Julius Vander in *The Professor* (1938), as offering to 'the masses of people' the 'vast relief of the practical enjoyment of hatred', is rooted in that novel's worries about coming war, and the grim realization of the Professor's revolutionary son that 'War' (even a revolutionary war) 'is a movement of masses in which the individual counts for less than his true value'. The war that's declared over the radio to be imminent in Louis MacNeice's play *Out of the Picture* (1937), is a movement of masses. 'Armies are reported to be massing on the Volga, the Rhine, the Rhone, the Danube', and so on; and the Announcer's Song has it that:

The foreground needs men in the mass,
Beneath a sky of bombs and gas.

And it's not surprising to discover MacNeice in *I Crossed the Minch* resisting the current masses rhetoric because of the goosestepping Caesar tendencies of 'Collective Man':

A world society must be a federation of differentiated communities, not a long line of robots doing the goosestep. In the same way the community itself must be a community of individuals. Only they must not be fake individuals—archaizers and dilettantes—any more than the community must be a fake community, a totalitarian state strutting in the museum robes of Caesardom.

Even more strongly, Aldous Huxley perceives the 'mass-murder' and 'mass-suicide' of warfare—as he describes them in *What Are You Going to do About It?*—to be direct consequences of fascistic mass-movements. In *Beyond the Mexique Bay* (1934) he gives explicit political edge to the earlier and more general disquiets of *Brave New World* about mass life, presenting the orgiastic crowd-swaying doctrines of Nationalism as the essence and inevitable precursors of War: 'up-to-date mob leaders' like Hitler and Mussolini, ministering to 'the nameless urge which men satisfy in the act of associating with other men in large unanimous droves', provide 'the intoxicating delight of being one of thousands bawling "*Deutschland, Deutschland über alles*",' and so engender violent horrors:

Bawling in mobs is almost as good as copulation; but the subsequent action generally leads to discomfort, extreme pain, and death all round.

And that a great deal of caution was advisable is clear from the tendency of Day Lewis's broadcast thoughts in 1938 on poetry as the provider of imaginative goals for the coming revolution.[173] This was socialist vision going deliberately beyond the football crowd into crypto-fascist revivalism:

I think we've got to have some sort of social, political—oh, call it what you like—some sort of movement corresponding to the religious mass emotion that created the drama, which is going to create a continuous stream of sympathetic feeling amongst great masses of the people. Your Cup Final creates a mass emotion, but it only lasts for forty minutes each way: I think you've got to have something that will create a mass emotion which will persist over a long period, and which, by binding people together, will create a kind of public the poet can appeal to.

Day Lewis thought he was hinting shyly (this was, after all, a BBC radio conversation) at Socialism; but his dark sayings kept sliding into dubiously fascistic formulae about leaders:

To recreate a public for poetry I believe you need not the curiosity of one individual about the doings of another [as expressed in newspapers], but some force which can make a mass of individuals aware of their common ties and interests: the force, for example, which drives people, either through a sense of social injustice suffered by themselves or through a feeling that others are not getting a fair deal under a certain system of society. That force is in embryo till it's led in various ways—it must be led, obviously—on political lines for one thing. And there are certain people who arise out of the mass of the people who find themselves able to do this leading. But I don't believe these political leaders can give to the mass movement an imaginative understanding (an imaginative compulsion) towards the goal which, rightly or wrongly, they're after. That, I believe, has to be done by working upon their emotions . . .

Artists will do this; they are to be the crowd-mastering politician's imaginative ally:

> I think the artist's business on such occasions is to give the mass an imaginative picture (if you like, a Utopia, or a Mountain of God like Isaiah did) of the world as it would be if the action had been taken . . . [compelling] the individual, and thus the mass, towards the goal by working on his imagination.

It was a perfect description of Leni Riefenstahl's cinematic jobs for Hitler. No wonder, perhaps, that Auden and Spender shied away from this kind of talk. *Forward from Liberalism* showed Spender still holding on to a classic individualism (at a place where he'd just suggested communist persecutors often turned into Fascists, 'especially since the persecuting type of mind is fascist'):

> In the last analysis, the only integrity is personal integrity. Therefore, whilst it is right to demand absolute loyalty from the individual to his group, it is wrong to try and transform his mind into a generalized, group mind.

'[V]ast masses of people', Spender agrees, have to be educated into Socialism but not in fascist crowds:

> Education in socialist ideas cannot be achieved by flattering speeches made to great crowds at election time, promising spectacular reforms. This is the method of fascist dictators who demand acquiescence from their people in time of peace, co-operation only on the battle-field. Great crowds are the unit of fascist organisation, which is mass hypnotism. The socialist unit is the small cell . . .

('The mass socialist movements in Germany and Austria' have easily, Spender glooms, 'been tracked down by the fascist gunmen.')

Auden tried to effect compromises, in his article 'Psychology and Art Today' (in Grigson's gatherum *The Arts Today*, 1935) as also in his poetry, between individual-slanted, Freudian solutions ('the man who sees the patient, or at most the family, in the consulting-room') and group-minded Marxist ones ('the man who studies crowds in the street'). But he never got rid of the fear that a mass-movement could go wrong: 'Leaders of all mass movements are aware of the fact that they do not flourish unless the unconscious is tapped, unless the "heart" is touched. It is the author of good and evil alike.' That was said in his piece on the Buchmanites' pseudo-mass-movement, 'The Group Movement and the Middle Classes', in Richard Crossman's *Oxford and the Groups* (1934). The Oxford Group's communal proceedings were widely criticized as ur-fascist—F. L. Lucas, for instance, objected that they breathed the Dictators' air of *Massenmenschlichkeit*—and they were commonly cited as a type of the mass-mind's queering tendencies (the 'groupishness' of Group Theatre, sneered Grigson in *New Verse* (December 1935), 'seems as false in its way as an Oxford Group').

Auden and Spender's preference for the clique, the small group, the cell as against the mass, reaped only expectable criticism from the more committed mass-men of the Communist Party. Montagu Slater wasn't unwarm towards Group Theatre and Auden and Isherwood's dramatic attempts ('Their literary group life has given contemporary poetry quite a new liveliness'), but in the end 'the small group' was failing 'the mass of the people': 'It comes back to the old problem, How can the poet's voice reach the people?'[174] Randall Swingler condoned writing for the 'small group of sympathetic friends' ('We want poetry that we can read at home'), but the poet must go on 'to find a community of interest with the majority of mankind . . . an active identification of himself with the mass, the majority struggling for self-realization'. 'The poetry of intimacy must not despise the crowd, nor the crowd jeer at the circle of friends', Swingler arbitrated. But he was evidently keenest on the enriching and 'fructifying' progression that he envisaged from the private towards the communal:

> We also want poetry that can be cried in the streets, from platforms, in theatres; that will be sung in concert-halls and in pubs and in market-places, in the country and the town. Poetry to bind many together in a deeper sense of community.[175]

If the Party men, the Slaters and Swinglers, had serious misgivings about crowds they hedged or kept them quiet, as Alick West did. He confessed later on, in his autobiography *One Man in His Time* (1969), that he'd damaged his argument in *Crisis and Criticism* by avoiding any mention of the attractions of the fascist crowd, the great alternative 'we': 'I remained silent even to myself':

> When I spoke of the uneasiness about 'I' and 'we', I said nothing about the fascist 'we' nor about its power to release collective emotion. This was the real intellectual and political issue, and the book was weakened because it did not face it.

As an organizer of the Left Book Club Writers and Readers Group West would frequently urge meetings to unite against Fascism:

> But when I spoke . . . on Koestler's *Spanish Testament*, I kept silent about the essence of the book: that when Franco's forces were advancing on Malaga, Koestler, against all political reason and duty, under 'a strange and uncomfortable fascination', as he put it, got out of the car that was taking him from Malaga into Republican territory and was next day arrested by the fascists. Why did he go back unless on him, as on me, fascism exerted an attraction?

Divided 'Communist' Grassic Gibbon had his crowds both ways: dividing positive and negative responses to the masses between two characters in J. Leslie Mitchell's *Stained Radiance*. As John Garland moves towards the novel's Anarchocommunist party, Storman grows away from it. He gets disillusioned by economic enslavement in the Soviet Union, the persistence there of a ruling-class (the Party), and opts for individualism:

> I can no longer believe in the saving of the world through the sinking of individuality in a common cause. Mob salvation is a proven lie. I can no longer believe that the common good is greater than the good of the individual. There is no common good.

He heads for the suburbs, Purley or Wallington, 'finished with politics and parties':

> Communism was slavery, the tyranny of the mob-mind upon the individual. He had realized it at last.

Such realizations came fairly easily to ex-Communists and non-Communists. But, intriguingly, even those English authors on whom Fascism exerted some attraction seem not to have been moved much by its crowd-appeal. Nor, curiously, was the

English Literary Right all that much more vociferously individualistic than were the disillusioned or reluctant Leftists, the fellow-travelling Liberals, the occasional anarchist. It was natural enough that the worries of MacNeice and Auden and Spender should be shared by bourgeois leftish-liberal Julian Bell (' "the masses" are crowds, and crowds are idiotic': 'Open Letter to C. Day Lewis'), by the anarchist Herbert Read ('The artist, the individual endowed with exceptional sensitivities and exceptional faculties of apprehension, stands in psychological opposition to the crowd—to the people, that is to say, in all their aspects of normality and mass action': *Art and Society*, 1939), by liberal Socialist Edwin Muir (who wrote to Spender in October 1937 that he agreed 'with the ends of Communism completely' but found the 'whole impulse of Left literature . . . in danger of being dehumanised, formalised, throttled by an automatic ideology, which denies humanity except in a great bulk, so huge that it has no immediate relation to our lives: the "masses", for instance, not as a collection of men and women, but as an instrument, dehumanised as an army, a single objective mass possessing the attribute of force, and able to act only as a force'), and by fingers-crossed pinko Geoffrey Grigson. In his own contribution to *The Arts Today* Grigson keeps up a steady antithesis between the artist and 'the masses', 'the mass of individuals', 'the masses of the people' who don't find art 'easy to apprehend or accept'. Betrayed by his audience, the artist retreats from the crowd: 'what Ortega y Gasset has called the revolt of the masses ("The characteristic of the hour is that the commonplace mind, knowing itself to be commonplace, has the assurance to proclaim the rights of the commonplace and to impose them wherever it will") drives him to produce art in which the affective and spiritual dispense with the attractive element.' And it was just the rebarbative, non-crowd-pulling kinds of writing that *New Verse*, initially not unfriendly to what the keener Leftist poets like Day Lewis and Charles Madge were up to, came increasingly to advocate. From the start *New Verse* assumed Ortega's analysis and deplored the state of affairs it described; in fact *New Verse*, Grigson declared in its first number in January 1933, had to come into existence precisely to provide the outlets for poets that the revolt of the masses was presently denying them:

> Poets in this country and during this period of the victory of the masses, aristocratic and bourgeois as much as proletarian, which have captured the instruments of access to the public and urge them to convey their own once timid and silent vulgarity, vulgarising all the arts, are allowed no longer periodical means of communicating their poems.

No wonder Spender's *Vienna* got deplored (*NV*, December 1934) as being less sensitive 'to words put together, or rhythm' and 'to the shape, substance, and sound of words' than 'to mass emotion and to the emotion following upon a recital of the incidents of mass emotion'. No wonder either that Grigson should pun aggressively on the name of John Dos Passos: William Coldstream's paintings, Grigson jeered (January 1939), 'resemble chunks of Dos Massos' (or 'a double number of *Left Review*').

But this was also the line Britain's literary Right took. On the whole, there was not much crowd mania for them either: the sort of English individualism that kept a lot of literary Leftists leery of crowds worked in similar ways for the literary Right. It was part, astonishingly, of Hitler's attraction for Henry Williamson that he judged the dictator (*Goodbye West Country*) to have eradicated the 'crowd hysteria' and 'mass panic' of Weimar Germany, astonishing not least because one of *Goodbye West Country*'s most rhapsodic peaks is Williamson's attendance at a Nürnberg Parteitag ('I had been carried away by mass-emotion'). The literary Right tended, of course, to go on and on about the horrors of socialist mass society in a way that the Marxist fellow-travellers, the reluctant Leftists, were too ashamed to, but there was a clear overlap of antagonistic rhetoric. After all Grigson was only rehearsing His Master's Voice; his Dos Massos aversions came stimulated by Wyndham Lewis. In the *Doom of Youth* Lewis had struck again his old individualist note ('nothing interests me at all outside *the individual*'), inveighing characteristically against the modern 'mass-midget person' and his 'mass-amusements of all sorts, from community-singing to mass sun-bathing'. Modern man, victim of the Jewish-Bolshevik conspiracy of the nigger- and jazz-loving enemies of the Paleface, the civilized White Man, 'is being imperceptibly thrust back to the condition of stone-age man. And the droning and rapping of the mass-music of primitive tribes is there to hypnotise him into acquiescence'. Proclaiming himself in *Men Without Art* (1934) 'The Deputy of the Party of Genius', Lewis stood against 'all mass-organisation . . . aristocratic or demotic', and against the declining cultural standards of the 'gang-mind'. And in 1939, in *The Hitler Cult*, he brought himself unambivalently to plug Hitler and Germany into his masses–hostility. In *Hitler* (1931) he had granted the stunning force of the pro-Hitler crowds just as in *Blasting and Bombardiering* (1937) he had dwelt on the 'pup-like intensity' of the war-time London crowds and had republished his article 'The Crowd Master' (from *Blast* No. 2, July 1915) as that book's chapter entitled 'The War-Crowds, 1914'. But in 1914 as in 1937—and in his contribution to the first number of Mosley's *British Union Quarterly* of 1937, ' "Left Wings" and the C3 Mind', those war-time 'mafficking' London crowds merge glibly into the '30s cinema crowds and the mob of Abdication watchers—Lewis was intent on his own aloofness from the masses. His persona Cantleman was 'a master *in* the crowd' but not of it: 'It was a triumph (as I saw it then) of the individualist principle. I believed a great deal in the individual. And I still prefer him to his collective counterpart.' Genuine Fascism stood, Lewis warned the crowd-prone Mosleyites, for smallness— 'for the small trader against the chain-store . . . the peasant against the usurer . . . the nation . . . against the superstate,' and so on. So when in 1939 events compelled an energetic back-tracking, Lewis had his approach more or less ready: *The Hitler Cult* rats on Hitler because (according to a spiritedly tortured bit of argumentative slipperiness) he's become too big, been spoiled by 'Jewish bankers and business men' working in tandem with the favourite 'barbaric mass music' of the idle rich, the musical opium of Wagner. Wagner has taught Hitler to feel big, and theatrical impresario Max Reinhardt has shown him how to look big:

> Floodlight all your *own* performances, surround yourself with a barbaric symbolism, conjure up a torch-lit scene in which to hold your million-headed corroborees—copy the technique of the Reinhardt Mysteries—such things *do* smack of the barbaric . . .
> This cult of the Kolossal is the symptomatic expression of the new barbarity of machine-age man. The vast face of the *Massenmensch*—the enormously magnified visage of the Little Man—is a degeneracy.

Here, of course, Lewis was just giving a new lilt to his own old tune, and to the characteristically anti-democratic theme of all rightist culture critics, who deplored the steadily more obvious manifestations of the lives and tastes of a rising population of the ordinary people who were sowing their little bungalows and mortgaged villas about Southern England and insisting on their rights to walk all over the country-side, to drive their newly acquired cars to every beauty-spot, to have shops offer them cheap mass-produced goods they could afford, and to be entertained *en masse* by the mass media of newspaper, wireless, and film. Lewis's distaste was what came gradually to dominate D. H. Lawrence's writing. The little Faber pamphlet of his poems, *Nettles (Criterion* Miscellany No. 11) which came out in 1930, the year of his death, keeps up the vigorous flow of Lawrence's developed crowd hatred: '13,000 People', have shown themselves a 'half-witted lot' over his exhibited paintings; doom threatens 'The Factory Cities' ('the industrial masses—Ah, what will happen, what will happen?'); 'The Cry of the Masses' is a death rattle:

> Trot, trot, trot, corpse-body, to work.
> Chew, chew, chew, corpse-body, at the meal.
> Sit, sit, sit, corpse-body, in the car.
> Stare, stare, stare, corpse-body, at the film.
> Listen, listen, listen, corpse-body, to the wireless.
> Talk, talk, talk, corpse-body, newspaper talk.
> Sleep, sleep, sleep, corpse-body, factory-hand sleep.
> Die, die, die, corpse-body, doesn't matter!

T. S. Eliot was happy to publish that kind of thing because he'd held for some time that the urban masses were dead: in fact in *The Waste Land's* 'Unreal City' section the crowd ('so many') flowing to work over London Bridge of a morning isn't only dead, it's in a Dantesque Hell. 'I had not thought death had undone so many' the poem's narrator is made to think, in words out of the *Inferno*. Nobody abhorred a collective more strenuously than Eliot. His articles in the *Criterion*, his book *The Idea of a Christian Society* (1939), his talks on the wireless,[176] his other essays and lectures, sustain a horrified rhetoric against mass-education, mass-production, mass-meetings, mass-identity, mass-civilization. How the contemptuous Arnoldian phrases ring repeatedly out: 'masses of the people', 'great mass of humanity', 'mass of the population', 'mass society organised for profit', 'the mass', 'bodies of men and women—of all classes—detached from tradition, alienated from religion, and suscep-tible to mass suggestion: in other words a mob', 'illiterate and uncritical mob', 'a state secularized, a community turned into a mob, and a clerisy disintegrated', 'herd-feeling'. Even in his dramatic piece *The Rock* (1934) with its flaunted cockneyfied proles and its apparent interest in the souls of the people at large, Eliot is bluntly hostile to 'mass-made thought': for mass man is probably the dupe of bolshevik agitators, and certainly sinful ('masses of my fellow creatures living without God in the world'; 'the ordinary deadly sins of ordinary men like the mass of men living to-day'). Even community singing came in (in *C*, June 1927) for heavy criticism from Eliot as a manifestation of a suspiciously 'Muscovite'-style British *Massenmensch*: it had

> more in common with what Matthew Arnold illustrated by 'bawling, hustling and smashing' and breaking the Hyde Park railings. We are already accustomed to seeing, from time to time, immense numbers of men and women voting all together, without using their reason and without enquiry; so perhaps we have no right to complain of the

same masses singing all together, without much sense of tune or much knowledge of
music; we may presently see them praying and shouting hallelujahs all together, without much theology or knowledge of what they are praying about. We cannot explain it.
But it should at present be suspect; it is very likely hostile to Art; and it may mark, and be
a means of hastening the disappearance of the English Individualist whom we have
heard so much about in the past, and his transformation into the microscopic cheese-
mite of the great cheese of the future.

Only a Christian community was tolerable for Eliot: and even among this elect he
predicted a still more privileged role for the gifted few, the clerisy, 'the consciously
and thoughtfully practising Christians, especially those of intellectual and spiritual
superiority'. The Christian intellectual, the Christian reader, was ranged by Eliot
(in his essay 'Religion and Literature', 1935) in self-consciously isolated opposition
to the secularist tendencies of most literature: 'the mass of contemporary authors' the
'mass movement of writers', and (for that matter) 'the mass of the contemporary
public'. Wyndham Lewis, Lawrence, Eliot—not to mention W. B. Yeats: the giants
of early modernism couldn't bear the idea of their fellow man grouped in the large
heeded numbers of the modern democratic world. And their bemoanings were
promptly repeated by the younger generation of rightist and rightish littérateurs.
Evelyn Waugh's anti-socialist tirade *Robbery Under Law: The Mexican Object-Lesson*
(1939)—sponsored by the financial empire of Cowdray-Pearson to advertise the
awfulness of its Mexican oil interests being nationalized—is infused by Eliotically
depressed antagonisms to monotonous modern life: all over the world, Waugh complains, you find 'the same firms in almost identical buildings displaying uniform
mass-products to uniform, mass-clothed, mass-educated customers'. Socialist
Mexico has succumbed, Waugh suggests, to the 'new sentimentality . . . which
infects the younger generations of every race', i.e. 'crowd-patriotism'. 'Hitherto the
instinctive reaction of a self-respecting man to a crowd has been one of revulsion';
now 'mob enthusiasm . . . demagogy . . . hysteria' rule:

> As traditional sources of intoxication have fallen into increasing disrepute, mass hyste-
> ria has grown. People find a masochistic relish in being jostled and stifled in a crowd
> and in surrendering their individual judgements. Instead of diversity of opinion, they
> prefer rival orthodoxies. 'How does so-and-so stand, Left or Right?' 'Well, it's hard to
> say exactly'. 'Ah, sitting on the fence. No contemporary significance'. They love a
> crisis because a sense of universal danger, real or imagined, draws them closer to
> the mob.

Just so, Laura Riding declared on behalf of her clique (Graves, Alan Hodge, Harry
Kemp) in *Epilogue* No. 4 (1938) that writers should 'adhere imperturbably to our
difference from the mass numbers'. And Dr and Mrs F. R. Leavis and the *Scrutiny*
group sustained a major, and, again, Arnoldian-Eliotic campaign through the '30s
and beyond against mass-civilization. The 'function of criticism' (the use of
Matthew Arnold's phrase in that circle was deliberate) was presented precisely as a
defence of 'minority culture' against 'mass civilisation'. Leavis's own *Mass Civilisation and Minority Culture*, the first in the Minority Pamphlets series (1930), set the
tone for his wife's *Fiction and the Reading Public* (1932), which, despite its occasional
sympathies with the socially lowly reader, was as fierce against the decline of popular
aesthetic sensibilities as anything coming from the *Scrutiny* corner, set the tone too

for his own *Culture and Environment: The Training of Critical Awareness* (1933, written with Denys Thompson, then Senior English Master at Auden's old school), a most influential book that helped form the convention of school English lessons that popular texts—films, best-sellers, an estate of bungalows, a Woolworths counter, above all commercial advertisements—should be axiomatically prime targets for a suspicious literary criticism. On the strength of such advocacies generations of schoolchildren have been trained in a literary criticism built to withstand the supposed weaknesses and errors of popular materials. ('It is plain'—Leavis and Thompson—that 'a modern education worthy of the name must be largely an education *against* the [modern] environment'.)

The revulsion from 'mass civilisation' had much in it, of course, of merely common-or-garden snobbery. That's clear as soon as one comes across Aldous Huxley on the 'New Stupidity' ('Universal education has created an immense class of what I may call the New Stupid': *Beyond the Mexique Bay*, 1934), or Arthur Pumphrey (i.e. Alan Pryce-Jones) in the mocking *Pink Danube* (1939) on armpits and toothbrushes ('The city of the future, towards which I must strive, is to be the symbol of one community. Your armpits are as good as mine; my toothbrush is at anyone's disposal'), or Rayner Heppenstall in his piece 'I am not in favour of the Working Classes' (in *New Road*, 1944) on the proletarian's love of false teeth ('Perhaps it pleases them to roll their tongue around each other's filleted mouths and playfully dislodge a chattering denture from the hardened gums'), or John Betjeman in the poem 'Slough' (already one of this book's blackest of *bêtes noires*) urging bomber pilots to make free with the mortgaged homes of ordinary people:

> Mess up the mess they call a town—
> A house for ninety-seven down
> And once a week a half-a-crown.
> For twenty years.

(particularly enraging to Betjeman, it seems, are these people's pantries filled with the cheap mass-produced eating stuffs of urban mass-man and mass-woman: 'Tinned fruit, tinned meat, tinned milk, tinned beans'). But, it has to be observed of this snobbery, as of so many other dubious '30s affairs, that it is not at all confined to the Right. The shivers of distaste from an Eliot, a Waugh, a Leavis, or a Betjeman do not actually seem all that far removed from Orwell's wry notes in *The Road to Wigan Pier* on 'the mass-production of cheap smart clothes since the war', the Depression's 'cheap palliatives' of 'fish-and-chips, art-silk stockings, tinned salmon, cut-price chocolate (five two-ounce bars for sixpence), the movies, the radio, strong tea, and the Football Pools'. When he inspected mass life closely the bourgeois Leftist seems to have found a snooty dismay striking him as readily as it did his non-Leftist opponents. In particular, most writers—Left, Right, and would-be neutral—agreed that a good deal was wrong with the mass media as currently constituted.

The '30s realized the new imperialistic power of the media of communication with an understandable sense of shock and excitement. For here was a colonization of the mind on a staggering scale. Newspaper sales, for example, had ballooned in the early '30s as the proprietors competed for the advertising revenue that had become their real source of income by bribing subscribers with insurance and then (after 1932) with gifts like silk stockings and sets of books (Raymond Williams has boasted that he still reads his Dickens in one of those blue-and-gold-bound newspaper sets that junk shops are still full of). Such investment paid off: the *Daily Herald* and *Daily Express* boosted their circulations to over 2 million. Orwell hugely pitied the canvassers he met in the North (as *The Road to Wigan Pier* shows), men hopelessly chasing meagre commissions in aid of the newspapers' 'swindling' offers of gifts.

Other writers dwelt rather on the political power of the newspaper boss who had a huge circulation. In Christopher St John Sprigg's novel *Fatality in Fleet Street* (1933), Lord Carpenter, governing director of Affiliated Publications (he has a 'famous Napoleonesque profile' and 35,563,271 readers) decides it's time to declare war on Russia, against the wishes of the Prime Minister. His planned newspaper campaign is foiled only by his murder (a fate many of their political opponents must have wished on Lord Rothermere of the *Daily Mail* who had supported Oswald Mosley, and Lord Beaverbrook of the *Daily Express*, notorious Empire protectionist).

If powerful newspapers had to be watched, how much more to be suspected were the newer media of radio and film. The wireless, commanded by the BBC's craggily wilful, puritanically conservative director, John Reith, was already an obsessing mass medium:

> Mr Craigan put wireless earphones over his head.
> 'You and yer wireless', Gates softly said, 'it's enough to make anyone that lives with you light 'eaded, listening like you might be a adder to the music'.

Mr Craigan, we're told in Henry Green's *Living*, 'listened to the wireless every night of the week except Mondays'. And if 'listening-in' had become prevalent across the classes, so had going to the cinema. '30s movie-goers liked to keep thrilling themselves with the statistics of their sheer numerosity. In 1934, declared Paul Rotha in his *Documentary Film*, over 18 million Britons were going to the cinema each week, and pushing 40 million pounds a year into cinema box-offices. Half-a-billion hours were spent 'by human beings' in the cinema every week; in eight weeks 'moving-picture audiences equal in number the total population of the globe': so *Night and Day* readers were informed (16 December 1937) in a note on Gilbert Seldes's *Movies for the Millions*. Even more than newspapers and the wireless, film had—as Auden claimed to the North London Film Society on the subject of 'Poetry and Film' (reported in *Janus*, May 1936)—become 'the art of the masses'. Worker and boss's son are equally keen in *Living*:

> A great number were in cinema, many standing, battalions were in cinemas over all the country, young Mr Dupret was in a cinema, over above up into the sky their feeling panted up supported by each other's feeling, away away, Europe and America, mass on mass their feeling united supported, renewed their sky.

In Greene's *It's a Battlefield* all central London holds its breath as the Queen enters a cinema: 'there was a traffic block for half a mile'; it was like Armistice Day.

Cinemas, from flea-pits to fairy-palaces, were everywhere. So were their stars. Their images would fall out of your fag packet in the shape of 'cigarette-cards'. Gossip about their lives was spread liberally across the newspapers and the rush of new film-fan magazines. Movie-talk was unavoidable. The huge '30s presence of cinema, as of newspapers and radio, guaranteed huge access to the pages of '30s writing. The mass media obsessed the mass-conscious writer.

Suddenly, fiction was cluttered with cinemas, wirelesses, newspapers. Going to the cinema, listening to the wireless, reading newspapers, were abruptly taken for granted as stock activities of fictional characters. Newspaper reports, newspaper tycoons and film moguls, radio announcers and film stars became obligatory to the up-to-the-minute novel or drama. Nor, if they wanted to be alert and socially observant, must poems miss out on such phenomena. Novelists and poets started to see themselves as reporters: *reportage* became a highly favoured literary mode; authors like Wyndham Lewis compiled whole books—his *Doom of Youth*, for instance—out of newspaper cuttings; Surrealistic technique continued the Dada-ist use of bits of newspaper for cut-up poems, collages of all sorts, the *objets trouvés* of its random craft. Writers began to envy film its formal properties; cinematic jargon—*montage, cutting, jump-cuts*—became fashionable for the efforts they sought to imitate from movies. And meanwhile the serious modern critical discussion of the media began in the pages of the Leavises and of Christopher Caudwell, in the Broadcasting Chronicle that featured regularly in the *Criterion*, in collections of essays such as Grigson's *The Arts Today* (where the documentary film-maker John Grierson wrote on 'The Cinema Today') and Day Lewis's *The Mind in Chains* (where Arthur Calder-Marshall dealt with 'The Film Industry', and Charles Madge with 'Press, Radio, and Social Consciousness'). The modern post-Leavisian feeling that theories of literary criticism must look welcomingly beyond the academy to take in popular stuff and the texts of the non-written media was getting grounded.

In general, the critical discussion was done in respectable earnest. Wyndham Lewis was perhaps the most glaringly loony exception. So given was he to nightmares of Jewish-Bolshevik conspirators in every department of British life that he believed the 'pink' press, radio, cinema, were responsible for Britain's 'redness'. 'Why are you English all so red?', Roy Campbell had been asked. The media were why. According to Lewis's *Count Your Dead: They are Alive! or a New War in the Making* (1937), the Gaumont-British newsreel company, *The Times*, and the BBC were all hotbeds of Communist, Jewish, anti-Franco, anti-German prejudice: an analysis that no doubt pleased Oswald Mosley whose own anti-semitism led him to similar conclusions, but must have surprised all those Leftists who kept finding the media so conservative and pro-Franco.

Historians have, of course, confirmed the Left's disgusted allegations of right-wing, anti-Socialist, anti-worker political bias in the BBC and the cinema companies.[177] But everyone, Left and Right, agreed with Lewis that the media as currently constituted would not do. Graham Greene even came close to agreeing with Lewis that film people were an ignoble gang of foreign, semitic 'gutter-people'. How can all those *emigré* film-people properly reflect 'our language and culture', Greene asked in the *Spectator*, 5 June 1936:

> Watching the dark alien executive tipping his cigar ash behind the glass partition in Wardour Street, the Hungarian producer adapting Mr Wells's ideas tactfully at Denham, the German director letting himself down into his canvas chair at Elstree . . . I cannot help wondering whether from this great moneyed industry anything of value to the human spirit can ever emerge.

An insistent nationalistic note of anti-Americanism, in fact, runs busily through a lot of the period's harshest objections to the new mass-communications. Mass-production began in American factories; Hollywood represents the Fordization of fictions and images and so of the British mind and the English language, rapidly filling up with the talk of the talkies (what is snootily put in its place in Anthony Powell's *From A View to a Death* as 'ill-imbibed patter culled from the talkies': the language of the ordinary people more equably described in T. E. Lawrence's *The Mint* as 'soaked . . . to the bones in years of picture-going', who resort in 'moments of emotion' to 'the melodrama of film-captions'). Chauvinism tainted objections as widely different as F. R. Leavis's conservative Little-Englandizing booklet *Mass-Civilisation and Minority Culture*, with its denigrating of the American way of 'mass-production and standardisation', and James Barke's socialist little-Scotlandizing *Major Operation* with its running fight against cinema and against the 'ideology of this dominant Americanised cultural institution' (the 'mightiest' cultural institution of all, Barke believed).

British critics were worried by the media's cultivation of 'the standardised' and the 'cheap response', by the way audiences were encouraged into mindless passivity. 'It is a question', wrote T. S. Eliot (in *C*, October 1927), 'of what happens to the minds of the thousands of people who feast their eyes every night, when in a peculiarly passive state under the hypnotic influence of continuous music, upon films the great majority of which have been confected in studios of the Hollywood type.' Attaching the morally dubious half-light of Conrad's *nouvelle* 'Heart of Darkness' ('We live in the flicker') to the goings-on in the 'flicks' or 'flickers', Eliot turned the mindless tube-train-travellers of 'Burnt Norton' (1936) into a kind of cinema audience:

> Only a flicker
> Over the strained time-ridden faces
> Distracted from distraction by distraction
> Filled with fancies and empty of meaning
> Tumid apathy with no concentration[.]

Aldous Huxley agreed with Eliot's moralized objection: the savage's reaction to the 'feelies' of *Brave New World* is schooled by his reading in some of the most rebarbatively anti-sexual passages in Shakespeare. What was the state of mind, enquired Mrs Leavis in *Fiction and the Reading Public*, of people who must switch on a wireless set as soon as they entered a room? 'Let's turn on the radio. Quick!' urges Lenina in *Brave New World*, appalled by the sight of the sea, and so appalling her author. The custodians of bookish culture certainly had the wind up. F. R. Leavis talked in *Mass Civilisation and Minority Culture* of a 'catastrophe', of the 'overthrow of standards', of 'the plight of culture': 'the currency has been debased and deflated', there is no longer any 'living tradition of poetry spread abroad, and no discriminating public'. 'Most people', lamented C. Day Lewis in his 'Revolutionaries and Poetry' essay (*LR*, July 1935), 'prefer watching Greta Garbo to reading Tennyson.' He was right.

Popular culture, it was widely agreed, was dope; newspapers, films, the radio, as well as popular novels and jazz musicians were all (as the Agitator puts it in Eliot's *The Rock*) 'dopin' the workers'. Actual dope, the 'morphia, heroin, and "snow" ' that Plomer's 'Mews Flat Mona' takes, was, of course, widespread enough in the '30s. Evelyn Waugh's little squib on sunbathing, 'This Sun-bathing Business', seems casually to assume that taking cocaine is as common as sunbathing on the

Riviera (girls' 'brown limbs give away their secret, like the scars of the cocaine taker').[178] Hundreds of crooners, to the BBC's horror, suggested in the words of Cole Porter that some people not only got kicks out of women and champagne, but also from cocaine. Brian Howard and his chums went about 'guzzling' opium ('morphinism' was the reason Howard's German boyfriend was expelled from England in 1935). During the Munich crisis Graham Greene was writing two books together, *The Power and The Glory* in the afternoons and *The Confidential Agent* in the mornings, on drug energies—his benzedrine 'breakfast'. Auden ended up needing daily benzedrine. Wyndham Lewis alleges (in *The Diabolical Principle and the Dithyrambic Spectator*, 1931) that the intellectuals' dyspeptic aloofness from ordinary mankind was not only fuelled by homosexuality: it was the 'drug habit' that separated them from 'the vulgar herd that does not drug'. The 'doper-drummer' of Oran, described with great relish in Lewis's *Filibusters in Barbary*, is intended to illustrate what all 'orientalism'—by which Lewis always means Jewishness and Slavic Bolshevism—comes to (all Jews, Orientals, Slavs inhabited, he charged in *Count Your Dead*, 'a sort of dope dream'):

> When they sang together in chorus, suddenly he would out-howl the rest—tossing his head about and plucking at his nose, beating feverishly upon his *agwal*. He made no pretence of handling a snuff-nut—he had a vesper-packet: he did not *go aside* to take cocain, but sniffed it up without stopping his performance. Tossing and twitching without remission, he did one solo recitative, which was one breathless howl of suffering. Stamping and twanging came in the others. With the same sad nasal howl they joined him, in one long wolfish outburst. And it seemed quite natural that the doped drummer should stagger at their heels within an inch of collapse . . . He writhed upon the flank of the indifferent sextet . . .
>
> The contortions of the doper-drummer, nursing his *agwal* like a restive brat, beating its bottom—now drooping over his instrument and twitching, now rolling his head from side to side upon his double-shouldered back—this studied epilepsy suggested the birth then and there of a new dance-form, invented for the harsh pentatonic howling . . . A distinctly 'Islamic Sensation', I think.

Dope pedlars and addicts appear as bits of fashionable debris in Connolly's *The Rock Pool*, in Powell's *Agents and Patients* ('Surely he'll bring his own supplies, won't he? He can't expect hospitality to extend as far as that'), and in Pumphrey's *Pink Danube*, much in the spirit in which Perceval is made to sniff coke in MacNeice's *I Crossed the Minch* and Cokey Minnie and Dopey Jim adorn Ostnia's Red Light District in *The Dog Beneath the Skin*, and the contemporary in Plomer's 'Epitaph for a Contemporary' 'took drugs'. A higher moral tone is taken about heroin trafficking in Eric Ambler's leftist crime novel *The Mask of Dimitrios* (1939) and in Rex Warner's novel for boys, *The Kite* (1936). Warner even ends with a bout of Marxizing on 'that evil white powder': the wicked Ostlich's trade in drugs began in post-war Austria, so an analogy with Nazism is hinted; a parallel with the arms trade is actually made ('Why should I draw the line at drugs, when others who supply equally humane commodities, like bombs and shells, are rolling in wealth?'); and heroin is defined as one more version of the opium of the people: 'And so long as there are other people who live poor, miserable, starved, and discontented lives, so long will there be a market for any poison which can give a few moments' distraction and an illusion of liberty.'

But even this lofty moralizing was outdone when it came to the obloquy heaped by all parties on the metaphorical dope of popular culture. For Marxists religion was still the people's opium, but the old drug was now joined by the cinema according to Jack in MacNeice's 'Blacklegs': 'Glamour girls in seven-&-elevenpenny stockings and drinking cocktails on swansdown sofas . . . a beautiful syrupy fairy-tale.'[179] John Grierson agreed ('Our cinema magnate . . . more or less frankly, is a dope pedlar'). Edgell Rickword wanted to add in detective stories ('that "other opium" of the bourgeoisie, the detective novel'), *Storm* magazine to include all popular fiction ('the reams of counter-revolutionary dope that are contained in other organs of popular fiction': note that 'other organs'). Caudwell would go still further and include jazz ('religion, jazz' and 'the detective novel' take 'vulgarised values and outraged instincts and soothe both in an ideal wish-fulfilment world'). *Red Stage* managed to reduce such allegations almost to absurdity by over-inclusiveness. ' "Sex-Appeal" is Dope for the Masses: Press, Cinema, Theatre, Radio and Church Poison' its headlines blared (June–July 1932):

> That is what we are up against. A capitalist state; capitalist control; capitalist propaganda making use of every thought to poison the thinker, or, better still, to keep him from thinking at all. Yards of sex stuff on the pictures. Thousands of girls' thoughts successfully diverted to follow the exploits of their favourite 'star'. Thousands of fellows' minds debased to think only in terms of capitalist sport. Poison, poison, poison. Anything will do to keep the workers [sic] mind occupied and keep him from seeing facts as they really exist.

Brave New World, published early in 1932, was clearly already having its effect with its scrabbling together of consumerist propaganda, mass-entertainment in feelie-palaces and dance-halls, revivalist religious 'orgy-porgy' and constant drug-taking as symptoms of how the modern world was going wrong and indications of how true totalitarian control might work. No doom-laden fiction of the period would over compete for imaginative horror with that scene in the Westminster Abbey Cabaret where Calvin Stoges and his Sixteen Sexophonists lull four hundred and one identical couples into warm, escapist, *soma*-holiday quietism, a mental and spiritual living death.

T. S. Eliot rejected, of course, the idea that religion was opium but agreed about popular fiction: 'Fiction, not religion (according to Marx's silliest epigram), is the opium of the people today' (and, he added, 'some other form of opium will be provided tomorrow, for some form of opium they must have, until you can give them either religion, or to each man a job in which he can be passionately interested, or both').[180] Hardened readers of popular fiction, Mrs Leavis agreed, had a dope habit.

It was widely accepted that escapism, the doper's dream, was bad. It could not satisfy for long:

> All was a tangle; reality was too hideous to look upon: it could not be shrouded or titivated for long by the reading of cheap novelettes or the spectacle of films of spacious lives. They were only opiates and left a keener edge on hunger, made more loathesome reality's sores.

That's Helen Hawkins in *Love on the Dole*, her harsh conclusions helped along by the counsels of her author Walter Greenwood. Their shared line was prevalent. In his article 'The Menacing Movie Machine' in *New Theatre* (September 1939) John Danvers Williams quoted the chairman of Paramount Films with zesty distaste

('Men and women, working all day at machines and ledgers, find the superficial entertainment created by Hollywood a very necessary antidote—a means of helping them to forget the boredom of their lives'). By contrast, Paul Robeson's rejection of commercial films ('I am no longer willing to identify myself with an organisation that has no regard for reality—an organisation that attempts to nullify public intelligence, falsify life, and entirely ignores the many dynamic forces at work in the world today') was what *New Theatre* approved of.

A few authors could be found to disagree. Henry Green's *Living* does not disapprove of Lily's cinema-inspired visions of emigration, nor the tawdry touches of beauty that movies intrude into her existence ('Tulips, tulips she remembered time of infinite happiness in a cinema when a film was on about tulips. Not about tulips, but tulips came in'). MacNeice approves of zoos, and zoos are just like cinemas: 'a dream-world that comes easy to one':

> And I think that many of the two million [annual London Zoo visitors] do feel themselves at home there—just as they feel themselves at home in the bedroom of Loretta Young or the racing car of James Cagney or a Shanghai Express or a Garden of Allah or a Lost Horizon.

MacNeice openly sought escapist pleasure in cinemas as in zoos. He went to the movies, his autobiographical volume *The Strings Are False* (1965) reveals, 'solely for entertainment' four or five times a week:

> The organist would come up through the floor, a purple spotlight on his brilliantined head, and play us the 'Londonderry Air' and bow and go back to the tomb. Then the stars would return close-up and the huge Cupid's bows of their mouths would swallow up everybody's troubles—there were no more offices or factories or shops, no more bosses or foremen, no more unemployment and no more employment, no more danger of disease or babies, nothing but bliss in a celluloid world where the roses are always red and the Danube is always blue.

Nor in his poetry did MacNeice try to disguise this particular affinity with the enjoyments of the masses; as in his 'Ode':

> Tonight is so coarse with chocolate
>   The wind is blowing from Bournville
> That I hanker after the Atlantic
>   With a frivolous nostalgia
> Like that which film-fans feel
>   For their celluloid abstractions
> The nifty hero and the deathless blonde
>   And find escape by proxy
> From the eight-hour day or the wheel
>   Of work and bearing children.

For MacNeice, as for Henry Green's characters, the flicks offered a forgivably speedy way out of Birmingham. But for most of their fellow authors that was just the movies' trouble. The Chorus in *The Dog Beneath the Skin*, scathingly described

> . . . the cinemas blazing with bulbs: bowers of bliss
> Where thousands are holding hands: they gape at the tropical vegetation,
>   at the Ionic pillars and the organ solo.

Better, Auden and Isherwood suggest, to 'Look left' and take in the import of the 'locked sheds and wharves by water'. The strongly left-wing novelists, Harold Heslop, James Barke, John Sommerfield, Alec Brown thought so too. The brief flash of sympathy towards escapism in Barke's *Major Operation*, akin to Marx's own actually condoning tone towards religion's softening role as 'the heart of a heartless world . . . the opium of the people', is very rare on the '30s hard-line Left:

> Jean . . . was not enamoured of the sophisticated love-making of the foolish Hollywood puppets. Nor did she enjoy the fantastic exploits of the celebrated Slim M'Gurk. But for nearly two hours she was able to forget she lived in Walker Street . . . able to forget its sounds, smells and meannesses. This constituted a break in her week darg: her seven-day battle with existence.

Such sympathy was rare enough anywhere. For D. and the girl Rose in Greene's *The Confidential Agent*, to go to the cinema is to be enclosed luxuriously in Hollywood's unreal and narcissistic self-regarding:

> They sat for nearly three hours in a kind of palace—gold-winged figures, deep carpets, and an endless supply of refreshments carried round by girls got up to kill: these places had been less luxurious when he was last in London. It was a musical play full of curious sacrifice and suffering: a starving producer and a blonde girl who had made good. She had her name up in neon lights on Piccadilly, but she flung up her part and came back to Broadway to save him. She put up the money—secretly—for a new production and the glamour of her name gave it success. It was a revue all written in no time and the cast was packed with starving talent. Everybody made a lot of money: everybody's name went up in neon lights—the producer's too: the girl's, of course, was there from the first. There was a lot of suffering—gelatine tears pouring down the big blonde features—and a lot of happiness. It was curious and pathetic: everybody behaved nobly and made a lot of money.

The happy ending is as unreal, as far from D.'s world of grim political actualities as it's conventionally predictable ('If *we* lived in a world, he thought, which guaranteed a happy ending, should we be as long discovering it?'). Indeed, the whole experience runs on predictable lines: 'He felt her hand rest on his knee. She wasn't romantic, she had said: this was an automatic reaction, he supposed, to the deep seats and the dim lights and the torch songs, as when Pavlov's dogs saliva'd.'

Fictions about getting rich viewed in palatially plush circumstances: as MacNeice (gone less sympathetic since he left Birmingham) put it in *Autumn Journal*, XVII, the cinema

> gives the poor their Jacob's ladder
> For Cinderellas to climb.

The 'loving/Darkness' of the picture palace was, as Day Lewis's 'Newsreel' has it, 'a fur you can afford'. Hollywood exuberantly capped the most escapist lavishness of the worst kinds of fictional wish-fulfilments about 'aristocrats and plutocrats': 'those Don Juans, those melting beauties, those innocent young kittens, those beautifully brutal boys, those luscious adventuresses. Hence', Aldous Huxley went on in his 'Writers and Readers' essay (in *The Olive Tree*, 1936), 'Hollywood, hence the beauty chorus. When I was last at Margate a gigantic new movie palace had just been ·opened. Its name implied a whole social programme, a complete theory of art; it was called "Dreamland". At the present time, the cinema acts far more effectively as the opium of the people than does religion.'

The social snobbery and needlessly fantastic luxury of films was a constant theme of Greene's film criticism; for example in the *Spectator* (6 December 1935):

> An English film as a rule means evening gowns by Hartnell, suitcases by Asprey. An excursion steamer to Margate (*vide The Passing of the Third Floor Back*) becomes a luxury liner full of blondes in model bathing-dresses. Even in the worst French films one is not conscious of this class division, the cafés and dance-halls are of the kind familiar to the majority of the audience.

And for their part, of course, the stars need never leave the 'luxury liner'. The excited Minty casts a baleful eye in *England Made Me* on the return home to Sweden of a Garbo-like star cocooned in moneyed adulation:

> A number of people (were they hired by the hour? Minty wondered) began to cheer . . .
> Minty had one glimpse of a pale haggard humourless face, a long upper lip, the unreal loveliness and the unreal tragedy of a mask like Dante's known too well. The movie cameras whirred and the woman put her hands in front of her face and stepped into a car. Somebody threw an expensive bouquet of flowers (who paid for that? Minty wondered) which missed the car and fell in the road.

Miss Clara de Groot (the World's Sweetheart) of MacNeice's drama *Out of the Picture* (1937) (her 'hands cost ten pounds a week per nail') and Miss Lou Vipond ('The star of whom the world is fond') in *The Dog Beneath the Skin* (played by 'a shopwindow dummy, very beautifully dressed') make the same point, if with a more exaggerated flashiness. Keeping up with Miss Vipond's style at the Nineveh Hotel is, as the infatuated Alan discovers, quite bankrupting:

> 20 cases of champagne,
> A finest pedigree Great Dane,
> Half a dozen Paris frocks,
> A sable fur, a silver fox,
> Bottles of scent and beauty salves,
> An MG Midget with overhead valves,
> 1 doz pairs of shoes and boots,
> 6 lounge, 1 tails and 3 dress suits,
> A handsome two-piece bathing-dress,
> An electric razor, a trouser-press,
> A cutter for cigars, two lighters,
> 10 autographs of famous writers,
> Berths and tickets in advance
> For a trip round Southern France:
> Add to this his bed and board.

And the Chorus adds, wisely: 'It's more than one man can afford.' It was more, too, than Scotland's morals could afford, according to James Barke. As the hikers at Loch Lomond in *Major Operation* strum the 'latest hot music' from America on their ukes, the narration meditates on the mass-meaning of the new mass-distractions, especially the role of Garbo:

> New times, new machinery. Mass production, mass culture, mass gutter journalism—what could Loch Lomond mean? . . . A sixpence gets a packet of fags and leaves

enough for a seat at the cinema. Then the wise-cracking of Hollywood gets into the brain and the blues rhythm gets into the blood for there's nothing else to keep them out. So America becomes the cultural centre of the world. Even the Jap can tell you about Charlie Chaplin and Greta Garbo . . .

And Garbo, the Swede, is somebody, while Flora MacDonald might be a skivvy in Milngavie.

You did not, however, have to be a thorough-going Marxist or Socialist-Realist to find the movie-stars' pandered isolation from ordinariness distasteful and cinematic fiction's cultivation of escapism worrying ('It is not that one wants English directors to "go" proletarian', said Greene in his complaint just quoted), though a sort of Marxist assumption that the imagination's food should be close to everyday reality does undergird the usual complaints, wherever they come from. What irked was organized falsifying. The cinema attracted bogus scoundrels like Powell's Chipchase or Maltravers (who 'might easily have been a better-class gangster figure of any period'). It stank of the most attractive vices ('How does he get them?' Chipchase wonders of the huge and beautiful girls who constantly accompany dwarfish, hook-handed Herr Direktor Roth of Berlin's Niebelheimnazionalkunstfilmgesellschaft: 'The hook?'). It dealt in crude malformations of history and reality (Waugh makes joyful play in *Vile Bodies* with the Wonderfilm Company of Great Britain's version of the Life of John Wesley—'the most important All-Talkie super-religious film to be produced solely in this country by British artists and management and by British capital'—in which Wesley and Whitefield fight a duel and Wesley in America is rescued from Red Indians by the Countess of Huntingdon disguised as a cowboy). It encouraged absurd self-deceptions of the sort Boot indulges in Waugh's *Scoop* (1938: 'he was going to do down Benito. Dimly at first, then in vivid detail, he foresaw a spectacular, cinematographic consummation, when his country would rise chivalrously to arms; Bengal Lancers and kilted highlanders invested the heights of Jacksonburg; he at their head burst open the prison doors; with his own hands he grappled with Benito, shook him like a kitten and threw him choking out of his path; Kätchen fluttered towards him like a wounded bird and he bore her in triumph to Boot Magna . . .'). For such self-delusions Orwell blames the movies in Wigan ('fantastically cheap there. You can always get a seat for fourpence, and at the matinée at some houses you can even get a seat for twopence. Even people on the verge of starvation will readily pay twopence to get out of the ghastly cold of a winter afternoon'):

> You may have three halfpence in your pocket and not a prospect in the world, and only the corner of a leaky bedroom to go home to; but in your new clothes you can stand on the street corner, indulging in a private daydream of yourself as Clark Gable or Greta Garbo, which compensates you for a great deal.

In *It's a Battlefield* the star-stricken employees of a Battersea match-factory are distracted by film fantasies from industrial accidents, low wages, long hours:

> Between the line of machines the girls stood with tinted lips and waved hair, fluttering an eyelid, unable to talk because of the noise, thinking of boys and pictures and film stars: Norma, Greta, Marlene, Kay. Between death and disfigurement, unemployment and the streets, between the cog-wheels and the shafting, the girls stood, as the hands of the clock moved round from eight in the morning until one (milk and biscuits at eleven) and then the long drag to six.

So absorbed in fact are they into their useful escape world that they have become narcissistically assimilated, as it were, to the objects of their own devotion:

> In the cloakroom Norma put on her hat, Greta brushed her hair, Marlene made up her face.

It is from just such day-dreams that Socialism and socialist novels strive to rescue the deluded in the period's most heavily left-wing fiction. Sommerfield's *May Day* has the little maid Jean come to her senses once she's out of the cinema and back in the open-air: swooning cinematically in the arms of 'rich and handsome young men' only gets girls like her into old-fashioned trouble. Barke's *Major Operation* has George Anderson's marital problems start in his actress wife's desires for another sort of man, 'the Modern-Man of Fiction: of Hollywood, in fact'. Anderson is better off in the Party, better off even when he's killed in Party action on the streets than enduring Mabel's Hollywood-fuelled waspishness and betrayals.

Anxiety over cinematic falsities focused particularly on the Newsreels. Wyndham Lewis, typically, suspected a manipulative Jewish blackening of Hitler (*Count Your Dead* mocks the claim to be 'British' of Isidore Ostrer, Chairman of Gaumont-British Pictures). The newsreels, Lewis thought, were much too kind to the 'Red Militia' in Spain. As often his voice was hollow and isolated. 'Watch those Newsreels' (*DW*); 'News-Reel Poison' (*Red Stage*); 'Banned! Newsreels You Must Not See' (*New Theatre*): leftist complaints piled up: about the influence of the Rothermere family at British Movietone News; and the concentration of newsreels on anodyne trivia like the laying of foundation stones and the launching of ships; and the dilution for British audiences of the American *March of Time* news dramatizations; and direct censorship pressures from government and the London Licensing Committee (scenes of the bombing of Shanghai were frowned on, Paramount's reel on the Munich crisis had to be cut); and about subtler right-wing editorial bias, particularly over Spain (second-hand Red-atrocity features like Gaumont's notorious 'Blonde Amazon' story, supported by shots taken from old and unrelated film-stock; Red-bashing voiced commentaries; Franco-favouritism that had Gaumont anticipating the Fall of Madrid in November 1936 with the General shown triumphantly ascending some steps in another city altogether (Burgos) and that featured Franco always heroically high on his podia; shots of smart nationalists made to contrast pointedly with scruffy government supporters; the identifying of Soviet weapons but not of Italian Capronis in film of aerial bombing; the showing of the homecoming of O'Duffy's fascist Irish Brigade but not that of the British International Brigade contingent). The Left's case had far more substance than Lewis's ravings:

> Not . . . shown at cinema
> To blackout Paramount with the facts like lights,
> The horror facts, the humans in the horror.

Thus George Barker in his *Calamiterror* (1937), a theme expanded in Day Lewis's poem 'Newsreel'. In the 'dream-house' where the news that pleases is as consoling as the fiction movies—

> There is the Mayor opening the oyster season:
> A society wedding: the autumn hats look swell:
> An old crocks' race, and a politician
> In fishing-waders to prove that all is well—

even the bits of horror that newsreels allow in, lose their power to worry:

> Oh, look at the warplanes! Screaming hysteric treble
> In the long power-dive, like gannets they fall steep.
> But what are they to trouble—
> These silver shadows to trouble your watery, womb-deep sleep?

The Left's case against the tranquillizing effect of newsreels had gained fundamental impetus from *Brave New World*'s hostile dealings with Feelietone News. It
got frequent endorsement from Evelyn Waugh. He kept reverting to the antics of the
newsreel men in Ethiopia, pushy, ludicrous, and immoral, faking battles with iodine
for blood and fireworks to simulate bombardments and always at a price. 'Those', he
gibed in *Waugh in Abyssinia* (1936), 'who had worked during the Chinese wars—
where, it seemed, whole army corps could be hired cheaply by the day and even, at
a special price, decimated with real gunfire—complained of the standard of
Abyssinian venality'. Nor were the newsreel fakers alone in manipulativeness; the
equally venal press corps shares the low standard. *Remote People* (1931) is scathing
about 'anxious journalists whose only hope of getting their reports back in time for
Monday's papers was to write and despatch them well before the event'. And
Waugh's assumption of widespread media lying was common enough. In *Autumn
Journal* (XIX):

> The doll-dumb file of sandwichmen
> Carry lies from gutter to gutter.

That every newspaper report must be taken with a pinch of salt is demonstrated
before the final scene of *On the Frontier*, when Auden and Isherwood's chorus,
representing 'the typical readers of five English newspapers', read out the news in
the five widely differing guises it assumes in (it might be) the respectable *Times*, The
reactionary *Mail*, the Liberal *News Chronicle*, the communist *Daily Worker*, and the
sensational inside-story mongering *Week* run by Claud Cockburn. In the same play
there is ample revelation that you can't trust the wireless either ('Lies on the air
endlessly repeated': *Autumn Journal*, XVIII). The frontier divides the Ostnia-
Westland room in two, but in both totalitarian Westland and monarchist Ostnia
domestic life is dominated by portraits of the respective heads of state, the Ostnian
king and the Westland Leader, and by the wireless-sets that stand beneath the
portraits. And when the leaders speak, their rhetorics are locked into a merely
apparent opposition and in fact share the same war-mongering direction. No wonder
the play's Leftists warn that 'It's mad to die / For what you know to be a lie':

> Don't believe them,
> Only fools let words deceive them.
> Resist the snare, the scare
> Of something that's not really there.
> These voices commit treason
> Against all truth and reason,
> Using an unreal aggression
> To blind you to your real oppression;
> Truth is elsewhere.
> Understand the motive, penetrate the lie
> Or you will die.

Warnings like this frequently came layered in a thick sense of absurdity. MacNeice's *Out of the Picture* protests against the trivialization of the news:

> The news that blows around the streets
> Or vibrates over the air
> Whether it is rape, embezzlement or murder
> Seems frivolous, if not farcical, without dignity.
> Whereas the actual fact before it becomes news
> Is often tragic even when commonplace.

'Slapstick may turn to swordplay': but it's the slapstick that sticks in the mind, as MacNeice's Radio-Announcer dons pince-nez to give a talk on Aristotle and beats a drum to signify news of Crisis. Likewise, the cabaret absurdities of *The Dog Beneath The Skin* rather diminish the undoubted wisdoms of its two Journalists ('The General Public has no notion / Of what's behind the scenes'). The Second Journalist's 'We have got the lowdown / On all European affairs' deflates him as speedily as 'Oh Mr. Marx, you've gathered / All the material facts' undoes any sense of *The Dance of Death* as a Marxist tract.

Waugh's journalists and newsreel men can be as farcical as the film makers of *Vile Bodies*; his contempt for the absurdities of Fleet Street is as amusing in its presentation in *Scoop: A Novel About Journalists* as it must have been galling to live through in Abyssinia (where, according to *Waugh in Abyssinia*, he had been pestered by his employers with requests like 'Require comprehensive cable good colourful stuff also all news' that clearly did much to fertilize his novel). Lord Copper of Megalopolitan Newspapers and the *Daily Beast* (in *Scoop*) is as little frightening as Lord Monomark of the *Daily Excess* (in *Vile Bodies*), despite the suicide of gossip-columnist Lord Balcairn after a particularly successful crescendo of lies about Lady Metroland's party ('Lie after monstrous lie bubbled up in his brain . . . Excitement spread at the *Excess* office'). Other versions of Beaverbrook—Day Lewis's Bimbo in *The Magnetic Mountain* ('As for you, Bimbo, take off that false face') and Adam's Beethameer (Beaverbrook spryly conjoined with Rothermere) in *The Orators'* 'Diary of an Airman' ('Beethameer, Beethameer, bully of Britain, / with your face as fat as a farmer's bum')—have their indignations eroded just as limitingly by their authors' wish for acts of mainly comic devastation.

None the less, '30s literature's succession of megalomaniac newspaper proprietors with their huge circulations, its chain of dictatorially amplified voices on the impersonal wireless ('orders / Out of a square box from a mad voice', as MacNeice describes them in *Autumn Journal*; the Leader of *On the Frontier*'s Westland 'isn't a man at all! He's a gramophone': to be played, the authors' Notes inform, 'very stiffly, like a newsreel photograph of himself. His platform voice is a trance-voice, loud and unnatural'), all witness to widespread fears. 'Shall we', protested Michael Roberts in the Preface to *New Country*, 'see our lives dictated more and more by the proprietors of Guinness, the Gaumont Palaces, Harrods and the *Daily Mail*? Shall we watch our children hypnotised . . .?' He'd clearly absorbed *Brave New World*'s disquiet about *hypnopaedia*. Even Graham Greene, fonder of stressing the seedily Orwellian underbelly of the newspaper world (the situation of Fred Hale, for example, the minor crook and murder victim, employed as 'Kolly Kibber', seaside publicity tout for the *Daily Messenger*, in *Brighton Rock*), deals with the press's strong persuasiveness as well as its spiritual grubbiness in *It's A Battlefield*. Crime reporter Conder is a double-crosser, a fake, a creator of false fictional lives for himself, a glosser of the truth ('Condemned to the recording of trivialities, he saw the only hope of a posthumous immortality in a picturesque lie which might catch a historian's notice as it lay buried in an old file'). Trivializing, blurring banner headlines lace through the novel ('Are you Insured? . . . Drama at Red Meeting . . . Mr MacDonald Presents Golf Trophy'). Newspapers, though, are widely read ('Men stood in their doorways and read *The News of the World* and spat. In Wardour Street and Shaftesbury Avenue they were reading the *Sunday Express* . . .'). And the man in the street (the condemned Communist's brother) believes (wrongly) in the press's power for justice: newspaper publicity might save his brother. But it's the police who most powerfully manipulate the press, trading little scoops to Conder for inside dope on the Communist Party. A few turns of the political screw, in fact, and you got to Lord Stagmantle, Press Baron of *The Ascent of F6*, with his *Evening Moon*'s gutter-press opportunism ('Splashed it all over the front page—nearly doubled our sales, last week! . . . We were out to smash the Labour Government, you know: and, by God, we did! . . . any stick's good enough to beat a dog with, you know!': which is the very embodiment of Spender's *mot* in *Forward From Liberalism*, that newspapers are less the voice of the people than ' "the ear of the people" through which the bosses shout their propaganda'. The memorable Chapter IX, 'The Broadcast', of Rex Warner's *The Professor* (1938) featured happenings widely felt to be only a step or so away from fruition. Even as the Professor, temporarily his country's Chancellor, is spouting liberal democratic platitudes for, as he supposes, his people's wireless consumption, the Chief of Police has occupied the Radio Station and is announcing his own fascistic take-over of power. Brave New World indeed!

Beggars, perhaps, could not afford to be choosers, but, given this weighty and almost unanimous chorus of disapproval, it comes at first as some surprise to discover the readiness with which '30s authors cooperated with the media they or their friends kept chastising. They found it as hard to keep away from the opium distribution centres of the press, cinema, and wireless as the socialist writers from the dope of detective stories. Doubtless there was something of a 'Why let the Devil have all the best tunes' spirit about the BBC wireless appearances of a Wyndham Lewis or a T. S. Eliot (as, later, of an F. R. Leavis): they could see themselves as going on the air to undo some of the more frequent broadcasters' damage. And, to be sure, Day Lewis ran into problems in 1935 over what the BBC's new General Talks Department construed as an unacceptably excessive mingling of politics with literary criticism (the talk became Part One, 'The Revolution in Literature', of his *Revolution in Writing* pamphlet), just as John Reith had his doubts about Janet Adam Smith's keenness for the *Listener* to publish such 'modern' stuff as Auden and his friends. Still, as we've seen, the literary pages of the *Listener* quickly became almost the house organ of the Auden group. And Auden was not averse to broadcasting about 'Gossip'. Nor were the newspaper bosses' shekels any less acceptable than the BBC's. Graham Greene, of course, had quickly abandoned *The Times* for the risks of freelance writing, and Evelyn Waugh survived only a very short while at the *Express* (three weeks in 1927) and the *Mail* (three months at £30 per weekly article in 1930), but the Communist Tom Driberg managed to combine being a Communist and an

Express gossip columnist from 1928 onwards (he was expelled from the Party in 1941; his 'William Hickey' column started up in 1933). The rapport between Beaverbrook and some Leftists like Driberg and A. J. P. Taylor (and Michael Foot) is one of the most interesting sidelights both on Beaverbrook and on mid-century English intellectual Socialism. Meanwhile, C. Day Lewis did weekly novel reviews for the *Daily Telegraph* and detective fiction reviews for *The Spectator*—so much for the 'scavenger barons' and their 'jackal vassals' ('Your pimping press-gang, your unclean vessels') of *The Magnetic Mountain*—and Charles Madge, Humphrey Jennings, and Humphrey Spender found employment on the *Daily Mirror*, and Montagu Slater worked for the *Liverpool Post* (1924–8), then for the London *Observer* and *Daily Telegraph* (until he quit to edit *Left Review* in 1934). H. B. Mallalieu too was a journalist. Geoffrey Grigson combined being books-page editor of the right-wing *Morning Post* with running *New Verse* . 'No doubt the Book Society is, looked at from one angle, a racket' (thus Day Lewis plaintively defending his position in it in a letter to the scathingly critical Grigson in 1937): 'but so are the "literary" pages of daily papers: one has either to stand right outside the racket, or else to use it (as you used your position on the *Morning Post*) to get the better stuff over to ordinary readers'. Be that as it might, Grigson helped finance *NV*'s high tone towards the mass media by selling off spare *Morning Post* review copies. And almost equally high-minded chums would besiege him for a cynical share of the *Post*'s rightist loot. 'Do you know of anyone', clamoured MacNeice early in 1934,

> Who will pay me any money quickly for anything—reviewing or otherwise? I must have some money as I want to go to Spain in a few weeks.[181]

That very same year Spender tried the same touch:

> You mentioned a few weeks ago that I might do the novel reviews for a short time for the M.P. Do you think this is possible, as I am trying to fix up work for the autumn? Or, failing that, I would be very glad of any odd reviewing.[181]

No wonder Day Lewis felt it necessary to warn the young writer (in 'Writers and Morals') that going on to a newspaper was a false start:

> Unless he is lucky, he will find himself in an atmosphere of the cynical, the insistent and the plumb-crazy, to which the Foreign Legion would seem by comparison an outing of the Girls' Friendly Society. Some young men have, in a more discreditable sense, 'found themselves' here and signed on for life: others, lacking the boa-constrictor's digestion, have been unable to swallow it and pulled out before it was too late.

No wonder, either, perhaps, that Spender's attitude was so mixed. In 1936 he was writing to Grigson:

> It is a pity that my brother who was Lensman on the Mirror has left it. However, I dare say Madge will find Humphrey Jennings and the various other bibliophils and scholars who prostitute their talents there, quite good company. He might even write the much needed book about Fleet Street.[181]

It was the same ambivalent story in regard to the movies. The cinema's decriers—writers like Greene and Waugh—frequently spent as much time inside movie-houses as its professed friends, like Green and MacNeice. Greene was, of course, paid to attend, as film critic for *The Spectator* (July 1935–May 1937), for *Night and Day* (July–December 1937), and again for *The Spectator* (June 1938–March 1940). It had been only a short step, as it were, from being the self-appointed film critic of *Oxford Outlook* the termly student magazine Greene edited; and from reviewing it was only another 'small step to script-writing':

> That also was a danger, but a necessary one as I had a family to support and I remained in debt to my publishers until the war came. I had persistently attacked the films made by Alexander Korda and perhaps he became curious to meet his enemy. He asked my agent to bring me to Denham and when we were alone he asked if I had any film story in mind. I had not, so I began to improvise a thriller . . .

It was for Evelyn Waugh a similarly slight move from addicted cinema-going ('What a futile thing this diary really is', he noted towards the end of a film-packed school summer holiday in 1920, 'it consists chiefly of "shops in the morning, cinema in afternoon" ') and acting as film reviewer for the Oxford student paper *Isis*, to writing and acting in short student films (including the one made in the summer of 1924 after his Oxford Finals in which 'Sligger' Urquhart, the Dean of Balliol, played by Waugh, attempts to convert the king of England to Roman Catholicism), to attempts at work in the film industry (in 1932 for Basil Dean's Associated Radio Productions, in 1936 for Alexander Korda: all aborted ventures, like MGM's idea for a version in 1947 of *Brideshead Revisited*). 'At that period', according to Anthony Powell, 'the ambition of most young novelists, many older ones too, was to find employment script-writing for a film-company.' Less lucrative was reading novels for film companies. London seemed 'full', Walter Allen recalls, of authors picking up pennies this way from MGM. They included H. E. Bates as well as Allen himself—who even did an abortive synopsis of Beckett's *Murphy*! Powell never had to sink that low, and he enjoyed some early success—more than Waugh—when he was hired on to the payroll at Warner Brothers' studio in Teddington to do 'treat-ments' (£15, rising to £20, a week). But a hopeful visit to Hollywood in 1937 not only failed to establish him in film-writing, it dried him up as a novelist (so that on his return to London he had to resort to fortnightly novel reviewing for the *Telegraph* as well as regular pieces on autobiographies and memoirs for *The Spectator*. To be fair to Powell, his strictures on journalism in *Venusberg* (1932) and on film-making in *Agents and Patients* were only mild and jocular ones. When Aldous Huxley plumped for script-writing in 1938 ('You will get enough to set you up for life', Anita Loos promised of MGM's interest in him as writer for a film about Madame Curie) he had not Powell's excuse: Huxley *had* savaged the people's opium. And most notably in *Brave New World*, so that his letter to Anita Loos of 13 October 1945 outlining a nasty Hollywood cheapo screenplay for that novel—involving 'a gorgeous blonde' and a pimply scientist—is heavy with an ironic sub-text about how far even the most intelligent of writers and critics can be, simply, bought off.

Huxley's ready abandonment of intelligent and delicate scruples was as notable, given the opinions current among the man's friends, as Isherwood's alacritous incursion into film employment with Gaumont-British and the director Berthold Viertel in autumn 1933. He was, as he admits, in *Lions and Shadows*, 'a born film fan', an habitué of cinema since childhood, a devoted member of the Cambridge Film Club, and even, by accident as an undergraduate, a film extra for a day (the

24 shillings he got was 'the first money I had ever earned in my life'). When he took up film work again, in Hollywood early in 1940, Isherwood not only had Viertel to introduce him to MGM, he also had the example of Huxley's getting £1,500 a week for a screenplay of *Pride and Prejudice*. 'The studio', bluffed Isherwood uneasily, 'is just an office I visit in the daytime.' His studio experiences may have given him material for fiction (in particular the novel *Prater Violet*, 1946), and enabled those knowledgeable comments on the Hankow film studios in *Journey to a War* (the set 'had none of that unnatural newness which is such a besetting vice of the English studios'), but Isherwood was clearly still as guiltily cagey as in 1934, when, faced with his friends' gibes over Viertel's *Little Friend*, he made excuses to Plomer: 'After all, if I sold myself, I did at least make them pay handsomely for me.' The paragraph Isherwood put into *Lions and Shadows* (1938) about the documentary merits of even very bad films ('Viewed from this standpoint, the stupidest film may be full of astonishing revelations about the tempo and dynamics of everyday life') was evidently part of this long-running attempt at self-justification.

> 'Chalmers [Upward] was inclined to laugh at my indiscriminate appetite for anything and everything shown on a screen. He pointed out, quite truly, that as soon as I was inside a cinema I seemed to lose all critical sense: if we went together, I was perpetually on the defensive . . .'

Only a perpetually jokey self-denigrator like Isherwood would have professed to the loss of '*all* critical sense'. Most of the others, even the regular compromisers with the mass-media, found appeasing the critical consensus a good deal harder.

10

Mass Observations

FOR most Leftists, of course, confronting the mass-media problem didn't stop with mere denunciation or defensive compromise. They believed they must try to out-manœuvre and subvert the currently unsatisfactory media of mass-communication, to capture the mass-audience for good writing and art, to discover their own mass-appealing subject-matter and, the ultimate challenge and goal, seek to transform bourgeois art and aesthetics in the process, creating new, non-bourgeois kinds of art for the awakened masses—'a revolutionary literature', as Cornford envisaged it in his 'Left?' article in *Cambridge Review*, Winter 1933–4, 'stronger and more various than any which preceded it'.

These particular ambitions developed, of course, only on the Left. The solution to the mass-audience problem sought by Rightists like Eliot and Leavis was retreat into tradition and the cultural citadel of the educated and tasteful few. Even on the Left, goodwill towards the potential mass-audience varied. 'I am writing', pink Cyril Connolly announces at the start of *Enemies of Promise*, 'for my fellow bourgeois':

> the way I write, and the things I like to write about make no appeal to the working class, nor can I make any bridge to them till they are ready for it. I write for people who have been to schools and universities, whose interests and whose doubts I share.

But concern to transcend the narrow binds of middle-class culture did spread wide among broad-Leftists, as Gavin Ewart's jaunty 'Election Song, 1935' indicates:

> We shall preserve the ancient sweetness,
> And books published by Faber,
> But our class-culture lacks completeness,
> So vote Labour.

And among those animated by the issue at least three major overlapping problems presented themselves: first, how a mass audience could be claimed for an undebased and undebasing art; second, how the masses, particularly the industrialized working-class, should best become the subject of art; third, whether new, 'proletarian' forms might be discovered to unseat the hegemony of the old bourgeois ones, new forms apt to the needs of all the people.

'We wish poetry to be popular' proclaimed Day Lewis in *A Hope for Poetry*. In fact, he went on in 'Revolutionaries and Poetry', popularity was the only hope for poetry's survival: 'If poetry is to survive as a means of communication, it must become necessary again to people.' Caudwell and others were continually envious of the Soviet poets who rejoiced in mass audiences:

> The increase [this is *Illusion and Reality*] of the poet's public can already be seen in the Soviet Union where poets have publics of two or three million, books of poetry have sales of a size unknown previously in the history of the world.

Back at home the readership for serious literature was pitifully thin. Allen Lane's Penguins and Pelicans were regarded as a very bright sign (*Left Review*'s last number,

May 1938, was full of praise for Penguins and carried an article by Lane, 'Books for the Million'): but the Penguin list when it emerged was scarcely a Marxist (nor even a particularly poetic) one. For its part, the Left Book Club's rapidly enlarging circle of committed buyers of Victor Gollancz's books (39,400 by March 1937, 44,800 by May of that year, 57,000 by September 1939), though in general distinctly cheering, offered little promise to imaginative writers: the Club's steady diet of political analysis, sociology and memoir allowed room for only a handful of novels, the odd drama (Odets's *Waiting for Lefty*, June 1937), and the occasional volume of verse (*The Left Song Book*, edited by Alan Bush and Randall Swingler, March 1938; *Poems of Freedom*, edited by John Mulgan, December 1938). Any serious dents that 'poetry' might make in the cinema-going audience would be by Patience Strong, whose first book (1937) had already sold over 100,000 copies by the time Humphrey Jennings talked to her on the wireless in June 1938. Her poems were published daily in a popular newspaper (she used to knock off her week's batch every Monday morning; 'rather like washing day', she said).[182] No wonder Day Lewis took up detective stories (nor that Victor Gollancz kept packing his ordinary list with popular fiction). By contrast with Miss Strong, Jennings noted, a highbrow poet was lucky to sell 800 copies a year (the 2,350 copies of Auden's *Look, Stranger!*, sold within three months of publication, made it a poetic smash hit). The Mass-Observers Tom Harrisson and Julian Trevelyan used, missionary-wise, to hand round in Bolton's streets and pubs photos of pictures of the town painted by Trevelyan, Humphrey Spender, William Coldstream, Graham Bell, Michael Wickham, but (despite encouraging signs like the working-class interest in Picasso's *Guernica* when it hung in the Whitechapel Gallery, and Trevelyan's little raft of self-taught painters, and the Ashington group of miners, a famous but faraway WEA art class which painted up in Northumberland) most of the workers, in Blackheath as in Bolton, stubbornly preferred the other kind of pictures.

The first step Spender envisages for art in 'a classless society' in *Forward from Liberalism* will be that 'the great art and literature of the past five hundred years will be appreciated and criticized by a far wider audience than ever before and will affect the lives of a great many people, who will not only have galleries and libraries thrown open to them, but also a background of education, leisure and moderate comfort'. The galleries and libraries were, of course, open as he wrote. Still, Spender was right to emphasize leisure and education as keys to the culture he valued. But what if, ran an insistent and rather illogical worry, galleries and libraries packed with educated workers should vitiate that culture's intrinsic merits? This still familiar bourgeois bogey—does more mean worse?—was bravely addressed by Louis MacNeice in Section III of *Autumn Journal*, one of the finest parts of that poem, 'August is nearly over, the people / Back from holiday are tanned.' The poet acknowledges as utterly 'lost and daft' the

> System that gives a few at fancy prices
> Their fancy lives
> While ninety-nine in the hundred who never attend the banquet
> Must wash the grease of ages off the knives.

The people he celebrates are locked into an unthinking round of work followed by beer, 'the gossip or cuddle', 'the solace / Of films or football pools'; but doesn't even the apparently concerned poet secretly enjoy his, and his cultivation's, distances from them?

> And now the tempter whispers 'But you also
>   Have the slave-owner's mind,
> Would like to sleep on a mattress of easy profits,
> To snap your fingers or a whip and find
> Servants or houris ready to wince and flatter
>   And build with their degradation your self-esteem;
> What you want is not a world of the free in function
>   But a niche at the top, the skimmings of the cream'.

Extremely honestly, MacNeice fields the accusatory point, albeit in a bout of stilted prosing:

> And I answer that that is largely so for habit makes me
>   Think victory for one implies another's defeat,
> That freedom means the power to order, and that in order
>   To preserve the values dear to the élite
> The élite must remain a few. It is so hard to imagine
>   A world where the many would have their chance without
> A fall in the standard of intellectual living
>   And nothing left that the highbrow cared about.

'Which fears', MacNeice rebukes himself,

> must be suppressed. There is no reason for thinking
> That, if you give a chance to people to think or live,
> The arts of thought or life will suffer and become rougher
> And not return more than you could ever give.

It was easy to sneer at fears of this sort as the 'individualistic problem', and suggest that the difficulties would evaporate if only you 'went over' wholeheartedly enough, did enough Party work, participated more actively 'in the class struggle' (Cornford's article 'Left?' seems to be recommending to Spender that a spot of political violence will cure his hesitations). But you can see this kind of worry persisting, even for Party activists, and settling especially about any suggestions for improving art's popularity. Take the sometimes canvassed proposal that writers could achieve a wider audience by striving for simplicity.

T. S. Eliot had committed himself in his influential essay on 'The Metaphysical Poets' (1921) to the idea that in a civilization of 'great complexity and variety' modern poets 'must be *difficult*'. 'The poet', he'd said, 'must become more and more comprehensive, more allusive, more indirect, in order to force, to dislocate if necessary, language into his meaning.' Poems like 'The Love Song of J. Alfred Prufrock', 'Gerontion' and *The Waste Land* amply illustrated his point. Eliot was proving the inevitability of literary modernism's retreat from the common reader. It was a retreat consciously repudiated by Michael Roberts. He preferred simplicity. The *New Signatures* poets, he declared, were in revolt against ' "difficult" poetry': 'to be effective, a poem must first be comprehensible.' Poets 'aloof from ordinary affairs' produce 'esoteric work which' is 'frivolously decorative or elaborately erudite'. This is bad writing, for 'A poet who does not expect to find an audience of the right intelligence, experience, and sensibility cannot write well'. So the *New Signatures* poems 'represent a clear reaction against esoteric poetry in which it is necessary for the reader to catch each recondite allusion. Even' (Roberts added) 'Mr Empson . . .' (a necessary after-thought, for even Mr Empson himself admitted his poetry was of the 'clotted kind'). Roberts returned to the point with more political edge in his Preface to *New Country*: as the writer

> sees more and more clearly that his interests are bound up with those of the working class, so will his writing clear itself from the complexity and introspection, the doubt and cynicism of recent years, and become more and more intelligible to that class and so help in the evaluation of a style which, coming partly from the 'shirt-sleeve' workers and partly from the 'intellectual', will make the revolutionary movement articulate.

The declaration is recognizably an adaptation for writers and critics of the Left's favourite 'going over' passage from Marx and Engels. Its advocacy caught on. As the idea travelled about the '30s, though, the focus of the desired simplicity didn't stay still. Caudwell's *Illusion and Reality* stuck closely to the language point ('The vocabulary of the bourgeois poet became esoteric and limited'). MacNeice in his *Modern Poetry: A Personal Essay* (1938) quoted the *New Signatures* Preface but he suggested that what was in question was rather a shift of attitudes and a matter of poetry's content:

> These new poets, in fact, were boiling down Eliot's 'variety and complexity' and finding that it left them with comparatively clear-cut issues . . . they were deliberately simplifying [the world], distorting it perhaps (as the man of action also has to distort it) into a world where one gambles upon practical ideals, a world in which one can take sides.

And kicking off from a repudiation of that key modernist declaration by Eliot, lots of Marxists reacted against almost any kinds of formal experiment in art: against surrealism; against novelty in the novel; against anything subversive of plain, orderly, positivist, surface realities. Their reaction led, in fact, to the communist doctrine of Socialist Realism, made Party dogma at the Soviet Writers' Congress in 1934. Many such congresses had their bogeys-of-the-day: at Kharkhov (1930) Henri Barbusse was the target; in Spain (1937) it was Gide. In 1934 the chief Western target was Joyce. Not that Karl Radek's sustained attack on Joyce at the Congress, for his privatizing and specializing of reality and language (*Ulysses* 'cannot be read without special dictionaries', it's written in 'some kind of Chinese alphabet without commas so that it cannot reach the masses of the people'), revealed a very close reading of *Ulysses* (it just wasn't 'a book of eight hundred pages without stops or commas': Radek was getting carried away by Molly Bloom's soliloquy). But the direction was clear enough: the political revolutionary was to approach art and letters with an aesthetically arch-conservative, retrogressive doctrine—as Herbert Read was not slow to point out in reply to Radek in *Art and Society* (1937). (Caudwell had his answer ready to *that* in *Illusion and Reality*: the uprising of artistic modernism was 'rebellious' and anarchic, rather than truly revolutionary.) And Socialist Realism had a profound effect in Britain: it helped to slow down literary experiment and to smash up modernism especially in the novel, thus pushing the novel back beyond Henry James into the arms of nineteenth-century bourgeois naturalism. Herbert Read complained about this tendency; Radek seemed to welcome it ('The proletarian revolution firmly takes its stand on the basis of that stormy reality . . . which has been created by monopoly capitalism')—despite the obvious irony, one among several, that the consolations of form provided by the Balzacian novel, with its implications of an always translucently comprehensible and readily arrangable reality, had for archetype the detective story, whose manifestations and implications were continually decried by every '30s Marxist.[183]

But the Socialist Realist case for anti-modernistic simplicity by no means simply swept the leftist board. Radek could shout down his opponents (one Wieland Herzfelde was put lengthily right over his 'very dangerous speech' in favour of Joyce). Ilya Ehrenburg's sturdily independent speech was omitted from the Congress proceedings published in Britain by Martin Lawrence as *Problems of Soviet Literature* (1935). But still Amabel Williams-Ellis gave Ehrenburg's honest reflections on the low quality of Soviet machine-obsessed art and the cultural retardation of the Soviet 'broad masses', as well as his refusal to be programmed ('I have no programme for a literary school, no recipes how to write a novel'), lots of space in her Congress report in the second number of *LR* (November 1934). Mayakovsky and Pasternak were difficult poets, Ehrenburg acknowledged, but:

> Is a writer to be reproached if he is not accessible to everybody? Songs for the accordion are easier than Beethoven . . . We have a right to be proud that some of our novels now reach millions and millions of people. In this respect we have gone far ahead of capitalist society. But still at the same time we must foster and care for those forms of literature which today still seem only the privilege of Soviet intellectuals and the heights of the working class, but which, tomorrow, in their turn, will reach millions. Simplicity is not primitivism. It means synthesis, not lisping.

Day Lewis ('Revolutionaries and Poetry') didn't think that simplifying your language in order to reach a wider audience would work short of the revolution and the creation of a classless society (and, of course, all Marxist thinking about this issue was dogged by the determinist problem: was art a cause or an effect of social change?). Walter Allen deplored the '30s fashion for fictions made up of simple sentences: he had counted, he said in his article praising Henry Green's Flaubertian attention to style (*Folios of New Writing*, Spring 1941), only one relative clause in all of Leslie Halward's story 'Arch Anderson'. Roy Fuller ('The Audience and Politics', *TCV* June–July 1939) echoed Eliot and warned against attempts at undue simplifying: on the one hand, the magazine *Poetry and the People* might seem full of crude, inept poems, and *Twentieth Century Verse*, on the other, of narrow and difficult stuff, but

> There is no chance of poetry becoming simple, or of the class struggle easing to permit the sheltered blossoming of a sectional poetry. Poetry has to express an increasingly complex reality: only after a series of fascist wars, when the material bases of life had become worn, might reality be simple and give rise to a simple poetry—perhaps a poetry of minnesingers.

And if simplicity was a problematic notion to hold in theory it was even more troublesome to handle in practice. From the start there was Michael Roberts having to explain away the undoubted difficulty of Empson. 'In Mr Empson's poetry there is no scope for vagueness of interpretation, and its "difficulty" arises from this merit.' (There was 'no more *necessary* connection between Mr Roberts's essay and the poems [of *New Signatures*] than the embrace of one pair of covers', George Barker sharply observed in the *Adelphi*, June 1932.) Roberts avoided editorial hairsplitting and contradiction next time round by refusing Empson access to *New Country*. Janet Adam Smith had things both ways in her 'Introduction' to *Poems of Tomorrow* (1935): her selections from Day Lewis and Spender could not be charged with 'obscurity'; on the other hand, 'the reason I, at any rate, find poems of Herbert Read, W. H. Auden and Charles Madge so satisfying is that they seem to show a greater wisdom, a deeper feeling, and a wider experience at the fifth reading than at the first'—which 'implies a certain amount of effort and patience' in readers. MacNeice had to admit in *Modern Poetry* that though Auden and Spender might have been reacting against Eliot, Pound, and the Symbolists in spirit, 'in the letter' their sometimes 'difficult syntax, difficult imagery, obsure allusions' proved them much akin to their modernist forebears. And he was right. It's to be doubted whether Auden was greatly interested in being simple: not only can he be densely allusive, but he was also interested in a wide range of verse forms, including some most strict, complex, and gratuitously decorative ones like the sestina. Auden's practice in many ways represented, in fact, a revival of poetic complexity, a rebuttal of the extremes of earlier bluntly put, unrhymed, 'free' verse represented by, say, D. H. Lawrence's 'Pansies'.

Clarity in poetry is, in any case, very difficult to achieve. The frequent blurs and opacities of Spender's verses, lacks of clarity usually worsened not bettered by repeated bosh-shots in revision, show up by contrast the skill in plain statements of Louis MacNeice. MacNeice kept on neatly reconnoitring an adroit route between what he dismissed as the dense intellectual word games, 'the purely cerebral jigsaw writing' of Empson ('arid and therefore unpleasant'), and the obscurely 'sloppy' verbal mulch of Dylan Thomas. But MacNeice, as little afraid of bookish allusions and private references as of investing widely in varieties of poetic kinds and schemes, was sponsored in his plainness more by his classicism than by the Marxism he so frequently expressed doubts about. He never lost faith with poetry, unlike those who let the logic of the 'simplicity'-call drive them away from it. Did, one wonders, Day Lewis falter and Rickword and Cornford fall nearly silent, not just because the strains of Party activism consumed their time and energies but because the crying up of simplicity made poetry—inevitably a more complex kind of utterance than prose—intrinsically a problematic, not to say guilty pursuit? And if the logic of 'simplicity' pointed away from poetry to prose (with the slab of prose that passed as a 'poem', Charles Madge's 'Bourgeois News', for instance, or his 'Landscapes' II, III and IV, as a compromise that could, at a pinch, look for sanction to T. S. Eliot's 'Hysteria', in *Prufrock and Other Observations*), the same logic further demanded of prose writers the most straightforward of plain styles.

This conclusion was one clear, practical translation by British theorists of the meaning of Socialist Realism. The 'realist thought, the plain "prose" thought', will require, declared Ralph Fox in *The Novel and The People*, 'a simplistic, realistic' prose style. Edward Upward's notorious 'Sketch for a Marxist Interpretation of Literature' in Day Lewis's *The Mind in Chains*, is really no more hard-line than some other essays in that anthology. Like Anthony Blunt's 'Art Under Capitalism and Socialism' it rejects the obscure and the esoteric (unless the writer wakes up into Marxism, Upward warns, he will only 'at best' manage something like Lawrence's

The Man Who Died or the parts of Joyce's *Finnegans Wake* that Upward has seen),
and spurns anything fantastic or surreal:

> a modern fantasy cannot tell the truth, cannot give a picture of life which will survive
> the test of experience; since fantasy implies in practice a retreat from the real world into
> the world of imagination, and though such a retreat may have been practicable and
> desirable in a more leisured and less profoundly disturbed age than our own it is
> becoming increasingly impracticable today.

So Upward has bidden farewell to the quasi-surrealism of Mortmere and 'The Railway Accident', and gone on to the luminously simple and declaratively Kafkaesque
short sentences of 'Sunday' and 'The Colleagues' (in *New Country*) or 'The Island'
(*LR*, January 1935). He was not unique in this. There are some finely lucid '30s
monuments to what such an objectively plain prose could achieve, with Orwell and the
Isherwood of *Mr Norris Changes Trains* and of the pieces that compose *Goodbye to
Berlin* as arguably its two best exponents. But there are dangers in a prose that tries to
be so translucent. One, certainly, is dullness: it requires the contrived colloquial vividness of an Orwell, or the luridly violent subject matter of *Goodbye to Berlin*, to keep one
for long in front of such flatly yawning intensities of plate-glass. Upward dulled himself, as it were, into silence at the end of the '30s, and when he returned to fiction as a
Communist Party renegade with *The Spiral Ascent* he had only this dull medium at
his command, and his trilogy simply *bores* you. Another danger, more arguably, is
stylistic anonymity. Cleverly, in *Enemies of Promise*, Cyril Connolly soldered
together bits of *Wigan Pier*, 'Sally Bowles' (*Goodbye to Berlin*) and Hemingway's *To
Have and Have Not* in order to demonstrate the merging identity of what he calls 'the
colloquial style'. And the paragraph that resulted is indeed disconcertingly homo-
genous. It's hard to spot the joins. 'This, then', Connolly triumphed, 'is the penalty
of writing for the masses. As the writer goes out to meet them half-way he is joined by
other writers going out to meet them half-way, and they merge into the same creature.' (This was not a charge that would worry all would-be meeters of the masses:
collective writing had as we shall see, its '30s advocates.)

The effort to make the lives of the masses the subject of literature was no freer of
difficulties. The ambition on the Left was clear. After the revolution, the 'new
culture of the emancipated working-class' (Ralph Fox's phrase in *Communism and a
Changing Civilisation*, 1935) would be as closely bonded to the lives of the then
dominant workers as was believed to be currently the case in the Soviet Union. The
Manifesto issued by the British Section of the Writers' International, to be taken as a
preludic intimation of *Left Review*'s intent, craves a new revolutionary relevance:

> Journalism, literature, the theatre, are developing in technique while narrowing in
> content; they cannot escape their present triviality until they deal with events and issues
> that matter; the death of an old world and the birth of a new.

And writers, especially novelists, could assist at the parturition of the new workers'
world by trying to put into their work now the new subjects and attitudes that they
anticipated the revolution would shortly make inevitable. Michael Roberts' Preface
to *New Country* couldn't be clearer:

> But how is all this to affect our writing? First, it will affect our subject-matter and our
> attitude to it. The novelist, since it is his business to write about people, is compelled

more obviously than the poet to choosing what class he shall write about. And because his audience is predominantly middle class, and because it was the class among which his means permitted him to live, his novels represented mainly the life and aspirations of that class. How many novels of any note have dealt with shirt-sleeve labour? But the white-collar class, with its pitiful trade-balance superiority to the manual labourer, is doomed; even its cultural superiority has vanished under the standardisation and the vulgarisation of the Press.

The novelist, therefore, must either write in a way which shows the fatuity and hopelessness of that class, or he must turn for his subject-matter to the working class, the class which is, he thinks, not utterly corrupted by capitalist spoon-feeding and contains within itself the seeds of revolution. And he will find, in the lower working class, the clearest symbols of those passions and activities he values, for they will be less confused and muddled by the intricacies of a crumbling system. And so even though he feels that his job is to arouse and clarify the vision of the class from which he sprang, he will look elsewhere for his material and in so doing he will give new life and value to his work.

Reality—and Socialist Realism only reinforced the point—was *out there* (it was a refusal to face reality, Anthony Blunt declared in the *Mind in Chains*, for the artist 'to turn his eyes inwards on himself rather than on the outside world'), especially out there in the working-class. The trouble is that for most middle-class authors, that is, most authors, working-class experience was a long way out there, and could remain remote despite the most energetic efforts at 'going over'. The prose pieces Roberts put in to *New Country* illustrate the problem precisely. Two stories by T. O. Beachcroft (an Oxford English graduate of the '20s), one about a half-mile race in which a worker beats a Cambridge Blue, the other about the Workhouse, and a tiny tale by John Hampson about a girl troubled by her brother's imprisonment, do finger working-class life, but only very gently. The fiction most knowledgeably and extendedly in touch with the life of the proletariat is, revealingly, by South-African born William Plomer and is about Africa. Edward Upward writes what he knows about, drama-tizing the case of an anguished bourgeois: his own schoolmasterly despising of the bourgeois prep-school system in the story 'The Colleagues' and, in the story 'Sunday', an illustration of how a worried bourgeois can cut through his neurotic self-absorption by going over. For his own part, Michael Roberts quite shuns the workers and indulges in a knockabout surrealist radio-show about the London literary world and Cambridge intellectuals. This may be intended to reveal the hollowness of thoughtful bourgeois life, but it does seem rather taken with its own cleverness and too cosily enamoured-by-half of its rumbustious cast—'Superior Leavis', 'Little Willie Empson', 'the fat-legged Bloomsbury wenches', and the rest. It is left to Isherwood to try, in 'An Evening At the Bay, 1928', a straightforward kind of reportage that grants the importance of getting to know and of revealing how ordinary people live, but that also acknowledges that that life will probably never be available to a man of Isher-wood's background other than externally. So documentary exactitudes are called upon to aid the outside, bourgeois observer of working-class life, as the best substitute available for the inside information he was short on. (In *Lions and Shadows* Isherwood tells how he became the detached reporter even of fellow-bourgeois holiday-makers at the Bay: it 'was my horrible, fascinating little aquarium, which I never tired of study-ing. I was passing through yet another of my pseudo-scientific phases of class hatred. For Chalmers' benefit, I took verbatim notes of the scraps of dialogue I heard on the beach—though nothing but gramophone records could have done justice to those special intonations and accentuations which seemed, to my hyper-sensitive ear, to convey the very essence of these people's lives.') In 'An Evening at the Bay', Isherwood had already achieved his aloof 'I am a camera' stance, with both the manner (distant) and the matter (jotted-down surface details) that it entailed:

> Philip sat down in the corner by the window. He could see the whole of the Bay. The ruins of the little esplanade, undermined and broken up by the sea five years ago. The bathing-huts perched on the shingle bank. The large stucco house with boat-shaped verandas in the hollow of the swampy meadow. The gorse-bushes and the sandy lane stretching away to the left. The boarding-houses and lodgings with notices: Teas. Non-Residents Catered For. Marine View. Ocean Villa. Beach View, where we were staying. Bathing-dresses hung from windows blazing in the sunset. The opposite cliff was lit bright orange. The roofs of the new bungalows gleamed. The monument caught the light at the top of the swollen Down. On the right, the rocks of the bar were slowly drowning in a stagnant tide.

Clearly, the demands of this kind of realism would tug continually away from fiction towards documentary; away from even very factual novels like, say, *Mr Norris Changes Trains* or Orwell's *A Clergyman's Daughter*, towards documentaries scarcely disguised as fictions (like the stories of *Goodbye to Berlin* or Jack Lindsay's *1649* ('Sections 39, 41, 43, 49, 66, 73, 82, 83, 97, 106 and Endpiece consist of real documents', boasts Lindsay's authorial note—they're extracts from Cromwell's letters, Fox's Journal, the Leveller Journal and the like), or novels with an extremely high documentary content like John Dos Passos's *USA*, or, in Orwell's case, to actual documentary accounts such as *The Road to Wigan Pier* and *Homage to Catalonia*. The increasing scepticism towards imagined worlds in the periodical *Fact* (founded April 1937) nicely illuminates the tendency. The inventions of novelists can scarcely compete with communist actualities, opined Storm Jameson in her 'Documents' Essay in *Fact* (No. 4), repeating an insistence of Ralph Fox's *The Novel and The People*: 'if he is a novelist, the writer is not likely to be able to create a revolutionary hero under the eyes of the living Dimitroff'. The factual 'Examples' published in *Fact*, No. 4—they include stories by Leslie Halward, James Hanley, Mulk Raj Anand, Arthur Calder-Marshall, and a poem by Spender—are headed by a straight extract from an HM Stationery Office publication, *Report of the Causes and Circumstances attending the Explosion Which occurred at Gresford Colliery, Denbigh* ('the verbal deposition of John Edward Samuel . . . shows', alleged Arthur Calder-Marshall, 'a command of language and vividness of description, similar to Hemingway or Dos Passos. This deposition is printed as it appeared in the report. If the reader will break it up into the short paragraphs necessary for clarity, the resemblances to American fiction will be even greater'). By December 1938 (*Fact*, No. 20) Spender was announcing that 'It will not henceforward be the policy of FACT to review novels, unless they derive from a basis of factual material such as might form a number of FACT itself.' Malraux's novel *Days of Hope* was 'factual' because 'Malraux has first lived his books and then written them from his experience'; so was Lehmann's *Evil Was Abroad* because he 'writes with an almost photographic accuracy'. By publishing only such factual literature *Fact* would perhaps help hasten the 'new life' of the revolution which, when it comes, according to *Forward from Liberalism*, cannot fail to produce a great mass of documentary literature, in which writers record the changes that are taking place in society, their effect on the group or the individual. Such books, which will have their propagandist effect, will be produced in great quantities (as they are now in Russia) and may even take the place of sensational journalism or réportage, in a world which has far more leisure than ours in which to contemplate the significance of what is going on. Such books will be diaries, political propaganda, novels. The novelist of realism and adventure will be more concerned with his subject and less with wondering whether he is a poet.

What, though, about the workers themselves? In theory, at least, the life that was a closed book to the middle-class author was wide open to them. Day Lewis was enviously eloquent (in 'Revolutionaries and Poetry') about the possibilities the proletarian poet enjoyed:

> He has a magnificent opportunity before him. He stands inside the workers: he can see at first hand and feel immediately a world which has been to literature so far Terra Incognita. To speak to the workers and for the workers he does not need, as bourgeois poets do, to learn a new tongue: he has only to make poetry of what is his native language.

But new proletarian poets of note were extremely hard to find in '30s England, which was one reason for the period's surge of interest in the country's ancient folk poetry, the Border Ballad, and the 'folkish' gusto of John Skelton's most rumbustious manner. (It wasn't the only reason: the fad for ballads and the Skeltonic didn't begin with Auden and Garrett's left-leaning anthology *The Poet's Tongue* (1935), nor with Auden's article on Skelton in Garvin's *The Great Tudors* (1935), nor even with Philip Henderson's *Complete Poems of John Skelton* (1931), but with rightist Robert Graves. And so antagonistic was Graves to Auden and besotted with Laura Riding—Auden's poems were 'either synthetic or wilfully plagiaristic' of Miss Riding—that he sought in a letter to *Left Review* (December 1935) churlishly to discredit the Auden obligation to himself that Montagu Slater had rather generously declared in his *LR* article 'The Turning Point' two months earlier. Graves also wrote the most slashing review of Henderson: 'An Incomplete Complete Skelton', *Adelphi*, December 1931.) By contrast, prose fiction writers of proletarian origins were strewn rather more thickly about the period.

Bloated claims about the existence of worker-writers and worker-writing were, expectably enough, very easy to come by on the period's Left. The 'desire to express ourselves in print is prevalent among working people', claimed Julius Lipton in a review in *LR* ('Worker Writers', May 1938) of *Seven Shifts*, edited by Jack Common ('a collection of stories by men on the job'), and of *Let Me Tell You*, a collection of stories by that son of a Birmingham pork butcher, Leslie Halward. And *LR* and Lehmann's *New Writing*, as well as shorter-lived organs of the Left such as *Storm*, painstakingly sought out working-class authors, particularly short-story tellers. (*Storm* began in February 1933; its second number was April 1933; it was subtitled *Stories of the Struggle*; the struggle for life was too hard for No. 3, which probably never got published.) *LR* and *Storm* ran writing competitions with book prizes and a promise of publication for the winners. The *Left Review* competitions attracted a lot of hostile criticism, but they none the less winkled out a deal of new fiction-making talent. So much so that Tom Wintringham felt licensed to declare that the nine winning entries (they were published in *LR* No. 6) of the competition to describe 'either an hour or a shift at work', not only proved the importance of proletarian writing as the expression of a class's revolutionary consciousness, but also rebutted any worries over its quality. The early *LR* manifesto had distinguished 'writers' from 'working-class journalists and writers who are trying to express the feelings of their class': the distinction was no longer valid, Wintringham thought. But this was an optimistic assessment of the nine little winning pieces—a couple of tiny quasi-fictions plus a lot of straight reportage (they included Julius Lipton's sweatshop account, 'By the Dancing Needles').

Denigrating was, of course, easy. T. S. Eliot made merry in the *Criterion* (April 1932) over Harold Heslop's speech at the Second International Conference of Revolutionary Writers at Kharkhov (it was published in a special number of the periodical *Literature of The World Revolution*, 1931). Just who were these proletarian authors that Heslop discussed, James C. Welsh, John S. Clarke, Joe Corrie? ' "The great disappointment is Joe Corrie" ': Heslop's regret got itself quoted with a cheap magisterial sneer (only possible because of Eliot's ignorance of Scottish life and culture: the Fifeshire miner Joe Corrie achieved considerable fame in Scottish radical circles with his plays about industrial working-class lives, particularly *In Time O' Strife*, published as a pamphlet in 1930, about the miners' lockout at the end of the 1926 General Strike). For his part, George Orwell alleged in his review of Philip Henderson's *The Novel Today* in 1936 that there was no real 'proletarian' literature at all: 'like all other "proletarian" literature', Alec Brown's 'huge wad of mediocre stuff called *Daughters of Albion*', was 'by a member of the middle classes' ('Mr Henderson is careful to explain' that ' "proletarian" literature' 'does not mean literature written by proletarians; which is just as well, because there isn't any'). It was a characteristically sweeping accusation, greatly modified in Orwell's later broadcast discussion, 'The Proletarian Writer' (1940). This time Orwell agreed that 'lots' of 'good books' had been produced under the banner of 'proletarian literature'. Among others, Jack London, Jim Phelan, George Garrett, and James Hanley received the Orwell seal of approval. The seamen Garrett and Hanley both featured in *LR*; both were encouraged and published by John Lehmann in *NW*. And there were other decent working-class writers he might have named: Leslie Halward, Willy Goldman, the coalminers B. L. Coombes and Sid Chaplin, from in and about *NW*; A. P. Roley who figured in *Storm*; and James Barke, and seamen/carpenter's mate/ stage-manager John Sommerfield, and the miners Harold Heslop, Lewis Jones, and Walter Brierley, not to mention the now rather famous Lewis Grassic Gibbon, all constellated about *LR*. And even this rather robust little clutch of names does not exhaust the possible list. Walter Greenwood, for instance, Salford-born proletarian jack-of-all-trades had with *Love on the Dole* (1933) made his name and fame as a novelist of working-class life before *LR* and *NW* commenced their missionary work. And there were other possibles, some of them a little hard, especially in the more-proletarian-than-thou '30s, to place exactly in terms of class origin. Were V. S. Pritchett and Walter Allen (both published in *NW*) proletarian? Was John Hampson? Was, for that matter, Leslie Halward, really? MacNeice says he discovered the so-called Birmingham School of novelists around 1936, but that he read them as being far less Prolet-cultish than Oxbridge Leftism might have assumed: Hampson was friendly with E. M. Forster, Walter Allen went to Birmingham University,

Halward's father sold pork chops. What class exactly—the question obviously niggled—are pork butchers? Again, if reading and writing had embourgeoisified you, as they tend to, did you still count? Orwell had his doubts ('The Proletarian Writer'): 'you can see what is really the history of a proletarian writer nowadays. Through some accident—very often it is simply due to having a long period on the dole—a young man of the working class gets a chance to educate himself. Then he starts writing books.' And even though his subject is working-class life his manner and language are middle-class, because like D. H. Lawrence the education he has sought, in school, evening class, college, or in private reading, has been 'not very different from that of the middle class'.

And, once more, Orwell was exaggerating a good point. A number of 'proletarian' authors did make it eventually as more-or-less professional men of letters. James Hanley, for instance, did (with a lot of help from John Cowper Powys, among others); so did Joe Corrie, Lewis Grassic Gibbon, and the Durham miner Sid Chaplin. Walter Allen ended up as a distinguished critic and a Professor of English Literature. John Sommerfield got into films, radio, and television. Walter Greenwood became a Labour Councillor (more bourgeois than which it was, according to lots of Communist authors, impossible to get) and a prolific novelist and playwright. And so on. But the struggle was too much for many budding working-class writers; most of them, encouraged into a certain productivity by a friendly response from *LR* or kindly letters from John Lehmann, fell back eventually into defeated silence. 'Men and women are free to write', are they, snarled Tom Wintringham in *LR* ('Who is for Liberty?', September 1935)?

> As to that, ask Kenneth Bradshaw, who writes in this issue of *Left Review*. After he had been roused to take his first chance (by one of those condemned 'patronising' *Left Review* competitions) he got a deserved request from well-known publishers: had he the manuscript of a novel for them to see? He answered describing the life he leads, as a youth unemployed for some time; conditions under which no writer could possibly do work needing time, care, persistence. Then a month or more ago we sent back to him, for shortening, the story we print on page 509; he answered that he had at last got a job. A few words about the job made it clear why he wanted us to 'cut' his story: he had not the time, the energy, to do anything after finishing such a day's work. He had not—in fact—the freedom.

John Lehmann writes very movingly in his autobiographical volume *The Whispering Gallery* (1955) of just such difficulties—lack of time or energy or privacy in cramped, crowded housing—for the handful of proletarians he sponsored. B. L. Coombes, whose Left Book Club *These Poor Hands: The Autobiography of a Miner Working in South Wales* (1939) is dedicated to Lehmann, 'who cheered me by publishing my first short stories and who encouraged me to write this book', pleaded eloquently for help in these matters from sympathetic bourgeois authors ('Below the Tower', *FNW*, Spring 1941):

> I wonder can any of the Leaning Tower writers conceive the terrific struggle a man of the working class must put up before he can 'get through' as a writer? It seems that every door is shut against him, that he has set himself a most hopeless task, and that writing must be in every pulse of his being if he can survive and express himself at last.
> There are months of study to be faced; continual practice in the use of words to be

maintained; and every day the lack of privacy or quiet, and the exhaustion of heavy work—work for the pay envelope—must be borne.

The bourgeois child's relatively easy acquisition of literacy and bookishness, the glut of school and college magazines eager to receive his veriest juvenilia, the publishing and editing elders ready to give the youthies' early poems and first novels an avuncular leg-up into the big publishing world, the leisure and domestic space and financial support taken for granted among the better-off, all contrast sharply with the fierce battle to write and study needed in working-class circumstances. The case of Julian Bell, only an averagely talented boy, but who was granted years of leisure by indulgent parents in which to conduct his increasingly less promising prentice work, and all amidst the active encouragements of a circle of indulgently benevolent friends and relations, couldn't be more different from the compelling fate of Arthur Gardner, the Derbyshire coalminer and part-time college student of Walter Brierley's novel *Sandwichman* (1937). His 'scholarship' gives him days off work for attendance at college, but it's time-off without pay. And his progress is plagued by hostilities and distractions. His girl-friend wants him of an evening to be necking or at the cinema with her rather than reading. She doesn't want studies to delay their marriage, and Arthur naturally loses her to a more leisured rival. What's more, his family wants to colonize the precious studying hours he has left to him after work. All the sons must help father dig, plant, and weed the vegetable garden. Indoors, dad insists on a loud radio (' "Put the news on Albert", he said, "Let's 'ear close-of-play scores" '). So does brother Sidney, given to crooning noisily along with his favourite broadcast singers. And when Arthur seeks escape from the communal din, he's blamed for consuming too much of the family budget ('Does 'e think the damned 'ouse is made for 'im and 'is studyin'?': 'burnin' two lights'). A mining official notices that he reads his notebook down the pit at odd quiet moments, and when he makes a mistake in operating the coal tubs he's immediately sacked. 'You can't work down a pit', they tell him, 'and study as well.' So he has already discovered. It's no surprise when he fails his college exams.

Clearly, the popularity of shorter fiction among 'proletarian' writers is not attributable simply to the fact that the outlets most sympathetically open to them were magazines that could only accommodate shorter stuff. Nor was it necessarily due to a lack of talent for longer *nouvelles* and novels. One can guess that it had a lot simply to do with the working man's lack of time, energy, and opportunity in which to write, and revise, at any great length.

Shortwindedness was not, of course, all that was wrong with the fiction of worker-writers. And some of them themselves were anxious not to be given too uncritical an acceptance. John Sommerfield complained at the 1935 *Left Review* Contributors' Conference of 'a sort of "snootiness" ' in the pages of the *Review*, 'as if to say, "Isn't it marvellous that the workers should be able to write at all?" ' Gushing enthusiasts, though, were prepared to overlook a great deal. The interests of Party solidarity and Party lines kept a lot of critics quiet or cagey. Lewis Jones was, claimed Dave Garman in his Foreword to *We Live* (1939), 'a forceful and original writer': 'he had', Garman felt convinced, 'progressed far enough in the practice of literature to make, with his two novels', *Cwmardy* (1937) and *We Live*, 'a real and vital contribution to the culture of our times'. Jones, a Tonypandy miner on the dole, a communist member of Glamorgan County Council, Welsh organizer of the National Unemployed Workers' Movement, leader of three South Wales hunger marches, a man who had been jailed for three months for making 'seditious' speeches during the 1926 lock-out, and who dropped dead in January 1939 at the end of a day of speaking to over thirty meetings in aid of Spain, deserved praise; not all of it, though, for his novels. Fine and moving as they recurrently are, they nevertheless do not altogether avoid the faults of their sort: triteness and melodrama of plot, sentimental class chauvinism about workers, urgent dogmatisms, as well as a tendency to make the workers, especially members of the Communist Party, into men and women of excessive heroism and unbelievably steely militancy.

It would be foolish, naturally, to pretend that all 'proletarian' fictions can be shunted together into one capacious category or that they were all written and read in the same way. There are big variations in kind and scope. Just as there were huge differences in even leftist readers' responses. For example, John Lehmann thought Grassic Gibbon's *Grey Granite* (1934), the third part of his trilogy *A Scots Quair*, an intensely authentic proletarian fiction, an 'extremely remarkable and courageous' effort by an 'intellectual of bourgeois upbringing', so to 'steep himself in the life of' the proletariat 'as to write of the workers from the inside' (*LR*, February 1935). Exactly a year later in *Left Review* the Scottish proletarian novelist James Barke, author *inter al* of *Major Operation* and *The Land of the Leal* (1939), not only endorsed Lehmann's doubts about Grassic Gibbon's class origins but forcefully deplored *Grey Granite*'s active ignorance 'of the day-to-day actualities of the workers' struggle'. Gibbon knew Scottish *peasant* life, but he 'had never been in intimate contact with the industrialized worker'. He was thus prevented 'from presenting a true and powerfully convincing picture of working-class life'. On another point, it was John Sommerfield who disagreed at the *LR* Contributors' Conference with George Garrett over militancy: 'It would be wrong to show the working class only as militant.' In respect of their aims, ambitions, and the theory of a proletarian or socialist realist fiction, 'proletarian novelists' themselves could evidently differ as sharply as their fictional practices could vary.

It would be silly, too, to assume that ordinary bourgeois novels by happily bourgeois authors or novels by bourgeois authors trying hard to sympathize with the working-classes were not frequently as bad as, and sometimes much worse than, novels by present or former proletarians. Even the tritest, the most sentimental or melodramatic productions of the working-man authors I have named scarcely outdoes in these departments a novel like Naomi Mitchison's *We Have Been Warned* or Alec Brown's *Daughters of Albion* (1935). None the less certain badnesses—and shared badnesses—that proletarian fictionists were heir to are readily discernible.

Their going idea of what shaped a satisfying story greatly resembles the most routinated procedures of popular fiction. Tears were to be openly jerked with the energy of the most melodramatic of Victorian fictions. Death-beds are strewn about as thickly as Dickens would have desired. In A. P. Roley's *Revolt*, for instance (1933: billed in *Storm* as 'A first-class thriller about the working-class movement in this country between the years 1919 and 1921. The first real proletarian novel to be written about the struggle in Britain—and to be published'), Kitty Maguire, pregnant by the novel's politicized railway-worker hero, is knocked over by a car in a post-demo mêlée and dies having confessed she was once the mistress of the IRA traitor and agent provocateur who earlier had her Jud put in jail, and having made her confession to a priest so that she can die married to Jud. 'My wife', he pronounces over her expired form: 'He stood rigidly for a moment, and then imprinted his farewell salute on her smooth brow.' One of the earliest scenes in Lewis Jones's *Cwmardy* has Big Jim's young daughter dying most spectacularly in the agonies of childbirth—pregnant by Evan the Overman's son, who had hard-heartedly cast doubts on his paternity by accusing Jane of prostitution. Deaths sudden, deaths gruesome (Lewis Jones loves those: Jane's face quickly turns dirty yellow and snarling, the malodour of death pervading the parlour where she is laid out; in *We Live* the local shopkeeper, depressed over the slump, slashes his wife's throat carefully and repeatedly with a razor then hangs himself from a hook in the ceiling), deaths down the pit, the deaths of good workers or sympathizers under the tyrannical hooves of police horses in political demonstrations or by fascist bullets in Spain, deaths natural and deaths unnatural: they make a grim backdrop to a kept-up tale of life's—and, of course, the plot's—relentless unfairness to the working-class, which is battered continually by deprivations and dangers, by unwanted acts of man and 'Acts of God', by pregnancies, storms, accidents, and malevolences of all sorts. And the social and political struggle turns inevitably into melodrama. Just as the plot's worst friend is melodrama, so the political enemy, especially the Boss, keeps turning out to be, in fact, a melodrama. Again and again the ogre is exposed in his worst colours as he's bearded in his lair by irate workers—frequently harsh-tongued women workers. Women storm the Unemployment Board office in *We Live*, led by tough little Mary Roberts (blithely heedless of her suspender-belt's having snapped) in order to force concessions from the weak but tyrannical puppet of a bad government. When in Harold Heslop's *Last Cage Down* (1935), Mrs Cameron and Mrs O'Toole visit the odious pit manager Tate they are 'sneered' at 'evilly' and subjected to rough sexual innuendo ('He was like that man Goebbels when he was ill-treating the Jews'). But they best him toughly: 'Mrs O'Toole was beautiful in her righteous anger. She typified the working class beating down its enemy, the class oppressor.' The starving spinners of Grassic Gibbon's *Cloud Howe* (1933: second part of *A Scots Quair*) are less lucky: they're met in the hall of the factory owner's house by the sound of laughter and a bottle smashing, and the sight of a girl 'without a stitch on, nothing but a giggle', exiting from a room pursued by their boss. '*That's where the cash goes we make in the Mills*', they think, and all they get from their lascivious employer is 'a lecture on the awful times'. And the Big Bad Boss was as bad collectively as he was singly—in the sort of wicked group exposed, for example, in the scene in John Sommerfield's *May Day* where the directors of Amalgamated Industrial Enterprises crudely and machiavellianly arrange to shut factories, precipitate slumps, and ruin individuals in order to preserve the immediate interests of their cash and their shares.

Conversely, workers are not just the salt of the earth, in a glowing series of 'She was Poor but she was Honest' vignettes, they're inevitably made the true heroes of the Life-and-Boss-melodramas they're trapped in. Nowhere does the Bigness cult run bigger than among these fictions. The Big Worker—almost the same Big Worker—pops up everywhere. It's true that not quite *all* the big proletarians are actually called Jim. There's Lewis Jones's big Union Leader Ezra, and Heslop's morally gigantic miner Joe Frost (in *Last Cage Down*), and James Barke's shipyard worker Big Jock MacKelvie (in both *Major Operation* and *The Land of the Leal*), and

Barke's giant peasant Tom Gibson and the historically big Communist John MacLean ('the greatest man in Scotland today', 'about as big a man as Lenin') in *The Land of the Leal*. But Lewis Jones's two novels evince at length their fondness for Big Jim Roberts the miner with whom *Cwmardy* opens, and *Last Cage Down* is majorly concerned with Jim Cameron, the miner's leader with the 'great body', and *Grey Granite* has Big Jim Trease, leader of the unemployed. And what Big Jim gets up to does very decidedly blur into what Big Joe or Big Jock or Big John is about.

Last Cage Down ends most apocalyptically with a fire in the pit, in which the Communist Joe Frost gallantly risks his neck to save the helpless manager Tate ('Tate was proving in this, his last extremity, to be true to his nature, the nature of the parasite. He would get on fine without him'), and both men are in turn dug out by Jim Cameron— his stature confirmed by his carrying Tate to safety and also acknowledging Communism to be correct ('the likes of Joe Frost, his party, the Marxists, the communists are right'). And the heroes of proletarian fiction don't just commonly prove to be epically big, they also, especially if they're Communists, as lots of them are, keep being proved doctrinally right. Not that '30s proletarian fiction is always uniformly dogmatic. Hollywood, for instance, though predictably denounced as an opiate and a debaser, is an escapism keenly understood by some proletarian novelists. Jock MacKelvie's wife is, like her husband, a Socialist, but she's allowed (as we've already seen) to forget her mean existence for a couple of cinematic hours a week. James Barke and Lewis Jones aren't nearly as sceptical as lots of Leftists were about jazz. The Partick contingent of the International Brigade is sent off in *Land of the Leal*, not just to the tune of revolutionary and Soviet songs but to the sounds of a jazz piano. The night before Len Roberts leaves for Spain in *We Live* his family not only sing 'tearful' Welsh ballads and 'the old hymns of the people', but Mary herself plays a lot of jazz tunes on the piano ('Did you know that one, mam?'; 'That was "The Blues", mam'). But beyond these occasional spontaneous relaxations lie the more usual hard lines of repetitive doctrinal aggressiveness and confidence.

The 'surest Left pleasure is argument', as Harry Kemp groaned in *The Left Heresy* (quoting a *DW* Personal Ad: 'If Fenchurch Trip Communist wants continue argument, write Box, etc.'); and most '30s proletarian novels keep plugging away at the same arguments. A Soviet future is the only contemplatable one. The working-class movement needs leaders, and only the Communist Party can provide the toughly motivated leadership that lasts. The Labour Party, the Independent Labour Party, and the Trades Unions are no good. Their officials are half-hearted distracters from the path of true Socialist progress. They back down, they won't prolong fights (they betrayed the 1926 General Strike), and they grow corrupt, turning into sleek frauds fattening themselves on the misled worker's contributed pence. As good as fascist themselves, they encourage workers into Mosleyite Fascism ('flinging' them, as Heslop's *Last Cage Down* has it, 'to the wolves of Fascism'). They are almost as unimaginably bad as the lily-livered frequenters of churches and chapels; in fact, repeatedly they're shown to be compromising politically as they consort in Christian worship with the managers and owners. For the Christian God comes off badly, again and again, as 'The gaffer's God' who's 'no good for the working-man'. Rarely is any affection afforded even the working-man chapel steward or preacher. Dai Cannon, the drunken chapel rhetor of the Lewis Jones novels, is a manifestation of a rare sympathy. And even in this case, much of the welcome Jones's novels grant him is because of his lack of preacherly virtues—he's a cheerfully incorrigible boozer. And his occasionally moving, and always sound-hearted rhetoric ('I stand like Moses for my people') does keep having to be reproved for political impracticality and for its inflated blurs ('at last the octopus is closing his tentacles about the living bodies of our women and children. Like a gloating vulture, the hireling of the company . . .'; 'what did he mean by octopus's testicles?' wonders Big Jim.)

Proletarian novels are generally keen on the current Party line (the Communist Party lads in *Major Operation* have 'got the line right and they're sticking to it') and whoever puts the case for the line is always made to win his arguments. Plot conspires with events to prove them continually correct. 'Not only did' Jock MacKelvie 'feel he was right: he knew he was right.' And 'certainty became a powerful weapon'. Would-be opponents fall back in awe of the Party debaters. 'Well, I've got to hand it to you, mate. You've got the position weighed up all right. There're a lot of things I never understood before I understand now': thus the man Duff faced with MacKelvie's witness in *Major Operation*. And should the line shift the arguments will change gear without a pause and the converts will go on being just as readily convinced. In the novels of the later '30s, like *We Live* and *The Land of the Leal*, where the talk is all of United Fronts, Party men now eagerly chum up with the only recently abhorred Labour Party and Trades Union politicos. Barke's *The Land of the Leal* makes much of a radical Church of Scotland Minister: the cause needs him as much as the Duchess of Atholl 'when innocent children are being bombed to a pulp of bloody rags in Spain'. And MacKelvie is still winning the argument: 'I never', admits the minister's brother, 'cottoned on to this United Front till you explained its significance tonight.' And, of course, many committed authors felt, and in the political circumstances understandably so perhaps, that they should sacrifice a good deal to getting the message thus bluntly across. At the end of 'Struggle or Starve!', in *Storm*, No. 1, Rhys J. Williams unrepentantly spelled out his little fiction's purpose:

> You say this is no story? It has no crisis, no climax like proper stories should. Sorry, mate; I'm no fiction artist! My point is this: if us unemployed blokes don't get together, and weld ourselves into a powerful army and tell the damned rulers of this country that we are not going to starve in silence—well, we'll be left to starve like swine in a famine.

It was understandably easy, then, for bourgeois readers, irritated by working-class and Communist Party chauvinism, to dismiss the proletarian fictionists as beneath critical contempt. 'We have our proletarian writers', scathed 'R.H.' in the *Criterion* (October 1938):

> Indeed we have had our proletarian writers for so long now that we are rather bored with them. But they have been all of a type. They were men discontented with their jobs and in love with literature (i.e. with middle class language and manners), secret readers of the classics and shy frequenters of a night-school, who suddenly found themselves adopted by left-wing intellectuals of the middle-classes on the look-out for proletarian art and to whom it was said that if they portrayed their own life faithfully (i.e. according to the middle-class, communist mythology) they would be acknowledged as the true, adorable first-born of the new dawn. So they followed, all those proletarian novels of mean streets and baton charges, the day-to-day struggle and scenes of incredible, blaspheming violence, and the proletarian writer was established, the great *nouveau frisson*, cheaper and better than the black gigolo in every conceivable respect.

But enthusiasm of the Day Lewis kind for the proletarian writers' inside story was not entirely unmerited. The proletarian fictionist was indeed 'inside the workers', and to a degree no mere bourgeois 'goer-over', however careful an observer and note-taking documentarist he might be, was likely to out-do. To be sure, there were evident drawbacks to knowing so much, to having so great a burden of inside testimony to bear. Only the occasional proletarian novelist—Grassic Gibbon is the most distinguished case—was actually liberated into the zone beyond factualism, where direct social data could enjoy a confidently freer imaginative play. Generally the proletarian writer got himself at least as tightly constricted within *Fact*-style practices as Isherwood or Orwell was. In any case, 'straight', metonymic reportage is probably easier to do reasonably well than more imagined, invented, metaphoric writing. Certainly *Storm* carried more documentary than fiction despite its *Stories of the Struggle* subtitle, and the *LR* competitions went in very speedily for reportage. *LR*'s first competition invited readers to rewrite from a different point of view a scene from Amabel Williams-Ellis's novel *To Tell the Truth*; the second required descriptions of 'either an hour or a shift at work': documentary had been arrived at in only two strides. Still, ungainly and blunt though much of its fact-producing admittedly was, the '30s proletarian novel did reveal working-class life, did put working-class regions of Britain, the humble parts of cities, and their ordinary denizens, very firmly on to the twentieth-century novel's map.

Here was a body of fiction living as knowledgeably within the masses as any Leftist might desire, a literature of the mass of industrial workers, of the large urban mass, of the streets: 'The stream of life: mass consciousness: class consciousness'. *Major Operation*, a celebration of Glasgow as Britain's 'Second City', and notably filled with the life of the crowd ('Pavement patrol: in the Second City as elsewhere. Listen to the lullaby: the shuffle of the thousand feet: the clangour of the street trams'), is characteristic of these novels' attempts to come alive with mass life, the energies and excitements of the large meeting of workers, of the procession of strikers or of the unemployed, of leftist demonstrators on May Day, of the 'Mass United Front Demonstration Against Hunger, Fascism and War'. And some loss of the old individualism of the sort of characters familiar to conventional realistic novels did indeed result from what Kemp booed at as 'social accountancy'—that is, interpreting 'people factually', deducing 'human types from the economic processes'. These novels are anxious to generalize, to observe valuable social and political truths of wide applicability. Harold Heslop's *Last Cage Down* is not alone in continually making D. H. Lawrence's jump from the particular miners in front of the reader to miners in the general. 'A miner's fight is always with the dreaded thing, the goaf'; 'A miner's life is a sordid story of gentle tappings at a glittering face of coal'; 'A northern fireplace is the most wasteful thing in the world'; 'The male miner feeds well on Sundays, feeds and sleeps'. If one smelly earth-closet in yet another Coronation Street seems rather like all the others these novels offer, the answer is that they *are* alike. It 'is an unwritten law amongst the proprietors of Darlstone's colliery villages', declares *Last Cage Down*, 'that one street must be called Coronation Street.' Each of them has only one privy for every eight houses. Open drains lead to cesspools at the bottom of everyone of these Coronation Streets.

Those writers who were within the cage of working-class life knew its prevalent samenesses. And if even a sympathetic bourgeois like E. M. Forster was exasperated by the repetition ('It's all poverty, exasperation, disease, and attempts to free one-self', he complained to John (Hampson) Simpson about working-class writers in a letter of January 1938), and the lack of differentiation between 'Ted at the table, Ed in the mine, and Bert at the works' (that was in another letter, to John Lehmann, in December 1940), these could be taken as a measure of wide social gaps and of the propaganda still needing to be done. In any case these writers wanted to show a common plight as a prelude to stressing that salvation lay in communal action. Their novels would function as a kind of anthropology (*Major Operation* deplores the fact that Scotsmen are more familiar with the sex life of the Trobriand Islanders, 'The pink glow of Malinowski's eroticism', than with Scottish history), an anthropology of that numerous British tribe of the crowded street, the smelly slum, of the Unemployment Exchange, the Dole and the Means Test, of manual work and the violent manual workplace, of scrimping and saving and trying to stay respectable in a world where the old and sick are heartlessly scrapped, the fit but uppity are laid aside by arbitrary managements, and little boys are dropped into the bowels of the earth to dig coal at 5.30 in the morning.

The twentieth-century British novel was being taken *Behind the Scenes* of working-class existence (one's reminded of John Sommerfield's 1934 Discovery Book for children about the theatre entitled *Behind the Scenes:* backstage visits and a stage-hand's strike get into his *May Day*). These scenes were repeated thousands of times in life, so why not rehearse them a few times in fiction? The rest of the country needed to be told what it was like to stand in an unemployment queue or fiddle about making rugs at an Unemployment Centre. Walter Brierley is very good, in *Means Test Man* (1935)—dedicated to 'John Hampson, with Gratitude'—and *Sandwichman* at registering the mortal blows offered to the self-respect of working-people, their deep sense of shame over being out of work and over the legalized snooping of the Means Test inspector. What do the children and the neighbours think about their lost respectability; how can a workless man hold up his head in a male-dominant community when his wife is in charge of the home and working hard whilst he hangs idly around? The world must learn too about capitalism's relentless driving of the working-man, about 'speed-up', 'rationalization', the new 'Fordized' modes of production ('Not a minute lost. Always going. The mine had been Fordized. They were turning out coal like they turned out cars, and almost as cheaply'). People must be told how blood gets on the coal (as Len Roberts berates the slumming coal-boss's son in *We Live*):

> See the blood on every pound-note you change, taste the battered bodies on every bit of food you eat, see the flesh sticking on the coal you burn. Aye, and when we refuse to work to keep you fat and idle, send your police in to baton us down.

Hence that chain of hideous working accidents, the numerous deaths in the pit, the ripping off of a boy's arm by a coal-cutting machine in *Cwmardy*, the scalping of a beautiful girl in a piece of unguarded shafting in *May Day* ('Her cap was snatched off and whirled along a belt, her hair and scalp following it, winding round the pulley': 'The shafting was wound with long blonde hairs. A tangle of blood and hair was wedged between the belt and the pulley-wheel'), the shipyard dismemberment in *The Land of the Leal* ('Christ! his leg was right through the rollers—above the knee. *Reverse the bastard.* Mary, Mother of God—that human blood could spurt like that!'), the loss of a hand in a drilling machine in Leslie Halward's 'The Money's All Right' ('Fred felt something tugging at the sleeve of his overall. He looked puzzled. Then turned deathly white. He tried to cry out, but his throat was paralysed. The grip tightened. He grabbed the spinning chuck with his free hand and felt the skin sear. He bit his lip').

The more sober reports of pit accidents and the merely factual statistics of boys killed and disabled down mines—such as were to be found in B. L. Coombes's Left Book Club *These Poor Hands*—must be given lurid fictional life for maximum impact. The reader must be made to feel how easy it was for workers to die and, conversely, how hard it was for them to live. Miners' wages, let alone the Dole or the Public Assistance money, were scarcely enough to live on. The money in these fictions is not all right. *Means Test Man* dwells movingly on Jack Cook's stratagems to save pennies (he avoids paying for a 2*d.* stamp by withdrawing £1. 19*s.* 11*d.* rather than the dutiable £2 from his tiny savings account at the Co-op) and his subsequent despair when a threepenny bit gets lost through an unreckoned-on hole in his pocket. No wonder the wives and mothers in proletarian novels emerge as heroic ekers-out of short commons. Jean MacKelvie and her sister natter over a 'cup of tea (which had become a meal by the addition of a slice of bread and margarine)'. One of the best bits of the week's diary that makes up *Means Test Man* features the Cook family's Saturday night shopping: their hunt round the Co-op shops for the cheap cuts of meat, their joining 'the 8 o'clock lot' who hang about fresh-food shops till closing time to get the fish and fruit reduced in price because it won't stay saleable through the Sunday and Monday closing. Mrs Cook rather fancies an 18*s.* 11*d.* dress she sees in a shop window: but she has only 16*s.* 11*d.* on which to keep the entire family clothed, fed, warmed, and illuminated for a whole week. Even fish-and-chips, the working-class's gourmet treat—granted an entire and bloatedly laudatory section of *Major Operation* (entitled 'Rhapsody of Fish and Chips', it likens the 'fish supper' to the cinema as an 'imaginative escape'; 'The fried fish supper shop is the restaurant of the workers')—even fish-and-chips were just quick, cheap, unhealthily greasy and fattening. 'You can't get much meat for threepence, but you can get a lot of fish-and-chips', as Orwell complained in *Wigan Pier*: this was the slide from the culinary Gold Standard on to 'a fish-and-chip standard', and one more instance of modern decline. 'Hey, you, wha'r'about me tea?' grouses an uncouth husband in *Love on the Dole* to his 'gassin' ' wife: 'Ah suppose it'll be chips 'n'fish agen, eh? Y'll have had no time t'do any cookin', eh?' However much James Barke rhapsodized, fish-and-chips hastened the rotting of health that the rest of the workers' bad diet promoted. No wonder the British proletariat was notorious for its false teeth (the char Aggie in Alec Brown's *Daughters of Albion*, 'showed her lifelong and prenatal under-nutrition in the false teeth she wore, those false teeth which are the most striking thing in England to the foreigner coming in from a more primitive and better fed country').

Here, evidently, in the pages of writers natively familiar with ill-fitting dentures and chip shops, dole queues and demonstrations of unemployed workers, sounds the authentic voice of the British masses. And it's a voice consciously of the provinces, the British regions to which history had consigned the industrial working-classes: South Wales (Lewis Jones), Liverpool (Hanley, Phelan and Roley), the Durham coalfield (Heslop and Sid Chaplin), the Nottinghamshire-Derbyshire coalfield

(Brierley), Birmingham (Leslie Halward), Manchester (Walter Greenwood), Glasgow (James Barke, Grassic Gibbon). It's a voice rising with calculated self-consciousness from below, from the Orwellian lower depths, from the basement kitchens in which Virginia Woolf's female domestics toil ('These voices', as she wrote in her Introductory Letter to *Life as We Have Known It, By Cooperative Working Women* (1931), 'beginning . . . to emerge from silence into half articulate speech, these lives . . . still half hidden in profound obscurity'), from the hidden, murky underworld of James Hanley's stokers (and of Malcolm Lowry's seamen in *Ultramarine*) and of Hanley's *Men in Darkness*, his *Grey Children* in the economically depressed 'special areas'. 'They have said their say, and I have no comment to make', declared Hanley in *Grey Children*, his factual report from South Wales, 'for I could not better their own words.' After several pages of transcribed naval talk, including a good deal about venereal disease, Lowry's *Ultramarine* strikes an older revolutionary note in a set of quotations and misquotations from Wordsworth's Prefaces to *Lyrical Ballads*: this crude stuff is 'the real language of men', the *echt 'Lingua communis'*. The proletarian novelists strove to give the country its people in their very own words—what Evelyn Waugh objected to (in a review in the *Tablet* in July 1943 of Robert Graves and Alan Hodge, *The Reader Over Your Shoulder*) as 'the uncouth tongue of the pit and the factory'.

Very often the result is words strange to the reader who speaks 'standard' English. Lewis Jones fills his novels with Welsh terms, *bach, mam fach, duw, muniferni*. Heslop packs in dialect technical terms: the miner 'must know', for example, 'when to kirve, when to knick, when to smash down the "caunch" '. *Last Cage Down* leaves the outsider guessing what *cracket* and the rest mean, or what a *goaf* actually is. (One of Heslop's novels is actually entitled *Goaf*, 1934.) *Love on the Dole* is kinder to bourgeois readers, glossing its dialecticisms (rather oddly) in the body of its text: ' "Y'll be quartered" (fined a quarter hour's wage for impunctuality)'; ' "Let 'em try clemmin" (going hungry)'; ' "There's a rare lot on 'em i'Weaste" (cemetery)'; and many, many more. Greenwood glosses his people's slang as well: ' "Ah'm havin' a threepenny treble" (a wager upon three horses in which the winnings from the first horse, if any, are re-invested in a compound manner upon the second-named and then upon the third-named animals)'. So, occasionally, does James Barke. Who is 'Sweet Fanny Adam', bourgeois Anderson wonders? 'It's an alliterative euphemism for a succinct and effective working-class vulgarism', MacKelvie informs him. *Major Operation* is full of slang (*fan mail, domino, bohunk, sanfairyann*), a lot of it popular vulgar Americanisms off the movies. The language of James Hanley's people comes even less purged of the unrespectable. His *Boy*, for instance, is not only violently regional but is packed with swearing and abusive vulgarisms (lots of *bleeding* and *bloody*, many *bastards*, much use of *sweet effay, eff you, you shower of crimps, got your gob full, Brownie*). *Boy*'s Captain Wood is 'a five-to-two, and you know what I mean by that'. In fact, we might not know for sure (it's probably rhyming slang for a Jew), but at least the proletarian novel was opening the period's ears to the way people actually spoke in the lower social depths (you might not like the 'dreadfully dirty words' of Hanley's 'The German Prisoner', declared Richard Aldington in his 'Introductory Note' to it, but if 'you were not ashamed to send men into the war, why should you blush to read what they said in it?'). It was a truth-telling effort paralleled by the tremendous blows for democratic lexicography that were also being struck in this period by Eric Partridge's *Slang Today and Yesterday; a History and a Study* (1933), and especially by his undermining of the respectable *Oxford English Dictionary* in his great *Dictionary of Slang and Unconventional English*, which first appeared in 1937. Between them, these novelists and Eric Partridge showed the harder face of the quest for the people's authentic voice whose gentler side was the period's rediscovery of John Skelton's poetry.

The proletarian novels sought deliberately to oppose the workers' voice to the voices of the undemocratic and hostile media. Here were the people's alternative headlines. 'ALL OUT ON MAY DAY; MARCH FROM RAG FAIR AT 12.30', sing the Party pamphlets in John Sommerfield's *May Day*: the people's reality against the police-inspired lies and the frivolous obsessions with airmen, film stars, and peers that dominate the daily papers' headlines scattered gibingly through the novel. Each of *Major Operation*'s multitude of sections is enticingly headed like a newspaper story (GAFFER'S CHAFF, PAVEMENT PATROL, LOVE ON THE ROCKS, A QUESTION OF BIRTH CONTROL), so that the novel itself becomes a displacement and an undoing of the bourgeois news stories it mocks ('King Sol to Reign at Coast and Country Today'). Even when the public prints reported working-class events such as mining disasters, they were short on knowledge and sympathy. Joe Frost could, we're told in *Last Cage Down*, 'look within those pages and find the number of men and boys entombed within the raging fury of the earth beneath, but he could not find their names, their own dear personalities'. The likes of Harold Heslop sought to make up the deficiencies of journalists 'unskilled in the travail of the mines'. And, even more obviously, this kind of fiction deployed the voices that were too rarely heard on the wireless.

BBC English was widely vilified as the voice of a ruling élite: 'Tomorrow the news will be broadcast in dialect', announced Roger Roughton's 'Animal Crackers in Your Croup', a poem ironically envisaging an impossibly distant future. BBC language, grouched Ralph Fox in *The Novel and the People*, was bloodless and blameless, 'the thin speech of the gentlemen of Portland Place', conditioned by 'fear of the truth of life': quite the opposite of the people's talk. Tate, the manager in *Last Cage Down*, has an 'impossible BBC accent'. You 'needn't be clever', meditates the narration of *Major Operation*, to get into the BBC: 'Just a nice voice, you know' is all you need: 'wethah fawcaust.' And as the BBC represses the voice of the masses, so the crowd-opposing police in *May Day* are the ones equipped with radio. No wonder that novel's Pat Morgan (communist sub-editor on the *Evening Mail* who gets the paper's rightist May-day editorial blacked) rejoices over the voices that interrupt the regular wireless programmes to urge 'Workers, all out on May Day. Demonstrate for a free Soviet Britain!' This cry 'rang in a million ears': for a snatched moment the BBC had become, as the proletarian novel endeavoured to be, the voice of the working-class:

> It's grand, he thought, remembering the slogans and the rain of leaflets, the Communist voices raised everywhere where there are workers, speaking low in factories amidst the songs of machines, hammering home the May Day slogans until their clangour sounded everywhere, until even the radio and the newspapers, the loudest instruments in the orchestra of suppression were forced to echo the undertone of a working-class motif.

Just so, Charles Madge ('Press, Radio and Social Consciousness', in *The Mind in Chains*) welcomed the man who shouted the name of Mrs Simpson during a BBC variety show broadcast: 'voicing the desires of millions of listeners to break down official reticence.'

Giving the workers a voice, reporting their lives—the effort that kept drawing together the products of the privileged outsiders like Orwell and Isherwood with those of the striving insiders—was doubtless quite revolutionary. Orwell himself acknowledged ('The Proletarian Writer') 'the vitalizing effect of getting working-class experience and working-class values on to paper'. Spender agreed. The 'mere statement of social realities today, if it goes far enough, both suggests a remedy and involves one in taking sides'. That was in the Auden Double Number of *New Verse*. 'Descriptive reporting', wrote Montagu Slater in *Left Review* (June 1935), 'has a particularly revolutionary impact. (We have even invented a jargon name for it, *reportage*.) Certainly to describe things as they are is a revolutionary act in itself.' But, much though this sort of thing got into *New Writing*, was it—we have to ask—actually new writing?

Again and again on the '30s Left 'a new culture of the emancipated working-class' (Ralph Fox) was envisaged. The new world of the emancipated masses would entail the breaking of the old bourgeois forms of art and literature, and the concomitant production of new forms. The post-revolutionary workers and peasants, argues Warner's *The Wild Goose Chase*, will be 'a new race of men', who 'when they have filled their stomachs, will want to create for themselves' new art: 'Words are dug from mines and grow in fields.' Meanwhile, according to the suggestion of Caudwell's *Illusion and Reality*, the art of the bourgeois revolutionaries 'starts to revolutionise, not merely its productive forces but its own categories'; the patterns of 'the old consciousness' begin to be broken up, anticipating the 'richer pattern' of the coming 'new consciousness'. Caudwell remained very cagey about how far the new enriching had gone. Montagu Slater, though, felt distinctly encouraged by the contents of *NW* No. 4. 'Human Nature Changes', his *LR* piece about it was boldly headed. 'Writing is "new" in so far as it expresses new ways of thought, of feeling, and of being—which is as much as to say a new form of human nature, and an extraordinary change.' And such newness characterized, he thought, Alfred Kantorowicz's piece about Madrid, Margot Heinemann's poem 'Grieve in a New Way', Forster's 'brilliant account' of the 1937 Paris Exhibition, and Leslie Halward's use of 'Birmingham patois'.[184] John Lehmann had only a little while before preferred to hedge more of his bets, but he still thought newness had more or less arrived:

> I not only believe that a new Socialist society will eventually discover new forms for its culture, but also that the Soviet Union is in process of making these discoveries; and . . . is likely in a very few years to give something entirely new to the world.[185]

Such claims continued to be made, but with an increasing stridency of tone. For the desired new literary forms simply kept failing to materialize in the desired fashion.

Bourgeois forms, bourgeois control of forms, were very hard to crack open. As Leftwingers might infiltrate the rightwing press, but make little impression, so the few leftist thrillers, for instance, did not make a revolution. Eric Ambler and Nicholas Blake could manage only small steps towards politicizing their chosen forms. A. P. Roley's *Revolt* (1933), that 'first-class thriller' that we've already heard

Storm praising as 'The first real proletarian novel to be written about the struggle in Britain—and to be published', and one mixing guns and morphine, detection and thriller violence with a tale of trade union agitation—made little public impression. The failure L. C. Walker confessed to in his *Storm* piece 'Writing a Thriller' is instructive. Unemployed, he had felt he could out-write the improbable crudities of a best-selling thriller he'd picked up. He assembled his characters, but quickly realized he could not carry on because the thriller's stereotypes caved in under the interfering weight of known social realities:

> I could not help seeing behind the millionaire the gaunt, starved figures of a hundred families such as mine. In the dilettante and his lady love I could see the accumulated leisure of a hundred working men and the essence of the beauty stolen from their wives. The Scotland Yard man set me thinking of a certain day at a corner by the fire station, when I saw the batons rise and fall in sickening cadence.

And so on: in all conscience he couldn't abide by the form's demands; so 'Where's my cap? I'm off with the chalking squad'. For his part, Stephen Spender's distress ('The Creative Imagination in the World Today', *FNW*, Autumn 1940) over the magazine *Poetry and The People*'s burly demands ('Our poets must break with the past; new verse forms, and new methods of direct, simple utterance must be found'; pseudo-revolutionary bourgeois poets, who 'have no experience or understanding of working-class struggle', 'must learn new methods of expression (and that implies a whole new orientation of temper which is not of course easy to achieve), or they must go') was understandable, even though he was himself one of *Poetry and The People*'s public sponsors. Spender knew very well by 1940 that 'in other countries' directives like these had resulted in the actual deaths of writers. Silencing writers, driving them to suicide, bumping them off were not ways of improving their writing any more than threats or exhortations, let alone mere wishes, had been guaranteed to produce the formal changes then desired.

On the whole, literary forms proved widely resistant to change. Even the 'proletarian writer' 'isn't really', as Orwell observed, 'creating an independent literature. He writes in the bourgeois manner, in the middle-class dialect. He is simply the black sheep of the bourgeois family, using the old methods for slightly different purposes.' For all that the novel was made to include proletarian dialect speakers, the larger grammar and ideolect of the form stuck more or less sturdily to the same old and received bourgeois style. This is the case even with, say, V. S. Pritchett whose story 'Many Are Disappointed', one of the best short stories *NV* ever published, comes vividly and memorably alive on the strength of the phrase that provides its title, uttered by the strugglingly respectable, tight-lippedly suffering woman who runs a pub that's marked on the holidaying cyclist's map but which is after all no pub at all. She's not exactly stuck in the Orwellian lower depths, in her restrained and washed-out misery (a British version of the rural Depression heroines Walker Evans photographed in America), but she's proletarian enough. Her run-down tea-house is almost as far from the cultural centre's usual notice as a Hanley stoke-hold, and she's representative of the many ordinary characters Pritchett's stories captured by unearthing and bringing their ordinary catchphrases to light. But radical though such fictional interests and practices make Pritchett's social emphases, his stories are, *formally* speaking, close kin to, say, Elizabeth Bowen's, whose social emphases could scarcely be more bourgeois. In form Pritchett's tales live easily, in fact, within the tradition of the English short story, which they are not at all inclined to fracture. Even Henry Green, whose influential and notable assault, especially in *Living*, on the conventional languages of narrative—the ditching of definite articles; the stubbily shortened, assertively direct statements; the sentence line sticking ruggedly to the twists and meanders of unsophisticated people's talk— amounted to the best adaptation in England of Hemingwayese and the nearest anybody got to establishing what might be described as Workers' Pidgin, even Green is heavily constrained by the broader usages of usual novel practice. And the same is more or less true of Hanley, Lowry, Orwell, Greenwood, Grassic Gibbon, and the rest of the seekers after a proletarianized fiction, whether 'proletarian' by birth or by adoption and inclination. Convention's demands kept proving terribly difficult to shift; even by the most extreme triers.

Bourgeois-born Alec Brown, notorious to readers of *LR* for his aggressively prickly demands (in No. 3, December 1934) for 'the proletarianization of our actual language' and author of the slogans 'LITERARY ENGLISH FROM CAXTON TO US IS AN ARTIFICIAL JARGON OF THE RULING CLASS; WRITTEN ENGLISH BEGINS WITH US' and 'WE ARE REVOLUTIONARY WORKING-CLASS WRITERS; WE HAVE GOT TO MAKE USE OF THE LIVING LANGUAGE OF OUR CLASS', appended a cocky and predictable 'Author's Note' to his turgidly lengthy *Daughters of Albion*. It repeated his allegation about literary language since 'Caxton and company', asserted that 'the proletariat is the only class which can put forward and realise the slogan of a literary English . . . which is the mass language of the people', and went on:

> For this reason one of the basic tasks of a proletarian writer or a writer who is allying himself to the proletariat (proletarianizing himself, in the deepest sense) is to study both the preceding spasmodic attempts to make a real literary English and the actual spoken speech, and then to try to write a normal English based solidly on spoken English. In my preceding novel, *A Winter's Journey*, I made a first conscious—and amazingly stilted—attempt to do my part. This new novel is in a language which I am confident is a step forward towards what is needed.

It's a surprising boast. One peruses his *A Winter's Journey: A Simple Country Tale* (1933), an obvious attempt to combine cashing-in on the Webb-Powys rurality stunt with a bit of Marxizing about landlords and the need for agricultural mechanization, for the promised proletarianizing of literary English, but in vain. For its part, *Daughters of Albion* makes repeated stabs at undermining standardized printing practices—every *aren't, don't* and *isn't*, and so on, appears as *arent, dont*, and *isnt*; but otherwise its style is not noticeably different from any other straightforward novel of its period. Couldn't Brown, one wonders, see that his style differed scarcely one jot from the conventional? Amabel Williams-Ellis was scarcely less myopic. The criteria for worker-writers evinced by her adjudication of *Left Review* competitions were clamantly ordinary, having to do with old-fashioned skill at manipulating 'point of view' and with sensitivity to experience. In the name of newness ('Workers are not simply individuals with five senses, whose writing must touch these senses; they are creators of a new social order, and their writing a part of that creating') Alick West complained at the *Left Review* Contributors' Conference of just this critical and practical old-fashionedness:

> The fact is of course that out of all this we get nothing new but something which is indistinguishable from the aesthetics current at the end of the nineteenth century.

West's complaint was widely applicable. The products of Socialist Realism and proletarian fiction (especially the Stalin-prize winning sort) were in form frequently very mouldy fig, cousins to Zola, as Zola was cousin to Balzac. Formal innovativeness in British fiction was generally left to the obviously bourgeois (like Virginia Woolf), to the non-radical (like Joyce), or to the dubiously politicized (like the surrealistic or surrealized Dylan Thomas, Hugh Sykes Davies, Lawrence Durrell, and Samuel Beckett). Where new notes were sounded at all in 'proletarian' fictions, they were either very cinematic (of which more anon), or, more oddly, in the keys of the suspect avant-garde: of Joyce the officially despised, Eliot whose pessimism and religiosity were found so wayward, and D. H. Lawrence the lost leader of the proletarian novel. The city-scapes of John Sommerfield's *May Day* keep reminding one of Eliot and Joyce. James Barke's *Major Operation* is explicitly Joycean in its allusions to *Ulysses* (Molly Bloom is mentioned several times; the 'Rebel Song played by the Springburn Unemployed Workers' flute band' is made to slur into 'Love's Old Sweet Song' of Leopold Bloom). Barke's techniques for rendering the flow of George Anderson's thoughts are straight out of *Ulysses*. The wide-spread fondness for using newspaper headlines cannot have been indulged in unmindfully of Joyce's techniques in Part 7 ('Aeolus') of *Ulysses*. The part of Orwell's *A Clergyman's Daughter* concerned with Dorothy's return to consciousness after her weird blackout seems not unakin to the Joycean, even to Virginia Woolf's, experiments in streams of consciousness. Walter Brierley's *Sandwichman* is not only set in Lawrence country, but pursues self-consciously Lawrence's route ('the wide valley of woods and fields of "Sons and Lovers" '; 'The bus moved on. He glanced down the street where Lawrence was born; a shabby street with flat-faced houses and a grimy chapel. All the adult students in his group at Trentingham were crazy over Lawrence'; Bob Peel, exhilarated by the pit's vivid, sweaty life, is said to have 'gone all Lawrence'). What's more, many of what can be praised as the characteristic strengths of '30s 'proletarian' fictions were generally discoverable already in the pages of earlier novels.

Mrs Gaskell's *Mary Barton*, Charles Kingsley's *Alton Locke*, Charles Dickens's *Hard Times*, D. H. Lawrence's early fictions, had all anticipated these novels in stirring the reader's sentiments and sentimentality on behalf of proletarians in hardship and economic struggle. Mrs Gaskell and Dickens had been connoisseurs of the harrowing death-bed a century or so before. It was stories of Lawrence that schooled the '30s pitman-author in the plot-tactic of the injured or dead miner carried to his home and his grave by his mates. The history of the nineteenth-century novel at large involves the schooling of British readers to accept provincial, working-class, dialect-speaking ordinary life as an interesting and proper object of fictional attention. Where Alec Brown rushed glibly in, George Eliot and the Brontës, Mrs Gaskell and Hardy, Arnold Bennett, Mark Rutherford and D. H. Lawrence had already resolutely trod. It could easily be maintained, of course, that the great provincial tradition was collapsing into the excessively soft hands of Winifred Holtby (*South Riding*, 1936), or A. J. Cronin (*The Citadel*, 1937), or Richard Llewellyn (*How Green Was My Valley*, 1939), or Howard Spring (*Fame is The Spur*,

1940). It was undeniably shameful and rightly angering that any part of the nine-teenth-century documentary task should need repeating; that *Love on the Dole* should have to explain and illustrate again the Lancashire verb *to clem* (to go hungry), whose meaning *Mary Barton* so painfully glossed and spelled-out in the later 1840s; that Britain's economic, class, and regional differences should still need recording and lamenting. What couldn't be denied, though, was that the politically radical '30s fiction was not as innovative as some of its friends wished. But then, some eager over-throwers of tradition had too little grasp of the history they were decrying. Harold Heslop, ranting against the current state of British writing at the Kharkov Congress ('Modern British bourgeois literature has sunk to a depth that is truly astonishing. It has reached a level of rottenness that can only be described as positively nauseating. It is the literature of dead people. It betokens the final phase of dying capitalism'), detected glimmers of hope in 'a new school of writers'. Some of them, 'especially James Hanley and Henry Green', were 'of proletarian stock'. It was perceptive of Heslop already in October 1930 to have spotted the radical promise of James Hanley, then a very new author. It was sharp, too, to link him with Henry Green, to the style of whose *Living* Hanley's own early style owes a great deal. And Hanley was indeed proletarian born. Henry Green, however, was an Old Etonian, and *Living* was in its way as much in a thoroughly established and by no means new tradition of exploratory, provincial realism as, say, Walter Brierley's *Sandwichman* with its explicit references to *Jude the Obscure*, and *Love on the Dole* with its eye continually on *Mary Barton*, were to be.

However discomfiting to '30s theory and practice, writing novels and poems kept on proving themselves solitary occupations, done by bourgeois, noticeably ex-bourgeois or painfully would-be bourgeois people, with results frustratingly marked by the bourgeois history and development of the genres involved, especially that of realist prose fiction. Other arts, like the cinema (still new and experimental) and theatre (still enjoying a European renaissance), responded rather better to the pressure for a collective art of the people. For a start, they were by nature collective arts requiring continuous collaboration from groups of creative personnel, and so in principle ripe to become arts not only for the crowd but of it. Here, as celebrated crowd scenes in Soviet movies or in German or Soviet agitational dramas proved, were exploitable mass arts (Okhlopov's 'mass action', at Irkutsk, May Day 1921, involved around 30,000 people; in 1920, a cast of 8,000 re-enacted at St Petersburg the taking of the Winter Palace, a re-enactment used in the making of Eisenstein's film *October*, 1927).

Nothing produced in England ever matched these continental crowd excitements. Group Theatre (founded in February 1932) did its best with its choruses and dancing, its masks, its stark theatricality and revue intimacy, to import something of the avant-garde European stylization, the punchy satire, the political edge, together with the idea of the acting commune: an opportunistic mixture, in other words, of what English visitors to Weimar Germany had gleaned from earlier-'20s expression-ism and later-'20s agitprop, from Piscator and Brecht and the like, or that refugees like Berthold Viertel brought with them over to England.

Group Theatre started out cautiously with Vanbrugh's *Provok'd Wife* (Sunday, 3 April 1932), followed by W. J. Turner's *The Man who Ate the Popomack* and an experimental reading of *Peer Gynt*, but it was soon mounting Eliot's *Sweeney*

Agonistes and Auden's *Dance of Death*, MacNeice's translation of the *Agamemnon* and his *Out of the Picture*, Auden and Isherwood's *Dog Beneath the Skin*, *Ascent of F6* and *On the Frontier*, Spender's *Trial of a Judge* and Cocteau's *The Human Voice*. Most of the directing was done by Rupert Doone, the Group's driving force (he had worked as a dancer with Diaghilev and with Reinhardt in Berlin). Tyrone Guthrie occasionally co-directed as did Doone's boyfriend the painter Robert Medley; Viertel directed the Cocteau. Robert Medley usually designed the costumes and masks (John Piper did *Trial of a Judge*, though); Benjamin Britten wrote most of the music. There were also classes in movement and voice and fencing; lectures (by T. S. Eliot, Herbert Read, Tyrone Guthrie, Arthur Elton, E. M. Forster, A. S. Neill, 'and many others'); summer schools; exhibitions. Group Theatre acquired its own rooms in Great Newport Street (March 1933), but it never had its own theatre: shows and lectures were usually Sunday-night affairs in Great Newport Street, or in hired theatres, with occasional larger commercial runs *(Dance of Death* and *Sweeney* at Westminster Theatre in the winter of 1935–6; *The Ascent of F6* at the Mercury Theatre in 1937).

And Group Theatre remained a London Club (subscription one guinea; 7s. 6d. for students and 'Worker Members' of Trades Unions)—almost a club for Old Boys of Gresham's school (Auden, Medley, Britten, Pudney). Its publicity, though, sounded radically communal ('We want our Theatre to become a social force a real cooperation between artists and audience and a company trained in common'; 'The Group Theatre is a cooperative. It is a community'). One typescript draft document (in the Berg collection: by MacNeice) talks of cooperation 'with the other Left organisations'. Spender (at one point the Group's 'Literary Director') had his usual changes of mind about Group: the hero of *F6* was 'a Fascist type'; Group would do better with 'a calm realism' than with its 'expressionism'. But, none the less, Spender heralded poetic drama as 'a way out of isolation and obscurity' and the loneliness of the 'single poem'. What's more, his two friends had 'solved' the 'most important' problem of 'creating a contemporary poetic drama', that 'of finding an audience', 'better than anyone for a generation'.[186]

But Group's handfuls of Sunday nighters were scarcely the masses: 'small Sabbatical assemblies' of bourgeois Lefties in 'juvenile beards, dark-blue shirts, and horn-rimmed spectacles, which are not the representative insignia of the working class', was how Ivor Brown saw them. (Interestingly, George Orwell's own dark-blue shirts have been taken as a deliberate imitation of *French* workmen's style.) Brown couldn't 'see much point' in audiences who 'either see the point of the propaganda already or see the point of nothing but their own importance'.[187] The amateurism of Group's acting was frequently criticized; more damagingly, Spender found Auden and Isherwood's plays 'undergraduate smoker' (he meant Oxford and Cambridge) stuff.[188] Group was run, in fact, very like an undergraduate society, with its programme cards, bottle parties, and its intimate revues before invited audiences. A comparison Auden cannot have found welcome. His declaration 'I WANT THE THEATRE TO BE' (aggressively capitalized throughout), in *The Dance of Death* and *Sweeney* programme for Group's Westminster Theatre season 1935–6, faithfully sounded Group's communal note. 'Drama began as the act of a whole community. Ideally there would be no spectators. In practice every member of the audience should feel like an understudy.' That is 'living drama'. But where is it to be found?

At the Musical Hall (so far so good; though with fans like T. S. Eliot, Music Hall clearly was not an exclusively radical taste). In the Christmas Pantomime (which was suspiciously childish, not to say bourgeois). And, finally, in 'the country house charade': a quite astonishing claim that helps measure Auden's and his London clubmates' distance from most of the population they purported to serve. The ambivalences were fundamental (they're nicely revealed in Robert Medley's memoirs written in old age, *Drawn from the Life* (1983), an account eager to claim Doone's European experience but to disclaim much Expressionist influence, proud of Brecht's personal interest—he saw and enthused over *The Dance of Death* and *Sweeney Agonistes*—but anxious to disown left-wingery).

There were no such ditherings about the Workers' Theatre Movement which was founded in 1926 by members of the Independent Labour Party and Communist Party and flourished in the period (1928–34) of the Communist International's so-called 'Left' turn when Party politics and culture were kept sectarianly isolationist and self-contained. It was unabashedly more populist than Group Theatre ever sought to be. At its zenith in the early '30s the WTM turned against dramatic realism and, typically, united small bands of amateur players throughout the country into mobile Agit-Prop groups on German lines (a WTM troupe toured the Rhineland in 1931; Germans came over to help inspire the English). They staged small sketches ('Meerut', 'War', 'The First of May', 'The Crisis', 'The Rail Revolt', 'The NUWM Sketch', 'Jimmy Maxton and the ILP'), agitational happenings and 'the organized shouting of mass slogans'. They operated on street corners, at demonstrations, at political meetings. Workers' Theatre Movement acquired the rhetoric current in the period's Leftist literary criticism. Agit-Prop theatre, declared the first WTM National Conference (June 1932), was a 'new form': consciously spartan (no scenery or make-up and little 'costume') because its actors were poor ('the property-less class is developing the "property"-less theatre'). Its performances were for the people and done by ordinary people among the people ('surrounded by and part of the crowd'). 'Collective Writing' was advocated. 'Mass speaking' and 'revolutionary mass singing' were practised. The 'task of the WTM is the conduct of mass working-class propaganda and agitation through the particular method of dramatic representation'. Once more, the usurpation of entertainment and truth by bourgeois theatre and other media (including 'Left'-bourgeois theatre: Workers' Theatre Movement had grown very hostile to Labour and Independent Labour Party stuff) was being challenged. Hence those characteristic names of WTM troupes (again on German lines), such as Red Megaphones and Red Radio. Hence, too, pieces like Tom Thomas's *Their Theatre and Ours* (1932) (with its jibes at Miss Greater Garbage and other agents of the capitalist entertainment industry) and the *Suppress, Oppress and Depress* sketch, which put the *Herald*, the *Suppress (Express)* and *Pail (Mail)* in the dock:

> News of current events affecting the lives of the workers is received by means of a megaphone at the back of the stage, and the papers in turn twist it, disguise it or ignore it as suits the occasion. Finally the workers, unable to get the truth and disgusted with the anti-working class methods of the boss press are driven to protest, and the need for the workers' own press in the form of the 'Daily Worker' is very logically brought out . . . a 'Daily Worker' screen obliterates the boss press, and the end is a description of the role of the 'Daily Worker' and its use as a weapon in the class struggle.[189]

Workers' Theatre Movement was, in fact, full of the best sort of popular theatrical energy. It helped transmit the political dynamic of German Agit-Prop theatre into England (*On the Frontier*, for instance, clearly owes much to *Suppress, Oppress and Depress*) and inspired a number of theatrically important people. Ewan MacColl was a member of the Salford Red Megaphones: with Joan Littlewood he would carry on a lot of Workers' Theatre Movement's spirit in the Manchester Theatre of Action.[190]

By then, alas, in the late '30s, WTM had become defunct, killed off by the Communist International's rejection of Left sectarianism for Popular Front collaboration and the consequent soft-pedalling of blatant Agit-Prop themes and techniques, as well as by the Party's nearly simultaneous turn towards old-fashioned, albeit 'Socialist' realism. WTM had been liable anyway to change. There were *Red Stage* contributors even in 1931–2 who advocated individual acting ('to use mass acting, mass talking and mass glaring all the time on every occasion becomes aggressively boring'). The handful of professional producers and actors attracted into WTM inevitably softened the cheerfully aggressive amateurism. There were producers ready to move indoors off the streets as soon as the Communist Party's quest for Popular Front respectability was launched. The West Ham United Front troupe (in March 1935) and Manchester Theatre of Action acquired permanent homes. In Central London the inevitable result was Unity Theatre.

Once housed, first in Britannia Street, WC1 (February 1936), then in a converted chapel in Goldington Street, opposite the old St Pancras Town Hall ('New Unity Theatre Opens Today: Built by Workers', *Daily Worker*, 25 November 1937), Unity quickly restored a lot of the theatrical conventions. It soon acquired professional staff, had a policy of commercially attractive 'well-made plays', used professional producers (such as André Van Gyseghem), got big names like Tyrone Guthrie, Michel St Denis and Sean O'Casey on to its Council, and its actors (people like Paul Robeson, Alfie Bass, Bill Owen, and Ted Willis) included names known or rapidly to be known on the West End stage. Unity never became, of course, an *entirely* diluted and compromised theatre. It encouraged working-class writers (like the taxi drivers Herbert Hodge and Buckley Roberts who wrote *Where's That Bomb?* about a socialist writer tempted to write pulp fiction and his Bolshie hero who won't accept compromise). It imported radical drama from overseas: Odets' *Waiting for Lefty*, Pogodin's *Aristocrats*, Brecht's *Señora Carrar's Rifles*, the first play by Brecht ever performed in Britain.[191] It had its own mobile groups; it encouraged some 'mass speaking' (with sensational effect, apparently, at the end of *Waiting for Lefty*, where the audience was whipped up into shouting 'STRIKE, STRIKE, STRIKE!!!') and the 'Mass Declamation' of poetry (like Jack Lindsay's 'On Guard for Spain!' and *Who Are the English?*); it used the American Living Newspaper methods in *Busmen* and *Crisis* (1938); it enjoyed great success with the satirical Munich Crisis pantomime *The Babes in the Wood* zestily guying Chamberlain as a wicked uncle in league with the robbers Hit and Muss.

But, for all this, Unity's abandonment of more obviously Agit-Prop activities for the better-made play ('with a broad, almost liberal content') was clear, at least on its indoor stage. (Harry Kemp, in *The Left Heresy*, jeeringly claimed Unity had to resort to liberal stuff because Leftists couldn't write plays good enough for the public stage.) In March 1938, Group Theatre's production of Spender's *Trial of a Judge* transferred to Goldington Street; in December 1937 *Night and Day*'s theatre critic,

Elizabeth Bowen, was to be heard praising, almost in the same breath, both Unity and Group as radicalizers of the London theatre: ironies indeed (had not Auden in his listed 'WANTS' praised 'undocumentary' theatre and *straight* Panto, and resisted the notion that drama should 'provide . . . news'). As war broke out, Unity was preparing for even closer identification with mainstream theatre. It had certainly gone a long way since the opening night in February 1936 when songs and dance and a 'massed chant' of Ernst Toller's *Requiem* for Karl Liebknecht and Rosa Luxemburg had been thought as apt a fare for indoors as they might be for the street.

Undoubtedly, as the Left Book Club Theatre Guild's organizer John Allen put it in *Fact* magazine No. 4 (Allen was one of Group Theatre's founding dogsbodies and stage manager for its *Dance of Death* season), Unity Theatre and the Left Book Club Theatre Guild did a lot towards 'training the actors, spreading socialist ideas and opinions, taking plays to people who would never dream of buying a ticket for a theatre, and enlivening political meetings'. They didn't, though, do all that much for English dramatic literature. Their best plays came from overseas. Some local stage-writing talent was discovered. Herbert Hodge's 'political cartoons' were reportedly a great success; so were Montagu Slater's *Stay Down Miner* (produced at Westminster Theatre and nationally in 1936 and published in 1937 as *New Way Wins*), and his *Easter*, put on in South London Town Halls and at the Phoenix Theatre in winter 1935–6 (Slater went on to write the libretto for Britten's *Peter Grimes*). But this amounted to much less than Group Theatre's little explosion of theatrical works from Auden and Co. As for formal newness, 'Mass Speaking' was only a revival and expansion of the ancient chorus. And, as Group's production of the *Agamemnon* in MacNeice's translation and of Eliot's *Murder in the Cathedral* no doubt reminded people, choruses had not only been about a long time they were also not specially socialist. They could, though, be approved of, as Jack Lindsay argued of spoken poetry in 'A Plea for Mass Declamation', for restoring the links between literature and 'social process' that had been smashed up by capitalist individualism. Restoring such links would, argued Lindsay, make for socialist 'newness'. 'Here, then, I make my plea for Declamation, Mass-recitation, as the initial and primary form of our new poetry. For there we get the most direct contact with the new audience'. 'Mass-recitation' of poems with a 'new' socialist content will reveal 'most nakedly the new orientation demanded of the poet', and will help the poet 'work out the new relativity of form and content' that will solve his aesthetic problems:

> Mass-declamation becomes the form of contact from which endless new developments can stem; the Thespis' cart, the miracle-play platform in the market-place, which is the prerequisite of a new depth of drama; the first statement of the new convictions of relationship without which the future's lyric can never be born.[192]

The trouble is that Lindsay's Mass-Declamations are long and dull. What's more, even though a crowd of reciters spoke them, allegedly with great dramatic effect, and even though, according to Lindsay, Unity people collaborated with him by suggesting alterations in 'On Guard for Spain!', Lindsay himself did all the writing. The old 'bourgeois' problem of the lone man with the pen would not go easily away. (Significantly, Robert Medley blames the formal weaknesses of *F6* and *On the Frontier* upon Auden and Isherwood's inability and/or reluctance 'to subordinate themselves to the imperatives of cooperative production'.)

There were some more strenuous attempts than Lindsay's—or Auden and Isherwood's—to implant a creative mass element into the resistant world of the written text. *The White Sea Coral*, a piece of Soviet 'collective writing' sponsored by Gorki, was hailed in *The Novel and The People* for pioneering collective socialist history. John Lehmann praised such 'collective prose epics' (he instanced *Those Who Built Stalingrad* with *The White Sea Coral*) as 'first advances towards new forms in literature'.[193] 'If something' of the writer's 'unnatural apartness' from people, 'can be broken down by writers working together, by their coming into relation with their fellow-men, they may between them, provide the conditions, the warmth, for a new literature'. Thus Storm Jameson in *Fact*, No. 4. But poets and novelists found such collectivism harder than film and stage producers, choreographers, or group historians might. The collective novel Naomi Mitchison publicized in *New Masses* ('We're Writing a Book', September 1936)—it was allegedly being written by her under the guidance of a Committee of King's Norton Proletarians—ground to a halt only a quarter done. (The group of 'Comrades' to whom her *We Have Been Warned* is dedicated didn't have the same creative function at all.) Lewis Jones's claim that *Cwmardy* was a collective work ('really collective, in the sense that my fellow workers had to fight the battles I try to picture, and also in the sense that I have shamefully exploited many comrades for incidents, anecdotes, typing, correcting and the multifarious details connected with writing') sounds merely like special pleading. Arthur Calder-Marshall's perception (*Fact*, No. 4) of the 'composite' method as the answer to the collective problem (the mass-conscious writer 'has something new to say and the old forms devised for the acute analysis of individual character are not suitable for his purpose') doesn't seem very convincing. Calder-Marshall instanced Dos Passos's *USA*, a documentary fiction whose 'Newsreel' sections assemble a collage of real headlines and news items more busily than any British novel of the period dealt in fictionalized ones. But, of course, for all the 'mass' nature of these news items, and so of this novel's collage of American life, once again one author on the old-fashioned individual plan, John Dos Passos himself, did the gathering and arranging.

Too often, clearly, 'mass' and 'collective' as applied to writing were wishful metaphors that signalled a political desire to impart a collectivity into writing that was inspired by and culled from arts more susceptible to mass-ness, but didn't necessarily guarantee any real shift by written texts into the democratic and collective realities so envied elsewhere. And in its approach to film and photography '30s writing tended only to prove over again how stuck with mere metaphors of newness it was.

The new small portable camera was widely perceived as democratic. Any 'ordinary person', claimed Auden in *Letters from Iceland*, 'could learn all the technique of photography in a week. It is *the* democratic art, i.e. technical skill is practically eliminated.' Humphrey Jennings enthused that photography was 'the system with which the people can be pictured by the people for the people'. The snapshot could be politicized whether in the straight truth-telling exposé of photo-reportage that came into England particularly from pre-Hitler Germany (Humphrey Spender and Bill Brandt had lived in Germany, Felix Man was German), that flourished in Depression America (with Walker Evans as its most famous exponent) and in the memorable 'picture-story' style of *Picture Post, Paris-Match, Life Magazine*, and the rest, and that became one of the mid-century's sharpest journalistic modes, or in the flamboyantly disturbing caricature techniques of photomontage which had, again, been given notable expression in pre-Hitler Germany. John Heartfield was just one more of the German refugee photographers who fetched up in England (in his case, in 1938). His 'political photomontage in the service of the proletarian movement' was singled out for especial praise by F. D. Klingender in *5 On Revolutionary Art* (1935).

Even more deservedly admirable from this angle was the documentary film. Not only was film easily copiable and exceptionally mobile ('a far easier piece of luggage for the social gospeller than a cry of players with all their bits and pieces': Ivor Brown); but documentary film's revelations about the lives of ordinary people were more powerfully truth-telling still than the single photograph or the 'picture essay', and the opportunities it offered the radical editor to make revelatory and accusive connections by deft cross-cuttings and montages were obviously more extensive than with the single photomontage image. In the cinema, Heartfield's montage power could, as the movies of Eisenstein and other socialist film-makers had shown, be enormously expanded. What's more, documentary film was Ours, it was British: the charismatic Scot John Grierson had shown the world how the political intentions and editorial wizardry of Dziga Vertov and Eisenstein could be applied to documenting on film the lives of ordinary working people. Inspired by Eisenstein's dynamic cross-cutting (he helped prepare *Battleship Potemkin* for American audiences), impressed by his recent acquaintance with Robert Flaherty (but resistant to the love of exotic travelogue subjects that Flaherty famously evinced in *Nanook of the North* and *Moana*), Grierson had returned from America to make his herring-fishermen film *Drifters* for the Empire Marketing Board. The London Film Society premiered *Potemkin* and *Drifters* together on the Sunday afternoon of 10 November 1929. (Eisenstein was present: 'Why', he said to Grierson after *Drifters*, 'you must know all about *Potemkin*'.) Grierson had the word for *Drifters* and its 'new and vital form': *documentary* (he was the first person to anglicize the French word *documentaire*, in 1926). And *Drifters*, sensational because of its realistic outdoor and workaday subject ('material on one's doorstep', Grierson called it), as well as the relentlessly Eisensteinian collocations of its images, became a byword for the period's cinematic truthtellers. Eager movie-men clustered about Grierson. He was a genius at drafting in unlikely seeming sponsors—the Empire Marketing Board, the GPO, the Ceylon Tea Propaganda Board. Despite the grouches and occasionally justified suspicions of some Socialists—commercial backers wouldn't 'willingly pay for an exact picture of the human life within their enormous buildings', observed Auden, in a review of Paul Rotha's book *Documentary Film*—these government agencies, and private firms such as the Gas Light & Coke Company, were remarkably liberal in financing exposé accounts of work, food, health, housing, schools and workers' leisure. And Grierson not only made movies, he trained and helped a remarkable band of others to do so: Edgar Anstey, Arthur Elton, Stuart Legg, Paul Rotha, Basil Wright, Humphrey Jennings, Harry Watt, William Coldstream. It was Grierson's men (and women) who branched out into the documentary film companies that sprouted freely in and about Soho Square: Strand Films, the Realist Film Unit, the Progressive Film Institute, Associated Realist Film Producers. It was their films—'the "documentaries" in which this country has long led the world'—that made green oases in the grim deserts of Graham Greene's film-reviewing days: 'the only important films being made in England today come from Mr Grierson's system of film units'.[194] These radical tacticians had stormed the central stronghold of this arch-capitalist mass medium. 'Grierson plus Flaherty Equals Marx', someone scrawled on the Empire Marketing Board Film Unit's cutting-room wall. Documentary art was mass art for and about the masses, making movies that would exploit and 'reveal' (Grierson's words again) 'the essentially cooperative or mass nature of society'.

No wonder literary men wanted to become camera-men. Samuel Beckett wrote to Eisenstein (1936) offering his services as unpaid apprentice. When Eisenstein failed to reply Beckett approaches Pudovkin—with equally poor result. Auden went to work at Grierson's GPO Film Unit, where he wrote the voice-over poem for the film *Night Mail* (1936). His poem 'O lurcher-loving collier' was recited by a female recitative chorus in the GPO film *Coal Face* (1935), for which Montagu Slater also wrote verse. Auden produced verse for an abandoned GPO project on negroes. He and Britten collaborated on *The Way to the Sea* (1937), a Strand Film about Southern Rail electrification. Realist Film Unit's *The Londoners* also had verse by Auden. Auden equipped himself with a camera. So did other writers. Cameras were a key element in the writers' passion for Germany and the new 'objective' arts of Germany (Michael Spender worked in Germany in the late '20s for the Leitz company, who made Leicas; he it was who advised his younger brother Humphrey to acquire a Leica). The camera was one of the would-be hero's most essential bits of kit. The Leica enthusiasts writing in *My Leica and I*, edited by Kurt Peter Karfeld (1937), amply illustrated the connection: 'The Leica in the Himalayas', 'The Wonder of Flight and the Wonder of Sight', 'The Leica in Motor Racing Photography', 'The Skier and his Camera', 'On to Dynamic Photography!' Michael Spender took his camera to the Himalayas; Auden took his to Iceland; Auden and Isherwood had cameras in China. In fact, photographs became the essential accompaniment to the reporting narrative: for example, in the original Left Book edition of Orwell's *The Road to Wigan Pier*, in Auden and MacNeice's *Letters from Iceland*, and in *Journey to a War*. 'If only we could get some photographs!' exclaims Isherwood in that last book's 'Travel-Diary'. *New Writing* got some for its New Series that began Autumn 1938 (they were 'chosen and arranged with the assistance of Humphrey Spender'). Such photographs were sometimes amateurishly blurred and carelessly composed (some of the *Wigan Pier* ones are particularly scrappy, certainly no one as skilled as Bill Brandt—as is sometimes suggested—took these pictures). Occasionally, too, the fashionableness of all this photography was mocked: *Letters from Iceland*'s batch of ironically captioned snaps includes some ribbing of Grierson: 'Epic, the Drifters Tradition' (three men in a rowing boat), 'The Corpse' (a dead whale). But the widespread presence of photographic materials on the dust-jackets and between the covers of '30s books testified to the literary world's thorough anxiety to get in on the filmic act. So did the period's eager incorporation of filmic terms into the literary vocabulary.

Most notoriously, Isherwood turned himself in *Goodbye to Berlin* into 'a camera with its shutter open, quite passive, recording, not thinking'. And this wasn't just because of the *Camera Eye* sections in Dos Passos's *USA*. Both authors doubtless knew the work of Dziga Vertov, Leninist film-maker, producer of the Soviet *Kino-Pravda* newsreels, 1922–5, leader of the *Kinoki* (Cinema-Eye) group, maker of the famous *Enthusiasm: Symphony of the Don Basin* (1931) and *Three Songs of Lenin* (1934), who had declared 'I am a cinema-eye—I am a mechanical eye. I, a machine, show you a world such as only I can see.' On this view, the more like photography writing could get, the better. At least Spender thought so. In Lehmann's *Evil Was Abroad*, Spender enthused, Lehmann 'writes with an almost photographic accuracy . . . one is conscious, as it were, of the photographer's lens. Personally, I think that this is all to the good.'[195] In Spender's view, photographs sometimes outdid words for graphic power. The writing of Hoyningen-Huene's book *African Mirage* was far less good, he thought, than Auden's or MacNeice's travel books, but above 'the dull plains of his prose, the photographs stand like mountains . . . the camera's eye singles out objects of amazing beauty'. Some of the photographs he found 'dramatically effective, in the manner of stills from the best Russian films'.[196] (Spender admired the Russian film *Turksib* so much that he tried to replicate some of its powerful railway-engine imagery in his poem 'The Express'.) For his part, Graham Greene, knowing nothing of Sweden, but wanting to set *England Made Me* there, 'visited it, like a camera-team, to take the necessary stills': the 'photographs' (he means mental impressions) 'I brought back from Sweden were, I think, reasonably accurate (*Ways of Escape*, 1980). A socialist literature, asserted Storm Jameson in that most important article 'Documents' in *Fact*, No. 4 (July 1937), needed documents: presented not journalistically ('visits to the distressed areas in a motor-car'), nor as facts done up into fictions (she cited A. J. Cronin's *The Stars Look Down*), but produced like documentary films:

> Perhaps the nearest equivalent of what is wanted exists already in another form in the documentary film. As the photographer does, so must the writer keep himself out of the picture while working ceaselessly to present the *fact* from a striking (poignant, ironic, penetrating, significant) angle. The narrative must be sharp, compressed, concrete. Dialogue must be short—a seizing of the significant, the revealing word. The emotion should spring directly from the fact. It must not be squeezed from it by the writer, running forward with a, 'When I saw this, I felt, I suffered, I rejoiced . . .' His job is not to tell us what he felt, but to be coldly and industriously presenting, arranging, selecting, discarding from the mass of his material to get the significant detail, which leaves no more to be said, and implies everything.

The professionalism of the Grierson-trained documentaries had evidently impressed Miss Jameson. Documenting writers, though, were too incompetently amateurish: 'Again the relevant comparison is with the documentary film. It takes a sharpened and disciplined mind to handle a mass of material in such a way that only the significant details emerge.' And if the key to this revolutionary documentation was montage (Grierson's followers were all, he said, 'masters of camera and, more importantly, masters of *montage*'), writers had better act like film-editors. Evelyn Waugh complained that *Letters from Iceland* was ' "cut" like a film' (and 'never for one moment' looked like a book). 'The rapidity of movement . . . the "quick shots", to use a film term, of things in particular aspects, have been much prized by many of the younger poets, whose obscurity . . . is often merely due to their "cutting" '. So MacNeice, in his *Modern Poetry* (1938). The young Malcolm Lowry, a great devotee of German Expressionist cinema, admits that he strove for a montage effect in his novel *Ultramarine*. John Dos Passos's *USA* trilogy (and Alfred Döblin's novel *Berlin Alexanderplatz*) went in for montage more strenuously than most Britons' fictions. But rapid jumping from shot to shot, image to image, scene to scene, became a favourite period device. Arthur Calder-Marshall's *Dead Centre* is built up entirely, and rather clumpingly, of such a montage of scenes and people. The Mass-Observation *May the Twelfth* volume (1937), edited by Charles Madge and Humphrey Jennings and published by Faber, strove hard and consciously to achieve a cinematic kaleidoscope of events on George VIth's Coronation Day: 'Close-ups and long shot, detail and ensemble, were all provided'. The second 'Berlin Diary' in *Goodbye to Berlin* shows the mode at its most clearly fluid and politically revelatory. No writer, of course, ever became a camera; writing is not photography; nor are poems and novels films. And, to be fair, not all '30s authors wanted their work to be confused with these arts of the masses. Evelyn Waugh is persistently ribald about photographers. They manufacture lies (some half-finished buildings in Addis in *Waugh in Abyssinia* will be presented 'as the ravages of Italian bombardment'). They're frequently Americans in funny clothes, like the cheeky one in *Remote People* who 'wore a violet suit of plus-fours, a green shirt open at the neck, tartan stockings, and parti-coloured shoes'. The camera cannot 'discriminate', argues *Zoo*'s Writer to its camera-admiring Reader: it's 'much too glib'. Auden, for all his work with Grierson, agreed with Paul Rotha (in *Documentary Film*) about the documentary's 'continued evasion of the human being': 'the private life and the emotions are facts like any others.' Auden's 'WANTS' for the theatre included wanting to keep drama unphotographic: 'The development of the film has deprived drama of any excuse for being documentary.' ('I am Not a Camera', he would be insisting by 1972, in his volume of poems *Epistle to a Godson*.)

Even those writers who were enamoured of metaphors from the celluloid world would have, and in the end had to recognize they were merely metaphors. And metaphors that legitimately pointed them only in certain directions. The camera analogy may have sanctioned a certain aloofness. The occasional would-be airman or aerial-photographer may have taken from it a certain licence to be unmoved. And Isherwood's Camera-I did, after all, claim to be 'passive . . . not thinking'. (Granville Hicks blamed Dos Passos's 'camera eye' for his aloofness, his refusal 'to draw conclusions'.)[197] But this metaphor never legitimated a wholesale abandonment of the author's/narrator's personality, or of authorial responsibilities amidst a density of facts and factualities. Commentators have worried, for instance, about the degree of invention and rearrangement in the so-called documentaries of Orwell, 'A Hanging', 'Shooting an Elephant', 'Such Such Were the Joys', *Down and Out*, and so on.[198] But '30s writers never supposed in any strength that camera-work did not edit reality: 'nothing can be more arbitrary than a photograph', as Herbert Read said in his *Art and Society* (1937). Nor did people in the '30s suppose that camera pictures, once taken, did not need editing—after all what was montage? 'Some day', Isherwood's Camera-I went on, 'all this will have to be developed, carefully printed, fixed'. Storm Jameson talked, likewise, of the camera 'angle', and of 'presenting, arranging, selecting, discarding . . . to get the significant detail'. 'I am getting a little tired of that word "documentary" ', snapped Graham Greene. 'It has a dry-as-dust sound . . . it carries a false air of impartiality, as much as to say "this is what is—not what we think or feel". But the best documentaries have never been like that . . . So the personal element—the lyrical and the ironic—is the important thing in the documentaries which the GPO are sending to the World

Fair.'[199] Documentary proper, according to Grierson, took film far beyond journalism into 'art'—'to arrangements, rearrangements, and creative shapings' of natural material. Paul Rotha went still further: documentary was 'The use of the film medium to interpret creatively and in *social* terms the life of the people as it exists in reality.' *Drifters* cheerfully incorporated scenes shot in a trawler-cabin 'set' built in Lerwick Fish Market. When herring couldn't be found near Lerwick, Grierson even tried filling his empty nets with fish bought from another ship! Some of Robert Capa's famous Spanish Civil War photographs seem obviously posed, staged for maximal propaganda effect.

Neither the film documentarists nor the writers who admired them were as simplistically strait-jacketed over what Grierson called 'the creative treatment of actuality' as some of the later worriers over whether Orwell's claims for truth-telling sort well with the editorial hand that lies evidently heavy on his documentings. 'Documentary' was never the problematic form that Orwell's biographer Professor Bernard Crick makes it. Nor, strictly speaking, were the successful innovations of the film and photography world easily replicable in literature. Just borrowing the terminology of Grierson and Co. didn't grant documentary writings the documentary film's newness nor massness. Stealing ideas from the film world certainly helped dynamize some '30s writing, but of course 'photographic realism' had been known to fiction at least since Zola, and devices for impressionistic jumpiness that the '30s tended fashionably to label 'montage' had been developing gradually through the century. Joyce and Virginia Woolf were both, of course, cinema-age authors, and susceptible to technical influence from film. But the point remains: the '30s effort to make writing analogous to photography and film only enhanced possibilities of form and style that were, by the '30s, no longer brand-new. Nor, mass-media oriented though they may have been, were those formal enhancements of interest only available to paid-up leftist poets and would-be proletarian novelists.

Evelyn Waugh sported a camera, and rather grimly, if the snaps in his *Ninety-Two Days* are anything to go by. MacNeice was quite right (in *Modern Poetry*) to associate '30s poetry's taste for cinematic 'cutting' with the 'American hustle' of Pound and Eliot. So was Day Lewis to refer Eliot's 'Prufrock' to 'film technique' (in *A Hope for Poetry*).

> Just as a film director will use a series of superficially unconnected 'shots' to express an emotional state or to carry the mind from one dramatic point to another, so the poet will employ a series of superficially unconnected images. I have known intelligent people, who rarely go to the cinema, completely incapable of following the plot of a film in which this technique was employed. Similarly, with post-war verse . . .

Many readers noticed the deft cinematic cutting between different people and families, between home and work, between the classes (sweating foundry workers followed, for example, by Mrs Dupret and her son talking about dances and how tiring they are), in Henry Green's *Living*. But the novels of Anthony Powell and Evelyn Waugh can be just as acquisitive of cinematic effects. Lushington's last view of the harbour as he sailed for home in Powell's *Venusburg* 'was the final and rather masterly shot of the reel'. In Miss Runcible's dream in *Vile Bodies* the road 'unrolled like a length of cinema film'. Waugh's comic novels—*Black Mischief* in particular—frequently rely on the blackly ironic effects of a montage of juxtaposed scenes. So do

Anthony Powell's. Part One of *Afternoon Men* (1931) is actually entitled 'Montage'. '*Montage* will do the rest', Maltravers airily informs the gull Blore-Smith about his cinematic plans in *Agents and Patients* (1936). Maltravers is particularly up on documentary theory:

> When I say that the relative importance of cutting will be even greater than when an ordinary commercial film is being made, you will have some idea of the weight that I attach to this side of our work. The juxtaposition of sharp contrasts will be all important.

Technical attempts among leftist authors at the mass-condition of photograph and film instantly shed much of their acquired air of apparent social and formal hopefulness when one discovers them niftily replicated, more or less, among the likes of Waugh (not to mention Eliot and Pound) and the politically less committed or even uncommitted Old Etonians.

As for the literary activity of Mass-Observation, which was the result of an interest in art for and of the masses that kept paralleling and mirroring Leftist preoccupations of the period: it didn't provide any easier route to radically new forms of mass art. Founded early in 1937 by a fusion (some would say confusion) of the interests of Tom Harrisson (old Harrovian, ex-Pembroke, Cambridge, ornithologist writer, and professional *enfant terrible* with some experience of exploring foreign parts), Humphrey Jennings (a painter who had made strong impressions on Cambridge undergraduate literary life, surrealism, and documentary film-making: he had been trained in Grierson's GPO Film Unit), and Charles Madge the Cambridge Communist and poet, Mass-Observation set out to document British life (so as to tap its mystiques and reveal its wide intrinsic worths), but also to educate the British (and not just, as the art-postcards and photographs in pubs and the painting in the streets proved, middle-class Britons). Mass-Observation's ambitions were big. It was the brilliant innovation of Harrisson and Madge deliberately to bring anthropology 'home' to Britain, to treat Britain as though it were no different in kind from any other *Savage Civilisation* (title of Harrisson's 1937 book about the New Hebrides: reissued in September 1937 by the Left Book Club) in which intriguing kinship systems obtained, and strange rituals and odd instances of religious faith occurred, and where a lot of dancing went on. A place, in fact, where Malinowski might feel at home (he did: contributing the article 'A Nation-Wide Intelligence Service' to Madge and Harrisson's *First Year's Work 1937–8 by Mass-Observation*). But, too, this was popular, democratic anthropology relying not just on special teams of observers (the main one was led by Harrisson in Bolton or 'Worktown', a subsidiary one by Madge and Jennings in Blackheath, London) but on a network of scattered observers, volunteers who agreed to describe daily life where they lived, recording their activities and feelings on the twelfth day of each month (Day Surveys that found their centre in *May the Twelfth*, the Coronation Survey published in Autumn 1937). It was anthropology equipped with the best of techniques for realistic visual reportage: Harrisson recruited *Daily Mirror* documentary photographer Humphrey Spender (Michael Spender had been with Harrisson on the 1933 Oxford Expedition to the New Hebrides as official photographer) to go and photograph Bolton and Blackpool with his Leica. What's more, the photographable realism was tempered by Freudianized wisdoms about the personal psychology of the observed and the observer: there was a survey of Fears and of Dominant Images, and observers were probed about their own dreams and fantasies, their hatred of their fathers, and such. Mass-Observation was as up to the minute as could be—characteristically there was an M-O Exhibition at the newsy Peckham Health Centre in November 1938. M-O was a delicious period rag-bag. Like leftist art, Mass-Observation aimed to undo the distorting untruths of BBC radio, cinema, bad popular literature, and the sort of newspaper that employed Madge and Humphrey Spender. Charles Madge (his *Mind in Chains* piece) regarded the good features of mass newspapers as rescuable: in particular the way they served 'as vehicles for the expression of the unconscious fears and wishes of the mass', and their 'mass-produced character' which was their 'supreme virtue'. Mass-Observation was to be a kind of truth-telling radio ('through M-O', declared Madge and Harrisson in their Penguin Series *Britain by Mass-Observation*, January 1939, 'you can already listen-in to the movements of popular habit and opinion. The receiving set is there, and every month makes it more effective'). The same volume claimed Mass-Observation as a scientific kind of detective story ('His squalid boarding-house will become for the observer what the entrails of the dog-fish are to the zoologist—the material of sciences and source of its *divina voluptas*. Not for nothing has the detective become a figure of popular admiration: his is a profession which calls for a scientific analysis of human motives and behaviour. In the detection which we intend to practise, there is no criminal and all human beings are of equal interest'). Mass-Observers-to-be Humphrey Jennings and Julian Trevelyan had both helped organize the esoteric 1936 Surrealist Exhibition, but Mass-Observation would generate, according to Kathleen Raine, a mass surrealist art. For Madge, she claims in *The Land Unknown* (1975), Mass-Observation 'was less sociology than a kind of poetry, akin to Surrealism. He saw the expression of the unconscious collective life of England, literally, in writings on the walls, telling of the hidden thoughts and dreams of the inarticulate masses'. Onlookers and participants also thought of Mass-Observation as the best sort of cinema. Spender's Bolton photos 'look like stills from a wonderful film', enthused his friend (and exhibitor at the Surrealist Exhibition) the painter John Banting.

Mass-Observation 'is more than journalism or film documentary' boasted *May the Twelfth*, 'because it has the aim in view not only of presenting, but of classifying and analysing, the immediate human world'. But the burden of analysis didn't fall on the individual observer: he just observed. Even here, though, Christopher Isherwood's camera-work would be outdone: for Mass-Observation had scores of cameras on the go. 'The Observers' (*First Year's Work*) 'are the cameras with which we are trying to photograph contemporary life. The trained observer is ideally a camera with no distortion'. 'Untrained observers' tell what society 'looks like to them': trained ones, what it is actually like. Isherwood lacked M-O training. So did Auden, and Day Lewis, and most of the realistic novelists. Therefore, it was claimed, Mass-Observation would outstrip them all in its grasp of the lives people actually led, particularly northern, urban, working-class people, and especially people in their mass aspects—responding to the coronation of George VI or the Munich Crisis, doing the Lambeth Walk, congregating in pubs, all-in wrestling halls and chapels, or on holiday spree during Lancashire's Wakes Weeks. In that article that we've already declared to be one of the period's best pieces of literary criticism (in the special Tom Harrisson Number of the Oxford student paper *Light and Dark*, February 1938:

Harrisson was repaying undergraduate editor Woodrow Wyatt for his M-O work in Bolton), the one in which Harrisson vigorously debunked the social content and contract of current poetry as illustrated in *The Year's Poetry* (1937), *Letters from Iceland*, *NW* (No. 4) and the Auden Double Number of *NV*, the Mass-Observer picked up Auden's declaration that 'Nothing is made in this town' (from the poem 'Dover'). 'Nothing', Harrisson jeered, 'is ever made in an Auden town'. Mass-Observation can easily do better than that. It knows, for instance, what people really eat, and that they're not obsessed, unlike the poets, by death and angels. In scouring 150 pages of poems Harrisson had found, for nourishment:

APRICOTS, BREAD (five), BRISKET, CAVIARE, LOTUS (two), NUTMEG, OPIUM, OYSTERS (two), PARSLEY, PLUM, PORRIDGE
(Bread is allegoric, never buttered or jammed, twice as 'loaves'). Two poets have fellows starving, five have 'em hungry. One vegetarian. Three meals are eaten: all breakfasts. There is a famine, a drought and a vomatorium . . .

Incidentally there are about seventy-five specified deaths, and forty-three actual uses of the noun or verb . . . no guns. Though Spender has an appointment with a bullet. Timeless poet, he naturally fails to keep it. The mutilation group of Ill-Health themes has forty-six members, twenty of which are in the bloodshed sub-group, ten nervous (including two St Vitus, five loony, three poetic sterility or impotence); there are ten permanent physical disabilities (including three dumb, one deaf-mute, four and a half blind). The death-ceremonial group scores forty-one, including ten graves, seven tombs, one sepulchre, one coffin, four corpses, two skeletons, two burials, one carrion, one churchyard. No insurance policies, graceful deceases, or flowers by request. It is all of any century, immortal stuff.

There are no colliers, industrial workers or machinery, unemployed, top-hats, cars or trams, cigarettes, pints, potatoes, toothache or Simpson stories. Nothing of 'ordinary life'. No attempt to present the majority facts, or any facts in a way that the majority can understand them. And no reason why there should be, in art designed for the intellectual few. Well, then, why should Auden and Coghill get mad at M-O? . . . They can keep all the ten thousand who understand. Sociology's field must be far wider, in the living, the many, the 'obvious', the now.

In his *NV* piece 'Poetic Description and Mass-Observation', Madge likewise stressed M-O's superiority in matters of writing. An observer's description of the rudeness of a conductor on a Midland Red bus (a favourite passage: it was quoted again in *First Year's Work*) was '(i) scientific, (ii) human, and therefore, by implication, (iii) poetic'. 'Mass-Observation is a technique for obtaining objective statements about human behaviour . . . Poetically, the statements are also useful. They produce a poetry which is not, as at present, restricted to a handful of esoteric performers.' In fact, anybody can, by 'taking up the rôle of observer', become a realist artist, 'like Courbet at his easel'.[200] The poets were dithering, waiting for 'some sort of social or political mass movement', but Mass-Observation was already in the thick of mass movements. Thus, gibed *Britain by Mass-Observation*, referring particularly to the wireless discussions between C. Day Lewis, Herbert Read, and Humphrey Jennings, 'the poet inverts the responsibility, actually suggests that a new sort of public has got to be CREATED and ignores the numerous existing mass movements, like the weekly football pool, All-In or the Lambeth Walk'.[201]

Nor did the Mass-Observers think 'proletarian' novelists much better than leftist poets. 'Any investigation of modern life', insisted Harrisson, 'will at once reveal the wide divergence between the English "proletarian novel" and proletarian life, between the conversations in Calder-Marshall's books and in actuality. Our forthcoming publications may show the difference.' (Of the contributors to *New Writing*, No. 4. he wanted 'more of Sommerfield'—who was one of his Observers; he thought 'Hanley is bogus . . . no better than the dog's breakfast'; and he liked 'Leslie Halward's stuff . . . the only one of these writers that ordinary working people round my way enjoy'.) The scepticism was common within M-O. Introducing four proley or, at worst, petit-bourgeois reports, in their article 'They Speak for Themselves: Mass Observation and Social Narrative', Madge and Jennings boasted along the same lines:

> The reports which are written for Mass Observation come largely from people whose lives are spent in a world whose behaviour, language, and viewpoint are far removed from academic science and literature. Sociologists and realistic novelists—including proletarian novelists—find it difficult if not impossible to describe the texture of this world. After reading hundreds of Mass Observation reports, we find that they tend to cover just those aspects of life which the others miss. Why is this? Because, we suggest, in these reports people are speaking in a language natural to them—their spelling, punctuation, etc., are their OWN—in spite of a uniform State education . . . [T]here is a general wish among writers to be UNLIKE the intellectual, LIKE the masses. Much 'pro-letarian fiction' is a product of this wish. But it is not enough for such fiction to be ABOUT proletarians, if they in their turn become a romantic fiction, nor even for it to be BY proletarians, if it is used by them as a means of escaping out of the proletariat.
>
> Mass Observation is among other things giving working-class and middle-class people a chance to speak for themselves, about themselves.[202]

Many writers refused to stomach this hail of derision. MacNeice's *I Crossed the Minch* was dismissive about 'Madge's lab boys':

Crowder:	There's MacNeice. Don't you think we ought to observe him?
Percival:	What for?
Crowder:	Mass Observation, you know.
Percival:	Oh, yes, Mass Observation . . .

Letters from Iceland's 'Last Will and Testament' also rather scathed Madge:

> Item we leave to that great mind Charles Madge
> Some curious happenings to correlate.

Gavin Ewart, passingly wounded by Harrisson in *Light and Dark* ('thank God for Gavin Ewart's being omitted from all four volumes'), appealed to Madge with knowing jeers in 'Cage Me a Harrisson' to tame his wilder colleague ('Jealous a bit perhaps of dear old Auden?'):

> O you must learn through loss of love or money:
> Harrisson's only useful when he's *funny*.
> It's hard to find oneself in the same boat
> With people who consistently misquote—
> And think of Science with a capital S,
> How Harrisson occasions her distress
> By baiting poets with his subtle gibes
> And 'loving the women' of those savage tribes.
>
> . . .

Think, you and Science have a world to win,
But Harrisson is the Dog Beneath the Skin—
O put the kennel up and forge the chain,
Let Mass-Observation be itself again![203]

The novelists' revenge was fiercer still. Graham Greene's Mr Muckerji of *The Confidential Agent* (1939) is formidably keen ('we mass observers are always on duty'), always prying, has no misgivings about the oddest of questionnaires ('If you would answer me just one question? How do you save money?') and is full of the right anthropological jargon about rituals and West African Tribes.

'What do you do', the manageress said, 'with all this information?'
'I type it out on my little Corona and send it to the organisers.'
'Do they print it?'
'They file it—for reference. Perhaps one day in a big book—without my name. We work', he said regretfully, 'for science.'

Muckerji is precisely the Mass-Observer stationed in *Britain by Mass-Observation*'s 'squalid boarding house': but in practice he's useless at detective work. A witness saw young Else being pushed from a window, but Muckerji decided she had been influenced by 'the papers' and got the 'wrong' window. Catching the landlady's murderous ally painting out signs of struggle at this 'wrong window' Muckerji blathers anthropologically on about death superstitions and rituals.

Not dissimilarly, if more clumsily, C. Day Lewis, as Nicholas Blake, exposes the ludicrous inadequacies of Mass-Observer Paul Perry in *Malice in Wonderland* (1940). Perry, 'who on principle approved of mass movements' and 'mass production' and is well up on his Malinowski and the customs of the New Hebrides, comes to observe the goings-on at Wonderland Ltd. Holiday Camp, place of horde pleasure, of the 'mass sound' of the campers' war cry, of 'mass emotion' ('the Wonderland visitors were being welded into a Wonderland Community—a great pleasure unit with a single voice'). But he not only encounters the plain man's scepticism ('the public doesn't want science to tell them what they know all ready'); as a detective he's unable to solve the mystery of the disruptive Mad Hatter. Solutions are left to Nigel Strangeways, a detective unencumbered by New Hebridean wisdoms. As for Paul, he's said to have been on 'a wild goose chase'. Shrewdly alluding thus to Rex Warner's novel and so to the leftist 'movement of masses', Day Lewis at once debunks the fruitfulness of M-O's observations, and questions the political wisdom of taking Butlin's Holiday Camps—or Lambeth Walks, or Cup Finals, or whatever Tom Harrisson preferred—as the most important 'mass movements' of the day.

If anything, though, writing itself proved more resistant to M-O's claims than even these writers did: at least 'The Oxford Collective Poem' turned out to be a feebly damp squib.

Even at their best, M-O's published bits of reportage never excelled what *LR*'s competitions elicited from that journal's readers. It's hard, in fact, to tell the difference between what Madge offered as new and special and what *LR*, *Fact*, *NW*, and the rest were also producing. Such flat, objective, documents were, of course, one of the period's commonest written products. Strenuously edited, as *May the Twelfth* was by Madge and Jennings with the assistance (*inter alia*) of Kathleen Raine, T. O. Beachcroft, and William Empson (the volume even had Ruthven Todd to do its index), such documents could become as occasionally dynamic as a Grierson film. But does *May the Twelfth* actually out-do *USA* or *Goodbye to Berlin*? And, more damningly, how much really of the masses were Mass-Observation's literary efforts? On inspection, all those impressive persons-of-letters working up *May the Twelfth* turn out not to have been the only ones of their kind around in the M-O enterprise.

The original fifty observers soon shot up, it was claimed, to over a thousand. And much was made of the proletarians they included: a Bolton spinner, a Huddersfield power-loom turner, a miner, a dockyard armament fitter, a Bolton coalman. And John Sommerfield, the 'proletarian' novelist did write the M-O report on the Pub. But so much fuss was afforded the Bolton coalman because—it's clear from the records in the M-O archive at Sussex University—his presence was so rare in the M-O ranks. Masses of the Observers were just the sort of people you would expect to respond to the original publicizing letters in the *New Statesman*: journalists, doctors, pharmacists, schoolteachers, headmasters; the occasional solicitor, vicar, university lecturer, and barrister; lots of housewives; clerks; a notable number of schoolboys; and a substantial throng of students, particularly from Oxford and Cambridge. If there was a Left rent-a-crowd in the '30s and early '40s it was heavily represented in M-O: with the usual period preponderance of public school accents, Oxbridge outlooks, and bourgeois and would-be bourgeois literariness. Jack Lindsay became an Observer. So did J. B. S. Haldane, and Theodora Bosanquet the literary editor of *Time and Tide*, and the author Gay Taylor, and Naomi Mitchison (with little Valentine Mitchison aged seven and the slightly larger Avrion Mitchison aged nine), and Bernard Spencer. Notably, M-O also provided an outlet for striving and ambitious would-be writers: B. L. Coombes; the young Walter Allen, who joined as a twenty-six-year-old journalist; Robert Melville the future art critic, then a commercial clerk in Sparkhill, Birmingham; C. H. Sisson, then an office worker in the Ministry of Labour; J. F. Hendry, then an Assistant Inspector of Taxes in Leeds; Eric Edney (future poet of the International Brigade), then a 'musical assistant' in Bulawayo. It's not at all surprising to discover that the Huddersfield power-loom turner was Fred Brown, author of *The Muse Went Weaving*. It was literary types who signed up; not least among the students involved. The schoolboys included P. N. Furbank; the undergraduates Boris Ford (a pupil of Leavis, at Downing College; future editor of the *Pelican Guide to English Literature*), Herbert Howarth (the future literary critic), George Woodcock (anarchist and future friend of Orwell), Kenneth Allott, Denzil Dunnett (a name familiar to perusers of Oxford literary magazines of the '30s), Alan Hodge. The Day Surveys for 12 March 1937 are full of the feelings of chaps from Christ Church College, Oxford, who were just finishing their examinations for Classical Moderations. On 12 February 1937 reports from the same sources had been preoccupied with the University Labour Club Memorial Meeting for John Cornford and the 'legendary' 'bizarrity' of Cornford's sex life. No wonder Stephen Spender announced his attention of going to Bolton to look over the M-O material with an eye to writing a play from it; in the company of Mass-Observers like those, Worktown would be home from home.

Among Mass-Observers the class problem was as acute as among film documentarists with whom they so commonly overlapped. As Auden astutely observed in his *Listener* review of Rotha's *Documentary Film* (1936) when he suggested that the going-over problem was not easily solved by the documentarists ('It is doubtful whether an artist can ever deal more than superficially . . . with characters outside his own class'): 'most British documentary directors are upper middle-class.' So no one should have been surprised that when a piece of collective writing was actually produced by Mass-Observation it should have been 'The Oxford Collective Poem'.

> Believe the iron saints who stride the floods,
> Lying in red and labouring for the dawn:
> Steeples repeat their warnings; along the roads
> Memorials stand, of children force has slain;
> Expostulating with the winds they hear
> Stone kings irresolute on a marble stair.
>
> The tongues of torn boots flapping on the cobbles,
> Their epitaphs, clack to the crawling hour.
> The clock grows old inside the hollow tower;
> It ticks and stops, and waits for me to tick,
> And on the edges of the town redoubles
> Thunder, announcing war's climacteric.
>
> The hill has its death like us; the ravens gather;
> Trees with their corpses lean towards the sky.
> Christ's corn is mildewed and the wine gives out.
> Smoke rises from the pipes whose smokers die.
> And on our heads the crimes of our buried fathers
> Burst in a hurricane and the rebels shout.

These extremely trite eighteen lines were published by Madge in *New Verse*, No. 25 (May 1937), to illustrate further the connection between Mass-Observation and poetry that he'd acclaimed in the magazine's previous number. Twelve Oxford undergraduates (did they include the Oxford Mass-Observers whose names are known: Allott of Saint Edmund Hall, Hodge of Oriel, Howarth and B. W. Watkin of Christ Church, Dunnett of Corpus, H. A. Copeman and Christopher Cox of Queen's, E. D. Clanfield of Exeter, Wyatt of Worcester?) each logged 'predominant images' of their day for about three weeks. (M-O was very keen on the 'predominant image'.) The group then selected the six most recurrent images. Then each person composed six pentameter lines each containing these common images, after which six of these lines were picked out by vote. Each person then composed a poem incorporating those six lines. Finally, the twelve poems were passed 'round the circle' for evaluation, and the winner (that's it, above) was decided by yet another vote.

The poem which emerged from this process was, enthused Madge:

> much more a collective account of the Oxford [*sic*] than of any single person in the group. It has the Oxford scene with its stone buildings, its situation in a valley, and its associations and history, with a moral. It has the sense of decay and imminent doom which characterises contemporary Oxford. It expresses a feeling of a [*sic*] responsibility together with a sense of that responsibility being neglected now and in the past. This reflection of the immediate scene is what is looked for in a collective poem.

The 'scene' most immediately reflected was, of course, the current literary one: there's lots of the death that Harrisson deplored, war is approaching; rebellion tinged with revolution is in the air ('red . . . the dawn'); but so also is Freud ('our buried fathers'); not to mention fashionable Surrealism ('waits for me to tick'). Here was nothing new. What's more, the poem, like all the lines and poems essayed by this poeticizing group, is in iambic pentameters, the traditional staple of English verse. The winner also has neat rhymes and orderly stanzas. The runner-up (it's preserved in the M-O archive) was in rhyming couplets; only one of the twelve final poems, Madge reported, actually dared to abandon rhyme. This collective literary mind was as conventional as could be. It was also highly bourgeois.

'Christ's corn is mildewed and the wine gives out': Christ Church meadow and undergraduate distress over too little wine was as far removed from Worktown's pubs as Auden's Christ Church Essay Society was from Oxford's speedway. 'The tongues of torn boots flapping on the cobbles', intrusive reminder of the poor, is probably a cinematic or newspaper image. So that the rhetoric of the mass in Madge's *New Verse* account of the poem's composition—*cooperated, collective poem, the mass, collective account*, removing *the traces of the individual*, ensuring *anonymity, mass-poem, the group*—(*mass-alteration* is talked of in Madge's archive records) and the slyly sub-Marxist talk of *synthesis* and the stress on redness (red clothes, red plums, red dress, red hair: the only complete list of original lines preserved in the archive is of the ones sparked by 'The red garment of a woman' image), all ring as gratingly and as bogus as, say, Tom Harrisson's effort to sound like a Lancastrian in his piece 'What They Think in "Worktown" ': 'We . . . in our town'; 'I and my fellow townsmen', 'us Northerners'.[204]

The refusal, in short, of this Collective Poem to be other than pretty conventional and dominantly bourgeois was characteristic of the defeat of similar literary aspirations in the period. The Mass-Observation poem had about as much of the masses in it as Group Theatre or, for that matter, a country house charade did.

How massive, one wonders, did 'a mass' have to be before it counted? *Storm*, No. 1, sold 1,200 copies: if only 10,000 copies could be sold, readers of *Storm*, No. 2 were assured, the paper would 'begin to fulfil its role as a magazine of mass revolutionary culture'. *Left Review*, though, had to be content with less than that target: it sold only about 3,000 copies per issue. Was that sufficient to make it (in Simon Blumenfeld's words in No. 3, December 1934) 'the voice of the inarticulate', the expression of 'the struggling, dark consciousness of the broad masses'? And though *Red Stage* occasionally claimed large crowds for Workers' Theatre Movement troupes, it would also gloat over very exiguous 'masses'. Its headline 'Our Theatre Awakens the Masses' yawned over a photograph of three overalled actors shouting down megaphones at a blank-faced street of council houses and a midget group of tiny tots. On that reckoning twelve Oxford undergraduates made quite a large mass. All the same, that Madge and the rest had to pretend or whistle consolingly in the dark like this, revealed a state of affairs sad for their popular literary hopes, and saddening for everybody who endorsed them. As dismaying as Tom Harrisson's brisk dismissal (in *Light and Dark*) of literature *tout court*. Madge's 'Mass Poetry' had been, he said, 'a horrible perversion': 'Madge and myself now work on a common programme and are no longer concerned with literature—he got rid of that in the Coronation Book.'

11

Seedy Margins

ONE of documentary film's acknowledged ancestors was the travel film. The French word *documentaire* on which Grierson calqued *documentary* meant *travelogue*. And the '30s kept this old association vividly alive as writers in droves took their cameras and their camera-eyes with them on their writing travels. 'The seeing eye—even "the camera eye"—is admittedly the first virtue of the travel writer', declared Granville Hicks in a *New Masses* review of John Dos Passos's travel book *Journeys Between Wars*.[205] Dos Passos was not just the author of the monumental *USA*; characteristically of the period, his documentings of home were continually interrupted by journeys *In All Countries* (his title of 1934). And what was true of Dos Passos was even truer of British authors.

Importance, creative innovativeness, the centres of art and politics were, it was widely felt among British writers of the '30s, sited away from Britain. They were abroad, elsewhere. 'England is a problem', Connolly jotted in his diary in 1929; 'There is no place in England for a serious rebel. If you hate both diehards and bright young people you have, like Huxley, Lawrence, Joyce etc. to go and live abroad'. When, in a sharp little sketch in the same diary, Mr C. Congoly the exile returns to London there's little satisfying for him to do except commit suicide. Back in London early in 1933 after extensive travels in Germany and Spain, Stephen Spender was soon once again seized up. 'It is very grey and yellow here . . . and I am beginning already to feel nausea at London', he wrote to Isherwood, and 'it is very difficult for me to write poetry or anything else'. England, Connolly wrote to his friend Noel Blakiston in 1929, is 'a great sleepy pear'. 'England hasn't got It and doesn't want to have', moaned Humphrey Jennings in Grigson's *The Arts Today*. The bemoaned shortcomings and diminishments of British culture run like an unhealed sore through that volume of Grigson's. 'It is dangerous living on an island, even if the island is on the edge of Europe . . . It is', declared Grigson himself, 'the cause of provincialism'.

> One could fight perhaps for England as a country, if one kept one's eyes on autumn beeches, a pond with cows drinking, and did not look five hundred yards to the right at the bungalows and five hundred yards to the left at the arterial road. But how one wearied at the constant, careful harbouring of small impressions.

So thinks Oliver Chant in Graham Greene's *The Name of Action*, ludicrously small-time for the intervention he makes in Trier politics, aptly put-down by the Dictator's wife for his 'infamous' sexual 'suggestion': 'Mr Chant, of—of South-West London—offering the post of mistress to the wife of the Dictator of Trier.' British intellectuals in the '30s weren't afflicted with chauvinistic pride. Heated nationalism was, of course, the mark of the Fascist, and Socialists were on principle internationalist. But, for whatever reasons, the British intelligentsia of this period was notably leery of patriotism. Orwell's aggression (in the 'England your England' section of *The Lion and The Unicorn*, 1941) towards those selling patriotism short (a 'generally

negative, querulous attitude', 'complete lack at all times of any constructive suggestion', 'irresponsible carping', 'emotional shallowness') is unpleasantly marred by
the venom of British bulldoggery's most recent Leftist convert. But Orwell's hoarse
exaggerations do have, as usual, a point:

> In intention, at any rate, the English intelligentsia are Europeanized. They take their
> cookery from Paris and their opinions from Moscow. In the general patriotism of the
> country they form a sort of island of dissident thought. England is perhaps the only
> great country whose intellectuals are ashamed of their own nationality. In left-wing
> circles it is always felt that there is something slightly disgraceful in being an English-
> man and that it is a duty to snigger at every English institution, from horse racing to
> suet puddings. It is a strange fact, but it is unquestionably true that almost any English
> intellectual would feel more ashamed of standing to attention during 'God Save the
> King' than of stealing from a poor box.

Unquestionably true? What *was* unquestionably true was that the intellectuals'
discontents were not confined to cooking and politics. Almost everything they cared
about seemed to be done better somewhere else. No wonder the better future of
politics and writing was axiomatically perceived as a *New Country*.

Abroad was so much more eventful than home. James Hanley found mid-'30s
South Wales Communists curiously blinded to Welsh dilemmas by their obsession
with Spain and Russia, where they thought the 'future for good or ill was being
forged'. Hanley wondered (in *Grey Children*) whether the Communist 'looks twice at
the figure of the harassed and poverty-stricken woman dragging her tired feet up her
own street, carrying food back to cook for her hungry family'. 'Why travel a thousand miles', he asked, 'when the problem is on their own doorstep?'

Travelling a thousand miles, if only in spirit, was simply one of the commonest of
'30s manœuvres. The struggle at home for work and a decent wage, for democracy
and Socialism against local oppression and local Fascists was important enough. But
the typical shout of '30's leftist marchers wasn't just 'Down With the Means Test!',
it was 'Down With the Means Test, Fascism and War!', for the political mind of the
period ran straight to events overseas. The most engaging struggles were not against
employers in factories here, or against landowners who wouldn't let the people hike
freely over the British countryside, nor even on the streets of London's East End
against Mosley's British Union of Fascists, they were taking place on the wheatfields
of Russia, at the great Dniepestroi hydro-electric installations, on the streets of
Berlin and Barcelona, around the Karl Marx Hof in Vienna, at the University City
in Madrid, in Italy and Abyssinia and China. *Evil was Abroad* (as John Lehmann's
title of 1938 had it). So also, though, was salvation and hope, in literature as in
politics.

Saving doctrines came from abroad. 'We in England are provincials for socialism',
wrote Montagu Slater.[206] And politicals of all colours found their nostrums and
examples in foreign places—in Rome, Moscow, Berlin. Socialist Realism was a
directive from Moscow. Surrealism was a flamboyant import, particularly from
France, its birth here a matter of translated manifestoes and of foreigners like Breton
and Dali stepping in like famous gynaecologists giving the local midwives a hand
with a difficult parturition. The most admired work in committed cinema and drama
was done elsewhere—in Russia, or Germany, or America. 'We need plays that cut

ice', declared Elizabeth Bowen in her *N & D* drama column, 'plays that get some-where, plays that are of our time; we are badly behind most other countries in this'.[207] The new photography was German. The camera that the inquisitive intellectual sported was either American (Kodak) or, more likely, a German Leica. (Michael Spender, his brother Humphrey recalls, 'believed that good lenses were made only in Germany'.) Even our greatest national contribution to form in the period, documentary film, never quite discarded its memory of a foreign etymology. The long shadows of the Russian Eisenstein and the American Flaherty kept falling across the British work. All of Grierson's and Tom Harrisson's apologetics couldn't rinse away their unease over their rebarbatively parochial subjects. It was easy to feel that observing the natives in Bolton would never match for excitement more exotic anthropological work on far-flung islands. Whatever your tastes, in fact, they seemed to be met better elsewhere. The dim Blore-Smith is taken in Powell's *Agents and Patients* to Paris for sex (to 'shake off a few inhibitions') and to Berlin 'for a little research work in the art of the cinema'. His mentors had the current idea. If you liked boys, Berlin was best. If you fancied nudism, almost anywhere else was warmer than Britain. 'I, for my part,' says a character in Spain in Rose Macaulay's novel *Going Abroad* (1934), 'like abroad. The sun shines and the sea is blue, the sands are hot.' For its part, American jazz was the hottest of the hot music. Hollywood offered would-be scriptwriters most money. Our folk-poetry seemed noticeably less exciting than American folk-songs, spirituals, and the blues. 'Only in America', insisted Auden in his Introduction to *The Oxford Book of Light Verse* (1938), 'under the conditions of frontier expansion and prospecting and railway development, have the last hundred years been able to produce a folk-poetry which can equal similar productions of pre-industrial Europe.' Other people's writers, in fact, carried so much more clout than ours. Characteristically, the arch-modernist of the '30s writing in English was James Joyce, more at home in Zürich or Paris than his native Dublin. The period's most famous 'English' poets were the Irishman Yeats and T. S. Eliot the naturalized American. *NW* and *TCV* were, from their inceptions—as the *Criterion* had been before them—internationally minded. The examples of greatness offered by a characteristic critical work like Ralph Fox's *The Novel and The People* are mainly foreigners: Malraux, Dos Passos, Erskine Caldwell, Céline, Bloch, Gide, Jules Romains. Again and again American novels were praised for their proletarian vigour, what Calder-Marshall described (in Grigson's *The Arts Today*) as their 'vulgarity and contact with the life of the people which English fiction needs if it is to be revitalized', 'an energy and courage in technical experiment which, may, by example refertilize English literature also'. For the literature of the air one looked to the Frenchman Saint-Exupéry; in the literature of action the American Hemingway and the Frenchman Malraux had overtaken T. E. Lawrence; the documentary novel was exemplified best by Dos Passos and Alfred Döblin. Invigoratingly for them, foreign writers had closer to their hands the epic themes and heroic inspirations, the excitements, horrors, and pressures that were felt to count. The English had to have such writers imported, translated. The writings of Freud and Kafka, for instance, inched their way only very slowly and in translation into the '30s English consciousness. Periodicals like *New Writing* did their best to erode the language barriers. The arrival of refugee European intellectuals after 1933 helped. The Hungarian art historian Frederick Antal taught Francis Klingender and Anthony Blunt about Georg Lukàcs, for instance. Practitioners like Brecht were actually around in London, if only occasionally. But still many authors—Brecht himself, writers like Mayakovsky and Pasternak, let alone foreign theorists like Lukàcs, or Ferdinand de Saussure, or the Russian Formalist critics, or the men of the Frankfurt School like Walter Benjamin, Theodor Adorno, Erich Fromm and Herbert Marcuse, or writers like Roman Ingarden or Jacques Lacan, and so on and on—had to wait until after the Second World War to make their proper impact in Britain. Severe censorship even kept some notable books written in English out of the home bookshops. Connolly's *The Rock Pool*, the uncensored version of James Hanley's *Boy*, Durrell's *Black Book*, Henry Miller's *Tropic of Cancer*, *Tropic of Capricorn*, and *Black Spring*, all had to be published in Paris (by the Obelisk Press). *Ulysses* was not published in England until 1936, and then only in a tiny limited edition. Like the complete *Lady Chatterley's Lover*, *Ulysses* had been a book to be smuggled unofficially past the customs officials at England's frontier ports. The 'police and customs authorities', complained Orwell in 'Inside the Whale', 'have so far managed to prevent me from getting hold of' *Tropic of Capricorn*. Orwell's correspondence with Obelisk Press was intercepted, and the police seized his copies of *Tropic of Cancer* and *Lady Chatterley* on orders from the public prosecutor. The Lawrence was returned ('the public prosecutor wrote and said that he understood that as a writer I might have a need for books which it was illegal to possess'), but not the Miller. Waugh's *Vile Bodies* opens most feelingly with hapless Adam Fenwick-Symes having his books and manuscripts confiscated at the English Customs: 'this book on Economics comes under Subversive Propaganda. That you leaves behind. And this here *Purgatario* doesn't look right to me, so that stays behind, pending inquiries. But as for this autobiography, that's just downright dirt, and we burns that straight away, see.'

No wonder so many writers looked abroad for themes and inspirations. Ralph Fox, for instance, wrote about China and Genghis Khan and Lenin, an exotic clutch of interests calmly taken as characteristic of the times by V. S. Pritchett (reviewing *A Writer in Arms*, the memorial volume for Fox, killed in Spain): Fox 'turned his eyes abroad in his quest for the epic subject . . . to the heroic clashes where the myriad anomalies of our English classes did not exist and where one could talk more vividly of oppressor and oppressed.' He was thus like 'Other English romantics, our Catholics, for example', who 'have turned with nostalgia to pre-reformation times, to contemplate a society more dramatically divided and free of the gradual greyness of the English scene where, as recorded in Mr George Orwell's latest book, there is more uproar about a Football Pool than about Fascism.'[208]

No wonder, either, that so many '30s writers spent so much of their time abroad, hastening to join what Paul Fussell has called 'the diaspora . . . of literary modernism'.[209] The younger ones had before them the tradition and example of the famous literary expatriates. The Mediterranean coastline is composed for Connolly's Naylor (*The Rock Pool*) into a sketch-map of notable writing names. 'Naylor looked up at the brown village high in the pine woods, and then across at Vence where the lights began to flicker in the sanatoriums. There Lawrence died.' 'All along the coast from Huxley Point and Castle Wharton to Cape Maugham, little colonies or lonely giants had settled themselves: there was Campbell in Martigues, Aldington at Le Lavandrou, anyone who could hold a pen in Saint-Tropez, Arlen in Cannes, and beyond, Monte Carlo and the Oppenheim country. He would carry on at Nice and fill the vacant stall of Frank Harris.' Connolly himself had gone to write in Sanary precisely to be near the Aldous Huxley he 'loved' (he and his wife felt rebuffed by the Huxleys, and Connolly's writing blocks got worse at the thought of Huxley typing hard across the bay). Sanary was the announced death-place of the famous (and bogus) travel-writer T. T. Waring, in Anthony Powell's novel *What's Become of Waring?* (1939). It was easy to see a connection between writing and travelling (in *The Destructive Element* Spender links together Auden's Airman, Rilke, and D. H. Lawrence as artists who travel).

Mostly, British writers stuck to Europe. More or less everywhere had, of course, some takers. Alec Waugh went all over the place ('He is always "just going" ', brother Evelyn wrote in a *Bookman* profile, June 1930, 'his luggage is invariably packed'). Adventurous chaps like Peter Fleming would go to China. The Far East attracted more academic types too—Edmund Blunden, Empson, I. A. Richards, Peter Quennell, Julian Bell—who taught in the universities of China and Japan. Later in the decade the Sino-Japanese conflict also drew Isherwood and Auden to China. Fleming and Evelyn Waugh, Huxley, Lowry, and Greene travelled in South America; Evelyn Waugh and Greene in Africa. A few people went to North America—Alistair Cooke, Malcolm Lowry, Anthony Powell, Aldous Huxley, and, right at the end of the decade Auden, Isherwood, and MacNeice (as well as Benjamin Britten and Peter Pears). John Cowper Powys lived in the USA from 1920 to his resettlement in London in 1935. But of his own journey to the USA, Easter 1939, MacNeice wrote: 'I visited the Great Unvisited.' For it was Europe, and the edges of Europe, that attracted British authors most. Connolly's Mediterranean, it might be, where Evelyn Waugh headed for his first travel book, *Labels: A Mediterranean Journal* (1930); where Anthony Powell went to write (he worked on *Afternoon Men* at Toulon, *Venusburg* at Sainte-Maxime, and used his Toulon recollections at some length in *What's Become of Waring?*); where Brian Howard went to pretend to write; the Mediterranean on to whose un-European fringes bolder spirits sometimes ventured—Wyndham Lewis (to Morocco for *Filibusters in Barbary*), Rex Warner (who went to teach in Cairo in 1929: whence his Egyptian poems and *The Kite*), George Orwell (convalescing in Morocco for six months from September 1938).[210] But it might equally be any one of Europe's multiplied faces that attracted the writing Briton: from the continent's southern edges, the Anglicized Med and less Anglicized territory of Spain (where Roy Campbell and Robert Graves retreated—Campbell to Toledo, Graves to Mallorca; where Ralph Bates agitated and wrote, and the Civil War of 1936–9 provided one of the period's most magnetic places); up to Europe's northerly reaches (the Hebrides of MacNeice's *I Crossed the Minch*, 1938; the Iceland of *Letters from Iceland*; the Finland Anthony Powell visited in 1924 with Archie Lyall, later the author of *A Guide to the 25 Languages of Europe*, and that Powell put into *Venusburg*, 1933; the Sweden Graham Greene visited specially for *England Made Me*); across to Europe's more exciting eastern reaches in the Soviet Union, which probably attracted more literary and political pilgrims, tourists, and spyers-out of the land in the '30s than any other country on earth. And, of course, there were the classic European cities of the tourist, the holiday-maker, the seeker after cultural novelties and releasing excitements. 'I shan't mind if they choose me' declares one of the villagers in *The Dog Beneath the Skin*:

> There's lots of places I want to see:
> Paris, Vienna, Berlin, Rome.
> I shouldn't be sorry to leave the home.

Paris, so near, especially by air from Croydon aerodrome, still had the magic of all the painters and writers who had for the last century or so made its name for progressive culture. It was still a notoriously sexy city (still the place where John Cowper Powys had once found his 'alluring, lust-drugging erotic bookshops'), still an artistic mecca. Julian Trevelyan had gone there in 1931 to avoid Cambridge Tripos exams, to paint and to sound sufficiently respectable for Malcolm Lowry's father to accept him in 1933 as the wayward Lowry's 'guardian'. (Lowry, a year older than Trevelyan, got married in Paris in 1934, with Trevelyan as best man.) George Orwell spent almost the whole of 1928 and 1929 in Paris, writing, giving English lessons, driven to skivvying in an hotel, accruing material for what became *Down and Out in Paris and London* (1933)—'A Scullion's Diary' was an early version's title. Paris was where David Gascoyne gravitated, to be introduced to Surrealism at the age of seventeen. Roland Penrose remembers the English Surrealist Group in London stemming from a meeting with Gascoyne in a Paris Street in 1935. Paris was where Henry Miller lived, to whom Lawrence Durrell, a youthful disliker of England living on Corfu ('Fuck the English, Henry . . . I hate England') sent his first fan letter in 1935 (over *Tropic of Cancer*). In 1937 Durrell himself moved to Paris (in 1939 he was, after a spell in London, back in Athens, for the British Council). The International Congress of Writers for the Defence of Culture was held in Paris (June 1935): the British delegates, led by Forster, included Huxley, James Hanley, Ralph Fox, John Strachey, Amabel Williams-Ellis. Forster found it 'an impressive affair', despite all the 'Congress-addicts who would travel any distance for their drug'. A bit of an addict himself, he was there again in July 1937, speaking at an *entretien* on 'The Immediate Future of Literature', organized by the League of Nations' Committee for Intellectual Cooperation (Paul Valéry and Gilbert Murray were chairmen), and reporting on the great Paris Exhibition for *NW*. Auden kept putting in brief Parisian visits; Brian Howard and Cyril Connolly had periods living there; it was one of the many places in which Isherwood tried to settle with his German boyfriend Heinz; it was the main staging-post for British volunteers wishing to participate in the Spanish Civil War; it was where, in June 1937, Nancy Cunard got the signatures of Auden and Spender, Brian Howard and Ivor Montagu for the circular letter that resulted in the pamphlet *Authors Take Sides*.

Paris was still a desirable and possible city for many British writers and artists. But its unique magnetism had already peaked in the '20s. Isherwood's short story 'The First Journey' has Isherwood on his first trip abroad, a school walking tour in the Alps in 1922, visiting Paris and spitting (metaphorically) on Napoleon's tomb. Chalmers (Upward) 'suggested that the only adequate comment was to spit. Mentally, we spat.'[211] Paris wouldn't really do for them, nor for their chums. It was displaced first by Berlin, capital of the Weimar Germany that was home of the New Realism (*die Neue Sachlichkeit*), the new photography, the Bauhaus. Eventually Paris's old magic was 'scattered', not quite 'among a hundred cities' (Auden's words in 'In Memory of W. B. Yeats'), but among a fair number: Moscow, Vienna,

Amsterdam, Madrid, Barcelona, Rome occasionally (Evelyn Waugh went there for his second honeymoon). It's entirely characteristic of the period that Samuel Beckett did not return as soon as he could to Paris, where he'd been *lecteur* at the Ecole Normale Supérieure, 1928–30, and closely associated with Joyce, but instead dallied between Dublin and London and Germany before going back to Paris and Joyce at the end of 1937. It was in the later '20s that droves of British artists, writers, and intellectuals started going to Germany, especially to Berlin: the spitters Upward and Isherwood, Brian Howard, Auden, Stephen Spender and his brothers Michael and Humphrey, Malcolm Lowry, Alan Bush, Francis Bacon, Wyndham Lewis, John Lehmann. 'I've rather taken to Berlin, lately', Raymond Mortimer told Evelyn Waugh in October 1928. Midway between 1918 and 1939 Berlin became one of the central *entre deux guerres* sites. 'Well, here I am' thinks Isherwood's suicidal Edward Blake, in *The Memorial* (1932), drunkenly in a street in Berlin:

> For it had suddenly struck him—how queer; ten years ago I wasn't allowed to come down this road. Now it's allowed again. And in ten or twenty years' time perhaps it won't be allowed. How bloody queer. In 1919 we were going to have bombed Berlin. Mathematically speaking, there's no reason why I shouldn't be dropping a bomb on myself at this very moment.

Germany was now the place to be: for artistic progressivism, but also because there sunshine and cocaine and sex, especially homosex, were up until Hitler's intervention in 1933 so freely available. Berlin was a mythic sodom, and a sodomites' mythic nirvana. The British homosexuals excitedly went there 'to live'. In doing so, of course, many of them were taking the first step towards confirming themselves as members of the period's large band of perpetually unsettled drifters: those whom England bored or discontented; the England-haters like Lowry and Durrell; the people who thought home had betrayed them (like Waugh, willingly homeless after the collapse of his first marriage at the end of the '20s, forever moving on; even after his second marriage in 1937, and the bid for roots in the purchase of Piers Court, he trailed off to Mexico gathering material for *Robbery Under Law: The Mexican Object-Lesson*, 1939). They were joining the restless ones who preferred the working conditions offered by foreign parts (John Lehmann edited *NW* between Vienna and London, tacking about from Paris to Prague, Amsterdam to Moscow), the revolutionaries and freedom fighters perpetually eager to be near or in the front-line of the crisis in Moscow or Vienna or Madrid. In fact, they'd fallen into step with the huge army of expatriates and refugees, from Russia and Italy, Germany and Austria, as well as from Britain itself, who milled unresting and unrested about the world of the 1930s. Indeed, the British homosexuals and their boyfriends—Isherwood and Heinz, Howard and Toni, Spender and Tony Hyndman, John Lehmann—perpetually on the go, always *déraciné*, forever moving on and moved on, now in a *pension* here, now in a hotel there, trying that city and the other island, shunting dislocatedly between Spain and Greece, Portugal and Germany, on the move from London to Paris to Amsterdam to Vienna, from Mlini to Venice, to Prague, to Tiflis, harassed by border guardians, pursued by officialdom (especially so in the case of Isherwood's Heinz—wanted for German military service), never peacefully stationary for long (there was a rare moment in summer 1935 when Isherwood and Heinz, Spender and Hyndman, Howard and Toni, Forster and Buckingham, were all together in Amsterdam), can be considered to stand as representatives of an extra-ordinarily restless era.

The settled writer, like C. Day Lewis, who retreated deeper and deeper into the English countryside and who didn't go to Berlin or Moscow or Spain, who made his very first overseas trip only in July 1938, to attend a rally in Paris 'For Peace and Against the Bombardment of Open Cities', at which he stuck close to his chum Rex Warner, refused to speak French, and wouldn't mix with the travel-enthusiastic Stephen Spender, Rosamond and John Lehmann, and Guy Burgess, stands out for his home-hugging rarity. Steve Hannay, the trendy young leftist Etonian abroad, knowingly mocked in *Pink Danube* (1939), the parodic picaresque by Alan Pryce-Jones ('Arthur Pumphrey'), is much more representative of the period: he

> had gone out to learn German at the age of eighteen, in Berlin. He had four hundred a year, and a few tasteful possessions: a Picasso drypoint, a book of photographs by Man Ray, *Le Potomak*, with a dedication from Cocteau himself, and a very expensive camera, with which to collect material for a possible article on workers' flats (from a strictly functional standpoint) for the *Architectural Review*. I imagined his luggage: cheap suitcases, but the ties from Charvet; blue shirts with pointed collars, and a bottle of sunburn oil wrapped in a page of the *Daily Herald*.
>
> There had been summer in Munich, and a romantic interest in Ludwig II; but amongst these handsome and serious young Germans a good deal of serious luggage had to be discarded. The sunbaths remained; and the photography. But Picasso had become too versatile, and Cocteau too amateur. Besides, their connection with Paris was as uncomfortable as the glimpse of a discarded lover . . . On the one hand there was a comfortable inheritance, a studio flat off Fulham Road, the Bentley; on the other, a visit to the Russian Caucasus. And behind that, dimly referred to or hurriedly concealed, a variety of objects: a printing-press in a Hamburg cellar, one table in an Ottakring café, leaflets scattered in the Underground, a bottle broken over somebody's head. Behind these objects, however, stood others, and in the background a variety of landscapes. The broken bottle might have been a champagne bottle; the Ottakring café blurred into a picture of Maxim's at Juan-les-Pins. Gorki and Olivier of the Ritz appeared hand in hand.
>
> But it would not do to say so. For Hannay was a most serious man . . .

Tourists, Waugh declared in *Labels*, 'must form part of our "period" as surely as gossip-writers or psycho-analysts'. We've already glanced (in chapter 5) at the piece called 'Poetry and Politics' in *New Verse* (May 1933) in which Charles Madge picked out two major categories of 1933 poets: the settled and the unsettled, 'those who have got jobs, mostly as schoolmasters' (he cited Auden), and those who 'cannot settle down', who 'are to be found with a knapsack in the Tyrol, or sitting in a café at Perugia'. It was a nice distinction. But as the period went on it became harder with confidence to sustain it, for almost all the notable British writers settled for at least some overseas travel. Travelling was normal. Indeed, Empson's poem 'Manchouli', which is about the question of what is normal, concerns itself precisely with thoughts on normality that the poet has while 'passing these great frontiers'. In his poem 'Autumn on Nan-Yueh' Empson says he feels a 'brother' to Virginia Woolf because of the only bit of her prose he can remember: it begins 'Thank God I left'.

The desire to travel abroad was widespread. It wasn't only writers who called on, or longed to call on the travel-agent's services (travel-agents like Elizabeth Bowen's 'world-minded' Emmeline in *To the North*, who 'knows what all trains are doing all over Europe'). 'It's all very well' for those who 'can travel', wails Mrs A. of *The Ascent of F6*, listening to a wireless talk about Sudoland. Her husband tries to console her with escapist dreams:

> Better luck will come our way:
> It might be tomorrow. You wait and see.
> But, whenever it happens, we'll go on the spree!
> From the first-class gilt saloon of channel-steamer we shall peer,
> While the cliffs of Dover vanish and the Calais flats appear,
> Land there, take the fastest train, have dinner in the dining-car,
> Through the evening rush to Paris, where the ladies' dresses are.
> Nothing your most daring whisper prayed for in the night alone—
> Evening frocks and shoes and jewels; you shall have them for your own.
> Rome and Munich for the opera; Mürren for the winter sports;
> See the relics of crusaders in the grey Dalmation ports;
> Climb the pyramids in Egypt; walk in Versailles' ordered parks;
> Sail in gondolas at Venice; feed the pigeons at St Mark's . . .

Mrs A., refusing to be comforted, sees instead only a future of rainy English boarding-houses. The moneyed, though, could enact such dreams with little trouble, as Portright, the *soi-disant* painter of MacNeice's *Out of the Picture*, announced standing rhapsodically in front of the Group Theatre's very fine Art Deco Tourist Agency:

> Look at all these lucky people—they're wealthy—
> Wagon-Lits—the smell of orange blossom . . .
>
> Money is a terrible temptation!
> Look at these tourist agencies where money will buy you distance,
> Miles and miles of distance for your money.
> O the Tourist Agencies with revolving doors and marble floors,
> Schedules of the White Star Blue Star Green Star Red Star Black Star lines,
> This way for the belles of Andalusia and the garlands of Tahiti
> And the steam of the Victoria Falls.
> This way if you want to see the wonders of Yellowstone Park,
> The tomb of Tutenkhamen or Gandhi sitting at his wheel,
> The roof of the Sistine Chapel, the Great Wall of China,
> If you want to commune with strange winds, the mistral or the Sahara sandstorm,
> Or dip your fingers in holy rivers, Ganges or Oxus—
> If only I could, if only I could—
> Oh for the shrill hustle of the Gare de Lyon
> The swallows flying south, the moneyed swallows.

The dreams of the unlucky stay-at-homes were sustained at second hand by films, wireless talks and travel books. Mr and Mrs A. eventually get sated by wireless travel talk ('we are tired of descriptions of travel'). But there was a huge audience for travel books. The thirst for news from elsewhere scarcely let up. Travel-book publishers and travel writers were numerous enough for Anthony Powell to be sure of making quite unesoteric fun when he put Judkins and Judkins, the firm that makes a good thing out of such literature, and T. T. Waring who lives splendidly by cribbing other authors' old accounts of journeys, into *What's Become of Waring?* Some writers like Peter Fleming and Robert Byron specialized in accounts of travel: travel was their métier. Others, like Auden, MacNeice, Isherwood, Waugh, Greene,

Lehmann, whose craft was writing, went in opportunistically for travel books as attempts at supplementing the poet's or novelist's less than good income. If D. H. Lawrence could do it so would they. There was ready money in travel books; publishers and editors were pretty free with advances to cover travelling expenses; it was worth hustling to get contracts. '[O]nce I have got to Spain, I suppose your Morning Post wouldn't like to pay me for any of those half-witted little articles people are always writing about what they think they see in those places. Say the Seville do at Easter or the Joy in the Face of the Masses as the result of the last Election': thus MacNeice to Grigson in February 1934.[212] 'Will you please', Waugh instructed his agent in 1932, 'take any orders for travel articles—far flung stuff impenetrable Guiana forests, toughs in diamond mines, Devils Island, Venezuela. Particularly require payment on embarkation if possible.' ('[L]et me have the cash, dough, tin, spondulicks, ready, oof, doings or whatever it is', Waugh wrote to him in 1934, after 'an article of mine about debunking the bush'—actually called 'Rough Life'—had appeared in *Virginia Quarterly Review*.) Professional writers with any *nous* carried their notebooks with them everywhere to facilitate the travel reports they might send home to *New Writing* or *Left Review*, to *Night and Day* or the *New Statesman* or even (in Lehmann's case) *The Geographical Magazine*. 'You're all the same', gibes Liesl to Arthur Pumphrey, 'English, a little money, a little Communism, a little notebook half-full always.'

Titles of books and articles frankly promised journeys: *A Superficial Journey* (Peter Quennell, 1932), *One's Company: A Journey to China* and *News From Tartary: A Journey from Peking to Kashmir* (1934 and 1936, both by Peter Fleming), *Forbidden Journey* (Ella K. Maillart, Fleming's chum, 1937), *Journey to a War* (Auden and Isherwood), *Lapland Journey* and *Hebridean Journey* (Halliday Sutherland, 1938 and 1939), 'Journey to Paris' (Ignazio Silone, *New Writing*, Autumn 1936), 'Journey to Iceland' (Auden in *Letters from Iceland*). The journeys were openly rooted in the journal, the notebook: *Labels: A Mediterranean Journal* (Waugh, 1930), *Hindoo Holiday: an Indian Journal* (J. R. Ackerley, 1932), *Filibusters in Barbary (Record of a Visit to the Sous)* (Wyndham Lewis, 1932), *An Escapologist's Notebook* (Cedric Belfrage, 1936), *Escape With Me! An Oriental Sketch-Book* (Osbert Sitwell, 1939), *European Note-Book* (Bernard Wall, 1939), and 'Notes on a Visit to Ireland' by William Plomer, 'A Berlin Diary' by Isherwood, 'A Madrid Diary' by Alfred Kantorowicz (all in *NW*). Abroadness abounded particularly in the little magazines: *NW* carried Orwell's 'Shooting an Elephant' as well as his 'Marrakech' and Isherwood's 'A Berlin Diary' and E. M. Forster's report from the Paris Exhibition ('The Last Parade') and Lehmann's 'Via Europe: Scenes from a Travel Sequence, 1934–5'. In fact the magazines abundantly confirmed the impression to be gained from titles like *Letters from Iceland* and Ewart Milne's *Letters from Ireland* that authorship had turned into a species of Overseas Correspondence. They were filled with mailed reports from Our Man (and occasionally Woman) of Letters in the foreign part. 'A Letter from Moscow' (André Van Gyseghem), 'Letter from Tiflis' (Lehmann), 'An Open Letter to Aldous Huxley' (Spender: 'Writing as I am from Central Europe') all fell through the *Left Review* letter-box. *Night and Day*'s short life was mightily sustained by this kind of foreign correspondence: 'Letter from China' (as well as 'Travel Note' and 'Learning Chinese') from Empson; 'Our Continental Correspondent. Salzburg in the Distance' and 'Letter from Ireland' (Elizabeth

Bowen); 'A Reporter in Los Angeles' (Anthony Powell); 'New York Letter' (Alistair Cooke); 'Paris Letter' (Stuart Gilbert).

Naturally, all this foreign experience made its way into novels and poems and plays. 'How strange to remember', recalled Isherwood (in *Journey to a War*) as he passed Tung-ting Lake, 'that in London, only three months ago I had placed a finger on it in the atlas, and said "I wonder if we shall even get as far as *here*?" '. The Ordnance Survey map, the charting of local terrains in the literature of English travel was swapped for the wider geography of 'the immense improbable atlas' (Auden's phrase in his poem 'Dover'). The literary atlas came in many versions. It might be the 'school atlas and gazetteer heavily annotated' that Waugh's Father Rothschild carries in his suitcase on the channel crossing with which *Vile Bodies* opens. It might resemble rather J. F. Horrabin's political and critical *Atlas of Current Affairs*, published by Gollancz (1934). But whatever kind it was, with so many writers playing the travelmaniac (and 'A map', Waugh declared in a *Daily Mail* piece about travel, 16 January 1933, 'and particularly one with blank spaces and dotted rivers, can influence a travelmaniac as can no book or play'), place-names dominate any catalogue of '30s creative writing. *Vienna, Pink Danube, Goodbye to Berlin*, Orwell's *Burmese Days*, Auden's 'Macao', 'Hongkong', 'Brussels in Winter', Elizabeth Bowen's *The House in Paris*, Grigson's 'And Forgetful of Europe (Mlini 1935 to 1937)', MacNeice's 'The Hebrides' and 'Leaving Barra', all the poems about Spain: such texts announce a literature living greedily off its authors' experiences of foreign places. The world atlas was what the youthful would-be writer expected to grow up into. 'But we have left school now', Charles Madge asserts at the end of his poem 'Letter to the Intelligentsia' in the *New Country* anthology: 'we turn the pages / Of a larger atlas; telegrams come in / From China, and the world is mapped on our brains.' The foreign setting became normal in the poems of Spender and Auden and Lehmann, in the novels and short stories of Waugh and Isherwood, Spender and Ambler, Greene and Powell, Elizabeth Bowen and Jean Rhys. They did indeed have the world on the brain. 'We're internationalists', insists Kate Farrant in *England Made Me* (1935), yet another of Greene's internationally travelled fictions, this time one full of globe-trotting businessmen and international finance ('US Rubber, US Steel . . . It no longer gave' Krogh 'any pleasure to think that soon a new company under his control would be quoted there, as already he was quoted in Stockholm, London, Amsterdam, Berlin, Paris, Warsaw and Brussels'). Internationalists thronged '30s fiction, not least Isherwood's *Mr Norris Changes Trains*.

> By the time we had reached Bentheim, Mr Norris had delivered a lecture on the disadvantages of most of the chief European cities. I was astonished to find how much he had travelled. He had suffered from rheumatics in Stockholm and draughts in Kaunas; in Riga he had been bored, in Warsaw treated with extreme discourtesy, in Belgrade he had been unable to obtain his favourite brand of toothpaste. In Rome he had been annoyed by insects, in Madrid by beggars, in Marseille by taxi-horns. In Bucarest he had had an exceedingly unpleasant experience with a water-closet. Constantinople he had found expensive and lacking in taste. The only two cities of which he greatly approved were Paris and Athens. Athens particularly. Athens was his spiritual home.

The list drags on as much as Norris's travellers' tales were prone to.

Naturally enough, the impedimenta of international travelling were different from the homelier variety's gear. The hiker's shorts and little rucksack, the walker's boots and stick, the family roadster, the bicycle and the undergraduate motorbike, so familiar to the literature of English travelling, gave way in this literature to the hulking international touring motor-car (Mrs Melrose Ape's 'travel-worn Packard, bearing the dust of three continents' that's seen being loaded on ship-board at the beginning of *Vile Bodies*, or Mr and Mrs Aldous Huxley's custom-built scarlet Bugatti with the passenger seat specially stretched for his tallness), gave way also to the passenger ship, the posh liner, the aeroplane, and the inter-continental train.

Some authors still managed shrewdly to recall the simpler continental trampings of an earlier age and the pre-lapsarian cycling holidays of the sort Tony Last remembers in Waugh's *A Handful of Dust* ('bicycled along straight, white roads to visit the chateaux; he carried rolls of bread and cold veal tied to the back of the machine'). H. V. Morton's title *In the Footsteps of the Master* (1934) struck the old-fashioned note. (It was followed not only by Morton's own *Footsteps* sequels—so that Hugh Kingsmill and Malcolm Muggeridge mockingly suggested that 'A Company has been registered under the name "Footsteps Ltd." to deal with the literary work of Mr H. V. Morton'[213]—but by Louis Golding's Jewish alternative, *In The Steps of Moses the Lawgiver*, 1937—the year Waugh was toying with going *In the Steps of Caesar*, or St Peter or St Patrick or St Francis Xavier. When Robert Graves was asked in 1941 to do an *In The Steps of Hannibal*, he replied grouchily that he 'didn't go in anyone's steps—H. V. Morton's, Christ's or anyone's'.) Consciously old-fashioned in the same way was the title of Robert Byron's *The Road to Oxiana* (1937, reminiscent, perhaps, of Huxley's *Along the Road: Notes and Essays of a Tourist*, 1925, and certainly recalling John Livingston Lowes's book about Coleridge's 'Ancient Mariner' and 'Kubla Khan', *The Road to Xanadu*, 1927, a book that generated A. H. Nethercot's critical sleuthing around Coleridge's 'Christabel', *The Road to Tryermaine*, 1939). And there was some actual long-distance foot-slogging (the cousinly African journey described by Barbara Greene in *Land Benighted*, 1938, and Graham Greene in *Journey Without Maps*, 1936, an arduous long walk), not to mention the precarious messing about in small boats by Peter Fleming (*Brazilian Adventure*, 1933) and Evelyn Waugh (*Remote People*, 1931, and *Ninety-Two Days*, 1934), or the horse-riding of Fleming in China, of Auden and MacNeice in Iceland, and of Waugh in South America ('That evening after supper Sinclair came to me leading an Indian and said, "Chief, do you want to see this boy's arse?"'. 'I misunderstood him'—a *horse* was intended—'and said no, somewhat sharply'). And the real mountaineers were as familiar to '30s travel books, with their knapsacks, boots and goggles, as they'd become commonplace in other '30s texts. But on the whole, for means of getting about the world in this particular writing of the '30s, it was the big travelling machine that became strikingly predominant.

Writing had definitely entered the era of aeroplanes and aerodromes. How recurrent Croydon aerodrome becomes and how characteristically knowing about air travel is Fred Hall in *England Made Me* (1935):

> he was no longer interested by the flight from Amsterdam; he knew the airports of Europe as well as he had once known the stations on the Brighton line—shabby Le Bourget; the great scarlet rectangle of the Tempelhof as one came in from London in the dark, the headlamp lighting up the asphalt way; the white sand blowing up round the shed at Tallin; Riga, where the Berlin to Leningrad plane came down and bright pink

mineral waters were sold in a tin-roofed shed; the huge aerodrome at Moscow with machines parked half a dozen deep, the pilots taxi-ing casually here and there, trying to find room, bouncing back and forth, beckoned by one official with his cap askew. It was a comfortable dull way of travelling . . .

Even more prevalent were ships: 'The Cargo Ship' on which Greene travelled to Liberia in *Journey Without Maps*; the ship that carried Auden and Isherwood to Hong-Kong (described by Isherwood in 'Escales'[214]); the Mediterranean cruise-ship Waugh travelled on free in exchange for the publicity he gave the shipping company in *Labels*; the 'Boats to Iceland', factually described in the 'For Tourists' section of *Letters from Iceland*; the Russian vessel on which Lehmann sails at the end of *Prometheus and the Bolsheviks*, loud with the Internationale, bright with a huge Soviet star in flashing red bulbs; the ships in Auden's China poems, 'The Voyage', 'The Ship' (entitled 'Liner' in Auden's notebook) and 'Passenger Shanty'; the channel steamers that ply 'in and out' of the harbour in Auden's 'Dover', that carry home Father Rothschild in *Vile Bodies*, bring D. to England and trouble in Greene's *The Confidential Agent*, transport Alan to Europe in the second scene of *The Dog Beneath the Skin* (written by Isherwood), start the narrator of Isherwood's *Lions and Shadows* on his way to Germany ('Dover quay, enveloped in clammy brown fog; the third-class steamer saloon crammed with soldiers going out to the Army of Occupation at Wiesbaden; two Cambridge undergraduates with enormous red wrists'), and that carry Karen to her assignation with Max in Boulogne in Elizabeth Bowen's *The House in Paris* ('On board, the vibration of the ship starting half woke her nerves. Looking back at England, she asked herself: Or, is this courage? and walked the deck still not knowing'); and all those other '30s fictional ships galore: the one that takes Tony Last to an explorer's awful fate in *A Handful of Dust*, the storm-tossed vessel in Richard Hughes's Conradian *In Hazard* (1938), the numerous ones in numerous James Hanley novels and stories, in John Sommerfield's tale of a youthful wanderer *They Die Young* (1930), and in Malcolm Lowry's *Ultramarine* (1933)—a novel consciously imitative of Richard Henry Dana's *Two Years Before the Mast* and Nordahl Grieg's *The Ship Sails On*, less consciously, perhaps, drawing on Hanley's *Boy* and Lowry's friend Sommerfield's sea story. Greene's Anthony Farrant 'had been in more ports' than his sister 'could count'. So, almost, had '30s writing. Ships were so useful. They had all the mere excitements of planes and trains. Like them they easily implied the awaiting adventure, the unknown thrill ahead. Like trains, ships provided helpfully enclosed worlds, rich microcosms of the world or the class system, a busy stage. 'There is something dauntingly world-wide about a ship, when it is free from territorial waters', muses Graham Greene of his voyage home from Mexico at the end of *The Lawless Roads*:

> Every nation has its own private violence, and after a while one can feel at home and sheltered between almost any borders—you grow accustomed to anything. But on a ship the borders drop, the nations mingle—Spanish violence, German stupidity, Anglo-Saxon absurdity—the whole world is exhibited in a kind of crazy montage.

Some novelists, richly experienced in ships and planes and trains, gladly exploited all three. Greene did, and Powell, and Waugh. Much of Waugh's *Scoop* (1938) is taken up simply with transporting William Boot jokily from Croydon to Paris by air, from Paris to Marseille by the Blue Train, and from Marseille to Ishmaelia by a series of slow boats. Like other travelled novelists Waugh was reluctant to waste any of his hard-won travel knowledge. (So unwasteful were Waugh and Greene that they openly used material twice, recycling travel-book stuff in their fictions: *Journey Without Maps* started Greene's West African phase; *The Lawless Roads* began what *The Power and The Glory* continued; the miserable fate of Last in *A Handful of Dust* was born in the grey adventures Waugh recorded in *Ninety-Two Days; Black Mischief* draws on Waugh's *Remote People* travels, *Scoop* on the *Labels* travels, on the Abyssinian material in *Remote People* and, particularly, on Waugh's journalistic exploits first recorded in *Waugh in Abyssinia*, 1936). And least wasted of all '30s travel experiences were trains.

Trains were what waited for you on the other side of the English Channel. Trains confirmed you in the continental adventures the Channel crossing had initiated you into:

> At Ostende the train stood waiting to leave for Warsaw and Riga: the undergraduates unscrewed a *Niet Spuwen* notice as a souvenir. I parted from them in the waiting-room at Köln, as an official marched down the platform carrying, like a sacred banner, the wooden sign-board announcing the arrival of the Berlin express.

Thus the narrator of *Lions and Shadows*. His awed juvenile excitements are passed on to Isherwood's William Bradshaw, whom the opening of *Mr Norris Changes Trains* finds entrained for Berlin. ('The word "abroad" caused both of us naturally to look out of the window. Holland was slipping past'.) Trains were magic; even, Empson confessed in his poem 'The Beautiful Train', one belonging to the Japanese invaders of China: 'So firm, so burdened, on such light gay feet' that 'I a twister love what I abhorr'.[215] The big international trains were implicit with incident, with political intrigues, strange meetings, sexual plottings, providing an opportuneful montage (to use Greene's metaphor) of the international scene through which they passed but also (like ships) as usefully enclosed as a vicarage drawing-room and thus ripe for all sorts of strange deaths and smart detective work. Agatha Christie's *The Mystery of The Blue Train* (1928) was opportunistically followed up by Greene's *Stamboul Train* (1932)—'for the first and last time in my life I deliberately set out to write a book to please, one which with luck might be made into a film.' (Greene couldn't afford actually to take the train to Istanbul, so he played daily a record of Honneger's railway music *Pacific 231*, to achieve the moods his Cotswold Cottage and his third-class trip to Germany failed otherwise to provide.) Agatha Christie's *Murder on the Orient Express* got up steam along the same length of fictional track in 1934. Getting into a train clearly invited as much trouble as getting into a thriller— and thrillers of the period frequently opened in stations or on trains. Eric Ambler's trans-European thrillers tended to get the action aboard a train as soon as decently possible: the boat-train to Paris, the train for Zovgorod, the Bâle-Paris express in *The Dark Frontier* (1936); the slow train to Linz (for Vienna) in *Uncommon Danger* (1937: its foreign correspondent hero had to forego the Night Orient Express from Ostend because he'd lost heavily at poker). Greene's *The Name of Action* (1930) opens with Oliver Chant on the express train to Trier ('I'll go to Trier', he'd volunteered, eager for the adventure the exiled republican Kurtz promised him; frisked at Trier station, 'With an excited pleasure Chant realized that he had been searched for arms'). Elizabeth Bowen's *To the North* (1932) begins with Cecilia Summers at Milan

Station boarding 'The Anglo-Italian express—Chiasso, Lucerne, Basle and Boulogne—leaves at 2.15: it is not a *train de luxe*'. Menace brews most enticingly for Henry Green's party-goers at their fog-bound international station.

And the end of a train journey could be as unexpected as a thriller's end. 'Everyone waited for the train to impale them on London . . . Past midnight, that other train would crash into the Gare du Nord': so thinks Karen in Bowen's *The House in Paris*, a novel full of boats and trains, that begins with a taxi 'skidding away from the Gare du Nord', and ends at the Gare de Lyons ('more daunting than the Gare du Nord: golder, grander. Henrietta discovered that half past six is 18.30 in Paris: clocks must be larger, she thought'). Auden added in his 'Gare du Midi' a horrified little Brussels footnote to the warm GPO recommendations of rail-travelling in 'Night Mail' ('A nondescript express in from the South . . . a face . . . a little case, / He walks out briskly to infect a city').

In Winifred Holtby's poem 'Trains in France' (1931) (in her *The Frozen Earth* volume, 1935), train noises at night are bestial, 'savage, shrieking', reminding her of First War trains carrying men to their death. Trains carried Auden and Isherwood flinchingly on their *Journey to a War* ('We protested that we weren't discouraged—that the air-raids would help to pass the time, and a night in the paddy-fields would provide excellent copy. Nevertheless, as we approached the Canton railway station, I began to cast nervous glances at the sky. It was a warm, windless evening—perfect weather for the Japs'). A train transports Ulsterman MacNeice inconclusively towards alien Dublin in 'Train to Dublin'.[216] The beginning, middle and end of train-journeys were, in fact, so widely felt to be so potentially thrilling and/or menacing—a perception reinforced by train movies such as *Shanghai Express* (with Marlene Dietrich), *Rome Express, Turksib*, and by the movie versions of train novels (they included *Stamboul Train, Murder on the Orient Express, The Lady Vanishes*), not to mention Spender's verse version of *Turksib* in 'The Express'—that Peter Fleming felt obliged to sweep his Eight-Days in the Trans-Siberian Express from Moscow into his relentless debunking (in *One's Company*):

> the dignity, or at least the glamour of trains has lately been enhanced. *Shanghai Express, Rome Express, Stamboul Train*—these and others have successfully exploited its potentialities as a setting for adventure and romance. In fiction, drama, and the films there has been a firmer tone in Wagons Lits than ever since the early days of Oppenheim. Complacently you weigh your chances of a foreign countess, the secret emissary of a Certain Power, her corsage stuffed with documents of the first political importance. Will anyone mistake you for No. 37, whose real name no one knows, and who is practically always in a train, being 'whirled' somewhere? You have an intoxicating vision of drugged liqueurs, rifled dispatch-cases, lights suddenly extinguished, and door-handles turning slowly under the bright eye of an automatic . . .
>
> You have this vision, at least, if you have not been that way before. I had. For me there were no thrills of discovery and anticipation. One hears of time standing still; in my case it took two paces smartly to the rear. As I settled down in my compartment, and the train pulled out through shoddy suburbs into a country clothed in birch and fir, the unreal rhythm of train-life was resumed as though it had never been broken. The nondescript smell of the upholstery, the unrelenting rattle of our progress, the tall glass of weak tea in its metal holder, the unshaven jowls and fatuous but friendly smile of the little attendant who brought it—all these unmemorable components of a former routine, suddenly resurrected, blotted out the interim between this journey and the

last. . . This small, timeless, moving cell I recognized as my home and my doom. I felt as if I had always been on the Trans-Siberian Express.

In a sense Fleming *had* always been on a train. So had '30s literature. 'And the trains carry us about', declares MacNeice in 'Train to Dublin', and it's only 'during a tiny portion of our lives' that 'we are not in trains'. Or if not in trains, at least in transit in some other mode of long-distance transport. Malcolm Lowry gradually came to see that his *Ultramarine* belonged to a *roman fleuve*, that unending form of the novel, and one that should be called *The Voyage that Never Ends*—a voyage beginning at birth (he changed the name of *Ultramarine*'s ship from *Nawab* to *Oedipus Tyrannus*) and lasting all one's life. '[T]his last tooloose-Lowrytrek': that's what he labelled his wandering life, in an SOS letter ('sinking fast by both bow and stern') sent to John Davenport from Mexico in 1936. 'This is the perfect Kafka situation but you will pardon me if I do not consider it any longer funny. In fact its horror is almost perfect.' And since Lowry's protracted trek was only typical of the restlessnesses of this period's throng of wanderers, it's scarcely surprising the traveller's way of seeing should so shape '30s writing.

'What the Tourist Saw'—that's one of the section headings of Plomer's volume of poems *The Fivefold Screen* (1932)—is what a great deal of '30s writing sees. Written, a lot of it, in transit, on boats and in trains, in waiting-rooms and hotel-bedrooms, those classic locations for journalistic jottings, such writing is in a hurry, ready with the too quick response, the superficial reckoning, the first impression. The 'world of geography'—we've already heard Brother George Every's powerful complaint about Auden's travelling concerns—'is the world of railroads, films, trousers and ginger-beer bottles, spread out in a network over the earth by the economic energy of industrialism. The real China, the real Europe, the real America, lie below that network, and only in and behind their history can we find the universal and lasting elements in human experience.'[217] Undeterred, the tourists zoomed on. 'My impressions of Moscow, as here set down, are fragmentary and superficial. They are not intended to be anything else. In the course of my life I have spent eight days in that city.' Peter Fleming's honesty about his own flashy impressionism was refreshing ('The recorded history of Chinese civilisation covers a period of four thousand years. The population of China is estimated at 450 millions. China is larger than Europe. The author of this book is twenty-six years old. He has spent, altogether, about seven months in China. He does not speak Chinese'), but it only won him more and more readers. A lot of people seem actually to have welcomed this detached rush, past, over, and through experience: the aloof airman's speedy fly-over, the Orwellian outsider's view from the train, the movie-camera's rapid accumulation of moving images (in cinemas, complained Eliot in his essay 'Marie Lloyd' (1923), the 'mind is lulled by . . . . continuous action too rapid for the brain to act upon'). 'The tourist in space or time' was actually assisted by such detached, roving observation, according to MacNeice in his 'Letter to Graham and Anne Shepard' in *Letters from Iceland*:

> We are not changing ground to escape from facts
> But rather to find them. This complex world exacts
> Hard work of simplifying; to get its focus
> You have to stand outside the crowd and caucus.

Those writers who craved, like T. S. Eliot, poetic fulcrum points, and sought to make their observation-post at 'the still point of the turning world' were sceptical of any advantage to be gained from the moving view-finder's mobile vision, but they were often themselves caught, none the less, in the general restlessness. Elizabeth Bowen's travel-agent Emmeline sends her clients 'flying' about the world, but has sudden longings 'to be fixed'. Her craving for stillness, though, is in vain. Evelyn Waugh deplored the 'sick hurry' of the Futurist World of *Vile Bodies*, with its liners, airships, motor-cycles, aeroplanes, and racing cars, its fast jazzed-up music, its speeding cinematic views, and the general spiritual unsettledness of a world of the short relationship, Hollywood historiography, instant gossip-column reputations and roving evangelists. 'It unrolled like a length of cinema film. At the edges was confusion; a fog spinning past: "*Faster, faster*", they shouted above the roar of the engine . . . "Faster", cried Miss Runcible, "faster".' But in both his writings and his living Waugh was himself caught fast in the dynamic of movies and moving. So was Graham Greene, for all *Stamboul Train*'s sneering dismissals of the coarse and popular novelist Quin Savory, who relishes 'the beauty of landscape in motion' in movies, desires to convey the 'sense of movement in prose', but whose new novel *Going Abroad* is doomed to view the Far East 'through the eyes of a little London tobacconist'. Just so Cyril Connolly, striving vainly in *Enemies of Promise* to prove a distance between, on the one hand, the fruitful 'flight of the expatriate which is an essential desire for simplification, for the cutting of ties, the writer "finding" himself in the hotel bedroom or the café on the harbour' and, on the other, the deadening 'trajectory of the travel addict, trying not to find but to lose himself in the intoxication of motion', was himself sucked none the less into the formal and personal ambivalences of the prevailing travel addiction. *Enemies of Promise*, combining as it does a nifty guide book to the sights of modern literature and a swift confessional trip around the sites of Cyril Connolly's own life and labours, has all the shallow excitement of the best '30s *tours d'horizon*. And like it or not, it was hard for some writers to avoid the touristic style, that morally and tonally awkward combination of staying obstinately oneself, detached, uncommitted, unmoved even, amidst a life of incessant movement, whilst continually spotting with eager brightness from the plane or the train the passing surfaces, oddities, ephemera.

> I give you the incidental things which pass
> Outward through space exactly as each was

writes MacNeice in 'Train to Dublin', beginning a pastiche-Audenic list. 'This', MacNeice confessed in *I Crossed the Minch*, 'is the book of a tripper, a person concerned with the surface.' The tourist, gibed Auden in Chapter XIV of *Letters from Iceland*, 'Letter to Kristjan Andreirsson, Esq.', 'sees nothing important':

I question whether the reactions of the tourist are of much value; without employment in the country he visits, his knowledge of its economic and social relations is confined to the study of official statistics and the gossip of tea-tables; ignorant of the language his judgement of character and culture is limited to the superficial; and the length of his visit, in my case only three months, precludes him from any real intimacy with his material. At the best he only observes what the inhabitants know already; at the worst he is guilty of glib generalisations based on inadequate and often incorrect data. Moreover, whatever his position in his own country, the social status of a tourist in a foreign land is always that of a rentier—as far as his hosts are concerned he is a person of independent means—and he will see them with a rentier's eye: the price of a meal or the civility of a porter will strike him more forcibly than a rise in the number of cancer cases or the corruption of the judicial machine.

But it was touristic travel that fed the Audenesque taste for the bright or quirky surface which turned numerous '30s poems and books into smartly cluttered galleries for the intelligent but rushed spectator. What Spender liked in *The Dog Beneath the Skin* was characteristic of a lot of writings in this magpie vein: that play 'succeeded', Spender claims in his *New Writing* piece on the Auden/Isherwood 'Poetic Dramas', 'by the persistence of high spirits and great energy. Although some of the visits of Alan, the hero, to different parts of Europe in search of the long-lost heir Francis . . . are like the visits of a schoolboy to an International Exhibition, the total effect of a long and exhilarating journey made and a large and varied scene created, does get across'. The quick snap-shot mode had the blights of slickness ('A bit slinky. Slick, at any rate'; 'As bright as paint and as cheap. Acrid and cheap like cigarette smoke', is how Sophie describes the style of the New York advertising photographs she posed for, in Rayner Heppenstall's novel *The Blaze of Noon*, 1939. 'They call it streamlined?' her blind lover wants to know). But the rhetoric did also enjoy the advantages of the energies that the Futurists so admired.

'30s travel rhetoric could serve all kinds of turns, but like so much in this period it did have a disconcerting way of turning ominous. It informed the Audenesque's show-off chirpiness, but also the still more desperate brightness of Waugh's sick travel-jokes. Railway trains were as ambivalent in the 1930s as they'd been for Dickens and Carlyle. Revolution still rushed (for some) cheeringly forward like a steam-train, its Carlylian 'ominous, ever-increasing speed' welcomed by Hugh MacDiarmid's 'Second Hymn to Lenin' as 'A means o' world locomotion'.[218] But so also did death, its *Dombey and Son*-style threatening generalized in Roger Roughton's little story 'The Sand Under the Door', whose hero Var is actually dead and on a journey towards an unfunny farm which, beneath the piece's Kafka-esque, surrealized, Upward-like enigmatics, sounds just like Death itself.[219]

The plentiful kinds of movement in MacNeice's poem 'The Hebrides', or (as it was called at first) 'On Those Islands'—the poem of which he was so enamoured that he had it appear on three separate occasions in 1938—marvellously mirror the insistent mobility of '30s experience and writing. MacNeice's island vision is a matter largely of coming and going. The islands have their 'comers', the tourists; their sons have gone 'to Toronto or New York' or 'go poaching'; men 'go out to fish' and dream, 'returned', of how the 'fish come singing' and 'rush' into their nets; a girl 'goes out to marry' a man who 'goes each year to the south to work on the roads'. For their part, the years 'Pad up and down'; the 'leaping' salmon travel up river 'with a magnetic purpose'; 'The sense of life spreads out'; drinkers 'spread their traditional songs across the hills'; 'whooping dancers' celebrate a wedding; 'The black minister paints the tour of hell'. But it's in the vicinity of that black Bible-obsessed 'tour of hell' that one starts to notice just how negative a lot of the poem's motions are: 'No one hurries'; 'no decent girl will go with' the tinkers; 'no train runs on rails'; the local folk-wisdom doesn't get uttered because 'All is known before it comes to the lips.' As in the period generally, travelling has taken the narrative with some promptness into troubling zones. The Hebrideans are at peace; 'the art of being a stranger with your neighbour / Has still to be imported'; but disaster and death have established a daunting hegemony over an ageing population of fisherfolk:

> On those islands
> Where many live on the dole or on old-age pensions
> And many waste with consumption and some are drowned
> And some of the old stumble in the midst of sleep
> Into the pot-hole hitherto shunned in dreams
> Or falling from the cliff among the shrieks of gulls
> Reach the bottom before they have time to wake [.]

Stumbling, falling: these are the dangerous motions that lead to death. '[O]nly Death / Comes through unchallenged in his general's cape.' At his coming, 'the whole of the village goes into three day mourning'.

Most '30s comings and goings had a way of turning out like this to have deadly consequences or connections. It's not at all surprising to find, for example, that one of the period's most striking examples of a kept-up dynamic rhetoric should be closely connected to a prevailing sense that travel is trouble. *The Death of the Heart* (1938), perhaps Elizabeth Bowen's finest '30s novel, vividly dynamizes the spaces through which its heroine Portia moves by repeatedly deploying slangy metaphors of louchely nifty movement to describe human behaviour. ' "Mr Quayne started skidding about" '. Portia and Irene 'had been skidding about' too. Quayne 'was always dangling round' after his wife. Major Brutt comes 'crashing round'. People 'bump back', they 'cruise through', 'buzz back', 'nip down', 'buzz along', 'blow in', and 'bounce about'. And *The Death of the Heart* has only intensified a rhetoric the earlier Bowen novels began. *The House in Paris* (1935), about another girl in transit—left 'in mid-air for the day in Paris', condemned to a 'flying glimpse of Paris'—makes a good deal of the way Henrietta's been bumped about: 'Bumped all over the senses by these impressions, Henrietta thought: If *this* is being abroad . . .' Several times in Henrietta's novel people are said to 'bump down'. In *To the North* (1932), yet another Bowen novel in which the women come and go ('You lead such a funny life, like a cat; always coming and going'), Julian feels that he and his sister have 'both been shanghaied together', Cecilia thinks Emmeline 'looks all over the place', and so on. And in all these novels, as in *The Hotel* (1927), this world of perpetually restless movement, of trains and taxis, aeroplanes and cars, of life as a series of non-stop voyages, and temporary abodes, in which young women get into bad 'hotel habits' ('To carry your bag about with you indoors is a hotel habit, you know'), where people are constantly dirtied by travelling ('the grime, ingrained in one till one is almost polished, of a transcontinental journey'), this jostled, rushed way of life is made to represent the general plight of women, always at the mercy of arbitrary family whims, cruelly predatory lovers and of capricious yearnings of their own. 'I'm sick of motoring', declares the emancipated girl Sydney in *The Hotel*: in almost the same breath as she protests against Victor Ammering's kiss, 'I hate being messed about.' When Markie tells Emmeline she leads a funny life, 'always coming and going', she replies, 'Don't all women?'

In a way, of course, Elizabeth Bowen shows the whole civilized world agog with unsettledness, all its bourgeois occupants condemned to reel through a cinematic blur of unsettling experiences ('This is Paris. The same streets, with implacably shut shops and running into each other at odd angles, seemed to unreel past again and again'). 'All ages are restless', declares Lady Waters in *To the North*, 'But *this* age . . . is far more than restless: it is decentralized. From week to week, there is no knowing where anyone is.' Emmeline's work at her chic travel agency is precisely to promote this decentering of the age's self: 'she began to talk rapidly, fully alive. "Our organisation is really far-reaching", she said. "We can tell anyone almost everything: what to avoid, what to do in the afternoons anywhere—Turkestan, Cracow—what to do about mules, where it's not safe to walk after dark, how little to tip. We have made out a chart of comparative dinner times all over Europe, so's people need not waste their evenings; we are just bringing out a starred list of places good out of season and manufacturing towns that sound awful where there is really something to see. We keep very much up to date.' Emmeline's zest for speed ('Such an exalting idea of speed possessed Emmeline that she could hardly sit still and longed to pace to and fro', as she waited for Markie and the Paris flight at Croydon), is central to her novel's prevailing vision of a totally dynamic existence: 'long cars nosing like sharks, vans whirring in gear, the high tottering buses'; 'shadowy trams crashed by'; 'the car, clearing Cirencester, tore with a slick wet sound up the open road to Farraways' (Lady Waters's place); 'An apologetic white dog coasted round the chair-backs';

> Markie's fingers tightened, blood roared in ears as the plane with engines shut off, with a frightening cessation of sound plunged downward in that arrival that always appears disastrous. Tipped on one wing, they appeared to spin over Le Bourget in indecision; a glaring plan of the suburb tilted and reeled . . . Earth rejoined the wheels quietly and they raced round the bleached aerodrome in a whirr of arrival. Then, grasping their small baggage, tipped like grain from a shovel, they all stepped, incredulous, out of the quivering plane.

'It seems a short time', says the Vicar, 'since motoring was in itself a pleasure.' Now, rather, it characterizes the uprootedness of Emmeline, 'stepchild of her uneasy century'. It's 'neurosis,' declares Lady Waters of Cecilia's 'perpetual rushing abroad and then home': she 'never seems to be happy when she is not in a train—unless, of course, she is motoring'. 'The map of Europe was never far from [Emmeline's] mind, crowds rushing from platform to platform under the great lit arches, Cecilia's face sleeping against the cushions as the Anglo-Italian express tore into France from Switzerland on the return journey.' This continual rushing is good for neither of them:

> On a long journey, the heart hangs dull in the shaken body, nerves ache, senses quicken, the brain like a horrified cat leaps clawing from object to object, the earth whisked by at such speeds looks ephemeral, trashy: if one is not sad one is bored.

No wonder Markie declares he 'can't live at top gear', and Emmeline in distress 'looks all over the place', and their relationship founders in a welter of speed and travel metaphors ('Travelling at high velocity he had struck something—her absence—head on, and was not so much shattered as in a dull recoil . . . oppressed . . . by sensations of having been overshot . . . outdistanced . . . a good way past him . . . apart . . .'). 'But how am I to end?' Emmeline asks Markie as they talk tetchily of not marrying. Her speedy life can only end badly. Her travel agency's slogan is 'Move dangerously' ('a variant of "Live dangerously", you see'). Emmeline's idea of a custom-inducing advert is 'a Handley Page looping the loop full of passengers'. And she drives Markie and herself to an ending in a spectacular crash on the Great North Road. They're victims of 'taut ungoverned speed':

> Head-on, magnetized up the heart of the fan of approaching brightness, the little car, strung on speed, held unswerving way. Someone, shrieking, wrenched at a brake ahead: the great car, bounding, swerved on its impetus. Markie dragged their wheel left: like gnats the two hung in the glare with unmoving faces. Shocked back by the moment, Emmeline saw what was past averting. She said: 'Sorry', shutting her eyes.

To the North is a kind of revival of *Vile Bodies*. What rescues it however from mere pastiche is its insistence that females particularly are the victims of this reeling age. Unsettled, moved on, shoved about, let down, Elizabeth Bowen's Sydneys, Cecilias, Emmelines, Portias, and Henriettas reveal their author's keen awareness that all this travelling is a sort of spiritual malaise, and one that reveals its malevolence especially because of what it does to girls and women. Henrietta's bumpy ride in *The House in Paris* is not simply a matter of being bumped about: she's an instance, we're told, of how girls like her keep going in for the male 'cads' who will hurt them: for 'they like', the narrative alleges, 'to have loud chords struck on them'. In transition between old repressions and new freedoms, these girls move about more freely than did their mothers; they meet men and have affairs; but they're still oppressed, for all that they frequently appear to be choosing to behave in the way that leads to trouble. Emmeline drove herself to her death. And this is, in fact, the grimly paradoxical mixture of freedom and non-freedom that Jean Rhys also recognizes as terrifyingly inevitable to her emancipated women.

Shuttling about in London, or between Paris and London, transient in hotels and lodging-houses, familiar with trains and boats and taxis, Jean Rhys's women are less thoroughly bourgeois, more desperately down-at-heel than Elizabeth Bowen's, but they're recognizably subject to similar distresses. The men met up with in Jean Rhys's *After Leaving Mr Mackenzie* (1930), or *Voyage in the Dark* (1934) or *Good Morning Midnight* (1939), tend to be oppressive, like Miss Bowen's males, but their oppression is intensified, less well packaged in bourgeois politenesses; they're more caddish, more casually violent, more prone simply to ditch their girl friends after one-night stands. 'I get along with men,' boasts Laurie in *Voyage in the Dark*, 'I can do what I like with them.' But in practice they do much of what they like with the women. And usually they love them, hurt them, and leave them. ' "What is it?" she asked. "You aren't going? You promised to stay with me".' ' "Goodbye", said Mr Mackenzie.' 'I've realised, you see', confides Marya of *Quartet* (1928), 'that life is cruel and horrible to unprotected people.' And women, stuck with what Marya despises as 'her idiotic body of a woman', are the unprotected, always being shoved out to start messily again, elsewhere. Anna of *Voyage in the Dark*, exiled in London from her native West Indies, is 'Just bumming around'. 'There was a man I was mad about. He got sick of me and chucked me. I wish I were dead.' 'He looked as if he were making up his mind not to see me again. (Opaque, their eyes look.)' She has to have an abortion and she's left, like Sasha of *Good Morning Midnight*, to haunt cinemas and to wander on ('you've got to walk around by yourself'). Sasha is another victim of this unsettled cinematic run of experiences ('My film-mind . . . ("For God's sake watch out for your film mind . . .")'), another female version of Eliot's

Jewish landlord in 'Gerontion' ('Spawned in some estaminet of Antwerp, / Blistered in Brussels, patched and peeled in London'), another member of Jean Rhys's female foreign legion ('La Légion, La Légion Etrangère . . .'), dodging about precariously from Amsterdam to Brussels to London to Paris, to more men and more goodbyes: 'But I knew it was finished. From the start I had known that one day this would happen—that we would say good-bye.'

Jean Rhys and Elizabeth Bowen were not unique. F. Tennyson Jesse's *A Pin to See the Peepshow* (1934) is another powerful fiction about a woman, called Julia, whose desire for sexual liberty is expressed in literal mobility ('Calais she adored . . . she was "abroad", that was what mattered. It wasn't England'), and who meets a bad end— she's judicially hanged—an end that's connected with the ambivalent freedom of the Etrangère condition (she works at a shop called *L'Etrangère*, and has illicit sex there with the lover Leo who's responsible for her husband's death and so for hers). But Rhys and Bowen provide particularly outstanding instances of '30s writing's mixed feelings about inevitable mobility, in which the sense of all travel's exciting promise, and not just travel for women, is muted by apprehensions about what that journeying might bring forth. Setting off was widely felt to be a voyage in the dark; as rousing but also as daunting as getting into the ghost-train at a fairground.

Getting away, though, was an imperative. '[L]eaders must migrate', as Auden's Poem No. XXIV (1930) put it, and migrate now: ' "Leave for Cape Wrath tonight" '. Or, again (*The Orators*, Ode No. V, 'To My Pupils'),

> All leave is cancelled to-night; we must say goodbye.
> We entrain at once for the North . . .

'I leave for Zovgorod tonight,' replied Eric Ambler's Henry Barstow in *The Dark Frontier* (1936), in answer to the Audenic travel imperative, and replied in the voice of Conway Carruthers, the fictional special-agent that Barstow would dearly like to be. Markie and Emmeline also answered Auden's call: 'We're going north':

> The cold pole's first magnetism began to tighten upon them as street by street the heat and exasperation of London kept flaking away. The glow slipped from the sky and the North laid its first chilly fingers upon their temples . . . Petrol pumps red and yellow, veins of all speed and dangerous, leapt giant into their lights. As they steadily bore uphill to some funnel-point in the darkness . . . this icy rim to the known world began to possess his fancy . . .
>
> He saw 'The North' written low, like a first whisper, on a yellow A.A. plate with an arrow pointing: they bore steadily north between spaced-out lamps, chilly trees, low rows of houses asleep, to their left a deep lake of darkness: the aerodrome.
>
> 'Hendon', he said. 'I wish we were still flying.'
> 'So do I', she said with an irrepressible smile.

It's a lust for travel that commands *To the North*, and the text of this travelling is striated with inevitable menace. 'You are all very much too anxious to leave England', says Lady Waters. She blames 'our naughty Emmeline's propaganda', susceptibility to which runs in the family. Cecilia wants to go to America:

> Strangers, the kindly touch of the unforeseen—it was high time she was abroad again. The heart is a little thing and one can coerce it; she would step up the cheerful gangway and go abroad. She saw the bright decks and gilt saloons, heard the bugles and silent throb of the liner steaming rapidly west: so one leaves behind one's little coffer of ashes.

And however brightly alluring the gang-plank, departure was a little death. Miss Pym in Bowen's *The Hotel*, watching Milton and Ronald depart for the Genoa train, 'said to herself, "Morituri te salutant" ':

> People on the verge of departure always seemed to her to be saying this; there was something about them fated and sacrificial that made her feel self-conscious on their behalf yet somehow rather exalted.

There's a dog in *To the North* who's said to have been 'desolated by too many departures' and now 'dared form no more attachments, looking at newcomers with a disenchanted eye: a nervy luckless little white dog that yearned for a sweet routine'. And as early as his poem 'Extract' (in *Oxford Poetry 1927*) Auden had been preoccupied with the problematics of parting:

> Consider, if you will, how lovers lie
> In brief adherence, straining to preserve
> The glabrous suction of goodbye . . .

('Too long the suction of goodbye' was the *Poems* (1928) version.) Graham Greene too, was much taken by the rich melancholics of such moments. Kate Farrant's implication in the crookery of Krogh's business arrangements in *England Made Me* is measured by the accepting calmness of her farewell to legality:

> This is the moment I've always been expecting, the moment when we leave the law behind, push out for new shores. It seemed curiously unimportant. One had always expected the drawn-out business of goodbyes, tears on wharfs, last sight of shipping.

Her brother Anthony and his pick-up from Coventry, though, who are made of less stern stuff, wallow harshly in the pains of these occasions:

> They kissed, and into their kiss crept the desperation, the hunger of departure, the sadness of railway stations; one was going and one was staying; a holiday was over; the fireworks were dead on the grass, other people were watching the pierrots and curling of an evening in their glass shelter on the front, and one had done nothing more complete than this kiss. You have my address, the slamming of doors, write to me, the waving of a flag, we'll meet again, and the smoke blowing between. She took her mouth away. 'So limiting', she said again uncertainly. He looked for her mouth and missed it; her cheek tasted salt. 'Oh, damn it bloody all', she said . . .

Greene is very adept at celebrating this particular sort of sad ceremonial (a variant, Farrant knows, of the repeated First War dramas of 'leave trains'). One of its best occasions in all '30s literature is in Greene's abandoned novel of 1936, 'The Other Side of the Border', when Colley, seedy refugee from the margins of many a Greene fiction, leaves Victoria Coach Station for the North, in a moment resonant with the skewed rhythms of Greene's many dismayed departings:

> Colley, standing in the great steel coach station, felt the familiar sadness and unrest of departure. A modern clock-face without numerals, a chromium milk-bar, a faint smell of petrol: they took him back nevertheless to the stone quays and the slap of water, the oil and the seagulls of his usual loneliness. He was only going a few hundred miles north, but that's how new places, new people, always made him feel . . . it seemed just as important as when a liner leaves the dock.
> It was too familiar. He wanted a drink. There was no one he had to kiss and grin at

through the glass. It had been just the same when he was seventeen and went off to Brazil to take the place of a clerk who had died of yellow fever. All the goodbyes were always said in Surrey (they became on each occasion more perfunctory), and even that first time he had found his way to the mail boat's bar while everyone else was throwing paper ribbons. And it was the same the time he went to Africa, except that then the bar had not been open; he'd lain down on his bunk with a comic paper . . . You'd think you'd get used to new jobs and going away from the places you know, but the loneliness repeats itself every time . . .

. . . Three brandies were enough to give him the courage of goodbye . . .

He hated the world—that was the permanent, the first article of his creed—you couldn't help drinking, you couldn't help moving on, but the emotion which made every departure a sad one suggested that somewhere—something—he had no terms for it—there existed— . . .

. . . All the jobs he had held up and down the seedy margins of strange continents had left impressions which came up at this time of year into his sour consciousness with an effect of sadness and for some reason beauty: the face of a stoker on a mail boat in '31 and of a dago child; they had mounted for a breath of air on to the steerage deck; he saw their patient lamp-lit faces as they sat side by side on a coil of rope and panted in the heavy night . . .

It was just like any other sailing at night: the lamps slipping away from you, the turned faces, swinging round by the high stone palace wall in Grosvenor Gardens like a dockside, and into the churn of small tugs and tramps by Hyde Park Corner. Even the young men at the back were momentarily silent as London dragged slowly backward . . .

. . . England was like a magnet which had lost its power. There was nothing any longer to hold you to it. It shook you off. Colley thought: I've stuck it long enough for a good reference, and now I may as well shoot it abroad again.[220]

Here there's none of the triumphalism of Spender's 'The Express', where leaving the station is for the train a matter of 'restrained unconcern' and majestic readying for hieratic power soon to be achieved in 'the elate metre of her wheels', the 'stream-line brightness', when 'she acquires mystery, / The luminous self-possession of ships on ocean'; and then, 'At last, further than Edinburgh or Rome', 'she moves entranced', 'like a comet through flame'.[221] Greene's sense of the dubiousness of thresholds is closer to the mixed feelings of Spender's 'The Port', in which the bourgeois families may be happy and 'well-lit' but the whores, male and female, ply their trade in the ill-lit urban 'maze', 'far' from laughter, 'parched and hard', where illumination is temporary and what is seen ('Where magic-lantern faces skew for greeting'; 'a harsh lightning'), is deterringly ghostly.[222] '[D]oorsteps, docks and platforms make you', as Elizabeth Bowen's Naomi realizes in *The House in Paris*, 'clairvoyant', and what you see at those margins may indeed be as grim as what packs the recollections of Greene's marginal man or fuels the despairs of the marginal, bisexual poet Stephen Spender. The ancient taboos of the threshold—the place where the excitement of leaping across into the new is chequered by the fear of being stuck on the border, suspended in no-man's-land, dangerously exposed, fearfully indeterminate ('In that day', warned the Old Testament prophet Zephaniah, speaking for Jahweh, 'I will punish all those that leap on the threshold')—are inevitably focused at very many of the '30s huge roster of thresholds.

Greene's foreign agent D. in *The Confidential Agent* is well aware of the ill-luck attendant on such places. The girl Rose has told him she hasn't 'got a people':

'You're unlucky', he said, 'You are in No Man's Land. Where I am. We just have to choose our side—and neither side will trust us, of course.'

Characteristically, trouble comes to Jean Rhys's women on the *landings* of houses—menacingly indeterminate places, that are between rooms in the way that landing-stages and piers are, in a way, suspended between countries. Men lurk there in the darkness, ghosts 'of the landing', to kiss, to hurt: 'damned men'. 'I cry', narrates Sasha of *Good Morning Midnight*, 'in the way that hurts right down, that hurts your head and your stomach. Who is this crying? The same one who laughed on the landing, kissed him and was happy. This is me, this is myself, who is crying. The other—how do I know who the other is? She isn't me.' Much of the horror of Greene's *Brighton Rock* is enacted appropriately on Brighton pier. The nightmare of being inescapably stuck on a pier, a prey to the disorientations and threats of the threshold, haunts Auden. 'Who will endure?' asked No. XXV of his *Poems* (1933):

> For no one goes
> Further than railheads or the ends of piers,
> Will neither go nor send his son
> Further through foothills than the rotting stack
> Where gaitered gamekeeper with dog and gun
> Will shout 'Turn back'.

Auden had acquired D. H. Lawrence's fear of being doomed to liminality, as an 'in-between' in terms of classes and countries, keen to trespass, to progress, to move out, but perpetually thrust back by gamekeepers, those daunting guardians of the barred gate and the fence:

> You cannot be away, then, no
> Not though you pack to leave within an hour,
> Escaping humming down arterial roads.

Auden's 'Roman Wall Blues' fingers its dyspeptic legionnaire's troubles lightly ('I've lice in my tunic and a cold in my nose') but there's no disguising he's in a really bad way:

> The rain comes pattering out of the sky,
> I'm a Wall soldier, I don't know why.

He is as locked into his disorientation as Henry Green's travellers. And feelings of dislocation are common on such thresholds. At one point in *To the North* Emmeline is said to have 'looked bewildered—like a gentle foreigner at Victoria, not knowing where to offer her ticket, to whom if, at all, her passport, uncertain even whether she has arrived'. *The House in Paris* ends with startling indeterminacy at the Gare de Lyons where Ray and Leopold, the precious little Jewish stepson Ray has 'stolen' (they'll 'put in time' at a hotel 'somewhere'), see Henrietta off, and Ray has one of Mrs Bowen's moments of unnerving threshold clairvoyance. 'Ray had not seen Karen's child in bright light before; now he saw light strike the dilated pupils of Leopold's eyes.' 'Where are we going now?' Leopold asks:

Where are we going now? The station is sounding, resounding, full of steam caught on light and arches of dark air: a temple to the intention to go somewhere. Sustained sound in the shell of stone and steel, racket and running, impatience and purpose, make the soul stand still like a refugee, clutching all it has got, asking: 'I am where?' You could live at a station, eating at the buffet, sleeping on the benches, buying your cigarettes, going nowhere next. The tramp inside Ray's clothes wanted to lie down here, put his cheek in his rolled coat, let trains keep on crashing out to Spain, Switzerland, Italy, let Paris wash like the sea at the foot of the ramp. And a boy ought to sleep anywhere, like a dog. But the stolen boy is too delicate. Standing there on thin legs, he keeps his eyes on your face. Where are we going? Where are we going now?

Auden and MacNeice poured an opportunistic smidgin of dismissive scorn on the fashionable taste for *poésie de départ* when they bequeathed (in 'Last Will and Testament') 'the painted buoy / That dances at the harbour mouth' (lame pun, of course, on painted boy) to Brian Howard. But this was in a letter coming from Iceland, and, as they admitted, they weren't really minded to shun such writing: 'sooner or later / We all like being trippers'. Earlier in *Letters*, in Part II of 'Letter to Lord Byron', Auden had revealed a similar, marginally qualified enthusiasm. He didn't, he said, care much for the new *poésie de départ* that was 'Centred round bus-stops or the aerodrome'; none the less

> give me still, to stir imagination
> The chiaroscuro of the railway station.

And travelling in '30s Europe did rather multiply these stirred Bowenite, Audenesque moments: for every border offered the threshold's challenge, with the ancient taboos now actualized, released from myth and fiction into the substantive shapes of inquisitive border-guards, deterring inspectors of one's identity and pretensions as these were embodied and inscribed in one's passport (newly introduced for all countries only in 1915). Europe, a kaleidoscope of frontiers and frontier-guards, kept bumping you relentlessly up against threshold anxieties. 'All these frontiers . . . such a horrible nuisance', exclaims Isherwood's worried Mr Norris as he and young Bradshaw approach the German border and 'the sinister' officials of the *Deutsche Passkontrolle* ('They were not unlike prison warders'). Travelling through any continent was now an encounter with these newly harshened frontiers.

But this was especially true of encounters with Europe: for in Europe the traveller was plunged into that troubling and continual drama of drawn and re-drawn post-war borders that would lead inevitably to War again. It's the drama that enlivens the sober pages of Horrabin's *Atlas of Current Affairs* with its various sections on 'Germany's Western Frontier' ('as a matter of fact, of course, frontiers here have been constantly shifting for centuries'), 'Germany's Eastern Frontier: the "Corridor",' 'Poland's Eastern Frontier' ('The Supreme Council at Versailles originally fixed the eastern frontier of Poland on a line . . . running roughly north and south from Brest-Litovsk with Polish Ukraine (Eastern Galicia) as an autonomous area under Poland's sovereignty; but . . .'), not to mention its thoughts on 'Austria-Hungary's War Losses' ('When the Austro-Hungarian monarchy was broken up by the 1919 Treaties its population of 51 millions was divided up between seven states'—'where' now, as Lehmann put it in his 'Via Europe' travel sequence in the first number of *New Writing*, 'the Danube flows down through a patchwork of frontiers and peoples'). 'Not even in those parts of the world where frontiers are arbitrarily set up or altered by all-conquering imperialisms did the War make greater changes than in Europe itself', declared Horrabin, with reference to the six new sovereign states created in 1919 (Finland, Estonia, Latvia, Lithuania, Poland, and Czecho-slovakia). Talk of Europe was inevitability talk about frontiers. In his 1937 poem 'Meeting by the Gjulika Meadow' Grigson talks of talking about Europe with a 'Slovene', who carries a 'red plant' picked near the snow 'under / The suspicious frontier', 'under the / Nervous frontier':

> And talked under the thunder
> About Europe, about dealing
> In furs, about thunder, about rain
> And the invisible trout
>
> In the silk-blue Sava, and
> About Europe again, and frontiers.[223]

Obsessed by Europe, British writers were consequently obsessed by frontiers. They perceived frontiers as running through every department of their lives. At the beginning of *A Hope for Poetry* Day Lewis has the 'boy Keats', offered as a type of the young poet, riding with his belle dame 'across the frontiers of fancy'. He would inspire others on 'their wild-goose chases'. And not just poetic ones. Going over to the workers was frontiersman work, a matter of taking (with Upward) a *Journey to the Border* (a part of which story appeared as 'The Border-Line', the title of one of D. H. Lawrence's short stories, in *NW*, No. 1), or engaging (with Rex Warner) in *The Wild Goose Chase*: the journey 'to the frontier miles away, where very few of us had ever been', towards, beyond, and across the frontier of class, 'to get into touch with the revolutionary movement'. 'Can one ever', Warner's character George wonders, 'be quite at ease this side of the frontier?' Petra's mother in Spender's *Trial of a Judge* is bothered by being an exile from Poland in a land where strangers are hated; but Petra's brother has other frontiers on his mind:

> The frontier which I cross is that
> Bombed impassable road within
> Easy reach of trained machine guns, which
> Divides banks and cathedrals from the slums.
> Look up where wealth's Gibraltar stares across
> The workers' salt undifferentiated, fretting sea.
> . . . I must be poor,
> To cross that frontier all I need declare
> Is I have nothing and I give my life
> To those with nothing but their lives.

For his part, T. S. Eliot thought the 'real' frontier of 1936 lay between secularism and anti-secularism: 'between those who believe only in values realizable in time and on earth, and those who believe also in values realized only out of time. Here again the frontiers are vague, but . . . only because of vague thinking.'[224] One way or another '30s authors were, as Auden put it in Ode No. V 'To My Pupils', obsessively 'frontier-conscious': 'They speak of things done on the frontier.' Frontiers were very much on their minds (*Journey to a War* was a mysterious title, thought G. W. Stonier in his *NS* review of it, and one 'hinting at imaginary war and frontiers of the mind').[225] It was precisely its interiorization of the frontier that made Upward's *Journey to the Border* so powerful, according to V. S. Pritchett:

By its title ... Mr Upward's book suggests yet another conducted tour over the well-worn macadam of New Country; but a salutary and excellent change has taken place ... he has stopped to think. What is all this O.T.C. talk about the frontier and the border? Where is the border? It is, his hero discovers, in his own mind; it may be the symbol of a certain political idealism, but is it not also that point in the Left-wing mind where hysteria about the future breaks out, where a day-dream conflict between Fascist types and Left-wing types lures the timid intellectual into imagining himself a terrified hero or a choking posturing martyr, living with unreal courage in an unreal situation?

Journey to the Border presented, Pritchett thought, 'a brilliant . . . study of a widespread contemporary state of mind'.[226]

A lot of people, of course, alleged that borders were a pushover. Debunkers abounded. Camp old Julius Carson takes Arthur Pumphrey's cash before he entrains for England: 'You'll only get into trouble with it at the frontier. And that would be fatal.' And, of course, there's no trouble, for this is 'not even a wild-goose chase'. Evelyn Waugh was keen in *Robbery Under Law* to take the mickey out of 'one of those warm-hearted little articles which used to appear in the early days of the Spanish Civil War' in the *New Statesman*. 'Its author—Mr Cyril Connolly, I think—was describing his emotions when he crossed from capitalist France into the free, proletarian air of Catalonia.' The Workers' State had no beggars; but 'If this is so, the Mexicans, in this matter as in much else, have got their Marxism a little mixed.' So much, the implication was, for that sort of going over—and for other kinds too. In *Ninety-Two Days* Waugh sets off to cross the Brazilian border, only to find there are no frontier frissons to be had, for 'there were no formalities of any sort in crossing the boundary; our horses waded through the shallow water, stretching forward to drink; half way over we were in Brazil.' ('The Indians have probably very little idea of whether they are on British or Brazilian territory—they wander to and fro across the border exactly as they did before the days of Raleigh.') In such passages Waugh was consciously tongue-in-cheek, trying to demystify the period's frontier obsessions, to prick a lot of literary persons' bubbles. So was Orwell. 'In the end we crossed the frontier without incident,' he declared at the end of *Homage to Catalonia*: 'we slipped through the barrier.' He was laconically determined to play down just those border excitements that many of his contemporaries would have stressed.

And the frontier debunkers had the aeroplane on their side. Train-travelling subjected you to border controls; aeroplanes hopped cheerfully over those earthbound frustrations. Seen 'from the enchanted windows of a passenger aeroplane', the countries of the Danube basin 'refused', according to Lehmann's *Down River*, 'to create images of hatred and division'. And internationalists of all sorts likewise challenged the old frontier enclosures. The Brotherhood of Man spoke against divisive borders from Geneva. When Mr Norris called frontiers 'a horrible nuisance' the 'thought crossed' Bradshaw's 'mind that he was perhaps some kind of mild internationalist; a member of the League of Nations Union'. Internationalism of a less mild kind was advocated by socialists of all sorts. Socialism had always seen itself united in an international struggle.

> And in the borderless world of the many
> States and separate power melt away

as Spender's Third Red explains this particular vision in *Trial of a Judge*. And one of the period's most recurrent leftist complaints was that big business interests, especially in the arms trade, had created an evil version of internationalism: money's power acknowledged no other sovereignty than its own. 'But nationality's finished. Krogh doesn't think in frontiers', Greene's Kate Farrant expounds in *England Made Me*. And this 'new frontierless world' is certainly not a kind of Socialism: ' "Oh no", Kate said. "That's not for us. No brotherhood in our boat".'

But long before that final note of *England Made Me*, the reader's doubts about this frontierlessness have mounted high. Even Kate Farrant occasionally yearns for the limiting realities of the frontier-bound world of England, 'the Bedford Palace, the apples they'd eaten', 'the old honesties and the old dusty poverties of Mornington Crescent'. Living with the new frontierlessness only brought distinctly home a sense of how the old-fashioned power of frontiers still prevailed. The new forces of internationalism hadn't abolished borders, they were merely redrawing them and, in many ways, worsening their associated fears. The aeroplane's power to abolish frontiers only increased the fears nations had of would-be foreign transgressors of their borders. It's precisely on deadly action at frontiers that *Biggles Goes to War* (1938) is concentrated—for all that Biggles is a-warring by plane. Orwell's allegedly smooth slipping across the Spanish frontier is followed not only by his account of how he hadn't effectively crossed it at all ('We thought, talked, dreamed incessantly of Spain': border crossing was tougher than he'd imagined it), but also by glum fears about the bombing planes that really can slip across: 'the roar of bombs' over 'southern England' makes *Homage to Catalonia*'s final sentence a prophetic sentence of airborne doom. Just so, *Down River* ends with the roar of aeroplanes over Lehmann's London apartment: what he had described happening in Vienna and Prague could happen here or anywhere. That misguided version of internationalism, Pan-Germanism, had the scant respect for frontiers that none of its victims could smile on. Under Hitler, Germany's frontiers had 'advanced steadily' down the Danube. Being Hungarian, for instance, after the Austrian *Anschluss* was suddenly terrifically frightening: now Germans were 'just beyond' the 'thin green-white-red barrier' at the frontier that Lehmann crossed in a bus:

> It was a moment when all the small countries of the Danube saw ominous clouds gathering in their sky, when the ancient Djinn of Pan-Germanism seemed to have emerged in a flash from the bottle of Versailles to mock their independence.

The totalitarian abolishers of frontiers, in fact, only made the world more conscious of frontiers. Like the antics of big multinational companies whose wealthy disregard for borders emphasized to the poor how stuck they were within poverty's confines, aggressive internationalism may have redistributed frontiers but it didn't get rid of them. In Spain, 'The bloody frontier', as MacNeice called it, the period's sense of the frontier grew bloodier. And communist C. Day Lewis rejoices precisely in the way Soviet frontierlessness is making new frontiers ('On the Twentieth Anniversary of Soviet Power', in *In Letters of Red*):

> Your republic, Soviet Russia, is not contained
> Between the Arctic floes and sunny Crimea:
> Rather, its frontiers run from the plains of China
> Through Spain's racked heart and Bermondsey barricades

> To the factory-gates of America. We say,
> Wherever instinct or reason tells mankind
> To pluck from its heart injustice, poverty, traitors,
> Your frontiers stand; where the batteries are unmasked
> Of those who would shatter Life sooner than yield it
> To its natural heirs, your frontiers stand: wherever
> Man cries against the oppressors 'They shall not pass',
> Your frontiers stand. Be sure we shall defend them.

Whatever debunkers and internationalist theory might say, the frontiers still stood, and so did their terrors. In *The Professor* (1938), Rex Warner's pacifically liberal professor is moved by warm comradeliness to all people ('he saw . . . with a kind of love the attractive power of a reason that permeated frontiers and dominated the interests of classes') but the enemy powers massed on his country's borders ('our own frontiers, with a million soldiers, with tanks, guns, and aeroplanes waiting on the other side') soon punch tyrannically through the weak humaneness of his political efforts. 'I wish to disclose', declared the professor's communist son, 'the horrible fact that there is no enclosed space in Europe.' There was no escaping the traumatic darkness of that sort of frontier. And even if one persisted in faith in some other kind of border, and endorsed the idea of Graham Greene's Mr Calloway (in 'Across the Bridge', 1938),[227] that 'life . . . begins on the other side', any kind of Upwardian *Journey to the Border*, indeed goings-over of all kinds, still remained difficult and troubling.

Julian Symons began his review of a rag-bag of volumes that included Auden and Isherwood's *On the Frontier* by quoting from that play Eric's pious hopes of a frontier-less future:

> But in the lucky guarded future
> Others like us shall meet, the frontier gone,
> And find the real world happy.

'There are lots of frontiers, and there will always be lots of frontiers', Symons declared; 'and the existence of the frontier bothers us all.'[228] Mr Norris's sweat at the German frontier ('I was amazed to see what a state he was in; his fingers twitched and his voice was scarcely under control. There were actually beads of sweat on his alabaster forehead') was widely shared. Even callous little Bradshaw catches something of this worry:

> By now the train had stopped. Pale stout men in blue uniforms strolled up and down the platform with that faintly sinister air of leisure which invests the movements of officials at frontier stations. They were not unlike prison warders. It was if we might none of us be allowed to travel any further.

Certainly, it was in the nature of frontiers that they proved eventful. Looking across into Tibet (in his travel book *First Russia Then Tibet*, 1933), Robert Byron perceived 'no gradual transition, no uneventful frontier, but translation, in a single glance, from the world we know to a world that I did not know'. No uneventful frontier. In *One's Company—A Journey to China* (1934), Peter Fleming describes how he felt an oppressive sense of frontier at the Russian–Polish border. The cheerfully un-socialist Old Etonian wasn't likely to find socialism's internationalist leanings endorsed there:

> A legend, blazoned in Russian across an arch through which the train passes to cross the Polish frontier, informs you that 'The Revolution Breaks Down All Barriers'. The truth of this is not immediately apparent.

He 'negligently disposed' on top of his suitcase's contents a letter from Stalin to his brother Ian (who had been Reuter's correspondent in Moscow) in order to smooth his transition. Malcolm Lowry had no such aids when he tried to cross from Canada into the USA in 1939: drunk, anguished about his woman, Margerie Bonner, incoherent, he was refused entry. Lowry, his biographer Douglas Day writes, 'always terrified of anyone in authority—especially in *uniformed* authority—from now on added "Turned Back at the Border" to his litany of fears and woes, and always hereafter commenced sweating and trembling days before he had to pass through the Customs of any country.'

And, of course, for anyone with anything to hide or to lose—the crook, the spy, the smuggler, the double agent, the homosexual, the political refugee—the frontier's ordinary worrisomeness increased dramatically. 'Will it be all right on the frontier?' anguishes Juliet about her husband Dick in Lehmann's *Evil was Abroad*: he's a leftist newspaperman going 'across the frontier' on special missions, toting illicit pamphlets between Czechoslovakia and Germany, and the like. In thriller after thriller Eric Ambler plays on precisely this sort of frontier drama. Professor Barstow of *The Dark Frontier* day-dreamed of wandering 'over dark frontiers into strange countries where adventure, romance and sudden death lay in wait for the traveller', only to be plunged into just such dramas across Europe's borders. At 2 a.m. at the Ixanian frontier the scruffy guards confiscated his camera: 'The episode at the frontier had impressed him . . . This was reality.' In *Uncommon Danger* (1937), just one of Ambler's stories hostilely concerned with the plots of international capitalism, penniless newspaperman Kenton starts getting into trouble when he agrees to carry to Linz a packet of what turn out to be Russian war plans, for an implausible mittel-European posing as a German Jew on the run from Nazi agents ('They will search me at the frontier, strip me, send me to a concentration camp, where I shall be whipped'). Crossing frontiers wasn't easy, Kenton was driven to reflect, contemplating the thick barbed wire between Austria and Czechoslovakia:

> According to their biographers, men like Lenin and Trotsky, Masaryk and Beneš, Mussolini and Bela Kun, to say nothing of their friends, had spent half their lives 'slipping across' frontiers with prices on their heads and no passports in their pockets. But perhaps the rising generations of frontier officials had read those biographies too.

'They cross frontiers unnoticed, and slip through the barbed wire of concentration camps', marvelled John Lehmann, about Europe's revolutionaries.[229] But the truth was more as Kenton experienced it. Orwell nearly failed to 'slip' out of Spain: we 'ended', he wrote to Rayner Heppenstall (31 July 1937), 'by slipping over the border with the police panting on our heels.' Mr Norris's limp palpitations and the self-inflation of Spender's Dr Mur (in the story 'Two Deaths', in *The Burning Cactus*: 'He rushed about Europe in the fastest and most luxurious trains or aeroplanes, smuggling Communist literature and obtaining fantastic information') might be ludicrous, but the realities they touched on were not. Mur's 'Austrian comrades' *were* 'shot at when crossing frontiers'. Many a political exile really experienced what Spender's Judge calls 'the emigrant distress on the frontier's rim'. Turned back at the Franco-Spanish border in September 1940 Walter Benjamin, for one (he had earlier escaped from Nazi Germany to Paris and was now on the run again), committed suicide rather than be taken into German custody. Bernhard Landauer in *Goodbye to Berlin*, always talking of 'going to Paris, or to Madrid, or to Moscow', must have got away from the Nazis 'somewhere safe abroad', thinks Christopher Issyvoo. But he only got as far as the camps. Isherwood's own travels with Heinz in the '30s were a recurrent drama of being turned back at borders—the Harwich Immigration Officers' refusal to let Heinz enter England in January 1934 because he was an obvious homosexual was only a foretaste of the nightmares to come. 'Could I get over the frontier?' asks flamboyantly queer President Schplitz (he picks up good-looking 'mounted gendarmes' for private masochistic pleasures in the presidential gym), camp centre of E. M. Forster's facetiously vulgar story 'What Does it Matter? A Morality'. 'Would an aeroplane get him across' Schplitz wonders.[230] Heinz and Isherwood must often have thought of parachutes. Would they—as Ambler's Nick Marlow and the Soviet Agent Zaleshoff (the master of disguise who, with the aid of an Italian railway-worker Comrade, saves Marlow from being 'shot while attempting to evade arrest') are made to wonder in *Cause for Alarm* (1938)—would they ever 'get across the frontier?' Marlow and Zaleshoff did (into Yugoslavia); Isherwood and Heinz did not. No wonder Forster was given to fantasies about turning the tables on customs officials. One of the late fragments published in his *Arctic Summer and Other Fiction* (1980) has a traveller quelling the frontier custodian with his own question, 'Have you anything to declare?' Not insignificantly the official's sexual identity is mixed: now he's 'he', now 'her'.

Frontier dramas of the kind Isherwood and Heinz endured in the '30s, the border traumas of Europe's political refugees, the increasing traumatizing of borders as the forces of reaction pushed their frontiers across Europe and into Abyssinia and Spain and China, all intensified the sense of border danger for writers who guiltily shared Isherwood's and Mr Norris's sexual preferences, and/or imagined themselves as undercover men or spies ('A novelist', Greene's pompous popular writer Savory keeps declaring in *Stamboul Train*, with the voice of the Mass-Observer, is by nature 'a spy'). But for all the danger, writers couldn't leave borders alone; the promise they held out for one's art was worth risking their dangers for. Graham Greene, for instance, a persistent crosser of border zones, opens *The Lawless Roads* with a grim prologue about the hellish border that rifted his early life—an affair of repeated border crossings between school and home ('One was an inhabitant of both countries: on Saturday and Sunday afternoons of one side of the baize door, the rest of the week of the other. How can life on a border be other than restless?')—but he follows it up with Chapter I ('The Border') and its opening paragraph about the prospective joys of the border:

> The border means more than a customs house, a passport officer, a man with a gun. Over there everything is going to be different; life is never going to be quite the same again after your passport has been stamped and you find yourself speechless among the money-changers. The man seeking scenery imagines strange woods and unheard-of mountains; the romantic believes that the women over the border will be more beautiful and complaisant than those at home; the unhappy man imagines at least a different hell; the suicidal traveller expects the death he never finds. The atmosphere of the border—it is like starting over again; there is something about it like a good confession: poised for a few happy moments between sin and sin. When people die on the border they call it 'a happy death'.

And many of Greene's most memorable characters are marginal people, exiles caught between England and abroad, fringe Catholics poised between belief and unbelief, petty crooks and transgressors confined to the edges of social or religious acceptability. Greene is superb at presenting the crisis moments when these hangers-about on borders run the risk of exposure: *The Power and The Glory* is full of them; there's a particularly fine one in Chapter VIII of the *Name of Action* where Chant tries to smuggle a cargo of weapons across the German border ('How far is it to the frontier?') and is held up by the schoolmasterly precisions of the corrupt Herr Muller ('patronising voice of a schoolmaster, welcoming back to work, with mockery, a stupid boy') and the unexpected interest of Herr Mann, the passport officer, who once lived in the Tottenham Court Road.

In one way, '30s writers' taste for 'the gossip of the frontiers' (that's a phrase of Greene's from *The Name of Action*) witnesses merely a need for adrenalin-racing adventurism, the sort of adult boys' stuff that's mixed in with all Audenesque politicking, and for that matter with the political vision of some rather more earnestly engaged writers (border crossings, the smuggling of a republican leader out of Spain into France, and the running of revolutionary arms into Spain from France, formally enclose the action of Ralph Bates's *Lean Man*). But what is also being sought hereabouts is precisely the illuminations of Elizabeth Bowen's threshold clairvoyance, of Auden's stirred imagination of the margin. Borders might trap you and catch you out; if you stayed there long you risked, with the aloof airman, isolation and disorientation. What's more the view from the border might actually be distorting and blurred. But on the frontier, again like the airman, you sometimes actually saw both sides of the question. And though this bifocalism might condemn you to mixed feelings and edgy ambivalence—ranging from the unchosen Going Over confusions of Edward Upward's political and mental borderlands and the analytical ambivalences of Auden and Isherwood's *On the Frontier*, to the deliberately settled-for fence-sitting of a T. S. Eliot—the continual possibility of imaginative strength and sharp perceptions to be acquired in frontier zones was felt to be worth staking a lot for. For 'myself', declared Waugh in *Ninety-Two Days*,

> and many better than me, there is a fascination in distant and barbarous places, and particularly in the borderlands of conflicting cultures and states of development, where ideas, uprooted from their traditions, become oddly changed in transplantation. It is there that I find the experiences vivid enough to demand translation into literary form.

The 'rhythm of goodbye is in my blood', declares the narrator of Stevie Smith's *Over the Frontier* (1938), 'and I am set again for foreign parts.' For all the given darkness of the frontier, that was where she, and hosts of others, most wanted to be:

> Going abroad tonight? The face lit up by the booking-clerk's window. Poetry of the waiting-room. Is it wise, the short adventure on the narrow ship? The boat-train dives accomplished for the hoop of the tunnel; over the derne cutting lingering, its white excreta. Too late: smelling the first sea-weed we may not linger. The waving handkerchiefs recede and the gulls wheel after screaming for scraps. Throb of turbines below water, passing the mud islands, the recurrent light. Past. Handrail, funnel, oilskins, them, His will. The lasting sky.

Frontiers weren't 'wise' (as that passage from Auden's *The Orators* acknowledges), but the 'poetry of the waiting-room' depended on them, and none more obviously so than Auden's most famous frontier-poem, 'Dover'. 'Dover is a border town', as Grigson declared in his piece on Auden in *New Verse*'s Auden Double Number. It was one of the most important English gateways to the continent. Isherwood's 'I' leaves for Germany from 'Dover quay, enveloped in clammy brown fog' at the end of *Lions and Shadows*. Greene's *The Confidential Agent* opens with agent D. arriving at Dover. Fleming's *Brazilian Adventure* (1933) ends with his arrival at Dover, whose lighted parlours spell home ('A light suddenly turned on in a parlour window projected on to the yellow blind the outline of an aspidistra. I took it as a hint. I said goodbye to the jungle. I bought an evening paper'). Dover focused all the customary anxieties of the frontier. 'In Dover we know the secrets of men's hearts', an immigration officer told Alick West in 1924, refusing a friend, somebody else's mistress, admission ('Why don't you admit that she is your mistress?'). And Dover had peculiar resonances for the period's homosexuals. It was the place where the homosexual authors, in particular, met up with their chums before going abroad, where they gathered to welcome returners. Ports were always good for homosexual pick-ups (as Spender's 'The Port' reveals), but Dover added four regiments of soldiers to its quotas of sailors: it simply pullulated with boys. And the sexual fringe-men—mindful of Max Beerbohm's parody of Oscar Wilde's excitement over 'the *frou frou* of tweed trousers' and so self-consciously alert to the 'frou frou' (as William Plomer put it) of military kilts in Dover's streets and pubs—liked hovering in this town, on the edge of England, a location aptly emblematic of their social and sexual marginality, their conscious dislocation. E. M. Forster stayed there frequently between 1936 and 1938. His homosexual airmen friends, L. E. O. Charlton and Tom Whichelo lived together there, permanently. William Plomer lodged for a whole year in Dover. J. R. Ackerley would go down for the summer and comb the pubs with Charlton for sex. Isherwood and Auden stayed there with some regularity. Auden found it a good place to work. Dover was as welcoming a home as England could provide; it became one of 'our' places for this homosexual coterie. 'Dover would like us', announces Mr A. in *The Ascent of F6*, anticipating a weekend break. But, of course, it wouldn't (we know) because he's too heterosexual: he and Mrs A 'depart for Hove' instead (a bit of the play that Forster told Isherwood he specially liked). Dover, in Auden's and Isherwood's view, was suitable only for people like them; until, that is, it turned sour, as borders were always prone to, and Ackerley got thrown out of his Dover flat for behaviour outraging his landladies (the 'Holy Ladies', he dubbed them). (Jocelyn Brooke alleged in *A Mine of Serpents*, 1949, that his Dover landlady told him, about Auden and Isherwood, that '*they* were a pair of scamps, if you like'.) 'Dover is, alas', Forster lamented after Ackerley's trouble, 'henceforward a Closed Port'.

It's just this agitating mixture of congenial possibilities and anticipated dreads, 'between', as Grigson spelled it out in *New Verse*, 'the known and the feared, the past and the future, and the conscious and everything beyond control, the region of society and the region of trolls and hulders (and Goebbelses)', that nourished the work of Auden and his friends ('Auden lives very much in this frightening border territory'), and that in particular nourishes Auden's poem 'Dover'. Dover, in Auden's presentation, is a place of 'regular life', a place for enduringly secure privateness (the lighthouses 'guard for ever the made privacy of this bay'), where the taboos of thresholds have been normalized, and their terrors tamed

> And the old town with its keep and its Georgian houses
> Has built its routine upon these unusual moments;
> The vows, the tears, the slight emotional signals
> Are here eternal and unremarkable gestures
>   Like ploughing or soldiers' songs.

The soldiers are not, in fact, altogether 'unremarkable'. Dover contains people who know 'what the soldiers want'; and that doubtless includes what Auden and Co. wanted. 'The soldier guards the traveller who pays for the soldier'—the purchasable, pretty, sexually ambivalent boys:

> Soldiers who swarm in the pubs in their pretty clothes,
> As fresh and silly as girls from a high-class academy.

So 'Some of these people are happy', as Auden and his friends were until the Dover landladies turfed them out. But happiness is not universal, because Dover is edgy with the border-town's double-edged invitations. What awaits the traveller 'without', in 'the immense improbable atlas'? Migrants are hopeful and curious:

> The eyes of the departing migrants are fixed on the sea,
> To conjure their special fates from the impersonal water.

But they have only to look around to see some of their kind returning tearful and 'beaten'. In Dover, 'Within these breakwaters English is spoken'; but outside

> The aeroplanes fly in the new European air,
> On the edge of that air that makes England of minor importance.

Travelling from this place has to do, as elsewhere in Auden, with the problems of growing up, escaping Mother, defeating Nanny, launching into the unconsoling aloofness of the alien adult world where love comes scarce. And Dover looks intimidatingly out into this dangerous great world:

> High over France the full moon, cold and exciting
> Like one of those dangerous flatterers one meets and loves
> When one is very unhappy, returns the human stare:
> The night has many recruits; for thousands of pilgrims
>   The Mecca is coldness of heart.

(In *Murder in the Cathedral*, Thomas à Becket is urged to set sail for France: 'leave sullen Dover', the Chorus advises.) Yet this dislocating Dover is where Auden, his kind, his period, want to be located. 'Dover', composed in August 1937, became Auden's flagship poem, the one chosen to open the Auden Double Number of *NV*. It's an evident reworking of Spender's 'The Port'. For his part, Spender adapted both 'The Port' and 'Dover' in his own 'Port Bou'.[231] And, of course, Roy Fuller's 'August 1938'—and this helps to show how powerful a period emblem Dover, the

English frontier port, was—is simply a calculatedly faithful reproduction of much of Auden's poem:

> Mapping this bay and charting
> The water's ribby base
> By individual smarting
> And walks in shifting sand,
> We note the official place;
> Dover with pursed-up lips
> Behind the purple land
> Blowing her little ships
> To danger, large and bland.

12

Somewhere the Good Place?

ANY typology of '30s travelling is bound to be a rickety construct. For travellers came in all sorts of shapes. Travelling was done for all kinds of reasons. Even the same individual might perceive his or her journeyings in several different ways. '30s travelling was, in fact, no less mixed an affair than other important activities of this ambivalent decade. You could go as a mere escapist holiday-maker, or a would-be adventurer, or a youthfully jokey sender-up of things abroad, or an earnestly political traveller, or a deliberately inward journeyer. But however you did your travelling you plunged inescapably into the rich equivocations and doubts that all border-crossings focused and incited.

It is nice simply to get away for a holiday, to journey in what Graham Greene (praising Paul Theroux's *The Great Railway Bazaar*, 1975) has called 'the fine old tradition of purposeless travel for fun and adventure'. Lots of '30s travel writers tried to eschew seriousness. Osbert Sitwell prefaces his *Escape With Me!* (1939) with shrewd disavowals of more earnest purposes. 'The volume which you now open, Gentle Reader, is above all, in the phraseology of the day, *escapist*.' It's intended just 'for amusement' and 'for a record and description, and was not created with instruction for its purpose.' Travel writers, Sitwell thought, should simply use their eyes and avoid political distractions: 'In consequence, this is neither a communist book about Iceland or the Faroe Islands, nor a fascist volume about Spain.' 'I am no soldier of a cause militant': Sitwell was just a pleasure-seeker. So were lots of others. The '30s were, arguably, a pervasively escapist age. Cedric Belfrage's *Away From It All: An Escapologist's Notebook* (1936) begins its wry account of trying to get 'away from it all' in travel, with a 'Portrait of A Man Seeking Hashish'. Belfrage confesses that he had himself peddled dope to his age: he'd worked in Hollywood and on British newspapers. Exotic travel and the books about exotic places were, his argument runs, simply extensions of people's general wish to shut their eyes to quotidian social and political realities: they were just more dope.

And '30s writing bears witness to an extraordinary craving to achieve escapist interludes, Shangri-La, peaceful parentheses within the extremely unpeaceful inter-war parenthesis. In his old age Graham Greene has confirmed his own lifelong tendencies, and generalized about all writers, indeed all people as escapists: 'I can see now that my travels, as much as the act of writing, were ways of escape.' He goes on in the preface to *Ways of Escape* (1980): 'As I have written elsewhere in this book, "Writing is a form of therapy . . . to escape the madness, the melancholia, the panic fear which is inherent in the human situation". Auden noted: "Man needs escape as he needs food and deep sleep".' And such reflections grow directly out of the escapology that reigned even on the '30s Left, where *escapism*, the opposite of commitment, was one of the most prominent taboos (nicely caught by Waugh's *Put Out More Flags* where the Comrades are frightened of using the label escapism even after the runaway poets Parsnip and Pimpernell are well and truly ensconced in the United States). In particular, writers sought places where they could, literally as well as metaphorically, be islanded from the period's distresses. Not quite to a man, but nearly so, they were like 'The Man', in the D. H. Lawrence story, 'Who Loved Islands'. 'We seek an island', declares Auden's sound-track for the Southern Rail documentary *The Way to the Sea* (1937), and it describes the island-searchers as embarking 'for the pleasant island, each with his special hope: / To build sand-castles and dream-castles'. Auden and MacNeice (and Grigson) went north to Iceland: 'a fancy turn, you know, / Sandwiched in a graver show'; 'Holidays should be like this' (thus MacNeice's 'Postscript to Iceland'). MacNeice sought the 'easy tempo' of Hebridean Barra. And there were also magnetic softer and more southerly islands. Auden, Isherwood, and Upward made a cult of the Isle of Wight, scene of their stories (Upward's 'The Island', and Isherwood's 'An Evening At the Bay') and their poems, such as Auden's 'Look, stranger, on this island now'.[232] 'Look, stranger' presents the island as securely bounded ('the small field's ending pause'), slow and peaceful, 'stable' and 'silent'—clouds merely 'saunter' 'all the summer through the water'. Violence ('the pluck / And knock of the tide') is kept outside. On this island, the more urgent world can be recalled and glimpsed, but for the moment it's far away ('Far off . . . the ships / Diverge on urgent . . . errands'). Nor was the Isle of Wight the end of southern island-seeking. Isherwood and Co. sought the sun more southerly still 'On Ruegen Island'. Isherwood, Heinz and Erwin Hansen (who had formerly been at the now closed Hirschfeld Institute) fled in 1933 from Hitler's repressively anti-homosexual Germany further southwards yet to the Greek island Francis Turville-Petre had leased for himself and a covey of homosexual chums (it got into Isherwood's *Down There On A Visit*, 1962). Isherwood's continual craving, in fact, for islands in the sun with boys reads oddly like the fantasies of Kuno in *Mr Norris Changes Trains*, the dream fed by boyish, 'nephew' literature, 'that we are all living together on a deserted island in the Pacific Ocean', and that Bill Bradshaw professed to find embarrassing. For his part Kenneth Allott achieved his own island, a 'gully of silence', in bed with a woman, according to his poem 'Christmas, 1938':

> Distinct from the world as we are like Robinson Crusoe,
> What do they matter, the snowclouds, the headlights, the morning?[233]

Fenced in, Allott's 'I' refuses to 'succumb to the nightmare . . . of Europe': 'This is our moment held in the heart of Niagara.'

What they were all after was Spender's 'still centre', 'that central calm' that Day Lewis's *A Hope for Poetry* describes the poet as 'seeking to find and establish', especially within his own coterie ('that tiny, temporarily isolated unit with which communication is possible'). This was the chummily safe enclosure that Auden was not alone in discovering within the 'made privacy' of Dover's 'bay' or within the insulating womb of a ship. Ships in general, and cruise-liners in particular (Auden's 'The Ship' was entitled 'Liner' in his notebook) were as safe as islands: 'The streets are brightly lit; our city is kept clean.' Tomorrow 'the test for men from Europe' would come (Auden was on his way to China), but today there was calm:

> It is our culture that with such calm progresses
> Over the barren plains of a sea; somewhere ahead
> The septic East, a war, new flowers and new dresses.

Enisled on shipboard, Auden had actually reversed the chronology of the yearnings of his 'Spain'—where, as so often in the period, the grim awfulness of Today (Today the Struggle) is alleviated by fantasies about pastoralia ahead, the holiday joy that must come tomorrow: 'Tomorrow the bicycle races / Through the suburbs on summer evenings.' That sort of craving enraged political hard-liners like Edgell Rickword, but it was expressed on every hand: in Randall Swingler's poem 'Interim' (the revolutionary underground life will eventually give way to a 'lover's summer', corn and grass: 'This will come again I know'); in Michael Roberts's 'They Will Come Back' (*Poems*, 1936), where 'They will come back, the quiet days':

> Through bombs, and teargas, through the acute
> Machine-gun rattling answer, strict
> Self-knowledge, dark rebellion, death
> In the shuttered streets, through barricades,
> And doors flung open in the wind,
> They will come back.

It's there, too, at the end of Day Lewis's *Noah and the Waters*, as the sunny future after the catastrophic revolutionary floodtide is eagerly anticipated:

> Down the hillsides then shall the waters tumble apace,
> Finding their level, wearing the sun on their wide shoulders,
> To wed the radiant valleys: that reconciled embrace
> Shall raise—taller than sunflowers and record crops—the race
> That Noah foresaw in the veiled face of the avenging waters.

Such yearnings for the 'happier times' that Day Lewis anticipates in Section 31 of *The Magnetic Mountain*, after conflict and struggle are over and 'When the land is ours', were fuelled, of course, by T. S. Eliot's images of religious calm, of the still small voice of divine revelation imposing itself against the din of a turbulent secular world: *Ash Wednesday*'s

> Against the Word the unstilled world still whirled
> About the centre of the silent Word

which turned into *East Coker*'s dealings with 'the still point of the turning world'. Already in *Poems* (1934) Spender was writing about 'the centre of the turning year' (No. XXXVIII). Eliot's religious havens weren't for Spender, of course, but there were other varieties of *The Still Centre* (that was the pointful title of Spender's 1939 volume of poems) to try out. And like lots of island-seekers Spender turned inevitably to holidays. Christmas ones, like Kenneth Allott's, would do. The boy in *The Backward Son* finds home at Christmas a delightful refuge from school terrors: 'he is hung at the tiny point of the whole universe, which is his centre of happiness'; 'Christmas Eve . . . The burning brandy centre of the cold world'; 'The centre of the winter, the centre of the night, the centre of the hols'. (The episode was much admired by John Lehmann—according to the second volume of his autobiography.) Better still, though, were summer hols:

The happiness of this endless summer day reaches back into your earliest memories and may go on for ever. Why do you have to work? Why does every human activity have to go on its destructive way without stopping even for a moment? Why can't you just stay where you are at the centre of this happiness in which there is neither age nor youth, but a permanent, still existence outside yourself in which you could sink deeper and deeper?

Holiday sensations 'penetrated deep through his senses like a system of rods going through him and fixing him in a centre where he knew he could find happiness and be away from the anxiety which made him nervous and round-shouldered and gave him diarrhoea whenever he thought of the headmaster's lessons'.

'30s writing is extraordinarily obsessed by holidays and particularly by the pastoral interlude of summer. Poems on this theme abounded: George Barker's 'Summer Idyll', his 'Poem on People', Grigson's 'And Forgetful of Europe (Mlini 1935 to 1937)', Bernard Spencer's 'Picked Clean From the World', H. B. Mallalieu's 'Holiday', and Peter Hewett's 'Summer Night' to name no others.[234] Holidays were notable new activities of the masses (holidays with pay increased in Britain in this period; there was a Holidays With Pay Act in 1938; the modern French habit of everyone holidaying in August began in the '30s). Mass-Observation coupled the 'seaside holiday' as a new fruitful object of interest with 'the Cup Final, the monster rally', and 'the hiking excursion'. The sun was delightfully escapist. The girls of the opening chorus of Auden's *The Dance of Death* (they come on daringly sporting 'two-piece bathing suits') recommend the beach as an escape-route from 'sad news' and 'self-examination':

> Europe's in a hole
> Millions on the dole
> But come out into the sun.

How much more pleasantly escapist, then, the hotter sun of Auden's 'Continong'—on the Riviera, on the Baltic, in Salzburg and Vienna, in France, Spain, Portugal, Italy, Yugoslavia, Greece. Michael Roberts's 'Hymn to The Sun' makes the point. In it, a Northumbrian dustman finds August in England 'Voy wawm'. But he has 'never known' the shimmering whitenesses of stones, roads and houses 'round Millevaches'.

> '*Fait chaud*', as each old woman said,
> going over the hill, in Périgord,
> prim in tight bonnets, worn black dresses, and content
> with the lilt of sunlight in their bones.[235]

Geoffrey Grigson's continuing obsession with islanded escapisms (see *The Isles of Scilly and Other Poems* (1946), and *Places of the Mind*, 1949) lured him gradually southwards. Once having experienced the Adriatic and Sardinia, he reports in his autobiography, 'never again would we go north' to Iceland. '[N]ude at bathing places', the unemployed in Spender's *Vienna* escape, as Spender and his kind were seeking to escape (restless sun-seekers, Spender and Tony Hyndman tried Mlini, near Dubrovnik, at Grigson's suggestion in 1934):

> Not saying, life is happy, unhappy is ill,
> Death is reward, law just, but only
> Life is life, body is body, a day
> Is the sun: there is left only beauty
> Of merest being, of swimming, of somehow not starving:
> And merest beauty has a sun-tanned body.

Full-time writers and travellers could, of course, seek their islands in the sun as they chose and the funds allowed. Workers, schoolboys, and schoolmasters had to wait for what Mrs A. in *F6* refers to as 'the fortnight in August or early September', when the factories and schools emptied ('Empty as a school in August': Auden). Schoolmaster Auden is thought to have visited Isherwood on Turville-Petre's island in August 1933. In August 1934 he went for a motoring holiday in Hungary with a couple of youths from the Downs school ('flirted with policemen' in Dover, he solemnly informed the school mag). It was in August 1936 that MacNeice and a party of boys from Bryanston School joined Auden in Iceland for the 'Hetty to Nancy' adventures of *Letters from Iceland*. A lot of '30s living and writing got packed into August. In August 1937 Auden and Isherwood worked in Dover on *On the Frontier* and planned their China journey. It was then, too, that Auden wrote his 'Dover'. And how often August occurs explicitly in '30s literature: Fuller's 'August 1938', Mallalieu's 'Welcome in August', Auden's 'August for the people and their favourite islands', MacNeice's 'August à la Poussin' and his *Autumn Journal*, III ('August is nearly over, the people / Back from holiday are tanned')[236]. August became the prime month of the holiday and sun: 'Long happy hours of summertime, oh darling Tilssen of those long August days, empty of all significance but the minute upon the minute, how happy I was then', exclaims the narrator in Stevie Smith's *Over the Frontier*, recalling the six months of her author's absence from the office spent at Tilssen-Pillau in Germany. It was a heat MacNeice's summer *Zoo* story professed not to care much for:

> we have still got to face the deplorable fullness of August . . . The air will become denser with petrol fumes, the arterial roads with traffic. Hundreds of thousands of shirts will stick to the backs of hikers. Girls with peeling noses will extract pebbles from their shoes. In the hot-houses of the big stations there shall be three suitcases or packets to each perspiring pair of hands. String will cut into the fingers, throats will be parched, ears deafened, eyes sore from the dust. Chocolate will run in the pockets and asphalt stick to the shoes. Some will neck, some will grumble, some will forget the salt. For myself, I prefer Monkey Hill.

August was indeed the people's month: a time off, a time for holidays and Zoos and conferences (the Soviet Writers' Conference occurred in August 1934), for the pleasures of the interlude (Mass-Observation was kept busy in August 1938 noting the open-air dancing phenomenon, when the London County Council opened its parks for this form of mass summer recreation).

But it was also in August that the insecurity of '30s interludes was revealed. Every August in that decade reminded you inevitably that it was in August that war broke out in 1914. August had a way of going wrong. In August 1931 Ramsay MacDonald formed his National Government. The Spanish Civil War broke out in July 1936 to become another August war. The attack on Hill 481 on the Ebro, described by William Rust in his *Britons in Spain* as the British Brigade's 'toughest action', culminated in 'the final and most furious' (and unsuccessful) 'assault of all' on 1 August 1938 ('somebody mentioned that it was Bank Holiday Monday . . . For many . . . there were to be no more Bank Holidays'). A few days later on 12 August 1938, Germany mobilized in the crisis that crescendoed in Chamberlain's Munich Agreement in September. A year later the end was even nearer. The sense of crisis was again acute: the Hitler–Stalin non-aggression pact was signed on 23 August 1939. A 'week of severe crisis has begun', wrote David Gascoyne on the 22nd; 'For me, the interior crisis continues, more intense than ever. Labyrinthine. Incoherent. Inarticulate.'

> Read the advertisements for seaside lodgings,
> Bungalows to let and Mediterranean cruises.
> This is the type of the English summer.
> But will this summer run true to type?
> The official announcers would never mention them
> But there are certain factors to be considered.
>   First, there is a war about to be declared.

Thus the Listener-In of MacNeice's *Out of the Picture*, seizing the microphone to broadcast a debunking talk entitled 'Summer is A-Comen In'. The holiday enclosures, like the fragile frontiers of Europe, just couldn't keep out the hostilities you were trying to evade by being or going within them. The would-be still centre had a way of filling up with the din of war. 'I have become', regrets Spender's Judge as the Fascists invade, 'The centre of that clamorous drum / To which I listened all my life.' Auden found Iceland heavy with Nazi implication:

> I caught the nine o'clock bus to Myvatn, full of Nazis who talked incessantly about Die Schönheit des Islands, and the Aryan qualities of the stock 'Die Kinder sind so reizend: schöne blonde Haare und blaue Augen. Ein echt Germanischer Typus'.

This had already been Grigson's experience. 'As Wystan Auden and Louis MacNeice found sometime after, it was in landscape a counterpoint of the time, a barren and intimidating introduction to war, a land by which one was dwarfed, which was at once hot with underground fires and cold with immense winds.' The place described in Grigson's prose-poem 'A Queer Country' sounds like that surrealistically menacing and Nazified Iceland: 'The geysir goes off when fed with a still-born infant or soap'; 'The death-rate much exceeds the birth-rate'; 'They . . . have very fine blonde hair.' The ship on which Graham Greene sailed home from Mexico (*The Lawless Roads*) filled up with Franco-ists, more and more of them coming into the open as the journey proceeded, donning their uniforms, saluting stiff-armed, bursting into the Falangist hymn after Mass. Another island shattered:

> The shadow of the Spanish war . . . one couldn't expect to escape it in a German ship calling at Lisbon.

Auden's birthday poem for Isherwood, 'August for the people and their favourite islands', written in August 1935, starts by celebrating an islanded August holiday ('Daily the steamers sidle up to meet / The effusive welcome of the pier') in which people 'live their dreams of freedom', by the sea, 'laid here in the sun'. It recalls conversations on the Isle of Wight nine years earlier ('that southern island / Where the wild Tennyson became a fossil'), and then a Baltic summer five years after that. So the poem enfolds islands within islands, holidays within holidays, but to little protective avail now because the outer world of rampageous immorality (Scandal, Falsehood, Greed) and defeated virtues ('all Love's wondering eloquence debased / To a collector's slang') has penetrated the dense holiday carapaces, smashing the poet's former detachments:

> Louder today, the wireless roars
> Its warnings and its lies, and it's impossible
> Among the well-shaped cosily to flit,
> Or longer to desire about our lives
> The beautiful loneliness of the banks . . .

The still centre, however apparently far off from quotidian distresses, however southern and sunny it was, could not in practice hold out against the grimly real. On his holidays Grigson was perturbed by hearing the intrusive voice of Goebbels on a wireless in sunny Spain, by seeing in Munich a railway sign for Dachau, by the Fascist Exhibition in Rome, by a sight of Goering in Dubrovnik: 'Black intimations pushed themselves . . . into the sun'; 'Enjoyment in these years . . . was haunted by the spooks of Europe.' This is the point of Isherwood's heavily ironic island story 'A Day in Paradise', first published in the Teachers' Anti-War Movement magazine *Ploughshare* (April–May 1935)—a magazine Edward Upward was closely associated with (the story was republished in *Exhumations*). The guidebook's picturesque holiday claims are mocked and refuted by a Spanish-speaking island's slums, sick children, a barbed-wire enclosed military zone, and communist and pacifist graffiti. And it's the pastoral island vision itself, Isherwood's narrative concludes, that's delusively at fault:

> And if, just for a moment, passing through the slums of our native town, we are reminded of that filthy lane down by the port; if we remember, for an instant, our first dismayed impression of the island before the guide-book began to reassure us, of a sinister, squalid place where human beings are living as no animals should be allowed to live; if, seeing a poster in the street, we think of *No Mas Guerra!* scrawled on the whitewash and wonder whether, perhaps, there is some truth in it; whether the struggle against hunger and war isn't more immediate, more universal than we dream: then quickly the image of the island, as the travel-bureau and the guide-book present it, will slide, like a brightly-coloured magic-lantern picture, between us and our real memories, and we shall repeat, as hundreds of others have repeated: 'Yes, indeed . . . it's a paradise on earth'.

Isherwood wrote, of course, as the man for whom island-hopping had turned nightmarish. But whoever (s)he was it was difficult even for the note-taking, material-gathering writer not to feel the tripper's guilt on an islanded holiday, the guilt of the bourgeois slummer using his/her 'long holidays' on C. Day Lewis's advice (in *New Country*) to 'investigate the temper of the people'.

> For Europe is absent. This is an island and therefore
> Unreal.

So Auden in 'Journey to Iceland', mindful no doubt of his earlier gibes in *The Orators* about 'Robert and Laura spooning in Spain', and the hostility in 'A Communist to Others' towards the Gravesian escapism:

> Unhappy poet, you whose only
> Real emotion is feeling lonely
>   When suns are setting;
> Who fled in horror from all these
> To islands in your private seas
> Where thoughts like castaways find ease
>   In endless petting.

The exotically islanded travel world was as much a falsely magic-lantern world as the escapist dreamhouse of the cinema, or the enclosed scenarios of escapist detective stories: 'This little world of suspended activity' as the Riviera Hotel is described in

Elizabeth Bowen's *The Hotel*, 'just a dream'. Grand Hotels and Luxury Liners, islands and the mass media were all guilty of providing delusive dreams of paradise. Hotels were, as we have seen, built to look like liners; BBC Broadcasting House was made to resemble the Queen Mary; Auden and Isherwood's Nineveh Hotel, Paradise Park of the rich, apotheosis of such fictional luxury hotels of the period as the ones in Vicki Baum's *Grand Hotel* and Arnold Bennett's *Imperial Palace*, is aptly the residence of Lou Vipond the film-star. Debunking escape, Cedric Belfrage not only visited the South Sea Islands, spoiled by traders and missionaries, but fetched up with appropriate hostility in Hollywood, the dream-factory itself (he dished out his sternly unescapist medicine again in his wry Left Book Club Hollywood narrative *Promised Land*, 1938). Freedom on Hollywood terms was no freedom: offered that sort of escape you needed, as the last section of Belfrage's *Away From It All* has it, to 'Escape from Escape'. That is presumably why '30s writing continued to regard its islands and hotels, as well as its cinemas, ambivalently. They were prisons as much as paradises. Elizabeth Bowen's *The Hotel* is as full of prison bars as of pleasures; Isherwood's fictional island contains an ominous wall and barbed-wire enclosure, just as his Ruegen Island becomes a resort of Nazis; Auden fears lest 'no one goes / Further than . . . the ends of piers'. The last of the China Sonnets has Auden—and us, the readers, swept into his collective 'we'—sighing

> . . . for an ancient South,
> For the warm nude ages of instinctive poise,
> For the taste of joy in the innocent mouth.

But 'we' are stuck now—oddly, both by choice and by necessity—among cold and mountainous hardships, sleeping in huts, 'A mountain people dwelling among mountains.' 'There is no change of place', in fact, as Auden wrote in 'Who Will Endure': escape attempts were merely a change of prisons.

It was a thought that came as readily to dyspeptic observers of the period's new holiday camps as it did to MacNeice among the summer Zoo visitors and to disillusioned travellers and fellow-travellers to Russia and to Spain. It's impressed vividly on the reader of *A Handful of Dust*. Getting away from England and Brenda only lands Tony in a crazy, South American prison of Dickens-reading: 'Du Côté de Chez Todd', the imprisoning chapter's title in the novel locks it firmly into a mirror relationship with the opening 'Du Côté de Chez Beaver'. What's more, the search for island, holiday, pastoral tended to have the escapist end up back where he'd started, once the dream bubble burst and he debouched from the cinema into the street or the end of August returned him to the old workaday reality. The 'hands of the clock move on to the dangerous morning' in Allott's 'Christmas, 1938'; Auden's Liner sails inevitably towards its testing 'Tomorrow'; one has to say 'Goodbye to the Island' (that's the title of one of Plomer's poems in *Visiting the Caves*, 1936); holiday '*Joie de vivre*' is 'contraband' as MacNeice's *Autumn Journal* 'people' re-enter the working world, 'to face the annual / Wait for the annual spree' and to be solaced until next year by the little pastorals of 'films or football pools', 'the gossip or cuddle'. In other words, the end of the holiday 'dream atmosphere' (as it's called at the end of Nicholas Blake's holiday-camp novel, *Malice in Wonderland*) brought a rude awakening back into the old inescapable horrors. Just as, curiously, the war-torn '30s break from First War horrors ended at the end of August. Poland was invaded on

1 September 1939. Two days later Britain and France declared war on Germany. Julian Symons has dwelt (in his *Notes from Another Country*, 1972) on the disquieting sense of an ending heavy on the last number of *New Verse* when it came out at the end of August 1939. Auden's poem 'September 1, 1939' celebrates the end of 'a low dishonest decade': death is now intruding conclusively into 'our private lives'. And it does so in 'this neutral air' of New York, another travellers' destination, another holiday refuge that is proving, like all the others, no refuge at all. It was a grim consummation, oddly anticipated a year earlier in MacNeice's *Autumn Journal*, a poem gloomy with the wars and bombs that have all succeeded August: 'summer is ending in Hampshire', 'August going out', 'August is . . . over', and 'the new valkyries ride'. On 28 August 1938 Forster wrote to Isherwood confirming this coincidence of public and private woes: 'this Dover muddle', Ackerley's expulsion by his landladies, has occurred as the 'Munich' Crisis mounts ('alternating between gloom and resolute cheerfulness'; 'in a queer stasis', Forster reported to Day Lewis, a month later, 30 October 1938). Earlier yet, the dog that bombs down into the Edenic Riviera sunshine of Anthony Beavis and Helen Ledgwidge in Huxley's *Eyeless in Gaza* (1936) does so on 30 August 1933: 'bang, like a sign from heaven, down comes the dog!' A sign of the inescapable problems implicit in travel and summer that were anticipated still earlier in Elizabeth Bowen's *To the North*, whose deadly conclusion makes an apotheosis of 'The Summer Rush'. In the chapter of that novel called 'Wet Summer', Emmeline's agency offers southern suns as the antidote to English rains and as relief to minds 'on which a world's apprehension, strain at home and in Europe, were gravely written'. But she and Markie make an anti-August journey, north towards the 'cold pole', eschewing the offer of still centres that cannot pacify but only emphasize the general decentering, and enacting in their smash-up the collision course that the '30s holiday-mongers all sooner or later felt they were on. It was a realistic cooling towards the travel solution, one might feel: a cooling shared by T. S. Eliot, of course, for whom salvation lay precisely in the cold comforts of Christianity. 'A cold coming we had it . . . The very dead of winter.' So begins 'Journey of the Magi' (1927), adapting Bishop Lancelot Andrewes' Christmas Day sermon of 1622: journeying to Christ would be done unescapologically in the freezing depths of a traditional British winter: 'Just the worst time of the year / For a journey.'

Despite all this, a lot of '30s writers tried to maintain a sense of the frivolity of travel. Desperate efforts were made to insulate travel writing as an island of fun amidst the torrents of earnest '30s literature. There were masses of 'Bright Young Baedekers'. That phrase was Brian Howard's in a contemporary self-debate about what he might write ('About Writing': published for the first time in Marie-Jacqueline Lancaster's biography). 'Everything drives me to write a book', complains Howard's alter ego Russell:

> Money. The consciousness of not keeping up my position as a clever young man. The necessity of not disappointing one's father too long. Yet—what? A travel book? Every young man I know writes a travel book . . . observing how ugly beauty spots are. That travel book! The clever young man's travel book. A list of his irritations.

'The clever young man's travel book' debunked abroad. Waugh's first '30s travel book *Labels* set the tone for a whole slate of Bright Young Baedekers. Its purpose was to investigate 'with a mind as open as the English system of pseudo-education allows, the basis for the reputations these famous places have acquired'. What this meant was jeering steadily at them. England is full of pseudery and rotten pongs, of course, but—and Waugh's observations of this kind became formulaic—abroad is far worse. Bogus is a key word. Paris is 'bogus'; Cocteau's art is 'the apotheosis of bogosity' (Brick Top, the lady cabaret owner is 'the least bogus person in Paris'). Virtually no foreign prospect pleases. Travelling itself is awful: 'A railway journey is always disagreeable to me'; the Monte Carlo Casino is 'like Paddington Station in the first weeks of August'. Waugh's abroad is choked with absurdities, where bubbles are pricked in disgruntled cataracts of increasingly hysterical prose. How come the Sphinx got its reputation? 'As a piece of sculpture it is hopelessly inadequate to its fame . . . It is just about as Inscrutable and enigmatic as Mr Aleister Crowley.' 'Nothing I have ever seen in Art or Nature was quite so revolting' as Mount Etna at sunset. Every 'single example of Mohammedan art, history, scholarship, or social, religious or political organisation' is derisory. And if Waugh feels Arabic art is rubbish, 'this feeling is intensified and broadened a hundred times in relation to everything Turkish.' In Constantinople Christian art has been degraded everywhere by 'vile Turkish fripperies'. Waugh's chauvinism is unbounded. He suffers this outrageous exposure to foreigners only for the sake of his travel book. It's the duty of 'the young men and women who manage to get paid to write travel books' to 'have as many outrageous experiences' as they can:

> To stand for hours in a draughty shed while a Balkan peasant, dressed as a German staff officer, holds one's passport upside down and catechises one in intolerable French about the Christian names of one's grandparents, to lose one's luggage and one's train, to be blackmailed by adolescent fascists and pummelled under the arms by plague inspectors, are experiences to be welcomed and recorded. But . . .

So the chauvinist retreats into edgily raw jesting: about deck tennis, for instance: 'Some played with such vigour and persistence that they strained their backs and arms, slipped on the deck and bruised their knees, chafed raw places in the skin of their hands, struck each other in the face, twisted their ankles, and sweated profusely.'

The traveller on the Waugh plan becomes a mere scribbler 'down in my notebook' of the potentially amusing or deridable foreign event, someone relentlessly driven to see jokes and to poke fun, to prove his breezy English superiority. And Waugh's amused revulsion and steely snobbery became pervasive among the journeying Old Boys. Peter Fleming picked up Waugh's grinningly bad habits; even Robert Byron, a more serious critic and historian than either Waugh or Fleming, didn't escape the contagion (and, intriguingly, Waugh was harsh towards both of them in reviews). One can't, it's implied, respect the USSR if, as Robert Byron alleges in *First Russia Then Tibet*, the only industry to meet its Five-Year-Plan target is the Leningrad spats factory, and if Muscovites are to be caught scrabbling, as they are, for torn camisoles and threadbare galoshes in the Moscow Flea-Market: 'One man, as we passed, thrust a single spat at us.' And what price the Acropolis (*Labels*) when it turns out to be off-white, in fact 'pale pinkish brown', like 'the milder parts of a Stilton cheese into which port has been poured'? And what about the baroque Portuguese-style statuary of Rio, hyperbolically acclaimed in Fleming's *Brazilian Adventure* (1933)?

Victory has got a half-Nelson on Liberty from behind. Liberty is giving away about half a ton, and also carrying weight in the shape of a dying President and a brace of cherubs. (One of the cherubs is doing a cartwheel on the dying President's head, while the other, scarcely less considerate, attempts to pull his trousers off.) Meanwhile an unclothed male figure, probably symbolical, unquestionably winged, and carrying in one hand a model railway, is in the very act of delivering a running kick at the two struggling ladies, from whose drapery on the opposite side an eagle is escaping, apparently unnoticed. Around the feet of these gigantic principals all is bustle and confusion. Cavalry are charging, aboriginals are being emancipated, and liners launched. Farmers, liberators, nuns, firemen, and a poet pick their way with benign insouciance over a subsoil thickly carpeted with corpses, cannon-balls and scrolls. So vehement a confusion of thought, so arbitrary an alliance of ideas, takes the reason captive and paralyses criticism. But you cannot help feeling that such vigour of conception is hardly calculated to make for stability in execution; the thing *must* be top-heavy. You flinch. You tend to cower. It is with a feeling of relief that you turn the corner at the square.

This is precisely the kind of indecorously anarchic excess that Evelyn Waugh is always finding in the lives and works of the foreigner, among whom even the most overtly civilized turn out to be little more than wogs when measured by the high standards of the British Public School. And the wogs have got away too long with their insolent belief that rude incompetences, pretend accomplishments and a thin layer of civilized polish are more than a push away from delight in the missionary's glass beads. Opposed by 'two sentries and a posse of comrades' who suspect 'a professional *saboteur* sent by the British Government to upset the paint-front in 1932', Robert Byron effects entrance to a Russian church turned paint-factory by offering 'them each a gold-flake'. Among the Brazilian Indians Fleming's party

> dispensed the riches of Woolworth's with an air of magnificence. Necklaces and mirrors and knives and forks, and some silly toys—little white horses which were supposed to draw a sort of bicycle bell on wheels: only of course the wheels wouldn't go round and the bell wouldn't ring.

Mr Winter, the prospector described by Waugh in *Ninety-Two Days*, recruited his native labourers with such trinkets. 'He had a great success shortly after Christmas with some mechanical mice, emerald green drawers, and a gramophone.' The mice turn up again, large, spotted green and white, on wheels, in *A Handful of Dust*, and, instead of enticing help for Last and his distressed colleague from the Indians, these crude gadgets scare them right off. Such trinket incidents are, of course, intended only to reflect adversely on the recipients of the 'trade goods': however they respond, positively or negatively, they prove their savage stupidity. Fancy being bribed by a Gold Flake cigarette! Fancy being scared by a 3*s*. 6*d*. wind-up mouse! Nothing is conceded to the native population. The Indians in *A Handful of Dust* admire the look of a Boat Race rosette. Their repeated spitting, including the Pie-wies' spitting on Dr Messinger's head to make him a blood-brother (it was rumoured, reported Waugh in *Remote People*, that the Abyssinian Church consecrated its bishops by spitting on their heads), is made to look akin to Mrs Last's spitting into her mascara. But these facts do not qualify the foreigner's savagery, they imply only deficiencies in some of the English. Foreign natives remain stubbornly subordinate to what these travellers imply are proper standards of behaviour and intelligence. In other words, the traveller on the Waugh and Fleming plan is no anthropologist: no author he, unlike Tom Harrisson, of a book entitled *Savage Civilisation*. He's on the look-out simply for freakish experience snappily to record in order to jazz up his prose into best-selling stuff. He will be distantly satirical, toughly unmoved, egregiously superior to what he goes through: the child, in other words, of an age of slummers and airmen, in which even heavy intellectuals like the formidably intelligent William Empson kept up a cultic affection for Anita Loos's *'Gentlemen Prefer Blondes'*, the novel about the American girl who remains completely *herself* while abroad (see Empson's poem 'Reflection from Anita Loos').

Robert Byron's desire to show up Russian barbarism is understandable, given his political hostilities. So are Waugh's kept-up attempts in *Remote People* (1931) and *Waugh in Abyssinia* (1936) to undo British sentiment about the so-called civilization of Abyssinia and other parts of Africa. And Waugh's fictional touch for the absurdities of comic foreigners is very powerful: his novel *Black Mischief* (1932) is hilarious. But Waugh's theme hereabouts is the serious business of cultural imperialism; he's confronting the terrible disruptions that have occurred throughout Africa by the crude layering of modern upon medieval, white upon black, Western Christianity upon African paganism, Boy Scouting, school blazers and Hollywood movies on to societies scarcely prepared for their adequate consumption.

> We passed [*Labels* again] a game of football, played enthusiastically upon an uneven waste of sand, by Egyptian youths very completely dressed in green and white jerseys, white shorts, striped stockings, and shiny black football boots. They cried ' 'ip–'ip–'ooray' each time they kicked the ball, and some of them blew whistles; a goat or two wandered amongst them, nosing up morsels of lightly buried refuse.

But it's scarcely an adequate response to this social and political mess to keep finding Africa merely an uproarious shambles, hilariously disgusting, a perpetual sick joke. Those soldiers, in *Black Mischief*, one more lot of victims of Emperor Seth's loony modernizations, who, unused to boots, cook and eat a thousand pairs of them are plunged, amusingly, into a lively Chaplinesque farce

> Cook-pots steaming over the wood fires; hand drums beating; bare feet shuffling unforgotten tribal rhythms; a thousand darkies crooning and swaying on their haunches, white teeth flashing in the fire-light.

One has only to recall, by way of (a large) contrast, George Orwell's dealings with colonialism in *Burmese Days* (1934), or how the native prisoner who is led out to his execution in Orwell's 'A Hanging' (1931) becomes startlingly a human being—he's shown not wanting, just as no white man would want, to wet his feet even on his way to the gallows, and is not allowed to remain simply a *darkie* (or, to quote other favoured bits of Waugh rhetoric, a *nigger* or a *yid*)—to perceive something of the limitations of Waugh's method.

In Waugh circles only the British will do. So *bogus* Britishness is derided with particular determination. 'Very Olde Scotts Whiskey', a local brew, offends Waugh in a bar in Harar (*Remote People*). Just so Fleming pokes fun at Japanese-made 'Queen George Old Scotch Whisky' (*One's Company*). And Americans were evidently felt to stand to Brits as those bogus tipples stood to the real Scotch. The tourists in *Labels* ('this whole ragtag and bobtail of self-improvement and uplift . . . bruised and upbraided by the thundering surf of education'), Professor W. in *Remote*

People ('an expert of high transatlantic reputation on Coptic ritual' who is continually wrong during Haile Selassie's coronation service), Robert Byron's Mr Boggins the sufferer from phlegm, the magazine authoresses Evangeline Crossfoot Putz and Delia Olssen Dufflebury mocked by Fleming in the American chapter of *One's Company*: no North American is spared. How dare they pose as sham Britons, our language misleadingly on their lips ('The tongue of Shakespeare and *The Saturday Evening Post* is good enough for us', boasts the American Minister in Powell's *Venusburg*)?

In fact the youthies care little enough for any other travellers, particularly travellers who are, like the Americans, more earnest than they. The literary tradition (epitomized by Maugham's classic short-story 'Rain', first published in 1934) that harshly scathed missionaries, and that feeds into the gentler scepticisms of Sylvia Townsend Warner's missions-undoing novel *Mr Fortune's Maggot* (1927), the unstudied reaction that crops up in passing as though it were the most natural of feelings in Orwell's *Wigan Pier* (that 'ass of an American missionary, a teetotal cock-virgin from the Middle West' who pities Orwell's complicity in imperialist violence), or in Isherwood's *Lions and Shadows* (where a mock examination answer declares 'One cannot be a missionary nowadays—as a class they have been too much exposed; and, indeed, the demoralizing effect of having to wear boots in the tropics must be very serious'), this tradition peaked sturdily in '30s travel books. Missionaries get repeated mockings from Waugh and Fleming, especially the Protestant and the American ones (after *Labels* and his conversion Waugh reserved a soft spot in his travel books for propagators of Catholicism). Missionaries catch Fleming out (*One's Company*), always spoon in hand as they begin to say Grace before they eat. A Seventh Day Adventist missionary turns down an alcoholic drink Waugh offers him (*Remote People*), with an 'Oh, no, thank you' whose 'four monosyllables contrived to express first surprise, then pain, then reproof, and finally forgiveness'. Fleming finds large lady missionaries peculiarly repellent ('unusually well-developed woman . . . clad only in a pair of very tight shorts and a dirty white blouse'; 'a great globular European woman . . . face like a boot . . . clad in dark blue shorts and a sorely tried blouse'). He jeers at their *Redemption Songs* (' "Oh Tsidkenu!" is a favourite one') and at a Mr Titherton's 'straight talks' to God: 'Miss Tackle, Miss P. Flint (*I know those two ladies, Our Father. Please look on them today. They're two of the very best, I can tell you).*'[237] You're not at all surprised when the same earnest troupe of comic butts is wheeled on again in Auden and Isherwood's *Journey to a War*: the secretly smoking and drinking Baptist medics, the fox-terrier who jumps for cake only when asked if he's an 'American Baptist', religious professionals irkingly familiar with the Almighty and His Will ('Leave it to Jesus'):

> This morning Auden went again to the hospital and returned in a state of delighted fury against the lady missionary. Hearing that we were off to the front she had said: 'Are you insured with Jesus? Jesus has positively guaranteed eternal life . . . This life' (holding up her thumb) 'is just a teeny span.' Auden wishes he had bitten it.

Nothing was sacred, and the more earnest modes of travel writing bit the same satirical dust as the earnest travellers. The longwindedly maintained joke of Fleming's *Brazilian Adventure* is continuously to refuse to be a 'Jungle Epic'. 'This book is all truth and no facts . . . probably the most veracious travel book ever written; and . . . the least instructive.' It's 'Jungle Lampoon' in fact, perpetually maintaining the 'atmosphere of caricature' by guying a whole roster of available styles of travel reporting. There's the 'Travel and Adventure' code:

> If Indians approached us, we referred to them as the Oncoming Savages. We never said, 'Was that a shot?' but always, 'Was that the well-known bark of a Mauser?' . . . We spoke of water always as the 'Precious Fluid', we referred to ourselves, not as eating meals, but as doing 'Ample Justice to a Frugal Repast'.

'From my youth up', Fleming boasts, 'I have lost no opportunity of mocking what may be called the Nullah (or Ravine) School of literature':

> Whenever an author thrusts his way through the *zareba*, or flings himself down behind the *boma*, or breasts the slope of a *kopje*, or scans the undulating surface of the *chapada*, he loses my confidence.

The arch-poseur, supreme in the part of undeterredly phlegmatic Englishman (pooh-poohing dangers, insistent that 'even quite a bald account of a bicycle ride out from Oxford to Boar's Hill' would impress on 'the inhabitants of the Matto Grosso' a sense of peril), Fleming is an expert in all the possible poses and their related styles: *Times* reporter ('I was filled with a kind of forlorn glee when I reflected that all this would one day have to be translated into that impressive, non-commital prose, with its slight technical flavour, in which the activities of explorers are recorded in *The Times*'), *Wide World Magazine* contributor ('one of the Indians, advancing with a shy smile, rubbed noses with me; I suppose he had been reading the *Wide World Magazine*'; 'The beginning of the next day was straight out of the *Wide World Magazine*'), *Boy's Own* novelist (a scrub fire reminds him of a prairie fire pictured in an adventure book he read as a child), popular journalist ('what on another man's lips would have seemed trite and intolerable patter about Romance and Adventure and the Great Open Spaces was on Harman's real and true and easily acceptable. Here at last was that mythical figure, the Adventurer; it was the World's fault, and not his, that the phrases in which he revealed himself were hollow and hackneyed and stank of the Sunday papers during the hiking season'). Fleming has seen through every sort of travel writing ('wringing hands, exchanging gruff facetiae, turning abruptly on our heels . . . The partings of explorers are sweet sorrow for the reading public . . . our expedition was meticulously loyal to the best traditions'). In *Brazilian Adventure*, what Fleming calls the 'atmosphere of caricature' is total. He professes to find that atmosphere 'congenial and comforting'. 'The private code of nonsense' he and his chums maintained was, he claims, 'our chief defence against hostile circumstance'.

Fleming admits that 'To anyone who did not think it as funny as we did it must have been an intolerably tiresome kind of joke.' It's difficult to see how any reader of *Brazilian Adventure* could keep such a sense of tiresomeness at bay. Harold Nicolson 'had imagined that the literature of negation could go no further than Aldous Huxley. Mr Fleming carries it further'. Auden and Isherwood, though, were only too eager to join in Fleming's kind of youthily brash iconoclasm. Prankish ('screaming with laughter . . . singing in high falsetto or mock operatic voices'), taking irreverent precautions against constipation, eager for camp amusement of the kind the photographers Capa and Fernhout (camera-man on *Spanish Earth*) had generated on the voyage out ('with their horse-play, bottom-pinching, exclamations of "Eh, quoi! Salop!" and endless jokes about *les poules*, they had been the life and soul of the second class'), another case of boastful naïvety ('We spoke no Chinese, and possessed no special knowledge of Far Eastern affairs'), Auden and Isherwood slipped alertly into the Fleming tradition. In fact, their chance meeting with him proved a glorious satiric opportunity. For Fleming himself was now the measure for iconoclasts to delight in falling short of. 'In his khaki shirt and shorts, complete with golf-stockings, strong suede shoes, waterproof wrist-watch and Leica camera, he might have stepped straight from a London tailor's window, advertising Gent's Tropical Exploration Kit.' Fleming was the cool master of thresholds ('Fleming supervised our departure with his customary efficiency. One saw his life, at that moment, as a succession of such startings-out in the dawn'). His style was the most handily mockable one:

> Laughing and perspiring we scrambled uphill; the Fleming Legend accompanying us like a distorted shadow. Auden and I recited passages from an imaginary travel-book called 'With Fleming to the Front'.

'Well', said Auden, when Fleming finally left them, 'we've been on a journey with Fleming in China, and now we're real travellers for ever and ever. We need never go further than Brighton again.'

Determined irony and iconoclastic jokiness carried deliberate outrageousness right into the meat of the youthies' travel books. It was not only the traditions of travel books and magazines that were being flouted: the traditional contract of mutual respect between writer and reader, the idea that readers should get their money's worth, was continually being put at stake. The readily available contract between publisher and writer had, of course, been honoured on the publisher's side. The writer had had *his* money's worth, had enjoyed his advance, his paid-for jaunt. But when it came to producing his manuscript he extended his high-spirited casualness into self-indulgent reader-enraging confessions about how awful he found the task of actually writing this stuff. Waugh's *Ninety-Two Days* begins with complaints about having to buckle down to the drudgery ('lugubrious morning', 'day of wrath', 'the coming, miserable weeks'). Opening the book they'd just purchased, readers presumably did not want to come on starters like 'Who in his senses will read, still less buy, a travel book of no scientific value about a place he has no intention of visiting?' But lots of the youthful Baedekers are filled with such obvious devices for filling the page, for fulfilling the contract with the publisher any old how. *Letters from Iceland, I Crossed the Minch, Journey to a War* are opportunistic rag-bags. The '30s travel book had ousted the novel as writing's loosest, baggiest, most monstrously capacious form. Evelyn Waugh was scarcely the man to complain loudest, but in his review of *Letters from Iceland* he professed anxiety about the decline of the travel-book at the hands of publishers over-eager to 'cater for a growing taste in the semiliterate public for vicarious locomotion'. Because Byron, Fleming, and Patrick Balfour are good at travel writing 'the legend has grown up that it is only necessary to send an author abroad to compel him automatically to composition; writers in need of a holiday find that it can always be obtained in exchange for a contract. How burdensome these contracts can become is evident in *Letters from Iceland*.'[238] (Waugh knew all about such contractual burdens, of course.) Auden and Isherwood have, Waugh sneers, flinched from Iceland ('a tough job' for a book); they've gone in mostly for 'making weight' with typographical excesses, photographs, 'rough Byronic verses', travel hints, and 'the letters with which they had whiled away the time during their tedious little holiday'.

Waugh omits to mention Auden's discussion of travel writing. It was too close, perhaps, to Waugh's own padding devices in *Ninety-Two Days*. Self-reflection as filler never come more blatantly:

> Every exciting letter has enclosures,
>   And so shall this—a bunch of photographs,
> Some out of focus, some with wrong exposures,
>   Press cuttings, gossip, maps, statistics, graphs;
>   I don't intend to do the thing by halves.
> I'm going to be very up to date indeed.
> It is a collage that you're going to read.

Nor was arch wandering from the point ever done so pointedly, culminating in a taunting jest about the entire randomness of the word *point* itself ('no other rhyme except . . .'):

> A publisher's an author's greatest friend,
>   A generous uncle, or he ought to be.
> (I'm sure we hope it pays him in the end.)
>   I love my publishers and they love me,
>   At least they paid a very handsome fee
> To send me here. I've never heard a grouse
> Either from Russell Square or Random House.
>
> But now I've got uncomfortable suspicions,
>   I'm going to put their patience out of joint.
> Though it's in keeping with the best traditions
>   For Travel Books to wander from the point
>   (There is no other rhyme except anoint)
> They may well charge me with—I've no defences—
> Obtaining money under false pretences.

'Beside the major travellers of the day' Auden can only offer disclaimers ('I am no Lawrence'; 'I am not even Ernest Hemingway'; 'I'm not like Peter Fleming an Etonian') and stick to jestful sparring with readers

> So this, my opening chapter, has to stop
>   With humbly begging everybody's pardon.
> From Faber first in case the book's a flop.
>   Then from the critics lest they should be hard on
>   The author when he leads them up the garden,
> Last from the general public he must beg
> Permission now and then to pull their leg.

Auden deserved Waugh's chidings. So did Waugh. Bloodied, MacNeice refused to bow to Waugh's criticisms. It must have vexed MacNeice to be sneered at as a mere filling-device himself. 'I suspect', Waugh smirked, 'that his name was inserted to give a rhyme for "peace". (Mr Auden everywhere has difficulty with his rhymes. How lucky that he did not take his former collaborator, Mr Isherwood, on this jaunt).' So, defiantly, MacNeice helped fatten out *I Crossed the Minch* with 'Hetty's' talk of *Memoirs from Greenland* and its unsympathetic reviewers ('Let 'em go' to Greenland, 'I say, and see what they can do with it'). The reviews include one in *Dawn and Twilight* by Evelyn Priest, 'the man who used to write novels'. MacNeice, for one, was determined to keep playing the travel book game in the egregiously cheeky way. Innocence abroad, all clowning, caricature, light verse and *je m'en foutisme*, would still try to keep up its boyish self-image in the actual travelling about, and in the writing afterwards about travelling about.

But, of course, in the same way as thresholds would keep proving so terrifying and the bounds of still centres turn out to be ineffectual against nasty intrusions, so the jokes kept changing into sick jokes. Travel would keep on putting on its serious face. The real desperateness of Fleming's adventures in *Brazilian Adventure* is perceivable even amidst the steady hail of debunking. Waugh cannot much alleviate the mordant rigours of the journey described in *Ninety-Two Days*. What's more, none of the seemingly light-hearted reflections on the difficulties of writing travel books ever sounds entirely un-anxious: they continually have a way of reminding one of the real problems of all forms, all arts of writing. 'Of all possible subjects', Auden declared in his Introduction (1946) to Henry James's *The American Scene*, 'travel is the most difficult for an artist.' Waugh's frequently annoying text-side chats with the reader of his travel books blend, almost despite themselves, into those few moments in his fiction that are, formally speaking, more obviously serious disrupters of traditional narrative's claims on realism: the two endings of *A Handful of Dust*, or the narratorial reflections on the nature of Paul Pennyfeather's existence in *Decline and Fall* (Part II, Ch. 2, 'Interlude in Belgravia': 'the shadow that has flitted about this narrative under the name of Paul Pennyfeather materialised into the solid figure of. . .'; 'In fact, the whole of this book is really an account of the mysterious disappearance of Paul Pennyfeather, so that readers must not complain if . . .'; and so on). In fact the formal self-regard of '30s travel books, which turns them into a distinctive body of anti-texts, does a lot to help fill out the period's otherwise meagre ranks (Joyce, Woolf, Beckett, Durrell, Stevie Smith) of self-conscious, self-mirroring writings. On inspection the travel book turns out a much less naïf affair than it at first appears: it was perhaps no more innocently abroad in the field of publishing than its propagators were innocent of the wide-world's wickedness.

No one pretended, of course, to greater innocence than Christopher Isherwood—boyishly small, cheerfully urchin-like, readily playing the nice nephew to every uncle in sight. Bill Bradshaw is contrived for us as the archetypal innocent abroad, surrounded by narrative discretions, a naïf who is meant, apparently, to be slow at seeing through Mr Norris, and to come unscathed through most of the footsy-playing, hand-fondling, and offers of not-so-platonic friendship to which he is readily subjected. Chris Issyvoo actually claims the innocent eye of the camera. But it's as difficult to believe in Bradshaw's innocence as to accept in practice the objectivity of cameras. Isherwood seemingly believed that he was keeping his own homosexual proclivities covered up in *Mr Norris*, that the reader would not spot Bradshaw's author's excited attention to the louche, kinked sexual life of Berlin and of Mr Norris and his contacts. But it's Bradshaw's pretend innocence, in fact, that so blunts and mutes *Mr Norris*; just as it's the pretence that Sally Bowles is somehow attractively forgiveable within her acknowledged degradation—preserved by her Englishness perhaps: Jackson-Bowles to her author's Bradshaw-Isherwood—that won't finally pass muster.

Ordinary readers of *Mr Norris* and *Goodbye to Berlin* weren't informed, of course, that Isherwood had travelled to Berlin specifically for permissive sex with boys, nor that he had naturally fetched up in lodgings next door to Hirschfeld's notorious sex-clinic—that black museum of sado-masochistic gear, hospice for sexual freaks, mecca of homosexuals. But readers of Wyndham Lewis's richly disgusted rhapsody on Berlin's decadent night-life in his *Hitler* (1931) would have been well equipped to read between Isherwood's cagier lines. *Berlin Westens* (Lewis enthuses), 'huge grimcrack West End of the luxury night-life wonder-town': 'thrown up by the War out of the earth's bowels, as it were, from sweated cellars, traps, and gutters. It established itself overnight in the Kurfürstendamm, Nollendorf Platz, Wittenberg Platz, Motzestrasse, Tauenzienstrasse':

> The final touch came about two years ago. It is the electrical drum-fire, the high-volted light-bombardment from all sides, that is the finishing stroke. A great campaign, with the popular label *Berlin im Licht!* was inaugurated by the *Asphaltpresse* . . . The spurious germanism of the colossal wagnerian *Vaterland* of Kempinski, along with a thousand other night-circuses, *Negertanz* palaces, *naktballeten, flagellation-bars,* and sad wells of super-masculine loneliness, shining dives for the sleek stock-jobbing sleuth relaxing, and so forth, did indeed most luridly light themselves up and flaunt their names in fashionable electricity, to such good effect that, although Berlin cannot emulate the perpendicular night-scenery of the wan cañons and search-lit altitudes of New York City, it yet does decidedly convey an air of heavy and louche brilliance, as of a really first-class *mauvais lieu.* No city has anything on it as regards the stark suggestions of being the Hauptstadt of Vice, the excelsior Eldorado of a sexish bottom-wagging most arch Old Nick sunk in a costly and succulent rut—and that is what Berlin wanted, if by Berlin is meant that gilt-edged limelit fraction that enjoys *Berlin im Licht.* Paris has nothing to show at all like *Berlin Westens* . . . In harmony with all this, gang-violence in Berlin abounds, the armed *Zuhälter* or ponce fattens and flourishes. Berlin can show what is probably the most oddly unlovely gunmen of the earth . . .

Going to Berlin, being in Berlin, could not possibly be innocent activities. And, sure enough, the fact of political realities, violence, the need to make discriminations between Nazis and Communists, keep intruding by *force majeure* into the Berlin experiences of Isherwood's most ardent would-be naïfs. Such realities had a way of dispelling the most determined frivolities. It was seeing the world, especially in Spain and China, that snapped Auden out of his most monstrously inhumane aloofnesses. The poems he wrote after his jaunt to Spain are suffused with a mounting sense, almost hysterically so, of the worryingness of the world, represented repeatedly in images of travelling into all kinds of danger. 'Spain 1937' itself, the sickly violent 'Miss Gee', 'Victor' and 'James Honeyman', 'Dover', the China-journey sonnets and poems ('The Voyage', 'The Sphinx', 'The Ship', 'Passenger Shanty', 'The Traveller'), the poems about foreign cities (including 'Gare du Midi'), 'Musée des Beaux Arts' (about the indifference of art and the world to the death of the young airman-craftsman Icarus), the sonnets and poems about authors in various kinds of exile ('Rimbaud', 'A. E. Housman', 'Edward Lear', 'Voltaire At Ferney', 'Matthew Arnold'), culminating in his great celebration of Yeats who has entered the final exile of death ('Now he is scattered among a hundred cities': alive only in memories, in 'the guts of the living') and his long meditation on accepting his own exile in New York, 'September 1, 1939': the world has no more powerful body of travel literature to show than this group of poems. And the tone is always earnest, personal, engaged, grown-up; the 'test' now is deadly serious, as in 'The Ship':

> It is our culture that with such calm progresses
> Over the barren plains of a sea; somewhere ahead
> The septic East, a war, new flowers and new dresses.
>
> Somewhere a strange and shrewd Tomorrow goes to bed
> Planning the test for men from Europe; no one guesses
> Who will be most ashamed, who richer, and who dead.

Each stanza of 'Passenger Shanty', the one attempt to jack up these travelling glooms into a more frivolous acceptability, is sucked back into the pervasive despairing:

> The beautiful *matelots* and *mousses*
> Would be no disgrace to the Ballets Russes,
> But I can't see their presence is very much use.

The committed travellers and the travellers in search of some religious or political commitment, were, of course, pretty earnest from the start. They were acting on certainties, or hunches, that what Auden kept calling (after Henry James) 'The Good Place' existed somewhere; as in 'The Voyage':

> And, alone with his heart at last, does the traveller find
> In the vaguer touch of the wind and the fickle flash of the sea
> Proofs that somewhere there exists, really, the Good Place,
> As certain as those the children find in stones and holes?

The location of the Good Place varied. The religious looked for it inside themselves, or in Canterbury or Rome; the political in Berlin, in Rome, in Spain, or in Moscow. The Rightist meccas had their takers, of course; there *was* right-wing travel in the '30s. Wyndham Lewis hung on to his admiration for Hitler and the peaceful intentions of Hitlerized Berlin until very late in the decade. He and Henry Williamson relished the idea of Hitler in power in Berlin. Lewis carried his anti-semitic Nietzscheanism with him to Morocco and displayed it gustily in *Filibusters In Barbary*. Robert Byron and Peter Fleming were natural anti-bolshevists. The Italian invaders of Abyssinia were peaceful men, Lewis thought (*Left Wings Over Europe: or, How to Make a War About Nothing*, 1936). They were simply liberators and engineers. Waugh, too, rhapsodized over Mussolini's engineers (*Waugh in Abyssinia*): the new road to Addis will, he thinks, spread Christian civilization along with the 'eagles of ancient Rome'. Waugh's Roman Catholicism had not only made him a more earnest traveller (a note attached to *Labels* disavows its unregenerate jests about what had since become his new-found Church), it had also helped bring him closer to Italian Fascism. It led him, too (as it led Graham Greene in *The Lawless Roads* journey), to visit Catholic-harassing socialist Mexico—the visit on which Waugh enjoyed the sponsorship of the Cowdray-Pearson empire, peeved at having its oil-interests nationalized. *Robbery Under Law: The Mexican Object Lesson* dutifully mingles rabid anti-socialism with its Catholic pieties. For his part, F. (*Bengal Lancer*) Yeats-Brown hacked his way through the *European Jungle* (May 1935) armed with all the crudely rightist sympathies that Christianity, anti-Semitism, anti-Communism, Mussolini-admiration and pro-Hitlerism could bequeath a political analysis: it was time Britain woke up to her Jewish problem;

Mussolini had shone with human-kindness when Yeats-Brown met him; it would be 'stark lunacy' to fight Germany; the Red Trojan Horse was laden with more enemies than you'd suppose; let's give up 'this Hitler-hate and Moscow-mindedness'. What's more, a handful of political travellers was on the Nationalist side in the Spanish Civil War. But not many. In Spain, as in the period generally, the majority of politically-minded journeyers were precisely 'Moscow-minded'. It was the journey to Spain that provided, as we shall see, the crucial testing of the enthusiasm for foreign socialism that prevailed among so many British writers. But that was an enthusiasm based in and focused most intently on the seeking and finding of its meanings in the distant but also immanent Soviet Union—spiritually the ever-present shadow across all leftist experiences of the Spanish Civil War.

David Caute's *The Fellow-Travellers* (1973) has charted huge swathes of enthusiastic left-wing tourism to the Soviet Union. Every leftist intellectual, trades-union official and writer worth his salt wanted to check up on Utopia. 'For the Experience of a Lifetime Visit USSR (Soviet Russia)'; 'Russia? Why Not Go and See For Yourself!': so ran the Intourist and The Friends of the Soviet Union ads in *LR*. And it wasn't only Leftists who responded. Travel agents Peter and Emmeline in *To the North* discuss 'getting more closely in touch with Intourist with a view to doing more about Russia'. Robert Byron (*First Russia Then Tibet*) felt the modern shaking of values and went East to 'discover what ideas, if those of the West be inadequate, can with greater advantage be found to guide the world'. Moscow, he exclaims, was

> the cynosure of an agitated world . . . the capital of the Union, the very pulse of proletarian dictatorship, the mission-house of Dialectical Materialism. I looked across the river. Before me stood the inmost sanctuary of all: the Kremlin.

Moscow was 'the internationalist Mecca' according to Wyndham Lewis (*Left Wings*). It was Bernard Shaw's 'self-patented Russian Elysium' according to Robert Byron. All those 'who require a "breather" after too prolonged an immersion in the fumes and fogs of "capitalism" ', gibed Lewis in *Filibusters*, 'take the beaten track to Russia'. He himself went elsewhere; but the others queued up for Soviet entry-permits. 'When the opportunity arose to visit the Soviet Union', Charlotte Haldane declared, speaking for her husband J.B.S. and herself in *Truth Will Out* (1949: quickly reissued by the Right Book Club), 'we eagerly accepted it'. 'The entire British Intelligentsia has been to Russia this summer': Kingsley Martin's claim (1932) was exaggerated, but only just.

To see the world aright you had, it was widely felt, to see Moscow. In books and articles about it, if they were all you could manage (and how they thumped by the ton off the presses), but preferably in person. 'More quickly than Moscow itself, one gets to know Berlin through Moscow': thus Walter Benjamin in his 'Moscow' essay of 1928. Acquaintance with the map of Russia shakes up your reading of the map of Europe:

> However little one may know Russia, what one learns is to observe and judge Europe with the conscious knowledge of what is going on in Russia . . . the stay . . . is so exact a touchstone for foreigners.

The USSR was 'the embodiment of the hopes of all mankind' declared Bukharin at the Writers' Congress in Moscow (1934); quoting him (*LR*, October 1935), Montagu

Slater fervently hopes 'we' can prove him right. Moscow was mythic ('one of those cities', Peter Quennell, wrote in *Action*, 3 December 1931, 'which you approach for the first time with trepidation. So many ideas centre around it that you have come almost to disbelieve in its real existence'). It was a focus of vision according to Gide (*Back From the USSR*, 1937).

> Who shall say what the Soviet Union has been to us? More than a chosen land—an example, a guide. What we have dreamt of, what we have hardly dared to hope, but towards which we have been straining all our will and all our strength, was coming into being over here.

John Cowper Powys is to be found in his *Autobiography* (1934) 'Pondering so often upon their experiment in Russia . . . that amazing country'. Clough Williams-Ellis went to Moscow repeatedly 'because I needed reassurance in a mad and menacing world'. The last chapter of *Forward From Liberalism* finds Spender looking to Russia, that 'great diagram of the classless society' depicted by Sidney and Beatrice Webb in *Soviet Communism: A New Civilisation?* (the question mark got dropped in their second edition). Auden never visited Moscow. He did, however, try and get Naomi Mitchison to talk about Russia to his pupils at the Downs School (November 1932). She was strong on the still centre that was the USSR. When, according to MacNeice in his autobiography, the 'armchair reformist'

> fancies an allegro movement he has—or had, rather—only to turn to Moscow. Naomi Mitchison after a fortnight or so in Russia gave a lecture at Birmingham University about the joy in the faces of the masses.

Gaiety, 'childlike happiness', 'sublimely grinning', 'cheerfulness': joy is just about the only thing in the faces of the Russian masses in Lehmann's *Prometheus And the Bolsheviks*. There was, seemingly, a lot to be joyous about. Charles Madge wrote of the ' "tranquility" of the USSR' that was 'being gradually imparted to the whole world'.[239] Lehmann made a great deal, like so many others, of freedoms guaranteed by the new Soviet Constitution. Writing conditions could not, seemingly, have been more idyllic—huge advances, massive royalties, indulgent publishers, eager and ballooning readership, packed theatres. And it wasn't just the problem of sales that the Soviets had solved: as Edgell Rickword solemnly informed readers of his Communist Party pamphlet *War and Culture: The Decline of Culture Under Capitalism* (1936):

> The Soviet Union is the only country where this problem of giving a meaning to art has ceased to exist. By their understanding of the work of Socialist construction, the writers, musicians, painters and poets have found all the inspiration they need for works which satisfy their ambitions and appeal to the masses of the people.

At least, in the USSR, the aesthetics of technics, and the art of the machine-age, of mass production, the modern city and the factory, seemed to have come into their own. What the Futurists had anticipated, and Gropius and the Bauhaus had started to perform—that union of *Art and Industry* (Herbert Read's title of 1934) that made (to steal T. S. Eliot's words from another context, his '*Ulysses*, Order, and Myth' essay of 1923) 'the modern world possible for art'—appeared to have reached a zenith in Russia. Russia was indeed where Urban Dionysia, as Brian Howard called it, seemed to be happening (the invitation-cards to Brian Howard's 24th birthday party (4 April 1929), his own 'Great Urban Dionysia', were done in so-called Futurist style, parodying the tabulated Blastings and Blessings of Wyndham Lewis's *Blast*). And Futurism thrived in Russia with a clear purpose, unsullied (as it seemed) by the taint of Marinetti's Fascism, uniting the new semantics of socialism with the machine age syntax of Marinetti's proclamation called 'Destruction of Syntax—Imagination without Strings—Words-in-Freedom' (1913):

> Futurism is grounded in the complete renewal of human sensibility brought about by the great discoveries of science. Those people who today make use of the telegraph, the telephone, the phonograph, the train, the bicycle, the motorcycle, the automobile, the ocean liner, the dirigible, the aeroplane, the cinema, the great newspaper . . . do not realize that these various means of communication, transportation and information have a decisive influence on their psyches.

But the Futurist does realize, and, even more, so does the Moscow lover. A man or woman of the machine—after all, Moscow could only be reached by train, or ship, or plane—(s)he was aptly placed to relish the new mechanical, cinematic aesthetic of touristic dynamism, fast machines, air-aces, speedway-stars, record-breakers, which was triumphing there. These were the arts, as Marinetti's 'Destruction of Syntax' has it, of 'Man multiplied by the machine. New mechanical sense, a fusion of instinct with the efficiency of motors and conquered forces'; 'The passion, art, and idealism of Sport. Idea and love of the "record" '; 'New tourist sensibility bred by ocean liners and great hotels (annual synthesis of different races). Passion for the city . . .'; 'The earth shrunk by speed. New sense of the world . . .'; 'Love for the straight line and the tunnel. The habit of visual fore-shortening and visual synthesis caused by the speed of trains and cars that look down on cities and countrysides. Dread of slowness . . .'

From Russia, the writers of an old mechanized society, whose ageing factories and eyesores of ruined industrial landscapes were short on glamour and seemed anyway unable to support the population at a decent standard of living, caught the magic of industrialization, of the Five-Year Plans, hydro-electric schemes, the building of new cities, new factories, new railway liners, and the ploughing up of virgin steppes. The Russian Revolution was thrillingly modern, mechanical, urban, scientific. 'Now it is made clear to every Communist', declared Walter Benjamin in his 'Moscow' essay, 'that the revolutionary work of this hour is not conflict, nor civil war, but canal construction, electrification, and factory building. The revolutionary nature of true technology is emphasized ever more clearly.' And to be truly modern revolutionaries writers must endorse the endeavours of the 'engineers of the human soul', such as Gorki or Lenin. Lenin had, in fact, been altogether absorbed into the machinery of his revolution. 'His mind like an oxy-acetylene flame': so Day Lewis enthused in his 1937 poem 'On the Twentieth Anniversary of Soviet Power'. (The year 1937 was a particularly enthusiastic pro-Soviet year; *Left Review* for instance, celebrated with a Twentieth Anniversary Soviet Number in November of that year.) Lenin featured as a train engine, 'A means o' world locomotion', in MacDiarmid's 'Second Hymn to Lenin'. He was a machine, an engine, a factory, the Tractor Driver of the Revolution, the Immortal revolutionary Tractor itself, according to Tom Wintringham's truly awful poem 'The Immortal Tractor' (1933), surging forward on a revolutionary trajectory: Petrograd–Leningrad–London:

Lenin was speaking. Careful, searching, keen,
His comrades heard him. Words became flame—
Not the white furnace-hunger, nor the light
Of guns that curse at night—
Flame at the heart of a vast machine,
Sparks small as steel can measure, strong alone
By striking thought, the mist of thought, to action:
Petrol to power. Words had grown
Electric, surely placed at the millionth fraction
Of time, to leap, explode, become
The pulse's drum,
The living, lifting, and life-giving factor
In the steel strength of the Immortal Tractor.

. . .

Lenin's words beat
With the rhythm of the factory, the red heat of the forge,
Explosive as the dynamite that cuts a mountain gorge.
Lenin is speaking, and the workers go
Through blood, mud, snow,
Through fear, lies, hate;
And the Cossacks hesitate . . .

. . .

Lenin is living—every word a spark
Driving the great Tractor through the desert and the dark,
The million-powered Tractor, plunging on to victory—

. . .

Lenin is speaking. All who hear him know
Here, too, a Tractor's building, and will grow;
Here, in the cities where the cold fog kills,
In the ploughless valleys, on the blank, bare hills,
'Mid the famine of the mines and the phthisis of the mills,
We are moulding, forging, shaping the steel of our wills
Into pinions, into pistons, crankshaft-web and crankshaft-throw
We are building Lenin's Tractor. It will grow.[240]

News from Magnetogorsk and the Dnieperstroi dam, Stakhanovite visions of Red tractors, steel mills, railway trains, and new electricity pylons, much of it on film, sustained a wave of determined British imitators. Kark Radek had at the Moscow Writers' Congress denounced interest in Joyce as right-wing defection from the required revolutionary devotion to the ideal of Magnetogorsk. But already in 1934, even in the Soviet Union, the materialism-machine aesthetic was being found wanting. Ilya Ehrenburg, for instance, was reported in *Left Review* (November 1934) as praising Soviet *tractorists* rather than their tractors, which were, after all, designed on American models (and were part of the officially sought-after 'Fordization' of Soviet industry and agriculture: 'Americanize yourselves' the dying Lenin is said to have exhorted). And among British writers were lots of ruralists and Conservatives who had never been much enamoured of industrialism anyway. Aldous Huxley's *Brave New World* (1932) quickly became the key dystopian fiction, expression of widespread Western disquiet over the triumph of machine-age materialism, now in Huxley's Year of Our Ford, and in particular anxieties over the application of scientific factory routines to human processes that were better left messy, domestic and slow. Ford was 'in his flivver' all right and that was the reason all was *not* right with the world. Culture, civilization, literature, religion, human individualism were all threatened by the ant-like future of mindless Yankee–Soviet progressivism. People would *need* to be drugged all the time to bear the mindless tedium of that islanded pastoralism gone too far.

And Huxley's objections were a continuation of the old kept-up tones of D. H. Lawrence (one of his last poems to be published, in *Nettles*, is the grim 'The Factory Cities') and of T. S. Eliot's *Criterion*. Huxley's criticism struck the same sort of note as F. R. Leavis in *Mass Civilisation and Minority Culture* (Eisenstein's film *The General Line* 'does not afford the comfort that we are sometimes invited to find' in the USSR: it only endorses the American 'triumph of the machine'). It was the same line as Q. D. Leavis's (neither Naomi Mitchison in *We Have Been Warned* nor Amabel Williams-Ellis in *To Tell the Truth*, rasped Mrs Leavis—both of those novelists, of course, a million miles from having to join in the factory-work they admired on their Soviet trips—were capable of questioning 'the machine' as 'an absolute value' or entertaining 'any doubts about machine-tending as the good life').

It was natural enough for Evelyn Waugh to join in this machine-bashing, and Robert Byron ('conditioned reflexes, Ford lorries, and abortion clinics'), and Osbert Sitwell ('Magnetogorsk, the Nuremburg Stadium and the Great West Road'), and J. R. R. Tolkien (lightning was preferable, he declared in his lecture 'On Fairy Stories', to 'electric street-lamps of mass-produced pattern'). 'The notion that motor-cars are more "alive" than, say, centaurs or dragons is curious', Tolkien declared and his letters kept up a constant barrage against the Soviet new towns, 'Americo-cosmopolitanism', the 'filthy squelch' thrown up by motor-cars, the bad '*magia*' of machinery. Jeers at Spender's 'The Express' and 'The Pylons' from the likes of Charles Williams (in his *Descent into Hell*, published by Faber's in 1937, a couple of with-it characters went to see a play called *The Second Pylon*—'got the most marvellous example of this surrealist plastic cohesion'), or from Wyndham Lewis (*The Roaring Queen* opens with a train carrying the critic-artist Shodbutt most Spenderianly out of a station: 'With appropriate fuss the courteous ten-thirty, panting and with some proud snorts, as of a self-conscious charger (bestrid by a world-bearing generalissimo), left the station'): these were only expectable. So was Eliot's mock pro-Soviet poem recited by the Redshirt chorus in *The Rock*:

> in the cities
> 　　on the steppes
> production has risen by twenty point six per cent
> we can laugh at God!
> our workers
> 　　all working
> our turbines
> 　　all turning
> our sparrows
> 　　all chirping
> all denounce you, deceivers of the people!

So was Peter Fleming's debunking of Russia's industrial pretensions ('Will the Russians transform themselves into robots? . . . The lavatory is under repair, the lift does not work . . . Only one match in three strikes'). More disquieting, though, for

Moscow-fanciers was Graham Greene's scathing attitude towards the pylon lovingly photographed in Grierson's documentary *The Voice of Britain*—the same one drooled over in Read's *Art And Industry*: if only, said Greene, 'Grierson had included a few shots from the damper tropics where the noise of the Empire programmes . . . is just plain wails and windy blasts from instruments hopelessly beaten by atmospherics. At enormous expense from its steel pylon at Daventry the B.B.C. supplies din with the drinks at sundown.'[241] More worrying still was George Orwell's siding in *Wigan Pier* with *Brave New World* ('probably expresses what a majority of thinking people feel about machine-civilization') and with Karel Capek's anti-robot-age *R.U.R.* (Gollancz's Preface to *Wigan Pier* devotes considerable space disingenuously to refuting Orwell's charge that socialists glorify industrialism: 'the words "Magnitogorsk" and "Dnieper" make Mr Orwell see red—or rather the reverse'). Not to mention Julian Symons's scepticism in the first number of *TCV* (his relief that Dylan Thomas was 'not a Pylon-Pitworks-Pansy poet'), or, later on, Rex Warner's *The Aerodrome*, whose modern machine world of planes, motorbikes, pylons, and robotic clinicism is roundly spurned.

None the less, a great many British writers remained for a long time undeterred. For them Sovkino's technics were precisely poetry's way forward. A 'new seam of richest material has been opened up' declared Day Lewis in the part of *A Hope for Poetry* that praises poetry's latest welcoming of factory, railway, and pylon. A. S. J. Tessimond's '*La Marche des Machines* (suggested by Deslav's film)' was singled out for praise by Michael Roberts in his *New Signatures* Preface. 'Beauty breaks ground, oh, in strange places', exclaimed Day Lewis in *From Feathers to Iron*, celebrating 'a grain-Elevator in the Ukraine plain'. 'When the first tractor came . . . how we cheered it': for a moment, anyway, the poet thought he was actually performing in a Soviet movie. Spender's 'The Express' was, of course, inspired by *Turksib* (Victor Turin's film celebrating the completion of the Trans-Siberian railway, which was completed before the railway line: much like those parts of the Five-Year Plan said to have been completed in four years). No doubt, too, there's something in it of Louis Aragon's poem *The Red Front* as translated by e. e. cummings, which celebrates the USSR as an engine 'Whose pistons go SS RR and SSR SSR SSR' and whose 'effective cinematic imagery' Spender liked (*NV*, May 1933). Spender's praise in his *New Country* (1933) essay 'Poetry and Revolution' for the opening of Beethoven's *Eroica* symphony seems also to have its eye on cinematography of the *Turksib*-Aragon kind:

> The symphony begins with the dignity of a huge locomotive and we know that nothing can stop it until a certain task has been performed. The effect created is rather the same . . . as that created in some Russian films by photography of machinery moving, in which the action of the machine is really a model of the action of the human will.

New Country was, in fact, full of this kind of thing: Richard Goodman's aeroplanes ('The Squadrons') and socialist tractors ('Ode in Autumn'); John Lehmann's 'thunder of engines in a glass-roofed terminus', his train 'In quiet curves descending from the pass', 'The future calling . . . Roar of machinery'; Rex Warner's 'Choruses from "The Dam" ' (shades of Dnieperstroi); Tessimond's 'Steel April':

> I regret your bohemia's aesthetic blindness to the lovely world
> of wavesmooth tyrranous cars and departures in the Golden Arrow
> . . .

> My civilisation . . .
> will lubricate, tune the engine, until it is silken-silent;
> will make new factories flowers of steel, not flower-façaded.

And, of course, there was Spender's train sonnet beginning 'At the end of two month's holiday' ('Like the quick spool of a film / I watched hasten away the simple green'; 'Real were iron lines'), and his townscape vignette:

> The morning road with the electric trains
> Flashing over the bridge, the power station,
> Then the road leading over the canal, then to large buildings
> Then to the public washhouse (where I went) and the school.

Nor did Spender halt this line of work with *Poems* (1933), which contained 'The Express', 'The Pylons' and 'In Railway Halls'. *The Still Centre* (1939) is still busily about what Auden called the 'strict beauty of locomotive': in 'View From a Train', 'The Midlands Express' and 'Houses at Edge of Railway Lines'. It was a long while before Spender snapped out of the atmosphere of Charles Madge's 'Instructions':

> We shall free the political prisoners, the impulse, the desire to be,
> Our joy shall be as strong as the wheels of Dnieprstroi
> Deep in the racing blood revolving and dissolving
> Hard lumps of pain, electrolysing slumps.
>
> Along our cables flowing and in our streets going
> Into the houses breaking and the doors banging and shaking
> Marching along with drums and humming high in the pylons comes
> Power and the factories break flaming into flower.[242]

And Day Lewis also kept up such enthusiasms right to the end of the decade: as late as April 1938 (in the Sonnet 'When they have lost', published in the Virginian magazine *The Lyric*) his vision of the future, after 'oppression' shall have been conquered, is 'a power-house': 'To warm men's heart's again and light the land.'[243]

Soviet electrification seemed, of course, much more electrifying than the English kind. The period's poetic pylons are more evidently at home in Russian movies than in the English landscape. One commentator who has seen the Strand Films documentary *The Way to the Sea* (1936) about the electrification of the Southern Region—electricity plus railways, an electrifying combination!—a film for which Auden wrote the sound-track, joins Paul Rotha, one of its makers, in stressing the film's ironic intentions and results. They urge us to not to take as straight such lines as 'The pylon drives through the sootless fields with power to create and to refashion'.[244] The implication is of an ironic tone necessary in the English context. And, certainly, when Michael Roberts ascends to the power stations in the volume *These Our Matins* (1930)—

> Far above, the insulators,
> Hiss and spark like commutators
> For I go where bees are humming
> And dynamic turbines drumming—

it's in an obviously foreign mountainscape, in a poem titled 'Les Planches-en-Montagnes'. But still, England did have its new National Grid. 'We're entering the Eotechnic Phase / Thanks to the Grid and all those new alloys': thus Auden in his 'Letter to Lord Byron'. And it was no accident that the volume of Stephen Spender's

Poems that contained 'The Pylons' came out in the same year, 1933, as the Central Electricity Board's first National Grid was completed. In 1933, too, Tristram Hillier painted his picture 'Pylons'—peopling a bare landscape not unreminiscent of Salvador Dali with stark poles, wires, insulators, and a mysteriously boxed-in transformer excitingly labelled 'DANGER'. (A reproduction of it features in Herbert Read's *Unit 1, The Modern Movement in English Architecture Painting and Sculpture*, 1934: a reproduction erroneously ascribed to Humphrey Spender in the Gloversmith volume *Class Culture and Social Change*.) And the British Grid's pylonic clutter could, with a pinch or two of optimism, be assimilated into the more obviously politically revolutionary Russian version. In any case, English poetry must not, it was thought, flinch from such enthusiasms, for speaking of such things was 'Speaking Concretely', as Tom Wintringham called it, and that was the real road leftist poetry 'must take':

> steel wire must be
> Inseparable from concrete, you from me,
> We from the durable millions. Then there's a road![245]

This was undeniably the road, especially the rail-road, the USSR had taken (little Niko in Lehmann's *Prometheus and the Bolsheviks* wants to be 'a Stakhanovite engine-driver' when he grows up: 'His great ambition is to drive Stalin back from his summer house by the Black Sea in record time to Moscow'), but such mechanical beauties could emerge wherever workers were solid. Helped on his revolutionary frontier-crossing by railwaymen in *Lean Men*, Ralph Bates's Francis Charing has just such a moment of railroad vision (which among other things nicely reverses the horrified railway-truck scene with the Arab mare in D. H. Lawrence's *Women in Love*):

> A line of light railway tip-trucks smeared with clay and cement, about whose gaunt outlines dark mists lazily creep, may not be material for poetry, yet . . . All the suffering of the human race, all the splendour of martyred faiths, the significance of the dumb hills and the shattered rocks and bespoiled valley seemed to be somehow expressed in the beauty he perceived in a slow wreathing of vapour through the cast-iron spokes of an ore trolley wheel.

(Ralph Bates, born in Swindon, home of the Great Western Railway's great repair-shops, had trained as a locomotive engineer.)[246]

Clearly, wherever it cropped up, this hard, materialist Soviet affair of things and aggregations of things, where value was thought to reside in mere lists and numbers—the statistical arrays of the Soviet Pavilion in Paris described by Forster ('Statistics, maps and graphs preach a numerical triumph'), the triumphs of the Soviet Five-Year Plan trumpeted in books like Maurice Hindus's *The Great Offensive* (1933: 105,800 tractors; 3,229,150 horse-drawn ploughs; 9,330 beet-diggers; 27,000 potato diggers); and so on—had its romanticizable side. It was, of course, a naïve romanticism, wrapped up in the youthies' sympathy with that small boy's longing to be a train-driver (Auden was full of envying praise in his *Daily Herald* article 'How to Be Masters of the Machine', 28 April 1933, for 'Engine or crane drivers'), and in the slummer's safe sense that factory work is not his/her own lot. How else explain the curious notion that factory workers would like to drop in on other people's factories, even Soviet ones, during their Workers' Travel Association holiday? *Seeing Russia for Yourself* with the

WTA meant visiting the Ford Truck Factory at Autostroy, and the Stalingrad tractor works. Only a white-handed bourgeois would think an outing to the Ford Works at Dagenham (WTA had many such visits organized) a good idea for a day off work. It was the bourgeois Gide who was excited (*Back from the U.S.S.R.*) by some holiday-ing French miners who relieved a Soviet shift (and who *all* turned out to be Stakhanovites). When they went to the movies, actual workers wanted Hollywood glitter, not hours of documentary about post offices and gas-works. Visiting Five-Year-Plan-land made for an odd kind of holiday, even by the standards of the most regimented holiday camp.

Evidently, the leftist visitor who shared Ilya Ehrenburg's belief that 'Our foreign visitors are making a trip in the car of time', found it hard to hang on to ordinary critical standards. It was easy, gibed Waugh with some point in *Robbery Under Law*, for 'credulous pilgrims' to socialist paradises to 'forget that we, too, have hospitals and infant schools and recreation grounds at home': 'earnest students of the Left Book Club kind . . . will stare entranced at a cot or a blackboard if they have been told that they represent proletarian progress.' ('You Wops insist too much on the machine', Wyndham Lewis claimed he told Marinetti in the early days of Futurism: Lewis couldn't work up any childish enthusiasm about 'Automobilism' because he was so used to machines in this old industrialized country.) In much the same spirit, double standards were applied to the known facts of Soviet terror: liquidations, show trials, camps, prisons, censorship, Stalin's secret police, the worship of Lenin and Stalin. In practice, there was no real excuse for anybody continuing to swallow the glossier glosings of the Intourist brochures.

In his 1938 review of Eugène Lyons's *Assignment in Utopia* Orwell calmly accepted Lyons's exposure of the frame-up Trotskyist trials; he implied he'd realized the bizarrerie of events in Russia for the past two years (in English terms, Orwell suggests, Stalin's drama amounted to Beefeaters being discovered to be Comintern agents, GPO officials 'drawing moustaches on postage-stamps' and a sweet-shop proprietress transported to Australia for illicitly 'sucking the bull's-eyes and putting them back in the bottle').[247] In *Back From the U.S.S.R.* (Paris, November 1936; London, April 1937) and *Afterthoughts: A Sequel to Back From the U.S.S.R.* (1938), Gide reneged with mounting intensity on his earlier communist convictions: he couldn't keep on turning a blind eye to Stalinist lies and repressions (and it's silly of George Watson to pretend in his *Politics and Literature in Modern Britain* (1977) that Gide rejected Stalin out of a kind of super-totalitarianism). It's clear from Edmund Wilson's letters that it was easy for any clear-eyed visitor to the USSR to endorse Gide's response ('all the trials have been fakes . . . intended to provide scapegoats and divert attention from more fundamental troubles . . . a complete double stand-ard of truth in the Soviet Union . . . all this hornswoggling of the masses in the name of whom everything is being done': 15 April 1937, to Malcolm Cowley). There was a lot wrong with Trotsky, Wilson thought, but the man did keep spilling the beans about Stalin. Malcolm Muggeridge agreed: Trotsky 'blows the gaff, as far as the Soviet régime is concerned', and Muggeridge's *Winter in Moscow* (1934)—a col-lection of ironic factual-fictions designed to tell truths Muggeridge had gleaned as a foreign correspondent, especially about the refusal of ideologically biased Western visitors to admit the harsher verities—would follow suit. Muggeridge treasured 'as a blessed memory the spectacle' of British seekers after the Fabian Fairyland going with radiant optimism through a famished countryside; wandering in happy bands about squalid, overcrowded towns; listening with unshakeable faith to the fatuous outpourings of obsequious Intourist guides; repeating, like schoolchildren a multiplication table, the bogus statistics and dreary slogans that roll continuously—a dry melancholy wind—over the fairyland's emptiness.

Certainly there was no reason why anybody in the '30s should remain deceived. Disillusioned notes had been struck very early. e. e. cummings's appalled description of Russia, including his memorable hatchet-job on the worship at Lenin's tomb, published in his *Eimi* (1933), dates from his Russian visit of 1931.

But you could, evidently, remain an intelligent writer and still profess convinced optimism about the Soviet Union as late as 1937—which was when Lehmann's *Prometheus and the Bolsheviks* appeared. 'Guts. Durability. Physique. He suffers from a dilated heart, but otherwise his physical strength and endurance are enormous. He is no high-strung neurotic . . . Patience. Tenacity. Concentration.' And so on. *He* could, of course, be Mosley as described by A. K. Chesterton, or Mussolini as described by Yeats-Brown. But in fact he is Stalin being looked up to by John Gunther in his *Inside Europe* (1936). And such persistent heroizing of the USSR and Stalin was often done by people turning a careful blind eye to reported events. Edward Upward travelled Intourist to Moscow around 1931, returned a crusader for the cause (Spender describes him on his return to Berlin as 'not unlike the smiling young Consomol hero who saves the boys in the reform school in one of our favourite films— *The Way Into Life*'), and thereafter (until the late '40s) Upward refused to countenance any public doubts. He simply did not think about the Trials, he told his friends. Others pooh-poohed the adverse reports. John Cornford's friend Pat Sloan, who lived in Russia 1931–6 ('let us visit the Soviet Union . . . and see the new life that they are building'), advanced on his Left Book Club readers in *Soviet Democracy* (1937) behind the Webbs's pro-Soviet barrage (an allegiance repaid by Sidney Webb's glowing notice in *Left Review*: 'no book better fitted to make the reader understand what it all amounts to'). Sloan dismissed criticisms with furious disingenuousness. 'Soviet imprisonment stands out as an almost enjoyable experience.' Useful work is being done in the penal settlements: nobody there need waste his skills. Soviet citizens are free to criticize, but they do not wish to; but even if they want to, they can't, because theirs is the government of the whole people. *Bon mots* like 'Democracy, therefore, is also dictatorship' abound. Sloan's agility with whitewash resembles D. N. Pritt's in his Penguin Special *Light on Moscow* (October 1939), as he exculpates the Russians for the Hitler–Stalin pact and, in later editions, joins Bernard Shaw in justifying the Soviet invasion of Finland. Spender could not hope to compete with the hard Party sophists. His *Forward From Liberalism* wallows in liberal contradictions over repression, trials, state terror, secret police, censorship. Not for him the brisk toughness of John Cornford, who loved the idea of revolutionary violence. Or of the Webbs, who consoled themselves with reflection on the breaking of eggs as a necessary part of the making of omelettes—much like the functionary in Muggeridge's *Winter in Moscow* whose defence of the crushing of the kulaks is telegraphed to Western newspapers:

bolsheviks determined harmonise agricultural economy with industrial development plan stop admittedly involves cruelty and casualties like other forms war but dash putting it brutally dash impossible make omelettes uncracking eggs.

Stalinists were adept with egg-cracking talk: it was perhaps a help to killers to see broken human skulls, Trotsky's for instance, as mere bits of egg-shell. This way of seeing Russian conditions was, in effect, Christopher Caudwell's line in *Illusion and Reality* on Soviet censorship. The fellow-travelling artist 'goes to Russia not so much to see if the people are free, but if the artists are "interfered with" by the authorities'; scientists 'go to Russia prepared to "sacrifice" everything, provided scientific theory is not interfered with'. But the people *are* free; so the 'unfree' artists and scientists cannot therefore be people. And, of course, they are not: they're the Revolution's broken egg-shells.

Given the popularity of the Intourist route in the '30s, it's surprising just how many writers, even fellow-travelling ones, did *not* join Wells and Shaw, the Webbs, Fleming and Byron, Upward, Sloan, Ralph Fox, Naomi Mitchison, Valentine Ackland and all the rest on the trek to Moscow. Auden, Spender, Isherwood, Day Lewis, MacNeice, Orwell, Greene, and Caudwell, all steered clear. Many of those who didn't make it to Moscow while their fellow-travelling impulses were still strong did, however, make the shorter journey to Spain instead. And it was encountering Stalinist treacheries and OGPU evils in Spain that brought home to many Moscow admirers a realization at last of what was going on in the more distant Workers' Paradise. Many ears grew deaf to Intourist patter for the first time in Spain. Even so, again, as we shall see, large numbers of those interested in the Civil War didn't actually visit Spain. Fellow-travelling, to Spain as to Moscow and the cause of the working class in general, was a journey that went on inside as well as outside your head, and was sometimes confined entirely within the mental, the spiritual sphere. The Intour—as the Soviet trip was known in the period—was also an inward tour.

It is impossible, as we've seen repeatedly, to disconnect the obsession in '30s literature with journeys, goings-over, goodbyes, thresholds and travel machines, from the problematics of the inner life. So prevalent was the connection that the experience of any actual journey could be made, and was made, to provide lively emblems of the mental and spiritual, political and psychological positions that authors and their characters had reached or were traversing. And nowhere are inner and outer bolted more firmly and extendedly together than in Graham Greene's *Journey Without Maps* (1936). The book is about a journey Greene undertook into the unmapped interior of Liberia. But, as he explained later (see *Ways of Escape*, 1980), 'The account of a journey—a slow footsore journey into an interior literally unknown—was only of interest if it parallelled another journey': the journey into the interior of the self. (It's easy to agree with that when one reads the parallel account of the same journey by Greene's cousin Barbara Greene, called *Land Benighted* (1938), a naïf narrative peddling a pitifully small handful of tricks, it attempts to play the woman as debunking Peter Fleming, at the same time as jutting out the stiff British upper lip, and playing up the jungle's dismaying distance from the Savoy Hotel, cosmetic counters, and healthy stuff like Ovaltine and Eno's Fruit Salts, and all in a tone of bright female upper-class gush (the porters 'were charming to us and we got to know them all so well') that proves how banal a travel book, and Africa, can be in unsubtle hands.[248])

For his part, Graham Greene works hard to make his text not only a summation, a grammar-book or lexicon, of his period's travelling possibilities, but a personal answer to the needs he believes all that travelling ministers to. Readers are alerted to Greene's desire for period relevance by his book's sections' self-conscious '30s headings: *Up to Railhead, Border Town, The Edge of 'Civilisation', The Forest Edge,* and *Music at Night* (Aldous Huxley published books entitled *On the Margin* and *Music at Night*), *To the Frontier, New Country.* The account is full of borders, border-crossings, customs-sheds, and Greene's border anxieties. He dreams of being fined heavily at the border of Sierra Leone and Liberia. 'Would one have trouble with the Customs at the frontier?' Crossing into Liberia is consciously to defy the taboos of the unknown threshold: 'No one in Sierra Leone had ever crossed the border.' So it's consoling to hear of the Dutchman who eludes capture on several frontiers ('a good story to hear there in the dark, near the borders of a country of which no one in Sierra Leone had been able to tell me anything'). For even the most unpatrolled of borders, of which Africa has an exciting plenty, locations genially reminiscent of pre-passport days ('natives pass freely to and fro; indeed with a little care it would be possible to travel all down West Africa without showing papers from the moment of landing. There is something very attractive in this great patch of "freedom to travel" '), are none the less troublingly palpable. The '30's *Angst* of frontiers is inevitable everywhere in Greene's text; any ease of passage is delusive:

> That afternoon we went for a walk into French Guinea. The border is the Moa River . . . The curious thing about these boundaries, a line of river in a waste of bush, no passports, no Customs, no barriers to wandering tribesmen, is that they are as distinct as a European boundary; stepping out of the canoe one was in a different country. Even nature had changed.

Here, too, then, it only requires a minor adjustment for the *farer* to turn into Auden's *fearer*; and, in fact, Auden's ' "O do you imagine", said fearer to farer' is one of Greene's book's epigraphs. Like Auden and Freud, Greene associates travel with growing up and away from father and mother. He sails from Liverpool: a cold going in 'January wind':

> People sat crammed together below deck saying good-bye, bored, embarrassed and bonhomous, like parents at a railway station the first day of term, while England slipped away from the port-hole, a stone stage, a tarred side, a slap of grey water against the glass.

As so often in the '30s, the wide world is seen as a schooling: the British functionaries are the prefects to Sierra Leone smaller boys (Greene 'had never been a prefect'; acting the prefectorial headmaster to a chief at one point, he can only do it ironically: '*He* couldn't tell my satiric self-criticism as the ghost of Arnold of Rugby addressed his head prefect through my lips'). Greene's mapping of Africa is, again characteristically, a mapping of the whole world. The sections titled *Dakar, Freetown, Sunday in Bolahun, Hospitality in Kpangblamai, Bamakama, Galaye, Monrovia* and the like, are preceded by *Via Liverpool, Madeira, Las Palmas.* Just so, Greene's travelling reminiscences reach out from West Africa to encompass Nottingham, the Tottenham Court Road, Riga, Berlin, and Paris. And journeying is choosing, between destinations and so between possible goals. 'Intourist provides cheap tickets into a plausible future' for those who preferred Moscow. Berlin means 'swastika banners . . . the

Sunday processions with drums and bugles and bayonets under the Brandenburg Gate, the demonstrations at the Tempelhof'. Nottingham is where he recalls being received into the Roman Catholic Church.

Greene is well aware that some journeys are trivial and trivialized. In Teneriffe he sees the movie of his own novel *Stamboul Train*, a 'cheap banal film', unsuccessfully vulgar, ballyhooed by dreadful Hollywood publicity. 'One had never taken the book seriously' (it was written fast, for money); and here was Hollywood's quest for the trivial attraction cheapening the filmic version of his train journey into a still less serious affair: 'Two Youthful Hearts in the Grip of Intrigue. Fleeing from Life. Cheated? Crashing Across Europe. Wheels of Fate.' By contrast, this present journey and this current travel book are deadly serious. Eschewing the opposed but like nostrums of totalitarian Berlin and Moscow, wryly dismissive of British Imperialism at its pink gin worst ('the country which has given them only this: a feeling for respectability and a sense of fairness withering in the heat'), angry at arrant American colonialism in the predatory shape of the Firestone Company of Ohio, perpetually rueful about liberal democratic ideals in the messy tatters of the Republic of Liberia, Greene is waiving the political quest in favour of the religious one. '[M]y journey represented a distrust of any future based on what we are.' Answers would come not from Berlin or Moscow but from an inspection and a transformation of the human heart.

Africa is dark, unmapped, and its 'shape, of course, is roughly that of the human heart'. Africa encapsulates all the darkness, pain and evil of the world: the girl weeping in a bar in Leicester Square; 'pain at every yard' among Berlin's swastikas; 'poor and pinched' Communists meeting in the Parisian slums; Greene's first memory—'a dead dog at the bottom of my pram'; his adolescent discovery of 'the pleasure of cruelty'; the would-be suicide he once saw, knife in hand; the pair copulating in a Parisian street, 'like two people who are supporting and comforting each other in the pain of some sickness'. Africa is the ultimate 'seedy level'. 'There seemed to be a seediness about' Liberia that 'you couldn't get to the same extent elsewhere, and seediness has a very deep appeal'. Africa is, then, what we recognize as Greeneland, the very atmosphere of Greene's fiction. It is Greene's subconscious. As is usual in his work, Greene laces his Christian sense of evil with Freudian notions of libido and id, repression and trauma. His African journey is crucially a plunging down ('going deeper', 'getting somewhere') into the meaning of his dreams, a travelling into the subconscious: just like psychoanalysis, in fact, whose method

> is to bring the patient back to the idea which he is repressing: a long journey backwards without maps, catching a clue here and a clue there, as I caught the names of villages from this man and that, until one has to face the general idea, the pain or the memory. This is what you have feared, Africa may be imagined as saying, you can't avoid it, there it is creeping round the wall, flying in at the door, rustling the grass, you can't turn your back, you can't forget it, so you may as well take a long look.

What makes *Journey Without Maps* so powerful—and one of the most important of all '30s texts—is the way it combines, as well as distinguishes, a galaxy of possible journeys. The traveller is seen as a type of the delver into the self, he is a psychoanalytic explorer, but he is also united with other sorts of seeker after truth, religious pilgrims, explorers, detectives, and those very conscious tanglers with the undergrowths of epistemology and hermeneutics—writers, particularly novelists, and their surrogate narrators:

> Freud has made us conscious as we have never been before of those ancestral threads which still exist in our unconscious minds to lead us back. The need, of course, has always been felt, to go back and begin again. Mungo Park, Livingstone, Stanley, Rimbaud, Conrad represented only another method to Freud's, a more costly, less easy method, calling for physical as well as mental strength. The writers, Rimbaud and Conrad, were conscious of this purpose, but one is not certain how far the explorers knew the nature of the fascination which worked on them in the dirt, the disease, the barbarity and the familiarity of Africa.

No wonder Conrad's 'Heart of Darkness' is repeatedly alluded to: from Greene's text's opening arrival at 'The tall black door in the narrow city street' which 'remained closed', and 'the usual blank map upon the wall' in the Liberian Consulate, onwards. For Conrad's narrator Marlow was on just such a heuristically combinatory trek, mirroring the general problem of epistemology (how to know, to make out) and its urgent artistic, narratorial application (how to tell, to make) in the physical search for moral and political truths in Africa. And as T. S. Eliot recruited the problematics of morality and of moral description from Conrad to afforce his own religious quests, so Greene builds on Eliot's Christian and Christianized journeyings. Eliot straddles Greene's book. At the end, a 'Waste Land' image ('while I was fishing in the dull canal / . . . round behind the gashouse') is offered as illustration of the mixed experience of Africa ('its terrors as well as its placidity'). Early on, Prufrock's urban wanderings become a model for Greene's urge to travel 'further back' into the primitive depths:

> Streets that follow like a tedious argument
> Of insidious intent
> To lead you to an overwhelming question . . .

Greene is taking the pilgrim Eliot, as he takes Conrad's pilgrim-narrator Marlow, seriously. He takes other earnest, religious travellers seriously as well. Missionaries come out of no '30s travel book, except the ones they wrote themselves, so well as out of *Journey Without Maps*. The parallel book *Land Benighted* is kinder than most of its kind to missionaries; it's especially moved by the German missionary wife whose 'kindness lies enshrined in my heart'. But cousin Graham went further. The repressed, prudish missionary of Somerset Maugham's 'Rain' and the innocent Catholic priests of Greene's schooldays begging for 'a new altar cloth' for tin churches in the bush, are, he says, nothing like the heroic Dr Harley, Methodist medical missionary ('In Liberia I discovered another kind of missionary'):

> a man with a body and nerves worn threadbare by ten years' unselfish work, cutting away the pus from the huge swollen genitals, injecting for yaws, anointing for craw-craw, injecting two hundred natives a week for venereal disease.

Here Christianity was as fascinating as Africa, as Greene's subconscious, face to face as it was with the seedy, with the darkness of Pinkie's nightmares and Spanish Catholicism's grim imaginings, realistic about the evil of the human heart, and nothing at all like the breezily naïf dealings with sin indulged in by those notoriously laughing Christian travellers of the period, the Oxford Groupers, all cheerfulness, Rhodes Scholar, Rowing Blue, and International Houseparty. Freetown irked Greene precisely because of its Buchmanite aversion to booze and dirty stories: 'it might have been inhabited by rowing Blues with Buchman consciences and secret troubles.'

'30s literature gives the Oxford Group, or Moral Rearmament movement as it became, a rough time for its 'sin and vermouth' snobbery, covert homosexuality, crude anti-Communism, 'spiritual nudism' and moral triviality.[249] Auden particularly savaged them in his savaging of Miss Gee. The medical students of the poem 'Miss Gee' 'began to laugh' as the surgeon 'cut Miss Gee in half' not least because they include the Groupers who were taught to laugh as a way of attracting converts. Then

> They hung her from the ceiling,
> Yes, they hung up Miss Gee;
> And a couple of Oxford Groupers
> Carefully dissected her knee.

They dissect her knee because that is what she knelt on to pray, an anti-erogenous zone, and her cancer has been caused by sexual repression incited by religion. Miss Gee, of course, had deep yearnings for men to touch her erogenous zones. She used to dream about the vicar. So it's tough that she should be dead before some men get round to touching her knee, and, worse, that the men touching her up are sworn to Absolute Purity so cannot progress beyond her knee. Groupers, not Gropers (the joke was Greene's), they are as repressed, as sick (in Auden's view) as she was. And the trouble is they went all over the world. The Chinese New Life Movement encountered in *Journey to a War* is clearly an oriental MRA; 'We've had an enormous success in Norway', one Grouper boasts in *The Confidential Agent*; *Pink Danube* ends with an ironic retiring from international socialism into international Buchmanite homosexualism; hearty Groupers plague the holidaymakers in Rose Macaulay's *Going Abroad* ('I suppose they are having house parties down the Basque coast'). They were the period's most trivializing religious pilgrims.

So by rejecting Buchmanism Greene was spurning easy religious travel, the cheapening consolations of the jovial Groupers, and siding with the bleaker pilgrims of the period. He was in 'transit' with Eliot on the 'transitory', renunciatory path of 'Ash Wednesday' (the poem was published as a whole in 1930). This was 'a cold coming', even in the heats of Africa, or of California (where both Huxley and Isherwood found religion in Hindu mysticism). A 'coming', of course, not a 'going', for like psychoanalysis and the Liberian journey this was a Prodigal Son's coming home, a return to the faith of the Christian father and mother: *The Pilgrim's Regress* as C. S. Lewis called it (1933). In 1930 Lewis wrote to Arthur Greeves, his Plymouth Brethren friend in Belfast, about the beauty of finding himself 'on the main road with all humanity . . . It is emphatically coming home, as Chaucer says "Returneth *home* from worldly vanitee".' Lewis had one of his key moments of revelation on, quite literally, his way home: on top of a bus going up Oxford's Headington Hill. There, as in Eliot's *Four Quartets*, metaphor and actuality converged. In those poems Eliot's literal journeys, to the Cotswold house Burnt Norton, the Somerset village East Coker, to Little Gidding and, in a sense, to the Eastern Seaboard of North America, become analogues of the spiritual quest. Going back home, to the childhood memory, the village one's ancestors came from, the place where one's political forebears (Charles I and Nicholas Ferrar) were located, stands for the theological journey, the regression from modernity and from 'progress,' back towards the safe sources of memory, revelation, conservative attitudes, the world of Mother Church and Father God. Cyril Connolly's hostile summary in *Enemies of Promise* was rough and ready ('joining a church implies regression for an intellectual writer, it is a putting on of blinkers, for a writer is hiding under the skirts of one of the great reactionary political forces of the world, and the poet drawn to the confessional and the smell of incense finds himself defending the garotte and Franco's Moors'), but it applied more or less to Lewis and Tolkien, Charles Williams and T. S. Eliot, Evelyn Waugh and Roy Campbell, as it did to the Polish Count Potocki de Montalk and his *Right Review*, to Eoin O'Duffy's *Irish Christian Front* which fought for Franco in Spain, to Frank Buchman (he thanked heaven in 1936 'for a man like Adolf Hitler who built a front line defence against the anti-Christ of Communism') and, later on, to Auden himself.

Connolly's point did not, though, impugn George Bernanos, the French Catholic novelist who was busy against Franco. Nor leftist sympathizer and Catholic Graham Greene. What's more, the Christian pilgrims, for all their stress on the completed journey, the voyage back after the voyage out, the discovery and recovery of the Grail, the vision within the stilled centre at the 'heart of light'—whether in *Four Quartets*, Tolkien's *The Hobbit* (1937), Lewis's *The Pilgrim's Regress*, his *Out of The Silent Planet* (1938), or any one of Charles Williams's many fictions in which people are on the move as pilgrims, time-travellers, explorers, grail-seekers (his *Shadows of Ecstasy*, 1933, is, by the way, oddly anticipatory of *Journey Without Maps* in its dealings with Africa and the Africa 'within')—the Christian pilgrims remained after all curiously unrested, unresolved, still in transit, in exile.

Conrad's Marlow found Africa unspeakable and indescribable; he returned essentially clue-less from the labyrinth, a defeated detective, without solutions. Kafka's people also found the world Conradianly puzzling. And Greene's Africa is not only Conradian but Kafka-esque. Its Secret Societies, for instance, won't give up their secrets:

> It is a curiously Kafka-like situation: headmasters who wear masks and turn out to be the local blacksmith . . . One reaches the village at the foot of the *Schloss*, to discover that almost anyone may be the master of the *Schloss*; his agents are everywhere . . . there is an atmosphere of force and terror . . . occasionally beauty . . . 'meaning behind meaning, form behind form'. I can imagine that after seven years of investigating this formal but Protean religion, one may still despair of an interpretation. Olga in Kafka's novel, it will be remembered, tried to construct 'out of glimpses and rumours and through various distorting factors' an image of Klamm.

So the puzzled quester may be heroic, but his readings of the world's text remain (in Conrad's own word) absurd. The half-caste tax-gatherer Greene meets in the place Darndo ('marked on no map'), oddly reminiscent in his 'dirty pyjamas' of Kurtz's Russian side-kick in 'Heart of Darkness', calls himself a detective. 'I am a detective.' But he makes howlers, misreads evidence, believes Greene to be, if not a missionary then 'a member of the Royal Family'. So he is patently absurd. But he helped Greene with food and carriers. There is no one in Liberia 'for whom I feel now a greater affection'. And Greene finds him, a man of the Coast, 'stuck away in a tiny village of a strange tribe', 'heroic'. It's an admiration based, no doubt, in the man's likeness to Greene himself, the traveller poised eventually, at the end of his journey, irresolute, on the threshold, 'in the cold empty Customs shed' at Dover, mindful of Rimbaud, the poet 'estranged' of Auden's poem, whose traces Waugh pursued so compulsively in Harar (*Remote People*).

Greene's Catholicism brought neither him nor his characters any permanent rest from travel: he and they stayed on the move, irresolute, exiled; 'You couldn't help moving on.' 'So you're going back to England?' Minty enquires of Kate Farrant at the end of *England Made Me*. 'No', she replies, 'I'm simply moving on. Like Anthony.' It was as though—shades of Lewis's bus and his memorable ride to Whipsnade Zoo in the side-car of his brother's motor-cycle—Greene couldn't ever quite get off the tram on which he was instructed into the faith:

> In Nottingham I was instructed in Catholicism, travelling here and there by tram into new country with the fat priest . . . The tram clattered by the Post Office: 'Now we come to the Immaculate Conception'; past the cinema: 'Our Lady'; the theatre . . .

But then, this restlessness was endemic to the orthodoxy of pilgrimage. The Christian is hopeful of attaining the biblical 'rest' that 'remaineth' 'to the people of God', the Good Place, Heaven, the Grail; he wishes to make a good end. As Auden's poem 'The Voyage' has it:

> And maybe the fever shall have a cure, the true journey an end
> Where hearts meet and are really true.

And the believer's prospective good end should determine the quality of his earthly life, for in the beginning with God, at conversion, 'is my end'. This faith was nourished for Eliot by his favourite Divine, Lancelot Andrewes, not least in Andrewes's Ash-Wednesday sermon of 1619 on which Eliot drew for his poem 'Ash Wednesday':

> Whether a way be good or no, we principally pronounce, by the end. If (saith *Chrysostome*) it be to a *Feast, good*; though it be through a *blind lane*: if to *execution*, not *good*, though through the *fairest street* in the City. *Saint Chrysostome* was bidden to a marriage dinner; was to go to it through diverse lanes, and alleys; crossing the high street, he mett with one ledd through it to be executed: he told it his Auditorie, that *Non quà, sed quò* was it.
>
> If then our *Life* be a way (as a *way*, it is termed, in all Writers both *holy* and *humane*, *via morum* no less, than *via pedum*); the end of this way is to bring us to our end, to our *sovereigne good*, which we call *Happiness*.

But, meanwhile, the pilgrim is still left still seeking a 'continuing city'. 'Which *happinesse*'—Andrewes again—'not finding heer, but full of flawes, and of no lasting neither, we are sett to seek it.'

So classic Christianity doesn't evade, but reinforces a sense of earthly irresolution and restlessness. The poet of 'Ash Wednesday' is thrust back into the 'transitory' process, turning and returning, climbing, wavering, conscious of life as protracted exile from heaven's solutions. 'And after this our exile', he ends Part IV, deferring the paradisal promise held out in the Roman prayer to the Virgin ('and after this our exile show unto us the blessed fruit of thy womb, Jesus'). The poet of 'Four Quartets' is still in the 'unstilled world' of the 'by-pass', the tube-train, and the bombing-plane. 'Against the Word the unstilled world still whirled.' Evelyn Waugh's conversion was in fact a prelude to his most restless years. The 'twelve years . . . on the move' that he recalls at the beginning of *Robbery Under Law*, the perpetual homelessness and dislocation, the writing in pubs and friends' houses ('falling leaves in the autumnal sunshine remind me that it will soon be time to start out again somewhere else'), aren't just symptoms of a Tony Last-like disgust over his first marriage's collapse and revulsion from England's Bright Young People, they are also an enactment of the age's moral and spiritual dissatisfaction that he describes in *Vile Bodies*, an emblem of his sharpened religious awareness of a still unstill world. Chapter 5 of *A Handful of Dust*, in which Tony Last takes to exploring South America, is entitled 'In Search of a City': 'whose builder and maker is God', the Bible went on; and, orthodoxly, Last cannot find his City anywhere on earth. Waugh warns the reader of *Robbery Under Law* that he 'was a Conservative when I went to Mexico and that everything I saw there strengthened my opinions. I believe that man is, by nature, an exile and will never be self-sufficient or complete on this earth'. But Last's plight, stuck with a maniac in the jungle, and Waugh's '30s homelessness, were not the less terrifying for being theologically so sound. Waugh's letters of the early and mid '30s strive for cheerfulness in the condition of 'professional tourist', but can't help sounding constantly like a litany of exiled distress: 'no address'; 'no possessions, no home'; 'the haphazard, unhappy life I've led up till now', 'dislocation . . . a pain which can't be shared', 'God how S [ad]'; 'How sad. How sad'; 'Now I must go. How sad, how sad.' Tony Last 'lay back in' his 'hammock sobbing quietly'. And how urgently and unashamedly Last and Waugh yearned for home, for the end of journeys! 'Bath, with its propriety and uncompromised grandeur'—this is the end of *Ninety-Two Days*—'seemed to offer everything that was most valuable in English life; and there, pottering composedly among the squares and crescents, I came finally to the end of my journey.' Until the next one, that is.

The '30s comprised an uprooted generation ('I'm beginning to know that all human life is really uprooted', Julian Bell wrote to his mother from China), a tribe of nomads who (in Day Lewis's description in *Starting Point*):

> tramped from town to town in search of work, who streamed away from country villages as from a hopeless wound, who fled on motor bicycles from the furies of their own confused and avenging guilt, who travelled abroad or bivouacked in London flats, made desperate by the impermanence of their world. They were all exiles.

What Elizabeth Bowen calls in *To the North* 'an anxiety to be elsewhere' was widespread. For the *expulsés*, of course, the sadness and discontents of exile were willy-nilly one's lot. Trotsky did not choose to become the period's most famous political exile. The droves of German intellectuals, writers, musicians, artists, photographers, psychoanalysts who fled Germany after 1933, the Jews of Europe, and later the Czechs and Poles, the French, and all the rest, who choked Europe's roads and railway stations and the receptive foreign cities, had exile forced upon them. 'Once we had a country', sing the German Jews of Auden's 'Song' (March 1939), but 'We cannot go there now, my dear, we cannot go there now.' MacNeice was of the same mind. His fine 'Valediction' to Ireland (in *Poems*, 1935) is about leaving

Belfast, 'my mother-city, these my paps', for good ('Farewell, my country, and in perpetuum'), but involuntarily, under necessity ('I must go east and stay, not looking behind'). This sense of regrettable compulsion made MacNeice reiteratingly sympathetic to the period's exiles and refugees (particularly in his poem 'The Refugees', but note also 'Brains and beauty festering in exile' in *Autumn Journal*, XVIII, and 'The guttural sorrow of the refugees' in the poem 'The British Museum Reading Room').

The non-Jewish Germans who stayed at home in Germany would not agree that there was any compulsion. Gottfried Benn's 'Answer to the Literary Emigrants' sees Klaus Mann and his kind as voluntary travellers, wrongheaded deserters ('You are writing to me from the neighbourhood of Marseille. In the little resorts along the Gulf of Lyons, in the hotels of Zürich, Prague, and Paris—you write—the young Germans who once admired me and my books are now sitting as refugees'). But we must see them rather as victims of a period whose characteristic note is enforced alienation. They are the unlucky denizens of a modernist world whose citizenry, whether it paid tribute to Marx or Freud or Jesus, to totalitarian fascist or leftist regimes, was ruthlessly condemned to the sorry and reluctant status of aliens. To be European in the '30s was to be like King Lear (so Spender reflected in his diary on 15 September 1939), driven out, deprived, 'and filled with madness from within and outside.' He was thinking of his own German friends, and of all his English friends separated from their German and Austrian boyfriends. 'The whole of Europe is filled with people who are violently separated from those they love.' For 'years', he continued to meditate on 30 September, the newspapers have been 'filled with photographs' of refugees, in Spain, in Poland. 'Most homeless of all, little shreds of matter from distant countries that have nothing to do with them are driven through their flesh.' He was referring to bullets.

Auden wonders in his poem 'In Memory of Ernst Toller' whether this particular refugee German writer, who committed suicide in exile in New York, died because he had the modern European sickness in his head: refugee sickness ('had the Europe which took refuge in your head / Already been too injured to get well?'). And the literature of our time would look extremely paltry were it thinned of exiles, whether one thinks variously of Modernism's exiled founders such as Conrad the Pole and Eliot and James the Americans in England and Joyce the Irishman in Zürich, Trieste, Paris, or of the literary refugees from the horrors of the First War (Graves on Majorca, David Jones in his 'dug-out' in Monksdene Residential Hotel in Harrow on the Hill, Ivor Gurney in his Gloucester mental asylum) or of a Nabokov, the White Russian in flight from Red Russia, or a Viktor Shklovsky, a Red in flight from Stalinists, or a Walter Benjamin and Thomas Mann running from Nazis. The list of the great exiled—withdrawing into foreign lands, into the private and protective spaces of their own heads, into the ultimate self-exile of suicide—is enormous, lending high colour to the larger claims of Marxists, Freudians and Christians that we are all exiles. The '30s Marxists believed this strongly; so did the Freudians. Auden, notably, kept returning to the Freudian belief that everyone was born into the exile's plight, exiled from the womb and subsequently from mother's arms, so that Freud's death in September 1939 in exile from Germany had peculiar force for Auden: 'an important Jew who died in exile', Auden calls him in 'In Memory of Sigmund Freud'. Beckett, again notably, also agreed: his story *L'Expulsé* (written in

1946, after Beckett's own experience of flight upon flight: first from Ireland then from the German invaders of France) epitomises his repeated handling of the Freudian myth. Something fundamental of this sort animates Cyril Connolly's repeated distress, in his diary, at the loss of 'Eden': an Eden that's much more than, though it obviously includes, the mere Eton. The '30s Christians, too, saw themselves as unwilling exiles in the Biblical tradition of believers, 'strangers and pilgrims' on the earth, their New Jerusalem still way ahead of them. Movingly, Auden drew on *The Wanderer* and *The Seafarer*, the Anglo-Saxon poems about Christian *peregrini* that expressed the religious experience of exile from heaven, to reflect his heap of perennial discontents about growing up, leaving home and mother, being a homosexual, and perhaps also being distant from God:

> Doom is dark and deeper than any sea-dingle.
> Upon man it fall
> In spring, day-wishing flowers appearing,
> Avalanche sliding, white snow from rock-face,
> That he should leave his house,
> No cloud-soft hand can hold him, restraint by women;
> But ever that man goes
> Through place-keepers, through forest trees,
> A stranger to strangers over undried sea,
> Houses for fishes, suffocating water,
> Or lonely on fell as chat,
> By pot-holed becks
> A bird stone-haunting, an unquiet bird.

The exile's portion is self-mocking dreams of an impossible home—as unlikely of attainment as heterosexual happiness has proved:

> There head falls forward, fatigued at evening,
> And dreams of home,
> Waving from window, spread of welcome,
> Kissing of wife under single sheet;
> But waking sees
> Bird-flocks nameless to him, through doorway voices
> Of new men making another love.[250]

Everyone, according to Auden, shared Arnold's plight ('Matthew Arnold'), ousted from the protective 'mother-farms', alienated from 'the father's fond chastising sky'. Every poet ('The Composer') knew the way writing connected but also hurtfully disconnected Life and Art ('Rummaging into his living the poet fetches / The images out that hurt'). But only Auden and his special sexual kind shared the redoubled exilings of the homosexual author, reflections on which help make the sonnets 'Rimbaud', 'A. E. Housman', and 'Edward Lear' among the strongest poems of the '30s and of Auden's entire career. 'Verse was a special illness of the ear'; and three poems, three careers, write in a single authorial tone:

> Now, galloping through Africa, he dreamed
> Of a new self, the son, the engineer,
> His truth acceptable to lying men.
>
> No one, not even Cambridge, was to blame;

—Blame if you like the human situation—
Heart-injured in North London, he became
The leading classic of his generation.

Deliberately he chose the dry-as-dust,
Kept tears like dirty postcards in a drawer;
. . .

In savage footnotes on unjust editions
He timidly attacked the life he led.

Left by his friend to breakfast alone on the white
Italian shore, his Terrible Demon arose
Over his shoulder; he wept to himself in the night,
A dirty landscape-painter who hated his nose.

Affection, Auden glooms in the Lear sonnet, 'was miles away'. Spender, too, despaired over the sets of exiles enfolding him. The letter to Grigson from Vienna about stifling 'in a bugger's world' (1934) unites sexual with geographical alienation. In 'The Uncreating Chaos', which is about the rise of the mad Corporal Hitler ('the tide of killers, the whip masters'; 'pathics with rubber truncheons'; 'Spontaneous joy in padded cell'), he throws political alienation into the laden balance of his despairs: 'you', the poem declares, not a normal man, 'ever a bride', are now claimed in marriage by 'the uncreating chaos', fears of which drive into further exile: 'anxiety' is 'a globe trotter'; 'our fear makes being migratory'. There's real feeling in E. M. Forster's letter to Isherwood, 30 April 1937, at the height of the Heinz trouble that was keeping Isherwood out of England and out of any real place of settlement: 'how are you? I don't suppose you are very well, what an endless run round.' Exile on these models was not only terrible, it was unavoidable as illness, as mortality itself.

Some people did, by contrast, seem actually to be choosing exile, freely and gladly, volunteering for expatriation among the workers or actual emigration overseas, or opting willingly for internal emigration, the turning in one's heart to heaven or Moscow as a religious or political intourist. But questions of freewill and determination are as notoriously hard to decipher or resolve in the '30s as anywhere, whether literal or metaphorical, religious, or political alienations are in question. Easier are questions of contentment and discontentment. Even among those who believed themselves to be cheerfully, so to say, cooperating with God's will or History's determination, dementing dissatisfactions are not hard to detect. Orwell seems wilfully hasty when he implies, in his sneers about Communism as 'the patriotism of the deracinated', that such deracination was an easy option. Whatever happened, even the most cautious of internal exiles were people afflicted by dissatisfactions and feelings of homelessness, of being *dépaysé* as Connolly put it in a gloomy moment in his diary, even in their own country. No reader believes that Eliot's 'Journey of The Magi' is describing a journey that is other than unsettling. So it is too with, say, Spender's Marxizing 'Exiles from their Land, History their Domicile'. This title, the best line in Spender's poem (from *The Still Centre*), touches his own condition acutely. The words of the historical exiles, he explains, those libertarian thinkers and poets who sided with history and were driven into foreign countries ('And all outside the snow of foreign tongues') are now coming home to 'us': they are 'exiles long returned'. But their awakened audience, Spender and other British Marxists, the would-be siders with history, are still exiled: the historical exiles' success, purposeful action, and revolutionary achievement, contrast shamingly with 'our own wandering present uncertainty', 'life's exile'.

The exiles kept looking forward, at least while the tyrannously expelling forces prevailed, or conviction lasted that Better Places were ahead. 'We have no home. Our bourgeois home is wrecked.' Thus Charles Madge, in 'Delusions'. But it did keep dawning on the travellers how prone their anticipations were to dissolve into the disappointments of mirage:

> Sometimes the sight of gently waving green
> Invites the weary traveller's footsteps on
> Refracted far across the waste between,
> But, one step more, the glancing palms are gone.[251]

Travellers and exiles of all sorts, writers of travel books, 'the analyst and the immigrant alike', would desire Henry James's 'Great Good Place', the place Eric and Anna dream of in *On the Frontier*:

> *Eric*: But in the lucky guarded future
> Others like us shall meet, the frontier gone,
> And find the real world happy.
> *Anna*: The place of love, the good place.

But the disappointments in the putative Good Place could be intense. 'None of us really get there', was the glum conclusion of H. G. Wells's efforts to find 'that Great Good Place' that his 'old elaborate-minded friend, Henry James' had imagined (it comes in the prelude to Wells's *Experiment in Autobiography*, 1934). Auden lists his own American disappointments in his 1946 Introduction to James's *The American Scene*: 'the unspeakable juke-boxes, the horrible Rockettes and the insane salads . . . the anonymous countryside littered with heterogeneous *dreck* and the synonymous cities besotted with electric signs . . . radio commercials and congressional oratory and Hollywood Christianity . . . all the "democratic" lusts and licenses.' In such cases it was not only home that was ('Letter to Lord Byron') 'miles away and miles away'. And being suspended between the disappointing past and the still tantalizing future, with (as Day Lewis put it in Section 30 of *The Magnetic Mountain*) 'Nothing to rest the eyes on / But a migrant's horizon', finding oneself where one's friends can't find one, *Address Not Known* (that was the projected title of the abandoned travel book about America commissioned from Auden and Isherwood by John Lehmann at Hogarth), perpetually a traveller, caught between termini, like a Beckett character fraughtly tensed between life and death, could hardly not induce despair. Christopher, the hero of John Sommerfield's *They Die Young* (1930) went out to see the world—Oxford, Soho, New York, South America—delighted at first to travel ('I am on a ship actually I Christopher am on a ship'), but increasingly certain he was getting nowhere: 'There are no places left for Christopher'. Returning home only made things worse:

> Yet now there was not any place left for him; in going to sea he had found no content, and now, returning to his old environment, there was only a place for the Christopher he had been, not for the prisoning hulk of a forlorn spirit he now was.

He chooses suicide: deliberately crashing a Bentley. It is not a reassuring parable for those who eventually abandoned their New World hopes, whether the defeated

Moscow lovers or W. H. Auden himself discovering in 1972 on his emigré's return that Oxford was no more possible a continuing city than New York had proved.

The suicide of Sommerfield's Christopher confronted being a nowhere man (such as John Nower, that early Auden character in *Paid on Both Sides* with the hintful name—picked up, as a matter of fact, from one of the effigies in Christ Church Cathedral, Oxford) and did so more honestly than the glib pretence that this condition was a good thing. 'Well, we're off again', grinned Isherwood in the boat-train on the way to Southampton and New York, 18 January 1939. 'Goody', said Auden. As unconsoling a making the best of a bad job, of settling for exile's regrets, as when Auden celebrates Edward Lear for successfully reaching, and then becoming, the island of 'his Regret' ('children swarmed to him like settlers. He became a land'), or when Greene extols the seedy level, or Robert Graves talks toughly about accepting the inevitability of exile in his poem 'The Cuirassiers of the Frontier' (hard-nosed rebuke to Auden's 'Roman Wall Blues': 'Here is the frontier, here our camp and place'; 'We, not the City, are the Empire's soul: / A rotten tree lives only in its rind'), or when Graves, again, in 'The Cloak' briskly faces the plight of the man dodging back and forth across the Channel, never actually landing in England, domiciled now in Dieppe ('exile's but another name / For an old habit of non-residence / In all but the recesses of his cloak'), or when Christopher Isherwood professedly takes to the unsettling impermanence of his Californian 'home'. 'Few of the buildings look permanent or entirely real. It is rather as if a gang of carpenters might be expected to arrive with a truck and dismantle them next morning.'[252] The more provisional, it seems, the better:

> What was there, on this shore, a hundred years ago? Practically nothing. And which, of all these flimsy structures, will be standing a hundred years from now? Probably not a single one. Well, I like that thought. It is bracingly realistic. In such surroundings, it is easier to remember and accept the fact that you won't be here, either.[253]

Orwell, at least, would have thought this a specious glossing of the exile's troubles: not least because Isherwood's was one more case of the homosexual's double exile in which, by choosing expatriation, in Berlin, California, wherever, Isherwood had only confirmed and doubled the outcast narrowness his sexual nature anyway thrust him into. Being a mere heterosexual exile, like Henry Miller, was, in Orwell's view, disabling enough. 'That is the penalty of leaving your native land', Orwell declared:

> It means transferring your roots into shallower soil. Exile is probably more damaging to a novelist than to a painter or even a poet, because its effect is to take him out of contact with working life and narrow down his range to the street, the café, the church, the brothel. On the whole, in Miller's books you are reading about people leading the expatriate life, people drinking, talking, meditating, and fornicating, not about people working, marrying and bringing up children; a pity . . .

There could scarcely be a sourer footnote to a period in which the hopes for what exile—especially 'going over' to the working-classes—might bring to literature had run, so often, so deludingly high.

13

Spanish Front[254]

IF there is one decisive event which focuses the hopes and fears of the literary '30s, a moment that seems to summarize and test the period's myths and dreams, to enact and encapsulate its dominant themes and images, the Spanish Civil War is it. This war granted a special summary and mirroring of the period, an intense experience of self-reflection and self-realization. Very speedily indeed, the right-wing military coup against the legitimate Republican government of Spain got itself perceived in this way. In Spain, according to John Lehmann, 'everything, all our fears, our confused hopes and beliefs, our half-formulated theories and imaginings, veered and converged towards its testing and opportunity'. Spain made an epoch. George Orwell, volunteer fighter on the Republican side, seriously wounded in the throat, passionate to tell the world his truth about what had gone on in the political conduct of the war, was of course a special case. But his preoccupation with Spain, and the way the war clarified and transformed his life, were not all that atypical. After his experiences there he would keep harping, in a variety of moods, on the intense hold Spain had achieved over his imagination:

> the devil of it is that at present I simply can't write about anything but Spain . . . This Spain business has upset me so that I really can't write about anything else.

> I hate writing that kind of stuff [the political discussions in *Homage to Catalonia*] and am much more interested in my own experiences, but unfortunately in this bloody period we are living in one's only experiences *are* being mixed up in controversies, intrigues etc. I sometimes feel as if I hadn't been properly alive since abt the beginning of 1937. I remember on sentry-go in the trenches near Alcubierre I used to do Hopkins's poem 'Felix Randal' . . . over and over to myself to pass the time away in that bloody cold, & that was abt the last occasion when I had any feeling for poetry.

> The Spanish war and other events in 1936–7 turned the scale and thereafter I knew where I stood. Every line of serious work that I have written since 1936 has been written, directly or indirectly, *against* totalitarianism and *for* democratic Socialism, as I understand it.[255]

'Is this a strange country?' demanded Jack Lindsay's 'Requiem Mass: for the Englishmen fallen in the International Brigade.' And the firm answer was 'No. I have recognised it'. For Spain had, as it were, long been anticipated. Revolution and inevitable counter-revolution; the coming apocalyse that had already enjoyed rehearsals in Germany, China, Abyssinia; the need for drawing the line and seeking a showdown with Fascism; the necessary immersion in the destructive element: all these had been recurrent and prophetic themes among left-wing intellectuals long before the Spanish generals tried to seize power on 18 July 1936. Spain was the final confirmation—if one were, by that time, needed—of the Crisis.

Fulfilled prophecies can unsettle startlingly—as the boy who cried 'Wolf' found when the wolf at last turned up. No small part of the dedicated battiness of Roy

Campbell's poetic tirades on the subject of Spain is, I think, due to the way his triumphalist claims to have forecast certain detailed aspects of the war's outcome, in particular the discomforting of the Left ('Who promised this before the war begun, / And drilled them with my pen before my gun / To dance in dudgeon what I wrote in fun'), seem to have loosened his hold on reality. But it's not at all surprising, in the Spanish atmosphere of wish and fear, fantasy and nightmare come true, that lies and distortion of truth were commonplace, that accusations and counter-charges of contorting reality and history flew busily between apologists for the contending sides, that myths flourished. Not surprising, either, that the truth about Spain is still a disputatious matter (there are, for instance, people who go on denying that fascist aeroplanes destroyed Guernica or that the Germans and Italians put large numbers of flyers and planes at Franco's disposal), still less that myths still flourish.

In the minds of too many readers of poetry the war is still a poet's war. In their vision of Spanish events, inspired by the memorable phrases of Auden's poem 'Spain', the poets are still 'exploding like bombs'. British poets did, of course, go to Spain: most of them, including Auden, Spender, MacNeice, Edgell Rickword, Valentine Ackland, Sylvia Townsend Warner and David Gascoyne, as short-term visitors, or reporters, solid with the Republican cause, anxious to do something, but not actually fighting. Some poets did fight and, what's more, survive—Tom Wintringham and Miles Tomalin notable among them. Clive Branson was taken prisoner before he could do much fighting. The Irishman Ewart Milne preferred to work, for the duration of the war, with a medical unit. For his part, Roy Campbell repeatedly professed to have fought on the Franco side, but much doubt shrouds his raucous boasts of gallantry in the anti-Republican field. Some poets, famously, not only fought but died in the Republican cause: John Cornford, Julian Bell, the Irishman Charles Donnelly, the critic and aviator Christopher Caudwell. Ralph Fox, the other name commonly included among the 'poets' was not, of course, a poet but an historian, critic, novelist, and member of the staff of the *Daily Worker* and *Left Review*. And mention of Ralph Fox instantly underlines what is the case: that among writers and artists who were active in Spain, poets had no monopoly. George Orwell, critic and novelist, was only marginally a poet. Ralph Bates served as an important functionary in the Secretaria d'Agitacio of the United Catalonia Socialist Party, and he was a novelist. Just so, Jason Gurney who was disabled in the fighting, was a sculptor. Felicia Browne, the first Briton to be killed fighting in Spain, was another militant non-writing artist (there was a posthumous exhibition of her drawings in Soho in October 1936, to raise money for a British medical Aid Unit in Spain).

It was relatively easy, of course, reading the *DW* or the *LR* and in an air thick with pro-Republican manifestoes by writers and artists, to believe that the war actually depended on the moral support of a Hemingway or a Paul Robeson, on *los intelectuales* who motored about Republican Spain in July 1937 as part of the Second International Congress of Writers for the Defence of Culture, on the graduates and poets ready to die as the latest Byrons of the barricades. Even the fighting men appear to have placed great faith in their bourgeois comrades. The recollections of the working-class volunteers assembled in Judith Cook's admirable *Apprentices of Freedom* (1979) are prominently full of memories and myths of Fox and Cornford, of George Nathan's pukka guards-officer style, all swagger-stick and curled moustaches.

The day he was killed Cornford 'had a bandage round his head—he looked like Lord Byron'. 'I remember Ralph Fox looking very dashing. He wore a black beret, black leather coat and a large revolver—very dashing. Then there was young John Cornford, a real romantic type, six foot tall, twenty years of age, and he looked like a bloody Greek god.' 'I remember when the shelling was getting us down seeing Clifford Wattis, the actor's brother, who looked a real English aristocrat, all peaches and cream, sitting down and *shaving* in a wee mirror.' But, of course, for every Cornford or Fox or Caudwell scores of ordinary working men, trades-unionists, out-of-work manual labourers, assorted socialists and activists who would never be famous, never write a book or make a headline in the *Daily Worker*, had volunteered to fight. It has been estimated that over 80 per cent of the 2,762 Britons in the International Brigades were working-class. One of the cruellest injustices to their memory and bravery has been literary people's readiness at losing sight of them in the 'poets' war' legend.

Myths flourished in Spain, especially politically expedient stories, distortions promoted for the reasons Orwell kept attacking, the idea that any lie was justified if it helped 'our', the Republican side win. The peculiarly mythic atmosphere was engendered not least by the current sense of apocalypse. In Spain the period was to discover its favourite convergence of personal and public crisis most tellingly expressed. Here leftist authors found themselves challenged to write out, in their lives as well as on their pages, the '30s Book of Revelation. For here, starkly confronting them, were the classic Last Things of Christian dogma: Apocalypse in its traditional four kinds—Death, Judgement, Heaven and Hell. They were the things, of course, that revolutionary history and revolutionary talk had always been about: the final testing moment when, all at once, under the constant possibility of death, the Hell of Fascism, of the Destructive Element, and the Heaven of the revolutionary pastoral, beckoned. Nothing could be so exciting as such End Times, nothing so fearful.

Spain, like revolution, was of course a commitment less into the hands of an angry God on a Biblical Judgement Day, than into the testing crucible of history's secularized last day, 'On the last mile to Huesca, / The last fence for our pride.' 'The last fight let us face': so Cornford, in the tones of the revolutionary's traditional refusal to defer embracing the ultimate end in the words of 'The Internationale'. 'Then Comrades, come rally, / And the last fight let us face': the comrades sing it inside and outside the court-room in Lewis Jones's *We Live* when Len faces trumped-up charges of unlawful assembly, assaulting the police and allied misdemeanours, after a workers' demonstration. And *We Live* ends with Len going to Spain, being killed there, and his wife running to keep up with a procession of welcome for local survivors of the fighting: 'the people's day isn't over', declares Mary, 'and I must be with them till the last, as our Len was.' *We Live* offers a typically receding vista of revolutionary ultimateness, but the last thing we're actually given is a journey to Spain. So it is in James Barke's *The Land of the Leal* and Day Lewis's *Starting Point*: going to Spain, the War in Spain, are felt to be sufficiently climactic for a left-wing fiction to end on. The finality of Spain and of revolutionary struggle is emphasized by the death of Len and of James Barke's Andrew. Barke in particular tries consciously to release into his politicized fiction the energies of the Biblical apocalyptic he's drawing on. His last chapter, 'Thy Will be Done', has Chrissie dreaming of

David being bayonetted in Spain as a voice utters Scripture ('Behold: the days are coming in which they shall say Blessed are the barren, and the wombs that never bore, and the paps which never gave suck!'): 'She saw the bayonet sink into his midriff . . . saw his hands tear at the barrel . . . saw the boot crushing down on his face.' The yearning for the revolutionary Beyond had led to an abrupt end on the revolutionary battle-field. The socialist Heaven had proved as elusive in Spain as elsewhere. Just as utopian revolutionary literature (not just in the '30s) is consistently weak on the details of life after the revolution, so what might lie beyond the struggle in Spain is left unspoken by all three of these novels. For their heroes, the dream of that Heaven has clearly collapsed once more into the old hellish dream of violence, the sickening endlessness of struggling for a utopian hope forever deferred. *To Struggle Is To Live*, as Ernie Benson's autobiography (1980) has it, quoting Engels; *No Home But the Struggle*, in the words of the title of the last part of Edward Upward's trilogy; 'But today', as Auden's 'Spain' has it, 'today the struggle.'

Not that certain glimpses of Heaven hadn't been briefly vouchsafed in Spain. Nearer at hand than Moscow, nearer than some British Leftists had dared hope (and much too near for right-wing comfort), a revolution fleetingly emerged in Spain. 'I have seen wonderful things and at last really believe in Socialism, which I never did before': so Orwell enthused to Cyril Connolly about what he had seen in Catalonia (June 1937). 'For several months', he declared, 'large blocks of people believed that all men are equal and were able to act on their belief. The result was a feeling of liberation and hope that is difficult to conceive in our money-tainted atmosphere . . . No one who was in Spain during the months when people still believed in the revolution will ever forget that strange and moving experience.'[256] On the Aragon front, breathing the classless 'air of equality' of the workers' militias, Orwell had experienced, he says in *Homage to Catalonia*, 'a foretaste of Socialism'. And *foretaste* is, not accidentally, the Authorized Version of the Bible's word for the Christian's earthly apprehension of the Heaven that is to come.

Orwell's famous rhapsody about the revolution in the public appearance of Barcelona (in the opening chapter of *Homage to Catalonia*)—the first town he'd ever been in 'where the working-class was in the saddle', an urban Eden colourful with revolutionary posters and slogans, loud with constant revolutionary music, a comradely place from which smart clothes, tipping, private motor-cars, servility, polite modes of address, and the customary insignia of bourgeois, priest, and officer had been banished—was to be echoed and re-echoed in the diaries and letters of many a visitor. Franz Borkenau, Cyril Connolly, Philip Toynbee, Ralph Fox all got excited about similar evidences of revolutionary style. W. H. Auden entered into their enthusiasms in Valencia.[257] The air of fantasy incredibly come true is ministered to not least by the way heterosexual observers are so frequently taken by the emancipated girls, so fetchingly butch in their trousers and side-arms—'You girls in overalls with young breasts of pride', drooled Jack Lindsay, at a distance, in his poem 'Looking At a Map of Spain on the Devon Coast' (August 1937).

And yet, if the exhilaratingly bright days of the Spanish revolution hadn't already been overcast by the dark forces of Fascism (that modern version of Hell for the 'new Quakers' of the Left, as Evelyn Waugh sharply put it),[258] none of these excited observers would have paid their visits. 'I recognised it immediately as a state of affairs worth fighting for', Orwell avowed. But accepting one's immersion in the destructive element of war and civil war, was, for many, the short road to feeling disconcertingly tarnished by the fascistic hellishness of any and all killing. Auden steeled himself, notoriously ('Spain' again), to

> the deliberate increase in the chances of death,
> The conscious acceptance of guilt in the necessary murder.

He had publicly accepted the necessity for killing as a means of preserving the Spanish revolution, in his 'Valencia' report ('at the gates of Madrid where this wish to live has no possible alternative expression than the power to kill'). But Auden's poems on the sickness of psychotic killers that followed rapidly on the publication of 'Spain' ('Victor', June 1937; 'James Honeyman', August 1937), as well, perhaps, as the admission within the poem 'Spain' itself that the war was an enactment of 'our fever', our 'fears' and 'greed', indicate clearly enough how sickened Auden had been by the Spanish acceptance of violence. The non-combatant Ewart Milne took a principled stand against accepting 'murder', early on in his poem 'Sierran Vigil (1939)': 'exalting not the self-evident murder.' Publicly, at least, Spender worked his way cagily towards the same sort of rejection. His *New Statesman* review of *John Cornford: A Memoir* grants the strength and vigour of Cornford's poems, and is warm towards Cornford's exemplary courage, 'The spirit of Cornford and some of his comrades.' But Spender's worries about violence are also clear: Cornford's poems are defective because 'violent and insensitive'; 'his vision of life is impatient and violent, it leaves too many questions unanswered, he burns out too quickly, rushing headlong to his death'.[259]

Orwell couldn't have come out any more forcibly against the horror of war than he did in *Homage to Catalonia* and his other reflections on Spain. To 'fight you have to dirty yourself. War is evil': his essay 'Looking Back On the Spanish War' (1943) is plain. Orwell knew how shocking it was to have a bullet penetrate his flesh ('a violent shock, such as you get from an electric terminal: with it a sense of utter weakness, a feeling of being stricken and shrivelled up to nothing'). He retained his horrified sense that the shooting of human beings was not an event to be contemplated or spoken of lightly. When he and his first wife visited Georges Kopp, POUM officer and comrade, in a Barcelona jail:

> Kopp seemed in excellent spirits. 'Well, I suppose we shall all be shot', he said cheerfully. The word 'shot' gave me a sort of inward shudder. A bullet had entered my own body recently and the feeling of it was fresh in my memory; it is not nice to think of that happening to anyone you know well.

'Not nice': a fierce understatement that, quite understandably, gave way to his famous outburst against the 'good party man' Auden's 'conscious acceptance of guilt in the necessary murder' ('To me, murder is something to be avoided . . . Mr Auden's brand of amoralism is only possible if you are the kind of person who is always somewhere else when the trigger is pulled. So much of left-wing thought is a kind of playing with fire by people who don't even know that fire is hot'). It was easy for the wounded veteran to feel the un-wounded, non-fighting poet had been as thoughtlessly aloof from the violent realities he was touching on as the unworthiest of propagandist journalists or canting politicoes. And aloofness was indeed Auden's continual failing, not least over Spain. But the intensity of Orwell's attack must owe something to the fact that he himself had once been prepared to accept the necessity of revolutionary violence, and was still tormented by the dilemma over 'just wars'. 'Those who take the sword perish by the sword': he agreed in 'Looking Back on the Spanish War' with the traditional Christian assertion. But he added that 'those who do not take the sword perish by smelly diseases': 'To survive you often have to fight, and to fight you have to dirty yourself.'

Orwell's fierceness and its particular slant are related to the feelings and strategies of *Homage to Catalonia*, in which, in the light of Orwell's disillusionment over the Spanish revolution's collapse, kinds of killing are almost forensically distinguished. Orwell has accepted the necessity of violence. 'When I joined the militia I had promised myself to kill one Fascist—after all, if each of us killed one they would soon be extinct—and', he regretted, 'I had killed nobody yet.' Orwell fires shots, throws bombs, tries to bayonet an enemy soldier, and he's 'glad to say' that the men endangered by a seemingly traitorous officer who 'mucked up' a surprise-attack 'shot him dead on the spot'. A specialist in brutal truths, Orwell delights in admitting he would love to see jingoistic cowards bumped-off ('Sometimes it is a comfort to me to think that the aeroplane is altering the conditions of war. Perhaps when the next great war comes we may see that sight unprecedented in all history, a jingo with a bullet-hole in him'). But Orwell carefully minimizes the killing he is himself actually engaged in by playing up the farcicality of his anarchist/POUM unit. He presents himself and his comrades as badly armed, mere children, people who can't shoot straight (Orwell shoots only to miss), who tend to inflict wounds on themselves rather than on the enemy. They're a harmless joke. 'In this war everyone always did miss everyone else, when it was humanly possible.' It's a 'cock-eyed war', a ' "comic opera with an occasional death" ', a 'bloody pantomime'—a charade amply illustrated by Orwell's 'comic memory' of chasing along a trench a naked man who was clutching a blanket about his vulnerable person, Orwell the while taking ineffectual prods at his shoulder-blades with a harmless bayonet. Orwell presents himself as a usually clownish, ineffective man of violence. In any case he's much more sinned against than sinning, in the end the victim rather than the aggressor, terribly wounded in the throat, his voice nearly lost for good. What's more, the one effective bit of killing he allows himself to have committed is narrated as a decent, soldierly scrap. It's arranged, in fact, as a microcosm of The Just War, an illustration of the necessity of killing Fascists lest they kill you. Orwell and his comrades attack a fascist trench. Men are being hit all about him; he himself puts his hand up to his cheek in an engagingly futile defence against bullets hitting him in the face. So when he finally succeeds in potting an enemy with a grenade ('Ah! No doubt about it that time'), his deliberate act of violence seems only the most justifiable gesture of self-defence, a perfectly proper personal translation of the general need to defend oneself against Fascism. And the whole book's constantly wry current of farce is called deftly upon to provide this particular episode's tonal frame: Orwell shows himself lengthily wrestling the pin out of his first grenade; he throws it and a second grenade hopelessly wide; and at the climax of the whole episode there comes the pantomime with the bayonet. And so the problem of killing's necessity is subtly allayed, comfortingly buried between a play of contingency, farce and blokish decency. In Orwell's book it's other people who tend to do the serious or the dirty killings, the 'necessary murders'.

And away from the clean or, alternatively, comical air of the Orwellian fronts the ordinary soldier was being systematically betrayed, let down, lied to by cynical Stalinist agents and politicians, who would willingly use the old repressive methods and agencies to get rid of the 'heroes' they once recruited and praised, as soon as it suited their political ends and ambitions. Orwell's theme revives the indignations of First World War fighting men against politicians and officers far from the front lines ('Base Details' Siegfried Sassoon called them). In speedily re-corrupted Barcelona, officers strut about wearing automatic pistols, which 'we, at the front, could not get . . . for love or money'. The Barcelona police have good uniforms and weapons: 'I suspect it is the same in all wars—always the same contrast between the sleek police in the rear and the ragged soldiers in the line.' There was none of Barcelona's political trouble in the brotherhood of the *Frontkämpfer*:

> In the next bed to me there was an Assault Guard, wounded over the left eye. He was friendly and gave me cigarettes. I said: 'In Barcelona we should have been shooting one another', and we laughed over this. It was queer how the general spirit seemed to change when you got anywhere near the front line. All or nearly all of the vicious hatred of the political parties evaporated. During all the time I was at the front I never once remember any PSUC adherent showing me hostility because I was POUM. That kind of thing belonged in Barcelona or in places even remoter from the war.

The Stalinists themselves were not unmindful of First War analogies, though with less sense than Orwell, perhaps, of their irony. One of Orwell's named targets in *Homage* is Ralph Bates, made to stand for the typically mendacious political agent, because he stated in *New Republic* 'that the POUM troops were "playing football with the Fascists in no man's land" '—an echo of the First War myth of Yuletide compromise between British and German soldiery—'at a time when, as a matter of fact, the POUM troops were suffering heavy casualties and a number of my personal friends were killed and wounded'. Base Details like these signalled the degeneracy of the Spanish War into pointless violence akin, in Orwell's mind, to the First War.

If anything, though, this violence was bloodier. Certainly no war literature flows redder with blood than the Spanish War's. The bombing of open cities, the machine-gunning of refugees on open roads, introduced new horrors and new senses of horror. Under The New Barbarism, as T. C. Worsley called it, women and children as well as soldiers became for the first time in history regular targets for professional soldiers' angry guns. And images of that blood shed so freely merged naturally with the traditional rednesses of Spain: the blood of bull-fights, matadors' red capes, Goya's scenes, the blood of the thousands of effigies of the crucified Christ worshipped with such fierceness in the peculiarly sado-masochistic adorations of Spanish Catholicism. The 'blood-caked plains where the Spaniards fight' (Ruthven Todd, 'Christmas Present 1937'); 'Blood over China . . . and blood again / from sunburnt lovely limbs in Spain' (Peter Hewett, 'Notes for a Letter'); 'bloody sandwich' (Nancy Cunard, 'To Eat To-Day'); the oars 'Dripping with blood' in Day Lewis's 'The Nabara'; the Asturian miners who 'swim in blood' in A. L. Lloyd's translation of Tuñón's 'Long Live the Revolution': the war's grisly varieties of bloodshed are made to feel quite at home amidst the overlapping reds of the cross and the bullfight.[260] Right-wing Catholic convert Roy Campbell plays up the 'bloody Christ' angle, especially in his volume *Mithraic Emblems* (1936):

Toledo, hammered on the Cross,
And in her Master's wounds arrayed
('Hot Rifles')

Close at my side a girl and boy
Fell firing, in the doorway here,
Collapsing with a strangled cheer
As on the very couch of joy,
And onward through a wall of fire
A thousand others rolled the surge,
And where a dozen men expire
A hundred myrmidons emerge—
As if the Christ, our Solar Sire,
Magnificent in their intent,
Returned the bloody way he went,
Of so much blood, of such desire,
And so much valour proudly spent,
To weld a single heart of fire.
('Christs in Uniform')

Less Christian and more Republican poets prefer bull-fighting images: the children 'forced into the mould of the groaning bull' in Spender's 'The Bombed Happiness', the flaunted 'red rag' of Laurie Lee's 'A Moment of War', or George Barker's 'Elegy on Spain'. Barker's Spanish poems are ablaze with redness of all sorts (he shares the Surrealists' proclivities for imaged violence), but nowhere more so than in his 'Elegy' dedicated 'to the photograph of a child killed in an air raid on Barcelona'. This poem was, in John Lehmann's view (in his *New Writing in Europe*, 1940), 'as fine as Auden's *Spain*, and in many ways more moving because it is closer and angrier, more *wounded* one might almost say'. More wounded indeed, with its flaring rhetoric of blood: 'Bloody time', 'robe of blood that love illumines', 'the bomb's red wink and roar', red skies, 'bleeding sea', all clustered about the images of the child whose 'remains' are 'Staining the wall and cluttering the drains', and of the bleeding hero whose blood feeds 'the stones that rise and call' *No pasarán!*, a hero assimilated to the bleeding bull of tortured Spanish humanity: 'O bold bull in the ring.'

This bloodiness is smeared all over Spanish War writing. In Herbert Read's 'Bombing Casualties in Spain':

These blench'd lips
were warm once and bright with blood
but blood
held in a moist blob of flesh
not spilt and spatter'd in tousled hair.

In the 'mashed spouting redness' that an International Brigader's leg becomes when it's struck by bomb fragments in Wintringham's prose piece 'It's a Bohunk'. In the blood that 'poured out of my mouth', 'dribbling out of the corner of my mouth', that 'bubbled out of my mouth', in Orwell's descriptions of his throat wound. And so on and on, in a shockingly vivid set of exemplifications of what being immersed in war's destructive element actually meant. Such horrors were bearable, perhaps, in freedom's cause. As George Barker grimly tried to encourage himself in 'Elegy on Spain':

This flower Freedom needs blood at the roots,
Its shoots spring from your wounds, and the bomb
Booming among the ruins of your houses, arouses
Generation and generation from the grave
To slave at your side for future liberation.

They were bearable, too, so long as the 'red rag' lain across the 'hero's' eyes and baptizing the sand of Madrid with his blood was identifiable with a Red Flag that stood for liberty. As Cornford has it in 'Full Moon at Tierz: Before the Storming of Huesca':

Swear that our dead fought not in vain,
Raise the red flag triumphantly
For Communism and for liberty.

But if Freedom was in practice being betrayed then what was the point of the Spanish shedding of blood? In that case the war was merely the First War's pointlessly destructive element continued by another name.

So the Orwellian Base Details were particularly chastening. For they were indeed daunting evidence of the Spanish Revolution's decline and fall. It's no accident that *Homage to Catalonia* sounds like many a Trotskyite polemic on the decline of the Russian Revolution under Stalin. Spain provided a microcosm of Stalinism in action, a handy index to the whole repressive bag of Stalinist tricks, plottings, lies, smears, disappearances, the knock on the door in the middle of the night, the backstage bumpings off, imprisonment without trial, torture, brain washing, suborned witnesses, show trials. It made many Britons uncomfortably aware of continental political realities. 'It is not easy', Orwell writes in *Homage*, 'to convey the nightmare atmosphere of that time':

In England . . . There is political persecution in a petty way; if I were a coal-miner I would not care to be known to the boss as a Communist; but the 'good party man', the gangster-gramophone of continental politics, is still a rarity, and the notion of 'liquidating' or 'eliminating' everyone who happens to disagree with you does not yet seem natural. It seemed only too natural in Barcelona. The 'Stalinists' were in the saddle . . .

'Good party man': the phrase was deliberately applied, then, to Auden in 'Inside the Whale'. Orwell thought Auden had schooled himself to accept the sort of violence that Orwell and, he implies, all decent fighters for an unStalinist revolution, were repelled by. Certainly it was precisely Orwell's refusal to accept the politics of liquidation and elimination that caused harder Communists to sneer at *Homage to Catalonia*. 'The value of this book', jeered John Langdon-Davies in the *Daily Worker*, 'is that it gives an honest picture of the sort of mentality that toys with revolutionary romanticism but shies at revolutionary discipline.'[261] It's noticeable how much readier to live with the necessary murder were the tough proletarians who went to Spain with years of hardship and street-fighting behind them, and the committed members of the Communist Party like Fox, Donnelly, Cornford, and Caudwell. Cornford belonged, of course, to the generation younger than Auden and Spender, a generation schooled in hard-eyed violence not least by Auden's early verse. In a *NS* review (2 July 1976) of Samuel Hynes's book on the '30s and of the National Gallery's 1976 *Young Writers of The Thirties* exhibition, Spender was at pains to stress this generation gap. He, like Orwell, and, eventually, Auden, had found it hard to shed their liberal inhibitions. Not so the younger boys. And Spender quoted a letter of Julian Bell's to illustrate the kind of political unsentimentality it took to stay undisillusioned in Spain:

> I can't imagine anyone of the New Statesman doing anything 'unfair' to an opponent
> . . . Whereas for my own part . . . I can't feel the slightest qualms about doing anything
> effective, however ungentlemanly and unchristian, nor about admitting to myself that
> certain actions would be very unfair indeed . . .

Army deserters and mutineers, IRA veterans, men familiar with the inside of British prisons and of boxing-rings (how many boxing champs, one notices, military and civilian, went to fight in Spain: Sam Wild, Joe Norman, Tony Maguire, Fred Copeman, Tommy Picton, Wilf Winnick): the sort of Spanish volunteer Judith Cook interviewed wasn't prone to a pasty writer's liberal worries. They'd punched up gamekeepers (Wilf Winnick was as renowned for his part in the Kinder Scout scuffles as for his action in Manchester's YMCA boxing ring). They'd tackled Mosleyites and police horses with zest and lead piping. 'Two ex-Clydeside workers told me with relish of wrenching heavy metal litter bins off lamp posts in Glasgow to go out and beat Mosley's men over the head with them.' 'Let's face it', said one witness, 'not all the International Brigaders were crusaders': they included 'the kind of men I've met since who just like a fight and are going to get in one anywhere—you find them all over the world'. Good in a corner, reputedly relied on as Shock Troops to be called on in sticky situations, such men 'didn't take prisoners'. On the Ebro, for instance,

> Tony Maguire, one of our boxing champions, got to the top of the parapet of a Spanish
> trench, and they called out in Spanish, 'Surrender!' 'Surrender be fucked', he replied
> and emptied his automatic into them.

No Orwellian pantomime, nor Spenderian wavering, there. This was the spirit rather of John Sommerfield in *Volunteer in Spain* (1937). Sommerfield warms to his automatic pistol—good really only for shooting 'a fair-sized dog at twenty yards' but useful against Fascists who after all were only swine ('we too would be shooting the swine pretty soon')—and stays impersonal about what he labels 'clay pigeon shoot-ing' in Madrid:

> you did not think that you were making widows and orphans, robbing mothers of their
> children. You fired at something, dark and moving and if you hit it, you felt 'good shot'.

Sommerfield was one of the small band of trusty Communist volunteers assembled by Cornford during his brief return to England in September 1937. *Volunteer in Spain*'s dedication 'to the memory of my friend, John Cornford' is apt, not least because Cornford amply endorsed Sommerfield's violent enthusiasms. Cornford fitted easily into the Party that deplored Orwell's distaste for 'revolutionary disci-pline'. The friends contributing to his *Memoir*—edited by Soviet apologist Pat Sloan—evidently thought it high praise to list his cravings for violence, his liking for films with shooting in them, for all-in wrestling, for the more ruthless bits of Revolutionary history (Bela Kun machine-gunning 5,000 prisoners, for example). Cornford had a reputation for rough, brawling aggression: 'to break up a Fascist meeting was perhaps his highest enjoyment':

he knew little of boxing science or rules, he told me, but he could *hurt* an enemy, and would as soon kick him as anything else. He would certainly have been formidable, with his temper and powerful, ungraceful build, in any hand-to-hand fight.

Margot Heinemann's poem 'For R. J. C. (Summer, 1936)' remembers Cornford this way, rebarbative, harsh, a human being bending humaneness to history's inhuman will: he 'Divides talk coldly with the edge of will', he 'Charmed no acquiescence: he convinced and led', he had 'the force that hurts'. 'Rightly', she writes, he had

> no work but to dissect
> Romance and prove it unromantic,
> Breaking the scenery with his conscious hands.

No wonder his friends knew him rather formally as John, not chummily as Johnny or Jack:

> No, not the sort of boy for whom one does
> Find easily nicknames, Tommy and Bill.

There was nothing like Spriggy (Caudwell) or Charlie ('Charlie Donnelly, small, frail': in Blanaid Salkeld's poem 'Casualties') with this hard-headed boy whose political aggressions throve amidst the aggressiveness of Republican Spain.

A recurrent word in Cornford's letters home to Margot Heinemann is *tough*. The fighting is tough ('There's a tough time ahead'; 'I do know we're in for a tough time'), and so the men must be tough ('Fred Jones, he was a tough . . . Kept his head in a tough time'; John Sommerfield, 'tough'). Of course he's pleased to bag a Fascist ('We got one and both [Frenchmen] said it was I that hit him, though I couldn't be sure. If it is true, it's a fluke, and I'm not likely to do as good a shot as that again'). He finds crediting bombed Fascists with humanity hard ('The comrades with me on the roof were shouting for delight as each bomb landed. I tried to think of the thing in terms of flesh and blood and the horror of that village, but I also was delighted'). And he accepts entirely the need for revolutionary terror. 'In order to prevent a Fascist outbreak, every night splits, unpopular bosses, and known Fascists are taken for a ride. Assisted by the militia, there is a peasant war raging in the countryside and thousands of Kulaks and landlords have been killed . . . There is a real terror against the Fascists.' Necessary murders, indeed. He's prepared, too, to justify similar treatment for Mosley, to domesticate the Spanish principle of liquidating enemies:

It's as if in London the armed workers were dominating the streets—it's obvious that they wouldn't tolerate Mosley or people selling *Action* in the streets.

On occasion, reading Cornford's letters, you do find yourself wondering exactly what separated such violence from the fascist sort resisting which was the supposed *raison d'être* of the Spanish struggle. Spender accused Roy Campbell and the Fascists of a boastful thuggery that flowered not just in words, nor even simply in old-fashioned rifles:

Here we have the Talking Bronco, the Brute Life armed with abusive words and, most unfortunately, not with Mr Campbell's Flowering Rifle, but with Flowering Machine Guns, Flowering He [i]nkels, Flowering Capronis.[262]

But Cornford was no stranger to the moral burdens of the Brute Life. And the Revd Tom's anti-fascist sermon in *Land of the Leal* ('And if the poet William Blake felt that a robin in a cage set all Heaven in a rage, how much more must our offence against the Almighty be when we torture and beat men to death in concentration camps and rain high explosives upon the children playing at school') could cut both ways. The Republic too had its secret prisons and practised torture; it too caged you in violence.

A great deal of Spanish War writing and activity concerns prisons. If, as was earlier argued, the '30s was in general caged, imprisoned in the destructive element, Spain and the writing about it only confirmed and amplified that caging. In Clive Branson's poems from and about the Franco jails he suffered in, in Wyndham Lewis's extremely odd novel *Revenge For Love*, in Ewart Milne's stories 'Escape' and 'The Statue', in Arthur Koestler's *Spanish Testament* (1937), in *Homage to Catalonia*, in Spender's embarrassed quest to get his boyfriend T. A. R Hyndman released from punishment for desertion from the International Brigade, prison is the theme. Of course Wyndham Lewis contrives it that Percy Hardcaster, fat and bogus Republican, should be felt to deserve his Spanish jail. And Branson's prison poems stir up proper anti-Franco responses; they speak for all good men, women, poets and writers silenced and suffering in the jails of the intolerant insurgents. But T. A. R Hyndman was imprisoned by his own side. So was the English airman in Milne's story 'The Statue' ('I don't understand why boys who have come out to help and are airmen get themselves into jail'). Milne's story includes a memorably grim visit to the International Brigade detention centre in Salamanca, where the guards from 'the army of the proletariat' behave more toughly than ordinary warders would and even the visitors find getting out hard. And *Homage to Catalonia* boils over with indignation against the nightmarish filling up of Stalinist jails with Orwell's POUM friends, loyal and heroic Republicans all. Orwell presumes the POUM leader, Andrés Nin, to have been murdered in a socialist jail. Orwell's POUM comrade, Bob Smillie, died of 'appendicitis' in a Valencia jail. Orwell's friend and commanding officer Georges Kopp was still in jail as Orwell wrote *Homage to Catalonia*; Orwell expects that he has been shot. (Happily, Kopp did eventually get away to England.) Orwell's friends and comrades, his wife, he himself, were all in danger from this new 'reign of terror' conducted malignantly, brutally by Stalinist secret policemen. Orwell and his wife got away from their Stalinist hunters by the skin of their teeth. Others were less lucky. Without being charged, without trial, men who had fought for liberty in the cause their persecuters still proclaimed, were captured to languish in dark dungeons, filthy, starving, untended, constantly threatened with the bullet in the brain:

> Smillie's death is not a thing I can easily forgive. Here was this brave and gifted boy, who had thrown up his career at Glasgow University in order to come and fight against Fascism, and who, as I saw for myself, had done his job at the front with faultless courage and willingness; and all they could find to do with him was to fling him into jail and let him die like a neglected animal.

It was only fitting, perhaps, that Zoo-watcher MacNeice should notice the dying animals in Barcelona Zoo ('the Zoo is macabre—a polar bear 99 per cent dead, a kangaroo eating dead leaves'),[263] for Spain was where, for many Britons, the shades of the prison-house closed—as they had earlier done for Alexander Zamyatin, and were shortly to do for Arthur Koestler—dispiritingly about the Revolution. So far from providing any escape from the caging insanities of the time, Spain had merely confirmed them. 'It was', writes Orwell of Barcelona soured by Stalinist infighting, 'as though some huge evil intelligence were brooding over the town. Everyone noticed it and remarked upon it. And it was queer how everyone expressed it in almost the same words: "The atmosphere of this place—it is horrible. Like being in a lunatic asylum".'

If Spain caught up and extended the '30s crisis rhetoric of violence it also bore crucially on the period's rhetoric of travel. The Civil War added Spain to the period's conscious map, its map of conscience. History has this way, and not just in the 1930s, of animating geography:

> We will call it Going-into-History
> And you all know History is a cruel country.

Thus Ewart Milne in 'Thinking of Artolas', a Spanish War poem characteristically filled with the place names of the war, Extramadura, Cordova, Jarama, the streets of Madrid and Barcelona

> In the Gran Via in the Colon we went into conference.
> All day the starlings on the Ramblas whispered,
> All day the dead air pacified the street,
> Fat pigeons swaggered on the Plaza Catalunya.

In the Spanish Civil War, '30s literature's obsessive map-making and map-reading found their extreme apotheosis. Other places and place-names had their day in this period so extraordinarily mindful of geography, but no body of '30s literature rings so insistently with the language of the map than writing about Spain. Auden's 'Spain', H. B. Mallalieu's 'Spain', Edgar Foxall's 'Spain and Time', Barker's 'Elegy on Spain', Jack Lindsay's 'On Guard for Spain!', Read's 'Bombing Casualties in Spain', Clive Branson's 'Spain. December 1936', Rex Warner's 'Arms in Spain', Bernard Gutteridge's 'Spanish Earth', Margot Heinemann's 'On A Lost Battle in the Spanish War', Spender's 'To A Spanish Poet', Roy Fuller's 'Poem (for M.S., killed in Spain)', J. C. Hall's 'Postscript for Spain': no other country preoccupied '30s poets this much. Indeed, at no time in English literature has one foreign country so obsessed our poets. And the place-naming, as in Milne's 'Thinking of Artolas' just quoted, was extraordinarily wide-ranging in its geographical particulars: in, for example, Cornford's 'A Letter from Aragon' and 'Full Moon at Tierz: Before the Storming of Huesca', in Jacob Bronowski's 'Guadalajara', John Lepper's 'Battle of Jarama', Spender's 'At Castellon' and 'Port Bou', Sylvia Townsend Warner's 'Barcelona', Wintringham's 'Barcelona Nerves', A. S. Knowland's 'Guernica', T. A. R. Hyndman's 'Jarama Front', Richard Church's 'The Madrid Defenders', Milne's 'Sierran Vigil', Miles Tomalin's 'After Brunete'. Like generals following the progress of battles, the writers were much occupied in poring over the map of Spain. Poems and prose by Britons absorbed the Spanish people's own intense interest in the progress of the war across their land. Even Auden, inclined to make a joke as a map arrests his eye in Valencia, is none the less impressed by this Spanish attention to a landscape so different from his own Yorkshire or Iceland, or for that matter from Sunny Devon ('Impressions of Valencia'):

In the centre of the square, surrounded all day long by crowds and surmounted by a rifle and fixed bayonet, 15 ft high, is an enormous map of the Civil War, rather prettily illustrated after the manner of railway posters urging one to visit Lovely Lakeland or Sunny Devon. Badajoz is depicted by a firing-party; a hanged man represents Huelva; a doll's train and lorry are heading for Madrid; at Seville Quiepo el Llano [*sic*] is frozen in an eternal broadcast.

By the time he wrote his poem 'Spain' Auden had stopped trying to buffer the map's impact with flippancies about Sunny Devon, and had put his finger on why the geography of Spain mattered so much

> On that arid square, that fragment nipped off from hot
> Africa, soldered so crudely to inventive Europe;
>   On that tableland scored by rivers,
> Our thoughts have bodies; the menacing shapes of our fever
>
> Are precise and alive.

The course of the war in Spain filled up the map of Spain with the locations of painful memories, awful fears, large and small disappointments: would Fascism break through there?; so and so died here; there were won or lost; here, there and there Fascism triumphed and the Revolution foundered. The map of Spain, in other words and in the words of the first published version of Spender's poem 'To A Spanish Poet', became 'The map of pain'. This was, in many ways, a prevailing '30s experience, as regions once claimed by ordinary civilians and tourists, travel brochures and songs, ignited into trouble spots possessed rather by soldiers and bullets. What was happening to ordinary travellers' Spain had happened all over the world:

> I looked at film or chatty travel-book
> sun on the Yugo-Slavian sheepbell valleys
> light on the wrinkled peasants of the Dolomites
> with chimes from mountains; and *Vienna in Springtime*
> whistled by newsboys while the bullets hissed
> over the trembling roofs of Floridsdorf:
> Virginia was a tune, not negroes burning
> Venice was moonlight, and no tortured moan.

But 'Now the old myth is dead', we can't be kidded any longer about the world's niceness. So Peter Hewett in his poem 'The Old Lie', in Hogarth's first *Poets of Tomorrow* volume (the title's quotation from Wilfred Owen's disillusioned First War poem—'The Old Lie: Dulce et decorum est / Pro patria mori'—is revealing). But this transformation of the map comes home to Hewett, as it came home to his period, most sharply in the case of Spain:

> At school I learned about the Spanish main
> Armada, Inquisition, and deadly suave
> swarthy ambassadors from Philip's court.
>
> Later, the radio background and the island hate
> (the North was frozen decks, the South was sexy,
> the East mysterious and the West was wild)
> drawled out their monotone; and in my teens
> wilted in heat with *Streets in Old Seville*

> or *Barcelona Knuts* on gramophone
> and love again on *Balcony in Spain*.

And yet my friends could fall, and blood could rain.

Poet after poet made the point. In George Barker's 'Poem on Geography', the features of tormented countries stare fearfully from the map, 'faces of fear and fire':

> Glaring up with gutted Guernica, Spain.
> The distorted masks and the faces contorted
> Geography is the shapes of pain.

MacNeice sensed the danger behind the bland mask of the tripper's map on his pre-war visit to Spain celebrated in Section VI of *Autumn Journal*:

> But only an inch behind
> This map of olive and ilex, this painted hoarding,
> Careless of visitors the people's mind
> Was tunnelling like a mole to day and danger.

Time was still modifying Roy Fuller's sense of the Spanish map's pronesses to transformation, its acquisitions of new meanings in the Spanish War, long after the War had finished ('Times of War and Revolution'):

> The years reveal successively the true
> Significance of all the casual shapes
> Shown by the atlas.
>
> The pages char and turn. Our memories
> Fail. What emotions shook us in our youth
> Are unimaginable as the truth
> Our middle years pursue. And only pain
> Of some disquieting vague variety gnaws,
> Seeing a boy trace out a map of Spain.

Closer to, the pain that rhymed so readily with Spain (the word Spain itself, of course, containing pain, spelling it out), this reiteratively mapped set of pains was less vague, if no more disquieting. Readers of maps, military officers, gunners, and such, commanded the map precisely to inflict pain. As in Charles Donnelly's finest poem 'The Tolerance of Crows':

> Death comes in quantity from solved
> Problems on maps, well-ordered dispositions,
> Angles of elevation and direction.

Spender not only fills up the map, but the instruments of map-reading, with pain, in his related pair of poems 'The Moment Transfixes the Space Which Divides' and 'A Stopwatch and An Ordnance Map'. In both a man has been shot; his watch has stopped at 5 o'clock, the moment of death; he and his 'comrade' are divided by a Spanish death, a distance that can only now be measured ('O the watch and the compasses!'; 'dividers of the bullet') and mapped ('A stopwatch and an ordnance map'). The map of a pair of human beings' pain has been absorbed into the map of Spain (the reference to Spender's broken affair with Spanish volunteer T. A. R. Hyndman is clear). And in an even more direct fashion many specially geography-conscious pieces of Spanish writing (though no piece of writing about Spain, of course, is *unconscious* of geography) are memorable because in the places celebrated their authors themselves observed and shared the pain and fear of war: texts such as Orwell's *Homage to Catalonia*, Esmond Romilly's *Boadilla* (1937), Hyndman's 'Jarama Front', Wintringham's prose piece 'Comrades of Jarama', the Jarama soldiers' song 'There's a Valley in Spain called Jarama', Cornford's poems:

> Where in the fields by Huesca the full moon
> Throws shadows clear as daylight's, soon
> The innocence of this quiet plain
> Will fade in sweat and blood, in pain,
> As our decisive hold is lost or won.

Journeyers to war must always be mindful of the contribution geography makes to fighting: violent displacing increases the perturbed sense of placing. It's only expectable that the International Brigaders and the other international volunteers should peddle geography so hard. Tom Wintringham, for instance, makes the link. In 'Monument' he envisages all of Spain contributing to a memorial tower of Babel, a cenotaph to the Internationals who fought for her. 'You will remember the free men who fought beside you, / enduring and dying with you, the strangers'. And the most meaningful memorial to the frontier-crossing volunteers will consist of bits of all the fronts they served on:

> Bring to the tower, to its building,
> From New Castille,
> From Madrid, the indomitable breast-work,
> Earth of a flower-bed in the Casa del Campo,
> Shell-splinters from University City,
> Shell-casing from the Telephonica.
> Bring from old Castille, Santander, Segovia,
> Sandbags of earth dug out of our parapets.

The drama of the many Spanish fronts depended on the drama of the Spanish frontier that preceded and enclosed it. The foreigner who fought in Spain had first to travel to Spain, and cross into Spain. Spanish Front meant Spanish frontier. So just as the '30s frontier consciousness grew, in part, from a First War front-line mindfulness, with its particular apparatus of taking sides, confrontations and fearful goings-over, so now the period's general mythology of the frontier was reconnected to its military meanings in Spanish wartime fronts. As MacNeice put it at the end of *Autumn Journal* Section VI, it didn't take long for the period's liberals and Leftists to perceive that Spain would:

> . . . denote
> Our grief, our aspirations;
> . . . that our blunt
> Ideals would find their whetstone, that our spirit
> Would find its frontier on the Spanish front,
> Its body in a rag-tag army.

The front-line against Fascism was now across the Spanish frontier. 'The front-line trenches of democracy', so runs the argument of the Welsh comrades in *We Live*, 'are now in Spain, not Cwmardy'. So in animating the '30s geographical theme, in inviting the interest and then the presence of foreigners, Spain became the apotheosis of the '30s travel drama and trauma.

Spain is about departure. 'It's farewell to the drawing-room's civilised cry': Auden, going off to Spain, wrote out the poem that was later called 'Danse Macabre' and the poem 'Lay your sleeping head, my love' on a couple of musical scores that Benjamin Britten had with him when the two friends met to say goodbye in a Lyons Corner House café on 8 January 1937. 'It is terribly sad & I feel ghastly about it', Britten scribbled in his diary. They talked 'over everything, & he gives me two grand poems—a lullaby, & a big, simple, folky Farewell—that is overwhelmingly tragic and moving. I've Lots to do with them'. Auden found 'the glabrous suction' of this goodbye to a fellow Old Boy, musical collaborator and, it seems, a yearned-for beloved, particularly sticky. He kept going through his Spanish threshold rituals, first with Britten on the 8th, then again (because of delays) on the 11th when friends saw him off at Victoria Station, then again in Paris with Brian Howard and Christopher Isherwood. He'd made Isherwood and Upward his executors. In Paris he had frontier worries already, for his luggage had been sent ahead in error to the Spanish border. 'This was a solemn parting, despite all their jokes', writes Isherwood in *Christopher and His Kind*. He and Auden went to bed together— perhaps, Isherwood worried, for the last time. In the last words of 'Danse Macabre', which Auden also showed to Isherwood:

> So goodbye to the house with its wallpaper red,
> Goodbye to the sheets on the warm double bed,
> Goodbye to the beautiful birds on the wall,
> It's goodbye, dear heart, goodbye to you all.

Isherwood and Heinz had already, in Brussels, seen Giles Romilly and Tony Hyndman off to Spain. For his part, Orwell took pains to put in a hail and farewell to Henry Miller in Paris before he entrained for Spain (Miller told him he was a fool for not getting into the whale of apoliticism with him, but presented him none the less with a warm corduroy jacket).

In some senses parting was easier for the bourgeois. Auden, Orwell, and Giles Romilly did at least own passports and luggage. Those proletarians who, unlike Hyndman, happened to be unaccustomed to travel, had none of these accoutrements. They must travel light from Victoria Station in the guise of mere weekend excursionists to Paris (for which no passport was required). And there were no troops of friends to see them off in their sketchy trippers' kit—old flannels and plimsolls it might be—only policeman seeking to stop their departure ('Victoria Station was thick like flies with Special Agents and detective men looking for people like us', Judith Cook was told).

Still, once entrained at Victoria those working-man volunteers were making a good shot at entering the '30s travel ethos. For Spain, like so much of '30s writing, was also about trains. The train journey across France is dwelt on repeatedly in volunteers' letters and memoirs. John Sommerfield rhapsodically assimilated his train's noises to the Soviet-Futurist machine aesthetic (*Volunteer in Spain*): 'and still in the undeviating music of the dancing pistons we heard only the sound of our own impatience, and saw the map of France stretching away in front of us with the

Pyrenees and Spain still far away.' French peasants seem to have spent much of their waking day saluting the passing train-loads of volunteers with approving clenched fists. A 'most wonderful experience', Ralph Fox called it. Orwell was much moved too:

> The train, a slow one, was packed with Czechs, Germans, Frenchmen, all bound on the same mission. Up and down the train you could hear one phrase repeated over and over again, in the accents of all the languages of Europe—*là-bas* (down there). My third-class carriage was full of very young, fair-haired, underfed Germans in suits of incredible shoddiness—the first *ersatz* cloth I had seen—who rushed out at every stopping-place to buy bottles of cheap wine and later fell asleep in a sort of pyramid on the floor of the carriage. About half-way down France the ordinary passengers dropped off. There might still be a few nondescript journalists like myself, but the train was practically a troop train, and the countryside knew it. In the morning, as we crawled across southern France, every peasant working in the fields turned round, stood solemnly upright and gave the anti-Fascist salute. They were like a guard of honour, greeting the train mile after mile.[264]

One of the few wholly unmisgiving bits of Auden's 'Spain' celebrates this journey and its variants:

> They clung like burrs to the long expresses that lurch
> Through the unjust lands, through the alpine tunnel;
>   They floated over the oceans;
> They walked the passes. All presented their lives.

'They walked the passes'. Getting to the Front meant crossing the frontier; to go and shout a defiant *No Pasáran*, They Shall Not Pass, to the Fascists, you had to get through passport control. And that being impossible without a passport, many volunteers had to walk the passes, crossing the Pyrenees by night, on foot.

Even if you had a passport, though, entering Spain brought you face to face with the frontier. The dangerously fraught barrier, the familiar Dark Frontier of the literary imagination, loomed there, larger in life than on the page, suddenly as threatening and as unfamiliar as could be. On one side, comfortable France (and England) and merely writing about action; on the other, self-exiling, danger, the chance of the permanent exile of death, the challenge of translating images of action into action itself. Would one go over successfully? Would one find oneself stuck, ignominiously irresolute, on the border? Sylvia Townsend Warner describes (in 'Waiting at Cerbère') waiting at Cerbère, on the French side, in the 'White village of the dead'. Even the sea catches its breath in the contagion of the traveller's excitement and fear ('the white mane / Of foam like a quickened breath / Rises and falls again'), for just ahead is the frontier:

> And above, the road
> Zigzagging tier on tier
> Above the terraced vineyards,
> Goes on to the frontier.

Having made it to the other side, but only just, to Port Bou, Spender describes himself (in 'Port Bou') as stranded, suspended, paralysed with fear lest bullets stitch him through, caught on the frontier while others, cheerful militia men in their lorry, go rebukingly on into battle ('over the vigorous hill, beyond the headland').

The Spanish frontier, then, raised again the '30s challenge to take sides, to go over to the side of the workers of the world, and in a strikingly personal way. Orwell assimilates the journey to Spain quite naturally to his personal plunge downward into the working classes: travelling through France is going 'down France'; Spain is '*là-bas* (down there)'. Going to Spain, declared Connolly in *Enemies of Promise*, was intimately involved in going over to the workers' side. Spain catalysed the going-over issue. Many previously non-communist volunteers joined the Communist Party whilst in Spain. 'Often a writer is unable to go over. He approaches the barrier, shies, and runs away . . . But these fears can be surmounted by a moment of vision . . . It is too early yet', Connolly goes on, 'to say whether writers have done anything for Spain, but it is clear that Spain has done a great deal for writers, since many have had that experience there, and have come back with their fear changed to love, isolation to union, and indifference to action.' The question posed by Alick West's *Crisis and Criticism*, whether the hapless bourgeois *I* would be submerged savingly into the hopeful workers' *We* found its sharpest expression over Spain.

The issue seemed crystal clear, at first anyway. Not to be on the Republican side was to settle for a bourgeois individualism and its tyrannies over against unity with the workers. Spender's contribution to a *News Chronicle* symposium on beliefs summed up the belief of most liberals and Leftists about Spain:

> the struggle that is going on in Spain today seems to me the dramatization of a struggle between property, nationalism and tyranny against internationalism, freedom of expression and the classless society which is taking place all over the world.
>
> I believe that the interests of artists lie with the democrats and not with the tyrants, because I have seen that in several countries where the workers have been crushed, freedom of speech has been crushed also.[265]

The struggle was just as oversimplified on the out-and-out Right. Roy Campbell's polarities neatly mirror the Left's. In 'A Letter from the San Mateo Front', he speaks just as bluntly, but for the other side of the Front-line. He offers an alternative future ('I've got the future in my bag'), claims for his part to speak for 'the "Spanish Worker" ', sets his (Right) hand against the Left's, Catholicism against Communism:

> The Communist, whose bungling Left we fight
> With this Right hand—in every sense the Right!

No Catholic or non-Communist Republicans for him—whereas, of course, there were Catholics in favour of the Republic, George Bernarnos certainly, Graham Greene probably; and there were many non-communist Leftists. But Campbell sought a crude, simple picture; just like the militant Left. And over Spain the Left lost its patience; neutrality might have been tolerable in the past, but it wouldn't do now. 'There is no middle course. He who is not for us is against us', declared the Revd Tom in *The Land of the Leal*. Orwell agreed with Koestler when he said that

> Anyone who has lived through the hell of Madrid with his eyes, his nerves, his heart, his stomach—and then pretends to be objective, is a liar. If those who have at their command printing machines and printer's ink for the expression of their opinions, remain neutral and objective in the face of such bestiality, then Europe is lost.[266]

Orwell couldn't, he declared in *Homage to Catalonia*, write objectively about the Barcelona May Days: 'One is practically obliged to take sides.' And this was a little gibe at the communist opponents of the POUM who were all for taking sides over Spain because, they argued, personal neutrality at such a time would merely mirror the fascist-aiding pusillanimity of the French and British governments over 'non-intervention'. The 'policy of "neutrality" will prolong for months and even years a war that could speedily be ended': Cornford's view in 'The Situation in Catalonia' (*Memoir*) was the consensus of the Left. And the right-wing sympathies of so-called neutrals were indeed clear: whether in T. S. Eliot's stodgy polemics in the *Criterion* or Percy F. Westerman's lively non-interventionist yarn for boys, *Under Fire in Spain* (1937). Hence the posture of the famous *Left Review* pamphlet of 1937, *Authors Take Sides on The Spanish War*, that was referred to in Chapter 2.

'It is clear to many of us throughout the world', declared the questionnaire signed, *inter al*, by Spender and Auden and sent from Paris in June 1937 'to the Writers and Poets of England, Scotland, Ireland and Wales', 'that now, as certainly never before, we are determined or compelled, to take sides. The equivocal attitude, the Ivory Tower, the paradoxical, the ironic detachment, will no longer do.' There were only two sides to this question, as to every front and frontier: 'Are you for, or against, Franco and Fascism?' And, of course (and 'of course' was a common reply from Republic supporters), the majority of the published replies were on the Republic's side (127 of them against a mere 5 'Against the Government'). What's more most published respondents appeared to accept the premise of a necessary taking of sides. The *LR* compilers did fiddle their results a little. They did put as many equivocators as they reasonably could into the pro-Republic camp (Barker: 'I am for the people of Republican Spain, for the people of China, for the people of England, for the people of Germany, etc.'; Grigson: 'I am equivocal enough to be *against* politically, and not *for*, to fear and distrust any mass in its own control; but for me Hitler, Mussolini and Franco are man-eating mass-giants issuing from mediocrity and obscenity'; Tom Harrisson: 'The equivocal attitude, the Ivory Tower, the ironic detachment, are words your letter uses to sway with superstitious feeling our immediate judgement. But even without them we must feel horror and terror and hate at that Franco'). Some equivocal replies that we now know were given (Joyce's refusal, for instance, to do other than acknowledge receipt of the questionnaire; Orwell's angry tirade against Spender and the 'bloody rot' of the questionnaire's assumptions), were not published. But we don't know precisely how many of these there were. No list of unpublished replies is given, nor of those from whom no reply had been received (silence was potentially a very significant response). Mightn't Henry Green or John Cowper Powys or Anthony Powell have equivocated? The tone of Anthony Powell's account from Hollywood of the showing of the Hemingway–Ivens film *Spanish Earth* (in *N&D*) is spry and distant enough to suggest equivocation at best. Graham Greene's much-overlooked *Spectator* essay on the intervention of Tennyson and other Cambridge radicals in an earlier Spanish conflict is certainly sarcastic about current interveners and about the declarations printed in *Authors Take Sides*.[267] For their part, Graves and Laura Riding and Hilaire Belloc and Roy Campbell and Yeats would in all probability have been as unequivocally pro-Franco as the small handful the *Left Review* pamphlet allowed to speak for him (Blunden, Waugh, Arthur Machen, Eleanor Smith, Geoffrey Smith).[268] But the Right's ranks must appear as thin as possible, so Ezra Pound's hostility ('You are all had. Spain is an emotional luxury to a gang of sap-headed dilettantes'), Robert Byron's incipient Francoism and

Vita Sackville-West's pronounced anti-Communism are all diluted into an uneasy Neutrality. And, what's more, that Neutrality was itself demeaned by an appended question-mark: 'Neutral?' For, above all, neutrality, equivocation—the hesitations of a liberal like Grigson seriously worried by Communism, an Orwell who had learned that Spanish politics were exceedingly messy, a pacifist like Vera Brittain, a Christian like T. S. Eliot, intent however un-neutrally on achieving the Christian Third Way—were not acceptable. If no other label would avail, then Vera Brittain and Eliot must be content with a begrudging 'Neutral?'

But the Left's extremist stand against equivocation, so interestingly akin to that of the totalitarian Right as represented by Roy Campbell, belied the problematics of the frontier, of going over, of the attempt to merge *I* in *We*, that were being experienced even behind the bold faces put on in *Authors Take Sides*. Undoubtedly there was a good deal of successful going-over in Spain. All through the period, foreign countries had proved easier to live in than Britain was for the bourgeois seeker of exile from his own class among the world's workers. Public School accents, bourgeois names and table-manners were less deterring signals to Berlin boys, Parisian Left-Bankers, Russian mechanics, than to Wigan miners. And to the familiar foreign indifference to British middle-class-ness was added the acceptance that came from being obviously, at the risk of life and limb, on the same side in a foreign cause. Here was, at last, a brotherhood of human beings, and it included the Dalston busdriver and the Wigan coalminer as well as the Spanish peasant, the Italian volunteer and the Bohunk. That's why *Homage to Catalonia* begins with the memorable Italian militiaman who was also celebrated in Orwell's poem 'The Italian soldier shook my hand', 'the flower', as Orwell called him in 'Looking Back on the Spanish War', 'of the European working class', the kind of working-man with whom Orwell had long been unsuccessfully seeking intimacy. 'It was as though his spirit and mine had momentarily succeeded in bridging the gulf of language and tradition and meeting in utter intimacy.' The Italian was for Orwell the potent emblem of a successful going over, his newly achieved comradeship, in Spain, with workers of all sorts, whether Scotsman Bob Smillie 'the grandson of the famous miners' leader' or the thousands of worker-soldiers on the Aragon Front ('The ordinary class-division of society had disappeared to an extent that is almost unthinkable in the money-tainted air of England'). It was the comradeship marvellously rendered by Wintringham's account of being attacked from the air, 'It's A Bohunk':

> I did not know how close I was to panic until the Bohunk suddenly gripped my right hand with his left. He jerked out a word that was a statement, not an appeal or a curse. What the word meant I did not know: but I knew that it meant the fear was shadowing us again. Yet I was already released from fear; stronger than the courage of the dead or of my woman was the physical grip of his square hand, the friendship of men who knew how to die and live.

In this atmosphere, it's been reported, Christopher Caudwell shed his bourgeois-veiling pseudonym, happy to be known as 'Spriggy'. George Orwell, we're told, once again passed as Eric Blair.

But this idyllic scene was not easy or final. It didn't last for Orwell. Nor even for Cornford. Cornford's proven Communism led his poetry readily into first-person plural pronouns: 'Our bullets . . . our decisive hold . . . our testing . . . we stood . . .

We studied . . . We plunge . . . Our fight . . . our guard . . . our dead.' Thus 'Full Moon at Tierz', a classic demonstration of how 'my private battle with my nerves', the problems of 'The love that tears me by the roots' and 'The loneliness that claws my guts', the individual's personal difficulties, are resolved because *I* is sunk in *We*: 'Fuse in the welded front our fight preserves'; 'Now with my party, I . . .' Yet Cornford elsewhere struggled—admittedly in the private 'Diary Letter from Aragon' to Margot Heinemann—rather un-communistically to salvage a sense of self in just the militia mass that so exhilarated Orwell:

> For days I've been shoved about from place to place, lost and anxious and frightened, and all that distinguished me personally from a unit in the mass obliterated—just a unit, alternately worried, home-sick, anxious, calm, hungry, sleepy, uncomfortable in turn—and all my own individuality, such strength as I have, such ability to analyse things, submerged. Now that's beginning to be different, I am beginning to adapt . . . now I, John Cornford, am beginning to emerge above the surface again and recognize myself and enjoy myself, and it feels good.

This is, on the face of it, surprising. *I*-problems were supposed to be for the others, according to West. It was, after all, the fairly unreconstructed bourgeois Connolly who objected to Upward's notion of going over as an 'immersion' in socialist work. And it was, above all, sensitive Stephen Spender, target of many a communist jeer about the inability of bourgeois poets to 'go over' heartily and cut through all that nervous-nellying about the fate of individuals in revolutions, whose Spanish experiences and poems left him dithering on the frontier, worrying at but never resolving the problems of the self. But then one recalls the phenomenon (mentioned earlier) of working-class British volunteers appearing actually to be cheered by class-consciousness, gladdened by the presence among them of the manifestly bourgeois self-hood of Fox and Cornford and the rest. This sheds, of course, a possible new light on Caudwell's and Orwell's happy reversion to 'Spriggy' and 'Blair'. It connects, too, with Charles Donnelly's poem 'Between rebellion as a private study and the public/Defiance', which celebrates the writer as volunteer, the intellectual turned into public man of action, who is then mythicized publicly as a hero, but of whom it is noted that all this is hollow and error-mongering, false to the real self which still survives, true and private. Notably, Donnelly's poem ends by using the type of T. E. Lawrence, the private, shadowy, bourgeois writer hero, as an exemplum:

> Master of military trade, you give
> Like Raleigh, Lawrence, Childers, your services but not yourself.

Meanwhile, Randall Swingler stuck to the Party line, toughly exposing what he described as Spender's (and Auden's) continuing bourgeois dilemma, their inability to ditch the old *I*, with its equally suspect pluralization the old *we* of the poetic coterie, for the *we* of the workers and history. Swingler claimed to detect this non-resolution of the 'going over' issue in Auden's 'Spain', and then went on:

> In the sonnets from China, the opposition between 'we' and 'history' becomes ever more definite and more tragic. 'We' have little to do with history. Therefore it will destroy us. The same theme can be found in Stephen Spender's poems of the Spanish War. Not one of them is about the Spanish War. They are all about Spender and his

detachment from history in the making in Spain. He objectifies his own rejection of the historical issue in the Deserter, the Coward, the Wounded getting a little respite from the war. To those who had any part in that war, whether fighting or actively support-ing, these poems mean very little.[269]

But Swingler is too stern. Of course Spender's Civil War poems are about the Spanish War: they dramatize with particular intensity the personal dilemma of the British bourgeois writer nervously confronting the challenge to lose himself in revolutionary action, in a literal Spanish exile that graphically realizes the internal exile that going over to the workers entailed for such a person. Spender proved to be ideological dynamite to the Communist Party that had recruited him in order to send him off to Spain in a blaze of Republican publicity.[270] For he could never forget that his political motives for going to Spain were mixed up with personal ones: in particu-lar with guilt over Tony Hyndman who, piqued because of Spender's marriage, had rushed off to fight; a guilt exacerbated when Hyndman, finding he had no stomach for this war, deserted from the International Brigade and got incarcerated for his pains; and a guilt brought painfully home to Spender as he tried to winkle Hyndman out of the hands of the tough British Commissars brutally ready with anti-homosexual inuendoes. 'I think I know exactly why you don't recognise the worth-lessness of this particular Comrade'; 'you know too many boys for your own good': T. C. Worsley's documentary novel *Fellow Travellers* (1971) sharply renders the Party hard men's unsympathetic response to the worried homosexual author. Spender's own Party, in other words, was on this occasion as illiberally hostile as the incorrigible rightists Roy Campbell and Wyndham Lewis, with their repeated satiri-cal play on the prominence of leftist and, by implication, homosexual bottoms. False bottoms are the recurrent motif of Lewis's novel *Revenge for Love*. Bottoms are also made to characterize the Republican volunteers in *Flowering Rifle*: they're always showing cowardly behinds to the advances of the grinningly heterosexual Francoists:

> Their 'Progress' is to shunt along a track
> Where 'Left' means left-behind and 'Front' means back
> When was a Front so definitely split
> As this fat Rump they have mistook for it,
> And shown us little else as we advance
> Our proper *Front* from Portugal to France:
> And if they're facing 'Front'-wards, I'll not quiz
> What must the tail be like, if that's the phyz?
> With them, for opposites we have to hunt—
> 'Backwards' 's the word, when Popular the 'Front'.

Campbell angrily repeats his gibes about sexual inversion in his reply to Spender's review of *Flowering Rifle*. 'Who only by inversion can exist / As perverts' he had written in the poem; now the charge was quite personal: 'It requires a really perverse and passionate craving for inversion to arrive where Mr Spender is with regard to events.'[271] 'The Sodomites are on your side, / The cowards and the cranks', Campbell had jeered in the poem 'Hard lines, Azaña!'—a nasty text which is yet another run of bad jests about fronts and behinds.[272] Now he boasted shamelessly about the prophetic accuracy of those dismally crude-minded verses.

Campbell knew very well the hurtfulness of his taunts. 'I was driven on by a sense of social and personal guilt which made me feel firstly that I must take sides, secondly that I could purge myself of an abnormal individuality by cooperating with the workers' movement': so Spender wrote later on in *The God that Failed*. And Spain, occasion of so many anti-homosexual rebuffs, overshadowed for Spender by the Hyndman affair, only made his sexual dilemma worse. There were, for instance, so many beautiful boys among the visiting intellectuals (*World Within World* dwells on the 'beautiful young man, the poet Octavio Paz'), and among the soldiery ('He was a better target for a kiss' is how the poem 'Ultima Ratio Regum' notoriously celebrated one of them in his death). Battle lent itself to being written about in sexual terms ('As though these enemies slept in each other's arms': 'Two Armies'); and tauntingly so for Spender, because it is this war that has condemned him, as 'The Room Above the Square' implies, to sexual loneliness and separation from Hyndman: 'Torn like leaves through Europe is the peace / That through us flowed.' Pervasive fears of being shot now merge (in 'Port Bou') recognizably into Spender's old anxieties about buggery:

> But my body seems a rag which the machine-gun stitches,
> Like a sewing-machine, neatly, with cotton from a reel;
> And the solitary, irregular, thin 'paffs' from the carbines
> Draw on long needles white threads through my navel.
>   (*NW* version of 'Port Bou'.)

The honesty of Spender's Spanish poems is what's compelling. There is for them no ready resort to the *we* expected by orthodox Party men. The potential bogusness of that *we* is signalled by the glib way it was available even to stay-at-home poets: 'Tell them in England, if they ask/What brought us to these wars', 'We came because . . .' (Day Lewis's 'The Volunteer'); 'We came from field and town . . . We poured out on the Ramblas . . . We who have forged our unity on the anvil of battle' (Jack Lindsay's 'On Guard for Spain!'). Spender was not one to go in for shams like that. 'All I can try to do', he asserted characteristically when faced with Picasso's painting *Guernica*, 'is to report as faithfully as possible the effect that this very large and very dynamic picture makes on me.'[273] And his *I*'s experience of a perpetual sense of wartime threat ('I have an appointment with a bullet', begins his poem 'War Photograph', apparently a commentary on Robert Capa's famous photo of a Republican fighter being shot) led not towards, but away from, the much canvassed workers' *we*. Left only with an irresolutely fraught sense of self, Spender seeks escape, but in the dire self-erasure of impersonality, not in the fruitful multiplications of the communal plurality of pronouns. 'A Stopwatch and an Ordnance Map' retells the story of 'The Moment Transfixes the Space Which Divides' by ruthlessly deleting the first-person pronouns: a drama between 'I' and a dead man becomes one merely between the dead man and 'his living comrade', 'another who lives on'. Even more starkly, 'Thoughts During an Air Raid' eventually got most of its first-person pronouns revised into impersonal 'ones', so that a poem originally obsessed by loss of 'the great "I"' and greatly perturbed by the self's reduction to a figure in a casualty list—

> . . . Supposing that a bomb should dive
> Its nose right through this bed, with me upon it?—

is flattened out into a comically unmoving concern about 'one'. Spender couldn't, in the end, bring himself even to go on naming the loss of 'I'. But this was a self-protectiveness that, for all its prevalent anxieties, at least put him in a good position to expose the claims of all those bogus *I*'s (Jack Lindsay's 'I rose from the bed of my young wife's body/at the call of Liberty' was hysterically untruthful, Spender thought[274]), and to lament the loss of Cornford's person and potential in the war that might easily have also claimed Spender himself. The Cornford *Memoir*, Spender averred,

> gives a portrait of a character so single-minded, so depersonalized, that one thinks of him, as perhaps he would wish to be thought of, as a pattern of the human cause for which he lived, rather than as an individual, impressive and strong as his individuality was.

Spender would, in fact, never concede that the '30s challenges Spain focused were easily to be met or soon to be resolved. His Spanish writing keeps suggesting that what 'going over' meant and what ways one was to take in 'going over', how, in sum one was to travel in the '30s, were questions that were not only clamantly raised but were also kept open, never to be cheaply or routinely closed, by the experience of Spain.

What sort of journey would one take to Spain? Would it be, for instance (and odd though the question might initially seem), a holiday trip? For all the obvious seriousness of the war prospect, elements of mere holiday did keep on intruding. Spain meant a journey southwards, towards the sun and, at least when the war broke out in July 1937, a trip in the summer holiday season.

> I had left Victoria Station with an old friend of my Cambridge days. We had caught the night boat to Ostend, bought a couple of ancient bicycles with shrivelled saddles, and pedalled south to Paris through the battlefields of Flanders. The ubiquitous monuments to the fallen only strengthened our resolution—for *that* war, organized by incompetents, had been but senseless slaughter in the mud, whereas we were on our way to fight in the sun on behalf of the working people of the world. I am sure that the Spanish sun played a great part in the formation of the International Brigades—we would have volunteered less eagerly for a struggle in the rain or snow.

Thus John Bassett, a British Communist, in *The Distant Drum: Reflections on the Spanish Civil War*, edited by Philip Toynbee (1976). Even for the communist volunteer, the utterly serious soldier, the sun was an unavoidably important element of Spain. Here was sunbathing as never before. 'We were sitting elegantly with the sun pouring in on us and into us, with a long, even warmth, that soaked through into our veins and beyond, deeper, and that was what we wanted; we didn't shut our inner bodies against the sun, as people who go sunbathing usually do, so that they only succeed in getting burnt like sausages for breakfast fried too hurriedly on the outside.' Thus the narrator of Ewart Milne's 'The Statue'. And what did these volunteers sing? Their music celebrated sun as well as victory:

> We came to sunny Spain,
> To make the people smile again,
> And to chase the Fascist bastards
> Over hill and over plain.

When this sunny war broke out it attracted, or distracted, holidaymakers. That famous pair of early British volunteers, the East End Communists and tailors Sam

Masters and Nat Cohen, were on a summer cycling holiday in the south of France when the war attracted them across the border. John Cornford, intent on an August holiday in France, went over the frontier to see what was going on in Spain. Some of the early volunteers were socialist sportsmen and women in Barcelona for the summer games of the People's Olympiad. For some professors and students the war remained a vacation exercise. Professor J. B. S. Haldane, complete with revolver, would drop in on the trenches during university vacations. In his Christmas holiday, 1936, Oxford student Philip Toynbee went to Spain as part of a student delegation. Louis MacNeice paid *his* visit to Barcelona during his Christmas vacation from teaching at London University, December 1938. No wonder Spain suggested, for some writers, postcards. Ruthven Todd's 'Poem (for C. C.)' is a postcard *to* 'sunny' Spain: 'I', he writes,

> Think of the sun's glare
> And twisted cactus plant
> Against this grey and heavy air
> And a mast's indolent slant;
> Wish I were where you are, or you here,
> To stop these minutes tapping at my ear.

No wonder, either, that the holiday visitors, the flying-visitors, whether famous film-stars, professors, and poets or merely curious undergraduates, frequently wore the irresponsible air of the mere whorer after temporary political excitements in exotic locations. MacNeice admits in his autobiography that 'my motives were egotistical; I was sensation-hunting, testing myself, eager to add a notch to my own history'. Philip Toynbee had the decency to realize the shame of his privileged excursion—no different, of course, from the slumming about Vienna's barricades indulged in by the temporary Leftists of *Pink Danube* ('oh, look, there's a bullet-hole, I do believe'):

> We went to the Casa de Campo and up to the second line . . . On into the front line trenches were mortar-bombs falling behind and in front of us, and I took one shot, through sandbags, at the flitting figure of a Moor between the ruins. Felt mildly afraid, but also exhilarated . . . A wounded man was carried in and I went and stood behind a wall, watching soldiers running doubled-up across a gap . . . two boys of about sixteen hitching bandoliers over their shoulders and walking resolutely down to the barricades. And what a shit I felt, leaning elegantly against the wall with a glass of wine in my hand and watching these others do the fighting . . .

'See Spain and see the world': the last stanza of Rex Warner's 'The Tourist Looks at Spain' was fiercely ironic about those who would remain trippers, disconnected observers, aloof passers-through. MacNeice's *Autumn Journal* mocked the tourist's frivolity:

> And the standard of living was low
> But that, we thought to ourselves, was not our business;
> All that the tripper wants is the *status quo*
> Cut and dried for trippers.
> And we thought the papers a lark . . .

But Rex Warner never got beyond the Communist Party recruiting officer in London, MacNeice persisted in his own touristicism, and Republican Spain filled up with journalists, visitors, mere observers, enthusiasts keen for other people's fighting. The touch-lines were crowded with the likes of Jack Lindsay, cheering from a distance (it's hard to stay mindful during 'Looking at a map of Spain on the Devon Coast (August 1937)' that the narrator is hundreds of miles from danger), or Auden finding today's monstrous Spanish struggle as hard to write about as he found it difficult to live with, or Spender perusing his own reluctance to get involved. 'That fighting was a long way off'; Bernard Spencer's reflection in the poem 'A Thousand Killed', published in *New Verse* early in 1936, was prophetic. Spender was even inclined to make a principle out of the trippers'—it was also the camera-man's, and even the airman's—position that he discerned in the point-of-view and execution of Picasso's *Guernica*:

> it is not a picture of some horror which Picasso has seen and been through himself. It is the picture of a horror reported in the newspapers, of which he has read accounts and perhaps seen photographs.
> This kind of second-hand experience, from the newspapers, the news-reel, the wireless, is one of the dominating realities of our time. . . . The flickering black, white and grey lights of Picasso's picture suggest a moving picture stretched across an elongated screen; the flatness of the shapes again suggests the photographic image, even the reported paper words. The centre of this picture is like a painting of a *collage* in which strips of newspaper have been pasted across the canvas.

Furthermore, the agreeably utopian aspects of Spain, its revolutionary joys, were easily attracted into the touristic aspects to suggest that desired '30s haven, the still centre, the escapist island refuge. Sternly orthodox men of the Left rebuffed the very idea. When John Cornford's father mentioned, in his contribution to the Cornford *Memoir*, the possibility that Cornford was escaping from 'personal responsibilities' in going to Spain (he meant, one assumes, his son's domestic tangle, his discarded mistress, his young child, his new infatuation with Margot Heinemann, his growing reputation already in his first year after graduation for sexual bizarreness), Pat Sloan edited the suggestion out. But that some volunteers were escaping was obvious to less dogmatic observers. James Hanley's *Grey Children* records the dyspeptic obser-vations of a John Jones down in South Wales:

> Why should they be rushing over to another country to fight when there's all the fight they want here? . . . There's more in it than just this going out to fight Fascism. Take a young chap out of work. It's romantic, dare-devil, it means travel, seeing new things and new people. After you've seen a bit more round here you'll understand what I mean.

Even the most determined escapers, though, quickly found Spain was no picnic. T. A. R. Hyndman, running away from the ruins of his liaison with Spender, got speedily fed up with the continual danger and the chaos of what his friend John Lepper called 'the biggest shambles in Europe', contracted an ulcer, sought to go home, was prevented by the Commissars ('I was becoming almost paranoid concern-ing the Communist Party'), deserted with the now nearly blind John Lepper, and was arrested to spend three terrible and terrified months in prison with Lepper: 'a steady progress through jails, camps, then more jails. Our ailments became worse, especially John's. He wept with pain. And the visits by commissars with their questions . . .' Only Spender's diligent efforts got them out.

Spender's Spanish poems show the extent of his sympathies for Hyndman, and others enduring his plight. These poems are acutely aware of how the craved-for ideal of the 'still centre' failed in Spain. 'Port Bou', a frontier-poem, is also a harbour poem—the Catalan name Port Bou celebrated the bull's-horn shape of its harbour enclosure—and a variant, it's clear, of Auden's 'Dover'. And like Dover, Port Bou lets the poet down; its offered refuge is only temporary and provisional:

> As a child holds a pet
> Arms clutching but with hands that do not join
> And the coiled animal stares at the gap
> To outer freedom in animal air,
> So the earth-and-rock flesh arms of this harbour
> Embrace but do not enclose the sea
> Which through a gap vibrates to the Mediterranean
> Where ships and dolphins swim and above is the sun.
>
> （*NW* version）

What's more, quiet, empty Port Bou, a place of pause and peace (in the 'smiling faces' of the lorry-load of militia men who creak to a halt there 'the war finds peace'; their 'terrible' machine gun is wrapped in a cloth and 'rests'), is suddenly infected with noise, the 'bang-bang-bang' (or 'pom-pom-pom') of an old man, and the sounds of the firing practice:

> The echo trails over like an iron lash
> Whipping the flanks of the circling hills.

In fact Port Bou had become so empty of noise precisely because firing practice was about to fill it with the clamour of war:

> Now Port Bou is empty, for the firing practice.

And the poet's brief enjoyment of the still centre ('And I am left alone at the port's exact centre') turns terrifyingly into what feels like the stormy centre of a target:

> I on the bridge, solitary as a target.

(Or, 'At the exact centre, solitary as a target'; or, again, 'The exact centre, solitary as the bull's eye in a target'.) And there the poet endures his trauma of imagining himself stitched through by bullets, at one with the dead man in 'The Moment Transfixes' ('He the dead centre . . .') and the coward of 'The Coward', 'killed, not like a soldier / With lead but with rings of terror' (original version).

More literally, Orwell's still centre, the glorious 'interregnum' (as he calls it) of the Aragon front ('quite different from anything that had gone before and perhaps from anything that is to come'), was intrusively smashed up by the bullet that hit him. 'Roughly speaking it was the sensation of being *at the centre* of an explosion': a 'violent shock', that's the prelude to the wave of shocks that succeeds it, the political and personal shocks endured as Orwell experiences a Stalinist terror at first hand, and so has his idealized myths of the Spanish War broken up for ever. And Orwell and Spender weren't alone. On every hand the holiday soured, as the Spanish journey was discovered to be really only a grimly ironic parody of the soft, escapist southwards journey. The sun turned wintry—became the 'winter sunlight' of 'Port Bou'. It gave way to the awful cold of Spender's poem 'Two Armies', where 'Deep in the winter plain . . . Men freeze and hunger', the Spanish winter that's heaped as a curse on the Fascists' heads in A. L. Lloyd's translation of José Herrera Petere's poem 'Against the Cold in the Mountains'. Even during the summer months, the sun's heat could be unbearable and deadly. 'Two Englishmen were laid low by sunstroke', Orwell laconically observes. Tom Wintringham in 'It's a Bohunk' takes eager refuge from the sun in a shallow pool's tepid water, hugging the shadow of a bridge ('My head and shoulders were in the shadow of the railway bridge; the sun could not catch much of the rest of me'): for 'Spain in August is hotter than hell'. What's more, Spain is 'less safe: Hell's underground'. In the cloudless summer sky, enemy planes had perfect visibility. Better, Wintringham hints, the dark of a First War trench than this bright exposure. For him, holiday had become out of the question. As it had for Orwell, who realistically ditched journalistic tripperdom early on ('I had intended going to Spain to gather materials for newspaper articles, etc.': 'but I had joined the militia almost immediately'). And for John Cornford: 'I came out with the intention of staying a few days, firing a few shots, and then coming home. Sounded fine, but you just can't do things like that. You can't play at civil war . . . It didn't take long to realise that either I was here in earnest or else I'd better clear out.'

But not, though, for Auden. His 'Spain' is miserable precisely because Spain has turned out to be no holiday. His poem is tensed unhappily between holidays: yesterday's 'brochure of winter cruises' and tomorrow's 'walks by the lake' and 'bicycle races / Through the suburbs on summer evenings'. And this poet, like many another tripper, flinched away, lest he too become one of 'the poets exploding like bombs'. Auden returned home disillusioned after only a few weeks in Spain. Returned, what's more, to teach for the summer term of 1937 at the Downs School that he'd left in 1935 to join the GPO Film Unit. Auden's period as political activist was over, and he'd chosen to mark it by re-entry into schoolmastering, deserting the destructive for the old youthful element.

Serious, non-holiday travelling was grown-up travelling. Egregiously boastful though he be, at least Roy Campbell realized the importance of not being boyish in Spain. His *Flowering Rifle* consigns the defeated leftist volunteers to permanent little-boyishness. They're the'Tomboys of the Summer Schools', happier in shorts, fit only for Campbell's classroom lessons on the rightness of the Right:

> As these found out, these gutless weary-willies
> Who but that I had called this dance of wowsers
> Would still be hiking in their sawn-off trousers
> Or climbing grapenut-trees in some green lane—
> But that I gave the rendezvous in Spain,
> And came to greet them, shouting from my mule,
> 'Woodley! Old Woodley! welcome Home to School!'

And it's notable how, on either side, the Bright Young Baedekers tended as a class to steer clear of Spain. Isherwood preferred the distractions of the Heinz business and waving goodbye from the safety of the station platform: facing 'the Test' would remain for him a matter of fantasy untranslated into serious action. Robert Byron was more or less Francoist ('Had I been a Spaniard when the rebellion broke out, I cannot say for certain that I would not have favoured it') but his practical efforts went no further.

Evelyn Waugh thought of journeying to Spain;[275] he joined in the discussions of the Roman Catholic Archbishop of Westminster's Spanish Association as to the best ways of helping the Francoist insurgents;[276] he replied to the *Authors Take Sides* questionnaire that 'If I were a Spaniard I should be fighting for General Franco'; but he also declared that 'As an Englishman I am not in the predicament of choosing between two evils'; and he did not go to Spain. Auden, of course, brightest of the Bright Young Baedekers did go. But the tone of his article 'Impressions of Valencia' goes awry precisely because it's hard to carry your readers with you, even if you're a famous youthful Baedeker, when you sustain a debunking brightness in the face of catastrophic human suffering. Children play 'with large brown eyes like some kind of very rich sweet'. 'The foreign correspondents come in for their dinner, conspicuous as actresses.' So far, so Audenesque and clever. But he can't stop:

> Altogether it is a great time for the poster artist and there are some very good ones. Cramped in a little grey boat the Burgos Junta, dapper Franco and his bald German adviser, a cardinal and two ferocious Moors are busy hanging Spain; a green Fascist centipede is caught in the fanged trap of Madrid; in photomontage a bombed baby lies couchant upon a field of aeroplanes.

And there's the rub. Only a monstrously unfeeling childish joke-machine debunks a dead baby. It's a refusal to be serious cognate with the way another Spanish tripper, Cyril Connolly in *Enemies of Promise*—a book that strove to be thought the whippiest Young Baedeker among guides to the '30s literary scene—offers a kind of comic rewriting of *Homage to Catalonia*'s enthusiasm for Barcelona's revolutionary rejection of tipping, servility, smart clothes. Often, says Connolly, going over to the workers

> will only be recognised by external symptoms, a disinclination to wear a hat or a stiff collar, an inability to be rude to waiters or taxi-drivers, or to be polite to young men his own age with rolled umbrellas, bowlers and 'Mayfair men' moustaches . . .

In tones like that real revolutionary changes of style are dissipated into merely schoolboy indignations and undergraduate bohemianism, into a matter of Audenic charades and funny hats.

The failure of the Audenesque in 'Impressions of Valencia', at one with Auden's failure to become more than a Spanish tripper and with his 'Spain's' disconcerted aloofness is important not least because so many literary volunteers had Auden's voice ringing in their ears when they went to Spain. A generation had been taught about frontiers, fighting, and crypto-Stalinist dying 'without issue' by Auden's early OTC verse. At school and at Cambridge Cornford dedicatedly learned not only to imitate Auden, borrowing his phrases, his scenarios, his toughness, but also to assimilate them to his own developing Stalinist preoccupations. The fifteen pages of Cornford's pre-Spain poems printed in Jonathan Galassi's selection of his writing, *Understand the Weapon: Understand the Wound* (1976), would never have been composed without Auden. And Cornford's later Spanish experiences were amply prophesied in those early borrowed Audenisms: 'skyline operations', 'fighting on the frontier', 'Mood the more / As our might lessens'. So it was hard for him to forget Auden's words amidst what might readily have been seen as their Spanish enactments. And surprise that 'Heart of the heartless world' should have Auden's adaptation of

Karl Marx intruding into Cornford's trench-notebook remembrance of Margot Heinemann doesn't last long. On reflection the Audenism seems completely apt. And Cornford wasn't alone in being mindful of Auden over Spain. '[T]he frosted windowpane of importance', 'the legalised titles of theft': Lindsay's 'On Guard for Spain!' knew its Auden too. So did Rex Warner's 'The Tourist Looks at Spain' ('All those of us who loved, who read the classics / who were pleased when with friends / . . . followers of football'; 'abject surrender of the cracked nerve'; 'the ferret's temper'). Charles Donnelly found it as natural as Cornford did to take his Audenesque equipment with him into Spain. The action envisaged in his 'Poem' ('Between rebellion as a private study') is Audenic action

> In a delaying action, perhaps, on hillside in remote parish,
> Outposts correctly placed, retreat secured to wood, bridge mined
> Against pursuit, sniper may sight you carelessly contoured.

So are the listed attributes of the heroic actor to whom teachers will 'make reference / Oblique in class':

> Man, dweller in mountain huts, possessor of coloured mice,
> Skilful in minor manual turns, patron of obscure subjects, of
> Gaelic swordsmanship and mediaeval armoury.

The difference, though, between Auden and Cornford and Donnelly was that whilst these two poetic disciples lived—and died—with the uneasy consequences of the translation of those Auden metaphors into the violent realities of Spain, Auden himself could not stomach that transition. They, so to say, grew up out of the boyish metaphors into the adult realities of battle. He went back a while longer to school. Cornford, of course, still had school on his mind. 'I think', he wrote to Margot Heinemann,

> the days spent in the village alone were the hardest I have yet spent in my whole life. It was the same loneliness and isolation as the first term in a new school, without the language and without any kind of distraction of something to do. All the revolutionary enthusiasm was bled out of me.

He had a dream about holding his own against the bullying school rugger captain ('One of the toughest people when I was small at school'). Both Giles and Esmond Romilly were glad to recall school: their OTC training came in extremely handy in Spain (at militarist Wellington College they'd both been keen anti-militarists). Esmond Romilly's *Boadilla* repeatedly draws analogies between school and battle experiences: next to this war, school was what the youthful Romillies knew best. Orwell, too, couldn't help remembering the school boxing instructor demonstrating 'how he had bayoneted a Turk at the Dardanelles', as he chased that naked Spaniard farcically along the trench. But all these men moved on, in their practice, beyond the OTC, beyond those Let's-Pretend Field Days where no one really got killed or maimed—affairs as little dangerous as an adrenalin-stirring read through *Paid On Both Sides*—to become actual soldiers on real and lethal battlefields. It was the dedicated trippers who hung back, settling once more for the OTC. At Tarragona, MacNeice helped unload a lorry-load of dried-milk sacks. He hadn't, he reports in his autobiography, sweated so much for years. 'I found myself in the mood of a schoolboy's O.T.C. rag'. And there, as far as action in Spain went, he was content to stay.

If Spain was a stumbling-block for the Bright Young Baedeker, it also tested the idea of adventurous, heroic travel. Spain held out the possibility of real action to writers who had long envisaged action, the prospect of journeying to a war whose cause looked good and brave and just enough at least to expunge those collective bad memories of the journey to fight in an inglorious, unheroic, soured First World War. Here was the Test, as Isherwood had imagined it. He, of course, didn't take up the challenge; he failed the test of Spain before he had even put himself to it (as the narrator of *Lions and Shadows* said of his motorbike: 'I . . . began to fail the Test almost before it had begun'). Lots, though, of those who did respond to the call to action were sustained by a strong conception of the epic nature of the journey they were undertaking. The hard Left vented no public doubts. To journey to fight in Spain was to embark on a modern odyssey. Everyone, it seemed, deserved a medal. Just to step into the volunteers' ranks was considered an heroic exploit, for the cause of Republican Spain was axiomatically heroic. The 'Spanish people', ran Harry Pollitt's appeal for Solidarity With the Spanish People (*DW*, 25 July 1937), 'workers, peasants, intellectuals, men and women alike, are fighting in unforgettable bravery and heroism.' That kind of company, that kind of talk, allowed no intrusive suggestions of holiday to blur the heroic analysis. 'He'd made no summer excursion there', insists H. B. Mallalieu of his 'Departed Hero'.[277] And the bravery unto death of so many fighters for Spanish 'democracy' was indeed impressive. Even Orwell granted that: John Sommerfield, he said, harshly, had written 'a piece of sentimental tripe' in *Volunteer in Spain*, but none the less he had 'fought heroically in the defence of Madrid' as part of that 'thin line of suffering and often ill-armed human beings standing between barbarism and at least comparative decency'.[278] The Communist Party was completely ungrudging. 'Poplar Pays Tribute to Heroes Killed in Spain' blared the *Daily Worker*: the heroes on that occasion included Caudwell.[279] 'The Making of a Modern Hero' was the *Daily Worker* banner for Hugh Slater's review of the Cornford *Memoir*.[280] 'This young hero' Spender called him, in his *New Statesman* review. *Poetry and The People* carried L. Kendall's poem 'To the Heroes of the International Brigade in Spain' in its number of October 1938. The 'heroic dead' among the volunteers are celebrated in volunteer Eric Edney's poem 'Salud!'[281] Big Jock MacKelvie reappears in Barke's novel *The Land of the Leal* to recruit for Spain: he

> showed how the International Brigade had played a magnificently heroic part in preventing the success of the internal Fascist hordes in Spain—he himself had fought in the Brigade and had been wounded on the Jarama front—but MacKelvie did not speak of his own courageous qualities of leadership. He spoke of the political and human significance of the entire Brigade. He told of Hans Beimler, the German Red, who had escaped from a Nazi concentration camp and who had become a leading figure in the Thaelmann battalion and who had died in action—of Austrians and Frenchmen and Italians (how the Garibaldi battalion had routed the Italian Fascists at Guadalahara)—of Americans, Canadians, Poles, Norwegians—and Scots. It was a deathless record of how the best and bravest elements of the common people of the old world and the new world had, together with writers and scientists and intellectuals, gone to the defence of the heroic Spanish people and had led the counter offensive against the Fascist hordes.

And this, the note of the *Daily Worker* and *Left Review*, was readily struck in other quarters. 'Last Letters of a Hero' was the *News Chronicle*'s headline for its gathering of bits of Caudwell's epistles.[282] The hero of Richard Church's poem 'The Hero' leaves a woman and child behind, but the poem allows itself few doubts that he was actually 'Blown with heroism into Spain'. Auden, Swingler, and Benjamin Britten joined forces in the choral piece *Ballad of Heroes*, Britten's Opus No. 14, first performed 5 April 1939 in London's Queen's Hall at the Festival of Music for the People in celebration of the International Brigades. George Barker fans his blazing rhetoric with repeated blasts of heroizing. 'O hero akimbo on the mountains of tomorrow' his poem 'O Hero Akimbo on the Mountains of Tomorrow' keeps exclaiming. Barker's 'Elegy on Spain' comes stuffed with heroic stuff: 'The hero's red rag', 'The pall of the hero', 'the human hero', 'sing a breath taken by heroes'. 'What you see in Spain's heroic ardour / is your own noblest self come true', assures Lindsay's 'On Guard for Spain!' Spender's version of Altolaguirre's poem 'Madrid' dwells on 'this immense field of heroism'.

Cavillers, of course, wanted to amend the straight *Daily Worker* version. The Right had its Alcazar, whose 'continued and heroic defence . . . showed that although Spain was in her death agonies, the dauntless spirit of the conquerors of Mexico and Peru still survived'—according to Percy F. Westerman's *Under Fire in Spain*. Roy Campbell kept on about the Alcazar, 'This Rock of Faith', whose story will startle even the Fiends in Hell, so that they'll re-tell

> How fiercer tortures than their own
> By living faith were overthrown;
> How mortals, thinned to ghastly pallor,
> Gangrened and rotting to the bone,
> With winged souls of Christian valour
> Beyond Olympus or Valhalla
> Can heave ten million tons of stone![283]

In Campbell's version, as in Wyndham Lewis's *The Revenge for Love*, it's the Left who are the cowards, perpetually fleeing from battles in a funk: 'First clenching fists then throwing up their hands.' In Campbell's poetic vignettes, as among his few friends, Roy Campbell himself keeps turning out to be the hero. *Right Review* looked forward in June 1939 to a tournament in which the hectoring champion Campbell would triumph in verse over the poetic enemies whose side was on the run in Spain: 'A certain Drivelling Toad has written a review, threatening Roy Campbell with reprisals in the form of heroic couplets by W. H. Auden. Those who have read this immense tirade [*Flowering Rifle*] will hardly imagine Campbell in abject flight!' 'Heroic couplets such as' Campbell's in *Flowering Rifle*, in fact, 'can be written only by a Hero'. And how Campbell agreed with the *Right Review*'s estimate. He was no wet quitter. Unlike, of course, Auden and Spender.

> And flawlessly this axiom has been kept
> What Auden chants by Spender shall be wept.

The truth, of course, about Campbell's own self-publicized exploits in Spain has now been established by Peter Alexander's biography (1982). He and his family, terrified and appalled by tough republican measures against Toledo's Carmelite monks (seventeen shot dead) and other suspected Franco sympathizers, scuttled back to England as soon as they were able (the boat from Valencia, 9 August 1936, carried Graves, Alan Hodge, and Laura Riding too). Campbell never fired a single one of those pro-Franco shots that he kept claiming. His only actual anti-Red action during the Civil War was in his own heated fantasies. Only once was he at a front, and that was on the motorized trip he made in July 1937 as a correspondant accredited to the Roman Catholic *Tablet*. Nevertheless, he very properly admired the besieged defenders of the Alcazar for being braver than Spender or Auden. Orwell himself acknowledged that R. Timmermans' book *Heroes of the Alcazar* had a case: 'There is no need because one's sympathies are on the other side to pretend that this was not a heroic exploit.'[284] After what Orwell had been through he wasn't eager to join in the lopsided communist chorus. *Homage to Catalonia* would pay tribute to an act of non-communist, indeed anti-communist bravery: ILP man John McNair's crossing Barcelona by night during the May troubles to bring the beleagured Hotel Falcón's POUM contingent some cigarettes. 'I shall not forget this small act of heroism.'

And there, of course, it was: the possibility of heroism had manifested itself even in Orwell's downbeat, suspicious account. And to less suspicious observers on the Left the much wished-for heroic union of poetry and admirable action did indeed seem to have become possible in Spain. In the persons of Fox and Cornford and the other writing fighters Britons seemed at last to have stepped into the Romantic tradition most recently headed by Malraux. John Lehmann, T. A. Jackson and C. Day Lewis gave the tribute to Fox that they edited the title *Ralph Fox: A Writer in Arms* (1937). *NW* No. 3 (Spring 1937) was 'dedicated to the memory of one of its most valued contributors, without whose creative advice and sympathy it might never have been more than a project: Ralph Fox, who was killed defending in action those ideas which inspired his finest writing'. In his Introduction to *Poems for Spain*, which he edited with Lehmann, Spender is more tentative about the results of this journey into action—after all Cornford and the rest were by then dead, and the poetic results of the war were not overpoweringly impressive—but the possibility of the writer as man of action, and of the poem born in action, does mightily animate him. Cornford's 'preoccupation with the life of political action' hasn't entirely ruined his poetry; the Spanish poets' merging of the 'life of literature in the life of action' is extreme, but necessary; and even the worst Spanish War poems (Spender implies) have the virtues of action:

> where the issues are so clear and direct in a world which has accustomed us to confusion and obscurity, action itself may seem to be a kind of poetry to those who take part in it. Therefore these poems often seem like hasty transcriptions into words of an experience expressed not in words at all, but in deeds.

The Spanish Republicans were eager for culture ('that passion for education and popular culture which goes with a fundamental revolutionary change in a nation's life' as Spender put it in 'Pictures in Spain'[285]). Their poets recited poems to huge receptive crowds (David Gascoyne was 'much impressed' by 'a free public poetry-reading by Rafael Alberti and his beautiful wife, also a poet, in a quite large Barcelona theatre, which was packed with a wildly enthusiastic and very largely working-class audience'). Poetry flowed torrentially through Republican Spain, inspired by the cause. So here, in Spain's bracing combination of poetry, action, and a mass audience, there seemed hope too for British poetry. And poetry did, indeed, flourish in the pages of the International Brigade paper, *Volunteer for Liberty*. Some notable British poets emerged and matured among the volunteers: Miles Tomalin, Clive Branson, and Ewart Milne. John Cornford's brief period in Spain produced three remarkable poems from a poet who had earlier found political life a distraction from poetry. In Britain and America poetry and poets acquired new feelings of esteem and authority as the poetic shock waves from Spain were felt in the little magazines, on propagandist platforms, in the volume *And Spain Sings*—the American book of translated Spanish poems prepared largely by Rolfe Humphries (1937), and in the Spender–Lehmann *Poems for Spain*. People who had never written poems before wrote one now: like T .A. R. Hyndman, whose 'Jarama Front' was published in *Poems for Spain*, or the volunteer William Teeley who sent a poem to *Poetry and the People* (September 1938), or the unknown author of the famous poem 'Eyes' (sent, it now emerges, to Spender by Jack ('Russia') Roberts, Welsh miner and Communist, a Commissar in the Brigade, who had jotted the poem down probably during the Battle of Brunete).[286]

Instead, then, of going on guiltily feeling poetry to be a shamefully isolated, privileged and bourgeois affair, leftist poets had a moment of feeling licensed to walk tall. 'In a world where poetry seems to have been abandoned, become the exalted medium of a few specialists, or the superstition of backward people', Spender wrote in *Poems for Spain*, 'this awakening of a sense of the richness of a tomorrow *with* poetry, is as remarkable as the struggle for liberty.' It didn't seem absurd in this context to refer to the English Romantics, to see Ralph Fox as a modern Byron (as Pollitt did in *A Writer in Arms*), to call David Guest a Shelley (as Vivian de Sola Pinto did in *David Guest: A Scientist Fights for Freedom*[287]), to recall Wordsworth's sonnet 'Indignation of a High-minded Spaniard' (as Spender did in *Poems for Spain*). For a second great age of writers in action and of committed poetry appeared to have opened on the Spanish scene. Spain was, in fact, packed with the enacted emblems of the Bigness Heroic.

The young man in Spender's poem 'The Uncreating Chaos' went 'to the Pole, up Everest, to war'. Spender was assuming, even in 1935, that the apotheosis of mountaineering would come in battle. And Spain proved him right. For Spain meant mountains—the Pyrenees the volunteers had strenuously to cross; the Hills the British were always having to assault or hold in the teeth of shot and shell: Pingarron, or 'Suicide Hill' at Jarama, Hill 481 on the Ebro, Mosquito Hill at Brunete, Hill 666; the mountainous terrains on the Aragon front so vividly described by Cornford in 'Diary Letter from Aragon'; the 'Mountain muscles' of the country itself celebrated in Barker's 'O Hero Akimbo on the Mountains of Tomorrow'. And these mountains were not the less heroic for being southern climbs. Spain also meant flying: the bravery of Malraux himself and of the flyers he helped organize, celebrated by Malraux in his novel *L'Espoir* (*Days of Hope*, 1938) and by Spender in his *Daily Worker* article 'Spain's Air Chief in a Mechanic's Blue Dungarees'. 'I have never felt so much at ease with a man of action', claims Spender's article: Colonel Hidalgo Cisneros is as much at home with his friend Alberti's poetry, as with his fellow-pilots. Under his guidance the Spaniards are coming into their air-heroic inheritance, a mass-heroic ('however brave an action, no pilot is ever singled out for special praise; one reads in the Spanish papers of "our heroic aviation", but never of heroes'): a version of Fox's claim that where the West had one T. E. Lawrence the Soviet

Union had produced a nation of heroic aviators. So Republican Spain inhabits the leftist heroic plane. In standing up to the encroachments of Goliath Franco it has expanded into immense moral stature. To enter Spain is to be elevated. It's 'above . . . Above' that the road 'Goes on to the frontier' in Sylvia Townsend Warner's 'Waiting at Cerbère', and Spender's militia men in 'Port Bou' go on 'Over the vigorous hill'. And so to leave Spain is to participate in a diminishment: 'as I flew down from the Pyrenees to a country where money still goes, I felt', wrote MacNeice in his *Spectator* piece 'Today in Barcelona', 'that my descent into this respectable landscape was not only a descent in metres but also a step down in the world.'

Not surprisingly, in the matter of Spain people began to quarrel with Wilfred Owen's rejection of English poetry's fitness to speak of heroes. ' "All a poet can today is to warn", the greatest of the English war poets, Wilfred Owen, wrote in 1918. That is true always of poetry written in the midst of great social upheaval; but the poets of the International Brigade have a different warning to give from that of the best poets of the Great War. It is a warning that it is necessary for civilization to defend and renew itself.' Thus Spender in *Poems for Spain*. One of Hugh MacDiarmid's most astonishing poems is his attack on Siegfried Sassoon, 'An English War-Poet': 'the members of the International Brigade / Were made of different stuff'; they were 'stamped' in Spain

> With a different bearing altogether
> Than even the best, the most anti-militarist, gained
> Of those who fought against Germany in the first Great War.

Madge, too, thought the dismays of the First War irrelevant to this age. He wrote of Cornford and Guest in the Guest Memoir:

> Their too short lives are a phenomenon of our time quite unlike the shortened lives of those who were killed in the Great War, because they did not perish in the confusion of Imperialist War, but with their minds made up by a logic from which there is no escape, and in a struggle of whose significance they were deeply conscious.

So the military virtues that the First War poets had taught the '20s and early '30s to debunk came back into fashion. First War veterans like Wintringham were now highly prized for the skills they brought to the Brigades. First War kit came in suddenly handy. John Cornford returned for his second spell in Spain carrying his father's old First War revolver; Hyndman and Giles Romilly togged up at the Army and Navy Stores with boots and underwear purchased on Colonel Romilly's charge-account; Mrs Romilly pressed on Hyndman the scarf she had knitted for her husband during the First War. Many, as well, put on First War mental dress. Cornford's 'spirit was not', insisted Spender in his review of the *Memoir*, 'a resurrection of 1914'. Others were less sure about that. Look back a bit, Orwell urged his readers in 'Looking Back on the Spanish War':

> just have a look at the romantic warmongering muck that our left-wingers were spilling at that time. All the stale old phrases! And the unimaginative callousness of it! The sang-froid with which London faced the bombing of Madrid! . . . here were the very people who for twenty years had hooted and jeered at the 'glory' of war, at atrocity stories, at patriotism, even at physical courage, coming out with stuff that with the alteration of a few names would have fitted into the *Daily Mail* of 1918 . . . the same

people who in 1933 sniggered pityingly if you said that in certain circumstances you would fight for your country, in 1937 were denouncing you as a Trotsky-Fascist if you suggested that the stories in *New Masses* about freshly wounded men clamouring to get back into the fighting might be exaggerated.

An usual, Orwell lays his point on a bit thick, but it's still a good point. Even in his own Spanish writing the valiant old British 'thin red line' of imperialism breathes hotly down the neck of the International Brigade's 'thin line' that he prized so greatly. (And, of course, Orwell was particularly valuable to the POUM militia for his old skills in drilling natives in small arms and general military discipline that he'd acquired in the Imperial Indian Police.) The Communist Party—and, it now appears, later arrivals in the International Brigade—just wouldn't accept Alec McDade's grumpy lines in the song 'There's a Valley in Spain called Jarama', about the waste of 'our manhood' in the Brigade's unacceptably long stint in the cruel Jarama trenches. Instead, the song got rewritten and happily sung as a celebration of 'our glorious dead'.[288] C. Day Lewis's 'The Nabara' was from the start much nearer the Party's line, enthusiastically cheered on by John Lehmann as (unconscious irony) 'a bigger achievement than his previous long heroic ode, on the Australian airmen'.[289] And 'The Nabara' sites itself unerringly in the fat tradition of jingoistic British verse. Just so, in the echoing lists of dead war-heroes so common in the period—at the end of Wintringham's prose piece 'Comrades of Jarama', for example; in Lindsay's 'Requiem Mass: for the Englishmen fallen in the International Brigade' ('Where is Ralph Fox of Yorkshire?'; 'Where is John Cornford of Cambridge?'; 'Where is Davidovitch of Bethnal Green?); in William Rust's *Britons in Spain* (1939)—one is reminded not just of the War Memorials to the Fallen that sprouted all over England after the First War (reaping the sneers of Isherwood and the generations of Armistice-Day-despising pacifist schoolboys) but of Tennyson's great echoing celebration of dead soldiery, 'Ode on the Death of the Duke of Wellington'.

Still, Orwell was also right in attributing Blimpish enthusiasms mainly to observers far from the actual fighting. (He made one of his clearest over-statements when he tried to incriminate Cornford, by alleging little difference between Henry Newbolt's 'There's a breathless hush' and Cornford's 'Full Moon at Tierz'.) Day Lewis and Lindsay were notorious stay-at-homes. The action dwelt on so Newboltianly in 'The Nabara' was derived by Day Lewis from an account in a book, G. L. Steer's *The Tree of Gernika*. The soldiers learned better. 'No wars are nice, and even a revolutionary war is ugly enough', wrote Cornford. 'Reality makes adults of us all', declared Miles Tomalin in his 'After Brunete':

> Never call heroes men of glory
> Who rake in bloody dung and climb stinking out.

T. A. R. Hyndman, disillusioned 'deserter', one of the many volunteers whose presence in Spain became involuntary, who were jollied along forcefully by distinguished visitors like Harry Pollitt, and Willy Paynter and Arthur Horner of the Mineworkers' Union, who were told they must stay in Spain to keep up the Brigade's and the cause's good image, and who were imprisoned if they tried to quit (all this, while the trippers dodged about Spain freely and with acclaim, and key Party men were being snatched back to safety in England), T. A. R. Hyndman also fell naturally into anti-heroic, First War tones.

> But he was dying
> And the blanket sagged.
> 'God bless you, comrades,
> He will thank you'.
> That was all.
> No slogan,
> No clenched fist
> Except in pain.

Orwell's experience in Spain only endorsed First War lessons ('Looking Back'):

> The essential horror of army life . . . is barely affected by the nature of the war you
> happen to be fighting in . . . The picture of war set forth in books like *All Quiet on the
> Western Front* is substantially true. Bullets hurt, corpses stink, men under fire are often
> so frightened that they wet their trousers . . . People forget that a soldier anywhere near
> the front line is usually too hungry, or frightened, or cold, or, above all, too tired to
> bother about the political origins of the war . . . the laws of nature are not suspended
> for a 'red' army any more than for a 'white' one. A louse is a louse and a bomb is a bomb,
> even though the cause you are fighting for happens to be just.

Homage to Catalonia acknowledges, and resists, the temptation to glory-mongering.
Orwell describes his own train, grisly with collapsed, vomiting, wounded men,
pulling into Tarragona station just as a train-load of fresh International Brigaders
pulled away, bound for death at Huesca. Those wounded who could, cheered:

> A crutch waved out of the window; bandaged forearms made the Red Salute. It was like
> an allegorical picture of war; the trainload of fresh men gliding proudly up the line, the
> maimed men sliding slowly down, and all the while the guns on the open trucks making
> one's heart leap as guns always do, and reviving that pernicious feeling, so difficult to
> get rid of, that war *is* glorious after all.

But the description that follows of the wounds on show at Tarragona hospital
('Under the muslin you would see the red jelly of a half-healed wound') goes a long
way towards keeping you disabused of those troop-train illusions. So, in their several
ways, do Wintringham's poems about being wounded and hospitalized ('The Splint'
and 'British Medical Unit—Granien'), and Cornford's lines about being afraid and
about the undignified burial of Ruiz, and those fine, subdued, mournfully angry
poems that grieve over the dead in Spain—Donagh MacDonagh's 'He is Dead and
Gone, Lady . . .' (about Charles Donnelly), Ewart Milne's 'Thinking of Artolas',
the verses—not the reassuring choruses—of Margot Heinemann's poem about John
Cornford 'Grieve in a New Way for New Losses', Roy Fuller's 'Poem (For M.S.,
killed in Spain)' (about his friend the Blackpool Communist, Maurice Stott):

> Now uncovered is the hero,
> A tablet marks him where his life leaked out
> Through grimy wounds and vapoured into air.

Nineteenth-century Imperialist verses and their Spanish War ilk seem disgustingly
mendacious in such honest, pained company. Somhairle Macalastair knew better
than Lindsay and Day Lewis that tub-thumping tones were rather the appropriate
vehicle of fascist sentiments: his poem against O'Duffy's Irish Blue Shirt crusaders,
'Battle Song of "Irish Christian Front": "Off to Salamanca" ' makes the Fascist
speak Kipling:

With the gold supplied by Vickers,
I can buy Blue Shirt and knickers,
Let the Barcelona Bolshies take a warning.
For I lately took the notion,
To cross the briny ocean
And I start for Salamanca in the morning.

And it's one of the great distinctions of Spender's Spanish writing that, though in terms of action he never rose beyond the tripper level, he none the less came round in his mind to the soldiers', and Wilfred Owen's, point of view. In *What I Believe*, and in his article 'Poetry' in *Fact* magazine No. 4, he stood up explicitly for Owen's opposition to 'propaganda about War heroes'. He wrote Owenite lines of verse ('Two Armies'):

> who can connect
> The inexhaustible anger of the guns
> With the dumb patience of these tormented animals?

He reported the soldiers' distrust of the publicists' versions ('Heroes in Spain'):

> People try to escape from a realization of the violence to which abstract ideas and high ideals have led them by saying either that individuals do not matter or else that the dead are heroes . . . But to say that those who happen to be killed are heroes is a wicked attempt to identify the dead with the abstract ideas which have brought them to the front, thus adding prestige to those ideas, which are used to lead the living on to similar 'heroic' deaths.
>
> Perhaps soldiers suspect this, for they do not like heroic propaganda. When I was at the Morata Front several men complained of the heroics in left-wing papers.

In this article Spender came nearest to voicing publicly the distaste he was expressing privately about the politicians' rhetoric leading would-be heroes to ugly deaths. About himself he had fewer hesitations. Admittedly, the blunt line 'I am the coward of cowards' took time to get into the poem 'Port Bou', but Spender's Foreword to *The Still Centre* (1939) was happy to accept the charge of unheroism:

> As I have decidedly supported one side—the Republican—in that conflict, perhaps I should explain why I do not strike a more heroic note. My reason is that a poet can only write about what is true to his own experience, not about what he would like to be true to his experience.
>
> . . . One day a poet will write truthfully about the heroism as well as the fears and anxiety of today; but such a poetry will be very different from the utilitarian heroics of the moment.

The echo of Owen's Preface was clear; and all the poet Spender felt he should do in this period of poetry's unreadiness for heroic speaking was to adopt the still small personal voice of the coward: 'I have deliberately turned back to a kind of writing which is more personal, and I have included within my subjects weakness and fantasy and illusion.'

Big Spender wasn't the only poet big enough to admit his exclusion from the heroic. '[N]o man here is heroic' insists Ewart Milne's 'Sierran Vigil'—a poem 'Speaking no good word for war / for heroics, for the kingly dust.' But it took special courage for Spender to admit the truth of Roy Campbell's charges of cowardice: after all Campbell was gunning for Spender and his friends by name, as well as for Spender's party at large.

> . . . these three hundred Red-Necks, thrilled and caught
> By Prophecy, on the live wires of thought,
> Brought here to learn why communists 'feel small'
> And we so perpendicular and tall
> (Like a Cathedral over Comrades' Hall).

And a footnote of Campbell's rammed home the insult with reference to Day Lewis's notorious confession, reinforcing destructive intentions with a destroying misquotation into the bargain: 'Day Lewis, the Rearguard poet, who "fee-foh-fummed" so ferociously in peacetime and then spent the war in an armchair, wrote this line: "Why when I meet a Communist, do I feel small?" ' But Spender would not be deterred from telling the truth about his own feelings, even in the face of Campbell's crowings. Self-abasement could scarcely cringe lower than in 'Port Bou'. There, Spender disclaims the action and elevation of the militiamen. Not for him their 'vigorous hill'. Unheroically belittled he has to 'look up' to them, above him in their hill-bound, war-aimed lorry. In fact, Spender turns out to be less heroic than anyone around in Port Bou. He's overtaken by the old man, eager for the military life if only of a grotesquely toothless kind ('With three teeth like yellow bullets, spits "bang-bang-bang" '); by the children, who 'run after'; even by the women who, slow and encumbered though they are, still opt for a more heroic elevation than the poet: 'Clutching their clothes, follow over the hill.'[290]

It was, to be sure, only the truth about himself that Spender was telling. In some cases, of course, the '30s images of the heroic had actually been translated into temporary reality. Malraux did fly his plane in defence of Madrid; Clem Beckett, the communist speedway ace did find his Futurist apotheosis in the awful dynamic of a moment's encounter with a bullet; Orwell did get wounded honourably in action; Fox and Cornford, Caudwell and Donnelly did die bravely on the field of battle; their host of working-class comrades, despite their lack of practice in the generally bourgeois pastime of mountaineering (Gary McCarthey and Bill Donaldson, experienced members of a Glasgow climbing club, were exceptional among the British proletarian volunteers), did toil up and over the Pyrenees, endure the cold in the mountains, charge heroically and repeatedly up fascist-held inclines; the frontiers of fear were crossed again and again; hosts of brave men proved that pressure to live within a time's revived cult of the heroic body can easily turn into the call to die because of it. 'Body awaits the tolerance of crows': even—especially—the body of the bravest of dead fighters. But none of these instances of bravery was ascribable to Day Lewis, or Isherwood, nor to MacNeice or Auden or Spender. It was just as well that Spender accepted this sooner rather than later.

What's more, Spender's sense of personal failure was soon being very widely shared. As the war ground on to defeat, as news of the Stalinist political betrayals got about, so the sense of glory, purpose, meaning, drained surely out of the cause. The death of Charles Donnelly scarcely rises to the meaningful in MacDonagh's 'He is Dead and Gone, Lady': 'Something'—that's all—'has been gained . . .' Milne's 'Thinking of Artolas' won't go even that far:

> I set them together, Izzy Kupchik and Donnelly;
> And of that date with death among the junipers
> I say only, they kept it: and record the exploded
> Spreadeagled mass when the moon was later
> Watching the wine that baked earth was drinking.
> Such my story, Sirs and Senoritas. Whether you like it
> Or pay a visit to the vomitorium, is all one . . .

In Spain, bristling with the enacted icons of the '30s Bigness Heroic, the arena of so many brave men, so much brave action, it got eventually to be widely accepted that the idea of the hero had shrunk back to its diminished, First War, Owenite proportions. That heroic map of the Spanish Republic's struggle—the 'enormous map of the Civil War', as described by Auden, 'surmounted by a rifle and fixed bayonet, 15 ft. high'—shrivelled up as the Republic's land-mass gradually disappeared into the fascist maw. Ewart Milne's downbeat prose piece 'The Statue' dwells on this sort of diminishment. The 'twenty-foot statue of the militiaman, a gigantic piece of work' which had once towered over Barcelona's Ramblas and Plaza Catalunya, has now been taken down. It reappears in the course of Ewart's narration, dragged along in procession. The narrator tries to whip up enthusiasm for its emblematic meanings:

> 'It wouldn't matter if a shell did get him', I said. 'He's always dying or being killed, he's used to it. He died when an anarchist, Durutti, died, and when an Englishman, Ralph Fox, died, and when millions of peasants died, and an Irishman, Connolly, died to mention but few. But there he is again, you see, always popping up like a daisy, that democratic flower—'

'I'm sure', adds Vita, the narrator's friend, 'he'd talk like García Lorca, tender and expressive and sensual for all his bigness and bayonet and village blacksmith arms . . .' But still the statue is a 'Moloch', a 'monstrosity', marked by a vulnerability to shells and fallenness. It's as much on the decline as the once-heroic airman, in fact: for this is the piece where Milne describes the English aviator, airman-hero of the Republican summer, of 'August, 1936', who 'brought down many Italian planes', but who is now held in a Republican prison hospital. And the piece ends with the unheroic effects of the ascendancy of the fascist bomber-pilot. The narrator broadcasts to the world over Barcelona radio

> People who go to Madrid and return always speak of the heroic spirit of the defenders of the city. It is one thing to be heroic when one's belly is full and quite another when one's belly is empty. It is one thing to be heroic at the front, and another to lie awake night after night, listening for the double explosion, for the deadly apples of the triple engined planes.

Aptly so, for the '30s bigness heroic was indeed brought low as the once heroically-promising Spanish war literature filled up with the noise of bombs that fell from aeroplanes on to helpless human flesh, and the world correspondingly filled up with fears of 'implacable aeroplanes' (Hernández, in the Spenders' translation), of the 'all-seeing, all-powerful' 'winged eyes' (Wintringham), of the 'winged curse' (L. Kendall). In every sense, Spain was transformed by the bombing plane; and along with all the consequential shifts and swerves of opinion the period's taste for heroizing was soured well nigh irrecoverably:

> Spain is not Spain, it is an immense trench,
> a vast cemetery red and bombarded:
> The barbarians have willed it thus.

(So Mr and Mrs Spender, translating Hernández's 'Hear this Voice'.) No wonder that Clive Branson, imprisoned at Palencia, thinking sorrowfully of home, should imagine (in his poem 'On Dreaming of Home') the Spanish failure of his own and others' dreaming and thinking, and the frustration of mind and will ('No good the map of the will alone'), as the bringing low, the crashing of an aeroplane. The disillusioned volunteer is a crashed aviator. Was it, then, one wonders, with an altogether unconscious irony that the device commonly used to help hospitalized Spanish fighters' wounded arms in healing was called an 'aeroplane splint': 'one of those huge wire contraptions' that Orwell's *Homage to Catalonia* mentions, 'nicknamed aeroplanes'?

The journey to Spain, which was supposed to accelerate the doubting leftist author into commitment and action, to seal him into the public mode, the public arena, ended by returning him sooner (in the case of a Spender or a Ewart Milne) or later (in the case of an Auden or Wintringham) to the personal, the inactive, the uncommitted or differently committed. The '30s event that challenged authors most directly with the public crisis, turned out to be also the way into the period's most private crisis. This public journey was also, as Greene had discovered his African trip to be, an inner journey, 'travel in the mind' as Ruthven Todd's poem 'It was easier' called it.[291] Spain's map was also Branson's 'map of the will', the map (in Lindsay's 'On Guard for Spain!') 'of your own fate'. Spain hastened the clarity of self-revelation.

> In Spain the veil is torn.
> In Spain is Europe. England also is in Spain.
> There the sea recedes and there the mirror is no longer blurred.

Thus Rex Warner's 'The Tourist Looks at Spain': and the onlooker, the tourist, was susceptible to inward crisis as much as if not more than the soldier. The problem of one's personal courage—would one go?—assailed the onlooker as harshly as the questions of how one would fight, and whether one could face the enemy again, afflicted the combatants. And the longer the war went on, the longer the commentator survived—and the trippers naturally lived longer than a lot of the soldiers did—the moral question got larger. Would one, could one, tell the truth? Propaganda demanded useful 'truth', it needed 'either angels or devils', to the ruination, as MacNeice perceived, of one's writing: 'in the long run a poet must choose between being politically ineffectual and poetically false.'[292] In direct ways writers suffered from the propagandists. The merchants of expedient visions rewrote displeasing lines: MacDonagh's 'He is Dead and Gone, Lady', had its pious subtitle, 'For Charles Donnelly, R.I.P.' lopped off when it appeared in MacDiarmid's communist magazine *Voice of Scotland*, and its restrained, even critical, final line, 'Something has been gained by this mad missionary' was utterly transformed into the more acceptably up-beat 'Time will remember this militant visionary'.

In his story 'An Incident of the Campaign' G. S. Fraser quoted a Brigader's unheroic, grousing song that blamed France, England, and Russia for their meagre help ('I am a poor sod of a Government man. / I fought till I blistered, I walked till I bust'):

A singularly ribald composition, it was said to be the work of a young English Communist who had fought in the International Brigade. For obvious reasons (it was co[a]rse, brutal, and unorthodox), it had not been included in his slim posthumous volume.[293]

To his immense moral credit, Orwell bluntly refused to accept such 'obvious reasons' and got himself maligned and ostracized on the Left for his truth-telling efforts. It was far easier to slink quietly away, to let one's enthusiastic tones leak slowly out and the damning confessions emerge, if at all, slowly and later. This was Auden's way, and the way of Brigaders who left the Party after the Spanish War—like Wintringham, or, for that matter, Fred Copeman who later became a Catholic, or Albert Cole, a founder member of the Liverpool Communist Party, or George Aitken another founder of the Communist Party in Britain, first manager of the *Daily Worker*, commissar of the British Battalion, then of the XVth International Brigade. And just how hard it was to come clean with oneself is shown by Spender's case. He was determined, he said in *Life and the Poet* (1942), not to be one of those intellectuals who wouldn't tell or accept the disillusioning truths of Spain, whether as an 'enthusiastic simpleton' or an 'adroit hypocrite'. His poems were bravely cowardly; his articles strove against the propagandists and the 'uncritical, heroic attitude towards the war';[294] he faced (in 'Heroes in Spain') the fact that volunteers might be politically mixed-up, and praised (in the same piece) the honesty of a volunteers' tears ('they seem to me a monument of personal honesty, of the spirit in which the best men have joined the International Brigade'). But the extent to which even his determination wilted is shown by the way *Life and the Poet* and his later autobiography *World Within World* are so much more revealing than anything he published in the '30s, about his self-disgust as a slumming and compromised intellectual and about his disquiet over the Writers' Congress's steady hostility to Gide's published defection. And some of his doubts stayed strictly confined to private correspondence. Nowhere did he wax so damning in public as in a long, private letter to Virginia Woolf.

The letter is a very unhappy text indeed, scribbled down in Cerbère on 2 April 1937 in the flustered emotion, the white heat of Spender's return across the frontier. In it Spender tells the story of his failed attempt to winkle T. A. R. Hyndman out of the Brigades. The letter registers Spender's shock at discovering that the Communists were running the Brigades with an iron hand. 'Lack of imagination', Spender labels it. He charges the recruiters at home with being unscrupulous and liars, sending out 'sensitive' and 'romantic' enthusiasts like Hyndman to be caught up in a harsh political machine from which they'll not be allowed to escape. Spain is a Hell, he alleges, of narrowness and political dogmatism which is frighteningly religious in its intensity. Virginia Woolf must warn Julian Bell off.

Spender's special interests, his guilt and despair over Hyndman, no doubt colour his attitude. None the less his letter has great force. It amounts to the private equivalent of Gide's *Retour de l'URSS*. But the striking aspect of the matter is that the letter is private. Revelatory and traumatic though the journey has been, and eager though Spender is to spread private disillusionment (successfully too: 'Your letter came in the nick of time', Virginia Woolf replied, 'to set' Julian Bell 'against the C.P.'), several more years, and another lengthy Spanish visit (to the Writers' Congress), were to elapse before Spain's full effects received any real measure of acknowledgement in Spender's published writings.[295]

14

Too Innocent A Voyage

SOME things must go on the credit side of the Spanish account. Spain made a lot of the Old Boys grow up and helped some bourgeois ditherers go over to the workers' side. It turned some writers into authentic men of action, and united their art with a people's hopes, at least for a while. But even these diminished returns were hedged about with Spain's wider sense of failure. It was art's loss, as well as a set of personal losses, that Cornford and Caudwell, Donnelly, Bell, and Fox were killed in action. The most positive result of Spain, the period's renewed perception of the destructive element's destructiveness, the relearning of Owen's First War lessons, was heavy with negativity and the denial of a period's aspirations. The Spanish frontier proved insurmountable or ultimately defeating for many would-be frontier-crossers; many failed even in Spain to 'go over'; their revolutionary fervours had flared up into life but had as quickly spluttered out in despair; they found they couldn't translate, couldn't export back home, the satisfying alliances between art and revolution that they'd glimpsed in Spain. *Poems for Spain* wasn't succeeded, as it were, by *Poems for Britain*. Remnants of the old leftist guard survived into the '40s; but the '50s and '60s would provide a new scene, require a new personnel. Some of the '30s crowd would join attempts at reviving the old manifestoes. Mulk Raj Anand, Charlotte Haldane, Pat Sloan, John Sommerfield, Randall Swingler, Sylvia Townsend Warner, Alick West all signed *Poetry and the People*'s appeal for funds in February 1940 ('We have the opportunity of becoming the first real poetry magazine, read by ordinary people in every town and village in Great Britain'). Some of these signatories would stick by the Communist Party through thick and thin. Others would peel gradually away in the later '40s (they included Edward Upward), yet others over Hungary in 1956, and so on. But the more famous fellow travellers had dropped almost uniformly away by the end of the '30s. The Hitler–Stalin Pact, and the bombing of Poland by Soviet planes, were, it's clear from memoirs like John Lehmann's, for many Party members as well as fellow travellers a traumatic extension of the already devastating trauma of Spain. Spender signed that 1940 appeal; but by 1939 he was not only dismissing Philip Henderson's *The Poet and Society* as a tiredly 'conducted tour through well-charted New Country', he was also gloomily describing poets who had joined the International Brigade as people who refused life in favour of exile in a 'violent environment', and at the same time firmly rebutting the idea that T. S. Eliot's pronouncements as 'Churchwarden' had any damaging effect on the importance of 'Ash Wednesday' or 'The Wasteland' ('If the political realists had the good sense and faith to read the lines instead of fretting at the acts and the opinions . . .').[296] By 1942 (*Life and the Poet*) he was attacking 'the attempt to make poetry serve a cause or interest . . . Poetry cannot take sides except with life'. He'd thrown in his weight with Connolly's new paper *Horizon*, whose first editorial declared in January 1940 a new separation of 'culture from life': 'and however much we should like to have a paper that was revolutionary in opinions or original in technique, it is impossible to do so when there is a certain suspension of judgement and creative activity.'

The '30s leftist Grails had failed their seekers, and that failure happened most substantively in Spain:

> The young men for whom the Spanish war had been a crusade in white armour, a Quest of the Grail open only to the pure in heart, felt as if their world had burst; there was nothing left but a handful of limp rubber rag; it was no good trying any more.

So MacNeice, in *The Strings Are False*. Utopian hopes formerly centred on Moscow collapsed as Moscow showed its ugliest face in Spain:

> These man-hunts in Spain went on at the same time as the great purges in the U.S.S.R and were a sort of supplement to them. In Spain as well as in Russia the nature of the accusations . . . was the same and as far as Spain was concerned I had every reason to believe that the accusations were false. To experience all this was a valuable object lesson . . .
> . . . I understood, more clearly than ever, the negative influence of the Soviet myth upon the Western Socialist movement.

Thus Orwell, declaring the Spanish origins of his own war against Communism and against all political distortions of language, in the Preface to the notorious Ukrainian edition of *Animal Farm*. And as individuals' Spanish journeys smashed in terminal disappointment, so did the period's other outings in optimism. 'Since the Spanish war people stopped believing they could change things', Berthold Lubetkin, famous in the period as king-pin of the architectural firm of Tecton, has since declared.[297] Spender's 'Coward' dies of fright in Spain and [in the *Still Centre* version]:

> To him, that instant was the birth
> Of the final hidden truth
> When the troopship at the quay,
> The mother's care, the lover's kiss,
> The following handkerchiefs of spray,
> All led to the bullet and to this.
> Flesh, bone, muscle and eyes
> Assembled in a tower of lies
> Were scattered on an icy breeze
> When the deceiving past betrayed
> All their perceptions in one instant,
> And his true gaze, the sum of present,
> Saw his guts lie beneath the trees.

By the end of the decade to celebrate Spain and the Spanish War was also to celebrate the demise of the hopes they once encapsulated. 'We have lived in Spain', declare a couple of Spaniards at the end of Sylvia Townsend Warner's novel *After the Death of Don Juan* (1938), as they search out Madrid on the map ('And there in the middle of it, like a heart, is Madrid'). But this historical fiction (set in the 1770s) also works busily to suggest that Spanish peasant/worker revolution has always been doomed to fail, as it does in this novel, in a welter of talk and vain promises of help from unreliable bourgeois sympathizers. Britten's *Ballad of Heroes*, performed in April 1939 in honour of the British Battalion of the International Brigade, also blends its positive notes into more negative gloomings. In its central scherzo it incorporates some of the verses of Auden's 'Danse Macabre'. The result (as Donald Mitchell points out) is a complex set of farewells.[298] We are reminded of the would-be soldier's farewell that Auden's poem started out as, but also of Auden's and Isherwood's then recent retreat from commitment and politics, in their voyage to America. The piece sounds, as well, as if it's Britten's own farewell to England (he and Peter Pears were about to depart for New York) and to political commitment (it was Britten's last politically committed work of the decade). The sense of endings lies heavy on the work, endings intimately of a piece with the International Brigaders' frustrated return home.

As those journeys America-wards proved, there were other Grails, other journeys to try, after the collapse of fellow-travelling hopes:

> Keen for exposure when you felt secure
> Far from the wolves that gnawed a stranger's door,
> Round what grim gas-house did you plan this tour,
> By what canal this 'Journey From a War'?

Indignation ran high, especially among those like 'Bill', author of that verse of 'Look Again, Stranger!',[299] whose personal fellow-travelling hopes hadn't altogether collapsed, but also on the Right, in the pages of Evelyn Waugh's *Put Out More Flags*, which relishes every moment of the Left's discomfiture at the apparent desertion of Parsnip and Pimpernell. ('What I don't see', declares a truculent Leftist in Waugh's novel, 'is how those two can claim to be contemporary if they run away from the biggest event in contemporary history'.) But since the Spanish journey, that particular self-exiling, had failed, Auden was willing to try another exile in the United States, there to accept, as other disillusioned refugees from '30s hopes did, the Christian hope—that other exile consolation. Isherwood, too, would find metaphysical consolations in America. And Connolly, for all his vexation in *Horizon* No. 2 in February 1940 over 'our best poet' and 'our most promising novelist' seizing 'the main chance', saving their own skins and abandoning 'the sinking ship', could only agree that they were right to give up European politics, the things that Spain had stood for, and 'by implication the aesthetic doctrine of social realism'. 'Are they right? It would certainly seem so'. For his part Orwell chose simply to come home, to retreat to England. And this was, like Spender's letter to Virginia Woolf, Orwell's *Retour de l'Espagne*, a turning of the '30s frontier myth inside out. The end of *Homage to Catalonia* deliberately recalls the tone of all those excited visitors to Moscow and Barcelona, all that '30s frontier-crossing, in order to debunk it. Crossing into France is easy: 'we slipped through the barrier', 'without incident'. It's not now a going over to the workers, but a going back to the bourgeoisie. Orwell recalls the earlier time when smartly dressed travellers were turned back by the Spanish anarchist border-guards ('it was looking like a proletarian that made you respectable'). Now he's glad to look conventionally 'respectable': 'to look bourgeois was the one salvation'. Tones with which Orwell once celebrated Barcelona's revolution—tones conventional among Moscow's trippers for recording the first sight of Soviet tractors or a People's Park—he now deployed to enthuse about the simple things of French and English bourgeois existence:

> I wonder what is the appropriate first action when you come from a country at war and set foot on peaceful soil. Mine was to rush to the tobacco-kiosk and buy as many cigars and cigarettes as I could stuff into my pockets. Then we all went to the buffet and had a cup of tea, the first tea with fresh milk in it that we had had for many months. It was

several days before I could get used to the idea that you could buy cigarettes whenever you wanted them.

And like Gordon Comstock accepting the aspidistra, Orwell returned to the paradisal England of his childhood. He knew his retreat was insecure, that the world was clamorous with danger, that the miserable industrial towns hadn't disappeared, that 'the roar of bombs' would disrupt this bourgeois, southern English utopia. But, in the meantime, England—even as it was—would do. And whatever should happen to this England the patriotism he now resumed was set to last him till he died. 'Down here'—he was ready now to redirect the downward plunge he had earlier made into Wigan and Spain—

> Down here it was still the England I had known in my childhood: the railway-cuttings smothered in wild flowers, the deep meadows where the great shining horses browse and meditate, the slow-moving streams bordered by willows, the green bosoms of the elms, the larkspurs in the cottage gardens; and then the huge peaceful wilderness of outer London, the barges on the miry river, the familiar streets, the posters telling of cricket matches and Royal Weddings, the men in bowler hats, the pigeons in Trafalgar Square, the red buses, the blue policemen—all sleeping the deep, deep sleep of England . . .

Of course the problems to which '30s intellectuals characteristically addressed themselves have not gone away. Unemployment, hunger, injustice, dictators, repression of the powerless, the destructive element of actual wars, and the arms race, the social divisions nurtured in the West by the hegemony of bourgeois capitalism and, in Britain still, by socially divided and divisive practices in education, the arts, and traditional culture: all these are still with us. And the '30s struggle to radicalize writers and to communalize writing still goes on. It's fashionable among some younger protagonists of this theoretical enterprise to dismiss the '30s theoreticians of that struggle (the name of Christopher Caudwell is commonly muddied). But the line of radical descent to our own day is clear. The workers' choirs assembled and encouraged by Alan Bush are still singing; Joan Littlewood's work has looked back with enthusiastic gratitude to the Workers' Theatre Movement in which she began; Julian Symons has agreed, more or less (in his essay 'Keeping Left in the Thirties', now in his *Critical Observations*, 1981), with Edgell Rickword's suggestion that in the '50s ' "the Weskers and Sillitoes were reading their elder brothers' copy of *Left Review*".' 'Or', Symons adds, 'their fathers' copy perhaps?'

And if resistance to the myths and structures of bourgeois power and entrenched authority hasn't died, if on the contrary it shows certain signs of lively hope, it's because it is now, in the '80s, returning the discussion to where the schoolmaster-poets, where Auden the one-time reviewer and maker of educational books, where the Leavises, who in the forties and fifties captured the critical ground vacated by the '30s Marxists, knew it must belong: in the class-room and lecture-room. Once more, in the vacuum left by the decline of the Leavises' influence, the debate has moved, armed now with insights from the newer Euro-Marxisms, into the sort of places '30s Marxists once flourished in, the old schools of university English (and their successors in the polytechnics and the newer departments of cultural studies). And some participants in the debate have clearly learned, whether they all put it to themselves like this or not, to shun the sort of self-deceiving fantasies about easy, just-around-the-corner futures, the emptily mass-minded delusions and quasi-totalitarian cravings that '30s radicals indulged in. There are plenty of new naïveties, many new false starts, but at least the prevailing tone is now more cautious, less heroizing, less self-romanticizing, than in the '30s, and the new critics—again unlike in the '30s, a period perpetually haphazard about theory—have realized that the ground of theory must be sternly besieged as part of the process of making the reading and writing of books (and of all their textual neighbours) of more interest and pertinence to all the people.

A difficult and intransigent struggle demands a realistic politics. The troubled '30s experience will not have been in vain if it drains expectations of unwary—even, as we've suggested, childish—utopianism. Progressive critics nowadays, of course, scarcely bother to consult Christian wisdoms. Perhaps they should. Where '30s Christians, and in particular the '30s Leftists who became Christians, or returned to Christianity, were wiser than most of their political rivals was in their acceptance of the persistent problem of evil, evil in the human heart that keeps skewing good human aspirations. Admittedly, the '30s Christians were an odd lot with enough cranky variants on the idea of evil to put even the most tolerant observer off. Their frequently strange concepts indicate noisily that just recognizing the fact of evil is not by itself a saving realization. C. S. Lewis's devils are as comically argumentative, as dangerously civilized as stiltedly ageing Oxford dons; Graham Greene's Catholic sense of evil, like Evelyn Waugh's, is too nearly Manichean and too sectarian (it's manifestly dotty of *Brighton Rock* to discount *totally* Ida Arnold's ordinary-person discriminations 'right and wrong' in favour of Pinkie's Catholicized 'good and evil'); for Charles Williams evil is too gaudily sensational, too relishingly sado-masochistic. Worse, most of the period's Christians equated the good too promptly with the *status quo* or with Fascism, and evil too completely with Marxism and Communism. They were driven too by their perturbed vision of humankind's sinfulness into anti-human, often inhumanly idealist, Platonic excesses, discarding and disregarding the body and sexuality with Eliot or the Oxford Groupers, over-dyspeptic with Greene and his relish for the perpetual dead-dog across the pathway of life, consciously aloof and superior with Waugh and his apparent dislike of most of his fellow beings (so that, hereabouts, Orwell's tendentious allegation, in his essay 'Lear, Tolstoy and the Fool' (1947), that Christians are always choosing 'the next' world in preference to this one and so are crucially self-debarred from feeling the full weight of human suffering and tragedy and the sensations of, in Orwell's phrase, 'a normal human being', has decided point). But for all that, the question of evil remained and remains. Waugh carries conviction when he praises Georges Bernanos's 'sense of sin' (in a *Night and Day* review of *The Diary of a Country Priest*, 28 October 1937) and tries to put wicked actions into theological perspective: 'We live in a world of authors who try to make our flesh creep by elaborating more and more perverse and bloody crimes, but beside sin, as the saint sees it, these exploits are only the naughtiness of children, while the naughtiness of some children may cry to heaven for vengeance.' Spender's *The Destructive Element* may be one of the most creakily cumbersome critical books this century, altogether too bleating an affair, but it does worry importantly about the state of literature's traditional moral concerns. If leftist critics want the political subject to flourish in fiction they must, Spender keeps suggesting, expand 'politics' to take in all the old domain of 'morals'.

His point carries weight. It coincides, more or less, with T. S. Eliot's vigorous doubts (in his 'Religion and Literature' essay of 1935) as to whether 'what I call Secularism'—the concern 'only with changes of a temporal, material, and external nature; . . . with morals only of a collective nature'—will do. In fact a recognition of individuals' propensity for evil—the recognition memorably developed after the Second War in the fictions of William Golding and Iris Murdoch, and that characterizes, explicitly in cases like Spender's as well as implicitly in lots of others, the disillusionments and shifts of position of so many '30s authors—such a recognition does seem indeed to be the moral maturation that T. S. Eliot claimed it was in his famous essay on Baudelaire (first published, intriguingly, as the original Introduction to Isherwood's translation of *The Intimate Journals* of Charles Baudelaire in 1930). Better, in Eliot's view, to go in absolutely for evil, than to deny or skirt its presence:

> Baudelaire has perceived that what distinguishes the relations of man and woman from the copulation of beasts is the knowledge of Good and Evil (of *moral* Good and Evil which are not natural Good and Bad or Puritan Right and Wrong). Having an imperfect, vague romantic conception of Good, he was at least able to understand that the sexual act as evil is more dignified, less boring, than as the natural, 'life-giving', cheery automatism of the modern world . . . So far as we are human, what we do must be either evil or good . . . Baudelaire was man enough for damnation . . .

Eliot ends his essay with a characteristic passage from his mentor and guru T. E. Hulme about 'man' being 'endowed with Original Sin' and 'essentially bad' (so that 'While he can occasionally accomplish acts which partake of perfection, he can never himself *be* perfect'). An apt place, I think, to end this consideration of the '30s failure to achieve the perfections it sought. For why else, at bottom, did going over and loving one's neighbour prove such problems? Why else did so many revolutionary hopes collapse so readily into savage repressions? Why else did people (as they still do) find it more difficult to feed the hungry than to gorge themselves, harder to share possessions and power than to keep them, harder to turn swords into ploughshares than to live fatly on the proceeds of arms trading, harder to open, as it were, the door of the classroom, the library, the museum to the ignorant, the unlearned and the deprived than to slam it shut in their face? In the sense that the '30s experience brought such unpleasing reflections home it was indeed a voyage—like so many voyages that happened in the '30s—out of Utopian notions of innocence into a kind of Christian awareness of sin: a voyage such as is represented, in fact, in Richard Hughes's prophetic novel of 1928, *A High Wind in Jamaica*; that is, a journey into the revelation of the hellish wickedness and bloodily murderous violence of a group of apparently innocent people—in this case actual children. An updating of James's *The Turn of the Screw* and *What Maisie Knew*, *A High Wind in Jamaica*'s more ironic American title is pertinent to my point: some American readers (at least) have known this book rather as *The Innocent Voyage*.

Or, to take another prominent '30s metaphor: when '30s hopes crashed down to earth, bringing the bigness heroic down with them, they were only upholding the Eliot/Hulme contention about sin. The romantic, Utopian, sin-denying attitude, Hulme wrote in his essay 'Romanticism and Classicism' (published in his *Speculations* in 1924, and again in 1936, under the editorship of Eliot's assistant Herbert

Read), this attitude 'seems to crystallise in verse round metaphors of flight . . . It is a question of pitch . . . a certain pitch of rhetoric . . . The kind of thing you get in Hugo or Swinburne.' 'In the coming classical reaction', Hulme went on, 'that will feel just wrong.' One doesn't have to accept Hulme's egregious right-wingery to see how vividly he put his finger on the '30s reaction when, confronted by manifest human intransigencies, all that old ambitious loftedness came to feel just simply no longer tenable.

Amherst-Konstanz-Merton Street 1980–6

NOTES

An endeavour has been made to keep most references in the body of the text. To avoid excessive congestion, though, some have been taken out, as follows.

In the Notes and Bibliography, where no place of publication is indicated, London should be assumed.

1. Thick With One's Spittle

1. 'The "Realism" Quarrel', *LR*, III, 3 (Apr. 1937).
2. Samuel Beckett, 'Intercessions by Denis Devlin', *transition* (Apr.–May 1938), reprinted in *Disjecta*, by Samuel Beckett, ed. Ruby Cohn (1983).

2. Vin Rouge Audenaire?

3. See Patrick Leigh Fermor's engrossing account of that journey, *A Time of Gifts* (1977).
4. Roger Poole, ed., *Fiction as Truth: Selected Literary Writings by Richard Hughes* (Bridgend, 1983).
5. See Jacques Derrida, 'Living On: Borderlines', in Geoffrey Hartman, ed., *Deconstruction and Criticism* (1980).
6. In 'The Eiffel Tower', in *The Eiffel Tower and Other Mythologies*, trans. Richard Howard (N.Y., 1979), Roland Barthes suggests the literary text is like a city interpretable, 'in its structure', from a vantage point like the Eiffel Tower. He has an unrealistically rigid idea of what any city like Paris—and so the literary and/or historical text—is actually like.
7. See Frank Budgen, *James Joyce and the Making of 'Ulysses'* (1934); reissued and supplemented (Bloomington 1960; Oxford 1972).
8. Geoffrey Grigson, 'Education in the Twenties', *NV*, no. 29 (Mar. 1938).
9. Harold Nicolson, *Diaries and Letters 1930–1939*, ed., Nigel Nicolson (1966), p. 153.
10. *PNW*, no. 18 (July–Sept. 1943).
11. No. XIII, in *Poems and Songs* (1939); now in *The Collected Ewart 1933–1980* (1980).
12. See Lawrence Lipking's powerful essay 'Aristotle's Sister: A Poetics of Abandonment', in *Canons*, ed., Robert von Hallberg (Chicago and London, 1983, 1984).
13. (30 Oct. 1936); in Stephen Spender, *Letters to Christopher*, ed., Lee Bartlett (Santa Barbara, 1980).
14. 'Writers and Leviathan', *Politics and Letters*, Summer 1948, in *The Collected Essays, Journalism and Letters of George Orwell*, ed., Sonia Orwell and Ian Angus (1968), vol. IV.
15. 'Freedom That Destroys Itself', *L* (8 May 1935).

16. E. M. Forster, *Selected Letters*, ed., Mary Lago and P. N. Furbank, vol. II (1985).
17. 'Words as Narrative', *TCV*, no. 1 (Jan. 1937).
18. 'On *The Human Predicament*' (from a broadcast), in Richard Poole, ed., *Fiction as Truth*.
19. 'My First Book', *The Author*, July 1981; in *A Moving Target* (1982).
20. (15 Mar. 1933). In 'Chronology and Documents' section, compiled by Mary-Lou Jennings, of her *Humphrey Jennings: Film Maker/Painter/Poet* (British Film Institute, 1982).

3. Destructive Elements

21. (To Lady Mary Lygon); *Letters*, ed., Mark Amory (1980).
22. (3 Aug. 1939); in *Letters to Christopher*, ed. cit.
23. (To Katharine Asquith, 23 Dec. 1935); in *Letters*, ed. cit.
24. Oswald Mosley, 'Crisis', *Action*, no. 1 (8 Oct. 1931); Alec Waugh, *Wheels Within Wheels: A Story of the Crisis* (1933); Storm Jameson, 'Crisis', *LR* (Jan. 1936); Alick West, *Crisis and Criticism* (1937); Martin Turnell, *Poetry and Crisis* (1938); Christopher Caudwell, *The Crisis in Physics* (1939); Allen Hutt, *This Final Crisis* (1935); Hilaire Belloc, *The Crisis of our Civilisation* (1937); John Strachey, *The Nature of Capitalist Crisis* (1935); Charles Madge and Tom Harrisson, 'What is a Crisis', Part (a) of 'Crisis', *Britain by Mass-Observation* (Jan. 1939).
25. I'm thinking particularly of John Stevenson and Chris Cook, *The Slump* (1977), but John Stevenson's efforts to deny or ameliorate the Slump pervade much of his writing. See Bibliography, section 27.
26. In *Class, Culture and Social Change*, ed., Frank Gloversmith (Brighton, 1980).
27. David Gascoyne, *Paris Journal 1937–39* (1978).
28. (1 Oct. 1938). *Letters*, ed., N. Nicolson and J. Trautmann, vol. 6 (1980).
29. *Selected Poems* (1940).
30. 'Buy this Book', *NV*, no. 22 (Aug.–Sept. 1936).
31. The poem appeared in *L*, 30 September 1936, when Dyment was only 22. The speaker is Dyment himself. The letter in the poem, to Dyment's mother, is printed in full in

Dyment's *The Railway Game: An Early Auto-biography* (1962).

32. Christina Stead, 'The Writers Take Sides', *LR*, I, 11 (Aug. 1935).

33. Spender, '*The Left Wing Orthodoxy*', *NV*, nos. 31–32 (Autumn 1938).

34. Day Lewis in *DW*, 1 August 1936.

35. Church was reviewing anthologies, including *The Faber Book of Modern Verse*, in *C*, XV, no. 61 (July 1936).

36. *C*, XVII, no. 66 (Oct. 1937).

37. 'Orwell and the Spanish War: A Historical Critique', in *Inside the Myth, Orwell: Views from the Left* (1984).

38. Berg MS, New York Public Library.

39. *LR*, III, 7 (Aug. 1937).

40. 'Torture and Death', *NS*, 10 December 1938 (Christmas Books Supplement).

41. Plomer's article 'Winner Take Nothing' (1934) is reprinted in *Then and Now* (1935).

42. 'Art and Literature: A Martian Reviews Our Books', *L*, 26 June 1935.

43. In his volume *One-Way Song* (1933).

44. *CPP*, no. 8 (Dec. 1936).

45. *Spectator*, 24 June 1938.

46. Stevie Smith's piece is collected in the admirable Virago gathering of her uncollected writings, *Me Again*, ed., Jack Barbera and William McBrien (1981).

47. 'An Open Letter to Aldous Huxley on his "Case for Constructive Peace" ', *LR*, II, 11 (Aug. 1936).

48. 'To Aldous Huxley', *LR*, III, 11 (Dec. 1937).

4. In The Cage

49. Review of Detective Stories, *C*, V, no. 1 (Jan. 1927).

50. Review of *The Beast With Five Fingers*, *C*, VIII, no. 30 (Sept. 1928).

51. *The Immaterial Murder Case* (1945); planned in collaboration with Ruthven Todd, but written entirely by Symons. See 'Ruthven Todd: Some Details for a Portrait', in Julian Symons, *Critical Observations* (1981).

52. *CPP*, no. 6 (Oct. 1936) and no. 9 (Spring 1937).

53. *TCV*, nos. 15–16 (Feb. 1939).

54. *NV*, no. 29 (Mar. 1938).

55. 'The Guilty Vicarage', *Harper's Magazine*, May 1938; reprinted in *The Dyer's Hand, and Other Essays* (1963).

56. *LR*, III, nos. 12 and 13 (Jan. and Feb. 1938).

57. 'Crime', *Epilogue*, II (Majorca, Summer 1938).

58. 'The Cinema', *Spectator*, 17 June 1938.

59. It was first published in *C*, XI, no. 42 (Oct. 1931).

60. 'Cage' appeared in *The Year's Poetry 1937*, compiled by D. K. Roberts and Geoffrey Grigson; as did 'The Cage', by Walter de la Mare.

61. 'The Landauers' was first published in *NW*, no. 5 (Spring 1938).

62. 'Men in Darkness', in *Men in Darkness: Five Stories* (1931).

63. Berg MS, New York Public Library.

64. Review of John Middleton Murry, *Son of Woman: The Story of D. H. Lawrence, C*, X, no. 41 (July 1931).

65. The lecture appeared in *Essays Presented to Charles Williams* (1947).

66. Swingler's poem is gathered in John Lucas, ed., *The 1930s: A Challenge to Orthodoxy* (Brighton, 1978).

67. 'Ill' appeared in *NV*, no. 24 (Feb.–Mar. 1937).

68. 'Escales', part of the Auden-Isherwood *Journey to a War* material published separately in *Harper's Bazaar*, October 1938: reprinted in Isherwood's *Exhumations* (1966).

69. 'A Sketch of the Past', written in 1938; first published in *Moments of Being: Unpublished Autobiographical Writings of Virginia Woolf*, ed., Jeanne Schulkind (Sussex, 1976).

5. Too Old At Forty

70. 'The Oxford Boys Becalmed', in Edmund Wilson, *The Shores of Light: A Literary Chronicle of the Twenties and Thirties* (1952).

71. Revised Foreword of *All The Conspirators*, 1958: reprinted in *Exhumations* (1966).

72. 'Unhappy Mr Coward', *L*, 10 March 1937 (Early Spring Book Supplement).

73. 'Lines Written When Walking Down the Rhine', *C*, X, no. 38 (Oct. 1930).

74. 'The Inheritor' first appeared in *TCV*, no. 11 (July 1938).

75. Excerpts from *The Prolific and the Devourer* are in *The English Auden*, ed., Edward Mendelson (1977).

76. 'Diary of Percy Progress—1', *N&D*, 15 July 1937 (reprinted in *Night and Day* selection, ed., Christopher Hawtree, 1985).

77. In the Berg Collection, New York Public Library.

78. 'Careers For Our Sons: Education; Truths About Teaching', *Passing Show*, 16 February 1929 (reprinted in *The Essays and Reviews of Evelyn Waugh*, ed., Donat Gallagher, 1983).

79. 'Poetry and Politics', *NV*, no. 3 (May 1933). The Auden poem was originally Part II of 'A Happy New Year' in *New Country*; afterwards published separately, as 'Now from my window-sill I watch the night'.

80. Gavin Ewart collected this song in his *Forty-Years On: An Anthology of School Songs* (1969).

81. See Maclaren-Ross's account of his encounter with Cyril Connolly in *Memoirs of the Forties* (1965), p. 63.

82. Another poetic trophy of Ewart's *Forty Years On* collection.

83. (8 July 1938); in *The Collected Essays* etc., vol. I.

84. *N&D*, 14 October 1937 (reprinted in *Night and Day* selection cit; and in Graham Greene,

The Pleasure-Dome: The Collected Film Criticism 1935–40, ed., John Russell Taylor (1972)).

85. (Dec. 1935): reprinted in Connolly's collection *The Condemned Playground, Essays: 1927–1944* (1945).

86. *Oxford Poetry 1921* was edited by Alan Porter, Richard Hughes and Robert Graves and contained poems by F. N. W. Bateson (he hadn't dropped the N., for Noel, at that stage), Edmund Blunden, Graves, Hughes, Edgell Rickword, L. P. Hartley, and John Strachey. *Oxford Poetry 1923* was edited by David Cleghorn Tomson and F. W. Bateson, with poems by Harold Acton, Bates, Lord David Cecil, Graham Greene, Michael Hollis, Lord Longford, and A. L. Rowse. *1924* was edited by Harold Acton and Peter Quennell (poems by Acton, T. O. Beachcroft, Greene, Brian Howard, Quennell, Rowse, James Sutherland); *1925* by Patrick Monkhouse and Charles Plumb (poems by Acton, C. Day-Lewis, Greene, Rowse, Sutherland); *1926* by Charles Plumb and W. H. Auden (poems by Auden, Day-Lewis, Driberg, Rex Warner); *1927* by Auden and Day-Lewis (poems by Auden, Day-Lewis, Driberg, Louis MacNeice, Geoffrey Tillotson, and Warner); *1928* by Clere Parsons and 'B. B.' (poems by Auden, Arthur Calder-Marshall, MacNeice, Spender, Tillotson); *1929* by MacNeice and Spender (poems by Calder-Marshall, R. H. S. Crossman, Douglas Jay, MacNeice, Graham Shepard, Bernard Spender, Spender); *1930* by Spender and Spender (poems by Crossman, Richard Goodman, Calder-Marshall, Goronwy Rees, MacNeice, Spender, Spender); *1931* by Spender and Goodman; *1932* by Goodman.

87. 'New Books', *N&D*, no. 20 (11 Nov. 1937).

88. 'The Joker in the Pack', in *The Dyer's Hand* (1963).

89. 'For Schoolboys Only', *N&D*, 8 July 1937 (reprinted in *The Essays, Articles and Reviews*, ed., D. Gallagher).

90. Berg Collection MSS, New York Public Library.

91. Upward had suppressed the story, like so much else, during his faithful Communist Party years, which came to an end in 1948.

92. 'The Background of Twentieth Century Letters', *Scrutiny*, VIII, June 1939: a review, in part, of Connolly's *Enemies of Promise*.

93. (5 Oct. 1936); *Letters to Christopher*, ed., Bartlett.

94. 'Present Discontents', *The Tablet*, 3 December 1938 (reprinted in *The Essays, Articles and Reviews*, ed., Gallagher).

95. Berg Collection MS, New York Public Library.

96. The letter is published in the 'Wise Stevie' chapter of Naomi Mitchison, *You May Well Ask: A Memoir 1920–1940* (1979).

6. High Failure

97. 'The Obelisk', like other fetishistic Forster writings kept fondly in the drawer, was not published until after his death: in *The Life to Come, and Other Stories* (1972).

98. 'The Uncreating Chaos', in *The Still Centre* (1939); it appeared first in *NV*, no. 17 (Oct.–Nov. 1935).

99. 'Adventures in the Air', *L*, 2 December 1936 (review of John Grierson's *High Failure*).

100. The first photographic number of *NV* was n.s., no. 1 (Autumn 1938).

101. Spender's Journal, 26 February 1975, in *London Magazine*, February–March 1976; now available in Spender's *Journals 1939–1983*, ed., John Goldsmith (1985).

102. The review appeared in *Now and Then*, 1934; it's reprinted in *The English Auden*.

103. Berg Collection MS, New York Public Library.

104. *TCV*, no. 8 (Jan.–Feb. 1938).

105. *L*, 2 December 1936.

106. 'The Last Parade', *NW*, no. 4 (Autumn 1937); reprinted in *Two Cheers for Democracy* (1951).

107. *Red Stage*, no. 4, March 1932.

108. See the report, by John Cournos, 'Russian Periodicals', *C*, XV, no. 58 (Oct. 1935).

109. The poem appeared in *A Time to Dance* (1935).

110. The poem appeared in E. Allen Osborne, ed., *In Letters of Red* (1938).

111. 'Cinema', *S*, 25 November 1938.

112. 'Dictator', *TCV*, no. 10 (May 1938).

113. David Hinton tries—and fails—to exculpate Arnold Fanck's 'mountain-film' genre from the charge of implicit Fascism in *The Films of Leni Riefenstahl* (1978).

114. The story was published for the first time in Marie-Jacqueline Lancaster, *Brian Howard: Portrait of a Failure* (1968).

115. Hughes's arresting conceit is to be found in his review of the *Letters* of T. E. Lawrence, ed., David Garnett, *NS*, 10 December 1938.

116. 'Twenty-Seven Sonnets', *NV*, n.s. I, 2 (May 1939); 'Auden as a Monster', *NV*, nos. 26–27 (Nov. 1937).

117. It is also reproduced in the Bartlett edn. of Spender's *Letters to Christopher*.

118. 'A Queer Country', *NV*, no. 21 (June–July 1936), and *Several Observations* (1939).

119. *L*, 11 May 1939.

120. Ceadel's essay is in *Class, Culture and Social Change: A New View of the 1930s*, ed., Frank Gloversmith (Brighton, 1980).

121. 'The Japanese Invasion of China', in *L*, 2 February 1938; 'Bomber', in Bronowski's *Overtures to Death* (1938); 'The Bombed Happiness', in *The Still Centre* (1939); 'Bombing Casualties', in *Poems for Spain*, ed., Spender and Lehmann (1939); 'Elegy on Spain', in Barker's *Lament and Triumph* (1940).

122. In *Old Lights for New Chancels* (1940).
123. 'Slough' appeared in *Continual Dew* (1937).
124. 'The Cinema', *S*, 8 May 1936.

7. Going Over

125. 'The Literature of Communism: Its Origin and Theory', *C*, VIII, no. 32 (Apr. 1929).
126. (1933–4); republished in *John Cornford: A Memoir*, ed., Pat Sloan (1938).
127. In *Julian Bell: Essays, Poems and Letters*, ed., Quentin Bell (1938).
128. There are some sharp comments on the problems of *going over* in Roy Fuller, 'The Audience and Politics', *TCV*, no. 18 (June–July 1939), (*The Poet and the Public*, a *TCV* Special Number).
129. In *Writing in Revolt*, *Fact*, no. 4 (July 1937).
130. *Romance and Realism, A Study in English Bourgeois Literature*, ed., Samuel Hynes (Princeton, 1970).
131. *Journey to the Border* was published by Hogarth in 1938; extracts appeared earlier in *NW*, including a part entitled 'The Border Line' in *NW*'s very first number, Spring 1936.
132. 'The Falling Tower', *FNW*, III (Spring 1941).
133. 'The Left Wing Orthodoxy', *NV*, nos. 31–32 (Autumn 1938).
134. There's more pronominal reflection in Alick West's piece, 'The "Poetry" in Poetry', *LR*, III, 3 (Apr. 1937).
135. In *NV*, no. 16 (Aug.–Sept. 1935).
136. Randall Swingler, 'Controversy', *LR*, I, 3 (Dec. 1934); John Lehmann, 'Via Europe', *NW*, no. 1 (Spring 1936); T. A. Jackson, 'Communism, Religion and Morals', in *The Mind in Chains*, ed., C. Day Lewis (1937); Maurice Hindus, *The Great Offensive* (1933).
137. 'André Malraux, The Path to Humanism', *NW*, n.s., no. 1 (Autumn 1938).
138. 'Address to Death' first appeared in *NS*, 7 May 1938 (Literary Supplement).
139. *NS*, 15 October 1932: reprinted in *The English Auden*.
140. As a matter of fact, as we learn from Orwell's *Wigan Pier* diary (in the *Collected Essays, Journalism and Letters* of Orwell, vol. I), he was not in a train when he observed this compelling scene. But his soured awareness of the perpetual train/bus/aquarium window between himself and the impoverished working-class makes this re-writing feel inevitable and to read as extremely truthful.
141. Todd, 'Do You Believe in Geography?', *NV*, n.s. I, 1 (Jan. 1939); Barker, 'Poem on Geography', *TCV*, no. 11 (July 1938); Madge, 'Landscapes', *The Disappearing Castle* (1937); Spender, 'The Landscape Near An Aerodrome', *Poems* (1933); 'The Landscape', *NS*, 4 December 1937 (Christmas Books Supplement).
142. Mallalieu, 'The Philologist', *TCV*, no. 17 (Apr.–May 1939); 'Follow the Map', *Poets of Tomorrow*, 1939; Todd, 'It was Easier', *TCV*, nos. 15–16 (Feb. 1939).
143. Spender's review is 'Poetry and Politics', *S*, 28 July 1939; Pritchett's is 'New Novels', *NS*, 12 March 1938.
144. 'Journey' appeared in *NV*, no. 8 (Apr. 1934).
145. For an illuminating account see John Lowerson, 'Battles for the Countryside', in *Class, Culture and Social Change*, ed., F. Gloversmith.
146. See Tolkien's lecture 'On Fairy Stories', in *Essays Presented to Charles Williams* (1947).
147. 'On Those Islands' was subsequently re-titled 'The Hebrides'. It actually appeared three times in 1938: in *C*, XVII, no. 67 (Jan. 1938); as the last chapter of *I Crossed the Minch*; and in MacNeice's volume of poetry, *The Earth Compels*.
148. Cyril Connolly, *Journal and Memoir*, ed., David Pryce-Jones (1983), p. 233.

8. Notes from the Underground

149. 'Apologia', *FNW*, IV (Autumn 1941).
150. 'An Artist of the Thirties', *FNW*, III (Spring 1941).
151. Connolly, 'A London Diary', *NS*, 16 January 1937; Spender Letter, 20 February 1937, in *Letters to Christopher*, ed., Bartlett.
152. 'Documents', *Fact*, no. 4, July 1937.
153. *DW*, 17 March 1937.
154. The discussion was printed in *L*, 19 December 1940.
155. 'The Fog Beneath the Skin', *LR*, I, 10 (July 1935).
156. 'Prometheus Unbound', *LR*, III, 6 (July 1937).
157. Berg Collection MS, New York Public Library.
158. (To Mrs Flora Strousse, 19 Feb. 1932.) *Letters*, ed., Grover Smith (1969).
159. See *The Road to Spain*, ed., D. Corkill and S. Rawnsley (Dunfermline, 1981).
160. The volume was edited by Margaret Llewellyn Davies and published by the Hogarth Press. The Introduction also appears as 'Memories of a Working Women's Guild', in Virginia Woolf's *Collected Essays* (1967), vol. 4.
161. In *Britain in the 30s* (1975), a limited edition of photography taken by Humphrey Spender for Mass-Observation (with introduction and commentary by Tom Harrisson).
162. Rayner Heppenstall in *Middleton Murry: A Study in Excellent Normality* (1934).
163. *L*, 8 February 1973.
164. For Spender on the *mono azul*: see 'Spain's

Air Chief in a Mechanic's Blue Dungarees', *DW*, 19 April 1937 (reprinted in *The Penguin Book of Spanish Civil War Verse*, 1980, ed., Valentine Cunningham).

165. *NV*, no. 20 (Apr.–May 1936).
166. See Thomas Sharp, *Town and Countryside: Some Aspects of Urban and Rural Development* (1933).
167. 'Stonebreaking', another one of Forster's closet fictions, was not published until 1980, in *Arctic Summer and Other Fiction*.
168. The September 1939 Journal, in Bartlett's *Letters to Christopher* (1980), is now also available in Spender's *Journals 1939–1983*, ed., John Goldsmith (1985).

9. Movements of Masses

169. 'High seriousness and Aston Villa', *N&D*, 18 November 1937.
170. *DW*. 19 September 1936.
171. *NV*, no. 5 (Oct. 1933).
172. *NS*, 22 April 1939.
173. 'The Poet Speaks', *L*, 9 June 1938: in 'The Poet and the Public' series of wireless interviews conducted by Humphrey Jennings.
174. 'The Turning Point', *LR*, II, 1 (Oct. 1935).
175. 'History and the Poet', *NW*, n.s., no. 3 (Christmas 1939).
176. See his four talks in 'The Modern Dilemma' series, *L*, 16 March–13 April 1932.
177. See, for example, Arthur Marwick, *Class: Image and Reality in Britain, France and the USA* (1980), and Anthony Aldgate, *Cinema and History: British Newsreels and the Spanish Civil War* (1979).
178. *Daily Mail*, 5 July 1930.
179. Berg Library MS, New York Public Library.
180. 'A Commentary', *C*, XI, no. 45 (July 1932).
181. Berg Library MS, New York Public Library.

10. Mass Observations

182. 'Poetry as a Best Seller', *L*, 30 June 1938 ('The Poet and the Public' series).
183. See Anthony Blunt's attacks on Picasso, and his defence of Balzac's kind of realism, in 'Picasso Unfrocked', *Spectator*, 8 October 1937, and Blunt's letters defending himself against Herbert Read and others, *Spectator*, 22 October and 5 November 1937. The whole debate is reproduced in *Spanish Front: Writers on the Civil War*, ed., Valentine Cunningham, (Oxford, 1986), pp. 213–20.
184. 'Human Nature Changes', *LR*, III, 13 (Feb. 1938).
185. 'Epic and the Future of the Soviet Arts', *LR*, III, 10 (Nov. 1937).
186. 'Fable and Reportage', *LR*, II, 14 (Nov. 1936); 'Poetry and Expressionism', *NS*, 12 March 1938; 'The Poetic Dramas of W. H.

Auden and Christopher Isherwood', *NW*, n.s., no. 1 (Autumn 1938).
187. 'Left Theatres', *NS*, 6 April 1935.
188. 'Fable and Reportage', *LR*, II, 14 (Nov. 1936).
189. *Red Stage*, February 1932.
190. There's a handy dossier of memories, theories and some WTM theatrical texts in the History Workshop volume, *Theatres of the Left 1880–1935: Workers' Theatre Movements in Britain and America*, ed., Raphael Samuel, Ewan MacColl, and Stuart Cosgrave (1985).
191. See *Fear and Misery of the Third Reich* and *Señora Carrar's Rifles*, transl. John Willett and Wolfgang Sauerlander, ed., John Willett and Ralph Manheim (Brecht, *Collected Plays*, IV, Part 3: *Plays, Poetry and Prose* of Brecht, ed., John Willett and Ralph Manheim, 1983).
192. *LR*, III, 9 (Oct. 1937).
193. John Lehmann, 'Epic and the Future of the Soviet Arts', *LR*, III, 10 (Nov. 1937).
194. 'The Cinema', *S*, 20 December 1935.
195. 'Novels', *Fact*, no. 20 (Nov. 1938).
196. 'African Photographs', *NS*, 7 May 1938.
197. In the American leftist periodical *New Masses*, 26 April 1938.
198. Most recently in the rather captiously nit-picking volume of essays edited by Christopher Norris, *Inside the Myth, Orwell: Views from the Left* (1984). For more serious attacks on manipulations and gaps in *Homage to Catalonia*, see Section IV of Claude Simon's novel *Les Géorgiques* (Paris, 1981: English translation by John Fletcher, 1987), and Anthony Cheal Pugh, 'Interview with Claude Simon: Autobiography, The Novel, Politics', *The Review of Contemporary Fiction*, vol. 5, no. 1 (1984), 4–13.
199. 'Cinema', *S*, 26 May 1939.
200. *NV*, no. 24 (Feb.–Mar. 1937).
201. The reference was to the 'Poet and the Public' talks: see n. 173, above.
202. *Life and Letters Today*, XVII, no. 9, Autumn 1937.
203. *TCV*, no. 11 (July 1938).
204. *L*, 25 August 1938.

11. Seedy Margins

205. *New Masses*, 26 April 1938.
206. 'The Turning Point', *LR*, II, 1 (Oct. 1935).
207. *N&D*, 16 December 1937.
208. 'Ralph Fox', *NS*, 1 May 1937 (Literary Supplement).
209. See his continually suggestive *Abroad: British Literary Travelling Between the Wars* (1980).
210. Orwell's article 'Marrakech' appeared in *NW*, n.s., no. 3 (Christmas 1939).
211. 'The First Journey' appeared first in *N&D*, 16 September 1937, and was later incorporated into *Lions and Shadows*.

212. Berg Library MS, New York Public Library.
213. 'Next Year's News: A Preview of 1938'. *N&D*, no. 18 (28 Oct. 1937).
214. *Harper's Bazaar*, October 1938 (reprinted in Isherwood's Exhumations, 1966).
215. 'And I a twister love what I abhorr' is a marvellously memorable line that greatly animates John Wain's novel *Hurry On Down* (1953).
216. 'Train to Dublin's appeared first in *NV*, no. 13 (Feb. 1935). (It appears to haunt a later, more famous train poem, Philip Larkin's 'The Whitsun Weddings': both poems have gauche young females 'preferring mauve'.)
217. *C*, XV, no. 58 (Oct. 1935).
218. *C*, XI, no. 45 (July 1932). The poem was dedicated 'To My Friends Naomi Mitchison and Henry Carr'.
219. *C*, XVII, no. 66 (Oct. 1937).
220. *Nineteen Stories* (1947).
221. 'The Express' is no. XXXII in *Poems* (1933).
222. 'The Port' is no. XV in *Poems* (1933).
223. 'Meeting by the Gjulika Meadow' first appeared in *NV*, no. 25 (May 1937).
224. 'A Commentary', *C*, XVI, no. 62 (Oct. 1936).
225. 'Auden and Isherwood', *NS*, 18 March 1939.
226. 'New Novels', *NS*, 12 March 1938.
227. *Nineteen Stories* (1947).
228. 'About Frontiers', *TCV*, no. 15–16 (Feb. 1939).
229. 'Three Sketches', *LR*, I, 4 (Feb. 1939).
230. The story is in Forster's *The Life to Come* gathering (1972).
231. 'Port Bou' first appeared in *NW*, n.s. 1, (Autumn 1938), as 'Port Bou—Firing Practice'.

12. Somewhere the Good Place?

232. 'The Island' appeared in *LR*, I, 4 (Jan. 1935), and 'An Evening at the Bay' in *New Country*.
233. 'Christmas, 1938' is in the very last number of *NV*, n.s. I, no. 2 (May 1939).
234. 'Summer Idyll', in *The Faber Book of Modern Verse*; 'Poem on People', *LR*, II, 16 (Jan. 1937); 'And Forgetful of Europe (Mlini 1935 to 1937)', *NV*, no. 30 (Summer 1938); 'Picked Clean from the World', *NV*, no. 28 (Jan. 1938); 'Holiday' and 'Summer Night', in *Poets of Tomorrow* (1939).
235. 'Hymn to the Sun' is in Michael Roberts's *Poems* (1936).
236. 'Welcome in August', *LR*, II, 13 (Oct. 1936); 'August à la Poussin', *NV*, no. 5 (Oct. 1933).
237. John Betjeman was full of gloomy affection for those melancholic Christians who knew and relished the Old Testament titles of Jehovah: 'To Jehovah Tsidkenu the praise!'; 'Jehovah Jireh! the arches ring'; 'Jehovah Nisi! from Tufnell Park'. See his poem 'Suicide on Junction Road Station after Abstention from Evening Communion in North London', *Continual Dew* (1937).
238. 'Bloomsbury's Farthest North', *N&D*, 12 August 1937.
239. 'Air Gun', *NV*, no. 18 (Dec. 1935).
240. *Storm*, February 1933.
241. 'The Cinema', *S*, 2 August 1935.
242. *NV*, no. 2 (Mar. 1933).
243. 'When they have lost', in *Best Poems of 1938*.
244. See Donald Mitchell, *Britten and Auden in the Thirties: the Year 1936* (1981).
245. 'Speaking Concretely: A Reply to C. Day Lewis', *LR*, I, 2 (Nov. 1934).
246. See Introductory note, 'The Author', in *The Olive Field* (reissued with New Introduction by Valentine Cunningham, Hogarth, 1986).
247. *New English Weekly*, 9 June 1938.
248. *Land Benighted* was reissued as *Too Late to Turn Back* (1981).
249. See, e.g., Arthur Calder-Marshall, 'More Frank than Buchman' in *The Old School*, ed., Graham Greene (1934); 'These Buchmanites', in R. C. Carr, *Red Flags: Essays of Hate from Oxford* (1933); C. Day Lewis, *Starting Point*; and Auden's contribution to R. H. S. Crossman, *Oxford and the Group*.
250. 'Doom is dark' appeared as no. II in Auden's *Poems* (1933).
251. *NV*, no. 20 (Apr.–May 1936); reprinted in *The Disappearing Castle* (1937).
252. 'Los Angeles', *Horizon*, October 1947 (reprinted in *Exhumations*).
253. 'California Story', *Harper's*, 1952 (reprinted as 'The Shore' in *Exhumations*).

13. Spanish Front

254. Most of the poems and a good deal of the prose referred to in this chapter can be found reprinted in my two anthologies, *The Penguin Book of Spanish Civil War Verse* (1980) and *Spanish Front: Writers on the Civil War* (Oxford, 1986).
255. Letter to Jack Common, October (?) 1937; Letter to Stephen Spender, 2 April 1938; 'Why I Write', 1946: all in *Collected Essays* etc., vol. 1.
256. Review of Spanish War books, *Time and Tide*, 9 October 1937.
257. 'Impressions of Valencia', *NS*, 30 January 1937.
258. 'Present Discontents', *The Tablet*, 3 December 1938.
259. 'The Will to Live', *NS*, 12 November 1938.
260. 'Christmas Present 1937', in Todd's *Poems* (1938); 'Notes for a Letter', in *Poets of Tomorrow* (1939); the rest in *The Penguin Book of Spanish Civil War Verse*.

261. *DW*, 21 May 1938 (Spain Weekend Supplement).

262. 'The Talking Bronco' (review of *Flowering Rifle*), *NS*, 11 March 1939 (Spring Announcements Supplement).

263. 'Today in Barcelona', *Spectator* 20 January 1939.

264. 'As I Please', *Tribune*, 15 September 1944.

265. *What I Believe*, by Fourteen Modern Thinkers (1937).

266. Review of *Spanish Testament*, *Time and Tide*, 5 February 1938.

267. Anthony Powell, 'A Reporter in Los Angeles—Hemingway's Spanish Film', *N&D*, 19 August 1937, and Grahame Greene, 'Alfred Tennyson Intervenes', *S*, 10 December 1937 (also in Greene's *Collected Essays* (1969)), are both reprinted in Cunningham, ed., *Spanish Front* (1986).

268. See, e.g., Hilaire Belloc's eulogy of Franco in *Places* (1942)—part of which is reprinted in *Spanish Front*.

269. 'History and the Poet', *NW*, n.s. 3 (Christmas 1939).

270. See 'Spender for Spain', *DW*, 19 February 1937.

271. Letter, ' "Flowering Rifles" ', *NS*, 8 April 1939.

272. It appeared in the first number of Oswald Mosley's *British Union Quarterly*, January–April 1937, and is reprinted in my *Spanish Front*.

273. 'Guernica', *NS*, 15 October 1938.

274. 'Poetry', *Fact*, no. 4, July 1937.

275. *The Diaries of Evelyn Waugh*, ed., Michael Davie (1976): 19 September 1936.

276. *Diaries*, 25 September 1936.

277. 'Departed Hero', *Poets of Tomorrow* (1939).

278. Review of *Volunteer in Spain*, 'Spanish Nightmare', *Time and Tide*, 31 July 1937.

279. *DW*, 14 April 1937.

280. *DW*, 14 October 1938.

281. In the International Brigade's paper, *Volunteer for Liberty*, May 1940.

282. *News Chronicle*, 28 June 1937. The article was issued as a fund-raising pamphlet, to send a Sprigg Memorial Ambulance to Spain.

283. 'The Alcazar', *Mithraic Emblems* (1936); republished as one of 'Three Poems from Toledo' in the first number of Oswald Mosley's *British Union Quarterly* of 1937.

284. Review of Spanish War books, *Time and Tide*, 9 October 1937.

285. 'Pictures in Spain', *S*, 30 July 1937.

286. This information was given to me in a letter, by 'Russia' Roberts's grandson Richard Felstead, author of *No Other Way: Jack Russia and the Spanish Civil War* (Port Talbot, 1981).

287. ed., Carmel Haden Guest (1939).

288. This point has been much pressed upon me by International Brigade correspondents annoyed at my earlier version of the alterations: I put them down to C. P. head-quarters interference, in the Introduction to my *Penguin Book of Spanish Civil War Verse*.

289. See his *New Writing in Europe* (1940).

290. This is the *NW* version: *NW*, n.s., 1 (Autumn 1938). The many versions of this poem up until 1955 are set out in my Penguin Spanish anthology. The latest version in *Collected Poems 1928-1985* (1985) effects some more.

291. 'It was easier': *TCV*, no. 15–16 (Feb. 1939); *Poets of Tomorrow* (1939).

292. 'The Poet in England Today', *New Republic*, 25 March 1940.

293. *Seven*, Summer 1938.

294. Not least in 'Spain Invites the World's Writers: Notes on the International Congress, Summer 1937', *NW*, Autumn 1937.

295. The letter from the Berg MSS, New York Public Library, is printed in full (for the first time) in my *Spanish Front*, pp. 307–9.

14. Too Innocent A Voyage

296. 'Poetry and Politics', *Spectator*, 28 July 1939.

297. See Christopher Dodd, 'Lubetkin and the Battle of Peterlee', the *Guardian*, 30 May 1981.

298. See his *Britten and Auden in the Thirties: the Year 1936*.

299. *Poetry and the People*, May 1940.

WORKING BIBLIOGRAPHY

Unless otherwise stated, assume University Press titles to be published where the university's name implies its location, Penguin books at Harmondsworth, Middlesex, and everything else in London.

1. Some Useful Author Bibliographies

Armitage, C. M., and Clark, Neil, *A Bibliography of the Work of Louis MacNeice* (Kaye and Ward, 1973; 2nd edn.; 1974).

Bloomfield, B. C., and Mendelson, Edward, *W. H. Auden: A Bibliography, 1924–69* (2nd edn.; University of Virginia Press, Charlottesville, 1972).

Funk, Robert W., *Christopher Isherwood: A Reference Guide* (G. K. Hall & Co., Boston, Massachusetts, 1979).

Gallup, Donald, *T. S. Eliot: A Bibliography* (Faber, 1969).

Gingerich, Martin E., *W. H. Auden: A Reference Guide* (G. K. Hall & Co., Boston, Massachusetts, 1977).

Handley-Taylor, Geoffrey, and d'Arch Smith, Timothy, *C. Day Lewis, the Poet Laureate: A Bibliography* (St James Press, Chicago, 1968). Introd. letter by W. H. Auden.

Higginson, F. H., *A Bibliography of the Works of Robert Graves* (Archon, Hamden, Connecticut, 1966).

Kirkpatrick, B. J., *A Bibliography of Virginia Woolf* (3rd edn.; Clarendon Press, Oxford, 1980).

— *A Bibliography of E. M. Forster* (2nd edn.; Clarendon Press, 1985). Foreword by E. M. Forster.

Lowbridge, Peter, 'An Empson Bibliography' *The Review*, nos. 6 and 7 (June 1963), 63–73 (William Empson Special Number).

Meyers, Jeffrey, and Meyers, Valerie, *George Orwell: An Annotated Bibliography of Criticism* (Garland Publishing, New York, 1977).

Munton, Alan, and Young, Alan, *Seven Writers of the English Left: A Bibliography of Literature and Politics, 1916–1980* (Garland Publishing, New York, 1981).

Pound, Omar S., and Grover, Philip, *Wyndham Lewis: A Descriptive Bibliography* (Archon Books, Dawson & Sons, Folkestone, 1978).

Rice, Thomas Jackson, *James Joyce: A Guide to Research* (Garland Publishing, New York, 1982).

— *Virginia Woolf: A Guide to Research* (Garland Publishing, New York, 1984).

Ricks, Beatrice, *T. S. Eliot: A Bibliography of Secondary Works* (Scarecrow Press, Metuchen, New Jersey, 1980).

Tolley, A. T., *The Early Published Poems of Stephen Spender: A Chronology* (Carleton University, Ottawa, 1967).

2. Periodicals of the Period used extensively

Action: The New Weekly of the New Movement (8 Oct. 1931 to 31 Dec. 1931).

Action: For King and People (21 Feb. 1936 to 6 June 1940).

The Adelphi (Oct. 1930 to   ).

Air Stories: Flying Thrills and Aerial Adventure (1935–40).

The British Union Quarterly (1937–40), continuation of *The Fascist Quarterly* (Jan. 1935 to Oct. 1936).

The Cambridge Review: A Journal of University Life and Thought (1879–   ).

Contemporary Poetry and Prose (May 1936 to Spring 1937).

The Criterion (Oct. 1922 to Jan. 1939).

Daily Worker (1930–66), continued as *The Morning Star*.

The Enemy: A Review of Art and Literature (Jan. 1927 to 1929).

Experiment (Nov. 1928 to Spring 1931).

Fact (Apr. 1937 to June 1939).

Folios of New Writing (Spring 1940 to Autumn 1941).

Horizon: A Review of Literature and Art (Jan. 1940 to Jan. 1950).

International Literature (1931–45), formerly *Literature of the World Revolution* (June to Oct. 1931).

Ireland Today (June 1936 to Mar. 1938).

Kingdom Come: The Magazine of War-Time Oxford (Nov. 1939 to Autumn 1943).

The Left Review (1934 to May 1938).

Light and Dark (Jan. 1937 to Feb. 1938).

The Listener (16 Jan. 1929 to   ).

The Mint: A Miscellany of Literature, Art and Criticism (1946–8).

The New English Weekly (21 Apr. 1932 to 8 Sept. 1949).

New Masses (May 1926 to   ).

The New Republic (1914–   ).

The New Statesman (1913–31); *The New Statesman and Nation* (28 Feb. 1931 to   ).

New Theatre (Aug.–Sept. 1939).

New Verse (1933–9).

New Writing (Spring 1936 to Christmas 1939).

Night and Day (1 July to 23 Dec. 1937).

Nude Life (Anglo-American Physical Training Company, Coalville, Leicester: Mar. 1932 to Mar. 1936).

Our Time (Feb. 1941 to 1949), continuation of *Poetry and the People*.

The Oxford Outlook: A Literary and Political Review (May 1919 to May 1932).

Oxford Poetry (1913–32).

The Penguin New Writing (1940–50).

Poetry and the People (July 1938 to Sept. 1940).

The Red Stage: Organ of the Workers' Theatre Movement (Nov. 1931 to 1932).

The Right Review (Oct. 1936 to Mar. 1940).

Seven (1938–47).

Spain at War (Apr.–Dec. 1938); continued as *Voice of Spain* (Jan.–Mar. 1939).

The Spectator (1828 to   ).

Storm: Stories of the Struggle: A Magazine of Socialist Fiction (1933).

The Sunbathing Review (1933–59).

Time and Space: the Workers' Chess League's Monthly (Aug. 1938 to   ).

Twentieth Century Verse (1937–9).

The Voice of Scotland: A Quarterly Magazine of Scottish Arts and Affairs (June 1938 to Aug. 1939).

The Volunteer for Liberty (Jan. 1940 to June 1946).

The War in Spain (1938–9).

3. Miscellanies and Anthologies

(For Spanish Civil War anthologies, see section 22, 'Spain'; and for Surrealism collections, see section 16, 'Surrealism'.)

Auden, W. H., and Garrett, John (edd.), *The Poet's Tongue: An Anthology* (Bell, 1935).

Bell, Quentin (ed.), *Julian Bell: Essays, Poems and Letters* (Hogarth, 1938).

Bowen, Elizabeth (introd.), *The Faber Book of Modern Stories* (Faber, 1937).

Carpenter, Maurice; Lindsay, Jack; and Arundel, Honour (edd.), *New Lyrical Ballads* (Editions Poetry London, 1945).

Day Lewis, C., and Fenby, Charles, *Anatomy of Oxford: An Anthology* (Cape, 1938).

Evans, Patrick; Durrell, Lawrence; Todd, Ruthven; Foxall, Edgar; Blakeston, Oswell; and Heppenstall, Rayner, *Proems* (Fortune Press, 1938).

Galassi, Jonathan (ed.), *John Cornford: Understand the Weapon, Understand the Wound: Selected Writings* (Carcanet, Manchester, 1976).

Grigson, Geoffrey (ed.), *New Verse: An Anthology* (Faber, 1939).

Grisewood, Harman (ed.), *David Jones: Epoch and Artist: Selected Writings* (Faber, 1959).

Grubb, Frederick (ed.), *Michael Roberts: Selected Poems and Prose* (Carcanet, Manchester, 1980).

Hawtree, Christopher (ed. and introd.), *Night and Day* (Chatto, 1985). Preface by Graham Greene.

Hewett, Peter; Mallalieu, H. B.; and Todd, Ruthven, *Poets of Tomorrow: First Selection* (Hogarth, 1939).

Leavis, F. R. (ed.), *A Selection from 'Scrutiny'* (2 vols.; Cambridge University Press, 1968).

Lee, Hermione (ed.), *Stevie Smith: A Selection* (Faber, 1983).

Lehmann, John (ed.), *Poems from New Writing* (John Lehmann, 1946).

— *English Stories from New Writing* (John Lehmann, 1951).

Lehmann, John, and Fuller, Roy, (edd.), *The Penguin New Writing: An Anthology* (1985).

Matthias, John (ed.), *Introducing David Jones: A Selection of His Writings* (Faber, 1980). Pref. by Stephen Spender.

Mendelson, Edward (ed.), *The English Auden: Poems, Essays, And Dramatic Writings, 1927–1939* (Faber, 1977).

Moore, Charles, and Hawtree, Christopher (edd. and introd.), *1936, as Recorded by the Spectator* (Michael Joseph, 1986).

Mulgan, John (ed.), *Poems of Freedom* (Gollancz, 1938). Introd. by W. H. Auden.

Munton, Alan (ed.), *Wyndham Lewis: Collected Poems and Plays* (Carcanet, Manchester, 1979). Introd. by C. H. Sisson.

North, Joseph (ed.), *New Masses: An Anthology of the Rebel Thirties* (International Publishers, New York, 1969; Seven Seas paperback, Seven Seas Publishers, Berlin, German Democratic Republic, 1972). Introd. by Maxwell Geismar.

O'Brien, Edward J. (ed.), *The Best Short Stories: 1934[–1939]* (Cape, 1934 (–9)).

Osborne, E. Allen (ed.), *In Letters of Red* (Michael Joseph, 1938).

Roberts, Denys Kilham, and Grigson, Geoffrey (edd.), *The Year's Poetry 1937* (John Lane The Bodley Head, 1938).

Roberts, Michael (ed.), *New Signatures: Poems by Several Hands* (Hogarth Living Poets, no. 24; Hogarth, 1932).

— *New Country: Prose and Poetry by the Authors of 'New Signatures'* (Hogarth, 1933).

— *The Faber Book of Modern Verse* (Faber, 1936). Reissued with introd. 'The Making of *The Faber Book of Modern Verse*' (from *Times Literary Supplement*, 18 June 1976), by Janet Adam Smith (Faber, 1982).

Skelton, Robin (ed.), *Poetry of the Thirties* (Penguin, 1964).

Smith, Janet Adam (ed.), *Poems of Tomorrow* (Chatto, 1935).

Sprigg, C. St. John (ed. and introd.), *Uncanny Stories* Nelson, 1936).

Then and Now: A Selection of Articles, Stories and Poems, taken from the first fifty numbers of 'Now and Then', 1921–35 (Cape, 1935).

Williamson, Henry, *As the Sun Shines* (Faber, 1941).

4. Poetry

Alberti, Rafael, *Selected Poems* (ed. and trans. Ben Belitt; University of California Press, 1966).

Auden, W. H., *Poems* (Faber, 1930; 2nd edn., 1933).

— *The Orators* (Faber, 1933).

— *Spain* (Faber, 1937).

— *Another Time* (Faber, 1940).

— *Collected Shorter Poems 1927–1957* (Faber, 1966).

— *Collected Longer Poems* (Faber, 1968).

— *Collected Poems* (ed. Edward Mendelson; Faber, 1976).

Barker, George, *Poems* (Faber, 1935).

— *Calamiterror* (Faber, 1937).

— *Elegy on Spain* (Contemporary Bookshop, Manchester, 1939).

— *Lament and Triumph* (Faber, 1940).

— *In Memory of David Archer* (Faber, 1973).

Betjeman, John, *Mount Zion; or, In Touch With the Infinite* (James Press, 1931).

— *Continual Dew: A Little Book of Bourgeois Verse* (Murray, 1937).

— *Old Lights for New Chancels: Verses Topographical and Amatory* (Murray, 1940).

— *Summoned by Bells* (Murray, 1960).

— *Collected Poems: Enlarged Edition* (ed. The Earl of Birkenhead; 3rd edn. of *Collected Poems*, Murray, 1970).

Bronowski, Jacob, *Spain 1939: Four Poems* (Andrew Marvell Press, Hull, 1939).

Campbell, Roy, *Mithraic Emblems: Poems* (Boriswood, 1936).

— *Flowering Rifle: A Poem from the Battlefield of Spain* (Longmans, 1939).

— *Talking Bronco* (Faber, 1946).

— *Collected Poems* (The Bodley Head, 1949).

'Caudwell, Christopher' (Christopher St John Sprigg), *Poems* (The Bodley Head, 1939).

— *Collected Poems 1924–1936* (ed. and introd. Alan Young; Carcanet, Manchester, 1986).

Church, Richard, *Collected Poems* (Dent, 1948).

Coward, Noël, *The Lyrics* (Heinemann, 1965).

Davie, Donald, *Collected Poems 1950–1970* (Routledge, 1972).

Day Lewis, C., *Transitional Poem* (Hogarth Living Poets, no. 9; Hogarth, 1929).

— *Collected Poems 1929–1933* (*Transitional Poem, From Feathers to Iron, The Magnetic Mountain*) (Hogarth, 1935).

— *A Time to Dance, and Other Poems* (Hogarth, 1935).

— *Noah and the Waters* (Hogarth, 1936).

— *A Time to Dance: Noah and the Waters and Other Poems, with an essay Revolution in Writing* (Random House, New York, 1936).

— *Overtures to Death, and Other Poems* (Cape, 1938).

— *Collected Poems 1954* (Cape, 1954).

— (trans.), *The Georgics of Virgil* (Cape, 1940). Reissued in *The Eclogues, Georgics and Aeneid of Virgil* (Oxford University Press, 1966).

Durrell, Lawrence, *Collected Poems 1931–1974* (ed. James A. Brigham; Faber, 1980).

Eliot, T. S., *Collected Poems 1909–1962* (Faber, 1963).

Empson, William, *The Gathering Storm* (Faber, 1940).

— *Collected Poems* (Chatto, 1955).

Ewart, Gavin, *The Collected Ewart 1933–1980: Poems* (Hutchinson, 1980).

Foxall, Edgar, *Water-Rat Sonata and Other Poems* (Fortune Press, 1940).

Fuller, Roy, *Collected Poems 1936–1961* (Deutsch, 1962).

— *As From the Thirties* (Trogara Press, Edinburgh, 1983).

— *New and Collected Poems 1934–1984* (Secker & Warburg, 1985).

Gascoyne, David, *Roman Balcony and Other Poems* (Lincoln Williams, 1932).

— *Man's Life is This Meat* (Parton Press, 1936).

— *Hölderlin's Madness* (Dent, 1938).

— *Collected Poems* (ed. and introd. Robin Skelton; Oxford University Press, 1978).

Golding, W. G., *Poems* (Macmillan's Shilling Series of Contemporary Poets; Macmillan, 1934).

Graves, Robert, *Poems (1914–1926)* (Heinemann, 1927).

— *Collected Poems 1975* (Cassell, 1975).

Grigson, Geoffrey, *Several Observations: Thirty-Five Poems* (Cresset Press, 1939).

— *The Isles of Scilly, and Other Poems* (Routledge, 1946).

— *Places of the Mind* (Routledge, 1949).

Gurney, Ivor, *Collected Poems* (ed. P. J. Kavanagh; Oxford University Press, 1982).

Gutteridge, Bernard, *Old Damson-Face: Poems 1934 to 1974* (London Magazine Editions, 1975).

Hoffmann, Dr Heinrich, *Struwwelpeter* (Blackie, 1903; repr. Pan, 1972).

Holtby, Winifred, *The Frozen Earth, and Other Poems* (Collins, 1935).

Jones, David, *In Parenthesis*. With a Note of Introduction by T. S. Eliot (Faber, 1937).

Lawrence, D. H., *Nettles* (Criterion Miscellany, no. 11; Faber, 1930).

Lear, Edward, *Nonsense Omnibus* (Warne, 1943).

Lee, Laurie, *The Sun My Monument*. (New Hogarth Library, vol. 13; Hogarth, 1944; 2nd imp., Chatto/Hogarth, 1961).

Lehmann, John, *The Age of the Dragon, Poems, 1930–1951* (Longmans, 1951).

Lewis, C. S., *Poems* (ed. Walter Hooper; Harvest/Harcourt Brace Jovanovich, New York and London, 1964).

Lewis, Wyndham, *One-Way Song* (Faber, 1933).

Lowell, Robert, *Day By Day* (Faber, 1978).

MacDiarmid, Hugh, *The Batle Continues* (Castle Wynd Printers, Edinburgh, 1957).

— *Complete Poems 1920–1976* (ed. Michael Grieve and W. R. Aitken; 2 vols., Martin Brian & O'Keeffe, 1978).

MacNeice, Louis, *Poems* (Faber, 1935).

— *The Earth Compels: Poems* (Faber, 1938).

— *Autumn Journal: A Poem* (Faber, 1939).

— *Collected Poems 1925–1948* (Faber, 1949).

— *Selected Poems* (sel. and introd. W. H. Auden; Faber, 1964).

— *Collected Poems* (ed. E. R. Dodds; Faber, 1966).

Madge, Charles, *The Disappearing Castle* (Faber, 1937).

Mallalieu, H. B., *Letter in Wartime, and other Poems* (Fortune Press, 1941).

Milne, Ewart, *Letter from Ireland* (Gayfield Press, Dublin, 1940).

Montalk, 'Count' Potocki of, *Abdication of the Sun* (Right Review, London, 1938).

Nichols, Robert, *Such Was My Singing: Selection of Poems 1915–1940* (Collins, 1942).

Owen, Wilfred, *Collected Poems* (ed. C. Day Lewis, with Memoir (revised) from *Poems* (ed. Edmund Blunden; Chatto, 1931, corrected 1933) Chatto, 1963).

Plomer, William, *The Fivefold Screen* (Hogarth, 1932).

— *Visiting the Caves* (Cape, 1936).

— *Selected Poems* (Hogarth, 1940).

— *The Dorking Thigh, and Other Satires* (Cape, 1945).

— *Collected Poems* (Cape, 1973).

Porter, Cole, *Cole* (ed. Robert Kimball, with biographical essay by Brendan Gill; Michael Joseph, 1972).

Pudney, John, *Open the Sky: Poems* (Boriswood, 1934).

— *Dispersal Point, and Other Air Poems* (John Lane, 1942).

Rickword, Edgell, *Twittingpan and Some Others* (Wishart, 1931).

Roberts, Michael, *These Our Matins* (Elkin Mathews & Marrot, 1930).

— *Poems* (Cape, 1936).

Sassoon, Siegfried, *Collected Poems 1909–1956* (Faber, 1956).

— *Selected Poems* (Faber, 1968).

Skelton, John, *The Complete Poems of John Skelton* (ed. Philip Henderson; Dent, 1931).

Smith, Stevie, *Collected Poems* (Allen Lane, 1975).

Spencer, Bernard, *Aegean Islands and Other Poems* (Editions Poetry London, 1946).

— *Collected Poems* (ed. and introd. Roger Bowen; Oxford University Press, 1981).

Spender, Stephen, *Nine Experiments, by S. H. S., Being Poems Written at the Age of Eighteen* (Printed at no. 10, Frognal, Hampstead, NW3 [by Spender], 1928).

— *Poems* (Faber, 1933; 2nd edn., Faber, 1934).

— *Vienna* (Faber, 1934).

— *The Still Centre* (Faber, 1939).

— *Collected Poems 1928–1953* (Faber, 1955).

— *Collected Poems 1928–1985* (Faber, 1985).

Symons, Julian, *Confusions About X* (Fortune Press, 1939).

— *et al.*, *Poems for Roy Fuller on his Seventieth Birthday* (Sycamore Press, Oxford, 1982).

Tessimond, A. S. J., *Not Love Perhaps . . .* (sel. and introd. Hubert Nicholson; Autolycus Publications, 1978).

Thomas, Dylan, *Collected Poems 1934–1952* (Dent, 1952).

Warner, Rex, *Poems* (Boriswood, 1937).

Warner, Sylvia Townsend, and Ackland, Valentine, *Whether a Dove or Seagull* (Chatto, 1934).

Warner, Sylvia Townsend, *Collected Poems* (ed. and introd. Claire Harman; Carcanet, Manchester, 1982).

— *Selected Poems* (Carcanet, Manchester, 1985). Afterword by Claire Harman.

Williams, Charles, *Taliessin Through Logres* (Oxford University Press, London, 1938).

— *Arthurian Torso: Containing the Posthumous Fragment of the Figure of Arthur* (Oxford University Press, London, 1948). With 'A Commentary on the Arthurian Poems of Charles Williams' by C. S. Lewis.

Yeats, W. B. *Collected Poems* (2nd edn., Macmillan, 1950).

5. Novels and Stories

Ambler, Eric, *The Dark Frontier* (Hodder & Stoughton, 1936).

— *Uncommon Danger* (Hodder & Stoughton, 1937).

— *Cause for Alarm* (Hodder & Stoughton, 1938).

— *Epitaph for a Spy* (Hodder & Stoughton, 1938).

— *The Mask of Dimitrios* (Hodder & Stoughton, 1939).

— *Journey into Fear* (Hodder & Stoughton, 1940).

'Aston, James' (T. H. White), *They Winter Abroad: A Novel* (Chatto, 1932).

Barke, James, *Major Operation: A Novel* (Collins, 1936).

— *The Land of the Leal* (Collins, 1939).

Bates, Ralph, *Lean Men: An Episode in a Life* (Peter Davies, 1934).

— *The Olive Field* (Cape, 1936; revised, New York, 1966; reissued, with introd. by Valentine Cunningham, by Chatto/Hogarth, 1986).

Baum, Vicki, *Grand Hotel* (= *Menschen im Hotel*; trans. Basil Creighton; Bles, 1930).

Beachcroft, T. O., *Collected Stories* (John Lane The Bodley Head, 1946). Incorporates *A Young Man in a Hurry* (1934), *You Must Break Out Sometimes* (1936), *The Parents Left Alone* (1940).

Beckett, Samuel, *More Pricks Than Kicks* (Chatto, 1934).

— *Murphy* (Routledge, 1938).

— *Watt* (Olympia Press, Paris, 1953; Calder, 1963).

'Blake, Nicholas' (C. Day Lewis), *A Question of Proof* (Collins, 1935).

— *Thou Shell of Death* (Collins, 1936).

— *Malice in Wonderland* (White Lion Publishers, 1937).

— *The Beast Must Die* (Collins, 1938).

Bowen, Elizabeth, *The Hotel* (Cape, 1927).

— *To the North* (Cape, 1932).

— *The House in Paris* (Cape, 1935).

Bowen, Elizabeth, *The Death of the Heart* (Cape, 1938).
—— *Collected Stories* (introd. Angus Wilson; Cape, 1980).
Brierley, Walter, *Means Test Man* (Methuen, 1935; reissued, with introd. by Andy Croft, by Spokesman, Nottingham, 1983).
—— *Sandwichman* (Methuen, 1937).
Brown, Alec, *A Time to Kill: Two Stories* (Cape, 1930).
—— *Green Lane or Murder at Moat Farm* (Cape, 1930).
—— *A Winter Journey: A Simple Country Tale* (Cape, 1933).
—— *Daughters of Albion: A Novel* (Boriswood, 1935).
Buchanan, George, *Entanglement* (Constable, 1939).
Calder-Marshall, Arthur, *Dead Centre* (Cape, 1935).
'Caudwell, Christopher' (Christopher St John Sprigg), *This My Hand* (Hamish Hamilton, 1936).
Charlton, L. E. O., *The Bush Aerodrome* (Oxford University Press, London, 1937).
'Cheyney, Peter' (Reginald Evelyn Peter Southose Cheyney), *This Man is Dangerous* (Collins, 1936).
Christie, Agatha, *The Mystery of the Blue Train* (Collins, 1928).
—— *The Murder at the Vicarage* (Collins, 1930).
—— *Peril At End House* (Collins, 1932).
—— *Lord Edgware Dies* (Collins, 1933).
—— *Murder on the Orient Express* (Collins, 1934).
—— *Death in the Clouds* (Collins, 1935).
—— *Murder in Mesopotamia* (Collins, 1936).
—— *Hercule Poirot's Christmas* (Collins, 1938).
Common, Jack (ed.), *Seven Shifts* (Secker & Warburg, 1938).
Connolly, Cyril, *The Rock Pool* (Obelisk Press, Paris, 1936; Oxford Paperback, with introd. by Peter Quennell, 1981).
Crawshay-Williams, Eliot, *Night in the Hotel* (Gollancz, 1931).
Cronin, A. J., *The Citadel* (Gollancz, 1937).
Davies, Hugh Sykes, *Petron* (Dent, 1935).
Day Lewis, C., *Starting Point* (Cape, 1937).
—— See also 'Blake, Nicholas' in this section.
Dos Passos, John, *The 42nd Parallel* (Constable, 1930); *Nineteen Nineteen* (Constable, 1932); *The Big Money* (Constable, 1936); published as trilogy, *USA* (Constable, 1938).
Durrell, Lawrence, *The Black Book* (Obelisk Press, Paris, 1938); first published in USA with new pref. by author (E. P. Dutton, New York, 1960); first published in Britain (Faber, 1973).
Faulkner, William, *Pylon* (1935); Collected Edition (Chatto, 1967).
Forster, E. M. *The Life to Come and Other Stories* (Abinger Edition, vol. 8, ed. Oliver Stallybrass; Edward Arnold, 1972).
—— *Arctic Summer and Other Fiction* (Abinger Edition, vol. 9, ed. Elizabeth Heine and Oliver Stallybrass; Edward Arnold, 1980).

Fox, Ralph, *This Was Their Youth* (Secker & Warburg, 1938).
Fuller, Roy, *The Second Curtain* (Verschoyle, 1953).
Garnett, David, *A Man in the Zoo* (Chatto, 1924).
—— *The Grasshoppers Come* (Chatto, 1931).
Garrett, George, *Out of Liverpool: Stories of Sea and Land* (Merseyside Writers Committee, Liverpool, 1982). Introd. by Jerry Dawson.
Gascoyne, David, *Opening Day* (Cobden-Sanderson, 1933).
Gibbons, Stella, *Cold Comfort Farm* (Longmans, 1932).
—— *Christmas at Cold Comfort Farm and Other Stories* (Longmans, 1940).
Grahame, Kenneth, *The Wind in the Willows* (1908; with illustrations by Ernest H. Shepard, Methuen, 1931).
'Grassic Gibbon, Lewis' (J. Leslie Mitchell), *Sunset Song* (1932); *Cloud Howe* (1933); *Grey Granite* (1934); published as trilogy, *A Scots Quair* (Hutchinson, 1946).
Graves, Robert, *'Antigua, Penny, Puce'* (Seizin Press, Deya, Majorca, and Constable, 1936).
'Green, Henry' (Henry Yorke), *Blindness* (Hogarth, 1926).
—— *Living* (Dent, 1929).
—— *Party Going: A Novel* (Hogarth, 1939).
—— *Loving, Living, Party Going: Three Novels* (Introd. by John Updike, Picador, 1978).
Greene, Graham, *The Man Within* (Heinemann, 1929).
—— *The Name of Action* (Heinemann, 1930).
—— *Rumour at Nightfall* (Heinemann, 1931).
—— *Stamboul Train: An Entertainment* (Heinemann, 1932).
—— *It's a Battlefield* (Heinemann, 1934).
—— *The Basement Room and Other Stories* (Cresset Press, 1935).
—— *The Bear Fell Free* (limited edn.; Grayson and Grayson, 1935).
—— *England Made Me* (Heinemann, 1935).
—— *A Gun For Sale: An Entertainment* (Heinemann, 1936).
—— *Brighton Rock: A Novel* (Heinemann, 1938).
—— *The Confidential Agent: An Entertainment* (Heinemann, 1939).
—— *The Power and the Glory* (Heinemann, 1940).
—— *The Ministry of Fear: An Entertainment* (Heinemann, 1943).
—— *Nineteen Stories* (Heinemann, 1947).
—— *Twenty-One Stories* (Heinemann, 1954).
Greenwood, Walter, *Love on the Dole: A Tale of Two Cities* (Cape, 1933).
Haldane, Charlotte, *I Bring Not Peace* (Chatto, 1932).
Halward, Leslie, *To Tea On Sundays* (Methuen, 1936).
—— *Let Me Tell You* (Michael Joseph. 1938).
—— *The Money's All Right, and Other Stories* (Michael Joseph, 1938).
'Hampson, John' (John Frederick N. Hampson Simpson), *Saturday Night at the Greyhound* (Hogarth, 1931).

Hanley, James, *The German Prisoner* (privately printed by the Author, Muswell Hill, n.d. [1930–1?]). Introductory note by Richard Aldington.
—— *A Passion Before Death* (privately printed, London, 1930).
—— *The Last Voyage* (William Jackson, 1931). Foreword by Richard Aldington.
—— *Boy* (1st edn. with asterisks, Boriswood, Oct. 1931; 2nd impression with asterisks deleted, Boriswood, Dec. 1931).
—— *Men in Darkness: Five Stories* (John Lane The Bodley Head, 1931).
—— 'A Changed Man' (*C*, XIV, no. 56 (Apr. 1935), 374–8).
—— *The Furys* (Chatto, 1935).
—— 'Aunt Anne' (*C*, XV, no. 60 (Apr. 1936), 418–20).
—— 'Day's End' (*NW*, 3 (Spring 1937), 166–77).
—— 'From Five till Six', (*C*, XVII, no. 68 (Apr. 1938), 432–42).
—— 'Seven Men' (*NW*, 5 (Spring 1938), 210–32).
Heppenstall, Rayner, *The Blaze of Noon* (Secker & Warburg, 1939; reissued with Author's Note, Allison and Busby, 1980).
Herbert, A. P., *The Secret Battle* (Methuen, 1919; Oxford Paperback, with introd. by J. Terraine, 1982).
Heslop, Harold, *Goaf* (Fortune Press, 1934).
—— *Last Cage Down* (Wishart, 1935).
Hilton, James, *Lost Horizon* (Macmillan, 1933).
—— *Goodbye, Mr Chips* (Hodder & Stoughton, 1934).
Holtby, Winifred, *South Riding: An English Landscape* (Collins, 1936).
Hughes, Richard, *A High Wind in Jamaica* (Chatto, 1929).
—— *In Hazard* (Chatto, 1938).
Huxley, Aldous, *Brave New World: A Novel* (Chatto, 1932).
—— *Eyeless in Gaza* (Chatto, 1936).
—— *After Many a Summer* (Chatto, 1939).
Isherwood, Christopher, *All the Conspirators* (Cape, 1928).
—— *The Memorial: Portrait of a Family* (Hogarth, 1932).
—— *Mr Norris Changes Trains* (Hogarth, 1935).
—— *Lions and Shadows: An Education in the Twenties* (Hogarth, 1938).
—— *Goodbye to Berlin* (Hogarth, 1939).
—— *Down There on a Visit* (Methuen, 1962).
—— *The Berlin of Sally Bowles* (i.e. *Mr Norris Changes Trains* and *Goodbye to Berlin*), with introd. by Isherwood (Hogarth, 1975).
Koestler, Arthur, *Darkness at Noon* (trans. Daphne Hardy; Cape, 1940).
James, Henry, *In the Cage* (Duckworth, 1898).
Jameson, Storm, *Women Against Men* (comprising *The Single Heart* (Ernest Benn, 1932), *A Day Off* (Nicholson & Watson, 1933), and *Delicate Monster* (Nicholson & Watson, 1937), with introd. by Elaine Feinstein; Virago, 1984).
—— *Love in Winter* (Cassell, 1935; reissued, with introd by Elaine Feinstein; Virago, 1984).

Jesse, F. Tennyson, *A Pin to See the Peepshow* (Heinemann, 1934; reissued by Virago, 1979).
Johns, W. E., *Biggles Goes to War* (Oxford University Press, London, 1938).
—— *Biggles in Spain* (Oxford University Press, 1939).
Jones, Lewis, *Cwmardy: The Story of a Welsh Mining Valley* (Lawrence & Wishart, 1937; reissued, with introd. by David Smith, by Lawrence & Wishart, 1978).
—— *We Live: The Story of a Welsh Mining Valley* (Lawrence & Wishart, 1939; reissued, with introd. by David Smith, by Lawrence & Wishart, 1978).
Joyce, James, *Finnegans Wake* (Faber, 1939).
Kent, Martin, *Flying on the Frontier* (Epworth Press, 1938).
Lawrence, D. H., *The Collected Short Stories* (Heinemann, 1974).
Lawrence, T. E., *The Mint* (1st, limited edn., Doubleday, Doran, New York, 1936; reissued Cape, 1955).
Lehmann, John, *Evil Was Abroad* (Cresset Press, 1938).
—— *In the Purely Pagan Sense: A Novel* (Blond & Briggs, 1976).
Lehmann, Rosamond, *Dusty Answer* (Chatto, 1927; reissued by Penguin, 1936).
—— *Invitation to the Waltz* (Chatto, 1932; reissued, with introd. by Janet Watts, by Virago, 1981).
—— *The Weather in the Streets: A Novel* (Colins, 1936; reissued, with introd. by Janet Watts, by Virago, 1981).
Lewis, C. S., *Out of the Silent Planet* (John Lane The Bodley Head, 1938).
Lewis, Wyndham, *The Apes of God* (Nash and Grayson, 1932 [1931]).
—— *The Revenge for Love* (Cassell, 1937).
—— *The Roaring Queen* (ed. and introd. Walter Allen; Secker & Warburg, 1973).
Lindsay, Jack, *1649: A Novel of a Year* (Methuen, 1938).
Linklater, Eric, *Poet's Pub* (Cape, 1929).
Llewellyn, Richard, *How Green Was My Valley* (Michael Joseph, 1939).
London, Jack, *The People of the Abyss* (1903; with introd. by Jack Lindsay, Journeyman Press, 1977).
Loos, Anita, *'Gentlemen Prefer Blondes': The Illuminating Diary of a Professional Lady* (Brentano's 1926).
Lowry, Malcolm, *Ultramarine: A Novel* (Cape, 1933; revised, 1963).
Macaulay, Rose, *Going Abroad: A Novel* (Collins, 1934).
Macdonell, A. G., *England, Their England* (Macmillan, 1933).
Malraux, André, *Man's Estate* (= *La Condition Humaine* (1933), trans. Alastair Macdonald first published in English (Methuen, 1934) as *Storm in Shanghai*; Methuen, 1948).
—— *Days of Contempt* (trans. Haakon M. Chevalier; Gollancz, 1936).
—— *Days of Hope* (= *L'Espoir* (1938), trans. Stuart

Gilbert and Alastair Macdonald; Routledge, 1938).

Mannin, Ethel, *Comrade, O Comrade: Or, Low-Down On the Left* (Jarrolds, 1947).

Manning, Frederic, *Her Privates We: by Private 19022* (Peter Davies, 1930).

Milne, Ewart, 'Gun Runner (An Incident in the Spanish Civil War)' (*Palantir*, ed. Jim Burns, no. 19; Preston, Lancs, 1982).

—— *Drums Without End: Short Stories Mainly About the Spanish Civil War* (Aquila Fiction, Isle of Skye, 1986).

Mitchell, J. Leslie, *Stained Radiance. A Fictionist's Prelude* (Jarrolds, 1930). *See also* 'Grassic Gibbon, Lewis' in this section.

Mitchison, Naomi, *We Have Been Warned: A Novel* (Constable, 1935).

Orwell, George, *Burmese Days* (Harper & Bros., New York, 1934; Gollancz, 1935).

—— *A Clergyman's Daughter* (Gollancz, 1935).

—— *Keep the Aspidistra Flying* (Gollancz, 1936).

—— *Coming Up for Air* (Gollancz, 1939).

—— *Nineteen Eighty-Four: A Novel* (Secker & Warburg, 1949).

Postgate, Raymond, *Verdict of Twelve* (Collins, 1940).

Powell, Anthony, *Afternoon Men* (Duckworth, 1931).

—— *Venusberg* (Duckworth, 1932).

—— *From A View to A Death* (Duckworth, 1933).

—— *Agents and Patients* (Duckworth, 1936).

—— *What's Become of Waring?* (Cassell, 1939).

Powys, John Cowper, *Wolf Solent* (Cape, 1929; with pref. by Powys, Macdonald, 1961).

—— *A Glastonbury Romance* (The Bodley Head, 1933; with new pref. by author, Macdonald, 1955).

—— *Weymouth Sands: A Novel* (New York, 1934); = *Jobber Skald* (John Lane, 1935; reissued, with introd. by Angus Wilson, by Rivers Press, Cambridge, 1973).

—— *Maiden Castle* (Cassell, 1937).

—— *After My Fashion* (Pan, 1980). Foreword by Francis Powys.

Powys, T. F., *Mr Weston's Good Wine* (Chatto, 1927).

Priestley, J. B., *The Good Companions* (Heinemann, 1929).

'Pumphrey, Arthur' (Alan Pryce-Jones), *Pink Danube* (Secker, 1939).

Rhys, Jean, *The Left Bank: Sketches and Studies of Present-Day Bohemian Paris* (Cape, 1927). Pref. by Ford Madox Ford.

—— *Quartet* (first published as *Postures*, Chatto, 1928; Deutsch, 1969).

—— *After Leaving Mr Mackenzie* (Cape, 1930).

—— *Voyage in the Dark* (Constable, 1934).

—— *Good Morning, Midnight* (Constable, 1939).

—— *Tigers Are Better-Looking: with a Selection from the Left Bank* (Deutsch, 1968).

Rhys, John Llewellyn, *The Flying Shadow* (Faber, 1936).

Riding, Laura, *Progress of Stories* (Seizin Press, Deya, Majorca, and Constable, 1935; enlarged, with new preface by Laura (Riding) Jackson,

Carcanet, Manchester, 1982).

Roley, A. P., *Revolt* (Arthur Barker, 1933).

Sayers, Dorothy L., *Lord Peter Views the Body* (Gollancz, 1928).

—— *The Five Red Herrings* (Gollancz, 1931).

—— *Have His Carcase* (Gollancz, 1932).

—— *Gaudy Night* (Gollancz, 1936).

Sherriff, R. C., and Bartlett, Vernon, *Journey's End: A Novel* (Gollancz, 1930).

Shute, Neville, *So Disdained* (Cassell, 1928).

Simon, Claude, *Les Géorgiques* (Editions de Minuit, Paris, 1981; trans. by John Fletcher into English, Calder, 1987).

Smith, Stevie, *Novel on Yellow Paper, or, Work it Out For Yourself* (Cape, 1936).

—— *Over the Frontier* (Cape, 1938; reissued by Virago, 1980).

Sommerfield, John, *They Die Young* (Heinemann, 1930).

—— *May Day* (Lawrence & Wishart, 1936).

Spender, Stephen, *The Burning Cactus* [stories] (Faber, 1936).

—— *The Backward Son: A Novel* (Hogarth, 1940).

Sprigg, C. St. John, *Crime in Kensington* (Eldon Press, 1933).

—— *Fatality in Fleet Street* (Eldon Press, 1933).

—— *Death of an Airman* (Hutchinson, 1934).

—— *The Perfect Alibi* (Eldon Press, 1934).

—— *The Corpse with a Sunburnt Face* (Nelson, 1935).

—— *Death of a Queen: Charles Venables' Fourth Case* (Nelson, 1935).

—— *The Six Queer Things* (Herbert Jenkins, 1937).

—— See also 'Caudwell, Christopher' in this section.

Stuart, Francis, *Pigeon Irish* (Gollancz, 1932).

—— *The Coloured Dome* (Gollancz, 1932).

—— *Try the Sky* (Gollancz, 1933). Foreword by Compton Mackenzie.

Symons, Julian, *The Immaterial Murder Case* (Gollancz, 1945).

Thomas, Dylan, and Davenport, John, *The Death of the King's Canary* (Viking Press, New York, 1977).

—— *The Collected Stories* (Dent, 1983).

Todd, Ruthven, *Over the Mountain* (Harrap, 1939).

—— *The Lost Traveller* (Grey Walls Press, 1943).

Tolkien, J. R. R., *The Hobbit, or There And Back Again* (Allen & Unwin, 1937).

Toynbee, Philip, *A School in Private* (Putnam, 1941).

Upward, Edward, *The Railway Accident, and Other Stories* (Heinemann, 1969). Introd. by W. H. Sellers.

—— *The Spiral Ascent: A Trilogy of Novels* (comprising *In the Thirties* (Heinemann, 1962), *The Rotten Elements* (Heinemann, 1969), and *No Home But the Struggle* (previously unpublished); Heinemann, 1977).

Waddell, Helen, *Peter Abelard: A Novel* (Constable, 1933).

Warner, Rex, *The Kite* (no. 8 in Tales of Action, by Men of Letters; ed. L. A. G. Strong;

Blackwell, Oxford, 1936).
— *The Wild Goose Chase: A Novel* (Boriswood, 1937).
— *The Aerodrome: A Love Story* (John Lane The Bodley Head, 1941).
Warner, Sylvia Townsend, *Mr Fortune's Maggot* (Chatto, 1927; reissued, with pref. by Sylvia Townsend Warner, by Virago, 1978).
— *The True Heart* (Chatto, 1929; reissued, with pref. by Sylvia Townsend Warner, by Virago, 1978).
— *Summer Will Show* (Chatto, 1936).
—*After the Death of Don Juan* (Chatto, 1938).
Waugh, Alec, *The Loom of Youth* (Grant Richards, 1917).
— *Wheels Within Wheels: A Story of the Crisis* (Cassell, 1933).
Waugh, Evelyn, *Decline and Fall: An Illustrated Novelette* (Chapman & Hall, 1928).
— *Vile Bodies* (Chapman & Hall, 1930).
— *Black Mischief* (Chapman & Hall, 1932).
— *A Handful of Dust* (Chapman & Hall, 1934).
—*Mr Loveday's Little Outing, and Other Sad Stories* (Chapman & Hall, 1936).
— *Scoop: A Novel About Journalists* (Chapman & Hall, 1938).
— *Work Suspended and Other Stories* (Chapman & Hall, 1943).
Wells, H. G., *The Shape of Things to Come: the Ultimate Revolution* (Hutchinson, 1933).
Webb, Mary, *Precious Bane* (Cape, 1924; with introd. by Stanley Baldwin, Cape, 1929).
Westerman, Percy F., *Under Fire in Spain* (Blackie, 1937).
White, T. H., *Earth Stopped; or, Mr Marx's Sporting Tour* (Collins, 1934).
— See also 'Aston, James' in this section.
Williams, Charles, *War in Heaven* (Gollancz, 1930; republished in the Dennis Wheatley Library of the Occult, Sphere Books, 1976).
— *Many Dimensions* (Gollancz, 1931).
— *The Place of the Lion* (Gollancz, 1931).
— *The Greater Trumps* (Gollancz, 1932).
— *Shadows of Ecstasy* (Gollancz, 1933).
— *Descent into Hell* (Faber, 1937).
Williams-Ellis, Amabel, *Volcano* (Cape, 1931).
Williamson, Henry, *The Beautiful Years* (Faber, 1929).
— *Dandelion Days* (Faber, 1930).
— *The Patriot's Progress: Being the Vicissitudes of Pte. John Bullock* (Bles, 1930).
— *The Dream of Fair Women* (Faber, 1931).
— *The Gold Falcon, or The Haggard of Love* (Faber, 1933).
— *The Sun in the Sands* (Faber, 1945).
— *A Fox Under My Cloak* (Macdonald, 1955).
— *The Phoenix Generation* (Macdonald, 1965).
Wodehouse, P. G., *Mulliner Nights* (Herbert Jenkins, 1933).
— *The Code of the Woosters* (Herbert Jenkins, 1938).
Woolf, Virginia, *Jacob's Room* (Leonard and Virginia Woolf, Richmond, 1922).
— *The Waves* (Hogarth, 1931).
— *The Years* (Hogarth, 1937).
— *Between the Acts* (Hogarth, 1941).

6. Theatrical Texts

Auden, W. H., *The Dance of Death* (Faber, 1933).
Auden, W. H. and Isherwood, Christopher, *The Dog Beneath the Skin, or Where is Francis?* (Faber, 1935).
— *The Ascent of F6, and On The Frontier* (Faber, 1948).
Brecht, Bertolt, *Fear and Misery of the Third Reich, and Señora Carrar's Rifles* (trans. John Willett and Wolfgang Sauerlander, ed. John Willett and Ralph Manheim; *Collected Plays*, vol. IV, part 3; Methuen, 1983).
Corrie, Joe, *In Time O' Strife* (Forward Publishing Co., Glasgow, 1930; repr. 7:84 Theatre Company, Edinburgh, 1982).
— *Three One-Act Plays* (Labour Bookshop, Glasgow, 1930).
Coward, Noël, *Tonight at 8.30: Plays* (3 vols.; Heinemann, 1936).
— *Plays* (4 vols., ed. Raymond Mander and Joe Mitchenson; Eyre Methuen, 1979).
Eliot, T. S. *The Rock: A Pageant Play Written for Performance at Sadler's Wells Theatre 28 May-9 June 1934 on behalf of the 45 Churches Fund of the Diocese of London; Book of Words by T. S. Eliot* (Faber, 1934).
— *Murder in the Cathedral* (Faber, 1935).
— *The Family Reunion* (Faber, 1939).
Famous Plays of Today (Gollancz, 1929). Includes R. C. Sherriff, *Journey's End* (1929).
MacNeice, Louis, *Out of the Picture: A Play in Two Acts* (Faber, 1937).
Odets, Clifford, *Three Plays* (Gollancz, 1936). Includes *Waiting for Lefty* (1935).
Samuel, Raphael, MacColl, Ewan, and Cosgrove, Stuart, *Theatres of the Left 1880–1935: Workers' Theatre Movements in Britain and America* (History Workshop Series; Routledge, 1985). Includes texts.
Sayers, Dorothy L., *The Man Born to be King: A Play-Cycle on the Life of Our Lord and Saviour Jesus Christ, written for broadcasting* (Gollancz, 1943).
Sherriff, R. C., *Journey's End: A Play in Three Acts* (Gollancz, 1929).
Slater, Montagu, *New Way Wins* (Lawrence & Wishart, 1937).
Spender, Stephen, *Trial of a Judge: A Tragic Statement in Five Acts* (Faber, 1938).
Thomas, Tom, 'The Workers' Theatre Movement: Memoir and Documents' (*History Workshop, A Journal of Socialist Historians*, no. 4 (Autumn 1977), 113–42; with 'Editorial Introduction' by Raphael Samuel, 101–12).
Woolf, Leonard, *The Hotel: A Play* (Hogarth, 1939).

7. Letters

Ackerley, J. R., *The Letters of J. R. Ackerley* (ed. Neville Braybrooke; Duckworth, 1975).

Branson, Clive, *British Soldier in India: The Letters of Clive Branson* (Communist Party, 1944). Introd. by Harry Pollitt.

Connolly, Cyril, *A Romantic Friendship: The Letters of Cyril Connolly to Noel Blakiston* (ed. Noel Blakiston; Constable, 1975).

Durrell, Lawrence, *Lawrence Durrell: Henry Miller: A Private Correspondence* (ed. George Wickers; Faber, 1963).

Forster, E. M., *Selected Letters* (ed. Mary Lago and P. N. Furbank: vol. I, *1879–1920*(Collins, 1983); vol. II, *1921–1970* (Collins, 1985)).

Garnett, David (ed.), *The White/Garnett Letters* (Cape, 1968).

Graves, Robert, *In Broken Images: Selected Letters 1914–1946*(ed. Paul O'Prey; Hutchinson, 1982).

Hitler, Adolf, *Hitler's Letters and Notes* (ed. Werner Maser (Econ, Düsseldorf, 1973); trans. Arnold Pomerans (Heinemann, 1974)).

Holtby, Winifred, *Letters to a Friend* (ed. Alice Holtby and Jean McWilliam; Collins, 1937).

Huxley, Aldous, *Letters* (ed. Grover Smith; Chatto, 1969).

Lawrence, T. E., *Letters* (ed. David Garnett; Cape, 1938).

Lewis, C. S., *Letters* (ed., with memoir, W. H. Lewis; Bles, 1966).

— *They Stand Together: Letters to Arthur Greeves (1914–1963)* (ed. Walter Hooper; Collins, 1979).

Lewis, Wyndham, *The Letters* (ed. W. K. Rose; Methuen, 1963).

Lowry, Malcolm, *Selected Letters* (ed. Harvey Breit and Margerie Bonner Lowry; Capricorn Books Edition, New York, 1969).

Miller, Henry, see Durrell, Lawrence, in this section.

Muir, Edwin, *Selected Letters* (ed. and introd. P. H. Butter; Hogarth, 1974).

Owen, Wilfred, *Collected Letters* (ed. Harold Owen and John Bell; Oxford University Press, London, 1967).

Powys, John Cowper, *Letters to Nicholas Ross* (sel. by Nicholas and Adelaide Ross; ed. Arthur Uphill; Bertram Rota, 1971; Village Press, 1974).

— *Letters to Louis Wilkinson 1935–1956* (ed. Louis Wilkinson; Macdonald, 1958; Village Press, 1974).

Rhys, Jean, *Letters, 1931–1966* (sel. and ed. Francis Wyndham and Diana Melly; Deutsch, 1984).

Spender, Stephen, *Letters to Christopher: Stephen Spender's Letters to Christopher Isherwood 1929–1939* (Black Sparrow Press, Santa Barbara, 1980).

Thomas, Dylan, *The Collected Letters* (ed. Paul Ferris; Dent, 1985).

Tolkien, J. R. R., *Letters* (ed. Humphrey Carpenter, assisted by Christopher Tolkien; Allen & Unwin, 1981).

Waugh, Evelyn, *The Letters* (ed. Mark Amory; Weidenfeld, 1980).

Wilson, Edmund, *Letters on Literature and Politics 1912–1972* (ed. Elena Wilson; Routledge, 1977).

Woolf, Virginia, *Letters* (ed. Nigel Nicolson and Joanne Trautmann: IV, *A Reflection of the Other Person: 1929–1931* (Hogarth, 1978); V, *The Sickle Side of the Moon: 1932–1935* (Hogarth, 1979); VI, *Leave the Letters Till We're Dead* (Hogarth, 1980)).

8. Diaries

[Anon], *A Soldier's Diary of the Great War* (introd. Henry Williamson; Faber & Gwyer, 1929).

Baudelaire, Charles, *Intimate Journals* (trans. Ch. [*sic*] Isherwood; introd. T. S. Eliot; Blackamore Press, London/Random House, New York, 1930).

Clarke, Tom, *My Northcliffe Diary* (Gollancz, 1931).

Gascoyne, David, *Paris Journal 1937–1939* (Enitharnon Press, 1978). With preface by Lawrence Durrell.

— *Journal 1936–37* (Enitharnon Press, 1980).

Nicolson, Harold, *Diaries and Letters* (ed. Nigel Nicolson; 2 vols., *1930–39* and *1939–1945* Collins, 1966–1967).

Pryce-Jones, David, *Cyril Connolly: Journal and Memoir* (Collins, 1983).

Sartre, Jean-Paul, *War Diaries: Notebooks from a Phoney War November 1939—March 1940* (trans. Quintin Hoare; Verso, 1984).

Spender, Stephen, *Journals 1939–1983* (ed. John Goldsmith; Faber, 1985).

Waugh, Evelyn, *The Diaries* (ed. Michael Davie; Weidenfeld, 1976).

Woolf, Virginia, *Diary* (ed. Anne Olivier Bell and Andrew McNeillie: III, *1925–1930* (Hogarth, 1980); IV, *1931–35* (Hogarth, 1982); V, *1936–41* (Hogarth, 1984)).

9. Autobiographies and Personal Testimonies

Ackerley, J. R. (ed.), *Escapers All: the Personal Narratives of Fifteen Escapers from War-Time Prison Camps 1914–1918* (John Lane, 1932).

— *My Father and Myself: A Family Memoir* (Bodley Head, 1968).

Ackland, Valentine, *For Sylvia: An Honest Account* (foreword, Bea Howe; Chatto/Hogarth, 1985; reissued Methuen, 1986).

Acton, Harold, *Memoirs of an Aesthete* (Methuen, 1948).

Allen, Walter, *As I Walked Down New Grub Street: Memories of a Writing Life* (Heinemann, 1981).

Barke, James, *The Green Hills Far Away: A Chapter in Autobiography* (Collins, 1940).

Bell, Julian (ed.), *We Did Not Fight. 1914–1918 Experiences of War Resisters* (Cobden-Sanderson, 1935).

Benson, Ernie, *To Struggle is to Live: A Working Class Autobiography*, vol. 2, *Starve or Rebel, 1927–1971* (People's Publications, Newcastle-upon-Tyne, 1980). Foreword by Jack Lindsay.

Berlin, Isaiah, 'Meetings with Russian Writers in 1945 and 1956', in *Personal Impressions* (ed. Henry Hardy; Hogarth, 1981).

Blunden, Edmund, *Undertones of War* (1928;

reissued in Penguin Travel & Adventure Series, March, 1937).

Brenan, Gerald, *A Life of One's Own: Childhood and Youth* (Hamish Hamilton, 1962).

— *Personal Record, 1920–1972* (Cape, 1974).

Brittain, Vera, *Testament of Youth: An Autobiographical Study of the Years 1900–1925* (Gollancz, 1933).

— *Testament of Experience: An Autobiographical Story of the Years 1925–1950* (Gollancz, 1957).

Bush, Alan, 'In My Eighth Decade', in *In My Eighth Decade and Other Essays* (Kahn & Averill, 1980).

Campbell, Roy, *Broken Record: Reminiscences* (Boriswood, 1934).

— *Light on a Dark Horse: An Autobiography* (Hollis and Carter, 1951; republished, with foreword by Laurie Lee, by Penguin, 1971).

Carr, Richard Comyns (ed.), *Red Rags: Essays of Hate from Oxford* (Chapman & Hall, 1933).

Christie, Agatha, *An Autobiography* (Collins, 1977).

Church, Richard, *Over the Bridge: An Essay in Autobiography* (Heinemann, 1955).

Cockburn, Claud, *In Time of Trouble: An Autobiography* (Hart-Davis, 1956).

— *I, Claud . . . : An Autobiography* (incorporating *In Time of Trouble* (1956), *Crossing the Line* (1958), and *View from the West* (1961); Penguin, 1967).

Coombes, B. L., *I am a Miner*; special issue of *Fact* magazine (no. 23, Feb. 1939).

— *These Poor Hands: The Autobiography of a Miner Working in South Wales* (Left Book Club, Gollancz, 1939).

Day Lewis, C., *The Buried Day* (Chatto, 1960).

Dodds, E. R., *Missing Persons: An Autobiography* (Clarendon Press, Oxford, 1977).

Driberg, Tom, *Ruling Passions* (Cape, 1978).

Dyment, Clifford, *The Railway Game: An Early Autobiography* (Dent, 1962).

Farson, Negley, *The Way of a Transgressor* (Gollancz, 1935).

Fuller, Roy, 'Poetic Memories of the Thirties', *Professors and Gods: Last Oxford Lectures on Poetry* (Deutsch, 1973).

— *Souvenirs* (London Magazine Editions, 1980).

— *Vamp Till Ready: Further Memoirs* (London Magazine Editions, 1982).

Grant Duff, Sheila, *The Parting of Ways: A Personal Account of the Thirties* (Peter Owen, 1982).

Graves, Robert, *Goodbye to All That* (Cape, 1929).

'Green, Henry' (Henry Yorke), *Pack My Bag: A Self-Portrait* (Hogarth, 1940).

Greene, Graham, *A Sort of Life* (Bodley Head, 1971).

— 'While Waiting for a War' (*Granta*, no. 17 (Autumn 1985), 11–29).

— ed., *The Old School: Essays by Divers Hands* (Cape, 1934; reissued as Oxford Paperback, 1984).

Greenwood, Walter, *There Was A Time* (Cape, 1967).

Grigson, Geoffrey, *The Crest on the Silver: An Autobiography* (Cresset Press, 1950).

Haldane, Charlotte, *Truth Will Out* (Weidenfeld, 1949; reissued by The Right Book Club, 1949).

Hanley, James, *Broken Water: An Autobiographical Excursion* (Chatto, 1937).

Hitler, Adolph, *Mein Kampf* (trans. James Murphy (Hurst and Blackett, 1939); trans. Ralph Manheim, introd. D. C. Watt (Hutchinson, 1969)).

Hodge, Herbert, *I Drive A Taxi*; special issue of *Fact* magazine (no. 22, Jan. 1939).

Hyde, Douglas, *I Believed: The Autobiography of a Former British Communist* (Putman's, New York, and Heinemann, 1950).

Isherwood, Christopher, *Christopher and His Kind, 1929–1939* (Eyre Methuen, 1977).

Jameson, Storm, *Journey From the North, Autobiography* (2 vols.; Collins & Harvill Press, 1969–1970; reissued by Virago, 1984).

Koestler, Arthur, *The Invisible Writing* (Collins/ Hamish Hamilton, 1954).

Lancaster, Osbert, *With An Eye to the Future* (Murray, 1967).

Lehmann, John, *The Whispering Gallery: Autobiography I* (Longmans, 1955); *I Am My Brother: Autobiography II* (Longmans, 1960).

— *Thrown to the Woolfs* (Weidenfeld, 1978).

Lehmann, Rosamond, *The Swan in the Evening: Fragments of an Inner Life* (Collins, 1967; revised edn. with new epilogue by the author, Virago, 1982).

Lewis, C. S., *Surprised by Joy: The Shape of My Early Life* (Bles, 1955).

Lewis, Wyndham, *Blasting and Bombardiering* (Eyre & Spottiswoode, 1937).

— *Wyndham Lewis the Artist: From 'Blast' to Burlington House* (Laidlaw and Laidlaw, 1939).

— *Rude Assignment: A Narrative of My Career Up-to-Date* (Hutchinson, 1950).

Lindsay, Jack, *Life Rarely Tells* (incorporating *Life Rarely Tells* (1958), *The Roaring Twenties* (1960), and *Fanfrolico and After* (1962); Penguin, 1982).

Lubbock, Percy, *Shades of Eton* (Cape, 1929).

Lunn, Arnold, *Come What May: An Autobiography* (Eyre & Spottiswoode, 1940).

Macartney, Wilfred, *Walls Have Mouths: A Record of Ten Years' Penal Servitude* (Left Book Club, Gollancz, 1936). Prologue, epilogue, and commentary by Compton Mackenzie.

Maclaren-Ross, Julian, *Memoirs of the Forties* (Alan Ross, 1965).

MacNeice, Louis, *The Strings Are False: An Unfinished Autobiography* (ed., and pref., E. R. Dodds; Faber, 1965).

Madge, Charles, 'Viewpoint' (*Times Literary Supplement*, 14 Dec. 1979, p. 119).

Martin, Kingsley, *Editor: A Second Volume of Autobiography 1931–45* (Hutchinson, 1968).

Medley, Robert, *Drawn From the Life: A Memoir* (Faber, 1983).

Mitchison, Naomi, *You May Well Ask: A Memoir, 1920–1940* (Gollancz, 1979).

Mitford, Jessica, *Hons and Rebels* (Gollancz, 1960).

Mortimer, Raymond, *Try Anything Once* (Hamish Hamilton, 1976).

Mosley, Diana, *A Life of Contrasts* (Hamish Hamilton, 1977).

Mosley, Oswald, *My Life* (Nelson, 1968).

Muggeridge, Malcolm, *In a Valley of This Restless Mind* (Routledge, 1938).

Muir, Edwin, *The Story and The Fable. An Autobiography* (Harrap, 1940; enlarged and revised as *An Autobiography*, Hogarth, 1954).

Nicholson, Ben, and Jones, David, 'Looking Back at the Thirties' (*London Magazine*, NS 5, no. 1 (Apr. 1965), 47–54).

O'Connor, Philip, 'The Thirties: A Rehearsal' (*London Magazine*, NS 5, no. 12 (Mar. 1966), 65–70).

Phelan, Jim, *The Name's Phelan: the first Part of the Autobiography of Jim Phelan* (Sidgwick & Jackson, 1948).

Plomer, William, *Double Lives: An Autobiography* (Cape, 1943).

Powell, Anthony, *To Keep the Ball Rolling: Memoirs*: I, *Infants of the Spring* (Heinemann, 1976); II, *Messengers of the Day* (Heinemann, 1978); III, *Faces In My Time* (Heinemann, 1980).

Powys, John Cowper, *Autobiography* (John Lane, 1934; reissued, with introd. by J. B. Priestley, by Macdonald, 1967).

Powys, Littleton, *The Joy of It* (Chapman & Hall, 1937).

Pryce-Jones, Alan (ed.), *Little Innocents: Childhood Reminiscences* (Cobden-Sanderson, 1932; reissued as Oxford Paperback, 1986).

Pudney, John, *Thank Goodness for Cake* (Michael Joseph, 1978).

Quennell, Peter, *The Marble Foot: An Autobiography 1905–1938* (Collins, 1976).

Raine, Kathleen, *The Land Unknown* (Hamish Hamilton, 1975).

Read, Herbert, *The Contrary Experience: Autobiographies* (= combined version of *The Innocent Eye* (Faber, 1935) and *Annals of Innocence and Experience* (Faber, 1940), Faber, 1963; reissued, with 'Personal Foreword' by Graham Greene, by Secker and Warburg, 1973).

Regler, Gustav, *The Owl of Minerva: Autobiography* (trans. Norman Denny; Hart-Davis, 1959).

Rhys, Jean, *Smile Please: An Unfinished Autobiography* (Deutsch, 1979).

Romilly, Giles, and Romilly, Esmond, *Out of Bounds* (Hamish Hamilton, 1935).

Rowse, A. L., *A Cornishman at Oxford* (Cape, 1965).

Sassoon, Siegfried, *The Complete Memoirs of George Sherston* (Faber, 1937).

Sitwell, Edith, *Taken Care of: An Autobiography* (Hutchinson, 1965).

Sommerfield, John, *The Imprinted: Recollections of Then, Now and Later On* (London Magazine Editions, 1977).

Spender, Stephen, *World Within World: An Autobiography* (Hamish Hamilton, 1951).

Stanford, Derek, *Inside the Forties: Literary Memoirs 1937–1957* (Sidgwick & Jackson, 1977).

Stuart, Francis, *Things to Live For: Notes for an Autobiography* (Cape, 1934).

Thomas, Dylan, *Portrait of the Artist As A Young Dog* (Dent, 1940).

Toynbee, Philip, *et al.*, *Leaving School* (London Magazine Editions, 3, 1966).

Trevelyan, Julian, *Indigo Days* (MacGibbon & Kee, 1957).

Uhlman, Fred, *The Making of an Englishman* (Gollancz, 1960).

Wain, John, *Sprightly Running: Part of an Autobiography* (Macmillan, 1962).

Waugh, Alec, *A Year to Remember: a Reminiscence of 1931* (W. H. Allen, 1975).

Waugh, Evelyn, *A Little Learning: The First Volume of An Autobiography* (Chapman & Hall, 1964).

Wells, H. G., *Experiment in Autobiography: Discoveries and Conclusions of a Very Ordinary Brain (Since 1866)*, vol. I (Gollancz/Cresset Press, 1934).

West, Alick, *One Man in His Time: An Autobiography* (Allen & Unwin, 1969).

What I Believe, By Fourteen Modern Thinkers (Frederick Muller, 1937).

White, T. H., *England Have My Bones* (Collins, 1936).

Williams-Ellis, Amabel, *All Stracheys Are Cousins* (Weidenfeld, 1983).

Williams-Ellis, Clough, *Architect Errant* (Constable, 1971).

Williamson, Henry, *The Wet Flanders Plain* (Beaumont Press, 1929; rev. edn. Faber, 1929).

— *Goodbye West Country* (Putnam, 1937).

Woolf, Virginia, *Moments of Being: Unpublished Autobiographical Writings* (ed. Jeanne Schulkind; Sussex University Press, 1976).

Worsley, T. C., *Flannelled Fool: A Slice of Life in the Thirties* (Alan Ross, 1967).

Wyatt, Woodrow, *Into the Dangerous World* (Weidenfeld, 1952).

10. Interviews

Allain, Marie-Françoise, *The Other Man: Conversations with Graham Greene* (trans. from French by Guido Waldman; Bodley Head, 1983).

Breit, Harvey, *The Writer Observed* (1956; Collier Books, New York, 1961).

Durrell, Lawrence, *The Big Supposer: An Interview with Marc Alyn* (trans. Francine Barker; Grove Press, New York, 1974).

Grigson, Geoffrey, 'A Conversation with Geoffrey Grigson' (*The Review*, no. 22 (June 1970), 15–26).

Penrose, Roland, 'Sir Roland Penrose In Conversation with Alan Young' (*PN Review*, IV, no. 4 (1977), 4–10).

Plimpton, George (ed.), *Writers At Work: The Paris Review Interviews* (2nd series (Secker & Warburg, 1963) includes Eliot, Huxley, Durrell; 3rd series (Secker & Warburg, 1967) includes Waugh; 4th series (Secker & Warburg, 1977) includes Isherwood, Auden; 6th series (Secker & Warburg, 1984) includes Spender).

Pugh, Anthony Cheal, 'Interview with Claude Simon: Autobiography, The Novel, Politics' (*Review of Contemporary Fiction*, V, 1 (1984), 4–13).

Scobie, W. I., 'The Youth that was "I": A Conversation in Santa Monica with Christopher Isherwood' (*London Magazine*, NS, 17, no. 1 (Apr.–May 1977), 23–32).

Upward, Edward, 'Back From the Border' (*London Magazine*, NS, 9, no. 3 (June 1969), 5–11).

11. Criticism from the Period and after by 'Period' Critics

(See also the periodicals listed in section 2, passim.)

'Actor', 'A Real Workers' Theatre Movement', (*Discussion*, III, no. 2 (Mar. 1938), 42–4).

Artists' International Association, *5 On Revolutionary Art* (Wishart, 1935).

Beckett, Samuel, and others, *Our Exagmination Round His Factification For Incamination of Work in Progress* (Faber, 1929).

Bell, Graham, *The Artist and His Public* (Hogarth Sixpenny Pamphlets, no. 5; Hogarth, 1939).

Beste, R. Vernon, 'On Mass Declamations' (*Our Time*, II, no. 11 (May 1943), 5–6, 29–30).

Blunt, Anthony, 'Picasso Unfrocked' (*S*, 8 Oct. 1937, 584).

Bodkin, Maud, *Archetypal Patterns in Poetry: Psychological Studies of Imagination* (Oxford University Press, London, 1934).

Bostock, J. Knight, *Some Well-Known German War Novels 1914-1930* (Blackwell, Oxford, 1931).

Brown, Ivor, 'Left Theatres' (*NS*, 6 Apr. 1935, 487–8).

Budgen, Frank, *James Joyce and the Making of 'Ulysses'* (Grayson & Grayson, 1934; reissued and expanded Indiana University Press, 1960, and Oxford University Press, 1972).

Campbell, Roy, 'Epitaph on the Thirties' (*Nine*, no. 5 (Autumn 1950), 344–6).

'Caudwell, Christopher' (Christopher St John Sprigg), *Illusion and Reality: A Study of the Sources of Poetry* (Macmillan, 1937; new edn. Lawrence & Wishart, 1946).

— *Studies In A Dying Culture* (John Lane, 1938). Introd. by John Strachey.

— *Further Studies in a Dying Culture* (ed. and pref. by Edgell Rickword, The Bodley Head, 1940).

— *Studies and Further Studies in a Dying Culture* (introd. Sol Yurick, Monthly Review Press, New York, 1971).

— *Romance and Realism: A Study in English Bourgeois Literature* (ed. Samuel Hynes; Princeton University Press, 1970).

Cole, G. D. H., *Politics and Literature* (Hogarth Lectures on Literature no. 11; Hogarth, 1929).

Comfort, Alex, 'The Darkness of Poetry' (*Kingdom Come*, III, no. 12 (1943), 28–30).

Connolly, Cyril, 'The Grigson Gallery' (*NS*, 16 Sept. 1939, 404).

Daiches, David, *Literature and Society* (Gollancz, 1938).

Day Lewis, Cecil, *A Hope for Poetry* (Blackwell, Oxford, 1934).

— (ed.) *The Mind in Chains: Socialism and the Cultural Revolution* (Frederick Muller, 1937).

Drapier, Colin, 'Wanted! New Poets for the Revolution' (*Poetry and the People*, no. 19 (July 1940), 23–4).

Eliot, T. S., 'A Note on Poetry and Belief' (*The Enemy*, no. 1 (Jan. 1927), 15–17).

— *Selected Essays* (Faber, 1932; rev. and enlarged, 1934; enlarged, 1951).

— *After Strange Gods: A Primer of Modern Heresy: The Page-Barbour Lectures at the University of Virginia 1933* (Faber, 1934).

— *Selected Prose* (ed. John Hayward; Faber, 1953).

— *Selected Prose* (ed. Frank Kermode; Faber, 1975).

Empson, William, *Seven Types of Ambiguity* (Chatto, 1930).

— *Some Versions of Pastoral: A Study of the Pastoral Form in Literature* (Chatto, 1935).

Forster, E. M., *Aspects of the Novel and Related Writings* (Abinger Edition, vol. 12, ed. Oliver Stallybrass; Edward Arnold, 1974).

Fox, Ralph, *The Novel and the People* (Lawrence & Wishart, 1937; pref. Mulk Raj Anand, Cobbett Press, 1944; pref. Jeremy Hawthorn, Lawrence & Wishart, 1979).

Graves, Robert, 'An Incomplete Complete Skelton' (*Adelphi*, Dec. 1931, 146–58).

Grigson, Geoffrey (ed. and introd.), *The Arts Today* (John Lane The Bodley Head, 1935).

Henderson, Philip, *The Novel Today: Studies in Contemporary Attitudes* (John Lane, 1936).

— *The Poet and Society* (Secker & Warburg, 1939).

Hulme, T. E., *Speculations: Essays on Humanism and the Philosophy of Art* (ed. Herbert Read; frontispiece and foreword Jacob Epstein, Routledge, 1924; 2nd edn., 1936).

Jackson, T. A., *Dickens: The Progress of a Radical* (Lawrence & Wishart, 1937).

Leavis, F. R., *Mass Civilisation and Minority Culture* (Minority Pamphlets, no. 1; The Minority Press, Cambridge, 1930).

— *New Bearings in English Poetry: A Study of the Contemporary Situation* (Chatto, 1932; rev, 1950).

— *Education & the University: A Sketch for an 'English School'* (Chatto, 1943; 2nd edn., 1948). The 2nd edn. reprints *Mass Civilisation and Minority Culture* as an appendix.

— *The Common Pursuit* (Chatto, 1952).

— ed., *Towards Standards of Criticism: Selections from 'The Calendar of Modern Letters' 1925-7*

(Wishart, 1933; reissued with new pref. by F. R. Leavis, Lawrence & Wishart, 1976).

Leavis, F. R. and Thompson, Denys, *Culture and Environment: the Training of Critical Awareness* (Chatto 1933).

Leavis, Q. D., *Fiction and the Reading Public* (Chatto, 1932).

— 'Lady Novelists and the Lower Orders' (*Scrutiny*, IV, no. 2 (Sept. 1935), 112–32).

— 'Class-War Criticism' (*Scrutiny*, V, no. 4 (Mar. 1937), 418–23).

Lehmann, John, *New Writing in Europe* (Pelican, 1940).

Lewis, C. S., *The Allegory of Love: A Study in Medieval Tradition* (Clarendon Press, Oxford, 1936).

Literature of the World Revolution, nos. 1, 2 (Moscow, 1931). Special number, *Second International Conference of the Revolutionary Writers*.

MacNeice, Louis, *Modern Poetry: A Personal Essay* (Oxford University Press, 1938).

— 'The Poet in England Today' (*New Republic*, 102 (25 Mar. 1940), 412–13).

— *The Poetry of W. B. Yeats* (Oxford University Press, 1941).

Powys, Llewellyn, 'T. S. Eliot: The Tutor-Poet' (*The Week-End Review*, VII, no. 167 (20 May 1933), 556–7).

Richards, I. A., *Principles of Literary Criticism* (Kegan Paul, 1925; 2nd edn., 1926; 3rd edn., 1928).

— *Science and Poetry* (Psyche Miniatures General Series; Kegan Paul, Trench, Trubner, 1926).

— *Practical Criticism: A Study of Literary Judgement* (Kegan Paul, Trench, Trubner, 1929).

— *Complementarities: Uncollected Essays* (ed. John Paul Russo; Carcanet New Press, Manchester, 1976).

Rickword, Edgell, 'Poetry and Two Wars' (*Our Time*, I, 2 (Apr. 1941), 1–6).

— (ed.), *Scrutinies, By Various Writers* (Wishart, 1928).

Robbins, Jack Alan (ed.), *Granville Hicks in 'The New Masses'* (New York, 1974).

Roberts, Michael, *Critique of Poetry* (Cape, 1934).

— *T. E. Hulme* (Faber, 1938).

Spender, Stephen, *The Destructive Element: A Study of Modern Writers and Beliefs* (Cape, 1935).

— *Life and the Poet* (Secker & Warburg, 1942).

Traversi, D. A., 'Marxism and English Poetry' (*Arena*, I, no. 3 (Oct.–Dec. 1937), 199–211).

Turnell, Martin, *Poetry and Crisis* (Sands: The Paladin Press, 1938).

West, Alick, *Crisis and Criticism* (Lawrence & Wishart, 1937).

— *Crisis and Criticism and Selected Literary Essays* (foreword, Arnold Kettle; introd. Elisabeth West; Lawrence & Wishart, 1975).

Williams, Charles, *Poetry at Present* (Oxford University Press, London, 1930).

— *The Figure of Beatrice: A Study in Dante* (Faber, 1943).

Wilson, Edmund, *The Shores of Light: A Literary Chronicle of the Twenties and Thirties* (W. H. Allen, 1952).

— 'The Oxford Boys Becalmed', in ibid.

Zhdanov, A., Gorky, M., Bukharin, N., Radek, K., Stetsky, A., *Problems of Soviet Literature: Reports and Speeches at the First Soviet Writers' Congress* (ed. H. G. Scott (Martin Lawrence, 1935); reissued as *Soviet Writers' Congress 1934: The Debate on Socialist Realism and Modernism* (Lawrence & Wishart, 1977)).

12. Miscellaneous Writing, Essays, Journalism, etc. of the Period; or by 'Period' Authors

Auden W. H., 'Everyman's Freedom' (*NS*, 23 Mar. 1935, 422–33).

— 'Criticism in a Mass Society' (*The Mint*, no. 2 (1948), ed. Geoffrey Grigson).

— *The Dyer's Hand, And Other Essays* (Faber, 1963).

— *Selected Essays* (Faber, 1964).

— *Forewords and Afterwords*, selected by Edward Mendelson (Faber, 1973).

Beckett, Samuel, *Disjecta: Miscellaneous Writings and a Dramatic Fragment* (Calder, 1983).

Bell, Clive, *Civilization* (Chatto, 1928; republished in one volume with *Old Friends* (Chatto, 1956) as *Civilization and Old Friends*, University of Chicago Press, 1973).

Benn, Gottfried, *Primal Vision: Selected Writings* (ed. E. B. Ashton; Boyars, 1976).

Betjeman, John, *An Oxford University Chest* (John Miles, 1938; reissued as an Oxford Paperback, 1979).

— *First and Last Loves* (ed. Myfanwy Piper, illus. John Piper; Murray, 1952).

'Caudwell, Christopher' (Christopher St John Sprigg), *The Crisis in Physics* (ed. and introd. H. Levy; John Lane The Bodley Head, 1939).

Connolly, Cyril, *Enemies of Promise* (Routledge, 1938).

— *The Condemned Playground: Essays: 1927–1944* (Routledge, 1945; reissued with new introd. by Philip Larkin, Hogarth, 1985).

Forster, E. M., *Abinger Harvest* (Edward Arnold, 1936).

— *Two Cheers for Democracy* (Edward Arnold, 1951).

Fox, Ralph, *Genghis Khan* (John Lane, 1936).

Golding, William, *A Moving Target* (Faber, 1982).

Graves, Robert, *But It Still Goes On: An Accumulation* (Cape, 1930).

Greene, Graham, *The Lost Childhood And Other Essays* (Eyre & Spottiswoode, 1951).

— *Collected Essays* (Bodley Head, 1969).

— *Ways of Escape* (Bodley Head, 1980).

Groddeck, Georg Walther, 'The Meaning of Illness' (in Groddeck, *The Meaning of Illness:*

Selected Psychoanalytic Writings, trans. Gertrude Mander; Hogarth, 1977).

Harrisson, T. H., *Letter to Oxford* (Reynold Bray, The Hate Press, Oxford, 1934).

Hemingway, Ernest, *By-Line: Selected Articles and Dispatches of Four Decades* (ed. William White, with commentaries by Philip Young; Collins, 1968).

Hughes, Richard, *Fiction as Truth: Selected Literary Writings* (ed. and introd. Richard Poole; Poetry Wales Press, Bridgend, 1983).

Huxley, Aldous, *Music at Night, And Other Essays* (Chatto, 1931).

— *Brave New World Revisited* (Harper, New York, 1958).

Isherwood, Christopher, *Exhumations: Stories, Articles, Verses* (Methuen, 1966).

Lewis, C. S. (ed.), *Essays Presented to Charles Williams* (Oxford University Press, London, 1947).

Lewis, Wyndham, *Men Without Art* (Cassell, 1934).

MacNeice, Louis, *Zoo* (Michael Joseph, 1938).

Mitchison, Naomi, *Comments on Birth Control* (Criterion Miscellany, no. 12; Faber, 1930).

— (ed.), *An Outline for Boys and Girls and their Parents* (Gollancz, 1932).

Orwell, George, *The Collected Essays, Journalism and Letters* (ed. Sonia Orwell and Ian Angus, 4 vols.; Secker & Warburg, 1968).

Partridge, Eric, *A Dictionary of Slang and Unconventional English* (Routledge, 1937; enlarged 1938; 5th edn. in two volumes, 1961).

Potter, Stephen, *The Muse in Chains: A Study in Education* (Cape, 1937).

Rickword, Edgell, *Essays and Opinions 1921-1931* (ed. Alan Young; Carcanet New Press, Manchester, 1974).

— *Literature in Society: Essays and Opinions (II), 1931-1978* (ed. Alan Young; Carcanet, Manchester 1978).

Riding, Laura, *The World and Ourselves* (= *Epilogue*, vol. 4; Chatto, 1938).

— ed., *Epilogue: A Critical Summary*, I (Autumn 1935, Seizin Press, Deya, Majorca; and Constable, London).

Smith, Stevie, *Me Again: The Uncollected Writings* (ed. Jack Barbera and William McBrien; Virago, 1981).

Sommerfield, John, *Behind the Scenes* (Nelson, 1934).

Spender, Stephen, *The Thirties and After: Poetry, Politics, People (1933-75)* (Collins, 1978).

Thomas, Dylan, *Quite Early One Morning: Broadcasts* (Dent, 1954).

— *Poet in the Making: The Notebooks of Dylan Thomas* (ed. Ralph Maud; Dent, 1968).

Warner, Rex, *English Public Schools* (Britain in Pictures: The British People in Pictures series, Collins, 1945).

— *The Cult of Power* (John Lane The Bodley Head, 1946).

Waugh, Evelyn, *Rossetti: His Life and Works*
(Duckworth, 1928 and 1931; reissued, with introd. by John Bryson, by Duckworth, 1975).

— *Edmund Campion* (Longmans, 1935).

— *A Little Order: A Selection from his Journalism* (ed. Donat Gallagher; Eyre Methuen, 1977).

— *The Essays, Articles and Reviews* (ed. Donat Gallagher; Methuen, 1983).

Woolf, Virginia, *A Letter to a Young Poet* (Hogarth Letters Series, no. 8; Hogarth, 1932).

— *Flush: A Biography* (Hogarth, 1933).

— *Three Guineas* (Hogarth, 1938).

— *Collected Essays* (4 vols., Hogarth, 1966-7).

13. Politics, Social Analysis, etc. from the Period

Bazeley, E. T., *Homer Lane and the Little Commonwealth* (Allen & Unwin, 1928).

Carr, E. H., *The Twenty Years' Crisis 1919-1939* (Macmillan, 1939).

Day Lewis, C., *We're Not Going to Do Nothing* (Left Review, 1936).

— 'A Reply' (in *Julian Bell, Essays, Poems and Letters*, ed. Quentin Bell; Hogarth, 1938).

Forster, E. M., 'The Freedom of the BBC' (*NS*, 4 Apr. 1931, 209-10).

Fox, Ralph, *The Class Struggle in Britain in the Epoch of Imperialism* (2 vols.; Martin Lawrence, 1932-1934).

— *Communism and a Changing Civilisation* (Bodley Head, 1935).

— (trans.), *Marxism and Modern Thought* by N. I. Bukharin *et al.* (Routledge, 1935).

Glover, Edward, *War, Sadism & Pacifism: Three Essays* (Allen & Unwin, 1933).

Golding, Louis, *The Jewish Problem* (Penguin Special, 1938).

Gorer, Geoffrey, *The Revolutionary Ideas of the Marquis de Sade* (Wishart, 1934). Foreword by J. B. S. Haldane.

Hannington, Wal, *The Problem of the Distressed Areas* (Left Book Club; Gollancz, 1937).

— *Black Coffins and the Unemployed* (special issue of *Fact* magazine; no. 26, May 1939).

— *Ten Lean Years: An Examination of the Record of the National Government in the Field of Unemployment* (Gollancz, 1940).

Heard, Gerald, *The Third Morality* (Cassell, 1937).

— *Man the Master* (Faber, 1942).

International Anti-Communist Entente, *The Red Network: The Communist International at Work* (Duckworth, 1939).

Jackson, T. A., *Dialectics: the Logic of Marxism* (Lawrence & Wishart, 1936).

Joad, C. E. M., *Guide to Modern Thought* (Faber, 1933).

Kemp, Harry; Riding, Laura, and others, *The Left Heresy in Literature and Life* (Methuen, 1939).

Langdon-Davies, John. *The Future of Nakedness* (Noel Douglas, 1929).

Levy, H., *A Philosophy for A Modern Man* (Left Book Club; Gollancz, 1938).

Lewis, Wyndham, *Paleface: or the Philosophy of the 'Melting-Pot'* (Chatto, 1929).

— *The Diabolical Principle and the Dithyrambic Spectator* (Chatto, 1931).

— *Hitler* (Chatto, 1931).

— *The Doom of Youth* (Chatto, 1932).

— *The Old Gang and the New Gang* (Desmond Harmsworth, 1932).

— 'Freedom that Destroys Herself' (*L*, XIII (8 May 1935), 793–4).

— *Left Wings Over Europe: or, How to Make a War About Nothing* (Cape, 1936).

— *Count Your Dead: They Are Alive! or A New War in the Making* (Lovat Dickson, 1937).

— ' "Left Wings" and the C3 Mind' (*The British Union Quarterly*, I, no. 1 (Jan.–Apr. 1937), 22–34).

— *The Mysterious Mr Bull* (Robert Hale, 1938).

— *The Hitler Cult* (Dent, 1939).

— *The Jews: Are They Human?* (Allen & Unwin, 1939).

Lunn, Arnold, *Revolutionary Socialism in Theory and Practice* (The Right Book Club, 1939).

Morton, A. L., *A People's History of England* (Gollancz, 1938; Lawrence & Wishart, 1945).

Nichols, Beverley, *Cry Havoc!* (Cape, 1933).

Ortega y Gasset, *The Revolt of the Masses* (Madrid, 1930, in Spanish; in English, Allen & Unwin, 1932).

Roberts, Michael, *The Modern Mind* (Faber, 1937).

Spender, Stephen, *Forward from Liberalism* (Left Book Club; Gollancz, 1937).

Spring Rice, Margery, *Working-Class Wives: Their Health and Conditions* (Penguin 1939; reissued, Virago, 1981).

Strachey, John, *The Coming Struggle for Power* (Gollancz, 1932; reissued with new preface, 1934).

White, Amber Blanco, *The New Propaganda* (Left Book Club; Gollancz, 1939).

Wintringham, Tom, *How to Reform the Army* (special issue of *Fact* magazine; no. 25, Apr. 1939).

14. Art, Architecture, Iconography, etc.

Bayer, Herbert; Gropius, Walter, and Gropius, Ise (edd.), *Bauhaus, 1919–1928* (Museum of Modern Art, New York, 1938; paperback reprint, 1975).

Blunt, Anthony, *Picasso's 'Guernica'* (Oxford University Press, New York, 1969).

Cooke, Catherine (ed.), *Russian Avant-Garde Art and Architecture* (AD Profile 47, *Architectural Design*, vol. 53, nos. 5–6, 1983).

Ede, H. S., *Savage Messiah* (Heinemann, 1931; new edition, with new authorial preface, Gordon Fraser, 1971).

Flint, Kate, 'Art and the Fascist Regime in Italy' (*The Oxford Art Journal*, III, no. 2 (Oct. 1980), 49–54).

Le Futurisme 1909–1916 (Musée Nationale d'Art Moderne, 19 Sept.–19 Nov. 1979; Editions des musées nationaux, Paris, 1973).

Gardiner, Stephen, *Le Corbusier* (Fontana Modern Masters, 1974).

Hillier, Bevis, *Art Deco of the 20s and 30s* (Studio Vista/Dutton Pictureback, 1968).

Hinz, Berthold, *Art in the Third Reich* (trans. Robert and Rita Kimber; Blackwell, Oxford, 1980).

Hobsbawm, Eric, 'Man and Woman in Socialist Iconography' (*History Workshop Journal*, no. 6 (Autumn 1978), 121–38).

Lewison, Jeremy (ed.), *Circle: Constructive Art in Britain 1934–40* (Kettle's Yard Gallery, Cambridge, 1982).

Lissitsky, El, *Russia: An Architecture for World Revolution* (trans. by Eric Dluhosch of *Russland, die Rekonstruktion der Architektur in der Sowjetunion* (Verlag Anton Schroll, Vienna, 1930); Lund Humphries, 1970).

Lucie-Smith, Edward, *Art of the 1930s: The Age of Anxiety* (Weidenfeld, 1985).

Mayakovsky: Twenty Years of Work (ed. David Elliott; Museum of Modern Art, Oxford, 1982).

Merker, Reinhard, *Die bildenden Künste im Nationalsozialismus: Kulturideologie Kulturpolitik Kulturproduktion* (DuMont Buchverlag, Köln, 1983).

Michel, Walter, *Wyndham Lewis: Paintings & Drawings* (with introductory essay by Hugh Kenner; Thames & Hudson, 1971).

Morris, Lynda, and Radford, Robert, *The Story of the AIA, Artists' International Association 1933–1953* (Museum of Modern Art, Oxford, 1983).

Neue Sachlichkeit and German Realism of the Twenties (Hayward Gallery, 11 Nov. 1978–14 Jan. 1979; Arts Council, 1978).

Penrose, Roland, *Scrap Book 1900–1981* (Thames & Hudson, 1981).

Read, Herbert, *Art Now: An Introduction to the Theory of Modern Painting and Sculpture* (Faber, 1933; revised, 1936, 1948, 1960).

— *Art and Industry: the Principles of Industrial Design* (Faber, 1934).

— *Art and Society* (Heinemann, 1937).

Richards, J. M., *An Introduction to Modern Architecture* (Pelican, 1940).

Rickaby, Tony, 'Artists' International' (*History Workshop Journal*, no. 6 (Autumn 1978), 154–68).

Alexander Rodchenko 1891–1956 (Museum of Modern Art, Oxford, 1979). Introd. by David Elliott.

Sembach, K.-J., *Into the Thirties: Style and Design 1927–1934* (trans. by Judith Filson of *Stil 1930* (Wasmuth, Tübingen, 1971); Thames & Hudson, 1972).

Tendenzen der Zwanziger Jahre. 15. Europäische Kunstausstellung unter den Auspizen des Europarates (Dietrich Reimer Verlag, Berlin, 1977).

Thirties: British Art and Design Before the War (Hayward Gallery, 25 Oct. 1979–13 Jan. 1980; Arts Council, 1979).

Tisdall, Caroline, and Bozzolla, Angelo, *Futurism* (Thames & Hudson, 1977).

Vorticism and Its Allies (Hayward Gallery 27 Mar.–2 June 1974; Arts Council, 1974). Introd. by Richard Cork.

Willett, John, *The New Sobriety: Art and Politics in the Weimar Period 1917–1933* (Thames & Hudson, 1979).

15. Film and Photography

Aldgate, Anthony, *Cinema and History: British Newsreels and the Spanish Civil War* (Scolar Press, 1979).

Auden, W. H., 'Poetry and Film' (*Janus*, May 1936, 11–12).

Barnouw, Erik, *Documentary: A History of the Non-Fiction Film* (Oxford University Press, 1974).

Barsam, Richard Meran, *Nonfiction Film: A Critical History* (Allen & Unwin, 1974).

Brandt, Bill, *The English at Home: 63 Photographs* (Batsford, 1936). Introd. by Raymond Mortimer.

— *A Night in London: Story of a London Night in 64 Photographs* (Country Life, London; Art et Métiers Graphiques, Paris; Charles Scribner, New York, 1938).

— *London in the Thirties* (Gordon Fraser, 1983).

Bill Brandt: A Retrospective Exhibition (The Royal Photographic Society National Centre of Photography, Octagon Gallery, 1981). Introd. by David Mellor.

The British Worker: Photographs of Working Life 1839–1939 (Arts Council, 1981). Introd. by David Englander.

Charles, Duncan (ed.), *Amateur Photography: A Practical Handbook for the Amateur* (George Newnes, 1935).

Delmar, Rosalind, *Joris Ivens: 50 Years of Film-Making* (British Film Institute, 1979).

Fielding, Raymond, *The March of Time, 1935–1951* (Oxford University Press, New York, 1978).

Greene, Graham, *The Pleasure Dome: The Collected Film Criticism 1935–40* (ed. John Russell Taylor; Secker & Warburg, 1972; Oxford Paperback, 1980).

Hardy, Forsyth, *John Grierson, A Documentary Biography* (Faber, 1979).

— (ed.), *Grierson on Documentary* (abr. edn.; Faber, 1979).

Hinton, David B., *The Films of Leni Riefenstahl* (Scarecrow Press, 1978).

Hodgkinson, Anthony W., and Sheratsky, Rodney E., *Humphrey Jennings—More Than a Maker of Films* (Clark University/University of New England Press, Hanover, USA, 1982).

Hogenkamp, Bert, *Workers' Newsreels in the 1920s and 1930s: Our History* (Pamphlet no. 68; Communist Party, n.d.).

Jennings, Mary-Lou, *Humphrey Jennings: Film-Maker, Painter, Poet* (British Film Institute/Riverside Studios, 1982).

Karfeld, Kurt Peter (ed.), *My Leica and I: Leica Amateurs Show Their Pictures* (Photokino-Verlag Hellmut Elsner, Berlin, 1937).

Korda, T. (ed.), *Photography Year Book 1935* (Cosmopolitan Press, 1935).

— *Photography Year Book: The International Annual of Camera Art*, II, *1936–37* (Cosmopolitan Press, 1936).

Kracauer, Siegfried, *From Caligari to Hitler: A Psychological History of the German Film* (Princeton University Press, New Jersey, 1947).

Leiser, Erwin, *Nazi Cinema* (Rowohlt Taschenbuch, Hamburg, 1968; trans. Gertrud Mander and David Wilson; Secker & Warburg, 1974).

Leyda, Jay, *Kino: A History of the Russian and Soviet Film* (3rd edn.; Allen & Unwin, 1983).

Lovell, Alan, and Hillier, Jim, *Studies in Documentary* (Cinema One Series, no. 21; Secker & Warburg/British Film Institute, 1972).

Macpherson, Don (ed.), *Traditions of Independence: British Cinema in the Thirties* (British Film Institute Publishing, 1980).

Man, Felix, *Felix H. Man, Pioneer of Photo-Journalism* (exhibition arranged by the National Book League and the Goethe Institut, 1977). Introd. by Tom Hopkinson.

Marris, Paul (ed.), *Paul Rotha: BFI Dossier Number 16* (British Film Institute, 1982).

Mellor, David (ed.), *Germany—The New Photography 1927–33: Documents and Essays* (Arts Council, 1978).

Modern British Photography 1919–39 (Arts Council, 1980). Introd. by David Mellor.

Photomontages of the Nazi Period: John Heartfield (Gordon Fraser, 1977).

Richards, Jeffrey, *The Age of the Dream Palace* (Routledge, 1985).

— and Aldgate, Anthony, *Best of British: Cinema and Society 1930–1970* (Basil Blackwell, Oxford, 1983).

Riefenstahl, Leni, *Kampf im Schnee und Eis* (Leipzig, 1933).

Rotha, Paul, *The Film Till Now: A Survey of the Cinema* (Cape, 1930).

— *Documentary Film* (Faber, 1936).

— *Documentary Diary: An Informal History of the British Documentary Film, 1928–1939* (Secker & Warburg, 1973).

Shudakov, Grigory (with Suslova, Olga, and Ukhtomskaya, Lilya), *Pionniers de la photographie russe sovietique* (Paris, 1983; trans. by Paul Keegan as *Pioneers of Soviet Photography*; Thames & Hudson, 1983).

Siepmann, Eckhard, *Montage: John Heartfield: Vom Club Dada zur Arbeiter-Illustrierten Zeitung: Dokumente—Analysen—Berichte* (Elefanten Press, Berlin, 1977).

Spender, Humphrey, *Britain in the Thirties: Photographs* (Lion & Unicorn Press, 1975). Introd. and commentary by Tom Harrisson.

— *Worktown: Photographs of Bolton and Blackpool Taken for Mass Observation 1937–8* (Gardner Centre Gallery, University of Sussex, Brighton, 1977).

Spender, Humphrey, *Worktown People: Photographs from Northern England 1937–8* (ed. Jeremy Mulford; Falling Walls Press, Bristol, 1982).

Stott, William, *Documentary Expression and Thirties America* (Oxford University Press, New York, 1973).

Syberberg, Hans Jürgen (ed.), *Fotografie der 30er Jarhe: eine Anthologie* (Schirmer/Mosel, Munich, 1977).

16. Surrealism

Ades, Dawn, *Dada and Surrealism* (Thames & Hudson, 1974).

— *Dada and Surrealism Reviewed* (introd. David Sylvester; Arts Council, 1978).

Audoin, Philippe, *Les Surréalistes* (Ecrivains de toujours, Seuil, Paris, 1973).

Benjamin, Walter, 'Surrealism' (in Benjamin, *One-Way Street and Other Writings*, trans. E. Jephcott and K. Shorter; New Left Books, 1979).

Breton, André, *What is Surrealism?* (trans. David Gascoyne; Criterion Miscellany, no. 43; Faber, 1936).

— *Manifestes du surréalisme* (Gallimard, 1973).

— *What is Surrealism? Selected Writings* (ed. and introd. Franklin Rosemont; Pluto Press, 1978).

Cardinal, Roger, and Short, Robert Stuart, *Surrealism: Permanent Revelation* (Studio Vista/Dutton Pictureback, 1970).

English and American Surrealist Poetry (ed. and introd. Edward B. Germain; Penguin, 1978).

Gascoyne, David, *A Short Survey of Surrealism* (Cobden-Sanderson, 1935).

Nadeau, Maurice, *The History of Surrealism* (trans. Richard Howard, introd. Roger Shattuck; Cape, 1968).

Ray, Paul C., *The Surrealist Movement in England* (Cornell University Press, Ithaca, New York, 1971).

Read, Herbert (ed. and introd.), *Surrealism* (Faber, 1936; reissued 1971).

17. Religion

Auden, W. H., 'Charles Williams: A Review Article' (*The Christian Century*, 73 (2 May 1956), 552–4).

Crossman, R. H. S. (ed.), *Oxford and the Groups* (Basil Blackwell, Oxford, 1934).

Dawson, Christopher, *Progress and Religion: An Historical Enquiry* (Sheed & Ward, 1938).

Eliot, T. S., 'The Modern Dilemma' series: 'Christianity and Communism' (*L*, 16 Mar. 1932, 382–3); 'The Search for Moral Sanction' *L*, 30 Mar. 1932, 445–6, 480); 'Building Up the Christian World' (*L*, 6 Apr. 1932, 501–2).

— 'A Lay Theologian' (*NS*, 9 Dec. 1939, 864, 866). Review of *The Descent of the Dove* by Charles Williams.

— *The Idea of a Christian Society* (Faber, 1939).

— 'The Significance of Charles Williams' (*L*, 19 Dec. 1946, 894–5).

— *Notes Towards the Definition of Culture* (Faber, 1948).

Foot, Stephen, *Life began Yesterday: A Book of Moral Re-Armament* (Heinemann, 1935).

King, Sidney A., *The Challenge of the Oxford Groups* (Allenson, 1933).

Leon, Philip, *The Philosophy of Courage, or The Oxford Group Way* (Allen & Unwin, 1939).

Lewis, C. S., *The Pilgrim's Regress: An Allegorical Apology for Christianity, Reason and Romanticism* (Dent, 1933).

Lewis, John; Polanyi, Karl, and Kitchin, Donald K. (edd.), *Christianity and the Social Revolution* (Gollancz, 1935).

Nichols, Beverley, *The Fool Hath Said* (Cape, 1936).

Noyes, Alfred, *The Unknown God* (Sheed & Ward, 1934).

The Oxford Group International Houseparty (Oxford, Aug. 1934).

Pike, James A. (ed.), *Modern Canterbury Pilgrims: The Story of Twenty-three Converts, and Why They Chose the Anglican Communion* (Mowbray, 1956).

Russell, A. J., *For Sinners Only* (Hodder & Stoughton, 1933).

Sayers, Dorothy L., *The Mind of the Maker* (Harcourt, Brace, New York, 1941).

Williams, Charles, *He Came Down From Heaven* (*I Believe: A Series of Personal Statements*, no. 5) (Heinemann, 1938).

— *Witchcraft* (Faber, 1941).

— *The Forgiveness of Sins* (Bles, 1942; Christian Challenge Series).

— *The Image of the City and Other Essays* (sel. and introd. Anne Ridler; Oxford University Press, London, 1958).

18. The Air

Davis, Capt. H. D., and Sprigg, Christopher St John, *Fly With Me: An Elementary Textbook on the Art of Piloting* (John Hamilton, 1932).

Fife, George Buchanan, *Lindbergh: The Lone Eagle: His Life and Achievements* (A. L. Burt, New York, 1927).

Garnett, David, *War in the Air: September 1939 to May 1941* (Chatto, 1941).

Grierson, John, *Through Russia By Air* (Foulis, 1933).

— *High Failure: Solo Along the Arctic Air Route* (W. Hodge, 1936).

Haldane, J. B. S., *ARP* (Left Book Club, Gollancz, 1938).

Langdon-Davies, John, *Air Raid* (Routledge, 1938).

Lewis, Cecil, *Sagittarius Rising* (Peter Davies, 1936).

Lindbergh, Charles A., *'We'* (Putnam's New York and London, 1927).

Monk, F. V., and Winter H. T., *Pilot and Plane* (Blackie, 1936).

Olley, Gordon P., *A Million Miles in the Air: Per-*

sonal *Experiences, Impressions, and Stories of Travel by Air* (Hodder & Stoughton, 1934).

Saint-Exupéry, Antoine de, *Night Flight* (= *Vol de Nuit* (Gallimard, 1931), trans. Curtis Cate, with acknowledgements to Stuart Gilbert's translation (Crosby Continental Editions, 1932); Heinemann Educational, 1971).

— *Vol de Nuit* (Gallimard, 1978). Préface d'André Gide.

— *Wind, Sand and Stars* (= *Terre des Hommes*, trans. Lewis Galantière; Heinemann, 1939).

Sprigg, Christopher St John, *The Airship: Its Design, History, Operation and Future* (Sampson Low, 1931).

— *British Airways* (A Nelson Discovery Book for Boys and Girls; Nelson, 1934).

— *Great Flights* (Nelsonian Library for Boys and Girls of All Ages and All Tastes; Nelson, 1935).

— *'Let's Learn to Fly!'* (Nelsonian Library for Boys and Girls of All Ages and All Tastes; Nelson, 1937).

Westerman, John F. C., *The Air Record Breakers* (Ward, Lock, 1937).

Wintringham, Tom, 'Empire Air Day 1936' (*LR*, II, no. 10 (July 1936), 502).

Yeates, V. M., *Winged Victory* (Cape, 1934; repr. with introd. by Henry Williamson, 1934; reissued with new pref. and a tribute by Henry Williamson, 1961).

19. The Face of Britain

Bertram, Anthony, *Design* (Penguin Special, 1938).

Betjeman, John, *Ghastly Good Taste, or, A Depressing Story of the Rise and Fall of English Architecture* (Chapman & Hall, 1932).

— *Devon. Shell Guide. Compiled with Many Illustrations and Information of Every Sort* (Architectural Press, 1936).

— *English Cities & Small Towns* (Britain in Pictures/The British People in Pictures series; Collins, 1943).

Blunden, Edmund, *The Face of England: In a Series of Occasional Sketches* (Longmans, 1932). Introd. by Sir John Squire.

— *English Villages* (Britain in Pictures series; Collins, 1941).

— *Cricket Country* (Collins, 1944).

Coe, Peter, and Reading, Malcolm, *Lubetkin and Tecton: Architecture and Social Commitment* (Arts Council, 1981).

Cohen-Portheim, Paul, *England the Unknown Isle* (trans. Alan Harris; Duckworth, 1930).

Collett, Antony, *The Changing Face of England* (Nisbet, 1926; reissued in The Traveller's Library, 1932).

Darley, Gillian, *Villages of Vision* (Architectural Press, 1975).

Dodd, Christopher, 'Lubetkin and the Battle of Peterlee' (*The Guardian*, 30 May 1981).

Ford, Charles Bradley (ed.), *The Legacy of England: An Illustrated Survey of the Works of*

Man in the English Country (The Pilgrims' Library series; Batsford, 1935).

Gaunt, William, *London Promenade* (The Studio, 1930).

Gibbs, Philip, *England Speaks: Being Talks with Road-Sweepers, Barbers, Statesmen, Lords and Ladies, Beggars, Farming Folk, Actors, Artists, Literary Gentlemen, Tramps, Down-and-Outs, Miners, Steel-Workers, Blacksmiths, The Man-in-the-Street, High-Brows, Low-Brows, And All Manner of Folk of Humble and Exalted Rank With A Panorama of the English Scene in this Year of Grace 1935* (Heinemann, 1935).

— *Ordeal in England (England Speaks Again)* (Heinemann, 1937; enlarged, 1938; reissued by The Right Book Club, 1938).

Grigson, Geoffrey, *Wild Flowers in Britain* (Britain in Pictures/The British People in Pictures series; Collins, 1944).

Hamilton, Cicely, *Modern England: As Seen By An Englishwoman* (Dent, 1938).

Hanley, James, *Grey Children: A Study in Humbug and Misery* (Methuen, 1937).

Hudson, Kenneth, *Food, Clothes and Shelter: Twentieth Century Industrial Archaeology* (John Baker, 1978).

Joad, C. E. M., *The Horrors of the Countryside* (Day to Day Pamphlets, no. 3; Hogarth, 1931).

— *A Charter for Ramblers, or the Future of the Countryside* (Hutchinson, 1934).

Keun, Odette, *I Discover the English* (John Lane The Bodley Head, 1934).

Lymington, Viscount, *Famine in England* (The Right Book Club, 1938).

MacNeice, Louis, *I Crossed the Minch* (Longmans, 1938).

Mais, S. P. B., *This Unknown Island* (Putnam, 1932).

Massingham, H. J. (ed.), *The English Countryside: A Survey of its Chief Features* (The Pilgrim's Library; Batsford, 1939).

Morton, H. V., *In Search of England* (Methuen, 1927).

— *In Search of Scotland* (Methuen, 1929).

— *In Search of Wales* (Methuen, 1932).

— *In Scotland Again* (Methuen, 1933).

— *I saw Two Englands: The Record of a Journey Before the War, and After the Outbreak of War, in the Year 1939* (Methuen, 1942).

Muir, Edwin, *Scottish Journey* (Heinemann/Gollancz, 1935; reissued with introd. by T. C. Smout; Flamingo paperbacks, 1985).

Nichols, Beverley, *Down the Garden Path* (Cape, 1932).

— *A Thatched Cottage* (Cape, 1933).

Oliver, Paul; Davis, Ian, and Bentley, Ian, *Dunroamin: The Suburban Semi and its Enemies* (Barrie and Jenkins, 1981).

Orwell, George, *The Road to Wigan Pier* (with Foreword by Victor Gollancz, and photographs, Gollancz, Left Book Club, Mar. 1937; reissued, First Part only, Left Book Club, May 1937).

Priestley, J. B., *English Journey: Being A Rambling*

But Truthful Account of What One Man Saw and Heard and Felt and Thought During a Journey Through England During the Autumn of the Year 1933 (Heinemann, 1934).

Prochaska, Alice, *London in the Thirties* (London Museum, 1973).

Renier, G. J., *The English: Are They Human?* (Williams and Norgate, 1931).

Roberts, Cecil, *Pilgrim Cottage* (Hodder & Stoughton, June 1933).

— *Gone Rustic* (Hodder & Stoughton, Apr. 1934).

— *Gone Rambling* (Hodder & Stoughton, 1935).

Sharp, Thomas, *Town and Countryside: Some Aspects of Urban and Rural Development* (Oxford University Press, 1933).

— *English Panorama* (Dent, 1936).

Thorns, David, *Suburbia* (MacGibbon & Kee, 1972).

Ward, Mary, and Ward, Neville, *Home in the Twenties and Thirties* (Ian Allan, 1978).

Williams-Ellis, Clough, *England and the Octopus* (Bles, 1928).

— (ed.), *Britain and the Beast* (Dent, 1937).

Williamson, Henry, *The Village Book* (Cape, 1930).

— *The Story of a Norfolk Farm* (Faber, 1941).

— (ed.), *Nature in Britain: An Illustrated Survey* (The Pilgrims' Library series; Batsford, 1936).

20. Mass-Observation

Calder, Angus, and Sheridan, Dorothy (edd.), *Speak for Yourself: A Mass-Observation Anthology 1937-1939* (Cape, 1984; Oxford Paperback, 1985).

Dudman, George, and Terry, Patrick, *Challenge to Tom Harrisson* (Oxford, n.d. [1938]).

Harrisson, Tom, 'Mass-Opposition and Tom Harrisson' (*Light and Dark*, II, no. 3 (Special Tom Harrisson Number, Feb. 1938), 8–15).

— 'Mass-Observation: A Reply' (*NS*, 12 Mar. 1938, 409–10).

Jennings, Humphrey, and Madge, Charles (edd.), *May the Twelfth: Mass-Observation Day-Surveys 1937*, by over two hundred observers (Faber, 1937).

Madge, Charles, and Harrisson, Tom, *Mass-Observation* (Mass-Observation Series, no. 1, foreword Julian Huxley; Frederick Muller, 1937).

— *Britain by Mass-Observation* (Penguin Special, Jan. 1939).

— (edd.), *First Years' Work 1937-8*, by Mass-Observation (Lindsay Drummond, Mar. 1938). With an essay, 'A Nationwide Intelligence Service', by Bronislaw Malinowski.

Stonier, G. W., 'Mass-Observation and Literature' (*NS*, 26 Feb. 1938, 326–7).

21. Abroad

Ackerley, J. R., *Hindoo Holiday: An Indian Journal* (Chatto, 1932).

Auden, W. H. (ed. and introd.), *The American Scene, together with Three Essays from 'Portraits of Places'* by Henry James (Charles Scribner's Sons, New York, 1946).

— and Isherwood, Christopher, *Journey to a War* (Faber, Mar. 1939).

— and MacNeice, Louis, *Letters from Iceland* (Faber, 1937).

Belfrage, Cedric, *Away From It All: An Escapologist's Notebook* (Gollancz, 1936).

— *Promised Land: Notes For a History* (Left Book Club, Gollancz, 1938).

Benjamin, Walter, 'Moscow' (in Benjamin, *One-Way Street and Other Writings*, trans. E. Jephcott and K. Shorter; New Left Books, 1979).

Brontman, L., *On Top of the World: The Soviet Expedition to the North Pole 1937* (ed. Academician O. J. Schmidt, Hero of the Soviet Union; Left Book Club, Gollancz, 1938).

Brown, Alec (trans.), *The Voyage of the Chelyuskin*, By Members of the Expedition (Chatto, 1935).

Byron, Robert, *First Russia Then Tibet* (Macmillan, 1933).

— *The Road to Oxiana* (Macmillan, 1937).

Citrine, Sir Walter, *I Search For Truth in Russia* (Routledge, 1938).

Dos Passos, John, *Journeys Between Wars* (Constable, 1938).

Dower, K. C. Gandar, *Amateur Adventure* (Rich & Cowan, 1934; reissued in Penguin Travel & Adventure series, 1939).

Farson, Negley, *Behind God's Back* (Gollancz, 1940).

Fermor, Patrick Leigh, *A Time of Gifts* (Murray, 1977).

Fleming, Peter, *Brazilian Adventure* (Cape, 1933).

— *One's Company—A Journey to China* (Cape, 1934).

— *News from Tartary: A Journey from Peking to Kashmir* (Cape, 1936).

— 'A Tent in Tibet' (*Times Literary Supplement*, 15 July 1939 (Summer Reading Supplement), v–vi).

Frank, Waldo, 'The Body of Lenin' (*Adelphi*, July 1932, 683–6).

Gibbs, Philip, *European Journey* (Heinemann/Gollancz, 1934).

— *Across the Frontiers* (Michael Joseph, 1938).

Gide, André, *Back from the USSR* (= *Retour de l'URSS* (Gallimard, Paris, 1936), trans. Dorothy Bussy; Secker & Warburg, 1937).

— *Afterthoughts: A Sequel to Back from the USSR* (= *Retouches à mon retour de l'URSS* (Gallimard, Paris, 1937), trans. Dorothy Bussy; Secker & Warburg, 1938).

Golding, Louis, *In the Steps of Moses the Lawgiver* (Rich & Cowan, 1937).

Greene, Barbara, *Land Benighted* (Bles, 1938; reissued as *Too Late to Turn Back: Barbara and Graham Greene in Liberia*, with introd. by Paul Theroux, by Settle Bendall, 1981).

Greene, Graham, *Journey Without Maps* (Heinemann, 1936).

— *The Lawless Roads* (Heinemann, 1939).

Gunther, John, *Inside Europe* (Hamish Hamilton, 1936).

Hamilton, Cicely, *Modern Germanies: As Seen By an Englishwoman* (Dent, 1931).

Harrisson, Tom, *Savage Civilisation* (Left Book Club, Gollancz, 1937).

Hindus, Maurice, *Broken Earth* (Cape, 1926).

— *Humanity Uprooted* (Cape, 1929).

— *Red Bread* (Cape, 1931).

— *The Great Offensive* (Gollancz, 1933).

— *Under Moscow Skies* (Gollancz, 1936).

Horrabin, J. F., *An Atlas of Current Affairs* (Gollancz, 1934).

Huxley, Aldous, *Beyond the Mexique Bay* (Chatto, 1934).

Johnson, Hewlett, *The Socialist Sixth of the World* (Gollancz, 1939).

Lambert, R. S. (ed.), *Grand Tour: A Journey in the Tracks of the Age of Aristocracy* (Faber, 1935).

Lehmann, John, *Prometheus and the Bolsheviks* (Cresset Press, 1937).

— *Down River: A Danubian Study* (Cresset Press, 1939).

Lewis, Wyndham, *Filibusters in Barbary (Record of a Visit to the Sous)* (Grayson and Grayson, 1932).

Macpherson, Aimee Semple, *I View the World* (Robert Hale, 1937).

Maillart, Ella, K., *Forbidden Journey: From Peking to Kashmir*, trans. Thomas McGreevy; Heinemann, 1937).

Mitchell, J. Leslie, *Persian Dawns, Egyptian Nights* (Jarrolds, 1932).

— *The Conquest of the Maya* (Jarrolds, 1934).

— and Gibbon, Lewis Grassic, *Nine Against the Unknown: A Record of Geographical Exploration* (Jarrolds, 1934).

Morton, H. V., *In the Steps of the Master* (Rich & Cowan, 1934).

— *In the Steps of St Paul* (Rich & Cowan, 1936).

Muggeridge, Malcolm, *Winter in Moscow* (Eyre & Spottiswoode, 1934).

Orwell, George. *Down and Out in Paris and London* (Gollancz, 1933).

Powys, John Cowper, 'Elusive America' (*The Powys Review*, no. 6, II. ii (Winter–Spring, 1979–80), 50–3; from *American Mercury*, VIII, Mar. 1927).

— 'Farewell to America' (*The Powys Review*, no. 6, II. ii (Winter–Spring, 1979–80), 54–63; from *Scribner's Magazine*, XCVII, Apr. 1935).

Pritt, D. N., *Light on Moscow: Soviet Policy Analysed* (Penguin Special, 1939).

Reed, Douglas, *Insanity Fair* (Cape, 1938).

Romm, Michael, *The Ascent of Mount Stalin* (trans. Alec Brown; Lawrence & Wishart, 1936).

Sitwell, Osbert, *Escape with Me! An Oriental Sketch-Book* (Macmillan, 1939).

Sloan, Pat, *Soviet Democracy* (Left Book Club, Gollancz, 1937).

— *Russia—Friend or Foe?* (Frederick Muller, 1939).

Smith, Howard, *Last Train from Berlin* (Cresset Press, 1942).

Spender, Michael, 'Expeditions Past and Present' (in 'Man Against Everest'; *L*, XIX (9 June 1938), 1213–16).

Wall, Bernard, *European Note-Book* (Sheed & Ward, 1939).

Waugh, Evelyn, *Labels: A Mediterranean Journey* (Duckworth, 1930).

— *Remote People* (Duckworth, 1931).

— *Ninety-Two Days* (Duckworth, 1934).

— *Waugh in Abyssinia* (Longmans, 1936).

— *Robbery Under Law: The Mexican Object-Lesson* (Chapman & Hall, 1939).

— *When the Going Was Good* (Duckworth, 1946). The author's selection from his travel books with 1945 Preface.

Webb, Sidney and Beatrice, *Soviet Communism: A New Civilisation?* (Longmans, 1935; reissued as *Soviet Communism: A New Civilisation*, 1937).

Yeats-Brown F., *European Jungle* (Eyre & Spottiswoode, 1939).

22. Spain

Acier, Marcel (ed.), *From Spanish Trenches: Recent Letters from Spain* (Cresset Press, 1937).

Atholl, Katharine, Duchess of, *Searchlight on Spain* (Penguin Special, June 1938).

Authors Takes Sides on the Spanish War (*LR*, 1937).

Belloc, Hilaire, 'The Salvation of Spain' (in Belloc, *Places*; Cassell, 1942).

Bernanos, George, *A Diary of My Times* (= *Les Grands Cimetières Sous la Lune* (Paris, 1938), trans. Pamela Morris; Boriswood, 1938).

Bessie, Alvah, *Men in Battle. A Story of Americans in Spain* (Charles Scribner's Sons, New York, 1939).

— (ed.), *The Heart of Spain: Anthology of Fiction, Non-Fiction and Poetry* (Veterans of the Abraham Lincoln Brigade, New York, 1952).

Bolloten, Burnett, *The Spanish Revolution: The Left and the Struggle for Power during the Civil War* (University of North Carolina Press, Chapel Hill, 1979).

The Book of the XV Brigade: Records of British, American, Canadian, and Irish Volunteers in the XV International Brigade in Spain 1936–38 (XV Brigade Commissariat of War, Madrid, 1938; republished Newcastle upon Tyne, 1975).

Borkenau, Franz, *The Spanish Cockpit: An Eye-Witness Account of the Political and Social Conflicts of the Spanish Civil War* (Faber, 1937; republished, with foreword by Gerald Brenan, by University of Michigan, 1963).

Brenan, Gerald, *The Spanish Labyrinth: An Account of the Social and Political Background of the Civil War* (2nd edn.; Cambridge University Press, 1950).

Brome, Vincent, *The International Brigades: Spain 1936–1939* (Heinemann, 1965).

Broué, Pierre, and Témime, Emile, *The Revolution and the Civil War in Spain* (trans. Tony White; Faber, 1972).

Browne, Harry, *Spain's Civil War* (Longmans, 1983).

Calmer, Alan (ed.), *Salud! Poems, Stories and Sketches of Spain by American Writers: A Literary Pamphlet* (International Publishers, New York, 1938).

Carr, Raymond, *The Civil War in Spain* (2nd edn. of *The Spanish Tragedy: the Civil War in Spain* (1977);'Weidenfeld, 1986).

— (ed.), *The Republic and the Civil War in Spain* (Macmillan, St Martin's Press, 1971).

— and Aizpurna, Juan Pablo Fusi, *Spain: Dictatorship to Democracy* (2nd edn.; Allen & Unwin, 1981).

Cook, Judith, *Apprentices of Freedom* (Quartet, 1979).

Corkill, D., and Rawnsley, S., *The Road to Spain: Anti-Fascists at War 1936-1939* (Borderline Press, Dunfermline, 1981).

Cortoda, James W., *Historical Dictionary of the Spanish Civil War, 1936-1939* (Greenwood Press, Westport, Connecticut, 1983).

Crome, Len, 'Walter (1897-1947): A Soldier in Spain' (*History Workshop Journal*, no. 9 (Spring 1980), 116-28).

Cunningham, Valentine, 'Saville's Row with *The Penguin Book of Spanish Civil War Verse*' (*The Socialist Register 1982* (Merlin Press, 1982), 269-83).

— (ed.), *The Penguin Book of Spanish Civil War Verse* (Penguin, 1980).

— (ed.), *Spanish Front: Writers on the Civil War* (Oxford University Press, 1986).

Edwards, Jill, *The British Government and the Spanish Civil War, 1936-1939* (Macmillan, 1979). Foreword by Hugh Thomas.

Felstead, Richard, *No Other Way: Jack Russia and the Spanish Civil War: A Biography* (Alun Books, Port Talbot, West Glamorgan, 1981).

Ford, Hugh, *A Poets' War: British Poets and the Spanish Civil War* (University of Pennsylvania Press, Philadelphia, and Oxford University Press, 1965).

Foss, William, and Gerahty, Cecil, *The Spanish Arena* (John Gifford, 1938).

Fraser, Ronald, *Blood of Spain: The Experience of Civil War 1936-1939* (Allen Lane, 1979).

Gerahty, Cecil, *The Road to Spain* (Hutchinson, 1937).

Gibson, Ian, *The Death of Lorca* (W. H. Allen, 1973).

Green, Nan, and Elliott, A. M., *Spain Against Fascism 1936-39: Our History*, Pamphlet no. 67 (Communist Party, n.d.).

Guest, Carmel Haden (ed.), *David Guest: A Scientist Fights for Freedom (1911-1938): A Memoir* (Lawrence & Wishart, 1939). Introd. by Harry Pollitt.

Gurney, Jason, *Crusade in Spain* (Faber, 1974).

Hemingway, Ernest, *The Spanish War* (special issue of *Fact* magazine (no. 16, July 1938)).

— 'The Heat and the Cold: Remembering Turning the Spanish Earth' (*Verve* (Paris, Spring 1938)).

Jackson, Gabriel, *The Spanish Republic and the Civil War 1931-1939* (Princeton University Press, New Jersey, 1965).

— *A Concise History of the Spanish Civil War* (Thames & Hudson, 1974).

Kantorowicz, Alfred, *Spanisches Kriegstagebuch: mit einem neuen Vorwort des Verfassers und einem Anhang bisher unveröffentlicher Dokumente und Briefe von Theodor Balk, Lion Feuchtwanger, Ernest Hemingway, Hans Kahle, Egon Erwin Kisch, Erika Mann, Wilhelm Pieck, Gustav Regler, Ludwig Renn* (Hamburg, 1979; Fischer Taschenbuch, Frankfurt am Main, 1982).

Koestler, Arthur, *Spanish Testament* (Gollancz, 1937).

Langdon-Davies, John 'ARP: Bombs over Barcelona' (*L*, 14 July 1938, 59-61, 93).

Last, Jef, *The Spanish Tragedy*, (trans. David Hadett; Routledge, 1939).

Lehmann, J., Jackson, T. A., and Day Lewis, C. (edd.), *Ralph Fox: A Writer in Arms* (Lawrence & Wishart, 1937).

Morrow, Felix, *Revolution and Counter-Revolution in Spain* (1936, 1938; Pathfinder Press, New York, 1974).

Muste, John M., *Say That We Saw Spain Die: Literary Consequences of the Spanish Civil War* (University of Washington Press, Seattle, 1966).

O'Brien, Kate, *Farewell Spain* (Heinemann, 1937; reissued, with new introd. by Mary O'Neill, by Virago, 1985).

Orwell, George, *Homage to Catalonia* (Secker & Warburg, 1938).

— *Homage to Catalonia, and Looking Back on the Spanish Civil War* (Penguin, 1966).

'Pitcairn, Frank' (Claud Cockburn), *Reporter in Spain* (Lawrence & Wishart, 1936). Introd. ('I Add My Witness') by Ralph Bates.

Powys, John Cowper, 'The Real and the Ideal, reprinted from *Spain and the World*, Supplement, May 1938' (*The Powys Review*, no. 3 (Summer 1978), 78-9).

Preston, Paul, *The Coming of the Spanish Civil War: Reform, Reaction and Revolution in the Second Republic 1931-1936* (Macmillan, 1978).

— (ed.), *Revolution and War in Spain 1931-1939* (Methuen, 1984).

Puccini, Dario (ed.), *Le romancero de la résistance espagnole: anthologie poétique bilingue* (Maspéro, Paris, 1976).

Romilly, Esmond, *Boadilla* (Hamish Hamilton, 1937; with introd. and notes by Hugh Thomas, Macdonald, 1971).

Rust, William, *Britons in Spain: A History of the British Battalion of the XVth International Brigade* (Lawrence & Wishart, 1939).

Sommerfield, John, *Volunteer in Spain* (Lawrence & Wishart, 1937).

Southworth, Herbert, 'The Divisions of the Left' (*Times Literary Supplement*, 9 June 1978, 649-50).

Spender, Stephen, and Lehmann, John (edd.), *Poems for Spain* (Hogarth, 1939).

Sperber, Murray (ed.), *And I Remember Spain: A Spanish Civil War Anthology* (Hart-Davis/MacGibbon, 1974).

Stead, C. K., 'Auden's "Spain" ' (*London Magazine*, NS 7, no. 12 (Mar. 1968), 41–54).

Steer, G. L., *The Tree of Gernika: A Field Study of Modern War* (Hodder & Stoughton, 1938).

Thomas, Hugh, *The Spanish Civil War* (Eyre & Spottiswoode, 1961; revised, Penguin, 1965).

Timmermans, Rudolf, *Heroes of the Alcazar* (Eyre & Spottiswoode, 1937). Introd. by F. Yeats-Brown.

Tisa, John (ed.), *The Palette and the Flame: Posters of the Spanish Civil War* (Collet, 1980).

Tomalin, Miles, 'Memories of the Spanish War' (*NS* 31 Oct. 1975, 542).

Toynbee, Philip (ed.), *The Distant Drum: Reflections on the Spanish Civil War* (Sidgwick & Jackson, 1976).

Trotsky, Leon, *The Spanish Revolution (1931–39)* (introd. Les Evans; Pathfinder Press, New York, 1973).

Weintraub, Stanley, *The Last Great Cause: The Intellectuals and the Spanish Civil War* (W. H. Allen, 1968).

Wintringham, Tom, *English Captain* (Faber, 1937).

Worsley, T. C., *Behind the Battle* (Robert Hale, 1939).

Weil, Simone, 'Lettre à George Bernanos (1938?)', (*Ecrits Historiques et Politiques* (Gallimard, Paris, 1960), 220–4).

23. Thirties-devoted Issues of Periodicals

Gulliver, deutsch-englische Jahrbücher, 4: *Die roten 30er Jahre* (Argument-Verlag, Berlin, 1978).

Renaissance & Modern Studies, 20 (1976), 'The Thirties—A Special Number'.

The Review, no. 11 (n.d.), 'The Thirties—A Special Number'.

Stamp, Gavin, *Britain in the Thirties* (AD Profile 24), *Architectural Design*, Double Issue (n.d.).

The Thirties Society Journal, no. 1, 1981– .

24. Period Studies (Literary)

Barnes, James J., and Barnes, Patience P., *Hitler's Mein Kampf in Britain and America: A Publishing History 1930–39* (Cambridge University Press, 1980).

Bergonzi, Bernard, *Reading the Thirties: Texts and Contexts* (Macmillan, 1978).

Carter, Ronald (ed.), *Thirties Poets: 'The Auden Group': A Casebook* (Macmillan, 1984).

Clark, Jon; Heinemann, Margot; Margolies, David and Snee, Carole (edd.), *Culture and Crisis in Britain in the Thirties* (Lawrence & Wishart, 1979).

Constantine, Stephen, ' "Love on the Dole" and its Reception in the 1930s' (*Literature and History*, 8: 2 (Autumn 1982), 232–47).

Croft, Andy, 'The Birmingham Group: Literary

Life Between the Wars' (*London Magazine*, NS 23, no. 3 (June 1983), 13–22).

Enkemann, Jürgen and Klaus, H. Gustav, ' "Let the people speak for themselves". Zur britischen Dokumentaristik der dreissiger und vierziger Jahre' (I, *Gulliver*, 4 (Berlin, 1978), 90–106; II, *Gulliver*, 6 (Berlin, 1979), 145–72).

Gloversmith, Frank (ed.), *Class, Culture and Social Change: A New View of the 1930s* (Harvester, Brighton, 1980).

Hynes, Samuel, *The Auden Generation: Literature and Politics in England in the 1930s* (Bodley Head, 1976).

Johnson, Roy, 'The Proletarian Novel' (*Literature and History*, no. 2 (Oct. 1975), 84–95).

Johnstone, Richard, *The Will to Believe: Novelists of the Nineteen-Thirties* (Oxford University Press, 1982).

—— 'Travelling in the Thirties' (*London Magazine*, NS 20, nos. 5–6 (Aug.–Sept. 1980), 90–6).

Klaus, H. Gustav, 'Politische Lyrik im "Thirties Movement" ' (I, *Gulliver*, 4 (Berlin, 1978), 29–54; II, *Gulliver*, 5 (Berlin, 1979), 105–25).

—— *Caudwell im Kontext: zu einigen repräsentativen Literaturformen der dreissiger Jahre* (Peter Lang, Frankfurt am Main, 1978).

Loreman, Jack, 'Workers' Theatre: Personal Recollections of Political Theatre in Greenwich during the 1920s and 1930s' (*Red Letters*, no. 13, Spring 1982, 41–6).

Lucas, John, ed., *The 1930s: A Challenge to Orthodoxy* (Harvester, Hassocks, 1978).

McMillan, Dougald, *Transition 1927–38: the History of a Literary Era* (Calder and Boyars, 1975).

Maxwell, D. E. S., *Poets of the Thirties* (Routledge, 1969).

Mitchell, Donald, *Britten and Auden in the Thirties: The Year 1936* (Faber, 1981).

Mulhern, Francis, *The Moment of 'Scrutiny'* (New Left Books, 1979).

Poole, Philip; Clark, Jon, and Margolies, David, 'The Workers' Theatre Movement' (*Red Letters*, no. 10, 2–10).

Smith, Timothy D'Arch, 'R. A. Caton and the Fortune Press', (*Times Literary Supplement*, 12 Sept. 1980, 1003–5).

Spender, Stephen, 'Creating in the Mind A Map' (*NS*, 2 July 1976, 19–20).

Symons, Julian, *The Thirties: A Dream Revolved* (Cresset Press, 1960; rev. edn., Faber, 1975).

—— *Notes from Another Country* (London Magazine Editions, 1972).

—— *The Angry 30s: Picture File* (Eyre Methuen, 1976).

Tolley, A. T., *The Poetry of the Thirties* (Gollancz, 1975).

Waterman, Ray, 'Proltet: The Yiddish-Speaking Group of the Workers' Theatre Movement' (*History Workshop Journal*, no. 5 (Spring 1978), 174–8).

Watson, Don, 'Busmen: Documentary and British political theatre in the 1930s' (*Media, Culture and Society*, 3, no. 4 (Oct. 1981), 339–50).

Wilson, Edmund, *The Thirties* (ed. and introd. Leon Edel; Macmillan, 1980).

25. Lives and Memoirs

Ackroyd, Peter, *T. S. Eliot* (Hamish Hamilton, 1984).

Aldington, Richard, *Lawrence of Arabia: A Biographical Inquiry* (Collins, 1955).

Alexander, Peter, *Roy Campbell: A Critical Biography* (Oxford University Press, 1982).

Angier, Carole, *Jean Rhys* (Lives of Modern Women series, Penguin, 1985).

Auden, W. H., *Louis MacNeice: A Memorial Address* (Faber, 1963).

Bair, Deirdre, *Samuel Beckett: A Biography* (Cape, 1978).

Bedford, Sybille, *Aldous Huxley: A Biography* (2 vols.; Chatto/Collins, 1973, 1974).

Bell, Quentin, *Virginia Woolf: A Biography* (I, *Virginia Stephen 1882–1912*; Hogarth, June 1972; II, *Mrs Woolf 1912–1941*; Hogarth, Oct. 1972).

Betjeman, John, 'Louis MacNeice and Bernard Spencer' (*London Magazine*, NS 3, no. 9 (Dec. 1963), 62–4).

Bowker, Gordon (ed.), *Malcolm Lowry Remembered* (Ariel Books, BBC, 1985).

Brittain, Vera, *Testament of Friendship: The Story of Winifred Holtby* (Macmillan, 1940).

Brown, Terence, and Reid, Alec, *Time Was Away: The World of Louis MacNeice* (Dolmen Press, 1974).

Carpenter, Humphrey, *J. R. R. Tolkien: A Biography* (Allen & Unwin, 1977).

— *W. H. Auden: A Biography* (Allen & Unwin, 1981).

Cavaliero, Glen, 'Sylvia Townsend Warner: An Appreciation' (*The Powys Review*, no. 5; II, i (Summer 1979), 6–12).

Chesterton, A. K., *Oswald Mosley: Portrait of a Leader* (Action Press, 1937).

Chitty, Susan, *Now to my Mother: A Very Personal Memoir of Antonia White* (Weidenfeld, 1985).

Cohen, Stephen F., *Bukharin and the Bolshevik Revolution: A Political Biography, 1888–1938* (New York, 1973; Wildwood House, 1974).

Coppard, Audrey, and Crick, Bernard, *Orwell Remembered* (Ariel Books, BBC, 1984).

Crick, Bernard, *George Orwell: A Life* (Secker & Warburg, 1980; revised, Secker & Warburg, 1981).

Davies, Margaret Llewellyn (ed.), *Life As We Have Known It* (by Cooperative Working Women, with introd. letter by Virginia Woolf; Hogarth, 1931; reissued, ed. Anna Davin; Virago, 1977).

Day, Douglas, *Malcolm Lowry: A Biography* (Oxford University Press, 1973).

Day-Lewis, Sean, *C. Day-Lewis: An English Literary Life* (Weidenfeld, 1980).

Durrell, Lawrence, 'Bernard Spencer' (*London Magazine*, NS 3, no. 10 (Jan. 1964), 42–7).

Eason, T. W., and Hamilton, R. (edd.), *A Portrait of Michael Roberts* (College of St Mark and St John, 1949).

Ellis, Peter Beresford, and Williams, Piers, *By Jove, Biggles! The Life of Captain W. E. Johns* (W. H. Allen, Comet Books, 1985).

Ewart, Gavin, 'Reputations—VIII: Cyril Connolly' (*London Magazine*, NS 3, no. 9 (Dec. 1963), 35–50).

Fernbach, David, 'Wintringham, Thomas Henry (Tom) (1898–1949)' (*Dictionary of Labour Biography*, VII, ed. Joyce M. Bellamy and John Saville (Macmillan, 1984), 255–64).

Finney, Brian, *Christopher Isherwood: A Critical Biography* (Faber, 1979).

Fitzgibbon, Constantine, *The Life of Dylan Thomas* (Dent, 1965).

Ford, Hugh (ed.), *Nancy Cunard. Brave Poet, Indomitable Rebel, 1896–1965* (Chilton Book Co., Philadelphia, 1968).

Forster, E. M., *Goldsworthy Lowes Dickinson and Related Writings* (Abinger Edition, vol. 13, ed. Oliver Stallybrass; Edward Arnold, 1973). Foreword by W. H. Auden.

Fox, Ralph, *Lenin: A Biography* (Gollancz, 1933).

— 'Lawrence the 20th-c. Hero' (*LR*, I, no. 10 (July 1935), 391–6).

Fuller, Roy, 'Coming into His Own' [on Connolly] (*London Magazine*, NS 13, no. 3 (Aug.–Sept. 1973), 22–6).

Furbank, P. N., *E. M. Forster: A Life*, (I, *The Growth of the Novelist (1879–1914)*; Secker & Warburg, 1977; II, *Polycrates' Ring (1914–1970)*; Secker & Warburg, 1978).

Garnett, David, *Great Friends: Portraits of Seventeen Writers* (Macmillan, 1979).

Glendinning, Victoria, *Elizabeth Bowen: Portrait of a Writer* (Weidenfeld, 1977).

Goodway, David, 'Charles Lahr: Anarchist, Bookseller, Publisher' (*London Magazine* (June–July 1977), 41–55).

Goldsmith, Maurice, *Sage: A Life of J. D. Bernal* (Hutchinson, 1980).

Green, Roger Lancelyn, and Hooper, Walter, *C. S. Lewis: A Biography* (Collins, 1974).

Grigson, Geoffrey, *Recollections: Mainly of Writers and Artists* (Chatto/Hogarth, 1984).

Grossman, Carl M., and Grossman, Sylvia, *The Wild Analyst: the Life and Work of Georg Groddeck* (Barrie & Rockcliff, 1965).

Gutteridge, Bernard, ' "You can't abdicate and eat it": A Note on Roger Roughton' (*London Magazine*, NS 23, no. 11 (Feb. 1984), 76–9).

Hart-Davis, Duff, *Peter Fleming: A Biography* (Cape, 1974).

Hitchman, Janet, *Such A Strange Lady: An Introduction to Dorothy L. Sayers (1893–1957)* (New English Library, 1975).

Hopkins, Kenneth, *The Powys Brothers: A Biographical Appreciation* (Phoenix House, 1967).

Hurd, Michael, *The Ordeal of Ivor Gurney* (Oxford University Press, 1978).

Isherwood, Christopher, *Kathleen and Frank* (Methuen, 1971).

Kennedy, Richard, *A Boy at the Hogarth Press* (The Whittington Press and Heinemann Educational, 1972).

Kirkup, James, 'Dear Old Joe' [on Joe Ackerley] (*London Magazine*, NS 15, no. 1 (Apr.–May 1975), 19–37).

Lancaster, Marie-Jacqueline (ed.), *Brian Howard: Portrait of a Failure* (Anthony Blond, 1968).

Lehmann, John, 'Friend of Promise' [on Connolly] (*Encounter*, May 1975, 77–9).

Lytton, Victor, Earl of, *Antony (Viscount Knebworth): A Record of Youth* (Peter Davies, 1935). Foreword by J. M. Barrie.

Mack, John E., *A Prince of Our Disorder: the Life of T. E. Lawrence* (Weidenfeld, 1976).

Mahon, John, *Harry Pollitt: A Biography* (Lawrence & Wishart, 1976).

Mann, Golo, 'W. H. Auden, A Memoir' (*Encounter*, Jan. 1974, 7–11).

Meyers, Jeffrey, *The Enemy: A Biography of Wyndham Lewis* (Routledge, 1980).

Mortimer, Raymond, 'Survivor of a Vanishing Species' [on Connolly], (*London Magazine*, NS 13, no. 3 (Aug.–Sept. 1973), 18–21).

Nicolson, Nigel, *Portrait of a Marriage* (Weidenfeld, 1973).

Noble, John Russell (ed.), *Recollections of Virginia Woolf* (Peter Owen, 1972).

Osborne, Charles, *W. H. Auden: the Life of a Poet* (Harcourt Brace Jovanovich, New York and London, 1979).

Pryce-Jones, David, *Unity Mitford: A Quest* (Weidenfeld, 1977).

Quennell, Peter, 'Cyril Connolly' (*Encounter*, May 1975, 77–9).

Read, Bill, *The Days of Dylan Thomas: A Pictorial Biography* (McGraw Hill, New York, 1964).

R. E. S., 'Eva Reckitt: Obituary' (*History Workshop*, no. 2 (Autumn 1976), 238–9).

Rolph, C. H., *Kingsley: The Life, Letters and Diaries of Kingsley Martin* (Gollancz, 1973).

Seymour-Smith, Martin, *Robert Graves, His Life and Works* (Hutchinson, 1982).

Skidelsky, Robert, *Oswald Mosley* (Macmillan, 1975).

Sloan, Pat (ed.), *John Cornford: A Memoir* (Cape, 1938; reissued Borderline Press, Dunfermline, 1978).

Stallworthy, Jon, *Wilfred Owen: A Biography* (Oxford University Press/Chatto, 1974).

Stansky, Peter, and Abrahams, William, *Journey to the Frontier: Julian Bell and John Cornford: their lives and the 1930s* (Constable, 1966).

— *The Unknown Orwell* (Constable, 1972).

— *Orwell: the Transformation* (Constable, 1979).

Sykes, Christopher, *Evelyn Waugh: A Biography* (Collins, 1975).

Thomas, Hugh, *John Strachey* (Eyre Methuen, 1973).

Toynbee, Philip, *Friends Apart: A Memoir of Esmond Romilly and Jasper Ridley in the Thirties* (MacGibbon & Kee, 1954).

Warner, Sylvia Townsend, *T. H. White: A Biography* (Cape, 1967).

Wilkinson, Louis ('Louis Marlow'), *Welsh Ambassadors: Powys Lives and Letters* (Chapman & Hall, 1936; reissued Village Press, 1975).

Williamson, Henry, *Genius of Friendship: T. E. Lawrence* (Putnam, 1941).

Worsley, T. C., *The Fellow-Travellers: A Memoir of the Thirties* (London Magazine Editions, 1971).

26. Useful Studies of Particular Authors

Allott, Kenneth and Farris, Miriam, *The Art of Graham Greene* (Hamish Hamilton, 1951).

Bonnar, Robert, 'James Barke—A True Son of the Soil' (in *Life and Literature of the Working Class: Essays in Honour of William Gallacher*, ed. P. M. Kemp-Ashraf and Jack Mitchell (Humboldt-Universität, Berlin, 1966), 185–92).

Bradbrook, M. C., *Malcolm Lowry, His Art and Early Life: A Study in Transformation* (Cambridge University Press, 1974).

Brower, Reuben; Vendler, Helen, and Hollander, John, *I. A. Richards: Essays in His Honour* (Oxford University Press, New York, 1973).

Brown, Terence, *Louis MacNeice: Sceptical Vision* (Gill and Macmillan, Dublin, 1975).

Calder, Jenni, *Chronicles of Conscience: A Study of George Orwell and Arthur Koestler* (Secker & Warburg, 1968).

Callan, Edward, *Auden: A Carnival of Intellect* (Oxford University Press, New York, 1983).

Carter, Ian, 'Lewis Grassic Gibbon, *A Scots Quair*, and the Peasantry' (*History Workshop*, no. 6 (Autumn 1978), 169–85).

Cavaliero, Glen, *John Cowper Powys: Novelist* (Clarendon Press, Oxford, 1973).

— *Charles Williams: Poet of Theology* (Macmillan, 1983).

Chace, William M., *The Political Identities of Ezra Pound and T. S. Eliot* (Stanford University Press, Stanford, California, 1973).

Craig, Cairns, *Yeats, Eliot, Pound and the Politics of Poetry: Riches to the Richest* (Croom Helm, 1982).

Croft, Andy, 'Returned Volunteer: the Novels of John Sommerfield' (*London Magazine*, NS 23, nos. 1 and 2 (Apr.–May 1983), 61–70).

Dodsworth, Martin, 'Empson at Cambridge' (*The Review*, nos. 6 and 7 (William Empson Special Number, June 1963), 3–13).

Duchêne, François, *The Case of the Helmeted Airman: A Study of W. H. Auden's Poetry* (Chatto, 1972).

Duncan, Alastair B. (ed.), *Claude Simon: New Directions: Collected Papers* (Scottish Academic Press, Edinburgh, 1985).

Fernbach, David, 'Tom Wintringham and Socialist Defense Strategy' (*History Workshop*, no. 14 (Autumn 1982), 63–91).

Freyer, Grattan, *W. B. Yeats and the Anti-Democratic Tradition* (Gill and Macmillan, Dublin, 1981).

Gill, Roma (ed.), *William Empson: The Man and His Work* (Routledge, 1974).

Grant, Michael (ed.), *T. S. Eliot: The Critical Heritage* (2 vols.; Routledge, 1982).

Haffenden, John (ed.), *W. H. Auden: The Critical Heritage* (Routledge, 1983).

Hamilton, Ian, 'Louis MacNeice' (*London Magazine*, NS 3, no. 8 (Nov. 1963), 62–7).

Gardner, Helen, *The Composition of 'Four Quartets'* (Faber, 1978).

Gross, Miriam (ed.), *The World of George Orwell* (Weidenfeld, 1971).

Hynes, Samuel (ed.), *Twentieth Century Interpretations of 1984* (Prentice-Hall, Englewood Cliffs, New Jersey, 1971).

Jameson, Fredric, *Fables of Aggression: Wyndham Lewis, the Modernist Fascist* (University of California Press, Berkeley, 1979).

Janvier, Ludovic, *Beckett par lui-même* (Ecrivains de Toujours, Seuil, Paris, 1969).

Jeffares, A. Norman, *A Commentary on the Collected Poems of W. B. Yeats* (Macmillan, 1968).

Johnson, Roy, 'Walter Brierley: Proletarian Writing' (*Red Letters*, no. 2, Summer 1976, 5–8).

Kojecky, Roger, *T. S. Eliot's Social Criticism* (Faber, 1971).

Kush, Thomas, *Wyndham Lewis's Pictorial Integer* (Studies in the Fine Arts, The Avant-Garde, no. 19; UMI Research Press, Ann Arbor, Michigan, 1981).

Law, T. S. and Berwick, Thurso, *The Socialist Poems of Hugh MacDiarmid* (Routledge, 1978).

Lee, Hermione, *The Novels of Virginia Woolf* (Methuen, 1977).

— *Elizabeth Bowen: An Estimation* (Critical Studies Series, Vision Press, 1981).

Lewis, Peter, *George Orwell: the Road to 1984* (Heinemann/Quixote Press, 1981).

Littlewood, Ian, *The Writings of Evelyn Waugh* (Blackwell, Oxford, 1983).

McDiarmid, Lucy, *Saving Civilization: Yeats, Eliot, and Auden Between the Wars* (Cambridge University Press, 1984).

McLeod, A. L., *Rex Warner: Writer, An Introductory Essay* (Wentworth Press, Sydney, 1964).

Margolies, David N., *The Function of Literature: A Study of Christopher Caudwell's Aesthetics* (Lawrence & Wishart, 1969).

Marsack, Robyn Louise, *The Cave of Making, The Poetry of Louis MacNeice* (Clarendon Press, Oxford, 1982).

Materer, Timothy, *Vortex: Pound, Eliot, and Lewis* (Cornell University Press, Ithaca, 1979).

Mendelson, Edward, *Early Auden* (Faber, 1981).

Mengham, Rod, *The Idiom of the Time: The Writings of Henry Green* (Cambridge University Press, 1982).

Meyers, Jeffrey, 'André Malraux: the Art of Action' (*London Magazine*, NS 14, no. 5 (Dec. 1974–Jan. 1975), 5–34).

— (ed.), *George Orwell: The Critical Heritage* (Routledge, 1975).

— (ed,), *Wyndham Lewis: A Revaluation: New Essays* (Athlone Press, 1980).

Murdoch, Iris, *Sartre, Romantic Rationalist* (Bowes & Bowes, Cambridge, 1953).

Norris, Christopher, *William Empson and the Philosophy of Literary Criticism* (Athlone Press, 1978). Postscript by William Empson.

— (ed.), *Inside the Myth: Orwell: Views from the Left* (Lawrence & Wishart, 1984).

Nye, Robert, 'Tatterdemalian Taliessin: A Note on J. C. Powys and his critics' (*London Magazine*, NS 12, no. 6 (Feb.–Mar. 1973), 75–85).

Powys, T. F., 'A Tribute to a Friend, Sylvia Townsend Warner: T. F. Powys's Foreword to *A Moral Ending*, 1931' *The Powys Review*, no. 3 (Summer 1978), 5–6).

Pritchett, V. S., 'Henry Yorke, Henry Green' (*London Magazine*, NS 14, no. 2 (June–July 1974), 28–32).

Pryce-Jones, David, *Evelyn Waugh and His World* (Weidenfeld, 1973).

Reilly, Patrick, *George Orwell: the Age's Adversary* (Macmillan, 1986).

Sandison, Alan, *The Last Man in Europe: An Essay on George Orwell* (ibid., 1974).

Shklovsky, Viktor, *Mayakovsky and His Circle* (trans., and ed. Lily Feiler; New York, 1972; Pluto Press, 1974).

Skelton, Robin (ed.), *Herbert Read: A Memorial Symposium* (Methuen, 1970).

Slater, Ian, *Orwell: the Road to Airstrip One* (W. W. Norton, New York, 1985).

Smith, David, 'Underground Man: the Work of B. L. Coombes, "Miner Writer" ' (*The Anglo-Welsh Review*, 24, no. 53 (Winter 1974), 10–25).

— *Lewis Jones* Writers of Wales series, (University of Wales Press, 1982).

— and Mosher, Michael, *Orwell for Beginners* (Writers and Readers Publishing Cooperative, 1984).

Smith, Rowland, *Lyric and Polemic: The Literary Personality of Roy Campbell* (McGill-Queen's University Press, 1972).

Snee, Carole, 'Walter Brierley: A Test Case' (*Red Letters*, no. 3, Autumn 1976, 11–13).

Speaight, Robert, *George Bernanos: A Study of the Man and the Writer* (Collins & Harvill, 1973).

Spender, Stephen (ed.), *W. H. Auden: A Tribute* (Weidenfeld, 1974).

Stannard, Martin (ed.), *Evelyn Waugh: The Critical Heritage* (Routledge, 1984).

Thompson, Denys (ed.), *The Leavises: Recollections and Impressions* (Cambridge University Press, 1984).

Thompson, E. P., 'Caudwell' (*The Socialist Register 1977* (Merlin Press, 1977), 228–76).

Watt, Donald (ed.), *Aldous Huxley: The Critical Heritage* (Routledge, 1975).

Whitehead, John, 'Auden: An Early Poetical Notebook' (*London Magazine*, NS 5, no. 2 (May 1965), 85–93).

Williams, Raymond, *Orwell* (Modern Masters series, Fontana/Collins, 1971).

— (ed.), *George Orwell: A Collection of Critical Essays: Twentieth Century Views* (Prentice-Hall, Englewood Cliffs, New Jersey, 1975).

Woodcock, George, *The Crystal Spirit: A Study of George Orwell* (Cape, 1967).

Zwerdling, Alex, *Orwell and the Left* (Yale University Press, New Haven and London, 1974).

27. Some Period Studies (Historical, Political, Social)

Aron, Raymond, *The Opium of the Intellectuals*, (trans. Terence Kilmartin; Secker & Warburg, 1957).

Benewick, Robert, *the Fascist Movement in Britain* (Allen Lane, 1972). Revised version of *Political Violence and Public Order* (1972).

Bleuel, Hans Peter, *Strength Through Joy: Sex and Society in Nazi Germany* (trans. J. Maxwell Brownjohn from *Das Saubere Reich*; Secker & Warburg, 1973.

Blythe, Ronald, *The Age of Illusion: Glimpses of Britain Between the Wars 1919–1940* (Hamish Hamilton, 1963; reissued with new preface, Oxford University Press, 1983).

Boyle, Andrew, *The Climate of Treason* (Hutchinson, 1979; revised, Coronet Books, 1980).

Branson, Noreen, and Heinemann, Margot, *Britain in the Nineteen Thirties* (Panther, 1973).

Briggs, Asa, *The History of Broadcasting in the United Kingdom* (4 vols.; Oxford University Press, 1961–79).

— *Governing the BBC* (British Broadcasting Corporation, 1979).

Cockburn, Patricia, *The Years of The Week* (Macdonald, 1968).

Collier, Richard, *Duce! The Rise and Fall of Benito Mussolini* (Collins, 1971).

Cook, C., and Ramsden, J., *By-Elections in British Politics* (Macmillan, 1973).

Cowling, Maurice, *The Impact of Hitler: British Politics and British Policy 1933–1940* (Cambridge University Press, 1975).

Crossman, Richard (ed.), *The God that Failed: Six Studies in Communism* (Hamish Hamilton, 1950).

Douglas, Roy, *In the Year of Munich* (Macmillan, 1977).

Gay, Peter, *Weimar Culture: the Outsider as Insider* (Secker & Warburg, 1969).

Graves, Robert and Hodge, Alan, *The Long Weekend: A Social History of Great Britain 1918–1939* (Faber, 1940).

Griffiths, Richard, *Fellow Travellers of the Right: British Enthusiasts for Nazi Germany 1933–39* (Constable, 1980; Oxford Paperback, 1983).

Grunberger, Richard, *A Social History of the Third Reich* (Weidenfeld, 1971).

Hamilton, Alastair, *The Appeal of Fascism: A Study of Intellectuals and Fascism 1919–1945* (with foreword by Stephen Spender; Anthony Blond, 1971).

Hamilton, Cicely, 'Sun Bathing' (*The Week-End Review*, vii, no. 165 (6 May 1933), 524).

Hennessy, R. A. S., *The Electric Revolution* (Oriel, 1972).

Henrey, Robert, *The Foolish Decade* (Dent, 1945).

Homberger, Eric, 'Exiles from the National Consensus: Unity Mitford and Harry Pollitt in the 1930s' (*Journal of European Studies*, VII (1977), 278–88).

Lewis, John, *The Left Book Club: An Historical Record* (Gollancz, 1970).

Lunn, Kenneth, and Thurlow, Richard C. (edd.), *British Fascism: Essays on the Radical Right in Inter-War Britain* (Croom Helm, 1980).

Lyons, Eugene, *The Red Decade: The Stalinist Penetration of America* (Bobs-Merrill Co., Indianapolis, New York, 1941).

Macintyre, Stuart, *A Proletarian Science: Marxism in Britain 1917–1933* (Cambridge University Press, 1980).

Marwick, Arthur, *Class: Image and Reality in Britain, France and the USA since 1930* (Collins, 1980).

Mowat, C. L., *Britain Between the Wars 1918–1940* (Methuen, 1955).

Muggeridge, Malcolm, *The Thirties, 1930–1940, in Great Britain* (Hamish Hamilton, 1940).

O'Donnell, James P., 'Charlie Chaplin, Adolf Hitler and Napoleon' (*Encounter*, June 1978, 25–33).

Pearse, Innes H., and Crocker, Lucy H., *The Peckham Experiment: a Study in the Living Structure of Society* (Allen & Unwin, 1943).

Pelling, H., *The British Communist Party: A Historical Profile* (Adam & Charles Black, 1958).

Robson, W. A. (ed.), *The Political Quarterly in the Thirties* (Allen Lane, 1971).

Saville, John, 'May Day 1937' (in Saville, J., and Briggs, Asa (edd.), *Essays in Labour History 1918–1939*; Croom Helm, 1977).

Scannell, Paddy, 'Broadcasting and the Politics of Unemployment 1930–1935' (*Media, Culture and Society*, 2, no. 1 (Jan. 1980), 15–28).

Schöffling, Klaus (ed.), *Dort wo man Bücher verbrennt: Stimmen der Betroffenen* (Suhrkamp Verlag, Frankfurt am Main, 1983).

Sellar, Walter Carruthers and Yeatman, Robert Julian, *1066 and All That: A Memorable History of England* (Methuen, 1930).

— *And Now All This, by the Authors of 1066 And All That* (Methuen, 1932).

Skidelsky, Robert, *Politicians and the Slump* (Penguin, 1970).

Stevenson, John, *Social Conditions in Britain Between the Wars* (Penguin, 1977).

— *British Society 1914–45* (Penguin, 1984).

— and Cook, Chris, *The Slump: Society and Politics during the Depression* (Cape, 1977).

Taylor, A. J. P., *The Origins of the Second World War* (Hamish Hamilton, 1964).

— *English History 1914–1945* (Clarendon Press, Oxford, 1965).

Walker, Martin, *Daily Sketches: A Cartoon History of British Twentieth-Century Politics* (Frederick Muller, 1978).

Webster, Charles, 'Healthy or Hungry Thirties?' (*History Workshop Journal*, xiii (1982), 110–29).

— 'Health, Welfare and Unemployment During the Depression' (*Past and Present*, 109 (Nov. 1985), 204–30).

Werskey, Gary, *The Visible College* (Allen Lane, 1978).

Wood, Neal, *Communism and British Intellectuals* (Gollancz, 1959).

Zeman, Z. A. B., *Nazi Propaganda* (2nd edn.; Oxford University Press, 1973).

Ziemer, Gregor, *Education for Death, The Making of the Nazi* (Constable, 1942).

28. Some Critical, Cultural, Historical Studies pertinent to the Thirties but also more widely relevant

Aaron, Daniel, *Writers on the Left* (1961; reissued as paperback, Oxford University Press, New York, 1977).

Baldick, Chris, *The Social Mission of English Criticism 1848–1932* (Clarendon Press, Oxford, 1983).

Barthes, Roland, *The Eiffel Tower and Other Mythologies* (trans. Richard Howard; Hill and Wang, New York, 1979).

Beauman, Nicola, *A Very Great Profession: The Woman's Novel 1914–1939* (Virago 1983).

Blackmur, R. P., *Language as Gesture: Essays in Poetry* (Harcourt Brace Jovanovich, 1952).

Canetti, Elias, *Masse und Macht* (Claasen, Hamburg, 1960; trans. into English by Carol Stewart as *Crowds and Power*, Gollancz, 1962).

Bradby, David and McCormick, John, *People's Theatre* (Croom Helm, 1978).

Caute, David, *The Illusion: An Essay On Politics, Theatre and the Novel* (Harper & Row, New York, 1971).

— *The Fellow-Travellers: A Postscript to the Enlightenment* (Weidenfeld, 1973).

Cavaliero, Glen, *The Rural Tradition in the English Novel 1900–1939* (Macmillan, 1977).

Cockburn, Claud, *Bestseller: The Books That Everyone Read 1900–1939* (Sidgwick & Jackson, 1972).

Cowley, Malcolm, *A Second Flowering: Works and Days of the Lost Generation* (Deutsch, 1973).

Enzensberger, Hans Magnus, 'Tourists of the Revolution' (in Enzensberger, *Raids and Reconstructions, Essays in Politics, Crime and Culture*; Pluto, 1976).

Foulkes, A. P., *Literature and Propaganda* (New Accents Series, Methuen, 1983).

Fraser, G. S., *The Modern Writer and His World* Verschoyle, 1953; revised edns., Pelican, 1964, 1970).

Fussell, Paul, *The Great War and Modern Memory* (Oxford University Press, New York, 1975).

— *Abroad: British Literary Traveling Between the Wars* (Oxford University Press, 1980).

Grant, Joy, *Harold Monro and the Poetry Bookshop* (Routledge, 1967).

Green, Martin, *Children of the Sun: A Narrative of 'Decadence' in England after 1918* (Constable, 1977).

Grubb, Frederick, *A Vision of Reality: A Study of Liberalism in Twentieth Century Verse* (Chatto, 1965).

Hall, John, *The Sociology of Literature* (Longmans, 1979).

Hamilton, George Rostrevor, *The Tell-Tale Article: A Critical Approach to Modern Poetry* (Heinemann, 1949).

Hamilton, Ian, *The Little Magazines: A Study of Six Editors* (Weidenfeld, 1976).

Hawthorn, Jeremy (ed.), *The British Working-Class Novel in the Twentieth Century* (Edward Arnold, 1984).

Hewison, Robert, *Under Siege: Literary Life in London 1939–45* (Weidenfeld, 1977).

Hollander, Paul, *Political Pilgrims: Travels of Western Intellectuals to the Soviet Union, China, and Cuba 1928–1978* (Oxford University Press, New York, 1981).

Howard, Michael S., *Jonathan Cape, Publisher: Herbert Jonathan Cape, G. Wren Howard* (Cape, 1977).

Howe, Irving, *Politics and the Novel* (Horizon Press, Meridian Books, New York, 1957).

Jarrell, Randall *The Third Book of Criticism* (Faber, 1975).

— *Kipling, Auden & Co., Essays and Reviews 1935–1964* (Farrar, Strauss and Giroux, New York, 1980).

Kermode, Frank, *The Genesis of Secrecy: on the Interpretation of Narrative* (Harvard University Press, Cambridge, Massachusetts, and London, 1979).

Klaus, H. Gustav (ed.), *The Socialist Novel in Britain: Towards the Recovery of a Tradition* (Harvester, Brighton, 1982).

— *The Literature of Labour* (Harvester, Brighton, 1985).

Klingender, F. D., *Marxism and Modern Art: An Approach to Social Realism* (Lawrence & Wishart, 1975).

Lindsay, Jack, *After the 'Thirties: The Novel in Britain and its Future* (Lawrence & Wishart, 1956).

Mander, John, *The Writer and Commitment* (Secker & Warburg, 1961).

O'Brien, Conor Cruise, *Writers and Politics: Essays and Criticism* (Chatto, 1965).

Quigly, Isabel, *The Heirs of Tom Brown: The English School Story* (Chatto, 1982; Oxford Paperback, 1984).

Reed, John R., *Old School Ties. The Public Schools in British Literature* (Syracuse University Press, New York, 1964).

Rühle, Jürgen, *Literature and Revolution: A Critical Study of the Writer and Communism in the Twentieth Century* (trans. and ed. Jean Steinberg; Pall Mall Press, 1969).

Smith, David, *Socialist Propaganda in the Twentieth-Century British Novel* (Macmillan, 1978).

Smith, Stan, *Inviolable Voice: History and Twentieth-Century Poetry* (Gill & Macmillan, Dublin, 1982).

Swingewood, Alan, *The Novel and Revolution* (Macmillan, 1975).

Symons, Julian, *Bloody Murder: From the Detective Story to the Crime Novel: A History* (Faber, 1972). American edn., *Mortal Consequences: A History—from the Detective Story to the Crime Novel* (New York, 1972).

— *Critical Observations* (Faber, 1981).

Watson, George, *Politics and Literature in Modern Britain* (Macmillan, 1977).

Wellek, René, *Concepts of Criticism* (ed. S. G. Nichols; Yale University Press, New Haven and London, 1963).

— *The Attack on Literature and Other Essays* (Harvester, Brighton, 1982).

Wiener, Martin J., *English Culture and the Decline of the Industrial Spirit, 1850-1890* (Cambridge University Press, 1981).

Williams, Raymond, *Culture and Society 1780-1950* (Chatto, 1958).

— *The Long Revolution* (Chatto, 1961).

Winegarten, Renée, *Writers and Revolution: The Fatal Lure of Action* (Franklin Watts Inc., New York, 1974).

Wohl, Robert, *The Generation of 1914* (Weidenfeld, 1980).

Woodcock, George, *The Writer and Politics* (Porcupine Press, 1948).

Worpole, Ken, *Dockers and Detectives: Popular Reading, Popular Writing* (Verso, 1983).

29. Some Literary Historical Pointers

Belsey, Catherine, 'Introduction: Reading the Past' (in *The Subject of Tragedy: Identity and Difference in Renaissance Drama*; Methuen, 1985).

Cohen, Ralph (ed.), *New Directions in Literary History* (Routledge, 1974).

Dollimore, Jonathan, 'Introduction: Shakespeare, Cultural Materialism and the new historicism' (in *Political Shakespeare: New Essays in Cultural Materialism*, ed. Jonathan Dollimore and Alan Sinfield; Manchester University Press, 1985).

Hartman, Geoffrey H., 'Toward Literary History' (in Hartman, *Beyond Formalism: Literary Essays 1958-1970*; Yale University Press, New Haven, 1970).

Jameson, Frederic, *The Prison-House of Language* (2nd printing; Princeton University Press, New Jersey, 1974).

— *Marxism and Form: Twentieth-Century Dialectical Theories of Literature* (Princeton University Press, New Jersey, 1971).

— *The Political Unconscious: Narrative As A Socially Symbolic Act* (Methuen, 1981).

Jauss, Hans Robert, 'Literary History as a Challenge to Literary Theory' (*New Literary History*, I (1969); reprinted, slightly altered, in *Towards an Aesthetic of Reception*, trans. Timothy Bahti; Harvester, Brighton, 1982). Introd. by Paul de Man.

McGann, Jerome J., *The Beauty of Inflections* (Clarendon Press, Oxford, 1985).

Moretti, Franco, 'The Soul and the Harpy: Reflections on the Aims and Methods of Literary Historiography' (in *Signs Taken for Wonders, Essays in the Sociology of Literary Forms*, trans. Susan Fischer, David Forgacs, and David Miller; Verso, 1983).

Norris, Christopher, *Contest of Faculties: Philosophy and Theory After Deconstruction* (Methuen, 1985).

Said, Edward W., 'The Text, the World, the Critic' (in *Textual Strategies: Perspectives in Post-Structural Criticism*, ed., and introd. Josué V. Harari (Cornell University Press, 1979; Methuen, 1980); revised as 'The World, the Text and the Critic', in Said, *The World, the Text and the Critic*; Faber, 1984).

Weimann, Robert, *Structure and Society in Literary History: Studies in the History and Theory of Historical Criticism* (expanded edn.; Johns Hopkins University Press, Baltimore, 1984).

Williams, Raymond, *Writing in Society* (Verso, 1984).

ACKNOWLEDGEMENTS

The author and publisher gratefully acknowledge permission to reprint copyright material as follows:

W. H. Auden: from *The English Auden: Poems, Essays and Dramatic Writings 1927–1939*, ed. by Edward Mendelson reprinted by permission of Faber & Faber Ltd., and Random House Inc. Copyright © 1977 by Edward Mendelson, William Meredith and Monroe E. Spears, Executors of the Estate of W. H. Auden. *The Dance of Death* reprinted by permission of Faber & Faber Ltd., and Random House Inc. (Copyright 1934 by The Modern Library). W. H. Auden and Christopher Isherwood: from *The Dog Beneath the Skin, The Ascent of F6,* and *On the Frontier* reprinted by permission of Faber & Faber Ltd. Unpublished manuscript letters in the Berg Collection, New York Public Library reprinted by permission of Edward Mendelson. Copyright Edward Mendelson 1987.

Julian Bell: 'Bypass to Utopia', from *Letters and Works of Julian Bell*, ed. Quentin Bell reprinted by permission of Chatto & Windus and the Hogarth Press.

Roy Campbell: 'Christs in Uniform', from *Mithraic Emblems* (1936) reprinted by permission of Francisco Campbell Custodio and Ad. Donker (Pty Ltd).

Clifford Dyment: 'I found the letter in a cardboard box', from the *Listener*, 30 September 1936, and *The Railway Game* (1962) reprinted by permission of Miss Irene Dyment.

Gavin Ewart: 'Journey', from *New Verse*, April 1934 reprinted by permission of Gavin Ewart.

Graham Greene: 'The Other Side of the Border', from *Nineteen Stories* (1947) reprinted by permission of Laurence Pollinger Ltd., and Viking, Penguin Inc.

Louis MacNeice: from *Autumn Journal* (1938) reprinted by permission of David Higham Associates Ltd.

Materials in the Mass-Observation Archive reprinted by permission of the Tom Harrisson Mass-Observation Archive, University of Sussex.

George Orwell: from *The Road to Wigan Pier* (1937) reprinted by permission of the estate of the late Sonia Brownell Orwell, Secker & Warburg Ltd., and Harcourt Brace Jovanovich Inc.

Geoffrey Parsons: 'The Inheritor' reprinted by permission of the author.

Stephen Spender: unpublished letters in the Berg Collection, New York Public Library reprinted by permission of the author. 'The Funeral', Copyright 1934 and renewed 1962 by Stephen Spender; from *Collected Poems 1928–1953* reprinted by permission of Faber & Faber Ltd., and Random House Inc.

Evelyn Waugh: from *Vile Bodies* (1930), Copyright 1930 by Evelyn Waugh, © renewed 1958 by Evelyn Waugh. Reprinted by permission of A. D. Peters & Co. Ltd., and Little, Brown, Boston.

INDEX

OXFORD

MORE OXFORD PAPERBACKS

Details of a selection of other books follow. A complete list of Oxford Paperbacks, including The World's Classics, Twentieth-Century Classics, OPUS, Past Masters, Oxford Authors, Oxford Shakespeare, and Oxford Paperback Reference, is available in the UK from the General Publicity Department, Oxford University Press (JN), Walton Street, Oxford OX2 6DP.

In the USA, complete lists are available from the Paperbacks Marketing Manager, Oxford University Press, 200 Madison Avenue, New York, NY 10016.

Oxford Paperbacks are available from all good bookshops. In case of difficulty, customers in the UK can order direct from Oxford University Press Bookshop, 116 High Street, Oxford, Freepost, OX1 4BR, enclosing full payment. Please add 10 per cent of published price for postage and packing.

THE COMPLETE POEMS OF KEITH DOUGLAS

Edited by Desmond Graham

Here is a book which, as the *Scotsman* remarked of the hardback edition published in 1978, 'anyone with a serious interest in poetry in English of this century will want to possess'. Keith Douglas, who was killed in Normandy three days after D-Day at the age of 24, is now recognized as perhaps the finest of the Second World War poets. This definitive edition contains 105 poems, among them three previously unpublished or uncollected pieces. The editor provides a preface describing the developing phases of Douglas's work; a chronology of the poet's life; and illuminating textual notes.

'Keith Douglas was a poet of quite staggering resources . . . His death robbed English poetry of one of the most exciting talents to appear since Auden.' Vernon Scannell, *New Statesman*

ELIOT'S EARLY YEARS

Lyndall Gordon

Described by Jonathan Raban in the *Sunday Times* as 'the most valuable single book yet published about Eliot', this unusual biographical study of T. S. Eliot's formative years opens a new perspective upon the career of one of our century's most influential poets and critics. Drawing on unpublished manuscripts (his Notebooks and early poems, his mother's poems, his wife's diaries), Lyndall Gordon traces Eliot's journey across the 'waste land' to his conversion to Anglo-Catholicism at the age of 38. Eliot's poetry has a strong autobiographical basis, and the author here shows us its essential coherence within the context of his life.

'essential for all serious students of Eliot' *Times Literary Supplement*

Oxford Lives